T0260838

Geophysical Monograph Series

Including

IUGG Volumes
Maurice Ewing Volumes
Mineral Physics Volumes

118 **Magnetospheric Current Systems** *Shin-ichi Ohtani, Ryoichi Fujii, Michael Hesse, and Robert L. Lysak (Eds.)*

119 **Radio Astronomy at Long Wavelengths** *Robert G. Stone, Kurt W. Weiler, Melvyn L. Goldstein, and Jean-Louis Bougeret (Eds.)*

120 **GeoComplexity and the Physics of Earthquakes** *John B. Rundle, Donald L. Turcotte, and William Klein (Eds.)*

121 **The History and Dynamics of Global Plate Motions** *Mark A. Richards, Richard G. Gordon, and Rob D. van der Hilst (Eds.)*

122 **Dynamics of Fluids in Fractured Rock** *Boris Faybishenko, Paul A. Witherspoon, and Sally M. Benson (Eds.)*

123 **Atmospheric Science Across the Stratopause** *David E. Siskind, Stephen D. Eckerman, and Michael E. Summers (Eds.)*

124 **Natural Gas Hydrates: Occurrence, Distribution, and Detection** *Charles K. Paull and Willam P. Dillon (Eds.)*

125 **Space Weather** *Paul Song, Howard J. Singer, and George L. Siscoe (Eds.)*

126 **The Oceans and Rapid Climate Change: Past, Present, and Future** *Dan Seidov, Bernd J. Haupt, and Mark Maslin (Eds.)*

127 **Gas Transfer at Water Surfaces** *M. A. Donelan, W. M. Drennan, E. S. Saltzman, and R. Wanninkhof (Eds.)*

128 **Hawaiian Volcanoes: Deep Underwater Perspectives** *Eiichi Takahashi, Peter W. Lipman, Michael O. Garcia, Jiro Naka, and Shigeo Aramaki (Eds.)*

129 **Environmental Mechanics: Water, Mass and Energy Transfer in the Biosphere** *Peter A.C. Raats, David Smiles, and Arthur W. Warrick (Eds.)*

130 **Atmospheres in the Solar System: Comparative Aeronomy** *Michael Mendillo, Andrew Nagy, and J. H. Waite (Eds.)*

131 **The Ostracoda: Applications in Quaternary Research** *Jonathan A. Holmes and Allan R. Chivas (Eds.)*

132 **Mountain Building in the Uralides Pangea to the Present** *Dennis Brown, Christopher Juhlin, and Victor Puchkov (Eds.)*

133 **Earth's Low-Latitude Boundary Layer** *Patrick T. Newell and Terry Onsage (Eds.)*

134 **The North Atlantic Oscillation: Climatic Significance and Environmental Impact** *James W. Hurrell, Yochanan Kushnir, Geir Ottersen, and Martin Visbeck (Eds.)*

135 **Prediction in Geomorphology** *Peter R. Wilcock and Richard M. Iverson (Eds.)*

136 **The Central Atlantic Magmatic Province: Insights from Fragments of Pangea** *W. Hames, J. G. McHone, P. Renne, and C. Ruppel (Eds.)*

137 **Earth's Climate and Orbital Eccentricity: The Marine Isotope Stage 11 Question** *André W. Droxler, Richard Z. Poore, and Lloyd H. Burckle (Eds.)*

138 **Inside the Subduction Factory** *John Eiler (Ed.)*

139 **Volcanism and the Earth's Atmosphere** *Alan Robock and Clive Oppenheimer (Eds.)*

140 **Explosive Subaqueous Volcanism** *James D. L. White, John L. Smellie, and David A. Clague (Eds.)*

141 **Solar Variability and Its Effects on Climate** *Judit M. Pap and Peter Fox (Eds.)*

142 **Disturbances in Geospace: The Storm-Substorm Relationship** *A. Surjalal Sharma, Yohsuke Kamide, and Gurbax S. Lakhima (Eds.)*

143 **Mt. Etna: Volcano Laboratory** *Alessandro Bonaccorso, Sonia Calvari, Mauro Coltelli, Ciro Del Negro, and Susanna Falsaperla*

144 **The Subseafloor Biosphere at Mid-Ocean Ridges** *William S. D. Wilcock, Edward F. DeLong, Deborah S. Kelley, John A. Baross, and S. Craig Cary (Eds.)*

145 **Timescales of the Paleomagnetic Field** *James E. T. Channell, Dennis V. Kent, William Lowrie, and Joseph G. Meert (Eds.)*

146 **The Extreme Proterozoic: Geology, Geochemistry, and Climate** *Gregory S. Jenkins, Mark A. S. McMenamin, Christopher P. McKay, and Linda Sohl (Eds.)*

147 **Earth's Climate: The Ocean–Atmosphere Interaction** *Chuzai Wang, Shang-Ping Xie, and James A. Carton (Eds.)*

148 **Mid-Ocean Ridges: Hydrothermal Interactions Between the Lithosphere and Oceans** *Christopher German, Jian Lin, and Lindsay Parson (Eds.)*

149 **Continent-Ocean Interactions Within East Asian Marginal Seas** *Peter Clift, Wolfgang Kuhnt, Pinxian Wang, and Dennis Hayes (Eds.)*

150 **The State of the Planet: Frontiers and Challenges in Geophysics** *Robert Stephen John Sparks, and Christopher John Hawkesworth (Eds.)*

151 **The Cenozoic Southern Ocean: Tectonics, Sedimentation, and Climate Change Between Australia and Antarctica** *Neville Exon, James P. Kennett, and Mitchell Malone (Eds.)*

152 **Sea Salt Aerosol Production: Mechanisms, Measurements, Methods and Models** *Ernie R. Lewis and Stephen E. Schwartz (Eds.)*

153 **Ecosystem and Land Use Change** *Ruth S. DeFries, Gregory P. Asner, and Richard A. Houghton (Eds.)*

Geophysical Monograph 154

The Rocky Mountain Region: An Evolving Lithosphere Tectonics, Geochemistry, and Geophysics

Karl E. Karlstrom
G. Randy Keller
Editors

American Geophysical Union
Washington, DC 2005

Library of Congress Cataloging-in-Publication Data

The Rocky Mountain region--an evolving lithosphere : tectonics, geochemistry, and geophysics / G. Randy Keller, Karl E. Karlstrom, editors.
p. cm. -- (Geophysical monograph ; 154)
Includes bibliographical references.
1. Orogeny--Rocky Mountains. 2. Geology, Structural--Rocky Mountains. 3. Geophysics--Rocky Mountains. 4. Core-mantle boundary. I. Keller, G. Randy (George Randy), 1946- II. Karlstrom, Karl E. III. Series.

QE621.5.R63R63 2004
551.8'2'0978--dc22

2004029017

ISBN 0-87590-418-1
ISSN 0065-8448

Printed in the United States of America.

CONTENTS

Passive Source Seismic Studies

Synthesis Papers

PREFACE

The Rocky Mountains provide a key region for understanding the evolution of the western North American continent and processes that shape continents in general.

As a result, the region has prompted intense and pioneering geologic investigations for over a century, offering scientists an exceptionally rich field laboratory in which to gather data and to make and test interpretations. The Continental Dynamics of the Rocky Mountain (CD-ROM) experiment (1995-2004), from which this book derives, follows in this tradition, motivated by three leading questions: how are continents initially formed and stabilized; how do old lithospheric structures and boundaries influence younger tectonic events; and how did processes related to the plate boundary affect the evolution of the Cenozoic Rocky Mountains? To successfully answer such questions requires integrated studies focused from the surface, through the crust, into the mantle, and with a four-dimensional approach that also encompasses the time dimension.

We have thus organized this volume into five sections. Section 1 presents papers with new data on the tectonics, structure geology, and regional geophysics of the area. Section 2 offers geochemistry, geochronology and xenolith studies. Sections 3 and 4 summarize results from CD-ROM active-source seismic experiments (a 900 km long seismic refraction profile and two > 200 km long seismic reflection profiles) and passive source seismic studies, respectively. The last section provides a synthesis of the evolution of the Rocky Mountain continental crust and lithosphere. We emphasize the importance of the collaborative approach throughout, and have enlarged the scope of the investigations to reflect the grand scale of the most important constructional and reconstructional processes that affected the region. Thus, the papers cover the large spatial and 2-billion-year time scales that characterize the evolution of the Rocky Mountain region.

CD-ROM results have been presented at numerous oral and poster sessions of the American Geophysical Union and the Geological Society of America. Results have also appeared in articles in scientific journals. A series of five CD-ROM workshops helped build a better collaborative approach. This book presents primary new papers that provide a data-rich compendium of the main geological and geophysical results of this collaborative experiment, and an overall synthesis of the lithospheric evolution of the Rocky Mountain region that will be a valuable reference work for years to come.

We wish to acknowledge the participants of the broader CD-ROM working group (listed below), all of who have contributed to the papers in this volume. Our efforts at collaborative science have been an important success of the overall experiment. Funding has primarily been provided by the NSF Continental Dynamics Program, and we thank Leonard Johnson for his interest and support of the project. We also thank the *Deutsche Forschungsgemeinschaft* for its support of the seismic profiling. We especially thank the numerous reviewers of the individual papers, listed in the individual acknowledgements, for providing rigorous peer review of the science.

Karl E. Karlstrom and G. Randy Keller

CD-ROM Working Group

Chris Andronicos, Nicholas Bolay*, Oliver Boyd*, Sam Bowring, Steve Cather, Kevin Chamberlain, Nick Christensen, Jim Crowley, Jason Crosswhite*, David Coblentz, Ken Dueker, Tefera Eshete*, Eric Erslev, Lang Farmer, Rebecca Flowers*, Otina Fox*, Matt Heizler, Gene Humphreys, Micah Jessup*, Roy Johnson, Karl Karlstrom, Randy Keller, Shari A. Kelley, Eric Kirby, Alan Levander, M. Beatrice Magnani, Kevin Mahan*, Jennie Matzal*, Annie McCoy*, Grant Meyer, Kate Miller, Elena Morozova, Frank Pazzaglia, Claus Prodehl, Adam Read*, Oscar Quezada*, Mousumi Roy, Hanna-Maria Rumpel*, Jane Selverstone, Anne Sheehan, Liane Stevens*, Colin A. Shaw*, Elena Shoshitaishvili*, Scott Smithson, Cathy Snelson*, Mike Timmons*, Leandro Treviño*, Amanda Tyson*, Stacy Wagner*, Xin Wan*, Paul Wisniewski*, Michael Williams, Huaiyu Yuan*, Brian Zurek*

(*denotes graduate students)

The Rocky Mountain Region: An Evolving Lithosphere
Geophysical Monograph Series 154

10.1029/154GM00

Introduction to: The Rocky Mountain Region—An Evolving Lithosphere: Tectonics, Geochemistry, and Geophysics

Karl E. Karlstrom

Department of Earth and Planetary Sciences, University of New Mexico, Albuquerque, New Mexico

G. Randy Keller

Department of Geological Sciences, University of Texas at El Paso, El Paso, Texas

The Rocky Mountains are the eastern, high elevation portion of the continental-scale Cordilleran mountain chain that extends along the western portion of the North American continent. This wide zone records the complex tectonics associated with a series of Mesozoic to Recent additions to and modifications of the continent. This volume focuses on the southern Rocky Mountain region, the area that extends from southern Wyoming, across the Colorado Rockies, to meld with the region dominated by the Rio Grande rift and Colorado Plateau. These various physiographic domains share a common tectonic heritage as a region that was uplifted from near sea level in the late Cretaceous, underwent regional uplift, compressive deformation, and lithospheric modification at appreciable distance from the western margin of the North American plate during the 70–45 Ma Laramide orogeny, then underwent a complex series of magmatic modification events, localized lithospheric extension, and late Cenozoic denudation and incision. The result is the high elevation and topographically rugged western U.S. orogenic plateau shown in Plate 1.

There has been strong scientific interest in the region of the Southern Rocky Mountains for over a century, but our knowledge of the deep lithospheric structure of the region has been improved greatly by a series of seismic experiments that began in the mid 1990s. New insights have been provided by seismic data that were acquired by a combination of earthquake and natural source recording efforts; such as the Deep Probe experiment (*Henstock et al.*, 2001; *Gorman et al.*, 2002), Rocky Mountain Front experiment (*Sheehan et al.*, 1995), and CD-ROM experiment (*Dueker et al.*, 2001; *Karlstrom* and the CD-ROM Working Group, 2002). Previous work on the geologic evolution of this region has been extensive and includes recent summaries related to the CD-ROM experiment (*Karlstrom*, 1998; 1999).

This volume provides a synthesis and integration of geological and geophysical results of the CD-ROM (Continental Dynamics of the Rocky Mountains) collaborative investigation. This effort has involved more than 22 investigators at 14 institutions and 21 graduate students. After an initial workshop in 1995, CD-ROM was funded from 1997 to 2002 by the Continental Dynamics Program of the National Science Foundation, with supplemental funding from the *Deutsche Forschungsgemeinschaft*. The experiment was designed as a fully interdisciplinary series of coordinated geologic and geophysical studies to understand the tectonic evolution of the lithosphere of the southern Rocky Mountain region. As described in the papers in this volume, the overall project produced a seismic refraction line from Wyoming to New Mexico (Plates 1 and 2), geologic studies of shear zones and Laramide structures, xenolith studies, and teleseismic and reflections lines across key tectonic boundaries within the longer transect, including the Cheyenne belt and Jemez lineament.

The southern Rocky Mountain region is one of the best places in the world to study continental evolution because this region preserves a unique record of the assembly, stabilization, maturation, and early stages of disassembly of a continent. The core of North America is underlain by high velocity (old, cold) lithosphere (blue in Plate 2) whereas many areas

The Rocky Mountain Region: An Evolving Lithosphere
Geophysical Monograph Series 154

10.1029/154GM01

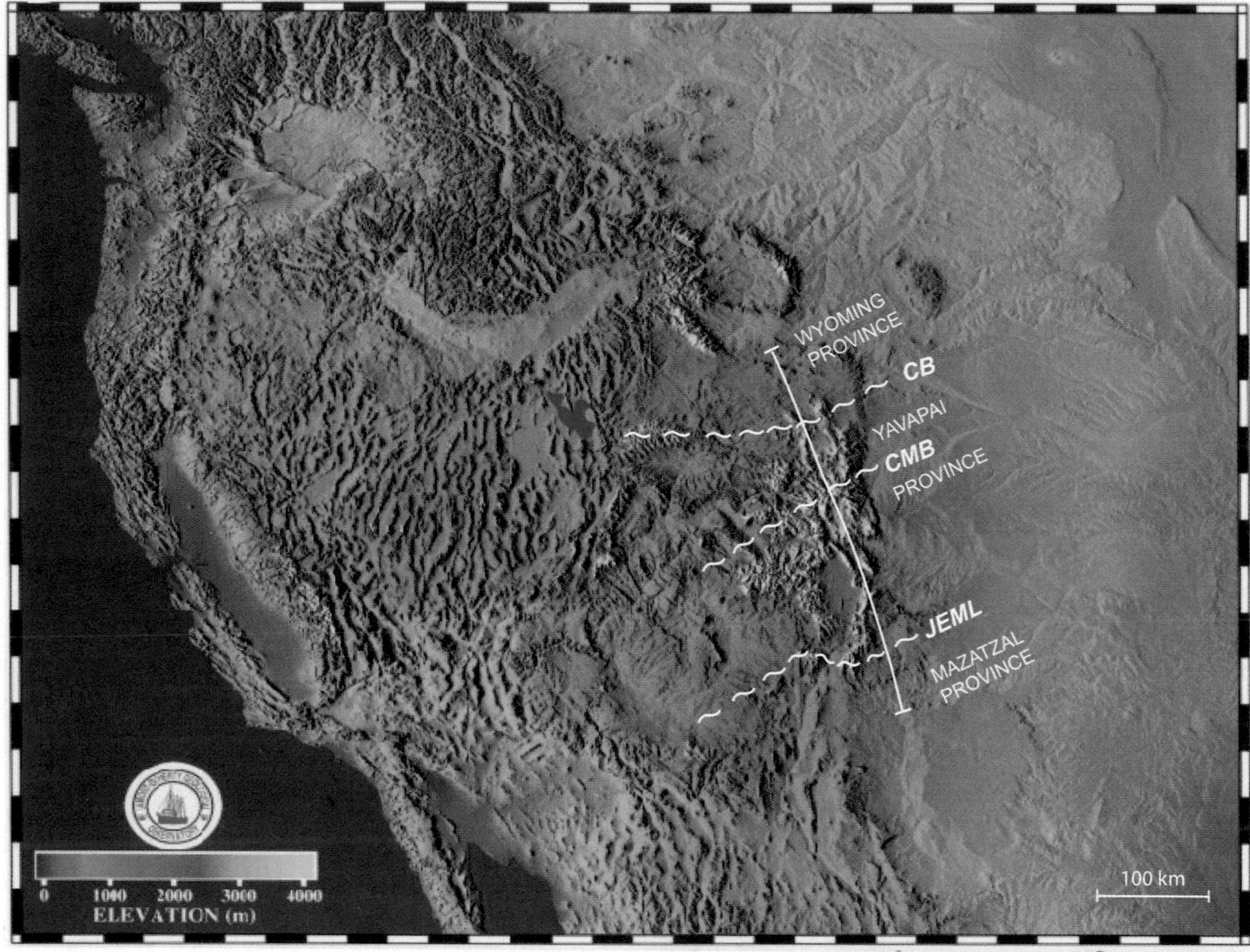

Plate 1. Digital topography in the western United States (from Simpson and Anders, 1992). The overall high plateau in the western United States records buoyant mantle. Subtle differences in Rocky Mountain topography appear to correspond to Precambrian lithospheric blocks, suggesting that different columns are responding differently (magmatically and isostatically) to present mantle reorganization. The CD-ROM transect crosses different Precambrian provinces and obliquely crosses the Rocky Mountains: *CB* = Cheyenne belt; *CMB* = Colorado Mineral belt; *JEML* = Jemez lineament.

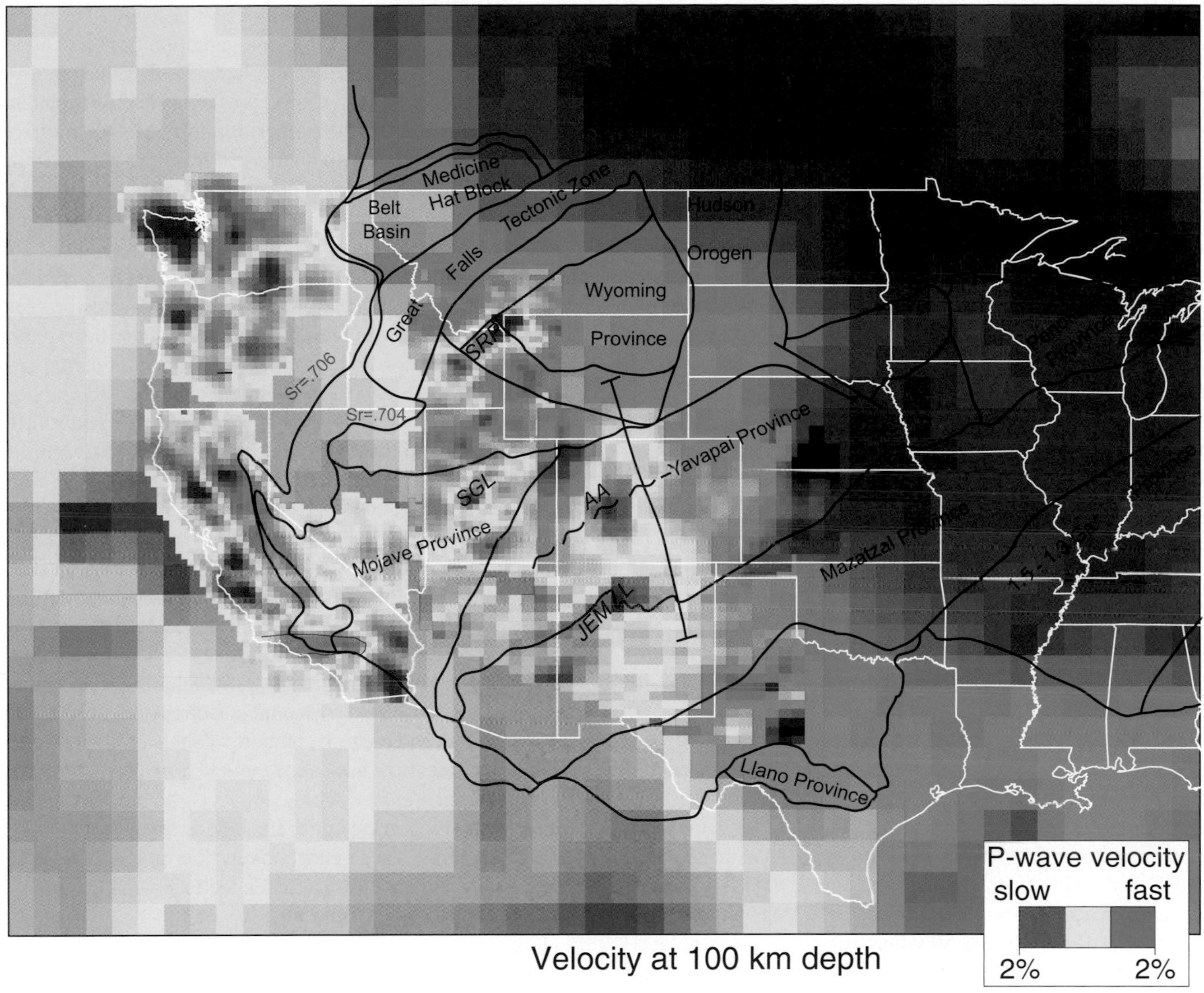

Plate 2. Composite image of the upper mantle seismic structure at 100 km depth beneath the greater North America region. Blue is high velocity mantle and red is low velocity mantle. The continental scale image is from the S-wave modeling of Grand (1997), with overlay of data from regional arrays (Humphreys and Dueker, 1994). Black lines are Precambrian provinces, showing some correspondence between crustal provinces and mantle velocity domains in the Rocky Mountain region: *SPR* = Snake River Plain; *SGL* = Saint George lineament; *AA* = Aspen anomaly; *JEML* = Jemez lineament. Location of CD-ROM transect is shown.

along the Pacific plate margin are underlain by low velocity (young, hot) mantle (red in Plate 2). This is one of the largest transitions in mantle velocity structure on Earth, some 14% variation in shear wave velocity from the core of the continent to the active plate margin. As shown in Plate 2, this same scale of velocity variation takes place over the shorter, 10-km, length scale of tectonic/magmatic provinces such as the Snake River Plain (*SRP*), St George lineament (*SGL*), Aspen anomaly (*AA*), and Jemez lineament (*JEML*). Part of the goal of the CD-ROM experiment was to investigate the origin of these velocity anomalies.

In the Southern Rocky Mountains, many excellent exposures of Precambrian rocks provide the ground truth needed to study how the old structures (in this case Archean and Proterozoic lithospheric provinces and sutures) are influencing younger lithospheric restructuring in response to Cenozoic sublithospheric processes. In addition, the region is an excellent field laboratory because different tectonic provinces record different aspects of the total history that involves rapid lithosphere accretion (1.8–1.6 Ga), long-term stability (1.6–1.4 Ga and 1.3– 0.53 Ga), crustal "overturn" (1.4 Ga), intracontinental deformation (Ancestral Rockies orogeny, Laramide orogeny), and early stages of extension (Rio Grande rift). The large amount of new data in the Southern Rocky Mountain region presented in this volume provide an opportunity to constrain and unravel aspects of Proterozoic, Laramide, and Neogene tectonism and their interactions, and the processes that have shaped the region during the past 2.0 Ga.

Continents are the long-term record keepers of plate interactions before 200 Ma. However, accumulated modification often obscures the record of older events. The southwestern U.S. is unique in that juvenile Proterozoic lithosphere was accreted as belts of volcanic arcs that were added to the south margin of an Archean cratonic core. These Precambrian provinces are also shown in Plate 2. These belts were then truncated in the Phanerozoic at a high angle so as to create a N-NW trending continental margin oblique to the NE-trending Precambrian accretionary structures (Plate 2). The result was the creation of a wide region where all stages of structural reactivation and continental modification are preserved. The Rocky Mountain deep interior of the Cordilleran orogen reveals two remarkable features: (1) much of the Precambrian assembly of the continent is still well preserved and well exposed over a large region; and (2) Cenozoic orogenesis is in a stage where the recently modified Precambrian continent strongly expresses both the older and the newer structures. Many of the papers in this volume address the hypothesis that the older lithospheric structures influenced younger tectonic and magmatic activity.

The scale of the investigations in the CD-ROM experiment is commensurate with the scale of the important constructional and reconstructional processes that affected the region, and the papers in this volume cover the large horizontal and vertical lengths and the billion-year time scales involved. At the same time, these papers feature the high level of resolution afforded by modern seismic and analytical techniques. An important overarching goal of this volume is a better understanding of processes of formation, stabilization, and reactivation and modification of continental lithosphere. The growth of continental crust from the amalgamation of island arcs is conceptually simple and often invoked. However, the processes involved in transforming accreted arcs to stable continental lithosphere with a thick lithospheric mantle are still not well understood. One result that emerges from the papers in this volume is a much better understanding of how dominantly oceanic tectonic elements (island arcs, back-arc basins, sea mounts, accretionary prisms) became amalgamated into "continental lithosphere". Accretion was followed by a series of underplating events that added to the mafic lower crust. Numerous magmatic differentiation events have been important in the evolution and stabilization of the continental crust and upper mantle, and the original mechanical and compositional anisotropies of the lithosphere have repeatedly influenced the tectonic and magmatic response of the continent to events around its margins.

The Laramide orogeny is an emphasis of several papers in this volume, and it probably represents the best-documented example of basement-involved foreland deformation on Earth. This impressive and enigmatic intracontinental restructuring and reactivation event was superimposed at high angles to the pre-existing lithospheric architecture. The mechanisms controlling basement-involved foreland deformation, including the role of basement structural anisotropy and lithosphere-asthenosphere interactions, remain subjects of great interest, as does the puzzle of how plate tectonic processes, both west of and underneath the Rockies, connect with the complex pattern of shortening in the upper crust. Several papers in this volume also address the geomorphology, thermochronology, and modeling studies of post-Laramide tectonism that gave rise to today's spectacular topography.

This volume is organized into five sections. The first nine papers present new data on the tectonics, structure geology, and regional geophysics for the area. A series of three papers deal with geochemistry, geochronology and xenolith studies. The CD-ROM project featured a 900 km long seismic refraction profile and two > 200 km long seismic reflection profiles. After an overview of controlled seismic techniques and previous studies, there are six papers that present and interpret the new CD-ROM data, and two papers that present results from older, but newly released, seismic reflection profiles. Passive source seismic studies have been featured by a series of experiments in the Southern Rocky Mountain region and seven papers present analyses of these data. The final two papers synthesize our present understanding of the evolution of the continental lithosphere in the Rocky Mountain region.

REFERENCES

Dueker, K., H. Yuan, and B. Zurek, Thick-structured Proterozoic lithosphere of the Rocky Mountain region, GSA Today, 11, 4–9, 2001.

Gorman, A.R., R.M. Clowes, R. M. Ellis, T. J. Henstock, G.D. Spence, G. R. Keller, A. R. Levander, C. M. Snelson, M. J. A. Burianyk, E. R. Kanasewich, I. Asudeh, Z. Hajnal, and K. C. Miller, Deep Probe: imaging the roots of western North America: Canadian Journal of Earth Sciences, 39, 375–398, 2002

Grand, S.P., Mantle shear structure beneath the Americas and surrounding oceans, JGR-B, 99, 11591–11621, 1994.

Henstock, T. J., A. Levander, C. M. Snelson, G. R. Keller, K. C. Miller, S. H. Harder, A. R. Gorman, R. M. Clowes, M. J. A. Burianyk, and E. D. Humphreys, Probing the Archean and Proterozoic lithosphere of western North America, GSA Today, 8, 1–5 and 16–17, 1998.

Humphreys, E.D., and Dueker, K.G., Physical state of the western U.S. mantle : Journal of Geophysical Research, v. 99, p. 9635–9650, 1994.

Karlstrom, K.E., Introduction to special issues: Lithospheric structure and evolution of the Rocky Mountains: Rocky Mountain Geology, v. 33, no. 2, p. 157–160, 1998.

Karlstrom, K.E., Introduction to special issues: Nature of tectonic boundaries in the lithosphere of the Rocky Mountains: Rocky Mountain Geology, v. 34, no. 1, p. 1–4, 1999.

Karlstrom and the CD-ROM Working Group, 2002, Structure and evolution of the lithosphere beneath the Rocky Mountains: Initial results from the CD-ROM experiment: GSA Today, 12, 4–10, 2002.

Sheehan, A. F., G. A. Abers, C. H. Jones, A. L. and Lerner-Lam, Crustal thickness variations across the Colorado Rocky Mountains from teleseismic receiver functions, J. Geophys. Res., 100, 20391–20404, 1995.

Simpson, D.W.; Anders, M. H., 1992 Tectonics and topography of the Western United States; an Application of digital mapping: GSA Today, 2, no. 6: 117–118, 120–121, 1992.

2D Laramide Geometries and Kinematics of the Rocky Mountains, Western U.S.A.

Eric A. Erslev

Department of Geosciences, Colorado State University, Fort Collins, Colorado

Kinematic hypotheses for 2D, cross-sectional Laramide deformation in the Rocky Mountains include (1) block-tilting models and (2) basement thrust models with (a) subcrustal shear during low-angle subduction, (b) lithospheric buckling, (c) thickened lower crust under Laramide arches, and (d) crustal buckling during detachment of the upper crust. Vertical tectonic models invoking vertical or normal faults are falsified by major Laramide thrust faults and pervasive minor faults indicating NE–SW to E–W horizontal shortening. Seismic tomography documenting 200-km-thick Rocky Mountain lithosphere contradicts models predicting wholesale removal of North American mantle lithosphere by subcrustal shear during low-angle subduction. Preliminary gravity and deep seismic interpretations indicate a lack of correspondence between the Moho and Laramide arch geometries. This apparent lack of major Laramide faulting and folding of the Moho under the Rockies contradicts block-tilting models invoking reverse faults cutting the Moho and lithospheric buckling models. In addition, the maximum distances between Laramide arch culminations are generally smaller than buckle wavelengths expected for cratonic continental lithosphere. Geophysical evidence for rootless Laramide arches is incompatible with models predicting pure shear thickening of the lower crust under individual arches but is consistent with models invoking distributed lower crustal thickening over the entire Rockies. Rocky Mountain crustal geometries are most consistent with initial crustal buckling followed by upper crustal detachment on subhorizontal thrust faults that root to the west. An originally thick Rocky Mountain crust may have facilitated delamination and detachment folding of the upper crust during both the Laramide and the Ancestral Rocky Mountain orogenies.

INTRODUCTION

The connections between foreland basement-involved deformation and plate tectonic processes remain poorly understood. The Laramide orogeny of the U.S. Rocky Mountains (Figure 1) is probably the best documented example of basement-involved foreland deformation on Earth yet there is no agreement on how the complex pattern of Laramide shortening in the upper crust was generated. Recent Laramide hypotheses (Figure 2) have invoked a range of mechanisms, from lithospheric buckling [*Tikoff and Maxson*, 2001] to extensive lateral intrusion of ductile crust [*McQuarrie and Chase*, 2000].

The scope of disagreement on the mechanisms of Laramide deformation is surprising considering the quantity and quality of available geologic and geophysical data for the Rocky Mountain region. The only major weakness of the Rocky Mountain data set relative to that of other orogens is the lack of current activity on Laramide structures – it is here that potentially analogous structures from the Andes [*Jordan and*

The Rocky Mountain Region: An Evolving Lithosphere
Geophysical Monograph Series 154

10.1029/154GM02

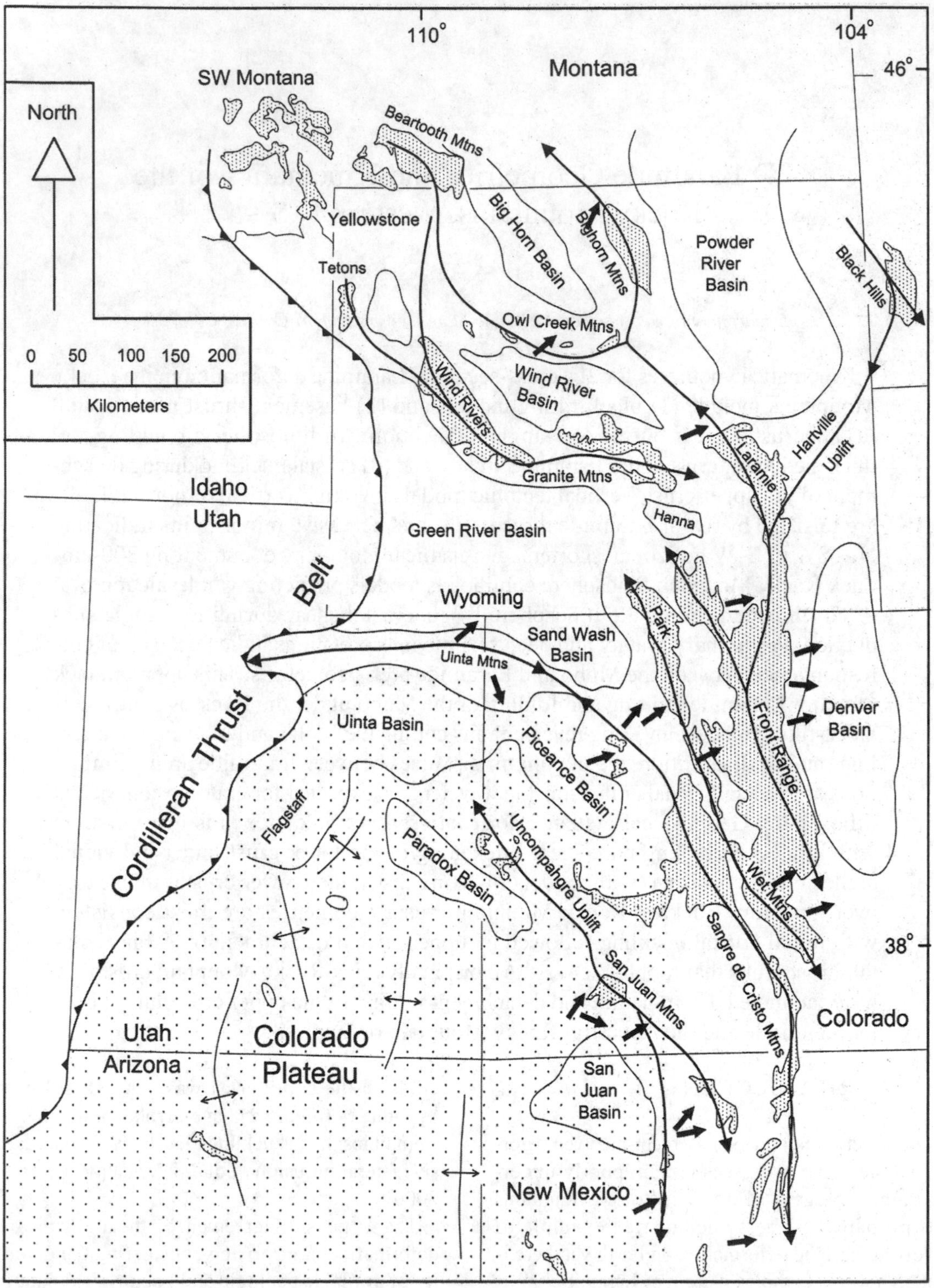

Figure 1. Tectonic map of Laramide arches in the Rocky Mountains. Arrows indicate average Laramide shortening directions derived from minor faults from Molzer and Erslev [1995], Gregson and Erslev [1997], Erslev [2001], DiUlio [2001], Erslev et al. [in press] and unpublished data.

Allmendinger, 1986] and the Tien Shan [*Nikishin et al.*, 1993] have distinct advantages. The purpose of this paper is to review kinematic models for Laramide deformation and test them using published geometric and kinematic constraints. This paper will focus on cross-sectional models for Laramide deformation and will leave an analysis of three-dimensional geometries and kinematics to a subsequent paper integrating GIS map analyses [*Koenig and Erslev*, in press] and minor fault studies [*Molzer and Erslev*, 1995; *Gregson and Erslev*, 1997; *Erslev*, 2001; *Erslev et al.*, in press]. It is important to note, however, that the papers cited above indicate generally NE–SW to E–W shortening and compression perpendicular to most NW–SE- and N–S-trending Laramide arches (see average shortening directions derived from minor faults in Figure 1). As a result, the cross-sectional models presented here probably are reasonable representations for the majority of the Laramide deformation. It is acknowledged that proposed 3D complexities due to transpression [*Wise*, 1963; *Sales*, 1968; *Chapin*, 1983; *Cather*, 1999a; *Karlstrom and Daniel*, 1993], slip partitioning [*Cather*, 1999b], reactivation of pre-existing weaknesses [*Blackstone*, 1990; *Karlstrom and Humphreys*, 1998; *Marshak et al.*, 2000; *Timmons et al.*, 2001; *Koenig and Erslev*, in press], possible rotation of the Colorado Plateau [*Hamilton*, 1988], out-of-plane contributions from the Cordilleran thrust belt [*Livaccari*, 1991] and multi-stage, multi-directional deformation [*Chapin and Cather*, 1981; *Gries*, 1983; *Bird*, 1998] play important roles, at least locally, in Laramide deformation.

This review of tectonic models attempts to encompass all reasonable, kinematically restorable models for the 2D (cross-sectional) kinematics of the Laramide orogeny. While it is more typical for papers to concentrate on a single, preferred tectonic model, orogens are commonly generated by multiple interacting mechanisms, with contributions from several end-member tectonic models. As a result, restricting discussion to one model, however dominant, can obscure the likely possibility that the Laramide orogen resulted from the complicated interplay of several mechanisms.

LARAMIDE CRUSTAL AND MANTLE GEOMETRIES

The volume of industry and academic data on Laramide deformation in the U.S. Rocky Mountains is staggering. This database can be an obstacle to regional syntheses because few people have had the time to be introduced to the breadth of Rocky Mountain data. Studies of upper crustal structural geometries and timing constraints revealed by Phanerozoic sedimentary strata have resulted in a variety of tectonic models that can adequately explain near-surface crustal geometries yet are highly divergent in their predictions of deeper geometries. But recent geophysical investigations of the deep structure of the Rocky Mountains such as the Rocky Mountain Front PASSCAL [*Sheehan et al.*, 1995], Deep Probe [*Snelson et al.*, 1998] and CD–ROM [*Karlstrom et al.*, 2002] experiments, when combined with earlier gravity [*Malahoff and Moberly*, 1968; *Hurich and Smithson*, 1982; *Hall and Chase*, 1989] and COCORP seismic data [*Smithson et al.*, 1979], can provide a clearer view of the deeper roots of Laramide deformation.

This paper is not intended to provide an exhaustive summary of Laramide near-surface geometries, for which other sources exist [e.g., *Brown*, 1988; 1993]. Rather, it will attempt to define the major geometric features of Laramide structures and consider them in the context of proposed models for Laramide deformation. Primary, or first-order Laramide features of the Rocky Mountains (Figure 1) include:

- anastomosing basement-cored arches, which split and join to form a web of basement highs,
- intervening, lens-shaped (in map view) foreland basins commonly containing synorogenic strata,
- major arch- and basin-bounding thrusts faults, and
- enigmatic syn-deformational igneous rocks (e.g., the northeast-trending Colorado Mineral Belt; see Larson and Drexler, 1987; Mutschler et al., 1987; Snoke, 1993 for reviews).

Laramide basement arches anastomose and change asymmetry along strike, suggesting that they are connected at depth [*Erslev*, 1993]. Large amplitude arches like the culminations of the Wind River, Bighorn and Front Range arches can be highly asymmetrical, with major thrust faults on one arch margin. Estimates of Laramide crustal shortening across the Rocky Mountains typically range from 8% [*Stone*, 1993] to 13% [*Brown*, 1993], in part depending upon which areas are included in a given analysis.

The geometry of the crust-mantle interface, or Moho, is much more poorly defined. Malahoff and Moberly [1967] provided the last comprehensive survey of the gravity field in Wyoming, which is the optimum location for studying Laramide crustal geometries because Wyoming has been minimally affected by post-Laramide rifting and igneous activity. They noted that Bouguer and isostatic anomalies are consistently positive over Laramide arches, indicating a lack local isostatic compensation. The arches' positive gravity anomalies are probably due to a combination of low-density sedimentary rocks in the flanking basins and anomalous masses at depth under the arches. Malahoff and Moberly [1967] concluded that "the crust-mantle contact under most of Wyoming is generally parallel to the sedimentary cover-basement contact". They used this interpretation to support a model using near-vertical, concave-downward upthrusts for Laramide arches. In their model, initial sharp breaks in the Moho caused by Laramide high-angle faults were smoothed out during post-Laramide time.

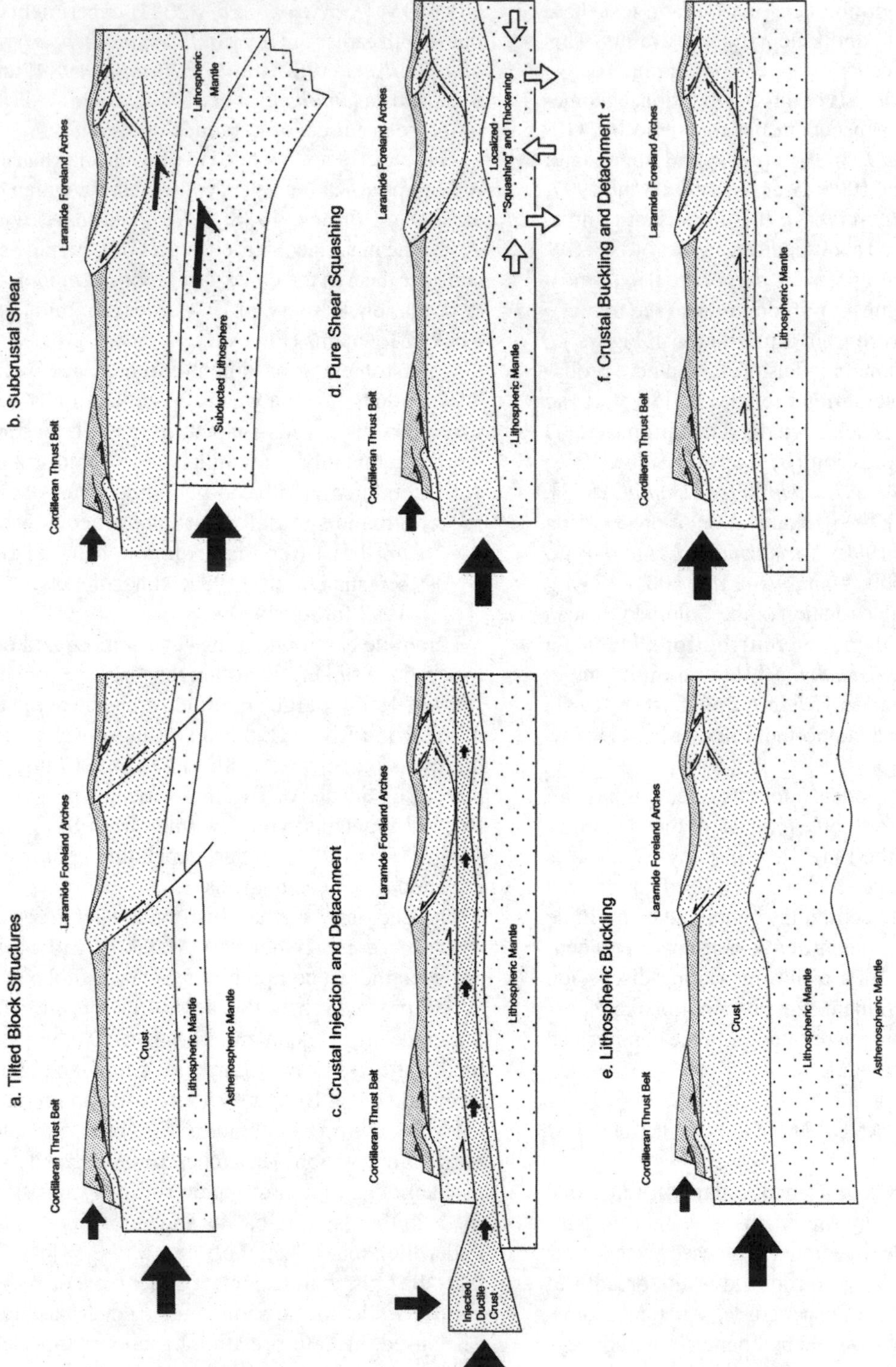

Figure 2. Schematic cross-sections illustrating possible kinematic linkages between Laramide upper crustal deformation and deeper lithospheric interactions.

The acquisition of a deep crustal seismic line across the southern Wind River arch by the COCORP consortium stimulated more detailed studies of the deep Rocky Mountain crust. Seismic results [*Smithson et al.*, 1989; *Lynn et al.*, 1983; *Sharry et al.*, 1986] document a major Laramide thrust fault below the Wind River arch with at least 22 km of slip. The fault appears to dip 45 degrees where it cuts the upper crust but shows no evidence of cutting the Moho. Hurich and Smithson [1982] tested this observation by using 1800 gravity measurements to generate high-resolution gravity profile through the Wind River arch. This profile has a minimum Bouguer gravity value in the eastern-most Green River Basin and a maximum Bouguer gravity value in the center of the Wind River arch. Their gravity modeling showed that 80% of this 85 mgal anomaly is due to density differences between the Precambrian and sedimentary rocks. Hurich and Smithson [1982] showed that models with a master fault cutting the Moho generated Bouguer anomalies that are too high in amplitude and broad in extent. A mid-crustal density anomaly, perhaps created by duplex thickening of the lower crust [Sharry et al., 1986], gave a good match to the gravity profile.

Hall and Chase [1989] modeled the Wind River arch and the Sweetwater arch to the southeast by defining a standard crust that would give a zero Bouguer anomaly at sea level, allowing the absolute gravity to be analyzed as well as the anomaly amplitudes. For both arches, their preferred gravity models showed thickening of the lower crust on listric thrusts that sole in the lower crust without disturbing the Moho.

More regional estimates of Rocky Mountain crustal thicknesses [see *Keller et al.*, this volume] have been offered by regional seismic refraction lines [*Prodehl and Lipman*, 1989; *Snelson et al.*, 1998] as well as teleseismic receiver function analyses [Sheehan et al., 1995; *Das and Nolet*, 1998; *Dueker et al.*, 2001] and surface wave studies [*Li et al.*, 2002] of seismic waves from distant earthquakes. The 1995 Deep Probe seismic refraction line stretched 2400 km from the plains of Alberta to the Colorado Plateau in western New Mexico [*Snelson et al.*, 1998]. It crossed the Rocky Mountains in Wyoming, traversing the Laramide Beartooth, Owl Creek, Wind River, and Uinta arches. Bouguer gravity along the transect showed positive anomalies across each arch, which were modeled by Snelson et al. [1998] without calling upon deflections of the Moho under the arches. Despite the fact that the Laramide uplifts and basins traversed by the seismic line have more than 9 km of structural relief on the top of the Precambrian basement, no analogous topography was interpreted on the Moho. They attributed a zone of 50-km-thick crust in NW Colorado to Proterozoic tectonism related to the Cheyenne Belt, the suture between Archean and Proterozoic crustal provinces. Similarly, no deviations in the Moho that correlate with Laramide arches were seen in the CD–ROM seismic refraction line [*Karlstrom et al.*, 2002], which extended from Rawlins in southern Wyoming to central New Mexico, where it crossed the Sangre de Cristo arch.

Published maps showing Rocky Mountain crustal thickness [*Li et al.*, 2002; *Keller et al.*, this volume] show a general increase in crustal thickness as the Laramide front is approached from the east. A fundamental problem is whether this thick Rocky Mountain crust was generated during the Laramide orogeny. Crustal thicknesses exceeding 52 kilometers are unusual for crust at sea level, which is where the Rockies were in the late Cretaceous. On the other hand, there is a lack of correspondence between local crustal thicknesses and individual Laramide structures. In fact, several areas of thickest crust are located in areas where there was minimal Laramide surface deformation (e.g., east of the Laramide front in the Denver Basin and eastern-most New Mexico). Within the Rockies, the most dramatic changes in crustal thickness occur across the E–W- to NE–SW-trending Cheyenne Belt, which marks the suture between Archean and Proterozoic crustal provinces near the Colorado–Wyoming state line. Recent detailed profiles across the Cheyenne Belt [*Dueker et al.*, 2001; *Crosswhite and Humphreys*, 2003] show that the belt not only corresponds to the change in crustal thickness between the dissimilar crustal provinces but is also marked by a welt of extra thickness paralleling the Cheyenne Belt and nearly orthogonal to Laramide arches.

The Cheyenne Belt welt, colorfully described as a mountainless root by Crosswhite and Humphreys [2003], suggests that Moho topography generated in the Proterozoic is still evident today. They suggested that the crustal welt, which they traced down to 62 km depth, was generated by dense eclogite emplaced in the mid-Proterozoic because it is not marked by either an isostatic gravity anomaly or a mountain belt. A consequence of this interpretation is that Laramide and post-Laramide processes did not restructure the Moho along the Cheyenne Belt in southern Wyoming. Elsewhere in the Rockies, post–Laramide modifications of crustal thickness appear to be limited to the Rio Grande Rift [*Keller and Baldridge*, 1999] and Tertiary volcanic centers like the San Juan Mountains [*Isaacson and Smithson*, 1976; *Li et al.*, 2002]. The lack of documentation of Moho topography created during the Laramide orogeny suggests that Laramide deformation had relatively little effect on the Moho's geometry even though significant upper crustal thickening is required by Laramide arch and basin geometries.

RESTORABLE CROSS-SECTIONAL MODELS FOR LARAMIDE DEFORMATION

One of the difficulties with comparing prior models of Laramide deformation is the diversity of ways used to represent the models. These range from conceptual maps to summaries of 4D (3D space and time) computer models. Since restorability is the primary criterion used in testing models for this compilation, I have attempted to summarize the models in conceptual, slip-perpendicular cross-sections (Figure 2) because this format provides the easiest way to evaluate the restorability and consequences of proposed deformation. It must be noted, however, that cross-section restoration cannot evaluate strike–slip models [e.g., *Karlstrom and Daniel*, 1995; *Cather*, 1999a] involving out-of-plane motions. An additional cautionary note is that the cross-sections in Figure 2 are not directly from the original author(s) of the models and should not be considered as being fully representative of their work. They represent my best rendition of the authors' concepts using a uniform structural geometry in the upper crust based on geometries from the Wind River and adjacent arches. These conceptual cross-sections show diverse structural geometries in the lower crust and mantle where I have tried to follow the dictates of models rather than specific information on Moho geometry.

Vertical tectonic models invoking regionally-extensive normal or vertical faults predict layer extension in the upper crust, which contradicts considerable evidence for layer shortening from seismic reflection profiles [e.g., *Smithson et al.*, 1979; *Stone*, 1993], well penetrations of major thrust faults [e.g., *Gries*, 1983], fold balancing [e.g., *Stone*, 1984; *Erslev*, 1986, 1993] and minor fault data (summarized in Figure 1). Vertical tectonic models with upthrusts, which are concave-downward faults with lower-angle thrusts at the surface and near vertical faults in the lower crust, cannot generate Laramide arch geometries by themselves. The horizontal shortening indicated by their shallow thrust geometries can't be matched by slip on near-vertical faults at depth because vertical faults do not generate horizontal shortening. In addition, slip on faults with concave-downward curvature would tilt the entire hanging wall toward the surface trace of the master fault. While this occurs immediately adjacent to the surface trace of many master faults, the back limbs of Laramide arches typically dip away from the surface trace of the master faults. This indicates concave-upward, listric faults that become shallower, not steeper at depth [*Erslev*, 1986], consistent with the Wind River seismic interpretations [*Smithson et al.*, 1979; *Sharry et al.*, 1986] and seismic interpretations from the analogous Sierras Pampeanas of Argentina [*Snyder et al.*, 1990].

If one extends the span of vertical tectonic models to include kinematic models where vertical displacements are roughly equivalent to horizontal displacements, then vertical tectonic models include reverse faulting models. Reverse faults cutting the entire lithosphere (Figure 2a) were proposed by Scheeval [1983] and McQueen and Beaumont [1989]. They explained Laramide structures of the Rockies as tilted block structures where the crust, possibly accompanied by the lithospheric mantle, tilted between basin-bounding thrust and reverse faults. This imbrication of lithospheric blocks must be accompanied by penetrative shear in order to cause uniform rotations between planar faults. The combination of domino-like tilting and penetrative shear would result in overall crustal thickening that could be used to explain the thick Rocky Mountain crust. McQueen and Beaumont [1989] calculated that their model is mechanically viable, requiring only 50 Mpa to produce a Wind-River-scale offset because erosion of the up-thrown tip of the blocks linked with sedimentation in the adjacent foreland basin can reduce the stress required for fault slip and block tilting. McQueen and Beaumont [1989] applied standard engineering theory [*Hetenyi*, 1946] for flexural beams to show that structures in Wyoming lie somewhere between short blocks, where rigid rotation dominates, and long blocks, where flexure is dominant and block length is limited to the flexural wavelength. Their initial elastic and elastic-plastic finite element models predict planar faults cutting the entire crust. In their discussion of the models, however, they noted that their models' prediction of crustal keels and Moho offsets are contradicted by the gravity signatures over Laramide arches, which suggest neither. McQueen and Beaumont [1989] concluded that Laramide faults may actually sole into the lower crust yet did not extend their mechanical analyses to test this geometry.

Bird [1984, 1988] directly linked Laramide deformation to shallow subduction of the Farallon and Kula plates indicated by the coincident magmatic gap west of the U.S. Rocky Mountains during the Laramide orogeny [*Coney and Reynolds*, 1977; *Dickinson and Snyder*, 1978; *English et al.*, 2003]. In a series of finite element models, Bird [1988] explored the consequences of basal tractions exerted on a rheologically layered continental lithosphere during low-angle subduction (Figure 2b). These computer models incorporated multi-layer rheologies for the lithosphere, with strong upper crust and upper mantle lithosphere and weaker lower crust and mantle asthenosphere. Some of the advantages of these models include separating the Laramide mechanical impetus from that of the Cordilleran thrust belt, which can explain the differences in their fault and fold trends, and their linkage with shallow subduction, which can explain the temporal coincidence between Laramide deformation and eastward migration of Cordilleran magmatism.

One difficulty with Bird's models is that the inclusion of a weak crustal layer in the models predicts the removal of continental mantle lithosphere from underneath the Rockies. The presence of intact North American lithosphere extending to considerable depth is clearly indicated by geochemical tracers in mantle-derived magmas of post-Laramide age [*Livaccari and Perry*, 1993] as well as deep seismic data indicating that North American Precambrian lithosphere extends to depths of 200 km over much of the southern Rockies [*Schutt and Humphreys*, 2001; *Dueker et al.*, 2001; *Humphreys et al.*, 2003]. In addition, Bird's models predict that shear in the lower crust should result in lower crustal thickening east of the Laramide arches. Crustal thickness estimates [see *Keller et al.*, this volume], however, show no evidence of a uniform welt of crustal thickening east of the Rockies. As a result, these computer models can be viewed as effective falsifications of the subcrustal shear hypothesis of Bird [1988]. Alternatively, however, it may just mean that additional processes are occurring that do not result in full delamination of the mantle under the Rockies [e.g., *Jackson*, 2002].

Another problem with subcrustal shear below the Rockies as a mechanism for Laramide deformation comes from focal mechanisms in the area of analogous low-angle subduction in the southern Andes. East of the Pampian Ranges, an active Laramide analog [*Jordan and Allmendinger*, 1986], Pardo et al. [2002] showed that focal mechanisms indicate that thrust earthquakes occur only down to 60 km depth. Deeper focal mechanisms, including those under the Pampian Ranges, indicate dip-parallel extension of the subducting slab. This suggests that thrust coupling between the two lithospheres occurs only to depths of 60 km, far to the west of the Pampian Ranges, and by analogy, far to the west of the Rocky Mountains.

In contrast to Bird's basal traction model, numerous authors have proposed that Laramide structures were generated during end loading of North American lithosphere during Cordilleran convergence. End loading of the North American lithosphere is appealing because it provides a single mechanism for Cordilleran thrust belt and Laramide deformation, and thus explains well-documented [see *Schmidt and Perry*, 1988] overlaps in deformation timing between these thrust belts. It also explains why similar basement-involved orogens are associated with continental collisions for which there is no evidence for shallow subduction [*Rodgers*, 1987]. In this way, it allows multiple mechanisms for basement-involved shortening in foreland regions, including continental collisions [*Burchfiel et al.*, 1989], microplate collisions [*Maxson and Tikoff*, 1996], and oceanic plateau collisions [*Murphy et al.*, 1999; *Saleeby*, 2003] as well as increased lithospheric coupling between converging plates during low-angle subduction [*Dumitru et al.*, 1991].

Egan and Urquart [1994] applied the conceptual framework of multi-layer continental rheologies to pure-shear foreland deformation (Figure 2d). They predicted that a weak lower crustal layer sandwiched between stronger upper crust and upper mantle layers would decouple deformation, with thrust faulting in the upper crust and ductile thickening in the lower crust and mantle lithosphere. Rather than connect detachments under the ranges into a master detachment, they hypothesized that Laramide fault slip and fold shortening in an upper crustal arch were locally compensated by ductile "squashing," forming a thickened crustal and lithospheric root in the lower crust and mantle lithosphere under each arch. This model of mirrored upper and lower crustal deformation is effectively falsified by earlier-cited gravity and seismic evidence against crustal roots under Laramide arches.

Kulik and Schmidt [1988] proposed a potentially more viable hypothesis where Laramide ductile deformation was distributed throughout the lower crust and mantle lithosphere under the Rocky Mountains, resulting in generalized crustal thickening and local detachment without discrete crustal roots. Unfortunately, the mechanism for such uniform thickening seems difficult considering the apparent present-day survival of the Proterozoic roots to the Cheyenne belt [*Crosswhite and Humphreys*, 2003]. In addition, it is not clear whether a ductile lower lithosphere contradicts geological and geophysical evidence for a strong North American mantle lithosphere [*Sales*, 1968; *Hall and Chase*, 1989; *Lowry and Smith*, 1995].

Humphreys et al. [2003] suggested that widespread ductile shortening and thickening under the Rockies may have been catalyzed by dewatering of the subducting Farallon plate. Earlier, Humphreys [1995] showed that Laramide hydration of the mantle lithosphere combined with post-Laramide asthenospheric upwelling can explain the post-Laramide ignimbrite flare-up in the Basin and Range, a volcanic event that did not occur throughout the Rockies. The question remains, however, if a slab under the 200 km thick North American lithosphere could have retained water such a long distance from the trench and then released enough water to cause widespread ductile thickening under the Rockies [*Gutscher et al.*, 2000]. In addition, it is difficult to understand why the southern Rocky Mountains experienced lithosphere-scale, hydration-catalyzed shortening without the Colorado Plateau to the west undergoing equivalent shortening.

McQuarrie and Chase [2000] proposed that the unusually thick crust and high Cenozoic elevations of the Colorado Plateau and Rocky Mountain region could be the result of the lateral injection of ductile crust. They proposed that ductile lower crust from the Cordilleran thrust belt might have been injected as a tabular, 5 to 10 km thick sill into the lower crust (Figure 2c). The mass balance of the hypothesis, however, appears marginal because thick crust and high elevations extend from the thrust belt all the way into the high plains, considerably east of major Laramide deformation. In addition, the fluidity of the crustal injector would have to be extremely high to maintain a nearly tabular channel of injection over such a distance. Analogous lower crustal movement proposed for the eastern Tibetan Plateau [*Royden and Burchfiel*, 1997] generated the very sharp topographic gradient from near sea-level elevations to the Tibetan Plateau. Analogous topographic gradients are not recorded in the Rocky Mountains. In addition, evidence for a strong Laramide lithosphere seems incompatible with this hypothesis, as are indicators of a strong lithosphere for the analogous Sierras Pampians of Argentina [*Jordan and Allmendinger*, 1986] like the region's crustal earthquakes at depths in excess of 30 km [*Smalley et al.*, 1993].

Tikoff and Maxson [2001] proposed that Rocky Mountain basement-involved deformation occurred due to buckling of the entire North American lithosphere (Figure 2e). Their model differs from earlier buckling models [*Fletcher*, 1984; *Kulik and Schmidt*, 1988] that proposed buckling limited to the upper crust. Tikoff and Maxson [2001] listed the following evidence for buckling of the continental lithosphere during the Laramide:

1) the regular spacing of Laramide Rocky Mountain and mid-continental arches,
2) their compatibility with analog experiments modeling continental lithosphere over asthenosphere using a 4 layer rheological package, and
3) the similarity of wavelengths between Laramide arches with wavelengths of buckled oceanic lithosphere in the Indian Plate.

Unfortunately, this innovative lithospheric buckling model predicts geometries not seen in the Rockies. Tikoff and Maxson [2001] used the experimental models by Martinod and Davy [1994] and Burg et al. [1994] as Laramide analogs but they show maximum deformation in the basins, where thrust faulting and folding accommodate synclinal crowding. In contrast, regional sections through the Rockies [e.g., *Stone*, 1993] show minimal deformation in the basins, with major thrust faults located on arch margins. This suggests that these analog models are, at the minimum, poor mechanical and geometric matches for Laramide deformation.

For an east–west profile at 40ºN latitude, Tikoff and Maxson [2001] cite an average spacing between Laramide arches of 190 km, which is nearly identical to the wavelength of folded, 60 Ma oceanic lithosphere in the Indian Ocean [*McAdoo and Sandwell*, 1985; *Zuber*, 1987]. In reality, Laramide arch spacings on this profile are trimodal, with spacings of 140 km on the Colorado Plateau, 80 and 60 km for the Rocky Mountains, and 300 and 230 km for midcontinental arches.

Determining the wavelength of Rocky Mountain arches is admittedly quite difficult. If the arches did develop by buckling, their anastomosing geometries suggest independent yet simultaneous initiation of buckling in adjacent domains, with folds growing laterally until they interfered with the folds of adjacent domains. If this is the case, a maximum wavelength can be established by measuring the maximum width between arch culminations that bound the elliptical foreland basins. This gives the following widths:

Bighorn Basin: 160 km
Wind River Basin: 110 km
Laramie Basin: 100 km
Hanna Basin: 70 km
North Park Basin: 80 km
South Park Basin: 100 km
Powder River Basin: 290 km
San Juan Basin: 120 km

The heterogeneous rheology of continental lithosphere makes it difficult to predict the expected wavelength of lithospheric buckling. Turcotte and Schubert [2002] show that elastic rheologies predict very large wavelengths but also require such large stresses (6.4 Gpa for 50 km-thick elastic lithosphere) that purely elastic buckling should not occur. When combined with the effects of sedimentation in synclines, models incorporating plastic rheologies, whose non-recoverable deformation is more realistic than elastic deformation, predict more reasonable stresses and buckling wavelengths [*McAdoo and Sandwell*, 1985; *Zuber*, 1987; *Martinod and Davy*, 1992; *Martinod and Molnar*, 1995]. These wavelengths match the observed wavelengths for the buckled lithosphere of the Indian Ocean, where the brittle-ductile interface is predicted to be 40 km deep.

Martinod and Davy [1992] suggested that buckling wavelengths typically are about 4 times the lithospheric thickness during compression of both coupled (e.g., oceanic lithosphere) and decoupled lithosphere (i.e., continental lithosphere with a ductile lower crust). Nikishin et al. [1993] suggested that surface undulations with an average wavelength of 360 km in central Asia might be due to buckling of the continental lithosphere. In their dynamic modeling of the Ferghana Valley of the western Himalayan orogen, Burov and Molnar [1998] concluded that biharmonic folding of crust (50–70 km wavelength) and lithosphere (200–250 km) can explain the uncompensated loads revealed by gravity analyses. In this case, the relatively short wavelengths for the continental crust were explained by Jurassic resetting of the lithospheric thermal structure in the Ferghana region.

If one compares the young (60 Ma), thin lithosphere (40 km) of the buckled Indian plate to the old (Precambrian), thick [>200 km; *Schutt and Humphreys*, 2001; *Dueker et al.*, 2001] Rocky Mountain lithosphere, the wavelength of buckling should be much greater in the Rocky Mountains than in the Indian plate. The maximum wavelengths between Rocky Mountain arches are, with one exception, smaller than wavelengths for the buckled Indian oceanic lithosphere. Phanerozoic thinning of the Rocky Mountain lithosphere might give smaller wavelengths, but this does not seem to be an option, especially in Wyoming and Montana. The 300+ km wavelengths between structures in the North American mid-continent are more compatible with the concept of continental lithospheric buckling than the wavelengths between Laramide arches.

In addition, current gravity modeling and seismic data for the Moho have not documented any major up-bowing of the Moho under the arches. As discussed earlier, gravity models through the Wind River and Granite Mountain arches were found to be incompatible with major uplift of the Moho under the arches [*Smithson and Hurich*, 1982; *Hall and Chase*, 1989]. An appeal to post-Laramide collapse [*Tikoff and Maxson*, 2001] can be made for these arches since they are cut by younger normal faults consistent with collapse [*Love*, 1970; *Steidtmann and Middleton*, 1991]. At this point in time, other arches with similar gravity anomalies and no evidence for post-Laramide collapse should be studied to further test this hypothesis.

Numerous authors have suggested that crustal detachment during top-to-the-east or -northeast shear [Figures 2f, Figure 3; *Lowell*, 1983; *Oldow et al.*, 1989; *Erslev*, 1993] generated both Laramide and Cordilleran thrust belt structures. From a regional point of view, this concept is appealing because it explains Laramide and Cordilleran thrust belt (or Sevier) deformation using a common tectonic mechanism. In addition, detachment of the upper crust can explain the incongruity between the fold- and fault-shortened upper crust and the apparently much less distorted Moho. Fletcher [1984] and Kulik and Schmidt [1988] hypothesized that the regular spacing of foreland arches and basins can be explained by initial buckling of the upper crust above a lower crustal detachment zone followed by fault break-through. Qualitatively, the arch wavelengths represented by the basin widths listed above seem compatible with crustal buckling and detachment, as are initial depth-to-detachment calculations [*Ehrlich*, 1999].

The continuity between Laramide foreland arches, which form a connected web of anastomosing basement highs [Erslev, 1993], can be explained by linking them with an underlying detachment zone. Connection via a common detachment zone also explains lateral changes in arch amplitude where, as arches decrease in amplitude (e.g., north plunge of Bighorn arch), other arches increase in amplitude (e.g., south plunge of Beartooth arch). In addition, the anastomosing pattern and laterally variable symmetry of Laramide arches are all characteristics of detachment folds [*Dahlstrom*, 1990; *Erslev*, 1993; *Epard and Groshong*, 1995]. Smaller arches [e.g., Laramie arch, *Brewer et al.*, 1982] commonly have less well-developed thrust faulting on their margins relative to larger arches [e.g., Wind River arch, *Smithson et al.*, 1979], which can be highly asymmetrical with major thrust faults cutting one flank. This suggests that folding of basement-involved foreland arches may have preceded emergent thrusting through arch flanks. This sequence of deformation is directly analogous to the sequence of detachment folding followed by fault-propagation folding seen in many analog experiments of fault-related folding [e.g., *Dixon and Liu*, 1991].

The geometry of larger, asymmetrical arches like the Wind Rivers can be modeled as rotational fault-bend folds [Figure 3, *Erslev*, 1986; *Stanton and Erslev*, 2001] where movement on a listric master thrust rotates a basement chip upward. This geometry can explain many second-order, basement-involved fault-propagation folds that are systematically arrayed with respect to the first-order arches and basins. In these structures, faulting dominates in the stronger, relatively homoge-

Basement-Involved, Second-Order Anticlinal Structures

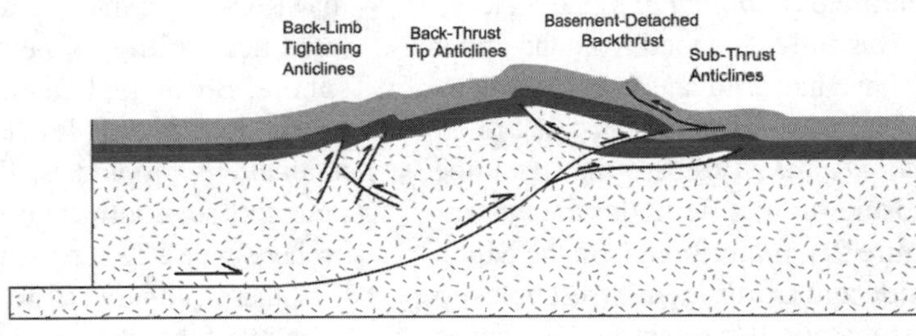

Restored Section

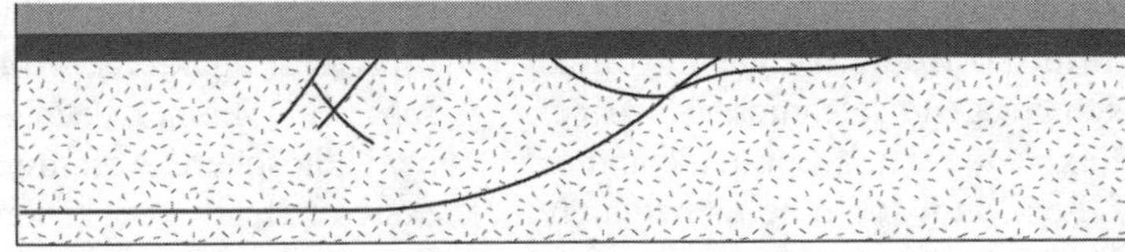

Figure 3. A geometric model for the development of first-order Laramide arches as detached fault-related folds and their associated second-order structures.

neous crystalline basement and folding dominates in the weaker, overlying Phanerozoic sedimentary strata [*Erslev*, 1991]. These second-order folds (Figure 3) include:

1) back-thrust tip structures: antithetic (opposite tectonic transport direction relative to the master fault), thin- and thick-skinned reverse and thrust faults with associated anticlines in the hanging wall of master thrust faults (e.g., Rattlesnake Mountain anticline south of the Beartooth arch),
2) sub-thrust splay structures: synthetic, low-angle fault-propagation and detachment structures which develop below master thrusts (e.g., Pinedale anticline southwest of the Wind River arch), and
3) back-limb tightening structures: variably-directed fault-propagation folds over higher angle thrust and reverse faults which develop on the backlimb of asymmetrical foreland arches (e.g., anticlines on the NE flank of Wind River arch).

One problem with crustal detachment models is their lithosphere-scale restoration, namely how do you generate shortening in the lower lithosphere equivalent to shortening in the upper lithosphere. One option is to detach at the base of the crust and extend the detachment to the west [*Oldow et al.*, 1989; *Erslev*, 1993] (Figure 3f) or east [*Paylor and Yin*, 1991], outside the limits of the section. Of these two options, extending the detachment to the west and invoking east-directed shear is more conceivable because the root of a major detachment system can be more easily concealed in the recently restructured Basin and Range than in the relatively undisturbed cratonic interior. This east-directed detachment faulting was probably driven by plate coupling to the west during Laramide "refrigeration" of the North American lithosphere by the down-going Farallon plate [*Dumitru et al.*, 1991].

If the mantle lithosphere was fully detached from the upper crust in the Rockies, it may have had equivalent shortening by unknown means under the main Cordilleran thrust belt (e.g., orogenic float model of *Oldow et al.*, 1989]. If the Laramide detachments root into lithospheric mantle under the eastern Cordilleran thrust belt (Figure 2f), major breaks in the Moho accounting for both Laramide and Cordilleran thrust belt slip would be predicted. Unfortunately, the Basin and Range in this area is so extensively overprinted by plutonism and extension that it is unlikely that any Laramide breaks in the Moho are preserved. In Idaho, however, large-scale (~150 km) thrust displacements have been inferred from changes in source regions for Late Cretaceous to Neogene volcanics [*Leeman et al.*, 1992]. And it might be expected that a Laramide and/or Cordilleran thrust break of the Moho would be immediately reactivated as normal faults once lateral compression was relieved in the mid Cenozoic. Reactivation of Laramide and Sevier crustal fault ramps is common [e.g, *Constenius*, 1996; *Kellogg*, 1999], and one could expect the same of faults cutting down through the Moho. It is tempting to speculate that the eastern margin of the main Basin and Range province records the position where the Laramide basal detachment(s) cut the Moho.

DISCUSSION AND CONCLUSIONS

One advantage of the Tikoff and Maxson [2001] model over detachment models is that their model ascribed the difference between Sevier and Laramide deformation to whether the mantle and crust were detached (Sevier) or coupled (Laramide). In criticizing detachment models, Tikoff and Maxson [2001] cited Cloetingh [1988] who stated that "transmission of stresses to the continental interior requires a rigid lithosphere and the absence of major detachments". As discussed above, however, modeling of lithospheric buckling using plastic instead of elastic deformation [*McAdoo and Sandwell*, 1985; *Zuber*, 1987; *Martinod and Davy*, 1992; *Martinod and Molnar*, 1995] has significantly reduced estimates of the stresses required for intracratonic deformation and detachment.

Mazzotti and Hyndman [2002] reported a potential modern analog of Laramide detachment in the northern Canadian Cordillera. In this area, GPS measurements and seismicity indicate that collisional shortening is being transferred by crustal detachment across the Cordillera to a foreland belt on its northern margin. This interpretation suggests that ductile lower crust, which in the case of the Laramide possibly resulted from originally thick Rocky Mountain crust and/or Laramide subduction dehydration [*Humphreys et al.*, 2003], can allow lower crustal delamination and strain transfer across large distances.

In fact, modeling by Burov and Diament [1995] suggests that Rocky Mountain crustal thicknesses may have necessitated decoupling of the upper crust from the mantle lithosphere. Burov and Diament [1995] compiled the effective elastic thicknesses (Te) of oceanic and continental lithosphere in an attempt to address the meaning of the diverse Te measurements for continental lithosphere. Unlike the oceanic lithosphere, where Te appears to correspond to the depth of the 600° C isotherm, continental Te values are bimodal. By modeling the strength of the lithosphere, they found that continental lithospheres with ages greater than 150 Ma have strengths dominated by the mantle. In these older lithospheres, they found that the degree of coupling in the lithosphere is determined by crustal thickness, with a critical thickness of 35–40 km separating thinner, coupled lithosphere from thicker, decoupled lithosphere. According to their analysis, crust with thicknesses currently seen in the Rocky Mountains and the Colorado Plateau should decouple and cause detachment tectonics.

This returns us to the genesis of the thick Rocky Mountain and Colorado Plateau crust. Recent analyses [*Bird*, 1984;

McQuarrie and Chase, 2000] have focused on how the Laramide and Cordilleran orogens might have thickened this crust because there is a first order match between the extent of thicker Rocky Mountain crust and the limits of Laramide deformation. Unlike the central Andean Plateau [*McQuarrie*, 2002], however, the thick crust of the Colorado Plateau has very little evidence of upper-crustal Phanerozoic deformation. Perhaps the correlation should be inverted – perhaps the extent of Laramide detachment was at least partially limited by the extent of already thick crust. The northern extent of Laramide deformation in Montana might have been predetermined by the thinning of the Rocky Mountain crust in this region [*Snelson et al.*, 1998] during the Belt orogeny, perhaps in combination with other changes in crustal strength (e.g., mafic intrusions associated with the Belt Basin). For the Rocky Mountain region, an anomalously thick Phanerozoic crust with a weak basal layer might help explain why the Pennsylvanian Ancestral Rocky Mountain structures of New Mexico, Utah and Colorado are so similar in geometry to Laramide structures even though they lack evidence (e.g., arc volcanics) of coeval subduction at depth. Perhaps anomalously thick crust in southwestern North American preconditioned the area for basement-involved deformation in both the Ancestral Rocky Mountain and Laramide orogenies by providing a weaker than normal lower crust. This crust then reacted in essentially similar ways to Laramide low-angle subduction [*Seleeby*, 2003] and Pennsylvanian continental convergence [*Dickinson and Lawton*, 2003].

In conclusion, the Rocky Mountains of the conterminous U.S.A. are an important locality for cratonic deformation, with classic examples of basement-involved thrust tectonics, plateau uplift, and later extension. Over the last quarter century, most of the major geometric and kinematic elements of upper-crustal, basement-involved deformation during the Laramide orogeny of the Rocky Mountains have been resolved by abundant industry and academic data. The overall structural geometry of first-order, anastomosing Laramide arches with second-order subsidiary structures can be adequately explained within the context of thick-skinned detachment folding. Prior hypotheses indicating a dominance of vertical motions during Laramide deformation have been falsified by extensive field observations of minor structures, seismic profiles, well data and structural modeling showing undisputable evidence for pervasive horizontal shortening.

The Laramide geometries and kinematics at deeper crustal and lithospheric levels are more poorly defined, but new data are allowing insights as to how Laramide upper crustal shortening was connected to Cordilleran plate processes to the west. Initial models of Laramide deformation by traction between North American lithosphere and the subducting Farallon plate predict too much thinning of the North American lithospheric mantle, which is over 200 km thick over much of the province. Preliminary evidence for a relatively planar Moho appears to contradict tectonic hypotheses invoking high-angle faults cutting the entire crust [*Scheeval*, 1983 and one of the models of *McQueen and Beaumont*, 1989), lower crustal thickening under individual arches [*Egan and Urquhart*, 1993], and lithospheric buckling [*Tikoff and Maxson*, 2001]. The injection of ductile crust from the thrust belt as a means to thicken the crust of the western U.S. [*McQuarrie and Chase*, 2000] is problematic due to the hypothesis' volumetric limitations, the lack of evidence for an abrupt topographic gradient analogous to that seen in the Himalayan modern analog, and the need for improbably fluid crustal rheologies relative to currently active Andean analogs with deep crustal seismicity. Instead, the combination of folded upper crust above a much more planar Moho suggests that Laramide arches formed by lower crustal detachment and buckle folding of the upper and middle crust during horizontal shortening.

The key question is how horizontal shortening, which is so apparent in the upper crust, was accomplished in the lowermost crust and lithospheric mantle. Widespread penetrative shortening of the lower lithosphere catalyzed by subduction hydration [*Humphreys et al.*, 2001] may be possible, although details of the kinematic linkage between the upper crust and the rest of the lithosphere have yet to be explored. This hypothesis could explain Laramide shortening in the southern Rockies without necessitating detachment of the relatively cool [*Helmstaedt and Schulze*, 1991] Colorado Plateau. Alternatively, originally thick Precambrian crust in the western and southwestern North American craton may have provided a weak lower crustal layer allowing detachment tectonics during both Laramide and Pennsylvanian orogenies. Rooting of the Laramide detachment to the west, in the area now occupied by the Basin and Range, seems consistent with the similar slip trends indicated by the main Cordilleran and Laramide thrust belts as well as the west-to-east progression of Laramide thrusting in Wyoming [*Brown*, 1988]. In addition, the overlap of thin- and thick-skinned thrusting in Wyoming suggests that the main Laramide and Cordilleran deformations were characterized by ENE-directed thrusting with minimal deformation in the lower plate. The discordance between the NE–SW trending Laramide magmatism associated with the Colorado Mineral belt may reflect fundamental differences in the deformation of the upper and lower parts of the Cordilleran lithosphere during Laramide deformation.

In reality, Laramide structures probably result from a combination of the 2D end-member models examined in this paper, as modified by additional 3D complications and constraints. A better understanding of basement-involved deformation will require more kinematic and geometric data from all levels in the North American lithosphere. Advances in structural balancing and mechanical modeling should allow full 4D (3D + time) analyses of the Laramide arches. More comprehensive gravity and deep seismic determinations of deeper lithospheric structure should finally determine how basement-involved deformation in the upper crust can be connected to plate tectonic processes.

Acknowledgements. This paper has benefited from insights from many people, including Steve Cather, Chuck Chapin, Robbie Gries, Ray Fletcher, Peter Hennings, Gene Humphreys, Karl Karlstrom, Randy Keller, Chris Schmidt, Art Snoke, Heather Stanton, Don Stone, and Chris Zahm. Detailed comments on this manuscript by Gene Humphreys, Steve Cather, and Chris Andronicos are gratefully acknowledged. This work is part of the CD–ROM project, funded by the Continental Dynamics Program of the National Science Foundation.

REFERENCES

Bird, P., Laramide crustal thickening event in the Rocky Mountain foreland and Great Plains, *Tectonics*, 3, 741–758, 1984.

Bird, P., Formation of the Rocky Mountains, western United States: A continuum computer model, *Science*, 239, 1501–1507, 1988.

Bird, P., Kinematic history of the Laramide orogeny in latitudes 35°–49° N, western United States, *Tectonics*, 17, 780–801, 1998.

Blackstone, D.L., Rocky Mountain foreland structure exemplified by the Owl Creek Mountains, Bridger Range and Casper Arch, central Wyoming, *Wyoming Geological Association Guidebook*, 41, 151–166, 1990.

Brewer, J.A., Allmendinger, R.W., Brown, L.D., Oliver, J.E., and Kaufman, S., COCORP profiling across the Rocky Mountain front in southern Wyoming, Part 1, Laramide structure, *Geological Society of America Bulletin*, 93, 1242–1252, 1982.

Brown, W.G., Structural style of Laramide basement-cored uplifts and associated folds, in Snoke, A.W., Steidtmann, J.R., and Roberts, S.M., eds., *Geology of Wyoming,* Geological Survey of Wyoming Memoir, 5, 312–371, 1993.

Brown, W.G., Deformation style of Laramide uplifts in the Wyoming foreland, in Schmidt, C.J. & Perry, W.J., Jr., (eds), *Interaction of the Rocky Mountain foreland and the Cordilleran thrust belt,* Geological Society of America Memoir 171, 53–64, 1988.

Burchfiel, B.C., Quidong, D., Molnar, P., Royden, L., Yipeng, W., Peizhen, Z., and Weiqi, Z., Intracrustal detachment within zones of continental deformation, *Geology*, 17, 448–452, 1989.

Burg, J.-P., Davy, P., and Martinod, J., Shortening of analogue models of the continental lithosphere: New hypothesis for the formation of the Tibetan Plateau, *Tectonics*, 13, 475–483, 1994.

Burov, E.B., and Diament, M., The effective elastic thickness (Te) of continental lithosphere: What does it really mean?, *Journal of Geophysical Research*, 100, 3905–3927, 1995.

Burov, E.B., and Molnar, P., Gravity anomalies over the Ferghana Valley (central Asia) and intracratonic deformation, *Journal of Geophysical Research*, 103, 18137–18152, 1998.

Cather, S.M., Implications of Jurassic and Cretaceous piercing lines for Laramide oblique slip faulting in New Mexico and the rotation of the Colorado Plateau, *Geological Society of America Bulletin*, 111, 849–868, 1999a.

Cather, S.M., Laramide faults in the southern Rocky Mountains—A role for strain partitioning and low frictional strength strike-slip faults?, *Geological Society of America Abstracts with Programs*, 31, A-186, 1999b.

Chapin, C.E., and Cather, S.M., Eocene tectonism and sedimentation in the Colorado Plateau–Rocky Mountain area, *Arizona Geological Digest*, 14, 175–198, 1981.

Chapin, C.E., An overview of Laramide wrench faulting in the southern Rocky Mountains with emphasis on petroleum exploration, in Lowell, J.D., ed., *Rocky Mountain foreland basins and uplifts*, Rocky Mountain Association of Geologists, 169–179, 1983.

Cloetingh, S., Intraplate stresses: A tectonic cause for third-order cycles in apparent sea level?, *Society of Economic Paleontologists and Mineralogists Special Publication*, 42, 19–29, 1988.

Coney, P.J. and S.J. Reynolds, Flattening of the Farallon slab, *Nature*, 270, 403–406, 1977.

Constenius, K., Late Paleogene extensional collapse of the Cordilleran foreland fold and thrust belt, *Geological Society of America Bulletin*, 108, 20–39, 1996.

Crosswhite, J.A., and Humphreys, E.D., Imaging the mountainless root of the 1.8 Ga Cheyenne Belt suture and clues as to its tectonic stability, *Geology*, 31, 669–672, 2003.

Dahlstrom, C.D.A., Geometric constraints derived from the law of conservation of volume and applied to evolutionary models for detachment folding, *American Association of Petroleum Geologists Bulletin*, 74, 336–344, 1990.

Das, T. and Nolet, G., Crustal thickness map of the western United States by partitioned waveform inversion, *Journal of Geophysical Research*, 103, 30,021–30,038, 1998.

Dickinson, W.R. and Lawton, T.F., Sequential intercontinental suturing as the ultimate control for Pennsylvanian Ancestral Rocky Mountains deformation, *Geology*, 31, 609–612, 2003.

Dickinson, W.R., and Snyder,W.S., Plate tectonics of the Laramide orogeny, in Matthews, V., III, (ed.), *Laramide folding associated with basement block faulting in the western United States*, Geological Society of America Memoir 151, 355–366, 1978.

DiUlio, A.A., Evidence for regional east-northeast shortening in the Wyoming-Idaho-Utah thrust belt, unpublished M.S. thesis, Colorado State University, 158 p., 2001.

Dixon, J.M., and Liu, S., Centrifuge modeling of the propagation of thrust faults, in McClay, K.R., ed., *Thrust tectonics*, Chapman and Hall, London, 53–70, 1992.

Dueker, K., Yuan, H., and Zurek, B., Thick Proterozoic lithosphere of the Rocky Mountain region, *GSA Today*, 11 (12), 4–9, 2001.

Dumitru, T.A., Gans, P.B., Foster, D.A., and Miller, E.L., Refrigeration of the western Cordilleran lithosphere during Laramide shallow-angle subduction, *Geology*, 19, p. 1145–1148, 1991.

Egan, S.S., and J.M. Urquhart, Numerical modeling of lithosphere shortening: application to the Laramide orogenic province, western USA, *Tectonophysics*, 221, p. 385–411, 1993.

Ehrlich, T. K., Fault analysis and regional balancing of Cenozoic deformation in northwest Colorado and south-central Wyoming, unpublished M.S. thesis, Colorado State University, Fort Collins, CO, 116 p, 1999.

English, J.M., Johnson, S.T., and Wang, K., Thermal modeling of the Laramide orogeny: testing the flat-slab subduction hypothesis, *Earth and Planetary Science Letters*, 214, 619–632, 2003.

Epard, J.-L., and Groshong, R.H., Kinematic model of detachment folding including limb rotation, fixed hinges and layer-parallel strain, *Tectonophysics*, 247, 85–103, 1995.

Erslev, E.A., Basement balancing of Rocky Mountain foreland uplifts, *Geology*, 14, 259–262, 1986.

Erslev, E.A., Trishear fault-propagation folding, *Geology*, 19, 617–620, 1991.

Erslev, E.A., Thrusts, backthrusts and detachment of Laramide foreland arches, in Schmidt, C.J., Chase, R., and Erslev, E.A., eds., *Laramide basement deformation in the Rocky Mountain foreland of the western United States,* Geological Society of America Special Paper, 280, 125–146, 1993.

Erslev, E.A., Multi-stage, multi-directional Tertiary shortening and compression in north-central New Mexico, *Geological Society of America Bulletin*, 113, 63–74, 2001.

Erslev, E.A., Holdaway, S.M., O'Meara, S., Jurista, B., and Selvig, B., Laramide minor faulting in the Colorado Front Range, *New Mexico Bureau of Mines and Mineral Resources Special Paper*, in press.

Fletcher, R.C., Instability of lithosphere undergoing shortening: A model for Laramide foreland structures, *Geological Society of America Abstracts with Programs*, 16, 83, 1984.

Gregson, J., and Erslev, E.A., Heterogeneous deformation in the Uinta Mountains, Colorado and Utah, in Hoak, T.E., Klawitter, A.L., and Bloomquist, P.K. (eds.), *Fractured Reservoirs: Characterization and Modeling*, Rocky Mountain Association of Geologists 1997 Guidebook, 137–154, 1997.

Gries, R.R., North-south compression of the Rocky Mountain foreland structures, in Lowell, J.D., and Gries, R.R., eds., *Rocky Mountain foreland basins and uplifts,* Rocky Mountain Association of Geologists, Denver, Colorado, 9–32, 1983.

Gutscher, M.-A., Spakman, W., Bijwaard, H., and Engdahl, E.R., Geodynamics of flat subduction: Seismicity and tomographic constraints from the Andean margin, *Tectonics*, 19, 814–833, 2000.

Hall, M.K., and Chase, C.G., Uplift, unbuckling, and collapse: Flexural history and isostacy of the Wind River Range and Granite Mountains, Wyoming, *Journal of Geophysical Research*, 94, 17,581–17,593, 1989.

Hamilton, W., Laramide crustal shortening, in Perry, W.J., and Schmidt, C.J., eds., *Interaction of the Rocky Mountain foreland and the Cordilleran thrust belt,* Geological Society of America Memoir, 171, 27–39, 1988.

Helmstaedt, H.H., and Schulze, D.J., Early to mid-Tertiary inverted metamorphic gradient under the Colorado Plateau: evidence form eclogite xenoliths in ultramafic microbreccias, Navajo volcanic field, *Journal of Geophysical Research*, 96, 13,225–13,235, 1991.

Hetenyi, M., *Beams on elastic foundation,* University of Michigan Press, Ann Arbor, Michigan, 247 p., 1946.

Humphreys, E.D., Post-Laramide removal of the Farallon slab, western United States, *Geology*, 23, 987–990, 1995.

Humphreys, E., Hessler, E., Dueker, K., Farmer, G.L., Erslev, E., and Atwater, T., How Laramide-age hydration of North American lithosphere by the Farallon slab controlled subsequent activity in the western United States, *International Geology Reviews*, 45, 575–594, 2003.

Hurich, C.A., and Smithson, S.B., Gravity interpretation of the southern Wind River Mountains, Wyoming, *Geophysics*, 47, 1550–1561, 1982.

Isaacson, L. B., and Smithson, S. B., Gravity anomalies and granite emplacement in west-central Colorado, *Geological Society of America Bulletin*, 87, 22–28, 1976.

Jackson, J., Strength of the continental lithosphere: time to abandon the jelly sandwich?, *GSA Today*, 12 (9), p. 4–10, 2002.

Jordan, T.E., and Allmendinger, R.W., The Sierras Pampeanas of Argentina: a modern analogue of Rocky Mountain foreland deformation, *American Journal of Science*, 286, 737–764, 1986.

Karlstrom, K.E., and Daniel, C.G., Restoration of Laramide right-lateral strike slip in northern New Mexico by using Proterozoic piercing points: Tectonic implications from the Proterozoic to the Cenozoic, *Geology*, 21, 135–142, 1993.

Karlstrom, K. E., and Humphreys, E. D., Persistent influence of Proterozoic accretionary boundaries in the tectonic evolution of southwestern North America: Interaction of cratonic grain and mantle modification events, *Rocky Mountain Geology*, 33, 161 – 179, 1998.

Karlstrom, K.E., Bowring, S.A., Chamberlain, K.R., Dueker, K.G., Eshete, T., Erslev, E.A., Farmer, G.L., Heizler, M., Humphreys, E.D., Johnson, R.A., Keller, G.R., Kelley, S.A., Levander, A., Magnani, M.B., Matzel, J.P., McCoy, A.M., Miller, K.C., Morozova, E.A., Pazzaglia, F.J., Prodehl, C., Rumpel, H.-M., Shaw, C.A., Sheehan, A.F., Shoshitaishvili, E., Smithson, S.B., Snelson, C.M., Stevens, L.M., Tyson, A.R., and Williams, M.L., Structure and evolution of the lithosphere beneath the Rocky Mountains: Initial results of from the CD–ROM experiment, *GSA Today*, 13 (3), 4–10, 2002.

Keller, G.R., and Baldridge, W.S., The Rio Grande rift: A geological and geophysical overview, *Rocky Mountain Geology*, 34, 121–130, 1999.

Keller, G.R., et al. – Moho summary - this volume.

Kellogg, K.S., Neogene basins of the northern Rio Grande rift—partitioning and asymmetry inherited from Laramide and older uplifts, *Tectonophysics*, 305, 141–152, 1999.

Koenig, N.B., and Erslev, E.A., Internal and external controls on Phanerozoic Rocky Mountain structures, U.S.A.: Insights from GIS-enhanced tectonic maps, in Raynolds, R. (ed.), *SEPM volume on the Cenozoic Rocks of the Rocky Mountains*, in press.

Kulik, D.M., and Schmidt, C.J., Regions of overlap and styles of interaction of Cordilleran thrust belt and Rocky Mountain foreland, in Schmidt, C.J., and Perry, W.J., Jr., eds., *Interaction of the Rocky Mountain foreland and the Cordilleran thrust belt,* Geological Society of America Memoir, 171, 75–98, 1998.

Larson, E. E., and Drexler, J. W., Early Laramide mafic to intermediate volcanism, Front Range, Colorado, *Colorado School of Mines Quarterly*, 83, 41–52, 1987.

Leeman, W.P., Oldow, J.S., and Hart, W.K., Lithosphere-scale thrusting in the western U.S. Cordillera as constrained by Sr and Nd isotopic transitions in Neogene volcanic rocks, *Geology*, 20, 63–66, 1992.

Li, A., Forsyth, D.W., and Fisher, K.M., Evidence for shallow isostatic compensation of the southern Rocky Mountains from Rayleigh wave tomography, *Geology*, 30, 683–686, 2002.

Livaccari, R.F., Role of crustal thickening and extensional collapse in the tectonic evolution of the Sevier-Laramide orogeny, western United States, *Geology*, 19, 1104–1107, 1991.

Livaccari, R.F. and F.V. Perry, Isotopic evidence for preservation of Cordilleran lithospheric mantle during the Sevier-Laramide orogeny, western United States, *Geology*, 21, 719–722, 1993.

Love, D., Cenozoic geology of the Granite Mountains area, central Wyoming, *U.S. Geological Survey Professional Paper*, 495-C, 154 p., 1970.

Lowell, J.D., Foreland deformation, in Lowell, J.D., ed., *Rocky Mountain foreland basins and uplifts,* Rocky Mountain Association of Geologists, 1–8, 1983.

Lowry, A.R., and Smith, R.B., Strength and rheology of the western U.S. Cordillera, *Journal of Geophysical Research*, 100, 17,947–17,963, 1995.

Lynn, H.B., Quam, S., and Thompson, G., Depth migration and interpretation of the COCORP Wind River, Wyoming, seismic reflection data, *Geology*, 11, 462–469, 1983.

Malahoff, A., and Moberly, R., Jr., Effects of structure on the gravity field of Wyoming, *Geophysics*, 33, 781–804, 1968.

Marshak, S., Karlstrom, K.E., and Timmons, J.M., Inversion of Proterozoic extensional faults: An explanation for the pattern of Laramide and Ancestral Rockies intracratonic deformation, United States, *Geology*, 28, 735–738, 2000.

Martinod, J. and Daviey, P., Periodic instabilities during compression or extension of the lithosphere 1. Deformation modes from an analytical perturbation method, *Journal of Geophysical Research*, 97, 1999–2014, 1992.

Martinod, J. and Davy, P., Periodic instabilities during compression of the lithosphere 2. Analogue experiments, *Journal of Geophysical Research*, 99, 12,057–12,069, 1994.

Martinod, J., and Molnar, P, Lithospheric folding in the Indian Ocean and the rheology of the oceanic plate, *Bull. Soc. Geol. France*, 166, 813–821, 1995.

Maxson, J., and Tikoff, B, Hit-and-run collision model for the Laramide orogeny, western United States, *Geology*, 24, 968–972, 1996.

Mazzotti, S., and Hyndman, R.D., Yakutat collision and strain transfer across the northern Canadian Cordillera, *Geology*, 30, 495–498, 2002.

McAdoo, D.C., and Sandwell, D.T., Folding of oceanic lithosphere, *Journal of Geophysical Research*, 90, 8563–8569, 1985.

McQuarrie, N., The kinematic history of the central Andean fold-thrust belt, Bolivia: Implications for building a high plateau, *Geological Society of America Bulletin*, 114, 950–963, 2002.

McQuarrie, N. and C.G. Chase, Raising the Colorado Plateau, *Geology*, 28, 91–94, 2000.

McQueen, H.W.S., and Beaumont, C., Mechanical models of tilted block basins: in Price, R.A., ed., Origin and evolution of sedimentary basins and their energy and mineral resources, *Geophysical Monograph*, 48, 65–71, 1989.

Molzer, P., and Erslev, E.A., Oblique convergence on east–west Laramide arches, Wind River Basin, Wyoming, *American Association of Petroleum Geologists Bulletin*, 19, 1377–1394, 1995.

Murphy, J.B., Oppliger, G.L., Brimhall, G.H., Jr., and Hynes, A., Mantle plumes and mountains: *American Scientist*, 87, 146–153, 1999.

Mutschler, F. E., Larson, E. E., and Bruce, R., Laramide and younger magmatism in Colorado, *Colorado School of Mines Quarterly*, 82, 1–45, 1987.

Nikinshin, A.M., Cloetingh, S., Lobkovsky, L.I., Burov, E.B., and Lankreijer, A.C., Continental lithospheric folding in central Asia (Part 1): constraints from geological observations, *Tectonophysics*, 226, 59–72, 1993.

Oldow, J.S., Bally, A.W., Ave Lallemant, H.G., and Leeman, W.P., Phanerozoic evolution of the North American Cordillera: United States and Canada, in Bally, A.W., and Palmer, A. R., eds., *The geology of North America, an overview,* Boulder Colorado, The Geological Society of America, The geology of North America, 1A, 139–232, 1989.

Pardo, M., Comte, D., and Monfret, T., Seismotectonics and stress distribution in the central Chile subduction zone, *Journal of South American Earth Sciences*, 15, 11–22, 2002.

Paylor, E.D., II, and Yin, A, A crustal scale kinematic model for development of Laramide structures in the Wyoming foreland, *Geological Society of America Abstracts with Programs*, 23, 5, A232., 1991.

Prodehl, C.P., and Lipman, P., Crustal structure of the Rocky Mountain region, in Pakiser, L.C., and Mooney, W.D., (eds.), Geophysical framework of the continental United States, *Geological Society of America Memoir*, 172, 249–284, 1989.

Rodgers, J., Chains of basement uplifts within cratons marginal to orogenic belts, *American Journal of Science*, 287, 661–692, 1987.

Royden, L.H., and Burchfiel, B.C., Surface deformation and lower crustal flow in eastern Tibet, *Science*, 276,788, 1997.

Saleeby, J., Segmentation of the Laramide slab – evidence from the southern Sierra Nevada region, *Geological Society of America Bulletin*, 115, 655–668, 2003.

Sales, J.K., Crustal mechanics of Cordilleran foreland deformation: a regional and scale model approach, *American Association of Petroleum Geologists Bulletin*, 52, 2016–2044, 1968.

Scheeval, J.R., Horizontal compression and a mechanical interpretation of Rocky Mountain foreland deformation, *Wyoming Geological Association 34th Annual Field Conference Guidebook*, 53–62, 1983.

Schmidt, C.J., and Perry, W.J., Jr., Interaction of the Rocky Mountain foreland and the Cordilleran thrust belt, *Geological Society of America Memoir*, 171, 582 p, 1988.

Schutt, D.L., and Humphreys, E.D., Evidence for a deep asthenosphere beneath North America from western United States SKS splits, *Geology*, 29, 291–294, 2001.

Sharry, J.R., Langan, R.T., Jovanovich, D.B., Jones, G.M., Hill, N.R., and Guidish, T.M., Enhanced imaging of the COCORP seismic line, Wind River Mountains, in Barazangi, M., (ed.), Reflection Seismology: A Global Perspective, *American Geophysical Union Geodynamic Series*, 13, 223–236, 1986.

Sheehan, A. F., Abers, G.A., Lerner-Lam, A.L., and Jones, C.H., Crustal thickness variations across the Colorado Rocky Mountains from teleseismic receiver functions, *Journal of Geophysical Research*, 100, 20,391–20,404, 1995.

Smalley, R., Jr., Pujol, J., Regnier, M., Chiu, J., Chatelain, J., Isaacks, B.L., Araujo, M., and Puebla, N., Basement seismicity beneath the Andean Precordillera thin-skinned thrust belt and implications for crustal and lithospheric behavior, *Tectonics*, 12, 63–76, 1993.

Smithson, S.B., Brewer, J.A., Kaufman, S., Oliver, J.E., and Hurich, C.A., Structure of the Laramide Wind River uplift, Wyoming, from COCORP deep reflection data and from gravity data, *Journal of Geophysical Research*, 84, 5955–5972, 1979.

Snelson, C.M., Henstock, T.J., Keller, G.R., Miller, K.C., and Levander, A., Crustal and uppermost mantle structure along the Deep Probe seismic profile, *Rocky Mountain Geology*, 33, 181–198, 1998.

Snoke, A.W., Geologic history of Wyoming within the tectonic framework of the North American Cordillera, in Snoke, A.W., Steidtmann, J.R., and Roberts, S.M., eds., *Geology of Wyoming*: Geological Survey of Wyoming Memoir, 5, 2–57, 1993.

Snyder, D.B., Ramos, V.A., and Allmendinger, R.W., Thick-skinned deformation observed on deep seismic reflection profiles in western Argentina, *Tectonics*, 9, 773–788, 1990.

Stanton, H.I., and Erslev, E.A., Rotational fault-bend folds: An alternative to kink-band models with implications for reservoir geometry and heterogeneity, *American Association of Petroleum Geologists Annual Convention Program*, 10, A191, 2001.

Steidtmann, J.R., and Middleton, L.T., Fault chronology and uplift history of the southern Wind River Range, Wyoming: Implications for Laramide and post-Laramide deformation in the Rocky Mountain foreland, *Geological Society of America Bulletin*, 103, 472–485, 1991.

Stone, D.S., The Rattlesnake Mountain, Wyoming, debate: a review and critique of models, *The Mountain Geologist*, 21, 37–46, 1984.

Stone, D.S., Basement-involved thrust-generated folds as seismically imaged in the subsurface of the central Rocky Mountain foreland, in Schmidt, C.J., Chase, R., and Erslev, E.A., eds., *Laramide basement deformation in the Rocky Mountain foreland of the western United States,* Geological Society of America Special Paper, 280, 271–318, 1993.

Tikoff, B., and Maxson, J., Lithospheric buckling of the Laramide foreland during Late Cretaceous and Paleogene, western United States, *Rocky Mountain Geology*, 36, 13–35, 2001.

Timmons, J.M., Karlstrom, K.E., Dehler, C.M., Geissman, J.W., and Heizler, M.T., Proterozoic multistage (ca. 1.1 and 0.8 Ga) extension recorded in the Grand Canyon Supergroup and establishment of northwest- and north-trending tectonic grains in the southwestern United States, *Geological Society of America Bulletin*, 113, 163–181, 2001.

Turcotte, D.L., and Schubert, G., *Geodynamics* (2nd edition), Cambridge University Press, 456 p., 2002.

Wise, D.U., An outrageous hypothesis for the tectonic pattern of the North American Cordillera, *Geological Society of America Bulletin*, 74, 357–362, 1963.

Zuber, M.T., Compression of oceanic lithosphere: an analysis of intraplate deformation in the central Indian Basin, *Journal of Geophysical Research*, 92, 4817–4825, 1987.

Eric A. Erslev, Department of Geosciences, Colorado State University, Fort Collins, CO 80523

Complex Proterozoic Crustal Assembly of Southwestern North America in an Arcuate Subduction System: The Black Canyon of the Gunnison, Southwestern Colorado

Micah J. Jessup[1], Karl E. Karlstrom[2], James Connelly[3], Michael Williams[4], Richard Livaccari[5], Amanda Tyson[6], and Steven A. Rogers[7]

The dominant orogenic fabric in Proterozoic rocks of the southwestern U.S. includes a series of NE-striking shear zones that are commonly interpreted as suture zones across which blocks of juvenile crust were assembled to the southern margin of Laurentia. New structural and geochronological data from southwestern Colorado suggest that fabrics related to assembly of tectonostratigraphic terranes in this area strike northwest. The NW-striking foliations represent deformation at ca. 10–20 km paleodepths (ca. 1.77–1.71 Ga), and are parallel to magnetic anomalies and to gradients in mantle velocity structure. The agreement between these data sets suggests that the NW-striking structures are important at lithospheric scale, extend to >100 km depth, and may record assembly of southwestern Colorado across NW-striking tectonic boundaries. Geochronologic data indicate that northwest (central Colorado)-and northeast (Cheyenne belt)-striking boundaries developed simultaneously during accretion of southwestern Laurentia between ca. 1.78–1.73 Ga. We propose that the Yavapai province at ca. 1.75 Ga may have involved a complex arcuate subduction system, with multiple arcs, analogous to that of the modern Banda Sea, in the Indonesia region.

INTRODUCTION

Models for forming juvenile continental crust commonly involve the complex collisions of oceanic elements and their amalgamation to the margin of an existing continent [*Hamilton,* 1979; *Snyder et al.,* 1996]. In the southwestern U.S., most models involve crustal growth from the northwest to southeast based on the dominant NE-striking orogenic fabric [*Reed,* 1987; *Karlstrom and Bowring,* 1988]. However, the Paleoproterozoic rocks exposed in southwestern Colorado add an important new element to these models; here the dominant fabric is NW-striking over a large region and geochronological data suggest that it formed at the same time that NE-striking shear zones were active elsewhere in the region. An

[1] Department of Earth and Planetary Sciences, University of New Mexico, Albuquerque, NM 87131; currently at: Department of Geological Sciences, Virginia Tech, Blacksburg, VA 24061.

[2] Department of Earth and Planetary Sciences, University of New Mexico, Albuquerque, NM 87131.

[3] Department of Geological Sciences, University of Texas, Austin, TX 78712.

[4] Department of Geosciences, University of Massachusetts, 611 North Pleasant Street, Amherst, MA 01003-9297.

[5] Department of Physical and Environmental Sciences, Mesa State, 1100 North Avenue, Grand Junction, CO 81502.

[6] Department of Earth and Planetary Sciences, University of New Mexico, Albuquerque, NM 87131; currently at: Department of Earth and Environment, Mount Holyoke College, South Hadley, MA 01075.

[7] Department of Earth and Planetary Sciences, University of New Mexico, Albuquerque, NM 87131; currently at: Department of Geological Sciences, University of Texas, Austin, TX 78712.

The Rocky Mountain Region: An Evolving Lithosphere
Geophysical Monograph Series 154

10.1029/154GM03

accurate model for assembly of the crust in the southwestern U.S. must account for coeval NW-and NE-striking zones [*Albin and Karlstrom,* 1991]. The goals of this paper are to describe, and provide time constraints on the NW-and NE-striking fabrics in southern Colorado. We highlight NW-striking aeromagnetic and mantle tomographic trends that coincide with the NW-trending surface foliations, and thereby delimit a ~200 km long and >100 km deep domain that may represent a major

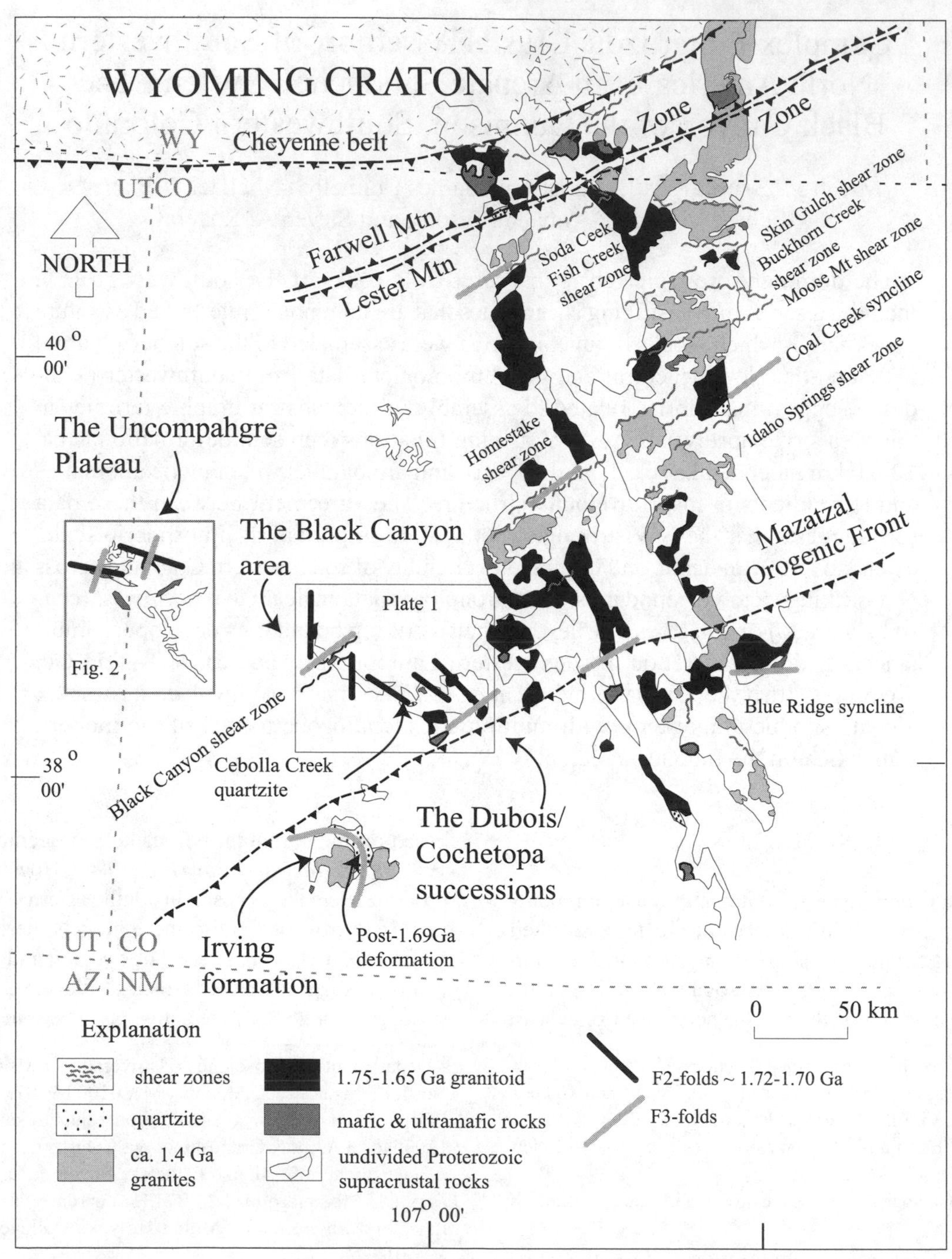

Figure 1. Proterozoic rocks exposed in Colorado. Boxes show locations of Plate 1 and Figure 2. Black bars represent the NW-trending F2 folds (~1.70–1.72 Ga) and gray bars depict NE-trending F3 folds. Granitic batholiths after *Reed et al.* [1987]. Shear zone locations after *Tweto* [1963].

lithospheric scale structural domain in southwestern Colorado.

This study seeks to unravel the ~350 m.y. long history and complex kinematics recorded by Proterozoic rocks of southwestern Colorado, specifically those exposed in the Black Canyon of the Gunnison [*Hansen and Peterman*, 1968]. New structural and geochronological work in the Black Canyon of the Gunnison presented here also leads to a better understanding of several aspects of the ca. 1.79 Ga-1.43 Ga history of the southwestern U.S., including ca. 1.4 Ga intracontinental tectonism related to A-type plutonism.

BACKGROUND

Proterozoic rocks in Colorado are part of the 1300-km-wide belt of juvenile crust that extends from the Cheyenne belt, Wyoming, to Sonora, Mexico (Figure 1). The area of the Cheyenne belt initially formed as a rifted margin (ca. 2.1 Ga) and then became the fundamental Proterozoic/Archean suture [*Karlstrom and Houston*, 1984; *Duebendorfer and Houston*, 1987; *Tyson et al.*, 2002]. The Proterozoic juvenile terranes south of the Cheyenne belt contain metasedimentary and metavolcanic rocks intruded by several generations of intrusive rocks [*Reed et al.*, 1987]. These rocks are believed to have formed in a complex series of arcs and oceanic terranes built on and tectonically juxtaposed with locally older (ca. 2.0–1.8 Ga) continental fragments [*Reed et al.*, 1987; *Karlstrom and Bowring*, 1988; *Bowring and Karlstrom*, 1990; *Ilg et al.*, 1996; *Shaw and Karlstrom*, 1999; *Selverstone et al.*, 2000; *Hill and Bickford*, 2001]. Uncertainty remains regarding the paleogeography and original geometry of tectonic blocks, and the location of possible sutures south of the Cheyenne belt.

We use the following terminology for provinces and events. The term "Yavapai orogeny" is used for the long and complex series of collisions involving oceanic tectonic elements (arcs, arc basins, oceanic terranes) and their assembly to North America (ca. 1.8–1.7 Ga). The resulting "Yavapai province" includes pre-1.70 Ga predominantly juvenile crust that extends from the Cheyenne belt south to central New Mexico and central Arizona and west to the Mojave province [*Karlstrom and Bowring*, 1988; *Karlstrom and Humphreys*, 1998]. This series of events was followed by crustal stabilization that resulted in a regional unconformity and deposition of mature quartzites (Plate 1) [*Williams*, 1987; *Soegaard and Eriksson*, 1989]. Scattered remnants of these quartzites extend well northward (into Colorado) from the sites of thickest preservation in northern New Mexico (Figure 1) [*Williams*, 1987]. The term "Mazatzal orogeny", refers to the deformation at ca. 1.65 Ga that resulted in the assembly of 1.68–1.65 Ga tectonic blocks to the southeastern margin of the Yavapai province along a complex, northeast-striking boundary [*Shaw and Karlstrom*, 1999]. After the Mazatzal orogeny at ca. 1.65 Ga there followed a ~200 m.y. period of cratonic stability and residence in the middle crust of the Proterozoic rocks that are presently exposed in the southwestern U.S. [*Bowring and Karlstrom*, 1990; *Williams and Karlstrom*, 1996]. This stability was disrupted by emplacement of ca. 1.45–1.35 Ga A-type granites and temporally associated regional transpressive deformation [*Nyman et al.*, 1994].

Southwestern Colorado provides excellent exposure of Paleoproterozoic rocks that range in age from ~1.8 to 1.68 Ga and record a sequence of events related to the growth, stabilization, and reactivation of juvenile lithosphere (Plate 1). The Dubois (ca. 1.77–1.76 Ga) and Cochetopa (ca. 1.74–1.73 Ga) successions are sequences of bimodal volcanogenic rocks intruded by two generations of calc-alkaline plutons (1.75 Ga and 1.72–1.70 Ga) and contain inherited zircons with ages of ~ 2.5 Ga and 1.87–1.84 Ga (Plate 1) [*Bickford and Boardman*, 1984; *Knoper and Condie*, 1988; *Bickford et al.*, 1989; *Hill and Bickford*, 2001]. Nd models suggest that these rocks represent juvenile material often interpreted as arc components [*DePaolo*, 1981]. The presence of inherited zircons supports models for the presence of older crustal material of ~2.0–1.8 Ga in the subsurface (Trans-Hudson/Penokean age) [*Hill and Bickford*, 2001; *Hawkins et al.*, 1996]. The Black Canyon area hosts quartz-rich metasedimentary rocks that are in contact with the Dubois and Cochetopa metavolcanic successions (Plate 1) [*Hansen*, 1981]. The nature of contacts between these three lithostratigraphic successions was obscured by intense polyphase deformation and may involve depositional contacts and/or tectonic juxtaposition of "blocks" [*Afifi*, 1981; *Shonk*, 1984]. Proterozoic rocks are also exposed in the Uncompahgre Plateau; these consist of migmatitic gneiss (ca. 1.74 Ga) intruded by a syn-kinematic monzogranite (1.72 Ga; Figure 2) [*Livaccari et al.*, 2001].

Regional ~1.4 Ga intracratonic tectonism is recorded by the emplacement of the Curecanti and Vernal Mesa monzogranites, several generations of pegmatite dikes (Plate 1) [*Hansen*, 1964; *Hansen and Peterman*, 1968; *Bickford and Cudzilo*, 1975; *Hansen*, 1981], and significant deformation and metamorphism [*Jessup et al.*, 2002a,b; *Jessup*, 2003]. The Curecanti monzogranite (1420 ± 15 Ma) is undeformed and was emplaced as sheets that crosscut the vertically foliated migmatitic gneiss host rock [*Hansen*, 1964; *Hansen and Peterman*, 1968]. The Vernal Mesa monzogranite (1434 ± 2 Ma; dated and discussed below) of the Black Canyon is interpreted as syn-tectonic [*Hansen and Peterman*, 1968; *Hansen*, 1981], but the Vernal Mesa of the Uncompahgre plateau (1430 ± 22 Ga) is interpreted as post-tectonic [*Livaccari et al.*, 2001]. The discrepancy between these two interpretations may reflect different levels of emplacement and/or partitioned deformation.

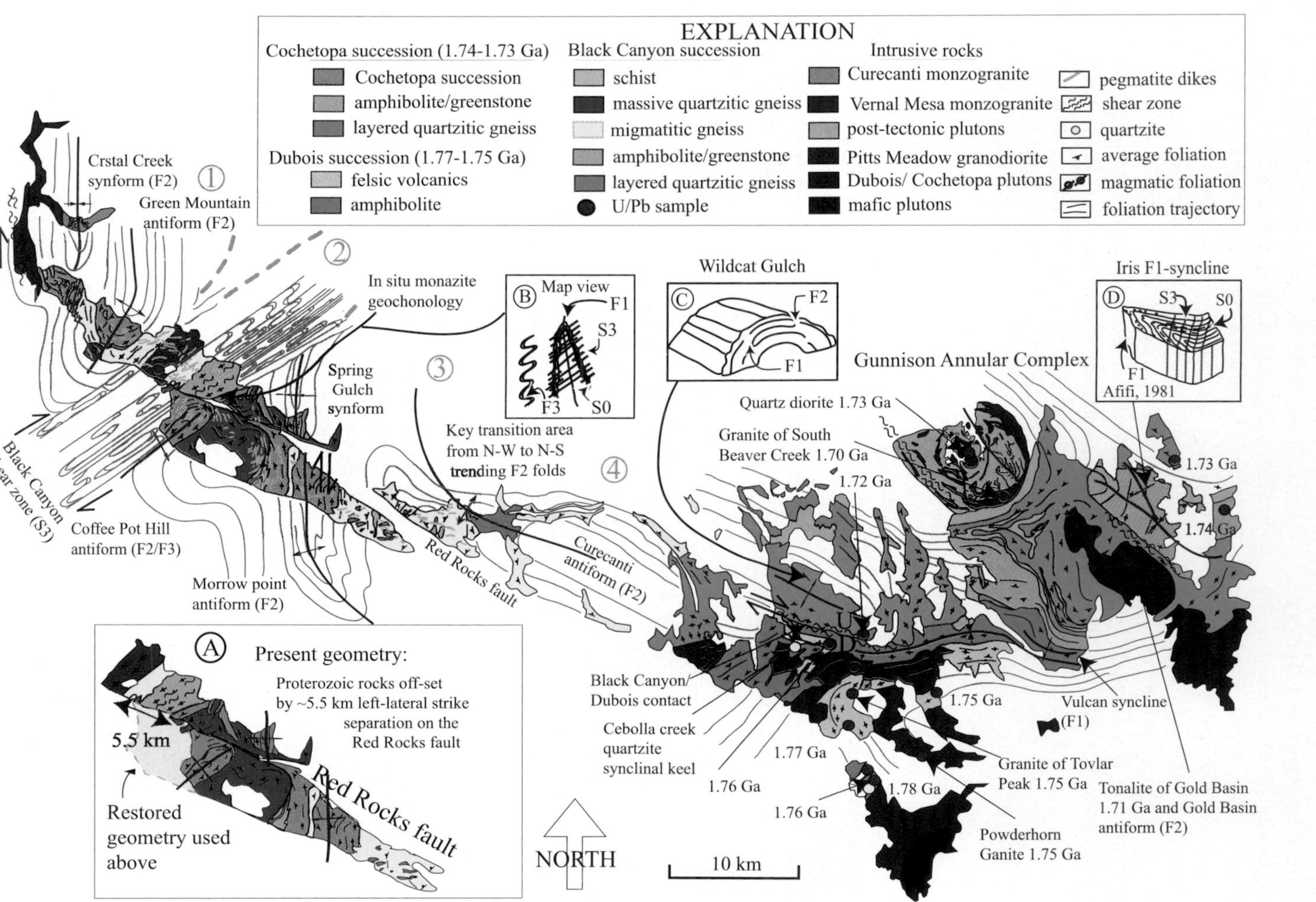

Plate 1. Generalized geologic map of the Black Canyon, Dubois and Cochetopa successions [*Hansen*, 1971, 1972; *Hedlund and Olson,* 1973, 1974a,b; *Olson,* 1975, 1976; *Olson and Steven,* 1976 a,b; *Lafrance and John,* 2001]. Geochronology after *Bickford et al.* [1989]. The Red Rocks fault is reconstructed for the structural analysis and foliation trajectory map. Inset A shows the position of the Proterozoic rocks on either side of the Red Rocks fault as they currently exist. Insets B, C, D show cartoons of key overprinting relationships.

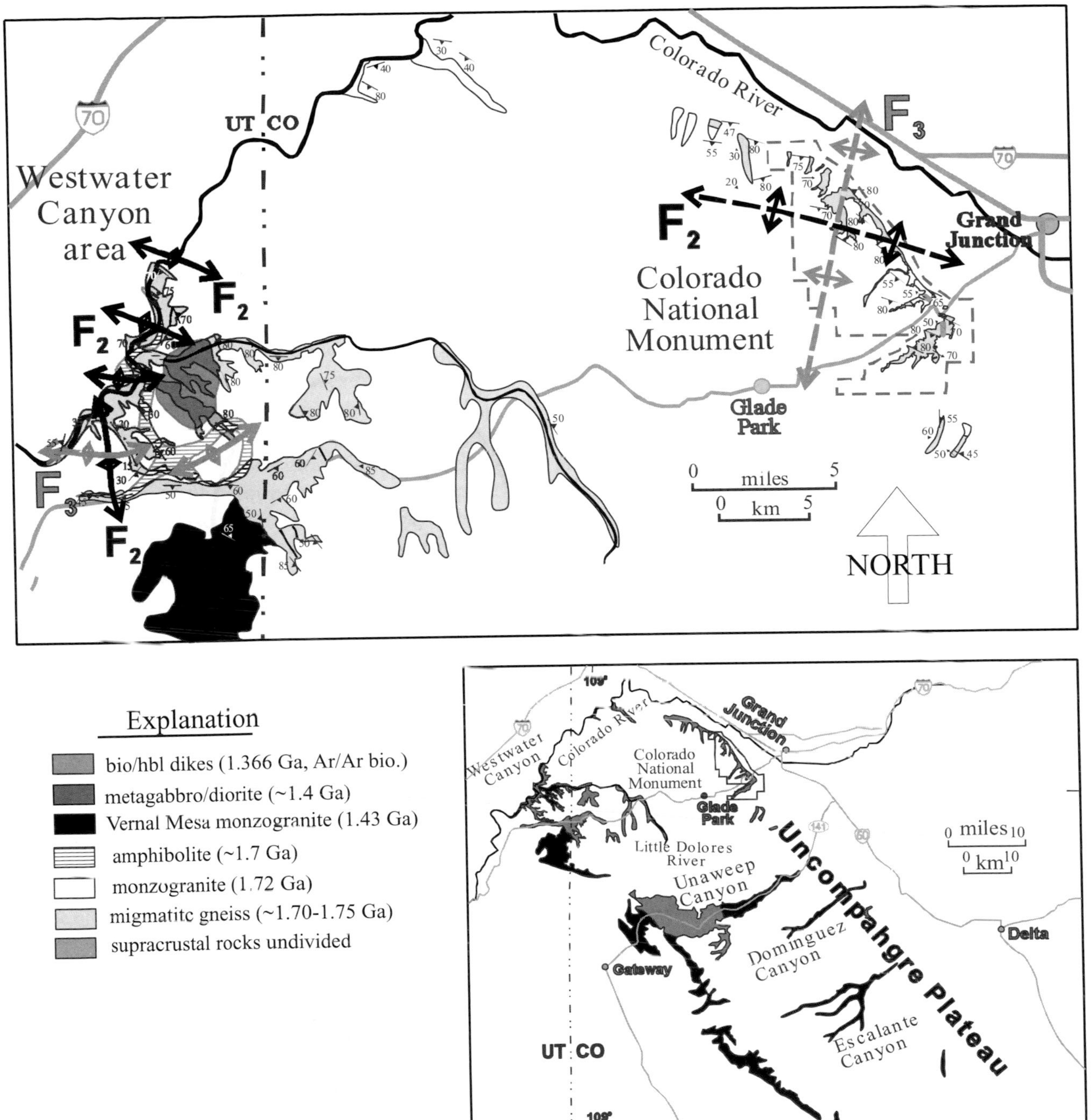

Figure 2. Generalized geologic map of the Proterozoic rocks exposed in the Uncompahgre Plateau. NW-trending F2 folds are tight to isoclinal with steeply dipping axial surfaces. The timing of this deformation is constrained by the syn-kinematic megacrystic monzogranite (1,721 ± 14 Ma). NE-trending F3 folds that result in basin and dome interference patterns refold the NW-trending folds. The Mesoproterozoic Vernal Mesa monzogranite is relatively undeformed (1,430 ± 22 Ma) [*Bickford and Cudzilo,* 1975].

Karlstrom and the CD-ROM working Group (Continental Dynamics of the Rocky Mountains), Structure and Evolution of the lithosphere beneath the Rocky Mountains: Initial results from the CD-ROM Experiment, *GSA Today*, 12, 4–10, 2002.

Knoper, M.W., and Condie, K.C., Geochemistry and petrogenesis of early Proterozoic amphibolites, west-central Colorado, U.S.A, *Chemical Geology*, 67, 209–225, 1988.

Lafrance, B., and John B.E., Sheeting and dyking emplacement of the Gunnison annular complex, SW Colorado, *Journal of Structural Geology*, 23, 1141–1150, 2001.

Livaccari, R.F., Bowring, T.J., Farmer, E.T., Garhart, Kimberly S., Hosak, A.M., Navarre, A.K., Peterman, J.S., Rollins, S.M., Williams, C.A., Kunk, M., Scott, R.B., Unruh, D., Proterozoic rocks of the Uncompahgre Plateau, western Colorado and eastern Utah, *Geological Society of America Abstracts with Programs,* 33, 44, 2001.

McCoy, A., The Proterozoic ancestry of the Colorado Mineral Belt: ca. 1.4 Ga shear zone system in central Colorado [M.S. Thesis]. University of New Mexico, Albuquerque, New Mexico, 2001.

North American Magnetic Anomaly Group (NAMAG), *Magnetic Anomaly Map of North America,* U.S. Geologic Survey, 2002.

Nyman, M.W., Karlstrom, K.E., Kirby, E., Graubard, C.M., 1994, Mesoproterozoic contractional orogeny in western North America; evidence from ca. 1.4 Ga plutons, *Geology*, 22, 901–904, 1994.

Olson, J.C., and Hedlund, D.C., Geology of the Gateview Quadrange, Gunnison County, Colorado, Geologic Quadrangle Map, *U.S. Geological Survey,* 1973.

Olson, J.C., Geologic map of the Spring Hill quadrangle, Gunnison and Saguache counties, Colorado, Geologic Quadrangle Map, *U.S. Geological Survey,* 1975.

Olson, J.C., Geologic map of the Houston Gulch quadrangle, Gunnison and Saguache counties, Colorado, Geologic Quadrangle Map, *U.S. Geological Survey,* 1976a.

Olson, J.C., and Steven, T.A., Geologic map of the Razor creek Dome quadrangle, Gunnison and Saguache counties, Colorado, Geologic Quadrangle Map, *U.S. Geological Survey*, 1976b.

Parrish, R.R., U-Pb dating of monazite and its application to geological problems. *Canadian Journal of Earth Sciences*, 27, 1431–1450, 1990.

Reed, J.C., Bickford, M.E., Premo, W.R., Aleinikoff, J.N., Pallister, J.S., Evolution of the early Proterozoic Colorado Province: constraints from U-Pb geochronology, *Geology*, 15, 861–865, 1987.

Selverstone, J., Hodgins, M., Aleinkoff, J.N., Fanning, M., Mesoproterozoic reactivation of a Paleoproterozoic transcurrent boundary in the northern Colorado Front Range: Implications for ~ 1.7 and 1.4-Ga tectonism, *Rocky Mountian Geology*, 35, 139–162, 2000.

Shaw C., Karlstrom, K., The Yavapai- Mazatzal crustal boundary in the Southern Rocky Mountains, *Rocky Mountain Geology*, 34, 37–52, 1999.

Shaw, C., Karlstrom, K., Williams, M., Jercinovic, M., McCoy, A., Electron-microprobe monazite dating of ca. 1.71–1.63 Ga and ca. 1.45–1.38 Ga deformation in the Homestake shear zone, Colorado: Origin and early evolution of a persistent intracontinental tectonic zone, *Geology*, 29, 739–742, 2001.

Shonk, M.N., Stratigraphy, structure, tectonic setting, and economic geology of an early Proterozoic metasedimentary and metavolcanic sequence, South Beaver creek area, Gunnison Gold Belt, Gunnison and Saquache counties, Colorado, [M.S. thesis] Colorado School of Mines, Golden, CO, United States, 1984.

Snyder, D.B., Prasetyo, H., Blundell, D.J., Pigram, C.J., Barber, A.J., Richardson, A., Tjokosaproetro, S., A dual doubly-vergent orogen in the Banda Arc continent-arc collision zone as observed on deep seismic reflection profiles, *Tectonics*, 15, 34–53, 1996.

Soegaard, K., and Eriksson, K.A., Transition from arc volcanism to stable-shelf and subsequent convergent-margin sedimentation in northern New Mexico from 1.76 Ga, *Journal of Geology*, 94, 47–66, 1986.

Spear, F., Metamorphic phase equilibria and Pressure-Temperature-Time paths, *Mineralogical Society of America,* 799 p, 1995.

Timmons J.M., Karlstrom K.E., Dehler C.M., Geissman J.W., Heizler M.T., Proterozoic multistage (ca. 1.1 and 0.8 Ga) extension recorded in the Grand Canyon Supergroup and establishment of northwest- and northeast tectonic grains in the southwestern United States, *Geological Society of America Bulletin*, 113, 163–180, 2001.

Tweto, O., and Sims, P.K., Precambrian ancestry of the Colorado Mineral belt, *Geological Society of America Bulletin*, 74, 991–1014, 1963.

Tyson, A.R., Morozova, E.A., Karlstrom, K.E., Chamberlain, K.R., Smithson, S.B., Dueker, K.G., Foster, C.T., Proterozoic Farwell Mountain-Lester Mountain suture zone, northern Colorado: Subduction flip and progressive assembly of arcs, *Geology*, 30, 943–946, 2002.

Williams, M.L., Stratigraphic, structural, and metamorphic relationships in Proterozoic rocks from northern New Mexico [Ph.D. thesis]: University of New Mexico, 138 p, 1987.

Williams, M.L., and Karlstrom, K.E., Looping P-T paths, HTLP metamorphism; Proterozoic evolution of the Southwestern United States, *Geology*, 24, 1119–1122, 1996.

Williams, M.L., Jercinovic, M.J., Terry, M., Age mapping and dating of monazite on the electron microprobe; deconvoluting multistage tectonic histories, *Geology*, 27, 1023–1026, 1999.

Wortman, G.L, and Bickford, M.E., Timing of arc accretion and deformation in early Proterozoic volcanogenic rocks, Central Colorado, *Geological Society of America Abstracts with Programs*, 22, A262, 1990.

James Connelly, Department of Geological Sciences, University of Texas, Austin, TX 78712; connelly@mail.utexas.edu.

Micah J. Jessup, Karl E. Karlstrom, Steven A. Rogers, and Amanda Tyson, Department of Earth and Planetary Sciences, University of New Mexico, Albuquerque, NM, 87131; micahj@unm.edu, ahj@unm.edu, micahj@unm.edu, micahj@unm.edu.

Richard Livaccari, Department of Physical & Environmental Sciences, Mesa State, 1100 North Avenue, Grand Junction, CO 81502; rlivacca@mesastate.edu.

Michael Williams, Department of Geosciences, University of Massachusetts, 611 North Pleasant Street, Amherst, MA 01003; mlw@geo.unmass.edu.

Signs From the Precambrian: The Geologic Framework of Rocky Mountain Region Derived From Aeromagnetic Data

Carol A. Finn and Paul K. Sims

U. S. Geological Survey, Denver, Colorado

Recently compiled aeromagnetic data greatly enhance our understanding of the Precambrian basement from the Rocky Mountain region by providing a means to (1) extrapolate known geology exposed in generally widely separated uplifts into broad covered areas, and (2) delineate large-scale structural features that are not readily discernable solely from outcrop mapping. In the Wyoming Province, Archean granite and gneiss terranes generate semi-circular bands of magnetic highs and lows, respectively, primarily reflecting Late Archean magmatic and deformation events that modified the older craton. In contrast, the subdued magnetic signature of the Paleoproterozoic crystalline basement of the Rocky Mountain region does not allow straightforward distinction of the Yavapai, Matzatzal and Mojave provinces. This is not the case for the Mesoproterozoic (~1.4 Ga) iron-rich granites. Although variable in magnetic expression where exposed and drilled, most are associated with highs. Many of these plutons intruded shear zones and therefore produce long, linear magnetic highs, particularly conspicuous in Arizona. A spectacular, high-amplitude magnetic potential high defines broad region of thick (~> 10 km) magnetite-rich granite, perhaps underlain by coeval mafic crust. In the east, this high corresponds to the Western Granite-Rhyolite Province. Based on the continuity of the regional magnetic high, we extend the western limit of the province from its current position in New Mexico to southeastern California. This implies that the province lay at the edge of the North American margin at the time of the late Proterozoic break-up of Rodinia and may be present in one of the conjugate rifted pieces.

INTRODUCTION

Updated aeromagnetic data from the Rocky Mountains greatly enhance our understanding of the Precambrian basement in the region. Magnetic anomaly data provide a means of "seeing through" nonmagnetic rocks and cover to reveal lithologic variations and structural features such as faults, folds, and dikes. Magnetic anomalies reflect variations in the distribution and type of magnetic, iron oxide minerals-primarily magnetite-in Earth's crust. As much as 90 per cent of the continental United States may be underlain by Precambrian rocks, yet only about 10 per cent are exposed [*Reed and Harrison*, 1993], making aeromagnetic data a powerful tool for mapping the geometry of basement rocks particularly where they are buried by younger sedimentary rocks. Because the younger strata generally have little magnetic character, they are often transparent in regional magnetic mapping. Consequently, observed magnetic patterns can be attributed to variations in magnetism of rocks in the Precambrian basement. However, veneers of younger volcanic rocks such as the Yellowstone volcanic field in northwest Wyoming and the San Juan volcanic field in southwest Colorado and adjacent areas are moder-

The Rocky Mountain Region: An Evolving Lithosphere
Geophysical Monograph Series 154

10.1029/154GM04

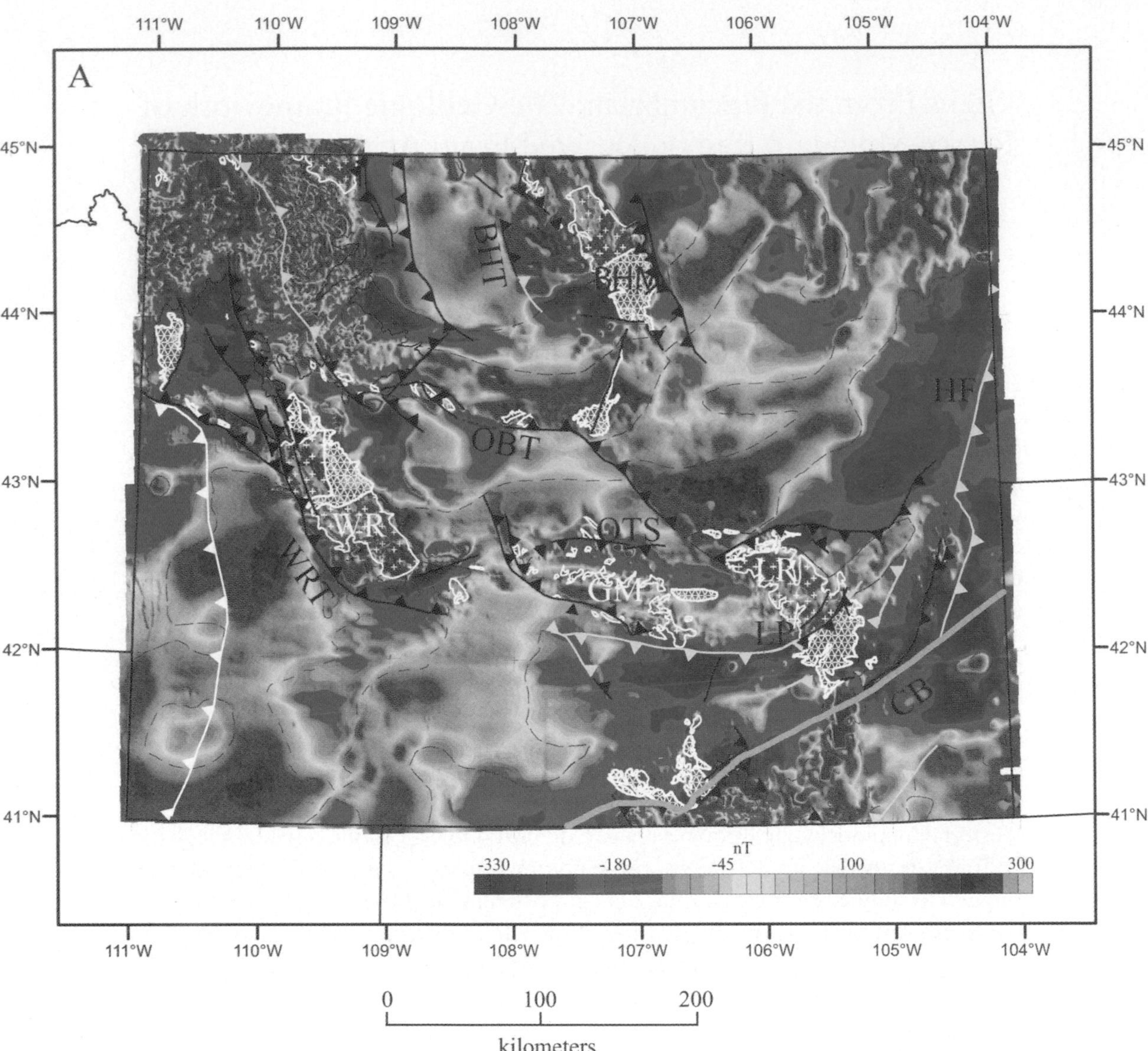

Plate 1a. Aeromagnetic Map of WY. Color shaded-relief image of a merged compilation of reduced-to-the-pole aeromagnetic data [*Kucks and Hill,* 2000] with outlines of mapped and inferred geologic units and faults. White plus pattern corresponds to Archean granites and cross-hatch pattern to Archean gneisses (see Plate 1b for legend). b) Precambrian basement map of Wyoming slightly modified from [*Green,* 1994; *Sims et al.,* 2001] based on interpretation of geologic mapping [*Blackstone,* 1989; *Green,* 1994; *Love,* 1985], aeromagnetic and drill hole information. New ages for the Sherman batholith [*Frost,* 1999] have been updated from the original map. Letters mark the following locations: BHT=Big Horn Thrust, BHM=Big Horn Mtns., CB=Cheyenne Belt, HF=Hartville Fault; LP=Laramie Peak; OBT= Owl Creek-Bridger Mountains Thrust; OTS= Oregon Trail Structure; WRT=Wind River Thrust.

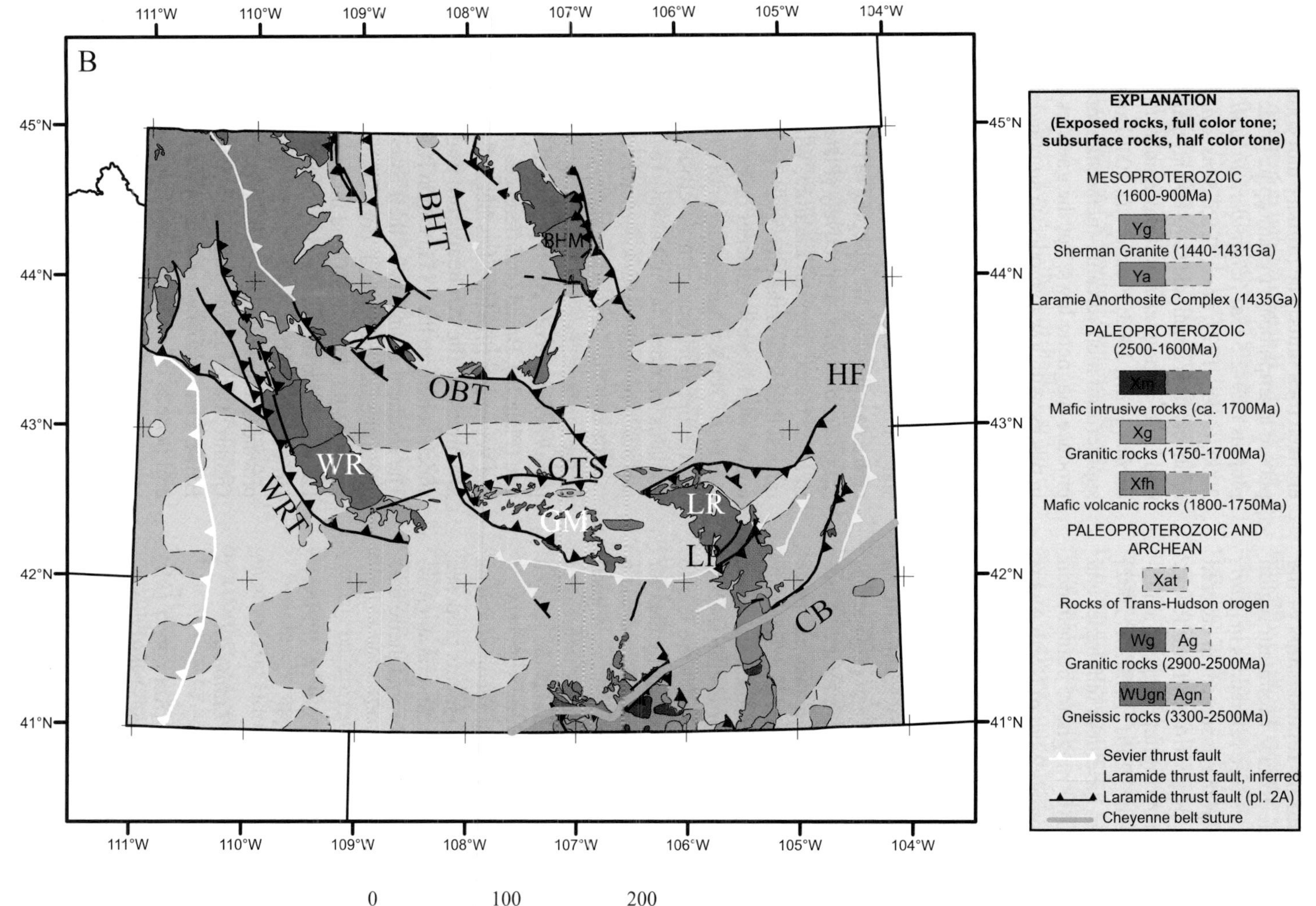

Plate 1b. Precambrian Geologic Map of WY. Precambrian basement map of Wyoming slightly modified from [*Green,* 1994; *Sims et al.,* 2001] based on interpretation of geologic mapping [*Blackstone,* 1989; *Green,* 1994; *Love,* 1985], aeromagnetic and drill hole information. New ages for the Sherman batholith [*Frost,* 1999] have been updated from the original map. Letters mark the following locations: BHT=Big Horn Thrust, BHM=Big Horn Mtns., CB=Cheyenne Belt, HF=Hartville Fault; LP=Laramie Peak; OBT= Owl Creek-Bridger Mountains Thrust; OTS= Oregon Trail Structure; WRT=Wind River Thrust.

Purucker, M., B. Langlais, N. Olsen, G. Hulot, M. Mandea, and Anonymous, The southern edge of cratonic North America; evidence from new satellite magnetometer observations, *Geophysical Research Letters*, *29* (9), 4 pp., 2002.

Ravat, T., D. Whaler, M. Pilkington, T. Sabaka, and M. Purucker, Compatibility of high-altitude aeromagnetic and satellite-altitude magnetic anomalies over Canada, *Geophysics*, *67*, 546-554, 2002.

Redden, J.A., Z.E. Peterman, R.E. Zartman, and E. Dewitt, *U-Th-Pb geochronology and preliminary interpretation of Precambrian tectonic events in the Black Hills, South Dakota*, 229-251 pp., 1990.

Reed, J.C., Jr., M.E. Bickford, W.R. Premo, J.N. Aleinikoff, and J.S. Pallisten, Evolution of the Early Proterozoic Colorado province-Constraints from U-Pb geochronology, *Geology*, *15*, 861-865, 1987.

Reed, J.C., Jr., and J.E. Harrison, Introduction to the Precambrian: Coterminus U.S., in *The Geology of North America*, edited by J. J.C. Reed, M.E. Bickford, R.S. Houston, P.K. Link, D.W. Rankin, P.K. Siims, and W.R.V. Schmus, Geological Society of America, Boulder, CO, 1993.

Ross, G.M., R.R. Parrish, M.E. Villeneuve, and S.A. Bowring, Geophysics and geochronology of the crystalline basement of the Alberta basin, western Canada, *Canadian Journal of Earth Sciences*, *28 512-522*, 1991.

Schruben, P.G., Raymond E. Arndt, and Walter J. Bawiec, Geology of the Conterminous United States at 1:2,500,000 Scale - A Digital Representation of the 1974 P.B. King and H.M. Beikman Map, *U. S. Geological Survey Digital Data Series 11, Release 2 (http://pubs.usgs.gov/dds/dds11/index.html)*, 2002.

Sears, J.W., and R.A. Price, New look at the Siberian connection; no SWEAT, *Geology*, *28* (5), 423-426, 2000.

Shaw, C.A., and K.E. Karlstrom, The Yavapai-Mozatzal crustal boundary in the southern Rocky Mountains, *Rocky Mountain Geology*, *34* (1), 37-52, 1999.

Shaw, R.D., P. Wellman, P.J. Gunn, A.J. Whitaker, C. Tarlowski, and M. Morse, Guide to using the Australian crustal elements map, *Australian Geological Survey Organisation, Record*, *1996/30*, 1996.

Shoemaker, E.M., Squires, R.L., and Abrams, M.J., Bright Angel and Mesa Butte fault systems of northern Arizona, in *Cenozoic Tectonics and Regional Geophysics of the Western Cordillera*, edited by R.B. Smith, and Eaton, G.P., pp. 341-367, Geological Society of America, Boulder, 1978.

Sims, P.K., V. Bankey, and C.A. Finn, Precambrian basement map of Colorado-A geologic interpretation of an aeromagnetic anomaly map, U.S. Geological Survey, 2001a.

Sims, P.K., and W.C. Day, Geologic map of Precambrian rocks of the Hartville uplift, southeastern Wyoming, U.S. Geological Survey, 1999.

Sims, P.K., C.A. Finn, and V.L. Rystrom, Preliminary Precambrian basement map of Wyoming showing geologic-geophysical domains, U.S. Geological Survey, 2001.

Sims, P.K., Z.E. Peterman, T.G. Hidenbrand, and S. Mahan, Precambrian basement map of the Trans-Hudson orogen and adjacent terranes, northern Great Plains, USA, U.S. Geological Survey, 1991.

Sims, P.K., H.J. Stein, and C.A. Finn, New Mexico structural zone-an analogue of the Colorado mineral belt, *Ore Geology Reviews*, *21*, 211-226, 2002.

Smith, D.R., J. Noblett, R.A. Wobus, D. Unruh, and K.R. Chamberlain, A review of the Pikes Peak batholith, Front Range, central Colorado-A "type example" of A-type granitic magmatism, *Rocky Mountain Geology*, *34*, 289-312, 1999.

Smithson, S.B., J. Brewer, S. Kaufman, J. Oliver, and C. Hurich, Nature of the Wind River thrust, Wyoming, from COCORP deep-reflection data and from gravity data, *Geology*, *6* (11), 648-652, 1978.

Southwick, D.L., Archean Superior province, in *Archean and Proterozoic geology of the Lake Superior region, U.S.A*, edited by P.K. Sims, and Carter, L.M.H., eds, pp. 115, U.S. Geological Survey Professional Paper 1556, Denver, 1993.

Sumner, J.S., Crustal geology of Arizona as interpreted from magnetic, gravity, and geologic data, in *52nd annual international meeting of the Society of Exploration Geophysicists*, edited by W.J. Hinze, pp. 164-180, Soc. Explor. Geophys., Tulsa, OK, Dallas, Texas, 1985.

Tarlowski, C., P. Milligan, and T. Mackey, Australian Geological Survey Organization Australia (AUS), 1996.

Torsvik, T.H., Enhanced: The Rodinia Jigsaw Puzzle, *Science*, *300* (5624), 1379-1381, 2003.

Tweto, O., Geologic Map of Colorado, U. S. Geological Survey Reston VA United States (USA), 1979.

Tweto, O., Rock units of the Precambrian basement in Colorado, *U.S. Geological Survey Professional Paper 1321-A*, 54, 1987.

Tweto, O., and J.E. Case, Gravity and magnetic features as related to geology in the Leadville 30-minute quadrangle, CO, *U.S. Geological Survey Professional Paper 726 C12-C29*, pp. 31, 1972.

U.S. Magnetic Anomaly Data Set Task Group, *Rationale and operational plan to upgrade the U.S. Magnetic Anomaly data base*, 25 pp., National Academy Press, Washington D.C, 1994.

Van Schmus, W.R., and M.E. Bickford, *Transcontinental Proterozoic provinces, Precambrian-Conterminous U.S.*, 171-334 pp., Geological Society of America, Boulder, Colorado, 1993.

Van Schmus, W.R., M.E. Bickford, and I. Zietz, Early and Middle Proterozoic provinces in the central United States, in *Proterozoic Lithospheric Evolution*, edited by A. Kröner, pp. 43-68, American Geophysical Union, Washington, DC, 1987.

Wellman, P., Block structure of continental crust derived from gravity and magnetic maps, with Australian examples, in *The utility of regional gravity and magnetic anomaly maps*, edited by W.J. Hinze, pp. 102-108, Soc. Explor. Geophys., Tulsa, 1985.

Wellman, P., Mapping of geophysical domains in the Australian crust using gravity and magnetic anomalies, in *Structure and evolution of the Australian continent*, edited by J. Braun, J. Dooley, B. Goleby, R.v.d. Hilst, and C. Klootwijk, pp. 59-71, American Geophysical Union, 1998.

Windley, B.F., Proterozoic anorogenic magmatism and its orogenic connections, *Journal of the Geological Society of London*, *150*, 39-50, 1993.

Wingate, M.T.D., S.A. Pisarevsky, and D.A.D. Evans, A revised Rodinia supercontinent: no SWEAT, no AUSWUS, *Terra Nova*, *14*, 121-128, 2002.

Wooden, J.L., and E. DeWitt, Pb isotopic evidence for the boundary between the Early Proterozoic Mojave and central Arizona crustal provinces in western Arizona, in *Early Proterozoic geology and ore deposits of Arizona*, edited by K. Karlstrom, pp. 27-50, Arizona Geological Society Digest, Tucson, 1991.

Yarger, H.L., Kansas basement study using spectrally filtered aeromagnetic data, in *The utility of regional gravity and magnetic maps*, edited by W.J. Hinze, M.F. Kane, N.W. O'Hara, M.S. Reford, J. Tanner, and C. Weber, pp. 248-266, Society of Exploration Geophysicists, Tulsa, 1985.

Zietz, I., Exploration of the continental crust using aeromagnetic data, in *Continental tectonics: Studies in Geophysics*, pp. 127-138, National Academy of Science, Washington, DC, 1980.

Zietz, I., P.C. Bateman, J.E. Case, M.D. Crittenden, Jr., A. Griscom, E.R. King, R.J. Roberts, and G.R. Lorentzen, Aeromagnetic investigation of crustal structure for a strip across the western United States, *Geological Society of America Bulletin*, *80*, 1703-1714, 1969.

Zietz, I., B.C. Hearn, Jr., and M.W. Higgins, Interpretation of an aeromagnetic strip across the northwestern United States, *Geological Society of America Bulletin*, *82* (12), 3347-3371, 1971.

Carol A. Finn and Paul K. Sims, U. S. Geological Survey, Denver Federal Center, PO Box 25046, MS 964, Denver, CO 80225

Low-Temperature Cooling Histories of the Cheyenne Belt and Laramie Peak Shear Zone, Wyoming, and the Soda Creek-Fish Creek Shear Zone, Colorado

Shari A. Kelley

Department of Earth and Environmental Sciences, New Mexico Institute of Mining and Technology, Socorro, New Mexico

The base of a fossil apatite fission-track (AFT) partial annealing zone (PAZ), which formed when the area now occupied by the central and southern Rocky Mountains was at sea level in Late Cretaceous time, has since been disrupted by Laramide and post-Laramide tectonism and denudation. New AFT data are used to identify this marker and to examine its disruption across Proterozoic boundaries in north-central Colorado and south-central Wyoming. The cooling history recorded by the AFT data in the Laramie Range is not strongly controlled by basement structures, but instead reflects either long-wavelength warping of the base of the PAZ during Laramide deformation or N-S variations in Paleozoic to Mesozoic sediment thickness across this range. In contrast, at least one structure associated with the Cheyenne belt in the Medicine Bow Mountains, the Rambler shear zone, influenced the Laramide cooling history of this range. The Rambler shear zone separates Laramide AFT cooling ages (60 to 79 Ma) to the northwest from >100 Ma AFT ages to the southeast. In the Sierra Madre, Wyoming, AFT ages from Archean rocks north of the Cheyenne belt and from Proterozoic rocks to the south are nearly equivalent (49–79 Ma); the Late Cretaceous PAZ is not preserved in this mountain range. Similarly, AFT ages north and south of the Proterozoic Soda Creek–Fish Creek shear zone the Park Range, Colorado are about the same (45–75 Ma). Thus, these shear zones apparently were not strongly reactivated during Laramide deformation.

INTRODUCTION

The Cheyenne belt (Figure 1), a Proterozoic crustal suture that separates Archean rocks to the north from Proterozoic rocks to the south, has long been recognized as integral to the assemblage of the North American continent [*Houston et al.*, 1968; 1989; *Karlstrom and Houston*, 1984]. The Proterozoic lithosphere in Colorado south of the Cheyenne belt was formed by accretion of diverse tectonic blocks against the Archean Wyoming craton between 1.78–1.65 Ga. The collision of the Green Mountain block composed of 1.76–1.78 volcanic arc rocks with the Wyoming craton led to the formation of the Laramie Peak shear zone [*Chamberlain et al.*, 1993]. Accretion of the Rawah block against the Green Mountain block was perhaps associated with the development of the Soda Creek–Fish Creek shear zone [*Foster et al.*, 1999] (Figure 1). Subsequent intracratonic events, particularly Laramide deformation, have reactivated the suture zones and large crustal structures that formed during lithospheric assembly in the area now occupied by the Southern Rocky Mountains [e.g., *Chamberlin*, 1945; *Houston et al.*, 1968; *Blackstone*, 1975].

The Rocky Mountain Region: An Evolving Lithosphere
Geophysical Monograph Series 154

10.1029/154GM05

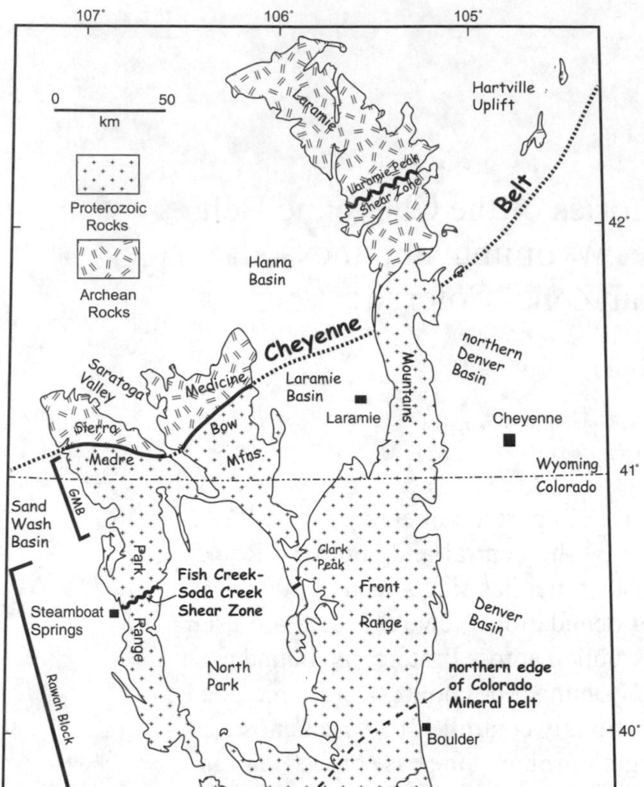

Figure 1. Index map showing the Laramide uplifts and Proterozoic structures examined in this study. GMB=Green Mountain block. After *Karlstrom and Houston* [1984].

Apatite fission-track (AFT) thermochronology has been used to examine the cooling histories of rocks recording Phanerozoic reactivation of Proterozoic boundaries in south-central Colorado and northern New Mexico [*Pazzaglia and Kelley*, 1998; *Kelley and Chapin, 2004*]. The base of an apatite partial annealing zone (PAZ) that developed in the Southern Rocky Mountain region prior to Laramide deformation has been used to constrain the geometry of Laramide structures and the amount of post-Cretaceous denudation across these features. AFT analysis demonstrates that Proterozoic boundaries like the Colorado Mineral belt [*Tweto and Sims*, 1963], the southern Yavapai- northern Yavapai boundary, and the Jemez lineament control the Laramide cooling history of the Front Range, Wet Mountains, and Sierra Nacimiento, respectively [*Pazzaglia and Kelley*, 1998, *Kelley and Chapin, 2004*]. In this paper, the influence of Proterozoic boundaries on the late Mesozoic to Cenozoic uplift and erosional history of the Rocky Mountains in north-central Colorado and southeastern Wyoming is examined using AFT thermochronology. The areas targeted for analysis include the Cheyenne belt in the Laramie Mountains, Medicine Bow Mountains, and Sierra Madre of Wyoming, the Laramie Peak shear zone in the Laramie Mountains, and the Soda Creek–Fish Creek shear zone in the Park Range, Colorado (Figure 1).

2. DEFINITION OF THE APATITE PARTIAL ANNEALING ZONE

2.1. Relationship Between Track Annealing and Temperature

Temperature and chemical composition control fission-track annealing in apatite [*Green et al*., 1986]. Figure 2a illustrates how fission-track age and track length changes as a function of increasing temperature in a drillhole. In a relatively stable geological environment where temperatures as a function of depth are at a maximum, AFT age and track lengths decrease systematically with depth [*Naeser*, 1979, *Fitzgerald et al*., 1995]. At temperatures less than 60 to 70°C, tracks that are produced by the spontaneous fission of ^{238}U are retained and annealing is relatively minor. The AFT ages of detrital grains in sedimentary rocks at shallow depths (temperatures < 70°C) are equivalent to or greater than the stratigraphic age of the rock unit and the mean track lengths reflect the cooling history of the source region. The AFT ages in basement rocks at shallow depth (temperatures < 70°C) reflect the cooling history of the basement prior to the period of stability, and the mean track lengths may be relatively long (>13 μm). In the temperature range known as the partial annealing zone (PAZ), which is between 60–70°C and 110–140°C, depending on the chemical composition of the apatite, the AFT ages and track lengths are reduced compared to the original AFT ages and lengths. Mean track lengths in the PAZ are typically 8 to 13 μm. Finally, at temperatures above 110 to 140°C, tracks that are formed are quickly annealed and the fission-track age is zero. If, after a period of relative stability, the crust is rapidly cooled during denudation related to a tectonic event, the fossil PAZ may be exhumed and the time of cooling can be constrained (Figure 2b). Furthermore, the paleodepth of the base of the PAZ can be estimated using certain assumptions about the thermal conditions (i.e., geothermal gradient, surface temperature, lithology of the removed section) prior to tectonism [*Fitzgerald et al*., 1995].

2.2. The Partial Annealing Zone in the Rocky Mountains

A fossil PAZ that developed during a period of relative tectonic stability at the end of Mesozoic time in western United States was disrupted by Laramide deformation and is preserved in several areas in the Southern Rocky Mountains [*Kelley and Chapin, 2004*]. The approximate geometry of the base of the fossil PAZ, which separates samples with AFT ages of 45 to 75 Ma with relatively long mean track lengths from samples with AFT ages >100 Ma and short mean track lengths,

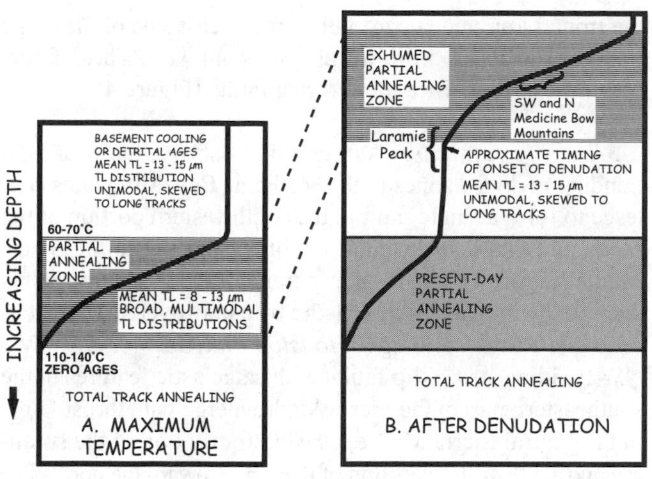

Figure 2. Diagram showing the relationship of fission-track age and length as a function of depth at time of maximum burial during the Mesozoic(a) and after denudation (b). Diagram described in the text. TL = track length. Representative portions of the AFT age-depth curve that are preserved in the Laramie and Medicine Bow Mountains are shown for reference.

is shown on Figure 2b. This exhumed PAZ, which was originally discovered in the Front Range by *Naeser* [1979] and further mapped in the Front Range and Wet Mountains by *Kelley and Chapin [2004]*, is also preserved in bits and pieces in the mountains of southeastern Wyoming.

3. GEOLOGIC SETTING

3.1. Proterozoic and Laramide Structures

3.1.1. Laramie Mountains. The Cheyenne belt is exposed in three Laramide-aged uplifts in southeastern Wyoming: the Sierra Madre and the Medicine Bow and Laramie Mountains. The easternmost range investigated in this study is the Laramie Mountains (Figure 3). Proterozoic deformation associated with the development of the Cheyenne belt in the Laramie Mountains has been masked by the intrusion of the 1.4 Ga Laramie Anorthosite complex and Sherman Granite along the suture. North of the Cheyenne belt, the Laramie Range is composed of Archean metamorphic rocks and the Archean Laramie batholith [*Condie*, 1969], while to the south, the range is underlain by Proterozoic granite, anorthosite, and metamorphic rocks. This mountain block, like others in the region, initially formed as the result of compressional deformation during the Laramide orogeny and is basically a long-wavelength, low-amplitude fold bound on the east side by a west-dipping thrust fault [*Blackstone*, 1975; 1996]. The Laramie Mountains are thrust over the western margin of the northern Denver Basin, which is a foreland basin (Figure 3). During the late stages of Laramide deformation, the Laramie Range was thrust northward over the south end of the Powder River Basin [*Gries*, 1983; *Stone*, 2002]. The Cheyenne belt is nearly coincident with the Wheatland–Whalen fault zone (Figure 3), which acted as a southeast-vergent thrust fault during Laramide deformation, and has since been re-activated as a zone of extension cutting rocks as young as Miocene [*Blackstone*, 1996]. Normal offset across this fault zone is down to the northwest. The Cheyenne belt divides the Laramie Mountains into two distinctive segments, each of which can be generally characterized as a doubly plunging anticline [*Blackstone*, 1975; 1996]. The southern Laramie Mountains merge with the northern Front Range of Colorado.

The Proterozoic-aged Laramie Peak shear zone (Figure 3) is located to the north of the Cheyenne belt in the northern Laramie Mountains [*Patel et al.*, 1999]. This significant Proterozoic boundary separates two blocks within the Archean crust [*Chamberlain et al.*, 1993; *Patel et al.*, 1999]. Based on interpretations of metamorphic mineral assemblages preserved across this structure, the block to the south of the Laramie Peak shear zone was uplifted about 10 km higher than the block to the north during 1.8 Ga collision along the Cheyenne belt. The Laramie shear zone is on trend with the Wyoming lineament [*Blackstone*, 1996]. This lineament, originally defined by *Ransome* [1915], separates NW-trending, SW-vergent Laramide-aged structures on the north side from N-trending, E-vergent structures on the south side of this feature, suggesting reactivation of this boundary during Laramide deformation.

3.1.2. Medicine Bow Mountains. The Cheyenne belt in the Medicine Bow Mountains (Figure 4) consists of several northeast-trending Proterozoic-aged mylonite zones separating intervening fault blocks. The northernmost mylonite zone in the Cheyenne belt is the Mullen Creek–Nash Fork shear zone, which separates Archean and Proterozoic metasedimentary and metavolcanic rocks sitting on Archean basement to the north from Proterozoic metasedimentary, metavolcanic and plutonic rocks to the south [*Houston and Karlstrom*, 1992]. South of the Mullen Creek–Nash Fork shear zone are the central mylonite zone, the southern mylonite zone, and the Rambler shear zone [*Houston and Karlstrom*, 1992]. All four mylonite zones merge into a single fault in the southwestern Medicine Bow Mountains. The mylonite zones have steep dips toward the south (60° to vertical) and are 50 to 400 m wide.

Blackstone [1983, 1987] discusses the Laramide structure of the Medicine Bow Mountains. The main frontal thrust fault, the Arlington fault (Figure 4), lies on the east side of the Medicine Bow Mountains and dips to the west. The south end of the Arlington fault merges with a splay of the Rambler fault system, where both brittle features and mylontic structures

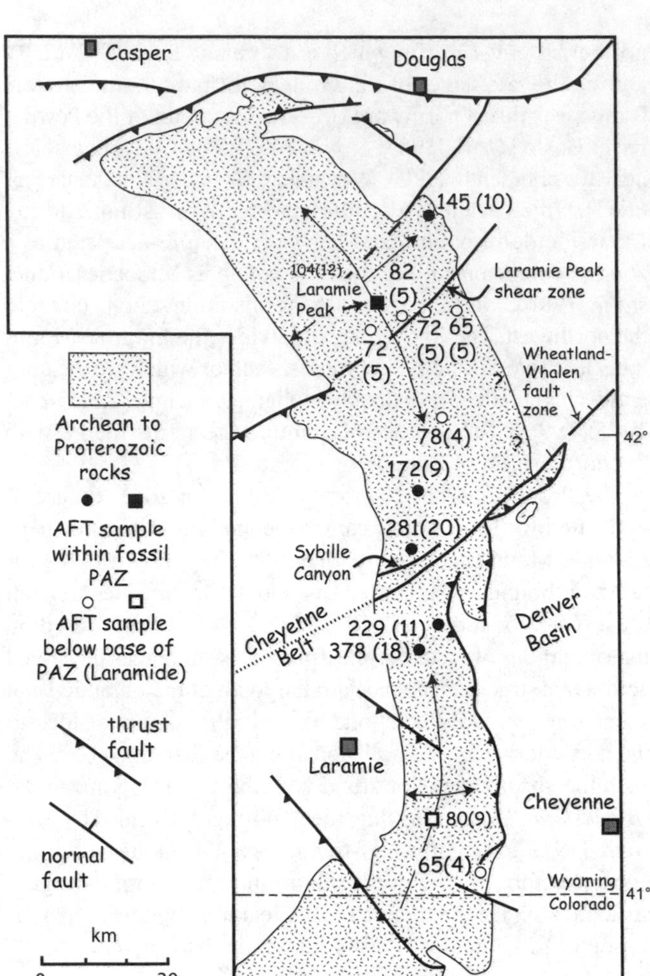

Figure 3. Generalized geologic map of the Laramie Mountains after *Blackstone* [1996]. The circles represent samples analyzed in this study. The squares mark the age-elevation traverse on Laramie peak and the sample from the summit of the range along Interstate 80 [*Cerveny*, 1990]. The approximate locations of the samples of *Cerveny* [1990] in Sybille Canyon are also shown. AFT ages, with the standard error of the age enclosed in parentheses, are shown adjacent to the sample localities (Table 1; see CDROM in back cover sleeve for sample numbers, elevations, and analytical data).

are exposed. The Corner thrust fault lies to the west of and overrides the Arlington thrust fault (Figure 4). The south end of the Corner fault merges with the southern mylonite zone. Brittle features overprint the mylonitic fabric of this structure, as well. The Middle Fork thrust fault just southwest of Centennial ties in with the main strand of the Rambler shear zone. The frontal thrust faults have dips of 17–30° toward the west. In the southern Medicine Bow Mountains in Wyoming, the frontal Laramide thrust faults on the east side of the range trend northwest and dip toward the southwest. These faults offset the earlier N-trending thrust faults (Figure 4).

3.1.3. Sierra Madre. As mentioned above, the four NE-trending mylonite zones in the Medicine Bow Mountains coalesce to form a single fault in the southwestern portion of the range and this fault extends a short distance into the Sierra Madre (Figure 4). The northeast trend of the Cheyenne belt is disrupted within the Sierra Madre by a north-vergent thrust of probable Proterozoic age [*Houston and Graff*, 1995]. Two NW-trending strike-slip faults with cataclastic features in the southeastern part of the Sierra Madre merge with thrust faults in the central Sierra Madre, causing translation of the southern and southwestern portion of the range toward the north. The NW-trending cataclastic faults are <40 m wide, are silicified, have hematite to limonite alteration, and at some localities, have copper and gold mineralization. In places the breccia zones appear to be intruded by granite, suggesting a Proterozoic age [*Houston and Graff*, 1995].

The Sierra Madre block appears to have been a part of the Medicine Bow Laramide uplift and is now separated from the Medicine Bow Mountains by a late Cenozoic NW-trending half-graben that may be the northern extension of the Rio Grande rift (Figure 4). The Independence Mountain thrust, an E–W trending thrust fault dipping toward the north, forms the south edge of the Sierra Madre, causing the Sierra Madre to override Mesozoic sedimentary units in North Park by as much as 7 km [*Blackstone*, 1977]. One of the principal NW-striking Proterozoic faults in the Sierra Madre appears to have been reactivated during Laramide deformation as the Independence Mountain thrust. Two phases of deformation and complex rotation of blocks characterize the Laramide tectonics of this area [*Houston and Graff*, 1995]. Several of the NW-trending faults in the Sierra Madre have been re-activated as normal faults in Cenozoic time, allowing the preservation of basins filled by Miocene Brown Park Formation within the range.

3.1.4. Park Range. The Soda Creek–Fish Creek (SC–FC) shear zone in the Park Range is a wide band of Proterozoic deformation that developed south of the 1.4 Ga Mt. Ethel pluton (Figure 5). The zone, which is about 4 km wide, consists of many discrete mylonites and shears with deformation restricted to bands only a few cm to 10 m wide. Some of the faults show left-lateral displacement, while others have right-lateral movement. The total apparent displacement across the Fish Creek–Soda Creek Shear zone is about 20 km of left-lateral motion [*Snyder*, 1978]. *Barinek et al.* [1999], using microstructural analysis, document north-side-down Proterozoic displacement in addition to left-lateral displacement.

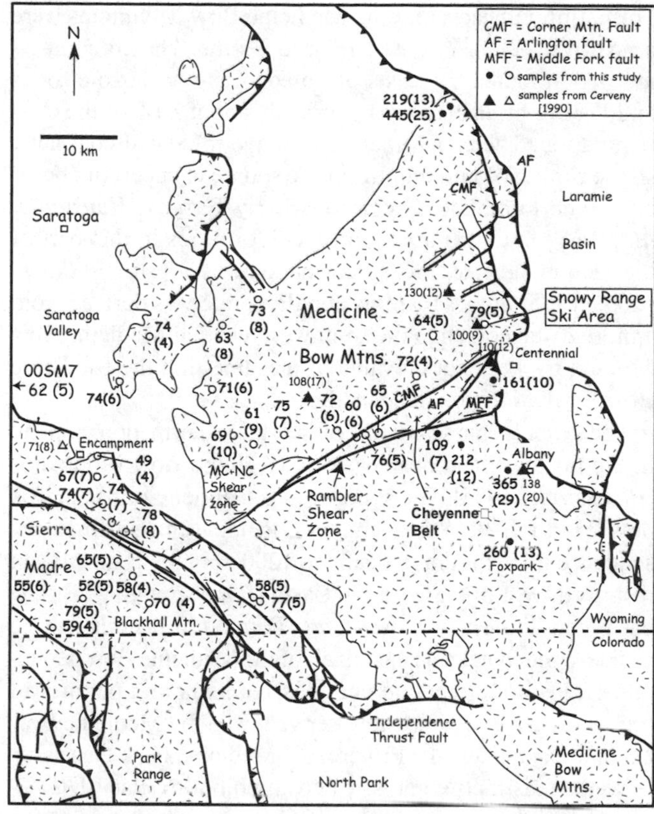

Figure 4. Generalized geologic map of the Medicine Bow Mountains and Sierra Madre after *Houston and Graff* [1995]. MC–NC = Mullen Creek–Nash Creek shear zone. Symbols defined in Figure 3 and in the upper right corner of the map.

Most of the displacement occurred prior to the emplacement of the 1.4 Ga Mt. Ethel pluton, but the last movement offsets 1.4 Ga dikes. *Foster et al*. [1999] note that the ages of the Proterozoic rocks are slightly older (1.78–1.76 Ga) to the north compared to the south (1.75–1.73 Ga). The rocks are more highly deformed to the south, but record higher pressures to the north. Some of the mylonite zones were reactivated during Laramide deformation [*Snyder*, 1978].

Examination of the simplified geologic map of the Park Range (Figure 5), which was compiled from the maps of *Snyder* [1980 a,b,c], indicates a difference in the style of Laramide deformation between the northern and southern Park Range that seems to be primarily controlled by the Mt. Ethel pluton, and to a lesser extent, by the Fish Creek–Soda Creek shear zone. North of the Mt. Ethel pluton, the Park Range is broken into at least four N-trending blocks, each of which is bounded on the west by a thrust fault. Folded Paleozoic to Mesozoic sedimentary rocks are preserved between each block. The east side of the largest block, which contains Mt. Zirkel, is bordered by a high angle fault that is likely a reverse fault [*Hail*, 1965]. At the latitude of the Mt. Ethel pluton, the Park Range is a single block bound on the west by a thrust fault zone. South of the Mt. Ethel pluton, the Park Range is essentially an anticlinal arch with Paleozoic sediments lapping onto the Proterozoic basement on the east side [*Hail*, 1968] and Miocene Browns Park Formation lapping onto the west side [*Snyder*, 1980c].

3.2. Synorogenic Sedimentary History of the Adjoining Basins

Figure 6 summarizes the relationships among the latest Cretaceous to Neogene sedimentary rocks preserved in the foreland (northern Denver Basin) and intermountain basins adjacent to the uplifts examined in this study. These sedimentary rocks and the unconformities between the sedimentary packages provide clues as to the timing and location of denudation associated with Laramide and post-Laramide deformation that can be tied to the AFT results.

The northernmost Denver Basin to the east of the Laramie Range contains rocks of latest Cretaceous age that mark the transition from marine deposition of Fox Hills Sandstone to marginal marine conditions recorded by the Lance Formation. However, synorogenic Paleocene to Eocene rocks that are preserved in the Denver Basin to the south between Denver and

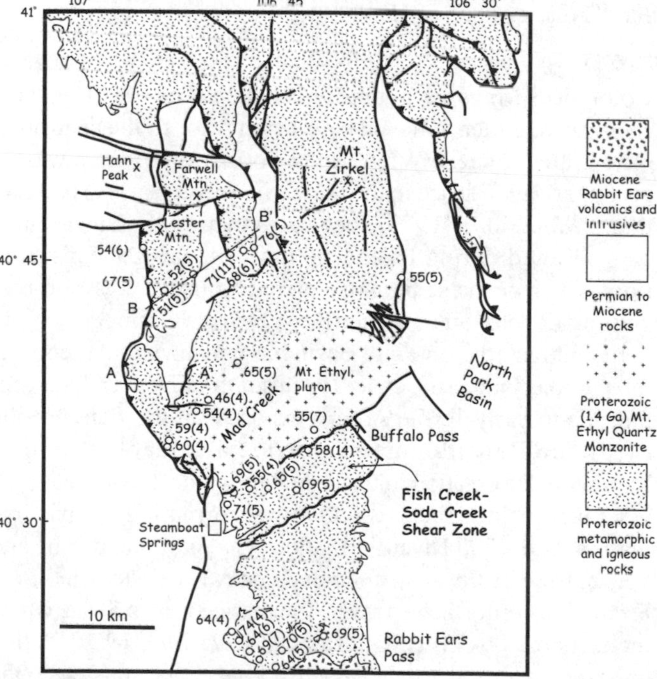

Figure 5. Generalized geologic map derived from Snyder [1980 a,b,c]. X marks key geographic features. Net Proterozoic left-lateral shear displacement across the Fish Creek– Soda Creek shear zone from *Snyder* [1978].

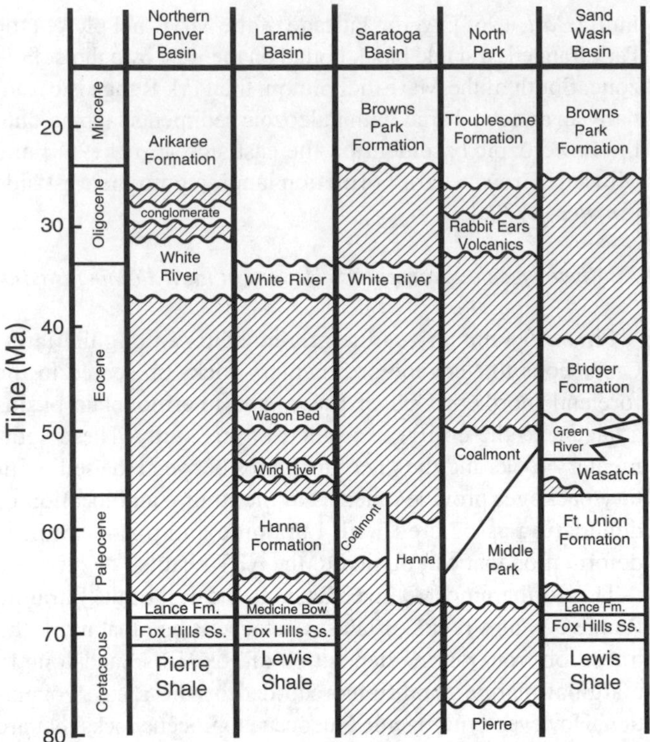

Figure 6. Correlation of stratigraphic units in basins bordering the areas of interest. Data based on *Lillegraven* [1993], *Montagne* [1991], *Hail* [1965; 1968], *Izett* [1968].

Colorado Springs are not preserved here. The early Cenozoic rocks were either removed by erosion prior to the deposition of the latest Eocene White River Formation or there was no accommodation space in the northern Denver Basin to preserve them. *Blackstone* [1975] speculates that the Laramie Range was elevated during Paleocene time, blocking the distribution of Paleocene sediments derived from the west and preventing deposition in the northern Denver Basin.

In contrast, the Laramie Basin has a history that is comparable to the southern Denver Basin south of Denver. Late Cretaceous to early Tertiary rocks exposed on the flanks of the Medicine Bow Mountains provide a valuable record of Laramide deformation in this area. From oldest to youngest the key units include the Medicine Bow Formation, Hanna Formation, and Wind River Formation. The marginal marine and non-marine shale, sandstone and coal of the Medicine Bow Formation conformably overlie the marine Lewis Shale, which includes the Fox Hills Sandstone [*Blackstone*, 1970]. Pollen preserved in coal is used to assign a Late Cretaceous age to the Medicine Bow Formation [*Houston et al.*, 1968]. *Knight* [1953] and *Houston et al.* [1968] suggest that the presence of gneiss and granite clasts in a 6-m thick conglomerate lens located about 157 m above the base of the Medicine Bow Formation indicates that the Medicine Bow Mountains were a positive feature in Late Cretaceous time. The conglomerate is dominated by clasts of Paleozoic to Mesozoic rocks and Precambrian clasts make up only about 10% of the clast population. The conglomerate is above the dated coal layer so the timing of initial uplift of the source is uncertain, but is bracketed to be Late Cretaceous-Early Tertiary. *Houston et al.* [1968] note that a lack of unconformities in this part of the section indicates local deformation was not strong. *Blackstone* [1975] argues that this conglomerate was derived from a more distal source to the southwest and that Medicine Bow Mountains were not elevated at the time that the Medicine Bow Formation was deposited.

The synorogenic sediments of the Hanna Formation crop out in the northern and northeastern Medicine Bow Mountains [*Blackstone*, 1975]. The synorogenic sedimentary package is only 244 m thick in the Laramie Basin, but this sedimentary sequence thickens dramatically to 1200 m in the Hanna Basin just northwest of the Laramie Basin [*Blackstone*, 1975]. The basal conglomerate of the Hanna Formation becomes more coarse-grained near the east flank of the Medicine Bow Mountains. This conglomerate contains boulders and pebbles of rocks that are unquestionably derived from the Medicine Bow Mountains, namely the Proterozoic Medicine Peak Quartzite, as well as distinctive gneissic and amphibolitic clasts [*Blackstone*, 1975]. *Houston et al.* [1968] note that there are two angular unconformities observed in the Laramie Basin, one between the base of the coarse-grained Hanna Formation and the underlying Medicine Bow Formation and one within the sequence of conglomeratic sandstone, shale, conglomerate, arkose, and coal of the Medicine Bow Formation. Thus, at least two episodes of deformation are preserved in the sedimentary record. Pollen collected from near the base of the unit and leaf fossils from a shale located 31 m above the base of the package yield Paleocene to Eocene (?) (Wasatch) ages [*Houston et al.*, 1968].

The Eocene Wind River Formation is earliest Wasatch in age, based on fossil evidence [*Princhinello*, 1971], and is dominantly composed of variegated claystones. The unit becomes somewhat coarser grained along the flanks of the Laramie Basin [*Blackstone*, 1975]. The Wind River Formation unconformably overlies the Hanna Formation and laps across the trace of the Arlington thrust fault in the northern Medicine Bow Mountains [*Blackstone*, 1975]. Based on these stratigraphic relationships, the Medicine Bow uplift was denuded primarily during Paleocene time.

The Sand Wash Basin to the northwest of the Park Range contains a record of the denudation of the Park Range and the Sierra Madre [*Tweto*, 1975]. The youngest marine unit of Cretaceous age is the Fox Hills Sandstone, which grades upward into the marginal marine, late Cretaceous Lance Formation, a unit equivalent to the Medicine Bow Formation in the Laramie Basin. The Lance Formation may contain reworked Cretaceous material derived from the Park Range [*Tweto*, 1975]. The Lance Formation is unconformably overlain by the Paleocene Fort Union Formation, which includes arkose and conglomerate with sources in the Park and Sierra Madre uplifts. The Eocene Wasatch Formation unconformably overlies the Fort Union Formation on the east side of the basin. Again, two episodes of Laramide deformation are found in the synorogenic sedimentary record. The fluvial Wasatch Formation derived from the Park and Sierra Madre ranges intertounges with the lacustrine Green River Formation. The middle to upper Eocene Bridger Formation rests conformably on the Green River Formation.

The Laramide-aged intermontane basins of south-central Wyoming (Saratoga Valley) and in north-central Colorado (North Park) have had complex histories compared to other Laramide basins. In the Saratoga Valley (Figure 1), the Paleozoic to Mesozoic sections were completely removed by erosion in the central part of the basin prior to deposition of the Paleocene Hanna Formation [*Montagne*, 1991]. Here the Hanna Formation sits on Proterozoic to Archean basement. Thus, in the earliest phases of Laramide deformation in this area, it appears that the Medicine Bow Mountains and the Sierra Madre were connected by an NE-trending arch that has since collapsed into the basin. According to *Montagne* [1991], the Hanna Formation occupied a syncline between the two mountain ranges and was subsequently mostly removed from what is now the Saratoga Valley prior to deposition of the Miocene Browns Park Formation.

Like the Saratoga Valley, North Park was uplifted in the earliest stages of Laramide deformation prior to subsiding to become a Laramide basin, although the amount of intrabasinal sediment removal was not as dramatic. The Pierre Shale in the North Park Basin is deformed and truncated by erosion prior to deposition of the Paleocene Coalmont Formation [*Hail*, 1965]. Equivalents to the Medicine Bow Formation (Laramie Formation in the southern Denver Basin, Lance Formation in the Sand Wash Basin), Fox Hills Sandstone, and the uppermost Pierre Shale down to the *Baculites eliasi* ammonite horizon [*Izett et al.*, 1971] are missing from North Park. Only ~1525 m of Pierre Shale is present in North Park, implying that ~ 920 to 1220 m was removed by erosion during Laramide time [*Izett et al.*, 1971]. *Hail* [1965] notes that the sandy nature of the Pierre Shale in North Park does not herald the beginning of the Laramide orogeny, but instead reflects the position of shorelines during a transgressive-regressive cycle in late Cretaceous time. Early Laramide folds in North Park were truncated by erosion before deposition of Coalmont Formation. The basal Coalmont Formation contains Precambrian detritus, indicating that the Paleozoic to Mesozoic cover in the Park Range had already been stripped off prior to deposition of the Paleocene unit. The sedimentary material was either never deposited in the western part of North Park or it was stripped off before the Coalmont Formation was deposited. In places in the narrow thrust belt along the east side of the Park Range in northwestern North Park (Figure 5), the Coalmont Formation lies on highly faulted and deformed Cretaceous units. The Coalmont Formation in turn has been offset by the Independence thrust fault.

Significant late Eocene paleovalleys, including the Downey Park and Toltec valleys in the Laramie Range and Kings Canyon in the Medicine Bow Mountains, are filled with thick sequences of latest Eocene to Oligocene White River Formation. This unit (Figure 6) is a tuffaceous sequence derived from distal volcanism in the Basin and Range province to the west that includes Colorado Mineral belt volcanic clasts [*Evanoff*, 1990]. The AFT results are generally insensitive to potential thermal effects White River deposition.

3.3. Previous Thermochronologic Analyses

Cerveny [1990] did a reconnaissance study of the denudation history of the mountains of southeastern Wyoming and northern Colorado using apatite fission track analysis. *Cerveny* [1990] analyzed six samples from vertical traverses in the Laramie Range, five samples from a road traverse across the Medicine Bow Mountains, one sample in the Sierra Madre, and five samples from a road traverse across the Park Range. The results of *Cerveny's* [1990] work are in general agreement, and in places supplement, the results of this study. The locations of his samples are noted on Figure 2 and his data are discussed in each appropriate section. *Kelley and Chapin [2004]* and *Cerveny* [1990] report AFT data for the northern Front Range. *Naeser et al.* [2003] recently presented AFT results from the Gore Range, which is located south of the Park Range. The results of these studies are incorporated into the interpretation of the data presented here later in the discussion.

4. METHODS

A total of 54 samples were collected in the Medicine Bow Mountains and in the Sierra Madre across the Cheyenne belt, and in the Park Range across the Soda Creek– Fish Creek (SC–FC) shear zone. In addition, Kevin Chamberlain of the University of Wyoming provided 24 apatite separates from the Sierra Madre, and the Medicine Bow and Laramie Mountains. The fission–track analysis methods used in this study are fully described in *Kelley et al.* [1992] and are briefly summarized here. Apatite was separated from the samples using standard heavy-liquid and magnetic separation techniques.

Apatite grains were mounted in epoxy, polished to expose the grains, and etched for 25 seconds in a 5 M solution of nitric acid to reveal the fission tracks. The grain mounts were then covered with muscovite detectors and sent to the Texas A&M Nuclear Science Center for irradiation. The neutron fluence was calibrated with age standards and Corning Glass. An apatite zeta value of 5516 ± 300 was determined using the CN-6 glass and the accepted age of 31.4 ± 0.5 for Durango apatite. Individual grain ages were calculated using the methods of *Hurford and Green* [1983] and the Chi-squared statistic [*Galbraith*, 1981] is applied to determine whether the individual ages belong to a single population. No chemistry was determined for the apatite, but unusually large etch pits, which indicate elevated levels of chlorine in apatite [*Donelick*, 1988], were not observed.

Confined track-length distributions in the apatite-grain mounts were determined using a microscope fitted with a 100-x dry lens, a drawing tube, and a digitizing tablet. Horizontal, well-etched, confined tracks (tracks completely enclosed within the crystal) in grains resting on their prismatic faces were measured. The orientation of the tracks with respect to the c-axis was also determined.

5. RESULTS AND INTERPRETATION

5.1. Laramie Mountains

The new AFT data from the Laramie Mountains were derived from apatite separates provided by Kevin Chamberlain at the University of Wyoming. These samples were gathered for U-Pb analysis and thus were not collected along age-elevation traverses that are preferred for AFT studies. *Cerveny* [1990] did collect samples along elevation traverses on Laramie Peak at the north end of the range, in Sybille Canyon just north of the Cheyenne belt, and from Interstate 80 near the south end of the range. The new AFT data, as well as the data of *Cerveny* [1990] are plotted on Figures 3 and 7. Two general populations of ages are depicted on Figure 7, those in the 65–82 Ma range with long mean track lengths of 13.5 to 14.0 μm (Table 1; see CDROM in back cover sleeve) associated with cooling during Laramide deformation, and those >100 Ma with short mean track lengths (Table 1; see CDROM in back cover sleeve) from within the PAZ. The data of *Cerveny* [1990] brackets the base of the fossil PAZ to be at a modern elevation of 2700 to 3100 m (~2900 m) on Laramie Peak and at ~1950 m in Sybille Canyon. The base of the PAZ is at or just below 2600 m in the southern Laramie Mountains along Interstate 80 [*Cerveny*, 1990]. The new data provide additional information on the geometry of the base of the PAZ in the Laramie Mountains. The two samples just south of the Cheyenne belt (Figure 3) yield ages that suggest that the base of the PAZ might be at an elevation ~250 m higher than it is north of the Cheyenne belt in Sybille Canyon (Figure 7). In the northern Laramie Mountains, the base of the PAZ has been flexed down about 1.1 km between Laramie Peak and the northeast margin of the range. The base of the PAZ has also been downwarped approximately 1 km between Laramie Peak and Sybille Canyon (Figure 7). The base of the PAZ dips north between the I-80 summit and just south of the Cheyenne belt, with a minimum elevation difference of 500 m between these two points.

Two samples key in understanding the relationship between the geometry of the PAZ and possible Laramide-aged reactivation of the Proterozoic Laramie Peak Shear zone and the Cheyenne shear zone are LR50 and LR32 (Table 1; CDROM in back cover sleeve), which are at about the same modern elevation. LR50 has an AFT age of 172±9 Ma and LR32 has an AFT age of 78±9 Ma (see Figure 3), indicating that the base of the PAZ lies between these two points and is dipping toward the south; note that no major structure lies between these two points. Thus, the base of the PAZ is undulatory from south to north, with a high near the Wyoming–Colorado line, a low near the Cheyenne belt, and a high-amplitude dome in

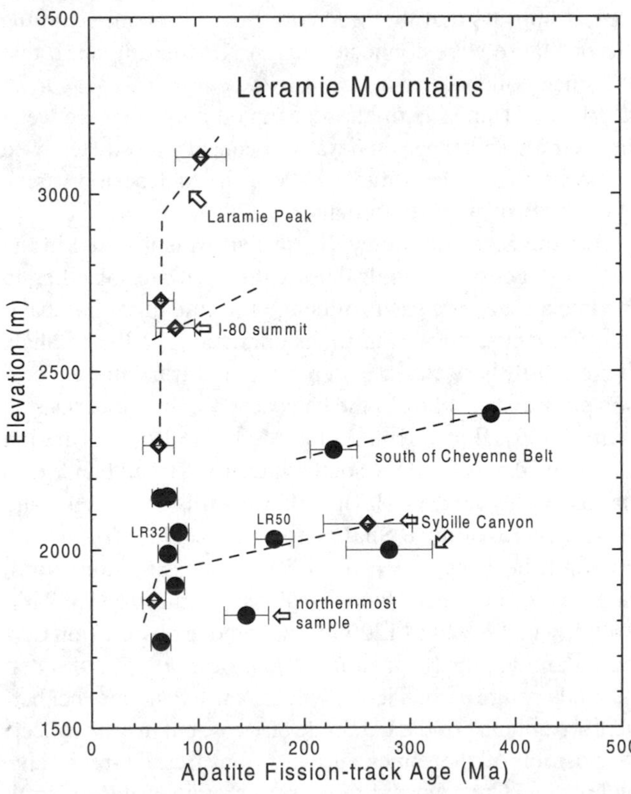

Figure 7. Age-elevation plot for the Laramie Mountains, including the data of *Cerveny* [1990], shown as open diamonds. Samples from this study are depicted as solid circles.

the Laramie Peak area. The cooling history in the Laramie Range during Laramide deformation does not appear to be strongly controlled by the Proterozoic structures, but instead reflects the formation of two doubly plunging anticlines, as envisioned by *Blackstone* [1996] (see Figure 3).

5.2. Medicine Bow Mountains

Sampling in the Medicine Bow Mountains was concentrated along a west to east traverse on Forest Road 500 between the North Platte River on the west and Albany, Wyoming on the east (Figure 4). Other samples were collected along Highway 130 and along forest roads to the north this traverse. Kevin Chamberlain provided the samples from the north end of the range. No attempt was made to do an age-elevation traverse in the Snowy Range because the Medicine Bow Quartzite that comprises the spine of the range does not contain significant quantities of apatite.

Cerveny [1990] reports five AFT ages from the Medicine Bow Mountains ranging from 100 to 140 Ma (triangles on Figure 4), with mean track lengths of 12.6 to 14.1 μm. Based on his general descriptions of the locales, there is overlap between the data sets on Centennial Ridge, near Albany, and at the Snowy Range Ski area (Figure 4); however, his data, in particular near the ski area, do not match the results presented here, which is surprising because the two data sets overlap elsewhere. The source of this discrepancy is unknown.

As was the case in the Laramie Mountains, the AFT ages from the Medicine Bow Mountains fall into two general age populations, the first group reflecting cooling during Laramide deformation and the second group associated with preserved pieces of the Late Cretaceous PAZ (Figures 4 and 8). Samples from the central part of the range recording Laramide cooling arc located northwest of the Rambler shear zone (Figure 4). These samples have AFT ages of 60 to 79 Ma and mean track lengths of 13.5 to 14.4 μm (Table 1; see CDROM in back cover sleeve). The relatively long mean track lengths, in addition to the observation that the ages are similar over a ~1.2 km elevation range (Figure 8), indicate rapid cooling during Laramide deformation. In the southeastern Medicine Bow Mountains, southeast of the Rambler shear zone, the AFT ages are 109 to 365 Ma, with mean track lengths of 10.6 to 12.1 μm, characteristic of samples within the PAZ. Note that the AFT ages from the Albany area do not correlate well with elevation (Figure 8); instead the map pattern (Figure 4) suggests that the PAZ might be tilted toward the east. The PAZ is also present on Centennial Ridge just southwest of the village of Centennial (Figure 4). Thus the block to the north of the Arlington splay of the Rambler shear zone appears to have been more strongly uplifted during Laramide deformation than the block to the south. At least 4 km of vertical motion across the Arlington fault in the central Medicine Bow Mountains is required to explain the AFT data and the 7 km of overhang near Centennial noted by *Blackstone* [1983; 1987], assuming that the fault dips about 30°. Furthermore, the preservation of the PAZ at the north end of the Medicine Bow Mountains is consistent with offset across the Arlington fault decreasing to the north [*Blackstone*, 1983; 1987].

5.3. Sierra Madre

Samples in the Sierra Madre were collected along two age-elevation traverses, one along Forest Road 409 in Archean rocks north of the Cheyenne belt, and a second on Blackhall Mountain in Proterozoic rocks south of the belt (Figure 4). A traverse on Willow Peak south of the Cheyenne belt was unsuccessful because the Proterozoic Sierra Madre granite does not contain apatite. *Cerveny* [1990] presents AFT results from an outcrop about 2 km southwest of Encampment (Figure 4); the AFT age at this point is 70.6±7.5 Ma with a mean track length of 13.7±2.1 μm, in general agreement with the results reported here.

The age-elevation plots for the profiles north and south of the belt indicate rocks on both sides of the Cheyenne belt

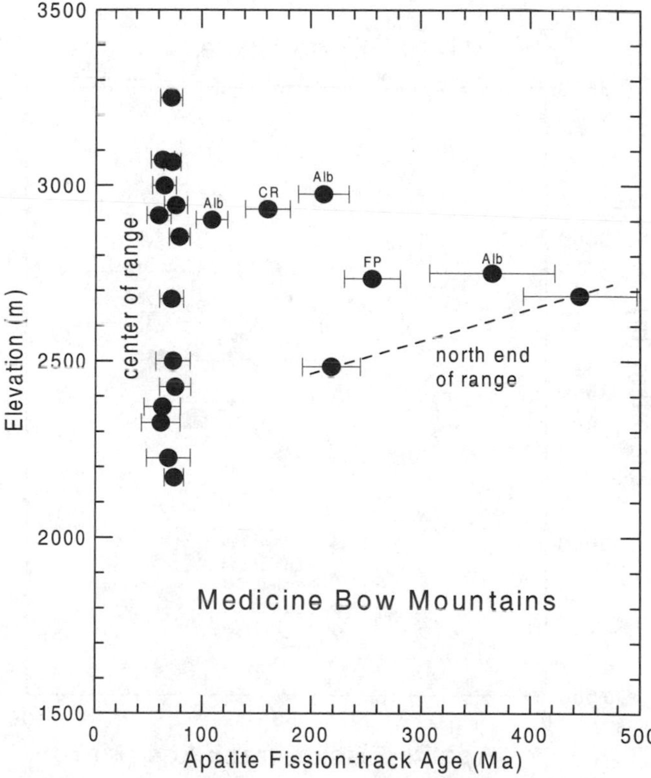

Figure 8. Age-elevation plot for the Medicine Bow Mountains, Wyoming. Alb = Albany, CR = Centennial Ridge, FP = Fox Park.

cooled during Laramide deformation (Figure 9). The Late Cretaceous PAZ is not preserved in the Sierra Madre. Although the AFT ages south of the belt are slightly younger than the ages to the north, the difference is not statistically significant. The mean track lengths for the Forest Road 409 and from 00SM07 in the western part of the range are 13.4 – 13.9 μm, consistent with the moderate cooling rate suggested by the slope of the profile in Figure 9. The mean track lengths of 13.8 to 14.2 to the south suggest somewhat more rapid cooling in this area. The limited data south of the projection of the Rambler shear zone into the Sierra Madre (Figure 4) indicate that this structure is not as profound a thermochronologic boundary as it is in the Medicine Bow Mountains.

5.4. *Park Range*

Samples were collected along four traverses in the Park Range. Three of the traverses, one along Highway 40 through Rabbit Ears Pass, one along the road through Buffalo Pass and one on a trail along Mad Creek, were sampled as a function of elevation (Figure 5). The fourth traverse was sampled at a nearly constant elevation through two fault blocks in the northern part of the range (Figure 5). *Cerveny* [1990] determined AFT ages of 61 to 65 Ma with mean track lengths of 13.4 to 14.4 μ from five samples at Buffalo Pass, comparable to the 55 to 71 Ma and 13.9 to 14.1 μm measurements from the current work in the same general area.

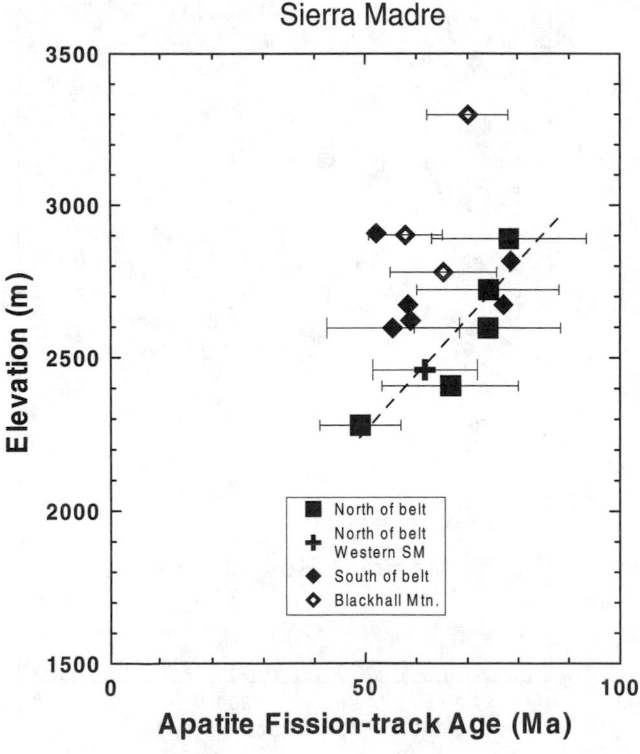

Figure 9. Age-elevation traverses from the Sierra Madre, Wyoming, showing profiles north and south of the Cheyenne belt.

The AFT ages from Rabbit Ears Pass south of the SC–FC shear zone range from 64 to 74 Ma and the mean track lengths are quite long at 14.1 to 14.9 μm, indicative of rapid cooling during Laramide deformation. The Miocene volcanism at Rabbit Ears Pass (Figure 5) did not affect the thermal history recorded by the AFT data. As mentioned previously, the AFT ages at Buffalo Pass within the SC–FC shear zone are 55 to 71 Ma. Samples from the Mt. Ethel batholith contained abundant fluorite, but also contained enough apatite for AFT analysis. The ages from Buffalo Pass show more scatter as a function of elevation compared to the Rabbit Ears Pass profile, but the data sets do overlap (Figure 10). Laramide cooling was rapid at Buffalo Pass, based on the track length data. The high elevation data collected so far in the Park Range suggest that the Late Cretaceous PAZ is not preserved here.

The AFT data from Mad Creek reveal a somewhat more complex picture of Laramide cooling north of the shear zone along the western margin of the range. Here the AFT ages are younger (46–60 Ma) and the mean track lengths are shorter (12.1 to 13.8 μm) than they are to the south. The AFT ages do not correlate well with elevation and the shortest mean track lengths are found in the middle of the profile (Figure 11). The geologic cross section constructed by *Snyder* [1980b] just to the north of this traverse provides some insight into the nature of the deformation of this block (Figure 5 and 12a). The basement has been folded, preserving outcrops of Mesozoic sedimentary rocks within the range. The younger AFT ages with short mean track lengths correspond to areas on line with the projection of the sedimentary remnants onto the Mad Creek profile. The areas beneath the downwarped sediments remained at elevated temperatures compared to the upwarped basement to the southwest and northeast during Laramide deformation. The area then completely cooled either during later stages of Laramide deformation or perhaps during post-Miocene time (note the presence of Browns Park Formation on this block; Figure 12a).

AFT data from Proterozoic rocks from two thrust-fault blocks south of Lester Mountain (Figures 5 and 12b) demonstrate that the eastern block cooled earlier (68–76 Ma) than the block to the west (51–54 Ma). The 67 ± 5 Ma on the west side of the western block is from the stratigraphically higher Triassic Chugwater Formation. This date is important because these data show that the 110°C isotherm (corresponding to a depth of ~4 km) was in the Mesozoic section ~200 m above the Great Unconformity prior to Laramide deformation.

One sample from the Mt. Ethel pluton on the east side of the range has an age of 55 Ma, similar to ages elsewhere in the range, but it has a short mean track length of 10.8 μm. This

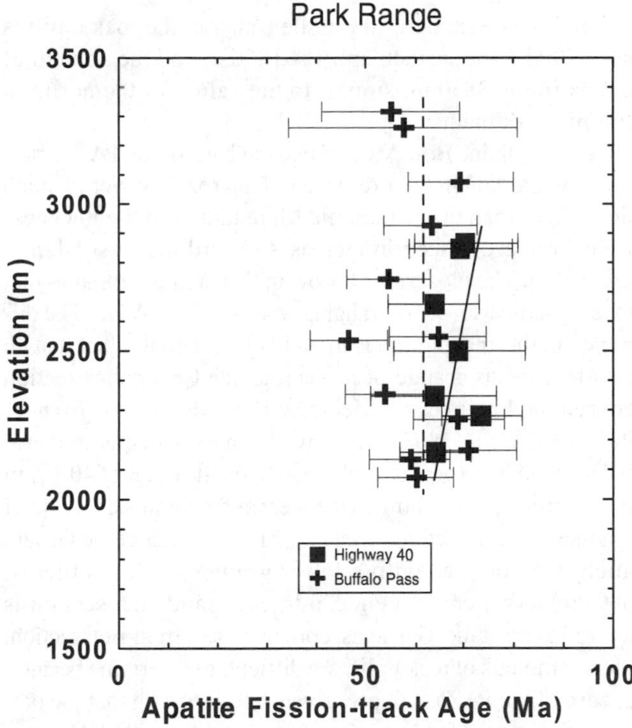

Figure 10. Age-elevation traverses from the southern (Highway 40–Rabbit Ears Pass) and central (Buffalo Pass) parts of the Park Range.

sample is near the high angle reverse fault that forms the eastern boundary of the range and it is close to a fluorite and uranium mine near Red Canyon. The significance of this age is uncertain.

Although the AFT ages north of the SC–FC shear zone are slightly younger than those to the south, the difference is not statistically significant. Thus, the shear zone apparently was not strongly reactivated during Laramide deformation, which is consistent with the minor amount of brittle deformation in the zone. The relatively undeformed 1.4 Ga Mount Ethel pluton seems to have been more important in controlling the location of Laramide thrust faults in the Park Range.

5.5. *Amount of Denudation Across the Region*

The amount of denudation, which is the result of both Laramide and post-Laramide processes, appears to increase toward the west between the Laramide Mountains and the Park Range. The base of the PAZ in the Laramie Mountains is preserved at modern elevations ranging from ~1900 to ~3000 m. The actual base of the PAZ is not exposed in the Medicine Bow Mountains, but the lowest sample within the PAZ is at a current elevation of ~2700 m in the southern part of the range and at 2500 m in the north. In contrast, Laramide cooling ages prevail at elevations of ~3300 m in the Sierra Madre and the Park Range; the PAZ has been removed by denudation.

Maximum estimates for the amount of section between the base of the Pennsylvanian section, which rests on Precambrian rocks, to the top of the Fox Hills Sandstone in the area ranges from 4152 m adjacent to the Medicine Bow Mountains [*Houston et al.*, 1968], 2883 m next to the Laramie Mountains in the northern Denver Basin [*Lowry and Crist*, 1967], and 2125–2415 m in North Park near the Park Range [*Hail*, 1965; 1968; *Snyder et al.*, 1987]. The North Park sections are truncated by late Cretaceous erosion [*Izett et al.*, 1971]. Using this information and the position of the base of the PAZ in the Laramie Range, and by making assumptions about the heat flow at the end of Cretaceous time, estimates of rock uplift, surface uplift, and denudation can be determined.

The temperatures in the Laramie Range at the end of Cretaceous time prior to Laramide deformation are estimated using the preserved sediment thickness, average thermal conductivity values [e.g., *Carter et al.*, 1998], and estimated heat flow values of 50 to 60 mW/m^2, which lie in the range of measured heat flow of 25 to 74 mW/m^2 for the Laramie Mountains [*Decker et al.*, 1980; 1988]. The estimated temperature-depth curve shown in Figure 13 shows that the 110°C isotherm, corresponding to the base of the PAZ, was at a depth of about 3800 m below sea level at the end of Cretaceous time, if a heat flow of 50 mW/m^2 is assumed. This heat flow value is consistent with the measurements closest to the northern Laramie Mountains [*Decker et al.*, 1988]. The model predicts that the base of the PAZ is within the Archean basement, as recorded in the AFT data of *Cerveny* [1990]. A higher heat flow of 60 mW/m^2 places the base of the PAZ in the Jurassic section (Figure 13), which is

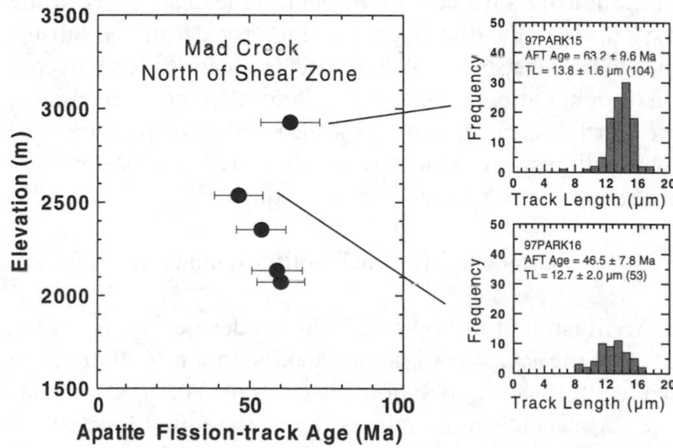

Figure 11. Age-elevation and track length data from the Mad Creek area, Park Range.

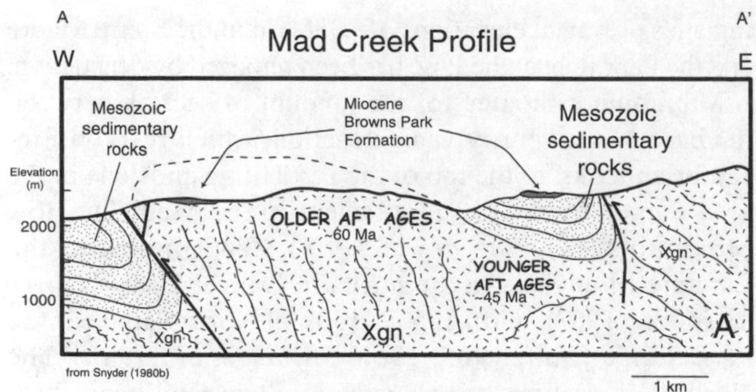

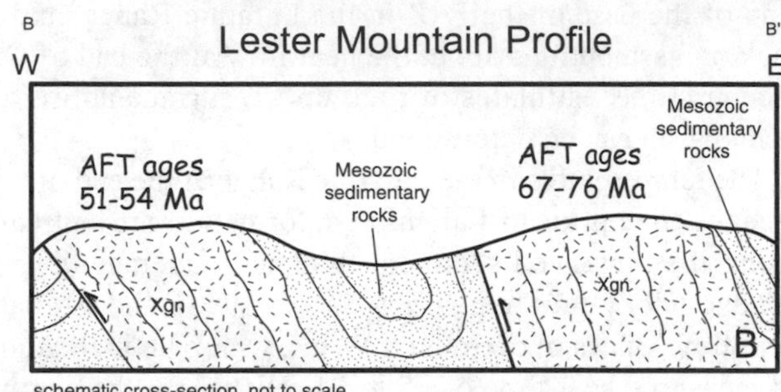

Figure 12. (a) Geologic cross section from *Snyder* [1980b] showing the relationship of folding to AFT age distribution along Mad Creek. (b) Schematic geologic cross section based on maps of *Snyder* [1980b,c] showing the difference in cooling history in two thrust sheets in the northern Park Range.

not compatible with observations. At this point, we can calculate the amount of rock uplift, surface uplift, and denudation following the methods of *Fitzgerald et al.* [1995]. Surface uplift and rock uplift are both tied to the same frame of reference, mean sea level. Surface uplift is the vertical motion of the Earth's surface with respect to mean sea level, while rock uplift is the displacement of the rock from the subsurface toward the surface with respect to mean sea level. In contrast, denudation is tied to a different frame of reference, the Earth's surface, and it is a measure of displacement of rock with respect to the surface. The three modes of motion are related through the following equation:

$$\text{surface uplift} = \text{rock uplift} - \text{denudation}$$

As illustrated in Figure 14, the modern elevation of the base of the PAZ on Laramie Peak is about 3000 m above sea level, thus the rock uplift is 6800 m. The modern average surface elevation in the northern Laramie Mountains is ~2400 m. Adjusting for the fact that sea level was about 200 m higher in late Cretaceous time than it is now, the amount of surface uplift is 2200 m. Consequently, the amount of denudation is ~3800 m. In Sybille Canyon, the rock uplift is only 5700 m, the surface uplift is 1850 m and the amount of denudation is 3850 m, similar to the value for the northern Laramie Mountains.

In the Medicine Bow Mountains, the base of the PAZ is preserved, even though the preserved Phanerozoic cover is much thicker here than in the Laramie Mountains and the thickness of the Cretaceous section increases toward the west [*Weichman*, 1961]. The modern heat flow in this area is similar to that in the Laramie Mountains, ranging from 41–68 mW/m^2. The difference in thermal regime is most likely due to the fact that an important facies change occurs in the late Cretaceous section between the Laramie and Medicine Bow Mountains [*Weimer*, 1961; *Steidtmann*, 1993]. The low-thermal conductivity Pierre Shale, which has a geothermal gradient on the order of 40°C/km through this section, thins to the west and contains a tongue of high thermal conductivity sandstone in the Mesa Verde Group, which is associated with a lower gradient on the order of 25°C/km. As a consequence, a thicker sand-rich section is required to reset the AFT ages, compared to a mud-rich section. Precise amounts of rock uplift are difficult to determine because the base of the PAZ is not preserved, but assuming that the preserved thickness of sediments nearly coincides with the base of the PAZ, and that the lowest elevation sample in the southern Medicine Bow Mountains is near the break-in-slope, then the amount of rock uplift is ~6800 m.

The lack of a PAZ at elevations of 3300 m in the Sierra Madre Range may imply greater rock uplift here than in the southern Medicine Bow Mountains. The modern heat flow values are quite low north of the Cheyenne belt in the Sierra Madre, on the order of 33–40 mW/m^2 [*Decker et al.*, 1988]. In contrast, the lack of a PAZ in the Park Range may be related to higher heat flow south of the Cheyenne belt. Figure 13 illustrates quite clearly that a slight increase in heat flow can dramatically influence the position of the 110°C isotherm. We have some evidence that the base of the PAZ was up in the Phanerozoic section in the Park Range (97Park30). The heat flow increases abruptly from 33–40 in the Sierra Madre to 80 to 103 mW/m^2 south of the Wyoming–Colorado state line [*Decker et al.*, 1988]. *Decker et al.* [1988] explained this step in heat flow as a difference in crustal heat production between the Archean and Proterozoic basement, with the Proterozoic basement being more radiogenic. This step in heat flow is likely related to Proterozoic lithospheric assembly processes rather than the modern difference in the thermal state of the mantle beneath Wyoming and Colorado [*Dueker et al.*, 2001]. *Decker at al.* [1988] argue that given the short half-width (~35 km) of the anomaly, the source of this step was has to be quite shallow, in the upper 5 to 10 km of the crust. Given these complications and the lack of PAZ preservation, no attempt is made to determine rock uplift in the Park Range.

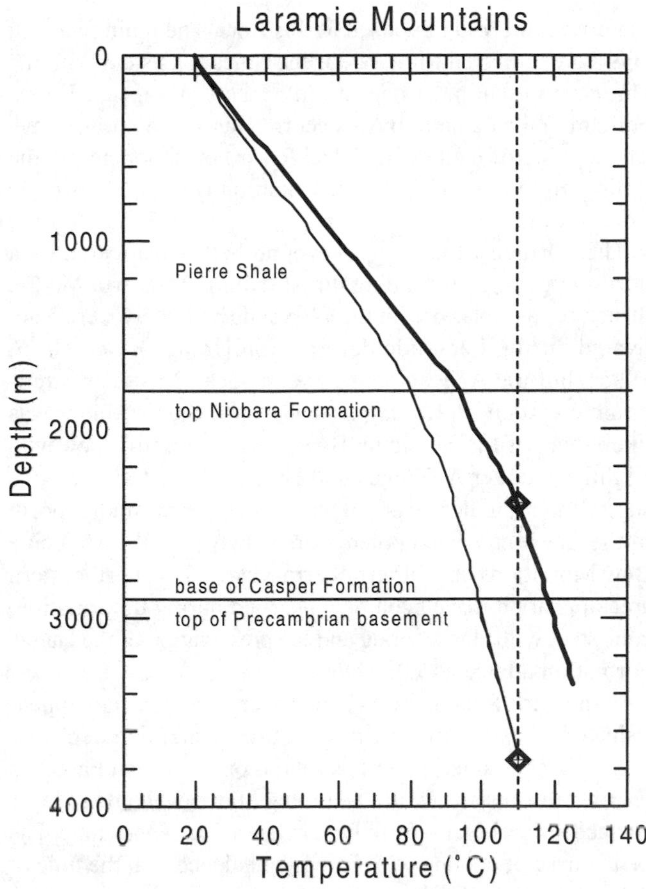

Figure 13. Calculated temperature-depth curve using the thermal resistance equation of *Bullard* (1939) assuming the sediment thickness of *Lowry and Crist* [1967] and the thermal conductivity estimates for certain lithologies of *Carter et al.* [1998]. The position of the 110°C isotherm (dashed line) and the base of the PAZ (diamond) are shown for heat flow values of 50 mW/m^2 (light line) and 60 mW/m^2 (heavy line).

5.6. Comparison to Other Thermochronologic Studies in the Region

The AFT data from the Laramie and Medicine Bow mountains, the Sierra Madre, and Park Range add to our understanding of Laramide cooling and deformation in the Southern Rocky Mountains and can be tied to results in the Front Range and the Gore Range. *Kelley and Chapin [2004]* have mapped the position of the base of the late Cretaceous AFT PAZ in the Front Range in Colorado. The base of the PAZ is missing from the northeastern part of the Front Range; all AFT ages north of Golden Gate Canyon on the eastern margin of the Front Range are Laramide cooling ages (45 to 70 Ma), similar to the ages in the Park Range and Sierra Madre. *Kelley and Chapin [2004]* have been able to document that, at a distance of ~30 km south of the Colorado Mineral belt, the base of the PAZ warps upward about 1 km in elevation. *Kelley and Chapin [2004]* suggest that the base of the PAZ south of the Mineral belt was warped up due to the thermal effects of Laramide plutonism. However, the base of the PAZ does not show symmetry across the Mineral belt. The base of the PAZ does not reappear along the eastern margin of the range until just south of the Cheyenne belt in the Laramie Range, more than 150 km north of the axis of the Mineral belt. This distance is much further north than expected for simple thermal effects related to the Laramide plutons in the Colorado Mineral belt (Figure 1). Similarly, *Cerveny* [1990] collected about five samples on Clark Peak, which lies on the southern end of the extension of the Medicine Bow Range into Colorado (Figure 1). *Cerveny* [1990] found an 81 Ma AFT age at the summit of Mt. Clark and AFT ages >100 Ma with short mean track lengths on the west side of Mt. Clark at low elevation. A west-dipping PAZ is present on the west side of the range; this area is ~100 km north of the axis of the Mineral belt. The thickness of the Pierre Shale approximately doubles across the Mineral belt and then thins northward across the Laramie Range [*Scott and Cobban*, 1965]. Consequently, variations in the thickness of the Pierre Shale may in part explain the distribution of the base of the PAZ in northern Colorado and southern Wyoming.

Naeser et al. [2003] recently completed an AFT analysis of the Gore Range 25 km south of Rabbit Ears Pass in the Park Range. In contrast to the 50–76 Ma Laramide cooling ages in the Park Range, the AFT ages in the Gore Range are 16 to 37 Ma on the west side of the range and are 5 to 19 Ma on the east side of the range. The AFT ages from the Gore Range reflect middle to late Tertiary cooling, which *Naeser et al.* [2003] associate with middle Tertiary heating and subsequent rift flank denudation related to the Rio Grande rift. The exact location of the transition between these two areas of contrasting thermal history is not known, but marked change in topography located near the Colorado River between the rugged Gore Range to the south and the flat-topped Park Range to the north is a likely candidate.

6. SUMMARY

The AFT data in the Laramie Range do not record Laramide reactivation of the Cheyenne belt. The emplacement of the 1.4 Ga Laramie Anorthosite complex and Sherman Granite along the suture effectively healed this zone of weakness, at least in the Laramie Range proper. Geologic evidence [*Blackstone*, 1996] indicates that this structure was reactivated to the northeast of the range during Laramide and post-Laramide deformation. Farther north, the cooling history across the Laramie Peak shear zone is similar on either side of this feature. Thus, in the Laramie Range, the AFT data do not reveal

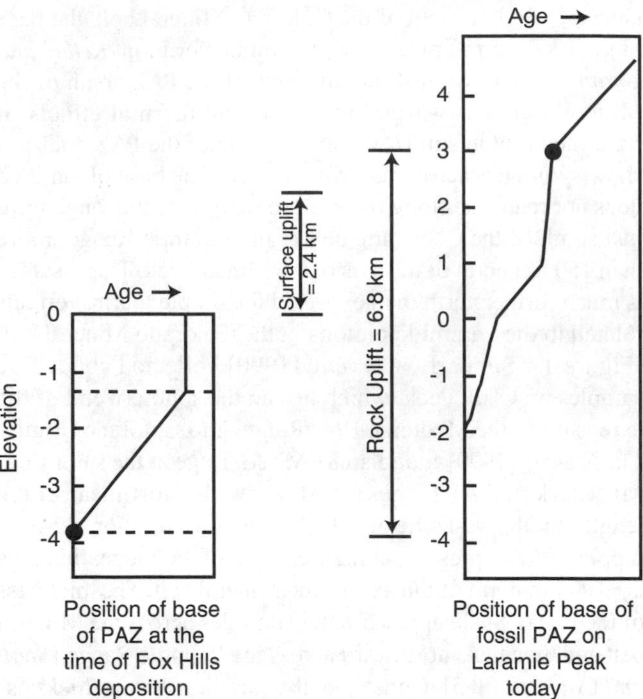

Figure 14. Plot of the estimated depth of the base of the PAZ at the end of Cretaceous time using the temperature curve for 50 mW/m^2 in Figure 13. The black circle show the position of the base of the PAZ in late Cretaceous time and today; the difference in elevation between these points is a measure of rock uplift. The modern elevation of the northern Laramie Mountains is derived from topographic profiles.

a significant difference in cooling history across the Proterozoic boundaries preserved in this mountain chain. Instead, the data are consistent with long-wavelength folding of the basement, with highs in the vicinity of Interstate 80 in the south and Laramie Peak in the north, with an intervening low along the Cheyenne belt, mirroring the topographic profile of the range. This interpretation is consistent with the observation of *Blackstone* [1996] concerning the dispersal of Paleocene sediments. The AFT ages can be used to infer that unroofing occurred 82 to 65 Ma. The AFT ages related to Laramide deformation here are generally older than they are to the west.

In the Medicine Bow Mountains, where the northeast-trending Cheyenne belt is perhaps best defined, the suture was definitely reactivated during Laramide deformation. The Rambler shear zone and the Arlington splay of the Rambler shear separate AFT ages of 60 to 79 Ma to the northwest from >100 Ma AFT ages to the southeast. The AFT ages suggest that the Medicine Bow Mountains began unroofing in latest Cretaceous time, perhaps contributing sedimentary clasts to the Medicine Bow Formation in the Laramie Basin. The basement was likely not exposed at this time. The main phase of cooling occurred during Paleocene time, consistent with the observation that most deposition in the synorogenic Hanna Formation the Laramie Basin occurred during Paleocene time. The lapping of the Eocene Wind River Formation across the Arlington fault implies that deformation was largely over by Eocene time.

The northeast trending Cheyenne belt is truncated by a north-vergent, east-trending thrust fault in the Sierra Madre. Structures associated with the Cheyenne belt likely were reactivated during Laramide deformation [*Houston and Graff*, 1995], but the AFT evidence is equivocal. Based on stratigraphic evidence [*Montagne*, 1991], the Sierra Madre was likely part of the Medicine Bow Laramide uplift; however, slightly younger AFT ages and shorter mean track lengths suggest that the northeastern part of the Sierra Madre, north of the Cheyenne belt, cooled more slowly than the Medicine Bow Mountains or southern Sierra Madre. This northeastern area appears to have been downwarped during deformation, consistent with slow cooling and the preservation of the Hanna Formation adjacent to this block.

In the Park Range, the SC–FC shear zone does not appear to have been strongly reactivated during Laramide deformation. AFT ages at the highest elevations on the Rabbit Ears Pass profile are consistent with the timing of removal of the Upper Cretaceous section in North Park in latest Cretaceous time. The main phase of cooling was during the Paleocene at the time of deposition of the Coalmont Formation. AFT and geologic evidence can be used to demonstrate that the basement was folded above thrust faults on the western margin of the range, and that the timing of thrusting appears to become younger toward the west. The undeformed 1.4 Ga Mt. Ethel pluton does seem to influence the style of Laramide deformation in the range. The Park Range is essentially a anticline to the south of the pluton and consists of a series of west-vergent thrust blocks north of the pluton.

Certainly, the idea of Laramide re-activation of Proterozoic structures related to lithospheric assembly in southern Wyoming and northern Colorado is not new, but this work serves to document more precisely which structures were important in controlling Laramide-aged denudation. The data also serve to emphasize the importance of basement folding during Laramide deformation. Finally, the AFT data from this project provide further documentation of a trend observed elsewhere in the Southern Rocky Mountains region. The AFT data from this study, for the most part, record only one phase of Laramide deformation in the area studied and do not indicate significant late Laramide or post-Laramide cooling events. In contrast, AFT data from the Front Range and Wet Mountains of Colorado record a mild middle Oligocene to early Miocene thermal event that is superimposed on a predominantly Laramide cooling history [*Kelley and Chapin, 2004*; *Naeser et al.*, 2003]. The middle Oligocene to early Miocene thermal event and subsequent cooling are the dominant events recorded by AFT data in the Gore Range [*Naeser et al.*, 2003], Rio Grande rift in southern Colorado and New Mexico [*Kelley et al.*, 1992] and the High Plains of northeastern New Mexico [*Kelley and Chapin*, 1995].

Acknowledgements. I wish to thank Tom Foster and his student, Mike Barinek, for their help in obtaining samples in the Park Range and Kevin Chamberlain for providing apatite separates. Charles Chapin provided many beneficial references and discussions, and he reviewed an early version of this manuscript. Discussions with Matt Heizler were also quite helpful. I especially want to thank Karl Karlstrom for the opportunity to participate in the CD–ROM project. James Steidtmann, John Murphy, Phillip Cerveny, and Charles Naeser reviewed the paper. This project was funded by the NSF Continental Dynamics Program grants EAR9614787 and EAR0003540.

REFERENCES

Barinek, M.F., C.T. Foster, and P.P. Chaplinsky, Metamorphism and deformation near the ~1.4 Ga Mount Ethel pluton, Park Range, Colorado, *Rocky Mountain Geology*, *34*, 22–33, 1999.

Blackstone, D.L., Jr., Structural geology of the Rex Lake quadrangle, Laramie Basin, Wyoming, *Wyoming State Geological Survey Preliminary Report 11*, 17 pp., 1970.

Blackstone, D.L., Jr., Late Cretaceous and Cenozoic history of the Laramie Basin region, in *Cenozoic History of the Southern Rocky Mountains*, edited by B.E.Curtis, Geol. Soc. Am. Mem. 144, 249–278, 1975

Blackstone, D.L., Jr., Independence Mountain thrust fault, North Park Basin, Colorado, *Univ. of Wyoming Contributions to Geology*, *16*, 1–16, 1977.

Blackstone, D.L., Jr., Laramide compressional tectonics, southeastern Wyoming, *Univ. of Wyoming Contributions to Geology*, 22, 1–37, 1983.

Blackstone, D.L., Jr., Northern Medicine Bow Mountains, Wyoming: Revision of structural geology, northeast flank, *Univ. of Wyoming Contributions to Geology*, *25*, 1–7. 1987.

Blackstone, D.L., Jr., Structural geology of the Laramie Mountains, southeastern Wyoming and northeastern Colorado, *Wyoming State Geological Survey Report of Investigations*, *51*, 128 p., 1996.

Bullard, E.C., Heat flow in South Africa, *Proceedings Royal Society London, Series A*, 173, 474–502, 1939.

Carter, L.S., S.A. Kelley, D.D. Blackwell, and N.D. Naeser, Heat flow and thermal history of the Anadarko Basin, Oklahoma, *American Association of Petroleum Geologists Bulletin*, *82*, 291–316, 1998.

Cerveny, P.F., III, Fission-track thermochronology of the Wind River Range and other basement cored uplifts in the Rocky Mountain foreland, Ph.D. dissertation, Univ. of Wyoming, 189 pp., 1990.

Chamberlain, K. R., S.C. Patel, B.R. Frost, and G.L. Snyder, Thick-skinned deformation of the Archean Wyoming Province during Proterozoic arc-continent collision; with Suppl. Data 9339, *Geology*, *21*, 995–998, 1993.

Chamberlin, R.J., Basement control in Rocky Mountain deformation, *Am. Jour. Sci.*, *243A*, 98–116, 1945.

Condie, K.C., Petrology and geochemistry of the Laramie batholith and related metamorphic rocks of Precambrian age, eastern Wyoming, *Geol. Soc. Amer. Bull.*, *80*, 57–82, 1969.

Decker, E.R., K.H. Baker, G.J. Bucher, and H.P. Heasler, Preliminary heat flow and radioactivity studies in Wyoming, *Jour. Geophys. Res*., *85*, 311–321, 1980.

Decker, E.R., H.P. Heasler, K.L. Buelow, K.H. Baker, and J.S. Hallin, Significance of past and recent heat-flow and radioactivity studies in the Southern Rocky Mountains region, *Geol. Soc. Am. Bull.*, *100*, 1851–188., 1988.

Dueker, K., Yuan, H., and Zurck, B., Thick Proterozoic lithosphere of the Rocky Mountain region, *GSA Today*, *11*, 4–9, 2001.

Donelick, RA , Etchable fission track length reduction in apatite: experimental observations, theory, and geological applications. Ph.D. Dissertation, 414 p., Renselaer Polytechnic Institute, Troy, NY, 1988.

Evanoff, E., Early Oligocene paleovalleys in southern and central Wyoming; evidence of high local relief on the late Eocene unconformity, *Geology*, *18*, 443–446, 1990.

Fitzgerald, P.G., R.B. Sorkhabi, T.F. Redfield, and E. Stump, Uplift and denudation of the central Alaska Range: a case study in the use of apatite fission-track thermochronology to determine absolute uplift parameters, *J. Geophys. Res.*, *100*, 20175–20191, 1995.

Foster, C.T., M.K. Reagan, S.G. Kennedy, G.A. Smith, C.A. White, J.E. Eiler, and J.R. Rougvie, Insights into the Proterozoic geology of the Park Range, Colorado, *Rocky Mountain Geology*, *34*, 7–20, 1999.

Galbraith, R .F., On statistical models for fission track counts, *Mathematical Geology*, 13, 471–478, 1981.

Green, P.F., I.R. Duddy, A.J.W. Gleadow, P.R. Tingate, and G.M. Laslett, Thermal annealing of fission tracks in apatite, 1. A qualitative description., *Chemical Geology (Isotope Geoscience Section)*, *59*, 237–253, 1986.

Gries, R., North-south compression of Rocky Mountain foreland structures, in *Rocky Mountain Foreland Basin and Uplifts*, edited by J.D.Lowell, Rocky Mountain Association of Geologists, 9–32, 1983.

Hail, W.J., Jr., Geology of the northwestern part of North Park, Colorado, *U.S. Geol. Surv. Bull.*, *1188*, 133pp., 1965.

Hail, W.J., Jr., Geology of the southwestern part of North Park, Colorado, *U.S. Geol. Surv. Bull.*, *1257*, 119 pp., 1968.

Houston, R.S., and P.J. Graff, Geologic map of Precambrian rocks of the Sierra Madre, Carbon County, Wyoming, and Jackson and Routt counties, Colorado, *U.S. Geol. Surv. Map I–2452*, 1995.

Houston, R.S., and K.E. Karlstrom, Geologic map of Precambrian metasedimentary rocks of the Medicine Bow Mountains, Albany and Carbon Counties, Wyoming: *U.S. Geol. Surv. Misc. Invest. Map I 2280*, 1992.

Houston, R.S., and 15 others, A regional study of rock of Precambrian age in that part of the Medicine Bow Mountains lying in southeastern Wyoming-with a chapter on the relationship between Precambrian and Laramide structure, *Geological Survey of Wyoming Memoir 1,* 167 p., 1968.

Houston, R. S., E.M. Duebendorfer, K.E. Karlstrom, and W.R. Premo, A review of the geology and structure of the Cheyenne belt and Proterozoic rocks of southern Wyoming, in *Proterozoic Geology of the Southern Rocky Mountains*, J.A. Grambling, and B.J. Tewksbury , Geol. Soc. Amer. Special Paper, 235, p. 1–12, 1989.

Hurford, A.J., and P.F. Green, The zeta age calibration of fission-track dating, *Isotope Geoscience*, 1, 285–317, 1983.

Izett, G.A., Geology of the Hot Sulphur Springs quadrangle, Grand County, Colorado, *U.S. Geol. Surv. Prof. Paper 586*, 79 p., 1968.

Izett, G.A., Cobban, W.A., and Gill, J.R., The Pierre Shale near Kremmling, Colorado and its correlation to the east and west*: U.S. Geol. Surv. Prof. Paper 684–A*, 19 pp, 1971.

Karlstrom, K.E., and R.S. Houston, The Cheyenne belt: analysis of a Proterozoic suture in southern Wyoming, *Precambrian Research*, *25*, 415–446, 1984.

Kelley, S.A., and C.E. Chapin, Denudational histories of the Front Range and Wet Mountains, Colorado, based on apatite fission-track thermochronology, in *Tectonics, geochronology and volcanism in the Southern Rocky Mountains and Rio Grande rift* edited by S.M. Cather, W.C. McIntosh, and S.A. Kelley, New Mexico Bureau of Geology and Mineral Resources Bulletin 160, 2004, in press.

Kelley, S.A., and C.E. Chapin, Apatite fission-track thermochronology of the Southern Rocky Mountains – Rio Grande rift - western High Plains provinces, New *Mexico Geological Society Guidebook 46*, 87–96, 1995.

Kelley, S.A., C.E. Chapin, and J. Corrigan, Late Mesozoic to Cenozoic cooling histories of the flanks of the northern and central Rio Grande rift,

Colorado and New Mexico, New *Mexico Bureau of Mines and Mineral Resources Bulletin 145*, 39 pp., 1992.

Knight, S.H., Summary of the Cenozoic history of the Medicine Bow Mountains, *Wyoming Geological Association Guidebook 8*, p. 65–77, 1953.

Laslett, G. M., W. S. Kendall, A. J. W. Gleadow, and I. R. Duddy, Bias in measurement of fission-track length distribution, *Nuclear Tracks*, 6, 79–85, 1982.

Lillegraven, J.A., Correlation of Paleogene strata across Wyoming – a users' guide, in *Geology of Wyoming*, edited by A.W. Snoke, J.R. Steidtmann, S.M. Roberts, Geological Survey of Wyoming Memoir, 5, p. 414–477, 1993.

Lowry, M.E., and M.A. Crist, Geology and ground-water resources of Laramie County, Wyoming, *U.S. Geological Survey Water Supply Paper 1834*, 71 pp., 1967.

Montagne, J., Cenozoic history of the Saratoga Valley area, Wyoming and Colorado, *University of Wyoming Contributions to Geology*, 29, 13–70, 1991.

Naeser, C.W., Fission-track dating and geologic annealing of fission tracks, in *Lectures in Isotope Geology*, edited by E. Jager, and J.C. Hunziker, New York, Springer-Verlag, 154–169, 1979.

Naeser, C.W., Bryant, B., Kunk, M.J., Kellogg, K., Donelick, R.A., and Perry, W.J., Jr., Tertiary cooling and tectonic history of the White River Uplift, Gore Range, and western Front Range, central Colorado: evidence from fission-track analysis, in press, 2003.

Prichinello, K.A., Earliest Eocene mammalian fossils from the Laramie Basin of southeastern Wyoming, *University of Wyoming Contributions to Geology*, *10*, 73–87, 1971.

Patel, S.C., B.R. Frost , K.R. Chamberlain and G.L. Snyder, Proterozoic metamorphism and uplift history of the north-central Laramie Mountains, Wyoming, USA, *J. Met. Geol.*, *17*, 243 – 258, 1999.

Pazzaglia, F.J., and S.A. Kelley, Large-scale geomorphology and fission-track thermochronology in topographic and exhumation reconstructions of the Southern Rocky Mountains, *Rocky Mountain Geology*, *33*, 229–257, 1998.

Ransome, F.L., Tertiary orogeny of the North American Cordillera and its problems, *Problems of American Geology*, New Haven, Connecticut, p. 287–376, 1915.

Scott, G.R., and W.A. Cobban, Geologic and biostratigraphic map of the Pierre Shale between Jarre Creek and Loveland, Colorado, *U.S. Geological Survey Map I–439,* 1965.

Snyder, G. L., Intrusive rocks northeast of Steamboat Springs, Park Range, Colorado, *U.S. Geol. Surv. Prof. Paper 1041*, 42 pp., 1978.

Snyder, G.L., Geologic map of the central part of the northern park Range, Jackson and Routt Counties, Colorado, U. S. Geol. Surv. Misc. Inv. Map I-1112, 1980a.

Snyder, G.L., Geologic map of the northernmost Park Range and southernmost Sierra Madre, Jackson and Routt Counties, Colorado, U. S. Geol. Surv. Misc. Inv. Map I-1113, 1980b.

Snyder, G.L., Geologic map of the northernmost Gore Range and southernmost Park Range, Grand, Jackson and Routt Counties, Colorado, U. S. Geol. Surv. Misc. Inv. Map I-1114, 1980c.

Snyder, G. L., L.L. Patten, J.J. Daniels, Mineral resources of the Mount Zirkel Wilderness and northern Park Range vicinity, Jackson and Routt counties, Colorado, U.S. Geol. Surv. Bull. 1554, p. 13–55, , 1987.

Steidtmann, J.R., The Cretaceous foreland basin and its sedimentary record, in *Geology of Wyoming*, edited by A.W. Snoke, J.R. Steidtmann, S.M. Roberts, Geological Survey of Wyoming Memoir, 5, 250–271, 1993.

Stone, D.S., Morphology of the Casper mountain uplift and related subsidiary structures, central Wyoming: implications for Laramide kinematics, dynamics, and crustal inheritance: *AAPG Bulletin*, *86*, 1417–1440, 2002.

Tweto, Ogden, and Sims, P.K, Precambrian ancestry of the Colorado mineral belt, *Geological Society of America Bulletin*, *74*, 991–1014, 1963.

Tweto, O., Laramide (Late Cretaceous-early Tertiary) orogeny in the southern Rocky Mountains, in *Cenozoic history of the Southern Rocky Mountains*, edited by B.F. Curtis, Geol. Soc. Amer. Mem. 144, 1–44, 1975.

Weichman, B.E., Regional correlation of the Mesaverde Group and related rocks in Wyoming, *Wyoming Geological Association Guidebook 16*, 29–33, 1961.

Weimer, R.J., Uppermost Cretaceous rocks in central and southern Wyoming, *Wyoming Geological Association Guidebook 16,* 17–28, 1961.

Shari Kelley, Department of Earth and Environmental Science, New Mexico Institute of Mining and Technology, Socorro, NM 87801

The Proterozoic Ancestry of the Colorado Mineral Belt: 1.4 Ga Shear Zone System in Central Colorado

Annie M. McCoy[1], Karl E. Karlstrom, and Colin A. Shaw[2]

Department of Earth and Planetary Sciences, University of New Mexico, Albuquerque, New Mexico

Michael L. Williams

Department of Geosciences, University of Massachusetts, Amherst, Massachusetts

A northeast-striking system of subvertical mylonites and ultramylonites, which formed in the Mesoproterozoic, provided a zone of weakness and conduit for the Paleocene to Oligocene magmatism and mineralization that are the Phanerozoic expressions of the Colorado Mineral Belt. The mylonites overprinted higher temperature Paleoproterozoic high-strain domains of similar orientation. Here, we distinguish the Phanerozoic Colorado Mineral Belt from a Proterozoic 'Colorado Mineral Belt shear zone system.' In each segment of the shear zone system, Mesoproterozoic mylonite strands, which are meters to tens of meters wide, overprint higher-temperature Paleoproterozoic high-strain domains, which are several kilometers wide. In situ electron microprobe monazite dating of the mylonites and higher temperature high-strain domains, and field studies of relative timing of shearing and pluton emplacement, show two main periods of shearing that each involve ~100 Ma of deformation. Higher temperature high-strain domains record pulses of deformation that occurred at 1.71-1.69 Ga, 1.67 Ga, 1.65 Ga, and 1.62 Ga. Mylonites record movement at 1.45 Ga synchronous with emplacement of the Mt. Evans pluton, at 1.42 Ga synchronous with emplacement of the Silver Plume pluton, and at 1.38 Ga synchronous with emplacement of the St. Kevin pluton. Post-1.38 Ga movements created ultramylonites. This shear zone system may be analogous to modern-day intracontinental zones of weakness like the Tien Shan of central Asia, which record both original assembly of tectonic blocks and recurrent reactivation during later plate convergence at a distant margin.

1. INTRODUCTION

The Colorado Mineral Belt shear zone system is defined here as a series of northeast-striking mylonitic and ultramylonitic shear zone segments that acted as a coherent shear zone system between 1.45 and 1.38 Ga. Each mylonitic shear zone segment overprints older Paleoproterozoic structures of similar orientation, and although these older structures are present throughout Colorado and are not unique to the Col-

[1] Presently at John Shomaker & Associates, Inc., Albuquerque, New Mexico

[2] Presently at Department of Geology, University of Wisconsin, Eau Claire, Wisconsin

The Rocky Mountain Region: An Evolving Lithosphere
Geophysical Monograph Series 154

10.1029/154GM06

orado Mineral Belt region, their presence suggests a common Paleoproterozoic ancestry to the shear zone system. Along the shear zone system, there is evidence for multiple episodes of reactivation throughout the Proterozoic and the Phanerozoic.

This paper presents a study of the structures, kinematic history, and timing of movement along each shear zone segment, and defines a 'tectonic fingerprint' for the shear zone system as a whole. The focus of this paper is to document the Proterozoic Colorado Mineral Belt shear zone system, and to illuminate the initiation and early evolution of this long-lived zone of weakness in the lithosphere.

2. AN INTRACONTINENTAL ZONE OF DEFORMATION FROM THE PROTEROZOIC TO THE PHANEROZOIC

The Colorado Mineral Belt is generally defined as a ~200 kilometer (km) long, northeast-striking zone in central Colorado marked by a concentration of Paleocene to Oligocene intrusions and related mineral deposits emplaced during and after the Laramide orogeny [Figure 1; *Tweto and Sims*, 1963; *Mutschler et al.*, 1987]. Mining along the Colorado Mineral Belt is famous for having produced billions of dollars worth of gold, silver, lead, zinc, molybdenum, tungsten, and fluorspar [*Tweto and Sims*, 1963]. The Colorado Mineral Belt is also defined by a pair of negative Bouguer gravity anomalies that are among the most negative in the United States [Figure 1; *Isaacson and Smithson*, 1976]. Modeling results by McCoy et al. [this volume] and Isaacson and Smithson [1976] suggest that the Colorado Mineral Belt anomaly may be explained in part by large, relatively low-density bodies in the crust, such as granitic batholiths, centered beneath the Colorado Mineral Belt.

The more southerly anomaly is centered on the San Juan Mountains, located along the southwestern extension of the Colorado Mineral Belt. The San Juan Mountains contain middle and late Tertiary (mainly Oligocene) magmatic centers, and its negative Bouguer gravity anomaly is similar to the more northerly anomaly of the Colorado Mineral Belt. Like the Colorado Mineral Belt, the San Juan Mountains show evidence for Proterozoic deformation along steeply-dipping structures [*Tweto and Sims*, 1963; *Baars et al.*, 1984]. In light of

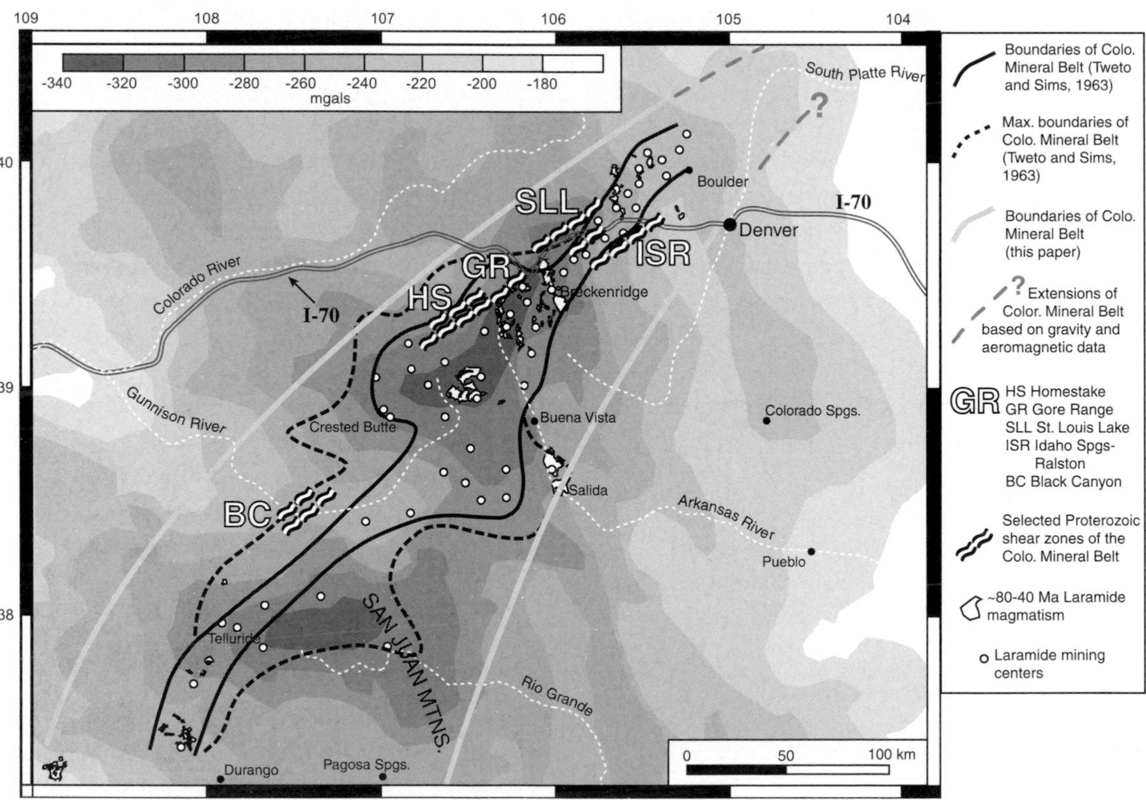

Figure 1. Map of Bouguer gravity data after Oshetski and Kucks (2001) and McCoy et al. (2001) with Colorado Mineral Belt shear zones, including Homestake, Gore Range, St. Louis Lake, and Idaho Springs-Ralston shear zones. Also shown is Black Canyon shear zone (Jessup et al., 2002) and Laramide magmatism and mining centers.

these similarities, we widen the boundaries of the Colorado Mineral Belt to include the San Juan Mountains [Figure 1].

The irregular geometry of Tweto and Sims' Colorado Mineral Belt boundaries [1963] does not appear to correspond with mapped structures. In this paper, we use smoother, more general Colorado Mineral Belt boundaries that include the negative gravity anomalies and major mining districts, and lie parallel to the northeast-striking Proterozoic structures [Figure 1].

A number of pieces of evidence suggest that magmatism and deformation of the Phanerozoic Colorado Mineral Belt were localized by the Proterozoic Colorado Mineral Belt shear zone system. First, apatite fission-track studies of Kelley et al. [2001] show that the Colorado Mineral Belt shear zone system coincides with a transition in Laramide-age structural style and timing of uplift in the Front Range. To the south of the shear zone system, Laramide structures are dominated by east-vergent thrusts and Proterozoic rocks have >100 Ma apatite fission-track ages. To the north of the shear zone system, Laramide structures are dominated by southwest-vergent back thrusts and Proterozoic rocks have 76–45 Ma apatite fission-track ages [*Kelley et al.*, 2001]. The discrepancy in ages and structural style across the shear zone suggest that movement along the zone juxtaposed deeper rocks to the north against shallower rocks to the south. Thus, the Colorado Mineral Belt shear system appears to have been reactivated during a protracted period of time in the Laramide, accommodating south-side down differential uplift and exhumation.

Second, stratigraphic studies by Allen [1994] documented multiple Paleozoic movements that directly reactivated at least one of the shear zone segments along the Colorado Mineral Belt shear zone system. The basal conglomerate unit of the Upper Cambrian Sawatch Quartzite thins across strands of the Homestake shear zone from north to south, suggesting that southeast-side up movement occurred along the zone during early stages of Sawatch Quartzite deposition [*Allen*, 1992]. The Homestake shear zone also coincides with the northern pinch-out of the Lower Ordovician Manitou Dolomite, suggesting southeast-side down reactivation of the Homestake shear zone and erosion prior to deposition of Middle Ordovician Harding Sandstone [*Allen*, 1993]. Subtle thickness and facies variations in the Upper Devonian Chaffee Group suggest further southeast-side up reactivations [*Allen*, 1993]. In all, variations in facies and thickness of Paleozoic strata indicate that the Homestake shear zone was reactivated at least four times between Cambrian and late Devonian time, and at least once after Early Mississippian time [*Allen*, 1994].

Third, tomographic images of Dueker et al. [2001] show a zone of anomalously low-velocity mantle imaged at depths exceeding 120 km that projects upward into the Colorado Mineral Belt, suggesting that this geologic feature is lithospheric in scale and coincides with a modern zone of anomalously low-velocity Proterozoic lithosphere [*Dueker et al.*, 2001]. This suggests continued reactivation of Proterozoic lithosphere compositional domains and/or interfaces during Cenozoic mantle reorganization in the western U.S. [*Karlstrom and Humphreys*, 1998].

3. THE COLORADO MINERAL BELT SHEAR ZONE SYSTEM

3.1. Previous Work

It has long been argued that the magmatism and mineralization of the Colorado Mineral Belt were localized along pre-existing weaknesses, as indicated by the presence of Proterozoic shear zones and plutons [*Tweto and Sims*, 1963; *Warner*, 1978]. We now know that this Precambrian ancestry involved focused deformation and magmatism at ~1.4 Ga and ~1.7 Ga [*Shaw et al.*, 2001], and molybdenite mineralization at ~1.4 Ga [*Sims and Stein*, 1999]. Although Tweto and Sims [1963] recognized Proterozoic shear zones in the Colorado Mineral Belt, they did not describe the variety of different fault rocks, or 'tectonites,' present in the shear zones, nor did they highlight the kinematics of the multiple Proterozoic movements in the shear zones. Tectonites such as cataclasite, ultramylonite, mylonite, and high-temperature striped gneisses, were grouped under the term 'cataclastic rock.' Moench [1964] defined two different tectonites along the Idaho Springs-Ralston shear zone, distinguishing the high-temperature gneisses from younger tectonites that resulted from cataclastic deformation of the previously foliated and deformed rocks. Re-examination and re-mapping of the Idaho Springs-Ralston shear zone, as well as other segments of the Colorado Mineral Belt shear zone system, according to a newer understanding of microstructures and kinematic indicators, is presented in this paper.

A recent detailed study by Shaw et al. [2001], which focused on the tectonites, kinematics, and timing of movement in the Homestake shear zone, a segment of the Colorado Mineral Belt shear zone system, sets the stage for this paper. Along the Homestake shear zone, Shaw et al. [2001] showed that transposition of a Paleoproterozoic low-angle S1 fabric was synchronous with granite intrusion and migmatization. This transposed S1 fabric was steepened during the formation of northeast-striking, subvertical S2 high-temperature high strain zones, which were reactivated during the formation of ~1.4 Ga mylonites and ultramylonites. Shaw et al. [2001] used in situ electron microprobe monazite dating to constrain the development or reactivation of S1 at 1700 +/- 7 Ma, movement along S2 at 1658 +/- 5 Ma and 1637 +/- 13 Ma, southeast-side down mylonitization at 1376 +/- 11 Ma, and southeast-side up ultramylonitization after 1376 Ma. Many features of the Homestake shear zone, as defined by Shaw et al. [2001], are common to each shear zone segment of the Colorado Mineral Belt shear zone system. Therefore, we use the Homestake study of Shaw et al. [2001] as a guide for comparison of the major segments of the Colorado Mineral Belt shear zone system.

3.2. Map Patterns and Overview of Shear Zone Geometry

The mylonite and ultramylonite shear zone segments of the Colorado Mineral Belt shear zone system lie along the northern edge of the Colorado Mineral Belt between Leadville and Golden, Colorado, and include the Homestake shear zone [Figure 1; *Shaw et al.*, 2001; *Allen*, 1994; *Tweto and Sims*, 1963], Gore Range shear zone [Figure 1; *Bergendahl*, 1969; *Tweto and Sims*, 1963], St. Louis Lake shear zone [Figure 1; *Taylor*, 1971; *Bryant et al.*, 1981; *Tweto and Sims*, 1963], and Idaho Springs-Ralston shear zone [Figure 1; *Graubard and Mattinson*, 1990; *Wells et al.*, 1964; *Tweto and Sims*, 1963; *Moench*, 1964; *Sheridan*, 1958]. The segments appear to represent en echelon shears, branches of a shear zone system, or even one continuous shear zone (if it was offset by post-Mesoproterozoic dextral fault motions) that extends at least 100 km in length.

At the northeastern extent of its exposure, the Homestake shear zone [Figure 1, Figure 2a] disappears under Phanerozoic cover east of the Eagle River in the northern Sawatch Range. Where Proterozoic rocks surface again, just east of Vail Pass, the Gore Range shear zone [Figure 1, Figure 2b] is directly along strike of the 044°, 79S Homestake shear zone. About 10 km north, and parallel to the Gore Range shear zone, several northeast-striking ultramylonite strands are present at Booth Lake. These strands are considered part of the Colorado Mineral Belt shear zone system, but are not discussed in detail in this paper.

The northeast extent of the Gore Range shear zone segment bends northward to an orientation of 030°, 76W just before it disappears beneath the Phanerozoic cover of the Blue River Valley. If projected across the Blue River Valley and Williams Fork Range at this orientation, the Gore Range shear zone connects with the St. Louis Lake shear zone segment [Figure 1, Figure 2c]. About 10 km east of the St. Louis Lake shear zone, several mylonite strands deform the Silver Plume granite at Berthoud Pass. In this paper, these strands are considered to be part of the St. Louis Lake shear zone because of similar orientations, shear sense, and timing of movement.

The Idaho Springs-Ralston shear zone segment [Figure 1, Figure 2d] does not occur along the strike of the St. Louis Lake shear zone, but is roughly aligned with the trend of the Homestake shear zone. Ancestral Rockies and/or Laramide movements along the north-striking Loveland Pass-Berthoud Pass fault system may have caused dextral strike-slip offset of tens of kilometers between the St. Louis Lake and Idaho Springs-Ralston shear zones.

As part of the current study, we also conducted reconnaisance mapping along the Montezuma shear zone described by Tweto and Sims [1963]. However, the Montezuma shear zone appears to contain only steeply dipping, high-temperature striped gneisses, which are common throughout central Colorado. The zone does not contain composite fabrics including mylonites or ultramylonites, nor does it show any evidence for simple shear, and therefore is not rightly defined as a 'shear zone' and is not addressed further in this paper.

3.3. General Characteristics of Shear Zone Segments

In each mylonitic segment of the Colorado Mineral Belt shear zone system, kilometer-wide mylonite zones contain multiple parallel mylonite strands that are one to tens of meters wide. The shear zones also contain ultramylonite strands that are typically narrower than the mylonite strands they overprint. Each mylonitic shear zone segment is northeast- to east-striking, with strikes ranging from 028° to 090°, although the northeast strikes dominate. The shear zone segments are subvertical and dip steeply to the northwest or southeast, with dips ranging from 74°NW to 66° SE. They contain steeply-plunging mineral stretching lineations.

Mylonites and ultramylonites of the Colorado Mineral Belt shear zone system overprint northeast-striking Paleoproterozoic high-strain domains that occur in biotite gneiss and migmatite along the Homestake and Gore Range shear zones [Figs. 2a and 2b]. Along the St. Louis Lake shear zone, the Paleoproterozoic high-strain domains occur in a tectonic melange composed primarily of amphibolite and granodiorite [Figure 2c]. Along Idaho Springs-Ralston shear zone, the high-strain domains occur in quartz monzonite along the southern limb of the Coal Creek synform [Figure 2d].

Although most of the shear zone segments do not appear to separate regions with distinctly different structures or metamorphic histories, they all show juxtaposition of different rock types that hint at the long tectonic evolution of the shear zone system. The shear zones locally follow Paleoproterozoic pluton margins [Figure 2], and show evidence for high temperature deformation synchronous with Paleoproterozoic pluton emplacement [*Shaw et al.*, 2001]. In the Homestake and Gore Range shear zones, relatively high-temperature high strain domains follow the southern margin of the Cross Creek batholith [~1675 [Rb-Sr], *Tweto and Lovering*, 1977]. In the St. Louis Lake and Idaho-Springs Ralston shear zones, mylonites follow the southern margin of the Boulder Creek batholith [1721 +/- 15 Ma [U-Pb SHRIMP], *Premo and Fanning*, 2000]. In these high-strain domains, biotite schist is

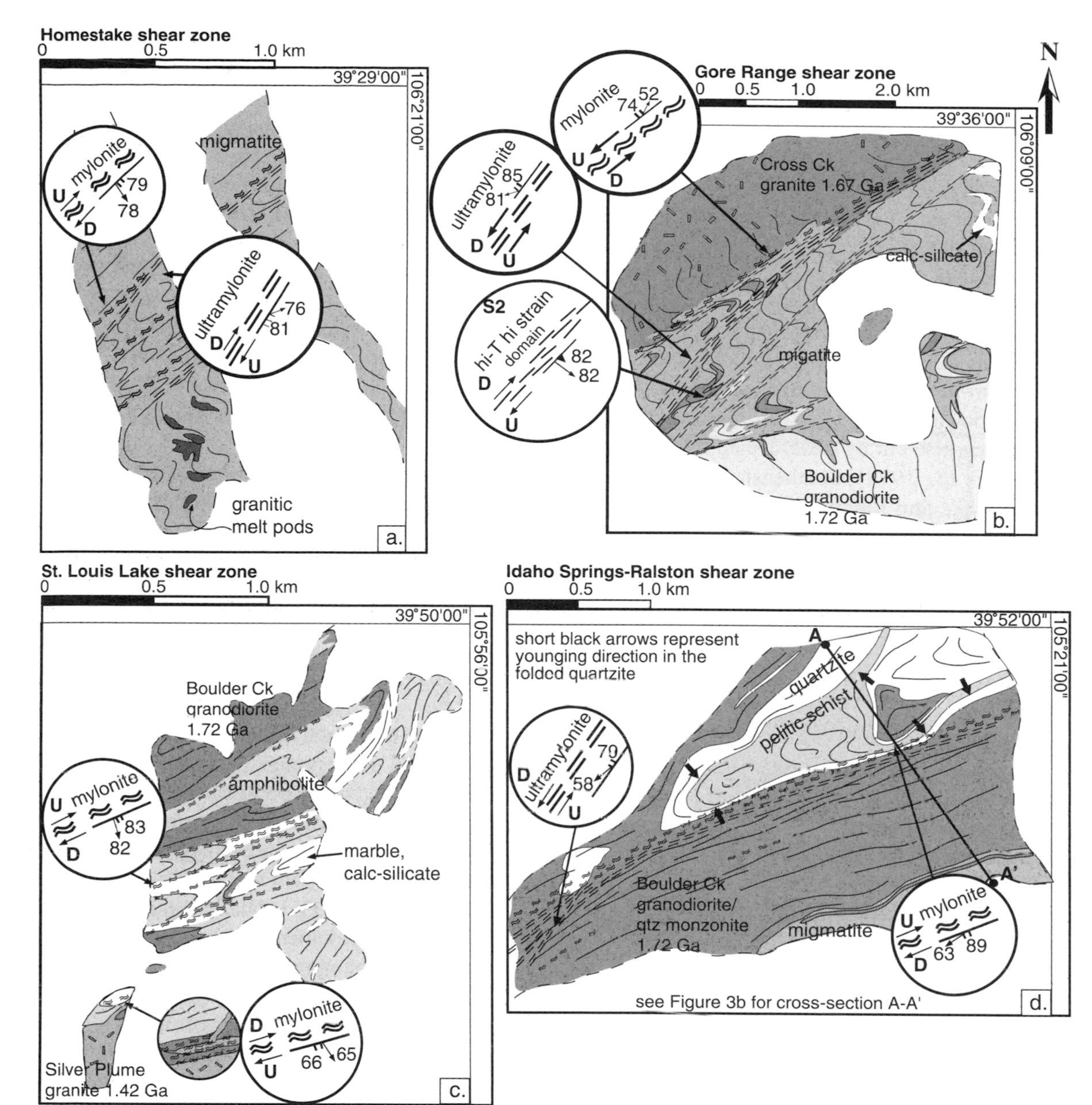

Figure 2. Maps of **a.** part of Homestake shear zone near the confluence of Homestake Creek and Eagle River, **b.** part of Gore Range shear zone south of Buffalo Mountain, **c.** part of St. Louis Lake shear zone at St. Louis Lake, and **d.** part of Idaho Springs-Ralston shear zone north of Ralston Creek in Golden Gate State Park.

interlayered and transposed with granite stringers and dikes. Late granite and pegmatite dikes that cut across the foliation suggest that granite intrusion outlasted deformation. Along the Homestake and Gore Range shear zones, there is widespread evidence for migmatization.

The St. Louis Lake shear zone overprints a tectonic melange containing boudinaged rocks of oceanic affinity. The melange contains marble, calc-silicates, possible metamorphosed chert, biotite schist, and amphibolite (including metamorphosed pillow basalts), with boudins of gabbro and ultramafic rocks, all interlayered on the mesoscopic to map scales [Figure 2c]. We interpret these rocks to be parts of a dismembered ophiolite complex tectonically emplaced along the Paleoproterozoic high-strain domains.

The Idaho Springs-Ralston shear zone deforms the Coal Creek Quartzite/pelitic schist sequence, which is one of several isolated Proterozoic meta-quartz arenites in central Colorado [Figure 2d; *Finiol*, 1992]. These quartzites are

may have been associated with the Yavapai orogeny or may have been a more local event coinciding with intrusion of the adjacent Cross Creek batholith.

1.65 to 1.62 Ga deformation dates coincide with the Mazatzal orogeny in southeastern Arizona [*Karlstrom and Bowring*, 1991], where subvertical, northeast-striking fabrics developed during formation of a continental margin batholith above a northwest- or north-dipping subduction system [*Selverstone et al.*, 1999] that has been projected from Arizona across southern Colorado [*Shaw and Karlstrom*, 1999]. 1.62 Ga deformation dates also correspond to U-Pb zircon dates of 1618 +/- 22 Ma for the Big Creek gneiss of the northern Front Range, 1627 +/- 4 Ma age for a quartz monzonite pluton in the Sierra Madre of northern Colorado, and emplacement ages for plutons in the Mount Tyndall area of the Wet Mountains, southern Colorado [*Premo and Van Schmus*, 1989; *Bickford et al.*, 1984].

Mesoproterozoic movements in the Colorado Mineral Belt shear zone system correspond in time with intrusions emplaced between 1.3 and 1.45 Ga along a belt that spans the southern margins of Laurentia-Baltica [*Nyman et al.*, 1994]. Although the plutons have been described as anorogenic, recent studies have shown evidence for substantial ~1.4 Ga deformation and metamorphism in the vicinity of many plutons [*Nyman et al.*, 1994]. In Colorado and New Mexico, ~1.4 Ga shear zones are moderately- to steeply-dipping and show evidence for synmagmatic deformation [*Nyman et al.*, 1994; *Kirby et al.*, 1995].

In the northern Front Range of Colorado, the northeast-striking, steeply-dipping Moose Mountain shear zone shows evidence for reverse-sense reactivation synchronous with intrusion of the ~1.4 Ga St. Vrain pluton [*Selverstone et al.*, 2000]. This shear zone is located 50 km north of, and is roughly parallel to, the Idaho Springs-Ralston shear zone. The relationship between the Moose Mountain shear zone, several other northern Colorado shear zones suspected to have been active at ~1.4 Ga, and the Colorado Mineral Belt shear zone system, is not fully understood. The northern Front Range shear zones have not been documented to extend southwest across Colorado, as the Colorado Mineral Belt shear zone system does.

4.3. Tectonic Significance of the Colorado Mineral Belt Shear Zone System

Metamorphic data indicate that Proterozoic rocks exposed in the Colorado Mineral Belt shear zone system were in the middle crust during tectonism, and deformation studies reveal a pattern of progressive overprinting of increasingly narrower, higher strain-rate/lower-temperature tectonites. This suggests that, at any given time, discrete structures at shallower crustal levels may grade into wider, more diffuse zones at deeper crustal levels. Greenschist-grade ultramylonites overprint wider zones of greenschist-grade mylonites, which overprint even wider, amphibolite-grade high strain domains. Each generation of tectonite appears to have caused grain size reduction and weakening along the shear zone system, leaving it prone to further reactivations.

4.3.1. Tectonic significance of Paleoproterozoic deformation along the Colorado Mineral Belt Shear Zone System. Major structural and metamorphic discontinuities have not been identified across the mylonite zones, S2 high-strain domains, or S1 domains. The Homestake, Gore Range, and Idaho Springs-Ralston shear zones separate plutons from metasedimentary rocks. The Homestake and Gore Range shear zones may have facilitated Paleoproterozoic pluton emplacement, while the Idaho Springs-Ralston shear zone appears to have developed after Paleoproterozoic plutonism and quartzite deposition.

Along the St. Louis Lake shear zone, the rock types within the melange are similar to the rock types in melanges identified within continent-arc collision zones [*Chang et al.*, 2000; *Polat and Kerrich*, 1999]. The presence of this possible fragment of oceanic melange, surrounded by granites and mica schists, suggests that the shear zone might have an ancestry as a lithospheric-scale structure that facilitated transport and tectonic juxtaposition of far-travelled rocks. The St. Louis Lake melange is similar to that described along the Moose Mountain shear zone in the northern Front Range [*Selverstone et al.*, 2000]. Both shear zones appear to have juxtaposed rocks from different structural levels during D2 and D3 (1.4 Ga) intracontinental steepening of the initially low-angle, continental assembly-related S1 fabrics.

The Paleoproterozoic structures, with inferred deformation dates from 1.7 to 1.62 Ga, developed during a time of regional tectonism that was likely associated with the collision of island arcs, and the welding of packages of arcs to the Archean Wyoming craton [Figure 6a]. Syn-tectonic plutons and batholiths of this period, such as the Cross Creek and Boulder Creek batholiths, do not appear to represent components of the initial magmatic arcs because they do not have isotopic signatures characteristic of arc plutons, and there is no evidence of andesites or adjacent suture zones [*Shaw and Karlstrom*, 1999; *Aleinikoff et al.*, 1993; *Reed et al.*, 1987].

4.3.2. Tectonic significance of Mesoproterozoic intracontinental deformation along the Colorado Mineral Belt Shear Zone System. In contrast to the D1 and D2 crustal assembly-related deformation that affected broad regions of Colorado, the Mesoproterozoic Colorado Mineral Belt shear zone system formed as a relatively narrow zone at a time when plate convergence was probably occurring some

1,000 km to the south, based on the proposed boundary between the Mazatzal Province and the Grenville Province near the present-day New Mexico/Texas border [Karlstrom and Humphreys, 1998]. At a great distance from this margin, the Mesoproterozoic mylonite system seems to have reactivated a zone of weakness related to Paleoproterozoic assembly. However, Mesoproterozoic mylonitization led to the development of a more focused belt of weakness and defined the trend that has influenced Phanerozoic deformation and magmatism along the Colorado Mineral Belt.

Intracontinental zones of deformation, located well away from plate margins, may be the loci of deformation that accommodate an important part of the observed plate convergence at the margin [Tien Shan of central Asia, Atlas Mountains of northern Africa; *Burov and Molnar*, 1998, *Yin et al.*, 1998], and may also record lithosphere/asthenosphere interactions. Intracontinental shortening in the Atlas Mountains has accommodated 17–45 percent of the total African-Eurasian plate convergence since the early Miocene [*Gomez et al.*, 2000; *Brede et al.*, 1992]. Late Cenozoic intracontinental shortening in the Tien Shan appears to have resulted in 20 to 40 km of shortening [*Yin et al.*, 1998].

The behavior of intracontinental zones of deformation at depth and over significant time intervals, and the processes controlling these zones, remain poorly understood. Intracontinental zones of deformation may have become weak due to anomalous heat, due to a pre-history that imparted compositional differences between the weak zone and the surrounding lithosphere, due to unusually thick crust, and/or due to the presence of major pre-existing mechanical weaknesses or domains of grain size reduction and strain-softening [*Burov and Molnar*, 1998; *Karlstrom and Humphreys*, 1998]. *Karlstrom and Humphreys* [1998] proposed that inheritance of Proterozoic structural grains throughout the southwest U.S. involves combinations of 'volumetric' inheritance (related to the density and fertility of compositionally different lithospheric blocks that influences isostatic and magmatic responses to tectonism) and 'interface' inheritance (related to mechanical boundaries that are zones of weakness and mass transport).

The Mesoproterozoic mylonites of the Colorado Mineral Belt shear zone system, as well as the Paleoproterozoic S2 high-strain domains they overprint, record primarily dip-slip movements on subvertical fault planes. The zones appear to have become subvertical during D2, with discrete mylonites reactivating broader S2 domains at middle crustal levels. Given these geometries, the shear zones appear to have caused large-scale 'jostling' of blocks [Figure 6a, Figure 6b]. These types of movements could occur along a flower structure or as part of a transpressive system, but we do not find extensive evidence for either horizontal stretching lineations or horizontal shear sense indicators of strike-slip movement.

Comparison of the Colorado Mineral shear zone system with younger analogues can shed light on the importance of intracontinental zones of deformation and the similarities among these zones, and can link the surface and shallow crustal level expression of such zones to their middle crustal analogues. One Cenozoic analogue is the Tien Shan of central Asia, where reactivation has taken place primarily along moderately- to steeply-dipping reverse structures [*Avouac et al.*, 1993; *Brookfield*, 2000]. Like the Colorado Mineral Belt shear zone system, the Tien Shan region records a complex tectonic history of continental assembly that occurred within a broad, diffuse zone containing slices of many different rock packages [*Allen and Vincent*, 1997]. In both areas, broad zones of assembly-related foliations were reactivated as narrower, more discrete zones of intracontinental deformation thousands of kilometers from the plate margin. Structures within the Tien Shan and the Colorado Mineral Belt shear zone system are adjacent to plutons, and there is evidence for early syn-plutonic deformation [Figure 7a, Figure 7b; *Brookfield*, 2000]. Although the magnitude of Precambrian offset across the Colorado Mineral Belt shear zone segments is difficult to determine, the Tien Shan faults have experienced several kilometers of offset during Cenozoic intracontinental reactivation [*Yin et al.*, 1998]. Most earthquakes on the Tien Shan faults have thrust solutions, indicating that this intracontinental zone of deformation is facilitating crustal shortening [*Yin et al.*, 1998].

The North Tien Shan fault is interpreted to have originated as a steeply-dipping 'back-stop' to a zone of lithospheric fragments assembled in the late Paleozoic [*Allen and Vincent*, 1997]. The fault was reactivated in the Cenozoic as a steeply dipping, north-directed thrust following the Paleozoic structural grain, but there is some evidence for dextral strike-slip movement [*Allen and Vincent*, 1997].

5. CONCLUSIONS

The Colorado Mineral Belt shear zone system is here defined as a Mesoproterozoic system of mylonites and ultramylonites that moved during a protracted period of orogenesis between 1.45 and 1.3 Ga. Although the shear zone system is Mesoproterozoic, the system overprints a broader, higher temperature, high-strain domain that records a >70 Ma Paleoproterozoic orogenic episode. Thus, the Colorado Mineral Belt shear zone system may have reactivated a more diffuse zone of weakness associated with continental assembly, and in doing so, it established the trend that controlled Phanerozoic deformation and localization of magmatic systems along the Colorado Mineral Belt.

The long history of deformation along the Colorado Mineral Belt shear zone system indicates that lithospheric zones

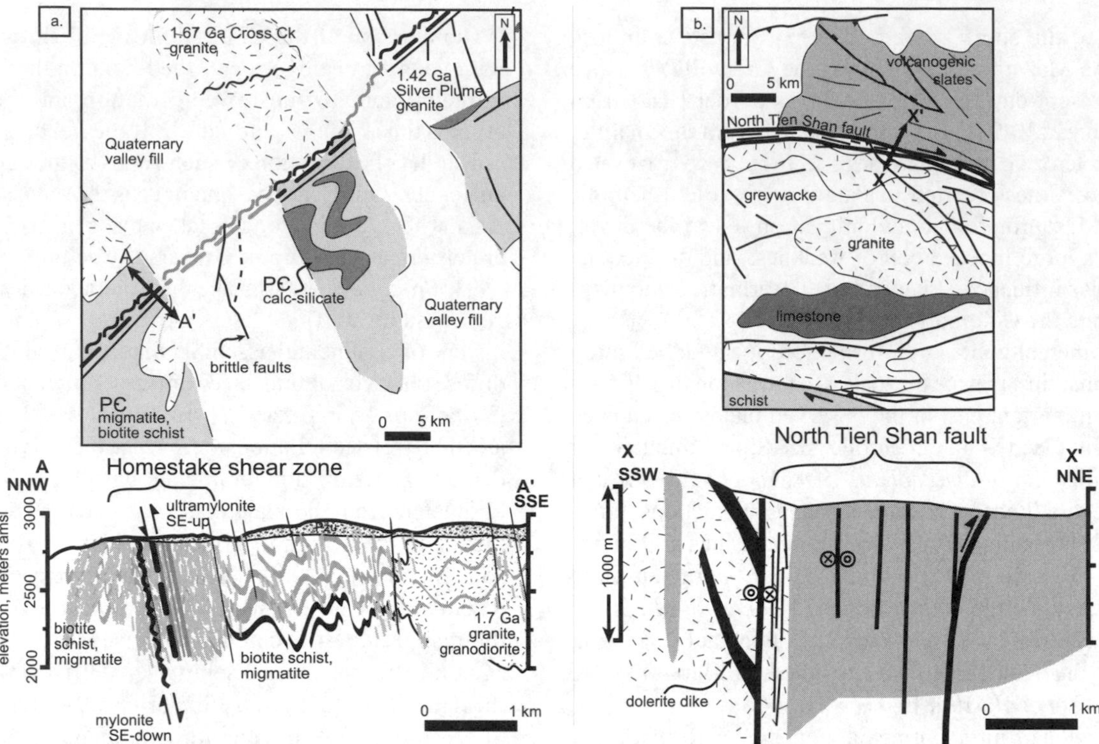

Figure 7. Comparison of maps and cross-sections of two intracontinental zones of deformation: Homestake shear zone (a., after Shaw et al., 2001), and the North Tien Shan fault of the Tien Shan (b., after Allen and Vincent, 1997).

of weakness, first established as diffuse zones of weakness during continental assembly and later reactivated as narrow intracontinental zones, may remain as loci of geologic processes for hundreds of millions of years. This study of the Colorado Mineral Belt shear zone system documents many characteristics of intracontinental tectonic zones including: 1. Origination of such zones in broad, subvertical domains of high strain and foliation intensification that have steepened what are inferred to be initially low angle sheet-like structures; 2. Reactivation of such zones as progressively narrower domains of increasingly higher strain-rate/lower-temperature grain size reduction at progressively shallower depths and lower temperatures, 3. Repeated emplacement of plutons and mineralization along the zones, and 4. Development of lithospheric-scale inhomogeneities such as negative gravity anomalies and slow mantle anomalies associated with the zone.

The Colorado Mineral Belt shear zone system is dominated by steeply-dipping structures with steeply-plunging mineral stretching lineations, indicating primarily dip-slip movements along subvertical zones. A kinematic model that accounts for these movements involves the 'jostling' of blocks, up and then down along the same zone of weakness, possibly facilitating pluton emplacement or an interplay between crustal shortening and crustal collapse. Such movements are observed in other intracontinental zones of deformation, such as the Tien Shan of central Asia, the Atlas Mountains of northern Africa, and the Laramide Rocky Mountains. Therefore, dip-slip movement along subvertical zones may be an important characteristic of long-lived intracontinental zones of deformation that have remained as weak zones in the lithosphere, experiencing multiple episodes of reactivation.

Acknowledgements. This study was made possible through the support of the NSF CD-ROM experiment funded by the Continental Dynamics Program, the Colorado Scientific Society's Ogden Tweto Memorial Grant, the Four Corners Geological Society's Masters Thesis Research Grant, and the University of New Mexico's Alexander and Geraldine Wanek Scholarship. I am grateful to Shari Kelley, Paul Bauer, and Daniel Holm for providing helpful reviews. Many thanks to Mike Spilde and Nelia Dunbar for guidance in the laboratory.

REFERENCES

Aleinikoff, J. N., J. C. Reed, Jr., and E. DeWitt, The Mount Evans batholith in the Colorado Front Range: revision of its age and reinterpretation of its structure, *Geol. Soc. Am. Bull.*, *105*, 791–806, 1993.

Allen, J. L., Influence of a basement shear zone on stratigraphy of the Cambrian Sawatch Quartzite, northeastern Sawatch Range, Colorado (abstract),

Geol. Soc. Am. Abstr. Programs, 24 (6), 1, 1992.

Allen, J. L., Cambrian-Mississippian reactivation history of the Homestake shear zone, central Colorado (abstract), *Geol. Soc. Am. Abstr. Programs, 25 (5)*, 2, 1993.

Allen, J. L., Stratigraphic variations, fault rocks, and tectonics associated with brittle reactivation of the Homestake shear zone, central Colorado, Ph.D. dissertation, Univ. of Kentucky, Lexington, 1994.

Allen, M.B., and S. J. Vincent, Fault reactivation in the Junggar region, northwest China: the role of basement structures during Mesozoic-Cenozoic compression, *J. Geol. Soc. of London, 154*, 151–155, 1997.

Baars, D. L., Tectonic significance of Proterozoic faults, San Juan Mountains, southwestern Colorado (abstract), *Geol. Soc. Am. Abstr. Programs, 16 (4)*, 214, 1984.

Barovich, K.M., Age constraints on Early Proterozoic deformation in the northern Front Range, Colorado, M.S. thesis, Univ. of Colorado, Boulder, 1986.

Bergendahl, M.H., Geologic map and sections of the southwest quarter of the Dillon Quadrangle, Eagle and Summit Counties, Colorado, scale 1:24,000, USGS Miscellaneous Geologic Investigations Map I-563, 1969.

Bickford, M.E., and S. J. Boardman, A Proterozoic volcano-plutonic terrane, Gunnison and Sailda areas, Colorado, *J. Geol., 92*, 657–666, 1984.

Bickford, M.E., R. D. Shuster, and S. J. Boardman, U-Pb geochronology of the Proterozoic volcano-plutonic terrane in the Gunnison and Salida areas, Colorado, in *Proterozoic geology of the southern Rocky Mountains*, edited by J. A. Grambling and B. J. Tewksbury, pp. 33–48, *Spec. Pap. Geol. Soc. Am. 235*. 1989.

Brede, R., M. Hauptmann, and H. Herbig, Plate tectonics and intracratonic mountain ranges in Morocco: the Mesozoic-Cenozoic development of the Central High Atlas and the Middle Atlas, *Geologische Rundschau, 81 (1)*, 127–141, 1992.

Brookfield, M.E., Geological development and Phanerozoic crustal accretion in the western segment of the southern Tien Shan (Kyrgystan, Uzbekistan, and Tajikistan), *Tectonophysics, 328*, 1–14, 2000.

Bryant, Bruce, L. W. McGrew, and R. A. Wobus, Geologic map of the Denver 1 degrees by 2 degrees quadrangle, North-central Colorado, scale 1:250,000, USGS Miscellaneous Geologic Investigations Map I-1163, 1981.

Burov, E. B., and P. Molnar, Gravity anomalies over the Ferghana Valley (central Asia) and intracontinental deformation, *J. Geophys. Res., 103 (B8)*, 18137–18152, 1998.

Chang, C.P., J. Angelier, and C. Huang, Origin and evolution of a melange: the active plate boundary and suture zone of the Longitudinal Valley, Taiwan, *Tectonophysics, 325 (1–2)*, 43–62, 2000.

Doe, B.R., and R. C. Pearson, U-Th-Pb chronology of zircons from the St. Kevin Granite, northern Sawatch Range, Colorado, *Geol. Soc. Am. Bull., 80*, 2495–2502, 1969.

Dueker, K., H. Yuan, and B. Zurek, Thick-structured Proterozoic lithosphere of the Rocky Mountain region, *GSA Today, 11 (12)*, 4–9, 2001.

Finiol, L. R., Petrology, paleostratigraphy, and paleotectonics of a Proterozoic metasedimentary and metavolcanic sequence in the Colorado Front Range, M.S. thesis, Colo. School of Mines, Golden, 1992.

Fraser, G. D., Coal Creek Quartzite, Jefferson and Boulder Counties, Colorado, *Geol. Soc. Am. Bull., 60 (12)*, 1960, 1949.

Gable, D. J., The Boulder Creek batholith, Front Range, Colorado, *USGS Prof. Pap., 1101*, 1-88, 1980.

Gomez, F., W. Beauchamp, and M. Barazangi, Role of the Atlas Mountains (northwest Africa) within the African-Eurasian plate-boundary zone, *Geology, 28 (9)*, 775–778, 2000.

Graubard, C. M., and J. M. Mattinson, Syntectonic emplacement of the ~1440 Ma Mt. Evans pluton and history of motion along the Idaho Springs-Ralston Creek shear zone, central Front Range, Colorado, *Geol. Soc. Am. Abstr. Programs, 22 (7)*, A245, 1990.

Hedge, C.E., A petrogenetic and geochronologic study of migmatites and pegmatites in the central Front Range, Ph.D. dissertation, Colo. School of Mines, Golden, 1969.

Hirth, G., and J. Tullis, The brittle-plastic transition in experimentally deformed quartz aggregates, *J. Geophys. Res., 99*, 11,731–11,747, 1994.

Isaacson, L. B., and S. B. Smithson, Gravity anomalies and granite emplacement in west-central Colorado, *Geol. Soc. Am. Bull., 87 (1)*, 22–28, 1976.

Jessup, M. J., K. E. Karlstrom, J. Connelly, R. F. Liviccari, Complex crustal assembly of southwestern North America involving northwest and northeast-striking fabrics: the Black Canyon of the Gunnison, southern Colorado, *Geol. Soc. Am. Abstr. Programs, 34 (4)*, 48, 2002.

Karlstrom, K.E., and S. A. Bowring, Styles and timing of Early Proterozoic deformation in Arizona: constraints on tectonic models, in *Proterozoic Geology and Ore Deposits of Arizona*, edited by K. E. Karlstrom, pp. 1–10, *Ariz. Geol. Soc. Digest 19*, Flagstaff, 1991.

Karlstrom, K. E., and E. D. Humphreys, Persistent influence of Proterozoic accretionary boundaries in the tectonic evolution of southwestern North America: interaction of cratonic grain and mantle modification events, *Rocky Mtn. Geol., 33 (2)*, 161–179, 1998.

Kelley, S. A., C. E. Chapin, and R. G. Raynolds, Influence of the Colorado Mineral Belt on the northwest margin of the Denver Basin (abstract), *Geol. Soc. Am. Abstr. Programs, 33 (5)*, A51, 2001.

Kirby, E., K. E. Karlstrom, C. L. Andronicos, and R. D. Dallmeyer, Tectonic setting of the Sandia Pluton: an orogenic 1.4 Ga granite in New Mexico, *Tectonics, 14 (1)*, 185–201, 1995.

McCoy, A.M., M. Roy, and G. R. Keller, Gravity modeling of the Colorado Mineral Belt, *Geol. Soc. Am. Abstr. Programs, 33 (5)*, 43, 2001.

Moench, R.H., Geology of Precambrian rocks, Idaho Springs district, Colorado, *USGS Bull., 1182-A*, A1–A69, 1964.

Montel, J., S. Foret, M. Veschambre, C. Nicollet, and A. Provost, Electron microprobe dating of monazite, *Chem. Geol., 131*, 37–53, 1996.

Mutschler, F. E., E. E. Larson, and R. M. Bruce, Laramide and younger magmatism in Colorado: new petrologic and tectonic variations on old themes, in *Cenozoic volcanism in the Southern Rocky Mountains*, edited by J. W. Drexler, and E. E. Larson, pp. 1–47, *Colorado School of Mines Quarterly, 82 (4)*, Golden, 1987.

Nyman, M.W., K. E. Karlstrom, E. Kirby, and C. M. Graubard, Mesoproterozoic contractional orogeny in western North America: evidence from ca. 1.4 Ga plutons, *Geology, 22*, 901–904, 1994.

Oshetski, K.C. and R.P. Kucks, Colorado aeromagnetic and gravity maps and data: a web site for distribution of data: *USGS Open-File Rpt., 00–0042*, web only at http://greenwood.cr.usgs.gov/pub/open-file-reports/ofr-00-0042/colorado.html, 2000.

Passchier, C. W., and R. A. J. Trouw, *Microtectonics*, 289 pp., Springer, New York, 1996.

Pattison, D. R. M., F. S. Spear, and J. T. Cheney, Polymetamorphic origin of muscovite + cordierite + staurolite + biotite assemblages: implications for the metapelitic petrogenetic grid and for P-T paths, *J. Metamorph. Geol., 17*, 685–703, 1999.

Polat, A., and R. Kerrich, Formation of an Archean tectonic melange in the Schreiber-Hemlo greenstone belt, Superior Province, Canada: implications for Archean subduction-accretion processes, *Tectonics, 18 (5)*, 733–755, 1999.

Premo, W. R., and C. M. Fanning, SHRIMP U-Pb zircon ages for Big Creek gneiss, Wyoming, and Boulder Creek batholith, Colorado: implications for timing of Paleoproterozoic accretion of the northern Colorado province, *Rocky Mtn. Geol., 35 (1)*, 31–50, 2000.

Premo, W. R., and W. R. Van Schmus, Zircon geochronology of Precambrian rocks in southeastern Wyoming and northern Colorado, in *Proterozoic geology of the southern Rocky Mountains*, edited by J. A. Grambling and B. J. Tewksbury, pp. 13–32, *Spec. Pap. Geol. Soc. Am. 235*, 1989.

Reed, J C. Jr., M. E. Bickford, W. R. Premo, J. N. Aleinikoff, and J. S. Pallister, Evolution of the Early Proterozoic Colorado province: constraints from U-Pb geochronology, *Geology, 15*, 861–865, 1987.

Selverstone, J., M. Hodgins, C. Shaw, J. N. Aleinikoff, and C. M. Fanning, Proterozoic tectonics of the northern Colorado Front Range, in *Geologic history of the Colorado Front Range*, edited by D. Bolyard and S. A. Sonnenberg, pp. 9–18, Rocky Mountain Association of Geologists, Denver, 1997.

Selverstone, J., M. Hodgins, J. N. Aleinikoff, and C. M. Fanning, Mesoproterozoic reactivation of a Paleoproterozoic transcurrent boundary in the northern Colorado Front Range: implications for ~1.7- and 1.4 Ga tectonism, *Rocky Mtn. Geol., 35 (2),* 139–162, 2000.

Selverstone, J., A. Pun, and K. C. Condie, Xenolithic evidence for Proterozoic crustal evolution beneath the Colorado Plateau, *Geol. Soc. Am. Bull., 111 (4),* 590–606, 1999.

Shaw, C. A., and K. E. Karlstrom, The Yavapai-Mazatzal crustal boundary in the Southern Rocky Mountains, *Rocky Mtn. Geol., 34 (1),* 37–52, 1999.

Shaw, C. A., K. E. Karlstrom, M. L. Williams, M. J. Jercinovic, and A. M. McCoy, Electron-microprobe monazite dating of ca. 1.71–1.63 Ga and ca. 1.45–1.38 Ga deformation in the Homestake shear zone, Colorado: origin and early evolution of a persistent itnracontinental tectonic zone, *Geology, 29 (8),* 739–742, 2001.

Shaw, C. A., L. W. Snee, J. Selverstone, and J. C. Reed, Jr., $^{40}Ar/^{39}Ar$ thermochronology of Mesoproterozoic metamorphism in the Colorado Front Range, *J. of Geol., 107 (1),* 49–67, 1999.

Sheridan, D. M., Map of bedrock geology of the Ralston Buttes quadrangle, Jefferson County, Colorado, USGS Mineral Investigations Field Studies map MF-179, scale 1:24,000, 1958.

Sims, P. K., and H. J. Stein, Re-Os ages for molybdenite record major Proterozoic crust-forming event in Colorado, *Geol. Soc. Am. Abstr. Programs, 31 (7),* 260, 1999.

Spear, F. S., Metamorphic phase equilibria and pressure-temperature-time paths, 799 pp., *Min. Soc. of Am. Mono.,* Washington, D.C., 1993.

Taylor, B., Precambrian geology of the Byers Peak area, central Front Range, Colorado, M.S. thesis, Colo. School of Mines, Golden, 1971.

Tullis, J., and R. Yund, The brittle-ductile transition in feldspar aggregates: an experimental study, in *Fault mechanics and transport properties of rocks: a festschrift in honor of W.F. Brace*, edited by B. Evans and T. Wong, pp. 89–117, *Academic Press International Geophysics SerAies*, *51,* San Diego, 1992.

Tweto, O. L., and T. S. Lovering, T.S., Geology of the Minturn 15-minute quadrangle, Eagle and Summit Counties, Colorado, *USGS Prof. Pap. P-956*, 1–96, 1977.

Tweto, O. L., and P. K. Sims, Precambrian ancestry of the Colorado Mineral Belt, *Geol. Soc. Am. Bull., 74,* 991–1014, 1963.

Warner, L. A., The Colorado lineament: a middle Precambrian wrench fault system, *Geol. Soc. Am. Bull., 89,* 161–171, 1978.

Wells, J. D., D. M. Sheridan, and A. L. Albee, Relationship of Precambrian quartzite-schist sequence along Coal Creek to Idaho Springs Formation, Front Range, Colorado: *USGS Prof. Pap., 0454-O,* 1–25, 1964.

Williams, M. L., M. J. Jercinovic, and M. P. Terry, Age mapping and dating of monazite on the electron microprobe: deconvoluting multistage tectonic histories, *Geology, 27,* 1023-1026, 1999.

Williams, M. L., and M. J. Jercinovic, Microprobe monazite geochronology: putting absolute time into microstructural analysis, *J. of Struct. Geol., 24 (6–7),* 1013–1028, 2002.

Yin, A., S. Nie, P. Craig, T. M. Harrison, F. J. Ryerson, X. Qian, and Y. Geng, Late Cenozoic tectonic evolution of the southern Tien Shan, *Tectonics, 17 (1),* 1–27, 1998.

K.E. Karlstrom, Department of Earth and Planetary Sciences, Northrop Hall, University of New Mexico, Albuquerque, New Mexico, 87131.

A.M. McCoy, John Shomaker & Associates, Inc., 2703-B Broadbent Parkway NE, Albuquerque, New Mexico, 87107.

C.A. Shaw, Department of Geology, University of Wisconsin-Eau Claire, 105 Garfield Avenue, Eau Claire, Wisconsin, 54702–4004.

M.L. Williams, Department of Geosciences, 233 Morrill Science Center, University of Massachusetts, 611 North Pleasant Street, Amherst, Massachusetts, 01003–9297.

Structure of the North Park and South Park Basins, Colorado: An Integrated Geophysical Study

Leandro Treviño and G. Randy Keller

Department of Geological Sciences, University of Texas at El Paso

The CD-ROM geophysical transect passes through the intermontane valleys of central Colorado called parks. However, relatively little is known about their structure. Geophysical exploration of the North and South Park basins is mostly of a proprietary nature and few of these investigations are ever published. In this study, we integrate gravity, geologic, seismic, and digital elevation data to determine the subsurface structure of the North Park and South Park basins of central Colorado and their structural setting. Digital elevation and gravity images and computer modeling of the gravity data show that the subsurface structure of these basins is not as simple as suggested by their surface expression. Modeling of gravity profiles constrained by drilling and geologic data shows that North Park and South Park are underlain by complex, fault-bounded, asymmetric basins. The gravity low over North Park is the result of basin fill of both Cenozoic and Mesozoic age that reaches a thickness of ~6 km. Very low gravity values found in the northern portion of the South Park are associated with the Colorado mineral belt. Modeling of two profiles south of this area indicates that a series of faults divide the basin. Particularly in the south, the basin is divided into sub-basins by a structural high. This study also provided constraints for the seismic refraction modeling that was part of the CD-ROM project.

1. INTRODUCTION

The intermontane basins of North Park and South Park are defined by the surrounding Laramide uplifts that expose Precambrian rocks. The physiographic expression of these features is very clear on the topographic map shown on Plate 1 (see CDROM in back cover sleeve) that was produced from digital elevation data obtained from the U. S. Geological Survey. The Park Range to the west, the Medicine Bow Mountains to the east, and the Rabbit Ears Range to the south define North Park. The Mosquito and Sawatch Ranges bound South Park to the west and the Front Range bounds it to the east. The topographic map shows that both of these basins are relatively flat compared to the surrounding region.

The Rocky Mountain Region: An Evolving Lithosphere
Geophysical Monograph Series 154

10.1029/154GM07

The North Park and South Park basins (Figure 1) are, at least in part, the result of horizontal compression that uplifted the southern Rocky Mountains during the Late Cretaceous-Early Tertiary (70–35 Ma) Laramide orogeny. However, their deep structure is poorly known and their tectonic origin and evolution are enigmatic. Summaries of the regional geology and tectonic evolution are found elsewhere in this monograph, but several structural events have affected the area containing these basins such as the Late Carboniferous (280–240Ma) Ancestral Rocky Mountains orogeny and the Neogene extension associated Rio Grande rift (~30Ma). In this study, we integrated gravity, drilling, seismic, elevation and geologic data to produce a variety of maps and models to elucidate the structural setting of the North and South Park basins. Our goal was to determine their deep structure and to facilitate the interpretation of the CD-ROM seismic refraction profile that passes through them [*Rumpel et al.*, this volume; *Snelson et al.*, this volume; *Levander et al.*, this volume].

2. BACKGROUND

The Rocky Mountains have been of great interest since the early exploration of the area. For example, *King* [1876, 1878] carried out the earliest geologic investigations in North Park. *Tweto* [1980 a, b] described the Precambrian geology and wrote tectonic descriptions of central Colorado. *Hail* [1965] described structural relationships in northwestern North Park. In the Delaney Butte-Sheep Mountain area (Figure 1), he mapped a series of imbricate, east-dipping thrust faults that form a resistant hogback composed of Mesozoic and Precambrian rocks. *Hail* [1965] also described the Walden and North Park synclines and a fault that is parallel to the North Park syncline known as the Spring Creek fault (Figure 1). *Behrendt et al.* [1969] combined gravity, magnetic, and a short (~30 km) seismic refraction profile in an integrated study of the North Park basin. One of the results of this study indicates that the steepest gravity gradients are associated with the Spring Creek fault (Figure 1). *Lange and Wellborn* [1985] carried out a seismic reflection survey in the North Park basin, and their results agreed with *Behrendt et al.* [1969] in showing that the North Park basin is an asymmetrical syncline, faulted on the northern end of the line. The sedimentary strata in the basin were described by *Lange and Wellborn* [1985]. A major unconformity separates the Precambrian basement from the overlying sedimentary sequence, and the Triassic Chugwater Formation, consisting of red shales, lies directly on this unconformity. Cretaceous units of marine and near shore sandstone and shale (Dakota Group, Morrison, Entrada, Niobrara, and the upper Cretaceous Pierre Formations, respectively) overlie the basal Chugwater Formation. The Tertiary Coalmont Formation consisting of alluvial and fluvial sandstones and shale caps the sequence.

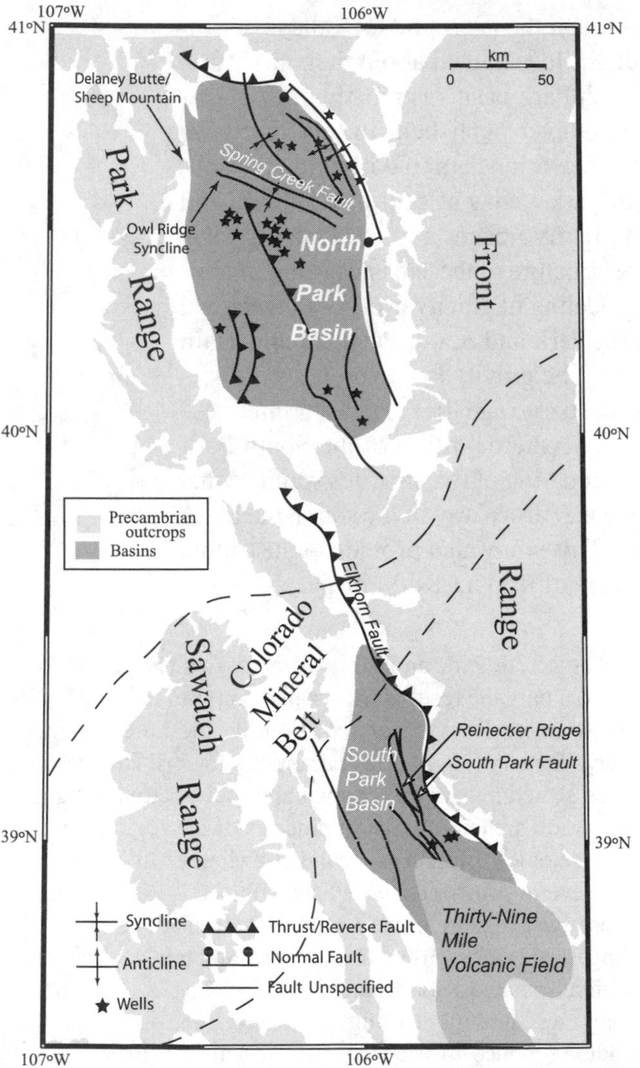

Figure 1. Tectonic map of the North and South Park region compiled from a variety of geologic studies that are referenced in the text. Stars indicate the location of wells that were used as geologic constraints in the geophysical interpretations.

Investigations into South Park geology and geophysics are abundant but are mostly in the form of unpublished Master's theses or Ph.D. dissertations [e.g., *Durrani*, 1980]. *Howland* [1935] described the bedrock geology of southern South Park while *Behre et al.* [1935] wrote on the bedrock geology of northern South Park. *Stark et al.* [1949] not only wrote about the structural geology of South Park but also its history. They recognized the complexity of the Laramide orogeny starting with the uplift and erosion of Cretaceous sediments, followed by outpourings of felsic volcanic rocks, and subsequent folding and thrust faulting. Block faulting and further erosion by the South Platte River, in the late Pliocene and later, defined South Park [*Stark et al.*, 1949]. *Sawatzky* [1964] in his description of southeastern South Park pointed to the Elkhorn fault as the main structural element present (Figure 1). The Elkhorn fault has displaced the Precambrian basement westward over Cretaceous and Tertiary sedimentary rocks [*Sawatzky*, 1964]. *Beggs* [1977], in a seismic reflection study, recognized that two basement controlled fault systems dominated the structure of central South Park. The subsurface of central South Park is an asymmetric, faulted, north-south trending syncline bounded by large basement uplifts associated with the Elkhorn thrust. Reinecker Ridge, a structural high, is located near the axis of the northern part of the South Park basin. Andesitic flows and breccias, conglomerates and immature sandstones and shales form Reinecker Ridge [*Sawatzky*, 1964]. Other minor structures are also present in this region [*Beggs*, 1977].

3. METHODOLOGY

The University of Texas at El Paso (UTEP) maintains an extensive gravity database, whose sources include governmental (U. S. Geological Survey, NOAA, NIMA, etc.), academic (University of Texas at Dallas, Purdue University, etc.), and commercial entities. Data for the study area were extracted from this database. A plot of the gravity data points shows that they are relatively well distributed in the study area (Figure 2). After removal of a small number of anomalous readings, the data were then gridded and a smoothing filter (10 km, low-pass, cosine-arch) was applied to produce Figure 3. By smoothing filter, we mean that a spatial domain low-pass operator (10 km) was employed to smooth the contours for display in this paper. In order to remove regional anomalies, a 20–100 km band-pass (cosine-arch) filter was applied using an algorithm in the Generic Mapping Tools (GMT) software [*Wessel and Smith*, 1998]. We tried several filters but found that the 20–100 km band-pass best enhanced the major features in the upper crust.

Computer models (2.5D) of the gravity data were produced using the approach of *Cady* [1980] along specific profiles of interest (Figure 3). The data used in the modeling are the Bouguer anomaly values not the band-passed values since

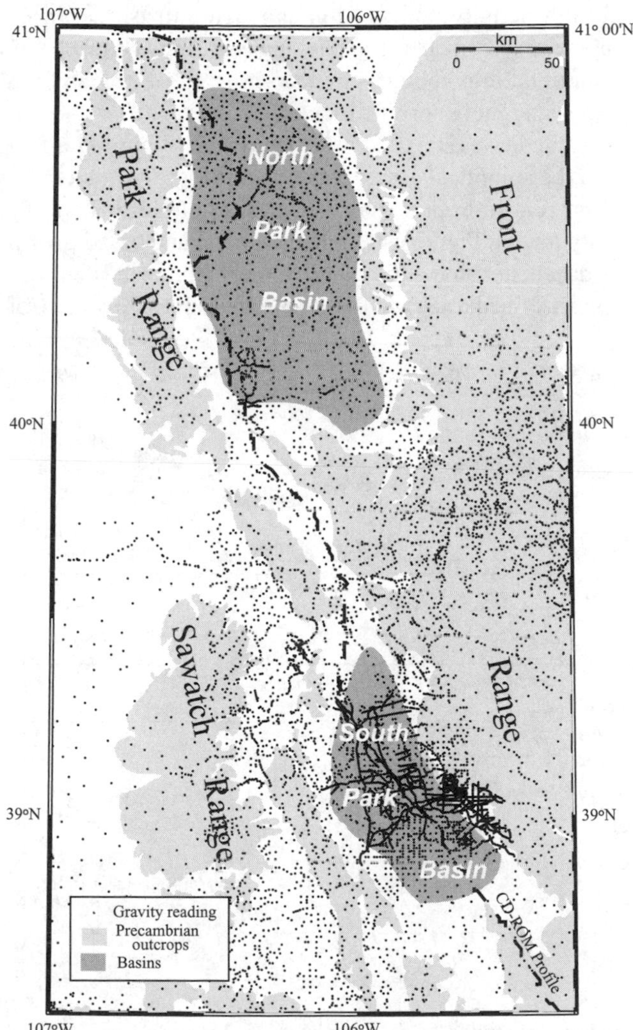

Figure 2. Plot of points in the area of the North and South Park basins where gravity measurements have been made.

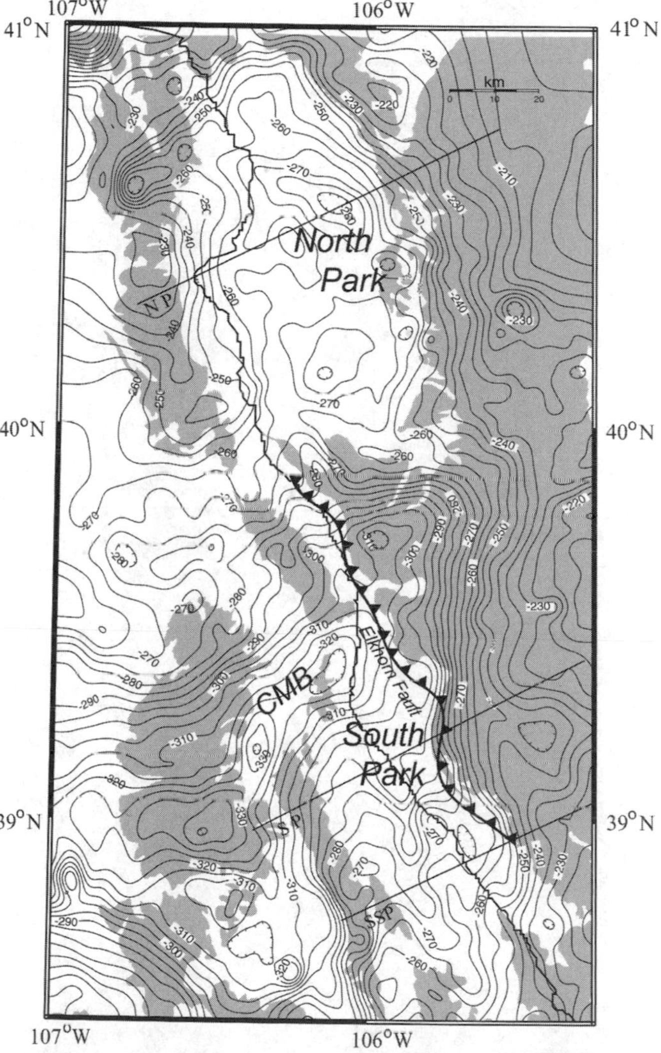

Figure 3. Bouguer anomaly gravity map, which as been smoothed for display purposes to remove wavelengths less than 10 km. Note the 50 mGal gradient between the South Park basin and the Front Range. Areas shaded in gray are outcrops of Precambrian rocks. The contour interval is 5 mGal. CMB – Colorado Mineral Belt. Lines crossing the map from southwest to northeast (NP, SP, SSP) are the profiles along gravity data were extracted and modeled.

this would imply interpolation of the data. These profiles were chosen so that they were perpendicular to the regional structural grain and features of geologic interest while tying to constraints such as wells. Modeling of these profiles was a trial and error process with geologic outcrops, drilling results, and seismic data as constraints. Thus, the models produced should be considered to be integrated geologic cross-sections of the upper crust.

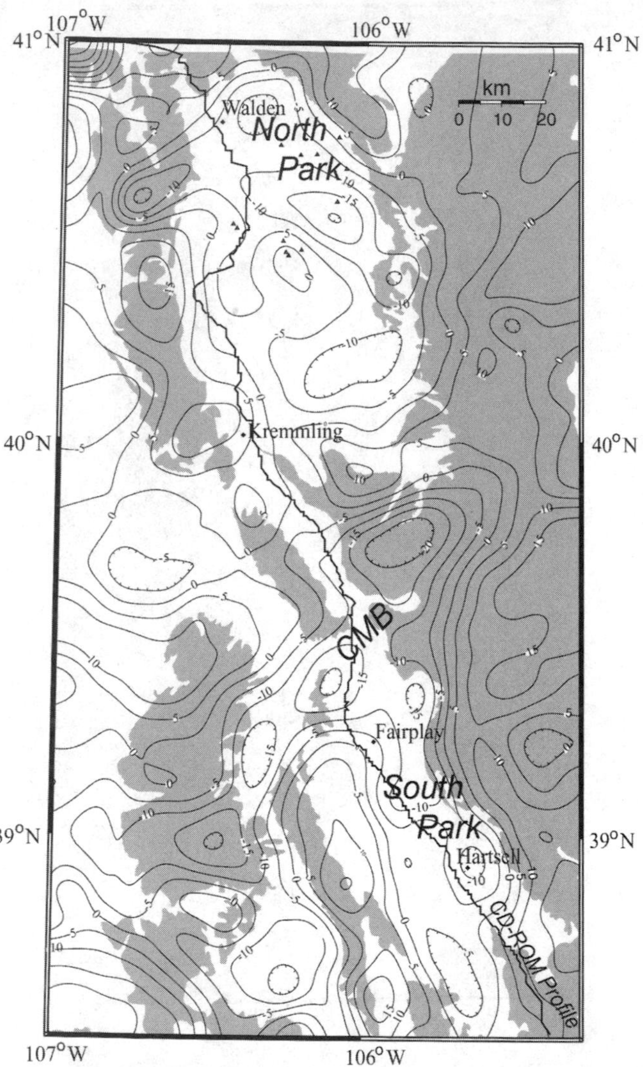

Figure 4. Filtered Bandpass (20–100 km) filtered map. The contour interval is 5 mGal. CMB – Colorado Mineral Belt. The regional effect has been removed from the data to produce map in order to emphasize upper crustal features. Areas shaded in gray are outcrops of Precambrian rocks.

4. RESULTS

4.1. Gravity Maps

A tectonic map (Figure 1) displays the main structural features related to the North and South Park basins. These structural components trend, for the most part, in a northwesterly direction. The Spring Creek fault parallels the Owl Ridge syncline in the North Park basin that is bordered by several faults.

The smoothed Bouguer gravity map (Figure 3) displays the integrated distribution of gravity anomalies from both upper crustal and deeper sources in the region. The gravity lows are associated with either basin fill or materials of lower density such as granitic bodies intruded into the regional metamorphic basement. Gravity highs are associated with structural highs and high-density bodies such as mafic intrusions. In the Bouguer gravity map, the gravity highs are mostly linked with the mountainous topography that consists of Precambrian granites and metamorphic rocks.

The bandpass (20–100 km) filtered map (Figure 4) is similar to the smoothed gravity map, and low gravity values are still present in both basins. In the case of North Park, the gravity low on Figure 4 is more extensive, and based on drill hole data better shows the extent of the region with significant amounts of sedimentary fill. The gravity low over the Col-

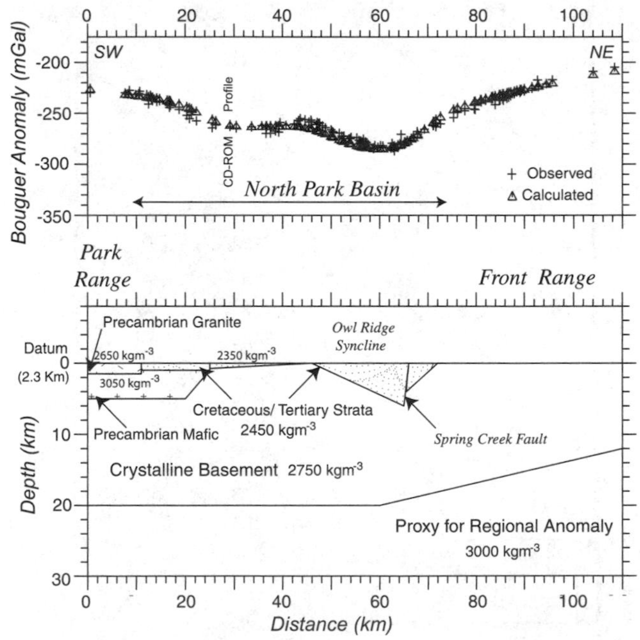

Figure 5. Integrated model of upper crustal structure for the North Park (NP-Figure 3). The numbers are densities that were assigned to each geologic unit.

orado mineral belt has a distinct linear northeasterly trend that is somewhat subdued in the Bouguer gravity map.

Two prominent gravity lows are present in the maps (Figures 3 and 4). In the case of North Park, a gravity low trends in a northwesterly direction parallel to the axis of the Owl Ridge Syncline (Figure 3). *Behrendt et al.* [1969] interpreted the steep gravity gradient in the North Park basin that strikes in a west-northwest direction, north of the axis of the syncline, as the Spring Creek fault (Figure 1). We attribute the gravity low in North Park to basin fill that *Behrendt et al.* [1969] described as Late Permian to Quaternary sediments but *Lange and Wellborn* [1985] showed were probably no older than Triassic. Older sedimentary rocks were either eroded by the Ancestral Rocky Mountains orogeny or they were never deposited [*Hail*, 1965].

A distinct northeasterly trending gravity low obscures the gravity signature of northern South Park. The lowest values in the basin are found on the northern fringe where the basin signature coincides with this linear feature. The Colorado mineral belt (CMB) strikes in a northeasterly direction, and a deep-seated feature associated with it appears to be the cause of this gravity low [*Brinkworth et al.*, 1968; *Case*, 1965; *Isaacson and Smithson*, 1976; *McCoy et al.*, this volume]. Case [1965], *Brinkworth et al.* [1968], and *Behrendt* [unpublished manuscript] have noted that the most negative gravity values on the continental United States are in the CMB. Furthermore, *Isaacson and Smithson* [1976], *Behrendt* [unpublished manuscript], and *McCoy et al.* [this volume] have modeled the Colorado mineral belt as being due to a steep sided batholith of granitic composition.

The gravity low associated with South Park is clearer to the south and has the form of two prongs and a sharp gravity gradient that defines the eastern structural boundary of the basin along the Front Range (Figure 4). The steep gravity gradient along the Front Range is coincident with the Elkhorn thrust fault but may also be influenced by changes in composition within the Precambrian basement between South Park and the Front Range. The gravity low in the middle of the South Park basin (Figure 3) is likely associated with the South Park fault that is located about the middle of the basin along the eastern boundary of Reinecker Ridge (Figure 1).

4.2. Gravity Models of the Basins

A series of 2.5D models was created to determine the structure beneath both basins. Surface features were constrained using the geologic map of Colorado [*Tweto*, 1979]. For initial constraints on depths of basins and strata present, the geologic cross sections constructed by *Tweto* [1989] and the available literature were consulted.

The North Park profile (NP, Figure 3) crosses several important structures in the basin such as the Owl Ridge syncline, the Spring Creek fault, and other minor faults. The 2.5D model for this profile (Figure 5) shows that the basin is an asymmetric syncline with a fault displacing the basin fill at the northern end of the line. The depth of the basin is approximately 6 km, and the sedimentary fill consists of both Mesozoic and Cenozoic strata overlying the Precambrian basement.

The northern South Park (SP, Figure 3) profile cuts across from the Mosquito Range on the southwest through South Park to the Front Range. Bodies of granitic composition bound the basin at either end of the line and the northeast end is cut by a series of faults (Figure 6). In the vicinity of Reinecker Ridge, Tertiary strata lie atop a Cretaceous unit cut by a normal fault. Another section of the Tertiary strata lies to the east of Reinecker Ridge along a thrust fault separating a bounding granitic body. Precambrian granitic rocks have been thrust westward, along the Elkhorn fault, over Tertiary and Cretaceous units. The granitic body at the east end of the profile masks the true width of the basin on the plot of observed/calculated gravity points. The basin appears much wider on the plot than it actually is.

The southern South Park profile (SSP, Figure 3) extends approximately from the Mt. Princeton batholith on the southwest to the Front Range on the northeast. This model shows that the basin is composed of two sub-basins separated by a structural high. The Upper Arkansas graben, a rift structure,

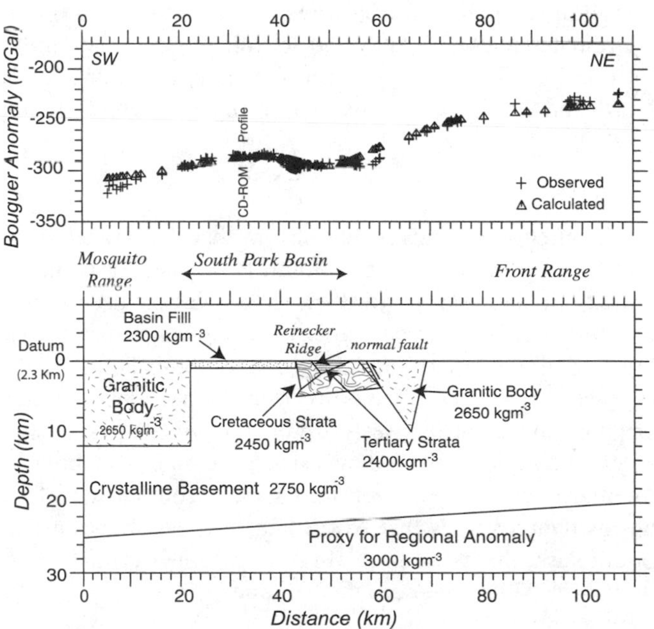

Figure 6. Integrated model of upper crustal structure for the central portion of South Park (SP-Figure 3). The numbers are densities that were assigned to each geologic unit.

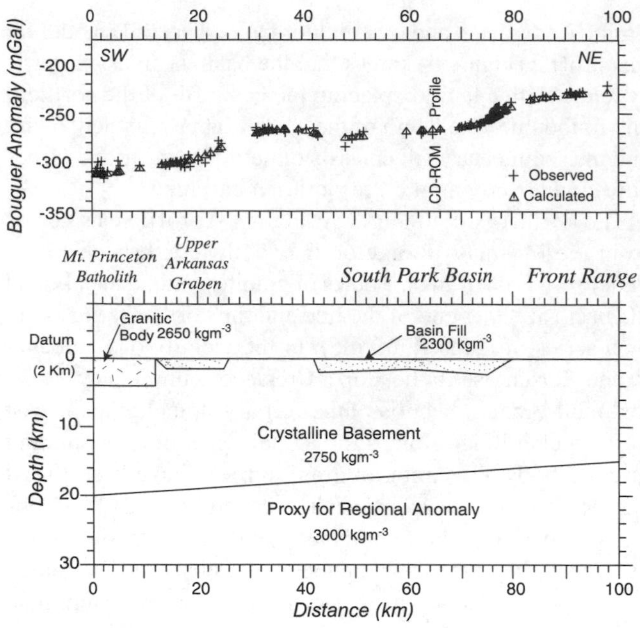

Figure 7. Integrated model of upper crustal structure for the southern portion of South Park (SSP-Figure 3). The numbers are densities that were assigned to each geologic unit.

is shown on the west of the basin, and appears to be ~ 2 km deep (Figure 7).

This study and the analysis of the upper crustal CD-ROM refraction along the profile (*Rumpel et al.*, this volume) were coordinated and the gravity and seismic models show similar relationships. The gravity and seismic models indicate both the North and South Park basins shallow in a southerly direction.

5. SUMMARY

Our integrated analysis has shown that the basinal structure beneath both North and South Park basins is complex. Gravity modeling of North Park shows that the deepest portion of the basin is an asymmetric syncline approximately 6 km in depth. Gravity modeling of the South Park profiles suggests that a series of faults divide the basin and that the eastern portion was thrust westward over younger units. Further south, the basin is divided by a structural high with thin sedimentary cover over the entire basin. A comparison of the basins demonstrates that North Park is a deeper basin than South Park; this is also seen in the new upper crustal velocity models [*Rumpel et al.*, this volume].

Acknowledgements. We thank reviewers Donald W. Forsyth, Mousumi Roy and Anthony Lowry for their helpful comments and suggestions that improved this paper significantly. Database and computational support as well as a research assistantship (LT) was provided by the Pan American Center for Earth and Environmental Studies, which is supported by NASA. Funding was also provided by the Continental Dynamics Program of the National Science Foundation (EAR-9614269) and the GEON project (EAR-0225670).

REFERENCES

Beggs, H. G., Interpretation of geophysical data from the central South Park Basin, Colorado, in *Exploration Frontiers of the Central and Southern Rockies,* edited by H. K. Veal, Rocky Mountain Association of Geologists, Denver, Colorado, 67–76, 1977.

Behre, C. H., Jr., I. T. Schwade, R. M. Dreyer, Bedrock geology of northern South Park. *Proceedings Geol. Soc. Am.* 66–67, 1935.

Behrendt, J. C., P. Popenoe, R. E. Mattick, A geophysical study of North Park and the surrounding ranges, Colorado. *Geol. Soc. Amer. Bull., 80,* 1523–1538, 1969.

Brinkworth, G. L., J. C. Behrendt, and P. Popenoe, Geophysical investigations in the Colorado mineral belt, *EOS Transactions AGU, 49*, 1968.

Cady, J. W., Calculation of gravity and magnetic anomalies of finite-length right polygonal prisms, *Geophysics, 45*, 1507–1512, 1980.

Case, J. E., Gravitational evidence for a batholithic mass of low density along a segment of the Colorado mineral belt, *Geol. Soc. Amer., Spec. Paper 82*, 26, 1965.

Durrani, J. A., Seismic investigation of the tectonic and stratigraphic history, eastern South Park, Park County, Colorado. Ph.D. thesis, 138 pp., Colorado School of Mines, 1980.

Hail, W. J., Jr., Geology of northwestern North Park, Colorado. *U.S. Geol. Surv. Bull. B-1188*, 133, 1965.

Howland, A. L., Bedrock geology of southern South Park, Colorado. *Proceedings Geol. Soc. Am., 1934*, 84, 1935.

Isaacson, L. B. and S. B. Smithson, Gravity anomalies and granite emplacement in west-central Colorado. *Geol. Soc. Amer. Bull.*, 87, 22–28, 1976.

King, C., Geological and topographic atlas: U. S. Geological Exploration 40th Parallel, 1876.

King, C. Systematic Geology: U. S. Geological Exploration 40th Parallel, vol.1, 803, 1878

Lange, J. K., and R. E. Wellborn, Seismic profile North Park basin, in *Seismic Exploration of the Rocky Mountains*, edited by R. R. Gries and R. C. Dyer, Rocky Mountain Association of Geologists and Denver Geophysical Society, Denver, CO, 67–76, 1985.

Levander, A., R., C. Zelt, and B. Magnani. Crust and upper mantle velocity structure of the southern Rocky Mountains from the Jemez Lineament to the Cheyenne belt, this volume.

McCoy, A., M. Roy, L. Trevino, and G. R. Keller. Gravity modeling of the Colorado mineral belt, this volume.

McCoy, A. M., K. E. Karlstrom, M. L. Williams, and C. A. Shaw, The Proterozoic ancestry of the Colorado mineral belt: ca. 1.4 Ga. Shear zone system in central Colorado, this volume.

Rumpel, H. -M., C. M. Snelson, C. Prodehl, G.R. Keller, Results of the CD-ROM project seismic refraction /wide-angle reflection experiment: The lower crustal and upper mantle, this volume.

Sawatzky, D. L., Structural geology of southeastern South Park, *Mountain Geol.* 1, 133–139, 1964.

Snelson, C. M., H. -M. Rumpel, G. R. Keller, K. C. Miller, and C. Prodehl, Regional crustal structure derived from the CD-ROM seismic refraction/wide-angle reflection experiment: The lower crustal and upper mantle, this volume.

Stark, J. T., J .S. Johnson, C.H. Behre, W.E. Powers, A. L. Howland, D. B. Gould, Geology and origin of South Park, Colorado, *Geol. Soc. Amer.*, 33, 188pp., 1949.

Tweto, O., Geologic Map of Colorado, scale 1:500,000, US Geological Survey, 1979.

Tweto, O., Tectonic history of Colorado, In *Colorado Geology*, edited by H. C. Kent, and K. W. Porter, 5–9, Rocky Mtn. Assoc. Geol., 1980

Tweto, O., Precambrian geology of Colorado, In *Colorado Geology*, edited by H.C. Kent, and K.W. Porter, 37–46, Rocky Mtn. Assoc. Geol., 1980.

Tweto, O., Geologic sections across Colorado, Miscellaneous Investigations Series-US Geological Survey, Report I-1416, 1 sheet, scale 1:500,000, 1983.

Wellborn, R. E., Structural style in relation to oil and gas exploration in North Park-Middle Park basin, Colorado, in *Exploration Frontiers of the Central and Southern Rockies,* edited by H. K. Veal, Rocky Mountain Association of Geologists, Denver, CO, 67–76, 1977.

Wessel, P. and W. H. F. Smith, New, improved version of the Generic Mapping Tools released, *EOS Trans. AGU*, *79*, 579, 1998

G. Randy Keller and L. Treviño, Department of Geological Sciences, University of Texas at El Paso, 500 W. University Ave., El Paso, TX, 79968.

Gravity Modeling of the Colorado Mineral Belt

Annie M. McCoy[1] and Mousumi Roy

Department of Earth and Planetary Sciences, University of New Mexico, Albuquerque, New Mexico

Leandro Treviño and G. Randy Keller

Department of Geological Sciences, University of Texas, El Paso, Texas

The Colorado Mineral Belt (CMB) is delineated as a belt of mostly Laramide mineralization within a broader zone of Laramide and older magmatism in central Colorado. This paper focuses on the profound negative Bouguer gravity anomaly that coincides with this broader zone of protracted magmatic activity, and is among the largest gravity anomalies in North America. Seismic studies suggest that parts of the CMB region are underlain by anomalously low-density crust and that the CMB lies within a broad zone of low seismic velocities in the uppermost mantle. This study explores simple distributions of subsurface mass deficits that can explain the CMB negative Bouguer gravity anomaly, and are constrained by geologic estimates of the extent of crustal plutonic bodies and by seismically-inferred crustal and upper mantle velocity anomalies. Specifically, we consider forward models that include (1) a low-density batholithic body in a 20-km thick upper crust with a density contrast of 150 kg/m^3, (2) a 25-km thick lower crust with a potential low-density body with a density contrast of 150 kg/m^3, and (3) an upper mantle with a low-density body with variable density contrast placed at variable depth. We discuss the viability of our first-order forward models and their consistency with seismic observations, and suggest refinements that can improve our understanding of the structure of the CMB.

1. INTRODUCTION

The Colorado Mineral Belt (CMB) is a northeast-trending belt of mainly Laramide mineralization in central Colorado within a broad zone that also includes older Proterozoic and younger Cenozoic magmatism [*Mutschler et al.*, 1987]. This zone extends more than 200 kilometers (km) across central Colorado and coincides with one of the most negative Bouguer gravity anomalies in the United States [*Case*, 1965; *Isaacson and Smithson*, 1976]. The Laramide (Late Cretaceous to earliest Oligocene) magmatism is typically expressed at the surface as relatively small, shallow intrusive bodies associated with major gold, silver, lead, and molybdenum deposits [*Tweto and Sims*, 1963]. Proterozoic rocks show evidence for magmatic episodes and shear zone movements around 1.7 billion years ago (Ga), 1.4 Ga, and 1.1 Ga focused along the CMB [*McCoy et al.*, this volume; *Shaw et al.*, 2001; *Nyman et al.*, 1994; *Barker et al.*, 1975].

A Bouguer gravity map of Colorado constructed using data of Oshetski and Kucks [2000] shows two major negative anomalies that are among the most prominent features in North America gravity [Plate 1; see CD ROM in back cover sleeve]. The northern negative anomaly is centered on the CMB and the southern anomaly is centered on the middle to late Tertiary (mainly early Oligocene) magmatic centers of

[1] Presently at John Shomaker & Associates, Inc., Albuquerque, New Mexico

The Rocky Mountain Region: An Evolving Lithosphere
Geophysical Monograph Series 154

10.1029/154GM08

the San Juan Volcanic Field [Plate 1]. Negative Bouguer gravity anomalies reach values of -340 milliGals (mGals) in the CMB and -337 mGals in the San Juan Volcanic Field.

Crustal thickness estimates based on teleseismic receiver functions indicate little crustal thickening from the Colorado Great Plains (49.9 km average thickness) to the Colorado Rocky Mountains (50.1 km average thickness), suggesting that the high topography of the Colorado Rocky Mountains is probably not simply compensated by an Airy-type crustal root [*Sheehan et al.*, 1995]. Rayleigh wave dispersion suggests greater variation in crustal thickness (40 to 45 km in the Colorado Great Plains and 48 to 55 km in the Southern Rocky Mountains; *Li et al.*, 2002], but is still consistent with the idea that the high topography of the Colorado Rocky Mountains must be supported by some combination of Airy and Pratt isostasy [*Li et al.*, 2002; *Snelson et al.*, this volume].

In this study, we explore Pratt and Airy-type forward models along northwest-trending profiles across the CMB in order to explain the observed negative Bouguer gravity anomaly [Plate 1]. Medium to short-wavelength (100 km) negative gravity anomalies can be adequately matched with inferred low-density structures in the crust, consistent with geologic mapping of crustal intrusions in the CMB. The longer wavelength (400 to 500 km) gravity low across the CMB is adequately explained by a low-density structure in the uppermost mantle or a regionally extensive low-density body in the lower crust. In our modeling we explore both possibilities and discuss the non-uniqueness of our results, which are somewhat constrained by geologic mapping and seismic studies.

We recognize that lower crustal features that explain the long-wavelength gravity low across the CMB are consistent with a study of Rayleigh wave dispersion from the Rocky Mountain Front experiment, which suggests low crustal densities below the Sawatch Mountains in the central CMB [*Li et al.*, 2002]. From the crustal structure in *Li et al.* [2002], we infer that the region of low crustal densities is roughly centered on the middle of our profiles and extends laterally about 100–200 km in the upper crust and 200–300 km in the lower crust [Plate 1]. However, to explain the 500 km scale gravity low surrounding the CMB, we find that a low-density body in the lower crust would need to be regionally extensive (>500 km), extending well beyond the CMB to the northwest and southeast. We conclude that current seismic interpretations of crustal structure in this region preclude such a low-density structure in the lower crust [see also *Rumpel et al.*, this volume; *Snelson et al.*, this volume].

On the other hand, low densities in the uppermost mantle are consistent with both P and S-wave travel-time tomographic results that show low seismic velocities (V_P and V_S) in the upper mantle (from 50 to 200 km depth) beneath the CMB region [*Lee and Grand*, 1996; *Lerner-Lam et al.*, 1998; *Dueker et al.*, 2001]. These results suggest that a low-velocity upper mantle anomaly underlies the Southern Rocky Mountains including the CMB, with varying estimates of velocity reduction [Plate 1]. For example, a 6 to 8 percent reduction in V_S at 75 to 125 km depth relative to western Kansas is presented in *Lee and Grand* [1996; Plate 1], a 4 percent reduction in V_P at a depth of 150 km relative to a regional average model is presented in *Lerner-Lam et al.* [1998], and a 2 percent reduction in V_P at depths of 50 to 250 km relative to a regional average is presented in *Dueker et al.* [2001]. Additionally, Rayleigh-wave dispersion also suggests up to 4 percent reduction in V_S relative to a regional model at 50 to 100 km depth below the Southern Rocky Mountains [*Li et al.*, 2002]. The region of low upper mantle seismic velocities coincides with large regional gradients in relative attenuation [*Boyd and Sheehan*, this volume], complicating the physical interpretation of this structure. Using combined models of attenuation and velocity structure, *Boyd and Sheehan* [this volume] interpret the western part of this low-velocity upper mantle region (with low attenuation) to be associated with moderately high temperatures, partial melt content, and/or anomalous upper mantle composition (e.g. the presence of phlogopite). They attribute the eastern part of the low-velocity upper mantle region (with high attenuation) to high temperatures associated with the Rio Grande Rift and its northward continuation [*Boyd and Sheehan*, this volume].

In the current study, we use simple gravity models as an added constraint on subsurface density structure, and highlight the issue of whether the large negative Bouguer anomaly centered over the CMB can be explained by low-density bodies in the crust and/or upper mantle. We interpret our models in combination with geologic data and the seismic results discussed above. In this study, we do not incorporate a detailed (10 to 50 km resolution), geologically-reasonable model of upper crustal structure, but we recognize that future refinements of this work will require such a regional model to adequately explain the shortest wavelength gravity anomalies.

2. METHODS

The observed Bouguer gravity values along northwest-trending profiles across the CMB (Profiles 1, 2, 3, and 4 in Table 1; see CDROM in back cover sleeve, Plate 1, and Figure 1) were extracted from a regional Colorado Bouguer gravity dataset [*Oshetski and Kucks*, 2000]. Gravity data within a 4-km wide swath were projected orthogonally onto these profiles using Generic Mapping Tools [GMT; *Wessel and Smith*, 1991]. The non-uniform coverage gravity data were also gridded using bi-cubic interpolation and projected onto the central profile line. Data extracted along each profile using both the swath-width method and the gridded data are plotted in Figure 1.

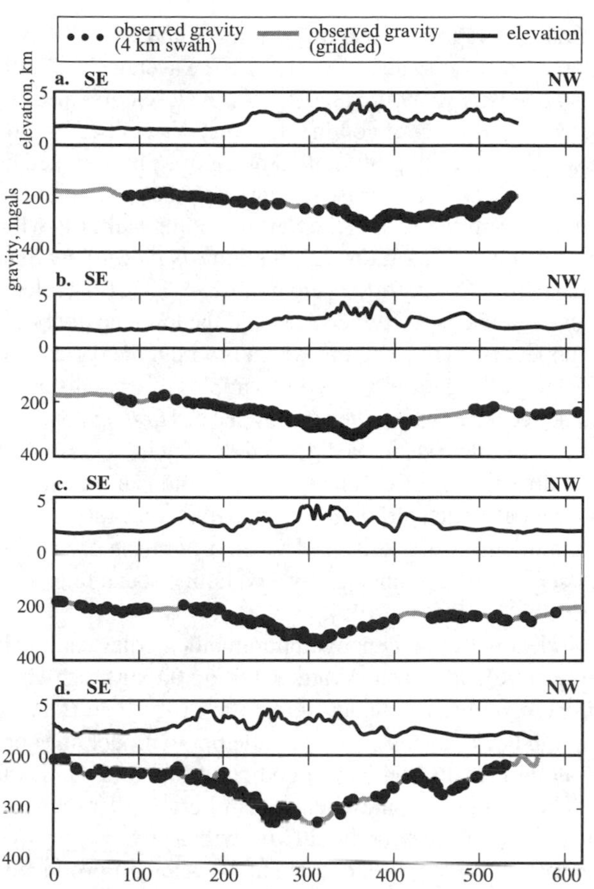

Figure 1. Observed gravity and elevation along Profiles 1 (in **a.**), 2 (in **b.**), 3 (in **c.**), and 4 (in **d.**). See text for details on the gridded gravity data and the swath widths used in this and subsequent figures.

A two-and-a-half dimensional modeling approach was used to determine the geometries and densities of subsurface bodies. We used forward-modeling software based on the Talwani method of calculating predicted gravity from bodies of arbitrary geometry [our code was written by D. Roberts, and is a rewrite of tal.25dgrav by S.F. Lai; *Talwani et al.*, 1959; *Cady*, 1980]. Matlab pre- and post-processors allowed for efficient organization and alteration of input models and graphical output. Input parameters included (a) the geometries of the modeled bodies defined by vertices with x,y coordinates in the two-dimensional planes of the profiles, (b) the densities of the modeled bodies defined with respect to an upper-crustal density of 2,750 kg/m^3, and (c) the distances to which the bodies are to be projected into and out of the plane of the profile (hence the two-and-a-half dimensionality of the model, although in the models presented here the in- and out-of-plane strike lengths are effectively infinite). In order to minimize boundary effects, typical lateral extents of bodies used in this study are >10,000 km).

3. OBSERVED GRAVITY ANOMALIES

The observed Bouguer gravity anomalies along the profiles [Figures 1a through 1d] are composites of several anomalies with varying wavelengths. First, a regional trend in the Bouguer gravity is present, with generally more negative values to the northwest [Figures 1a and 1b]. This regional trend is likely a reflection of gradually thickening crust from southeast to northwest, or a gradual reduction in average subsurface densities, or a combination of both. If we attribute this trend completely to a thickening of the crust, the amount of thickening required to explain the regional trend is no more than 4 or 5 km over a distance exceeding 500 km, which is well within errors of seismic Moho determinations. We do not attempt to remove this longest-wavelength regional trend in the current study; instead, we maintain a crust of constant thickness and focus on modeling the 500 km or shorter wavelength anomalies that are described below.

The longest wavelength anomaly we model is the 400 to 500-km wide anomaly with a shape that resembles a shallow dish with steep shoulders (shoulders are 20 to 50-km wide) centered on the CMB. Additionally, a shorter-wavelength (200 km) anomaly that coincides with the boundaries of the CMB as defined by Tweto and Sims [1963] appears to represent a shallower low-density body in the crust. Finally, the very short-wavelength (several km) Bouguer gravity variations probably represent density variations associated with lithologic changes in the upper and middle crust; these are not included in detail in this study, but should be incorporated into future, more detailed analyses of the CMB gravity.

4. FITTING A MODEL TO THE OBSERVED ANOMALIES

Unlike the Himalayas and the central Andes [*Zhao et al.*, 1993; *Zandt et al.*, 1994], the negative Bouguer gravity anomaly of the CMB cannot be explained by Airy-type isostasy. Profile 3 shows the predicted gravity anomaly if high topography were supported by an Airy-type crustal root [Figure 2]. The most negative predicted gravity value coincides with the most negative observed Bouguer gravity value and the highest topography along the profile. As pointed out by *Li et al.* [2002], the overall coincidence of the most negative Bouguer anomalies with the highest elevations in Colorado suggests the dominance of local compensation, although we recognize that regional compensation mechanisms may also play a role in supporting high topography [e.g., *Bechtel et al.*, 1990]. In this paper, we focus our attention on local compensation using a combination of Pratt and Airy-type isostasy. The predicted gravity anomaly in Airy-type compensation has more negative values along the sides of the CMB than observed and less

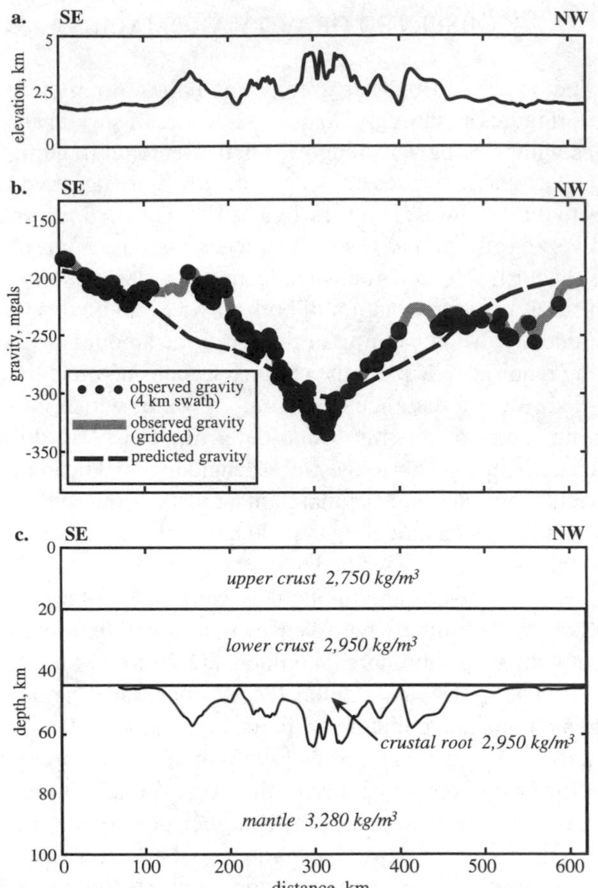

Figure 2. a. Elevation, **b.** observed and predicted gravity, and **c.** density model along Profile 3 based on Airy-type isostasy.

negative values at the center of the CMB. This is consistent with seismic studies that find a lack of a thick crustal root beneath the Colorado Rocky Mountains [*Sheehan et al.*, 1995; *Li et al.*, 2002], but also find anomalously low densities in the crust, suggesting the importance of Pratt-type isostasy [*Li et al.*, 2002].

Our gravity models are non-unique solutions with built-in assumptions about the densities of the Earth's crust and mantle. The assumptions used in this study include: an upper-crustal background density of 2,750 kg/m^3; a lower-crustal background density of 2,950 kg/m^3; an upper-mantle background density of 3,280 kg/m^3; and a batholithic crustal body with a density of 2,600 kg/m^3. The upper crust is 20-km thick, in accordance with the Conrad discontinuity indicated by seismic refraction studies [*Snelson et al.*, this volume; *Rumpel et al.*, this volume], and the lower crust is 25-km thick, giving a total crustal thickness of 45 km consistent with seismic observations [*Sheehan et al.*, 1995; *Li et al.*, 2002]. The assumed upper-crustal density is similar to that determined for Precambrian rocks in the Sawatch Range and Elk Mountains of central Colorado [2,710 kg/m^3, *Isaacson and Smithson*, 1976; 2,760 kg/m^3, *Tweto and Case*, 1972]. The average background crustal density is about 2,860 kg/m^3, in close agreement with the average density of continental crust determined by Christensen and Mooney [1995] of 2,830 kg/m^3. The assumed density of a body in the crust is similar to that determined for Tertiary granitic rocks in the Sawatch Range and Elk Mountains of central Colorado [2,630 kg/m^3, *Isaacson and Smithson*, 1976; 2,620 kg/m^3, *Tweto and Case*, 1972]. The density contrast between the upper crust and the low-density body in the upper crust is 150 kg/m^3, which is within the typical range of 50 to 180 kg/m^3 measured in samples of granites and the country rocks in which they are emplaced [*Bott and Smithson*, 1967]. The increase in background density in the lower crust is an attempt to reflect the depth-variation of densities for continental rocks, although we recognize that future refinements of this work will need to incorporate more realistic density-depth relationships for crystalline rocks [e.g. *Christensen and Mooney*, 1995].

Models with low-density upper mantle bodies can explain the relatively long wavelength (400 to 500 km) dish-shaped observed Bouguer anomaly across the CMB [Plate 2; see CD ROM in back cover sleeve]. We attempt to model this anomaly along Profile 3 with (1) a body in the mantle that is thick (200 km) and has a small density contrast of 20 kg/m^3 [Plate 2b] and (2) a thinner body (20 km) with a larger density contrast of 100 kg/m^3 [Plate 2c]. The wavelength produced by the thinner body with the higher density contrast is a better fit to the wavelength defined by the shoulders of the shallow dish-shaped observed anomaly. The wavelength form of these shoulders also places some constraints on the horizontal width of the model body. As the horizontal width is decreased [Plate 2e], the modeled anomaly develops a shorter wavelength.

We also vary the depth to the low-density body, equivalent to including a high-density (high-velocity) mantle lid, and investigate the trade-offs between density contrast, depth, and thickness of the low-density body below the mantle lid. The fit to the dish-shaped anomaly, particularly the relatively steep shoulders, is reasonable for relatively thin mantle lids (e.g., for a 100-km thick upper mantle body of density contrast 40 kg/m^3, the allowed mantle lid thickness is about 50 km; Plate 2f), but degrades for greater mantle lid thicknesses. Based on these results, we argue that the shape of the dish-shaped anomaly allows for low-density bodies in the upper mantle, either immediately below the Moho or in the 75 to 150 km depth range, but probably not deeper.

Alternatively, we model the shallow dish-shaped anomaly with a low-density body in the lower crust [Figure 3b] or a Moho with varying depth [Figure 3c]. A body in the lower crust with a density of 2,800 kg/m^3 and a thickness of about 10 km provides a good fit of the observed anomaly. Rayleigh-wave dispersion [*Li et al.*, 2002] and seismic refraction observations [*Snelson et al.*, this volume; *Rumpel et al.*, this volume] show a range in Moho depth of about 10 km, from depths of 40 to 50 km below the CMB, and this Moho topography may explain part of the shallow dish-shaped anomaly. Figures 3b and 3c show that both a low-density body in the lower crust and Moho topography provide good fits of the observed gravity and illustrate the fundamental non-uniqueness of gravity modeling, suggesting that further seismic work is required to determine the relative importance of low densities in the lower crust, low densities in the uppermost mantle, and/or crustal thickness variations. We note that to explain the broad anomaly with a low-density body in the lower crust we require a laterally-extensive (> 500 km wide on Profile 3; Figure 3) region of low densities, which is inconsistent with the crustal velocity structure from Rayleigh wave dispersion [*Li et al.*, 2002]. Thus, we confine the rest of our work to models that incorporate low densities in the mantle together with batholithic structures in the upper crust, consistent with recent seismic findings [*Dueker et al.*, 2001; *Li et al.*, 2002; *Lee and Grand*, 1996].

The shoulders of the shallow dish-shaped anomaly represent steep gradients in gravity values and are asymmetric; the northwestern shoulder has more negative Bouguer gravity values (even after removal of the longest wavelength regional trend). A model body that is thicker at the northwestern end provides a good fit of the anomaly's asymmetric geometry (see following section). This geometry may also represent lateral variations in density contrasts between the body and the mantle. Below, we present models that explore the most realistic geometries for subsurface density bodies that explain the observed Bouguer gravity along Profiles 1 through 4.

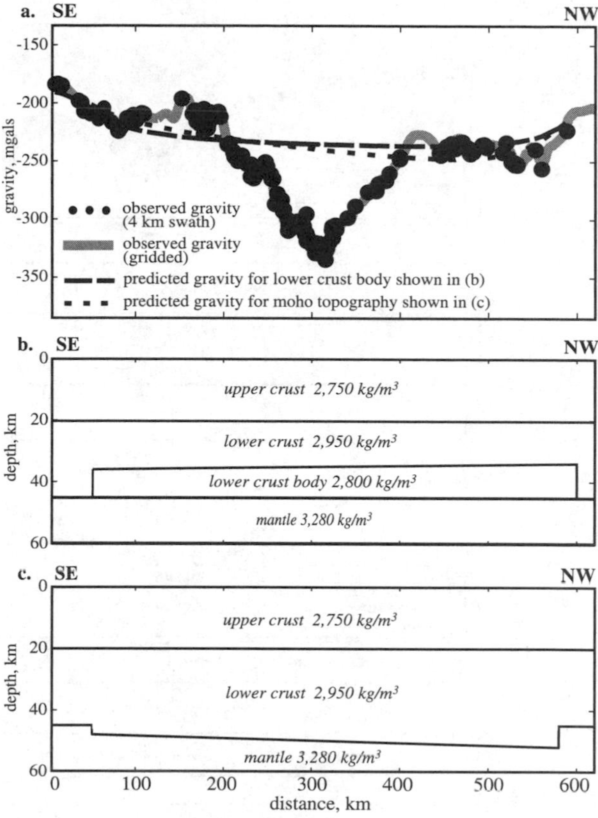

Figure 3. Attempts to match observed and predicted gravity along Profile 3 (**a.**) with **b.** a body in the lower crust that has a density of 2,800 kg/m^3, and **c.** topography on the Moho.

5. RESULTS OF FORWARD MODELING

All crustal models that fit the observed 200 km-wide gravity low centered on the CMB share some common features [Figure 4]. The low-density body in the crust extends from shallow crustal depths (several km below land surface) to about 20-km depths in each profile. The crustal batholithic body is about 150 to 200-km wide near the surface, but narrows to 10 to 20-km wide in the middle crust.

The longer wavelength low (500 km-wide) across the CMB is modeled with a low-density body in the upper mantle that is asymmetric and is thicker at its northwestern endpoint than in its central part. In the southern part of the CMB, along Profiles 3 and 4, the low-density body in the upper mantle is thicker at both endpoints than in its central part [Figure 4f and 4h]. The body in the mantle is typically 10 to 20-km thick in its central part, and is 400 to 500-km wide. Although we recognize the trade-off between the depth extent of the low-density body in the upper mantle and the density contrast between this body and the surrounding mantle [Plate 2], the wavelength and shape of the observed anomaly around the CMB leads us to favor a shallow upper mantle body with greater density contrast.

5.1 Profiles 1 Through 3

Along Profiles 1 through 3 [Profile 1 is the northeastern-most profile; Plate 1], the regional dish-shaped anomaly is modeled with a mantle body with a density of 3,180 kg/m^3, which represents a density contrast of 100 kg/m^3 (assuming crustal and mantle background densities as described in *Fitting a Model to the Observed Anomalies*) [Figure 4a through 4f]. In all cases, this body is more than 400-km wide and asymmetric [Table 2; see CDROM in back cover sleeve]. In Profile 3 [Figure 4f], the required body is slightly wider than in Profiles 1 and 2 [Figure 4b and 4d]. In all profiles, the northwestern end of the body is thicker than the central part of the body. In Profiles 3 and 4 [Figure 4f and 4h], in the southern part of the CMB, both ends of the body are much thicker than the central part of the body [Table 2; see CDROM

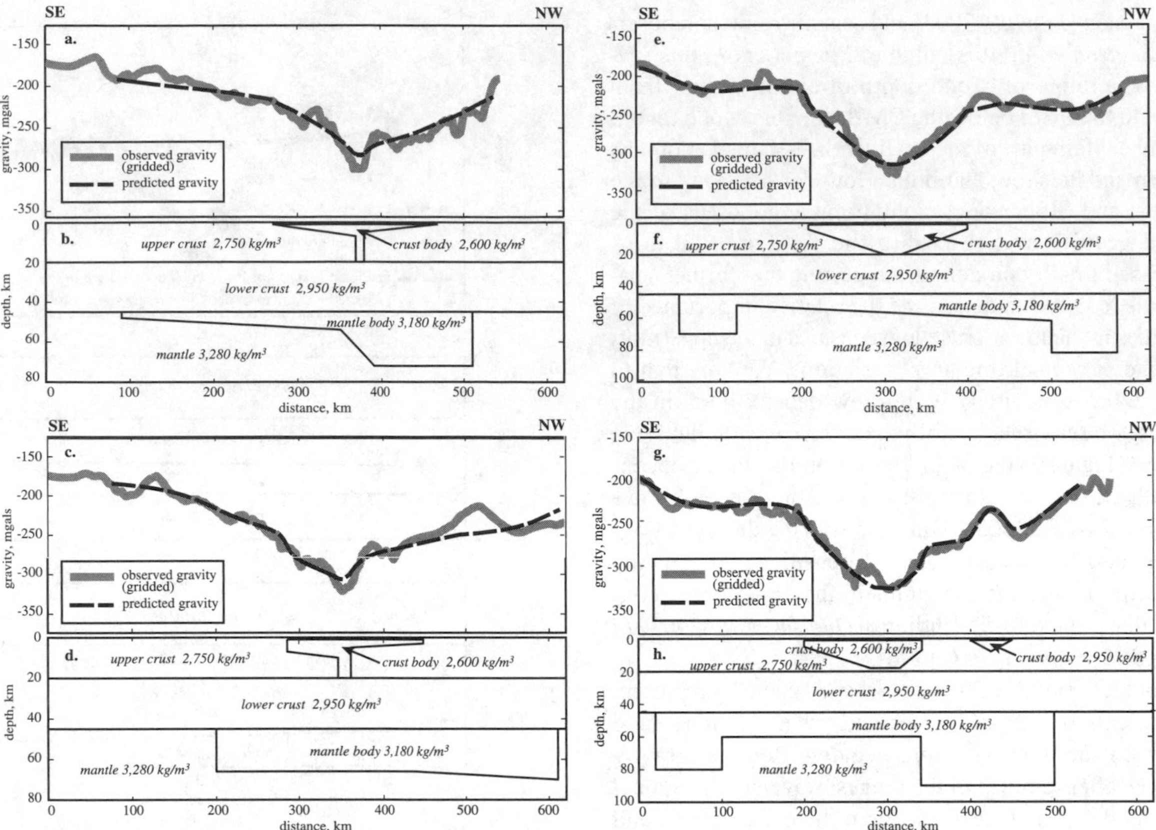

Figure 4. a. Observed and predicted gravity along Profile 1 based on a density model (**b.**) that includes a body with a density of 2,600 kg/m3 in the upper crust (2,750 kg/m^3) and a body with a density of 3,180 kg/m^3 in the mantle (3,280 kg/m^3). **c.** Observed and predicted gravity along Profile 2 based on the density model in **d.** (densities as in **b.**). **e.** Observed and predicted gravity along Profile 3 based on the density model in **f.** (densities as in **b.**). **g.** Observed and predicted gravity along Profile 4 based on the density model in **h.** (densities as in **b.**).

in back cover sleeve]. In contrast, Profile 2 requires an upper mantle low-density body that is almost uniformly thick (20 km) across the entire region [Table 2; see CDROM in back cover sleeve].

The shorter, 200-km wavelength anomaly was modeled in Profiles 1 through 3 with a body in the crust with a density of 2,600 kg/m^3 [Table 2; see CDROM in back cover sleeve; Figure 4b, 4d, and 4f]. In all three cases, the body is 10–20 km wide in the middle crust and 165–200 km wide near the surface. It is about 20-km thick in its central part and has lateral 'wings' that taper and pinch out at shallow depths.

5.2 Profile 4

Along Profile 4 [southwesternmost profile; Plate 1; Figure 4g and 4h], the results for the upper mantle body are similar to Profiles 1 and 3 above, in that the body is 480-km wide and asymmetric. The body is 35-km thick at its southeastern end and is 45-km thick at its northwestern end. As in Profile 3, the body's endpoints are more than twice as thick as the central part of the body.

The short-wavelength anomaly was modeled with a body in the crust with a density of 2,600 kg/m^3 [Figure 4h]. The body is 35-km wide in the middle crust and 142-km wide near the surface. It is 17-km thick in its central part, and about 5-km thick at its endpoints. This body has the winged geometry seen in crustal bodies in Profiles 1, 2, and 3, and there is a general thickening of the crustal body in Profiles 3 and 4, in a southwestern direction along the CMB.

A short-wavelength positive anomaly located southeast of the CMB was modeled with a relatively high-density body in the upper crust. This body is 50-km wide and 7-km thick and has a density of 2,950 kg/m^3. This high-density upper crustal body is one possible interpretation of this gravity feature. The gravity feature coincides closely with outcrops of the Early Cretaceous-age Dakota Sandstone and Purgatoire Formation [*Tweto*, 1979], sedimentary rocks that are not likely to generate positive density contrasts in the upper crust. There are relatively high-density ultramafic rocks exposed at the northern end of the Sangre de Cristo Mountains near Poncha Pass [*Shaw, C.A.*, written communication], and the positive anomaly may represent another occurrence of ultramafic rocks along the trend of the Sangre de Cristo Mountains. We suggest that further work needs to be done to investigate other (possibly deeper) density anomalies that might explain this gravity feature.

6. DISCUSSION

The negative Bouguer gravity anomaly of the CMB is similar to that of other parts of the U.S. Cordillera [e.g., the Southern Sierra Nevada, *Wernicke et al.*, 1996], in that it cannot be explained by simple Airy-type isostasy. In the Southern Sierra Nevada, for example, Wernicke et al. [1996] explained the negative Bouguer gravity anomaly through a combination of lateral changes in crustal and mantle densities and proposed asthenospheric upwelling beneath the high mountains. Previous work on the CMB Bouguer gravity anomaly in the vicinity of the Sawatch Range and the Elk Mountains attributed the anomaly to a granitic batholith [*Isaacson and Smithson*, 1976]. Isaacson and Smithson [1976] determined a maximum depth to the top of the low-density body to be 12 km and thus concluded that the negative anomaly is in the upper crust. Tweto and Case [1972] showed similar findings for the Sawatch Range near Leadville, Colorado. Isaacson and Smithson [1976] noted the close correlation between the negative Bouguer gravity anomaly and outcrops of Cretaceous and Tertiary granitic stocks.

In the current study, our models define two major contributions to the regional gravity anomaly across the CMB. First, a low-density batholithic body in the crust, which is wide and sheet-like in the upper crust, with a narrower root extending into the middle crust. The depth to the top of this body in our models is 1 km. We find that this crustal body thickens to the southwest, and is thickest in Profiles 3 and 4 in the southwestern part of the CMB. Additionally, along each profile, we model a regional dish-shaped anomaly (over 500-km wide) that may be explained by a low-density body in the uppermost mantle. This body is over 400-km wide in all parts of the CMB, and generally thicker at the endpoints than in the central part (although we note that in Profile 2 this body may be almost uniformly thick across the entire region). In all cases, the best-fit model has an asymmetric body in the upper mantle, and is thicker and/or has a higher density contrast at the northwestern edge of our study area. We recognize that the changes in thickness in the modeled upper mantle body may equivalently be represented by lateral changes in density rather than thickness.

Consistent with the trend toward more negative Bouguer anomalies from northeast to southwest along strike of the CMB, our gravity models require either a thickening of the inferred low-density bodies (and/or an increase in density contrast) in both the crust and upper mantle. Some of the increase in mass deficit could be explained by effects not included in our study, such as a thickening of the crust from 46 km in the northeast (Profile 1) to 52 km in the southwest (Profile 3) along the central CMB, inferred from Rayleigh-wave dispersion [*Li et al.*, 2002]. By comparing the mass deficits required for Profiles 1 and 3, we find that thickening the crust by 6 km can generate the necessary mass deficit, provided the thickening occurs over a 100 km wide region surrounding the CMB. Note, however, that the along-strike variations in batholithic bodies in our models are consistent with surface geologic constraints; the region of surface outcrop of crustal intrusive bodies (and associated mineralization) broadens from northeast to southwest along the CMB [*Mutschler et al.*, 1987; *Tweto and Sims*, 1963]. Thus, we conclude that the crustal density variations represented by the batholithic bodies in our gravity models represent a significant along-strike variation in the CMB, but recognize that crustal thickening may also play a role. Inferred mass deficits in the upper mantle also show a similar trend, consistent with an up to 2 percent reduction in V_S from northeast to southwest along the CMB at 75 to 125 km depths in the upper mantle [Lee and Grand, 1996] and similar along-strike variations in V_P e.g., *Dueker et al.*, 2001]. Additionally, inferred variations in relative attenuation show a trend from high attenuation in the northeast to low attenuation (but decreasing V_S) in the southwest along the CMB [*Boyd and Sheehan*, this volume].

The along-strike variations in subsurface mass deficit noted above therefore appear to be robust and constrained by corresponding trends in seismic velocities and surficial geology. While this coincidence is encouraging, the geophysical analysis above places no constraints on the time-development of the subsurface density structure and its relation to the protracted history of the broad zone of magmatism and deformation that includes the CMB. To further refine this work and interpret the structure of the zone, we need to incorporate detailed geologic knowledge of the deformation and magmatism in the region and upper-crustal density variations important to resolving the short-wavelength gravity features across the area. To refine our understanding of the size and shape of a large, crustal-scale batholith beneath the CMB and a regional-scale upper mantle low-density body, we need further work that combines seismic and gravity observations. For example, geologic maps and upper-crustal seismic velocity models combined with velocity-density scaling relationships [e.g., *Christensen and Mooney*, 1995] will be useful to determine the short-wavelength (10 km) variations in density and gravity. These will in turn highlight the nature of the more regional (over 200-km scale) gravity features. Specifically, the velocity data from the CDROM refraction line [Plate 1] will be used with empirical velocity-density scaling to constrain density structure in the crust along the CMB. In addition, better resolution in seismic tomography models might help resolve the size and shape of the zone of low densities in the uppermost mantle (e.g., resolve whether the body in the uppermost mantle does change appreciably in thickness and/or density across the region).

Acknowledgments. This study was made possible through the support of the NSF CDROM experiment funded by the Continental Dynamics Program NSF/EAR-00340. The authors thank reviewers Anne Sheehan, John Geissman, Holly Stein, and Paul Sims for providing helpful reviews in a timely manner, and editor Karl Karlstrom for his comments and suggestions.

REFERENCES

Barker, F., D.R. Wones, W.N. Sharp, and G.A. Desborough, The Pikes Peak batholith, Colorado Front Range, and a model for the origin of the gabbro-anorthosite-syenite-potassic granite suite, *Precamb. Res., 2,* 97–160, 1975.

Bechtel, T., Forsyth, D., Sharpton, V., and R. Grieve, Variations in effective elastic thickness of the North American lithosphere, *Nature, 343*, 636–638, 1990.

Bott, M.H.P., and S. Smithson, Gravity investigations of subsurface shape and mass distributions of granite batholiths, *Geol. Soc. Am. Bull., 78 (7)*, 859–877, 1967.

Cady, J.W., Calculation of gravity and magnetic anomalies of finite-length right polygonal prisms, *Geophysics, 45 (10),* 1507–1512, 1980.

Case, J.E., Gravitational evidence for a batholithic mass of low density along a segment of the Colorado Mineral Belt, in *Geol. Soc. Am. Abstracts for 1964, Spec. Pap., 82,* 26, 1965.

Christensen, M.I., and W.D. Mooney, Seismic velocity structure and composition of the continental crust: a Global View, *J. Geophys. Res., 100 (7)*, 9761–9788, 1995.

Dueker, K., H. Yuan, and B. Zurek, Thick-structured Proterozoic lithosphere of the Rocky Mountain region, *GSA Today, 11 (12),* 4–9, 2001.

Isaacson, L.B., and S.B. Smithson, Gravity anomalies and granite emplacement in west-central Colorado, *Geol. Soc. Am. Bull., 87*, 22–28, 1976.

Lee, D.K., and S.P. Grand, Upper mantle shear structure beneath the Colorado Rocky Mountains, *J. Geophys. Res., 101,* 22233–22244, 1996.

Lerner-Lam, A.L., A. Sheehan, S. Grand, E. Humphreys, K. Dueker, E. Hessler, H. Guo, D. Lee, and M. Savage, Deep structure beneath the Southern Rocky Mountains from the Rocky Mountain Front Broadband Seismic Experiment, *Rocky Mtn. Geology, 33 (2),* 199–216, 1998.

Li, A.B.; D.W. Forsyth, and K.M. Fischer, Evidence for shallow isostatic compensation of the Southern Rocky Mountains from Rayleigh wave tomography, *Geology, 30 (8),* 683–686, 2002.

Mutschler, F.E., E.E. Larson, R.M. Bruce, Laramide and younger magmatism in Colorado: new petrologic and tectonic variations on old themes, *in Cenozoic volcanism in the Southern Rocky Mountains,* edited by J.W. Drexler and E.E. Larson, pp. 1–47, *Colo. School of Mines Quart., 82 (4),* Golden, 1987.

Nyman, M.W., K.E. Karlstrom, E. Kirby, and C.M. Graubard, Mesoproterozoic contractional orogeny in western North America: Evidence from ca. 1.4 Ga plutons, *Geology, 22,* 901–904, 1994.

Oshetski, K.C. and R.P. Kucks, Colorado aeromagnetic and gravity maps and data: a web site for distribution of data: *USGS Open-File Rpt., 00–0042*, web only at http://greenwood.cr.usgs.gov/pub/open-file-reports/ofr-00-0042/colorado.html, 2000.

Shaw, C. A., K.E. Karlstrom, M.L. Williams, M.J. Jercinovic, and A.M. McCoy, Electron-microprobe monazite dating of ca. 1.71–1.63 Ga and ca. 1.45–1.38 Ga deformation in the Homestake shear zone, Colorado: Origin and early evolution of a persistent itnracontinental tectonic zone, *Geology, 29 (8),* 739–742, 2001.

Sheehan, A.F., G.A. Abers, C.H. Jones, and A.L. Lerner-Lam, Crustal thickness variations across the Colorado Rocky Mountains from teleseismic receiver functions, *J. Geophys. Res., 100 (10),* 20391–20404, 1995.

Talwani, M., J.L. Worzel, and M.G. Landisman, Rapid gravity computations for two-dimensional bodies with application to the Mendocino submarine fracture zone [Pacific Ocean], *J. Geophys. Res., 64 (1),* 49–59, 1959.

Tweto, O.L., and J.E. Case, Gravity and magnetic features as related to geology in the Leadville 30-minute quadrangle, Colorado, *USGS Prof. Pap., 726-C*, pp. 1–31, 1972.

Tweto, O.L., and P.K. Sims, Precambrian ancestry of the Colorado Mineral Belt, *Geol. Soc. Am. Bull., 74,* 991–1014, 1963.

Tweto, O.L., Geologic Map of Colorado, *USGS*, scale 1:500,000, 1979.

Wernicke, B., R. Clayton, M. Ducea, C.H. Jones, S. Park, S. Ruppert, J. Saleeby, J.K. Snow, L. Squires, M. Fliedner, G. Jiracek, G.R. Keller, S. Klemperer, J. Luetgert, P. Malin, K. Miller, W. Mooney, H. Oliver, R. Phinney, Origin of High Mountains in the Continents: The Southern Sierra Nevada, *Science, 271*, 190–193, 1996.

Wessel, P., and W.H.F. Smith, Free software helps map and display data, *AGU EOS Trans., 72*, 441, 1991.

Zandt, G., A.A. Velasco, and S.L. Beck, Composition and thickness of the southern Altiplano crust, Bolivia, *Geology, 22 (11),* 1003–1006, 1994.

Zhao, W., K.D. Nelson, J. Che, J. Guo, D. Lu, C. Wu, X. Liu, L.D. Brown, M.L. Hauck, J.T. Kuo, S. Klemperer, and Y. Makovsky, eep seismic reflection evidence for continental underthrusting beneath southern Tibet, *Nature, 366 (6455),* 557–559, 1993.

G.R. Keller, and L. Treviño, Department of Geological Sciences, University of Texas at El Paso, El Paso, TX, 79968-0555.

A. M. McCoy, John Shomaker & Associates, Inc., 2703-B Broadbent Parkway NE, Albuquerque, NM, 87107.

M. Roy, Department of Earth and Planetary Sciences, Northrop Hall, University of New Mexico, Albuquerque, NM, 87131.

Paleomagnetic and Geochronologic Data Bearing on the Timing, Evolution, and Structure of the Cripple Creek Diatreme Complex and Related Rocks, Front Range, Colorado

Jason S. Rampe and John W. Geissman

Department of Earth and Planetary Sciences, University of New Mexico, Albuquerque, New Mexico

Marc D. Melker

Cripple Creek and Victor Gold Mining Company, Victor, Colorado

Matthew T. Heizler

New Mexico Bureau of Mines and Mineral Resources, Socorro, New Mexico

Paleomagnetic data, combined with high-precision $^{40}Ar/^{39}Ar$ age determinations, show that the Cripple Creek Diatreme complex and surrounding rocks of the Colorado Front Range, have experienced about 10 degrees of north-side down tilting since the early Oligocene. We report data from 69 paleomagnetic sites collected from all representative rock types in the district. Overall, the demagnetization response based on both alternating field and thermal methods, directional data, and field tests indicate that most rocks in the Cripple Creek district carry geologically stable magnetizations. Group directions are of both normal (D = 1.1°, I = 63.9°, N = 24 sites, α_{95} = 4.2°, k = 50.3) and reverse (D = 165.4°, I = -67.4°, N =28, α_{95}= 3.7°, k = 55.4) polarity. Eight $^{40}Ar/^{39}Ar$ age determinations were obtained from hornblende, K-feldspar, phlogopite, and groundmass concentrates. These new data demonstrate that initial igneous events are represented by relatively felsic rocks (phonolite), emplaced at 31.59 ± 0.32 Ma, followed by mafic to ultramafic intrusions at about 31.12 ± 0.04 Ma, tephriphonolite at 31.15 ± 0.11 Ma, phonotephrite; and lamprophyre at 30.41 ± 0.21 Ma. Igneous activity ceased for about 2 m.y. and a final episode of ultramafic activity is represented by the 28.38 ± 0.21 Ma lamprophyre Cresson Pipe. The combination of high precision geochronology and paleomagnetic data indicate that magmatism occurred between polarity time scale chrons 12R and 10R. Such deformation could have been accommodated by motion along faults that were active in part of a northwest-directed transtensional setting during mid-Tertiary and younger extension.

The Rocky Mountain Region: An Evolving Lithosphere
Geophysical Monograph Series 154

10.1029/154GM09

INTRODUCTION

Throughout the Front Range of Colorado, Laramide contraction (about 70 to at least 40 Ma, if not younger) resulted in considerable topographic relief, and ultimately exposed a complex array of Precambrian crystalline rocks over a broad area. Laramide-style structures have been subsequently modified by Oligocene to present day extensional tectonism related to the Rio Grande rift, which extends from southern New Mexico to north-central Colorado. Evaluation of specific components of Laramide and younger deformation within the many Precambrian-dominated uplifts in Colorado, as well as adjacent areas, is difficult because the Precambrian rocks preserve complex structural histories, and recognizing the effects of Laramide and younger deformation is limited in the absence of cover Phanerozoic strata. In parts of the Front Range, for example, latest Cretaceous to mid-Tertiary igneous intrusions contain structures that clearly show a component of deformation that post-dates their emplacement. In some areas, structural features have been shown to be active following Laramide deformation and have been correlated with modest magnitudes of local crustal deformation. For example, paleomagnetic results from the Urad/Henderson intrusive complex (28.7–27.6 Ma) located in the northeast Colorado mineral belt, were interpreted as indicating that the complex has been subjected to moderate northeast side-down tilting of 15 to 25° [*Geissman et al.*, 1992].

The Cripple Creek Diatreme complex, Cripple Creek, Colorado, provides another example of a mid-Tertiary intrusive complex affected by a complex array of post emplacement structural features [*Lindgren and Ransome*, 1906; *Loughlin and Koschmann*, 1935; *Koschmann*, 1949] and the magnitude of post-emplacement deformation in the area is not known. The complex consists of volcanic breccias (the primary rock type in the district) that are cut by lamprophyre and aphanitic phonolite dikes and irregular porphyritic phonolite to phonotephrite intrusions. The igneous rocks were emplaced within a few kilometers of the paleosurface, between about 32.5 and 28.2 Ma, based on initial geochronologic work in the district [*Kelley et al.*, 1998] and further supported by data reported here. Gold exploration in the Cripple Creek district began in 1891 and has produced more than 22 million ounces of gold and silver making it the largest of the Rocky Mountain gold districts. Current mining activity relies on low-grade bulk disseminated deposits mined in an open-pit fashion, creating excellent, three-dimensional exposures of all representative rock types and the structures that affect them. This paper uses a combination of paleomagnetic, field structural, and geochronologic techniques to contribute to interpretations of post diatreme emplacement deformational history and provide insight into possible heterogeneity of post-Laramide deformation of the Precambrian core of the Front Range of Colorado. Specifically, paleomagnetic investigations were undertaken to identify components of crustal tilting and/or vertical axis block rotation that that may have affected the area during the transition from Laramide to Rio Grande rift tectonism or within the extensional stress regime related to Rio Grande rift development.

GEOLOGIC SETTING

The Cripple Creek district lies in the southern Rocky Mountains in central Colorado, about 70 km east of the northern part of the Rio Grande rift [Figure 1]. The district is bounded to the west by ≥1.7 Ga medium-grade biotite schist, to the south and east by 1.7 Ga granodiorite [*Wobus et al.*, 1976], to the southwest by 1.43 Ga. Cripple Creek Quartz monzonite [*Hutchinson and Hedge*, 1968], and to the north by the 1.08 Ga. Pikes Peak Granite [*Unruh et al.*, 1995] [Figure 2].

In the Cripple Creek area, the youngest recognized Precambrian event is intrusion of the ca. 1.08 Ga Pikes Peak

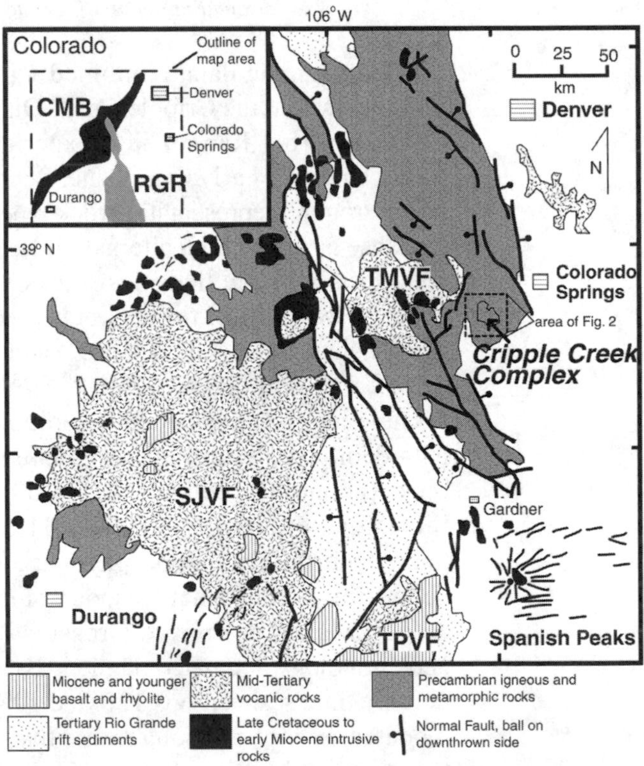

Figure 1. Distribution of latest Cretaceous to early Miocene intrusive rocks, central Colorado [after *Lipman*, 1981]. RGR = Rio Grande rift, CMB = Colorado Mineral belt, SJVF = San Juan volcanic field, TMVF = Thirtynine Mile volcanic field, TPVF = Taos Plateau volcanic field.

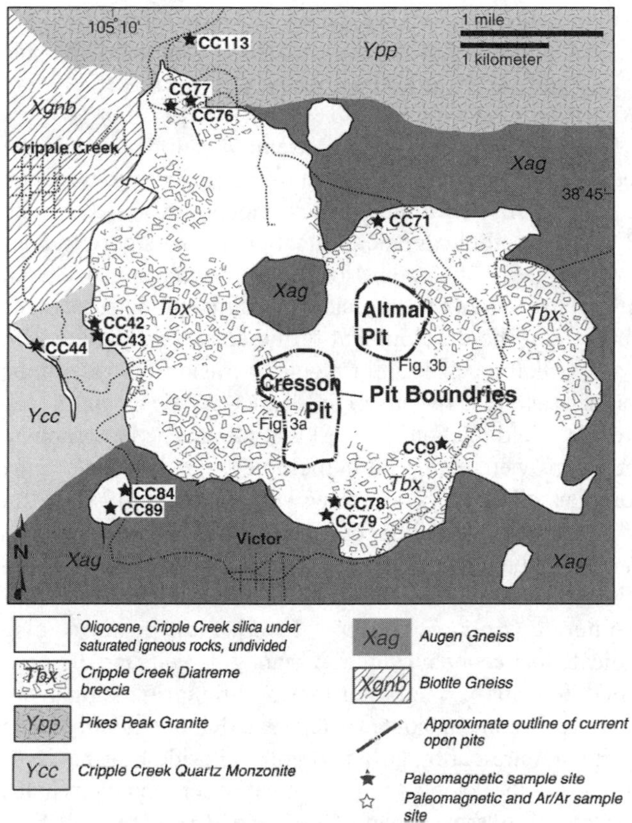

Figure 2. Simplified geologic map of the Cripple Creek diatreme complex and surrounding Proterozoic rocks, showing distribution of some of the paleomagnetic and geochronologic sampling sites.

batholith complex [*Unruh et al.*, 1995]. Cambrian to Late Devonian marine and non-marine sedimentation was interrupted by periods of non-deposition and erosion with subtle tectonic activity and little igneous activity in most of Colorado [*Ross and Tweto*, 1980]. Shallow marine sedimentation persisted into the Mississippian, but was interrupted by local and regional uplift [*Wallace*, 1990]. Beginning in the Early Pennsylvanian, deformation resulted in considerable local topographic relief and non-marine, coarse-grained sediments accumulated adjacent to uplifts [*De Voto*, 1980]. Together these processes formed the ancestral Front Range highland in central Colorado, as part of the north-northwest trending mountains and basins of the Ancestral Rocky Mountains [*Tweto*, 1975; 1980]. Between Permian and mid-Cretaceous time, marine and low-relief nonmarine depositional environments characterized Colorado with little evidence of substantial topographic relief [*Maughan*, 1980].

From the latest Cretaceous to early Tertiary, deformation in this area involved northeast-directed contraction [*Yin and Ingersoll*, 1997; *Erslev*, 2001; *Wawrzyniec et al.*, 2002] that formed north to northwest-trending basement-involved Laramide uplifts and associated basins within the Cordilleran foreland east and northeast of the Colorado Plateau. In the Front Range area, deformation began in the latest Cretaceous, at about 75 Ma, and continued into the early Tertiary, until at least 45 Ma, although recent work suggests that a contractional regime may have persisted to 35–30Ma [*Erslev*, 2001; *Wawrzyniec et al.*, 2002]. Overall, the locus of maximum Laramide shortening in the foreland progressed from north to south [*Dickinson et al.*, 1988].

Contemporaneous igneous activity was almost completely confined to the northeast-trending Colorado Mineral Belt, where shear zones of Precambrian ancestry intersect Laramide uplifts. Numerous small stocks and related volcanic or volcaniclastic rocks were emplaced along preexisting structures between 70 and 40 Ma [*Epis and Chapin,* 1975]. In the Cripple Creek area, Laramide deformation resulted in a structural high, with near complete erosion of Paleozoic and Mesozoic strata, and broad exposure of Proterozoic rocks [*Koschmann*, 1949]. Apatite fission-track (AFT) studies of Front Range Proterozoic rocks provide dates ranging from about 170 to about 45 Ma at low elevations from the south side of Pikes Peak and a 27 Ma date about one km south of the Cripple Creek diatreme complex [*Kelley and Chapin*, 2002]. Modeling of track lengths along with the age of the samples having relatively young AFT dates from south of Pikes Peak indicates rapid cooling and uplift of the crystalline basement between about 20 and 10 Ma [*Kelley and Chapin*, 2002]. The 27 Ma AFT result obtained near Cripple Creek is attributed to thermal effects of diatreme formation and hydrothermal activity [*Kelley and Chapin*, 2002].

Widespread erosion surfaces developed after the main pulse of Laramide deformation throughout south-central Colorado and fluvial deposits of the Eocene Echo Park alluvium accumulated in structurally controlled basins adjacent to erosional surfaces [*Epis and Chapin*, 1975; *Chapin and Cather*, 1983], only small remnants of which remain in the Cripple Creek area [*Wobus et al.*, 1976; *Epis et al.*, 1976]. In central Colorado, calc-alkaline and alkaline magmatism began in the late Eocene (~36 Ma), and continued into the early Miocene (~ 19 Ma) [*Epis and Chapin*, 1975].

In the early to late Oligocene, the tectonic regime in the southern Rocky Mountains began to transition to west- to northwest-directed extension or transtension due to intracontinental rifting [*Lipman*, 1981; *Chapin and Cather*, 1983; *Wawrzyniec, et al.*, 2002]. Most associated rift structures trend north-south and form a series of right stepping en echelon horsts and grabens. The most important of these extensional features, the Rio Grande rift, began in south-central New Mexico at about 27 Ma, and extends into north-central Colorado [*Lipman et al.*, 1978]. The timing of initial extension in much of southern and central Colorado remains relatively poorly known [*Ingersoll*, 2001].

DISTRICT GEOLOGY

As noted above, the Cripple Creek diatreme complex lies at the junction of several different Precambrian rock suites and at the intersection of north-northwest-trending dextral and northeast-trending sinistral strike-slip fault zones. The Cripple Creek diatreme outcrops in a roughly elliptical-shaped structure with a northwest trending long axis that is exposed over about 18 km² [Figure 2]. Several interpretations for the development of the complex have been proposed over the past century. An initial hypothesis involved diatreme formation by explosive volcanic activity [*Cross and Penrose*, 1895; *Lindgren and Ransome*, 1906; *Loughlin and Koschmann*, 1935]. Koschmann [1949], however, concluded that diatreme development and related structures were controlled by northwest-trending near vertical faults that were formed by intermittent subsidence, rather than as the result of explosive volcanism. More recent hypotheses for diatreme formation involve volcanic and brecciation events followed by prolonged subsidence, with attending lacustrine and fluvial sedimentation, interrupted by further brecciation, intrusion, K-metasomatism, diatreme development (e.g., the Cresson lamprophyric breccia pipe), and gold and gold-telluride mineralization [*Thompson et al.*, 1985; *Thompson*, 1992; *Pontius*, 1996].

The diatreme is divided into three major sub-basins- north, east, and south (largest, Figure 3, see CDROM in back cover sleeve). These sub-basins are separated by shallowly buried Proterozoic schist and granite "highs" that trend east-west (dividing north from south basins) and northwest-southeast (separating south from east basins) [*Koschmann*, 1949]. Also, an isolated high of Proterozoic granodiorite is exposed slightly northwest of the center of the district. The main complex is surrounded by a discontinuous ring of phonolite and phonolite breccia capping topographic highs [Figure 2].

The Cripple Creek breccia, the most volumetrically important rock type in the district, consists of poorly sorted volcaniclastic sediments, tuffs, angular to subangular clasts of Proterozoic igneous and metamorphic rocks, and associated silica-depleted alkaline rocks ranging in size from <1 cm to meter scale [*Loughlin and Koschmann*, 1935]. The Cripple Creek diatreme basin is filled with over one km of variably stratified and poorly- to infrequently well-sorted detritus of volcanic and nonvolcanic origin. Complete stratigraphic sections of nonvolcanic sediments include conglomerates, arkose, shale (with fossil leaves and bird tracks), and lacustrine limestone [*Loughlin and Koschmann*, 1935]. Detrital strata show evidence of channel deposits. The upper part of the section is interbedded with volcanic breccia. Precambrian rocks immediately to the southwest of the complex are capped by rhyolite and andesite flows related to central Colorado volcanic rocks and sandstone beds of early to mid-Tertiary age [*Koschmann*, 1949; *Thompson et at.*, 1985].

Previous assessments of diatreme development and the sequence of magmatic events in the complex were based on field relations and limited geochronologic work (K-Ar and $^{40}Ar/^{39}Ar$ isotopic age determinations). Cross-cutting relations suggest that diatreme formation, magma emplacement, and lacustrine and volcaniclastic sedimentation took place throughout the development of the complex [*Koschmann*, 1949]. Alkaline dikes and irregular intrusions, mostly in the southern and eastern parts of the district, cut the Cripple Creek breccia. Field relations suggest that ultramafic lamprophyre intrusions were emplaced in multiple stages. Outside of the complex, dikes, plugs, and flows crop out within 12 to 15 km of the diatreme and either cut and overlie Precambrian rocks or overlie the upper Eocene to lower Oligocene Wall Mountain Tuff and the lower Oligocene Tallahassee Creek Conglomerate [*Epis and Chapin*, 1975]. Field relations also indicate that economically important gold vein and disseminated deposits formed after mafic-ultramafic magmatism. Early veins contain biotite, K-feldspar, dolomite, fluorite, and pyrite and are cut by gold and gold-telluride bearing veins throughout the complex, with the greatest concentrations in the upper levels of the complex [*Loughlin and Koschmann*, 1935].

Previous geochronologic studies suggest that magmatism and related thermal activity spanned several million years in the early Oligocene. Two phonolite samples have K-Ar dates of 28.6 ± 0.7 and 30.1 ± 0.7 Ma [*Wobus et al.*, 1976]; the type of sample dated (e.g. groundmass concentrate or mineral separates) as well as location were not specified. McDowell [1971] reported K-Ar dates of 34.3 ± 1.0 and 34.7 ± 1.3 Ma from aegirine-augite separates from tephriphonolite samples from the eastern part of the district. K-Ar determinations from Wobus et al. [1976] and McDowell [1971] have been corrected according to the conversion factors of Dalrymple [1992]. Recently, Kelley et al. [1998] undertook a $^{40}Ar/^{39}Ar$ geochronology study, to more precisely defining the timing and sequence of igneous and hydrothermal activity in the complex. $^{40}Ar/^{39}Ar$ dates were obtained on biotite from tephriphonolite and a biotite-pyrite-quartz vein, adularia from an altered phonolite and adularia-quartz vein, and sanidine from phonolite. Tephriphonolite, altered phonolite, and whole rock determinations from trachyandesite and phonotephrite were interpreted by Kelley et al., [1998] to indicate a complex magmatic and hydrothermal history. Phonolite, tephriphonolite, and trachyandesite were argued to have been emplaced over a span of about 1 Myr from 32.5 to 31.5 Ma [*Kelley et al.*, 1998]. They also suggested a younger episode (30.9 ± 0.1 Ma) of phonolite emplacement outside of the complex. Based on dates from tephriphonolite, trachyandesite, and phonotephrite, two phases of intermediate to mafic magmatism were inferred to have occurred, the first from about 32.5 to 31.5 Ma and the second at about 28.7 Ma [*Kelley et al.*, 1998]. Step-heating $^{40}Ar/^{39}Ar$ analysis of hydrothermal biotite and K-feldspar from veins or disseminated deposits yield values between 31.3 ± 0.1 and 29.6 ± 0.1 Ma for early hydrothermal and 28.8 ± 0.1 and 28.2 ± 0.1 Ma for late hydrothermal episodes [*Kelley et al.*, 1998]. The approximate age of inception of volcanism and brecciation in the district is based on field relations involving breccia overlying Tallahassee Creek Conglomerate, which contains clasts of Wall Mountain Tuff (K-Ar age determinations of 35 – 36.6 Ma) [*Epis and Chapin*, 1975], and phonolite flows overlying Thirtynine Mile Andesite (34 Ma K-Ar age determinations) [*Epis and Chapin*, 1975]. Therefore, upper ages of about 35 Ma can be placed on the Cripple Creek breccia and of about 34 Ma for emplacement of phonolite flows and intrusions.

METHODS

This report summarizes and interprets results from 69 paleomagnetic sample sites in the mid-Tertiary Cripple Creek diatreme in order to assess the amount of post-Laramide tilting and/or rigid block rotation. Methods used for sampling, measurement of remnant magnetizations, and demagnetization follow standard procedures. ^{40}Ar-^{39}Ar geochronology was performed at the New Mexico Geochronology Research Laboratory (see methods section in CDROM in back cover sleeve).

PALEOMAGNETIC DATA

Paleomagnetic data from within and surrounding the Cripple Creek district are typically of high quality and suggest that most of the materials sampled are capable of carrying geologically stable and interpretable magnetizations. Progressive demagnetization generally yields an excellent response, with well-defined trajectories of magnetization decay and typically low dispersion, at the site level, of isolated magnetization components [Figure 4a,b; see CDROM in back cover sleeve]. NRM intensities range over four orders of magnitude, from about 0.03 mA/m for aphanitic phonolite to 2.73 A/m for phonotephrite. In turn, bulk susceptibility values range from 9.5 x 10^{-2} to 1.00 x 10^{-6} (SI volume) [Figure 5a; see CDROM in back cover sleeve]. About 13 percent (9 sites) of the samples demagnetized yielded uninterpretable results and therefore were rejected from further analysis. Of these unacceptable sites, eight sites have high dispersion at the site level [e.g., site CC71, (Table 1; see CDROM in back cover sleeve)]. Samples from one site (CC69) in aphanitic phonolite have very low NRM intensities (typically less than 0.1 mA/m). The low magnetic intensities provided random NRM directions and could not be demagnetized to resolvable intensities. Site CC55 was not included in mean calculations as it yielded a southeast declination and moderate positive inclination (D = 121.5, I = 57.5), the importance of this result is hard to evaluate.

Results of progressive alternating field (AF) and thermal demagnetization techniques, utilized independently and in conjunction to isolate and identify all magnetization components in rocks sampled, indicate that both magnetite and higher coercivity phases are present as principal magnetization carriers [Figure 5b,c; see CDROM in back cover sleeve].

Demagnetization treatment for all materials collected initially involved progressive AF treatment to peak fields of up to 120 mT. Subsequent thermal demagnetization was utilized for two purposes. The first was to assess whether similar magnetization components were isolated in both AF and thermal treatments. The second was to complete the specimen demagnetization when AF demagnetization only partially isolated components of the NRM. Incomplete demagnetization occurred because hematite was the principal remanence carrier or because the remanence was dominated by magnetite of higher coercivity, possibly single-domain particles, in which case we applied thermal demagnetization followed by AF treatment. For the majority of the sites, samples responded favorably to AF treatment [Table 1; see CDROM in back cover sleeve], with 80 to 90 percent of the NRM randomized [Figure 4a,b; see CDROM in back cover sleeve]. Most specimens responded, at least partially, to alternating field demagnetization and exhibited a uniform decay of a single magnetization component. For some sites, AF treatment had no effect in that there was no appreciable change in intensity and/or direction [Figure 4; see CDROM in back cover sleeve]. Some sites exhibited multiple component behavior with both north-seeking and positive inclination (normal polarity) and south-seeking and negative inclination (reverse polarity) magnetizations that were well defined and isolated completely in AF demagnetization (e.g., D = 29.7°, I = 72.5°, N = 10, α_{95} = 9.2°, k = 28.4 and D = 173.6°, I = -64.1°, N = 5, α_{95} = 3.1°, k = 595, in aphanitic phonolite site CC89) [Figure 4a, see CD ROM in back cover sleeve.] Thermal demagnetization of Cripple Creek breccia or other highly altered rocks, especially at temperatures above about 400°C, resulted in spurious behavior with respect to both direction and intensity; continued treatment was abandoned. Samples with specimens that were demagnetized with both AF and thermal methods typically either show internally consistent behavior [e.g., specimens from sample CC40A, Figure 4b; see CDROM in back cover sleeve] or yield significantly different directions depending on the demagnetization approach used [e.g., sample CC33n, Figure 4a; see CDROM in back cover sleeve].

Field tests to assess paleomagnetic stability allow estimates of the timing of magnetization acquisition. Specifically, breccia and baked contact tests were used to determine if the rocks sampled retained magnetizations that existed prior to a specific geologic event or were dominated by relatively young viscous or thermoviscous remanent magnetizations (VRMs or TVRMs). A contact test [*Everitt and Clegg*, 1962] is considered positive if the host rock has a unique magnetization to that of the cross-cutting intrusion and the host rock's remanence is progressively overprinted by that of the intrusion [Figure 6; see CDROM in back cover sleeve]. Nine baked contact tests were conducted on lamprophyre, aphanitic phonolite, and phonotephrite dikes, six of the tests yielded positive results. The other three were inconclusive in that the host or cross-cutting intrusion have the same polarity and essentially the same direction of magnetization. A breccia test [*Irving*, 1964] is positive if the magnetizations of each clast are random relative to one another and therefore distinct from that of the host material. Two breccia tests were carried out on lamprophyre breccia pipes, both of which provided "positive" results. For example at site, CC29, 10 independent phonolite breccia clasts within in the Cresson lamprophyre pipe were sampled. The population of magnetizations from the individual clasts is highly dispersed (α_{95} = 46.4°, k = 2.0, and R = 5.60). According to the test of Watson [1956], the population of magnetizations is random to between 90 and 95 percent confidence. Positive contact or breccia tests in either of these cases indicate that the host rock or breccia clasts retained pre-existing magnetizations (possibly thermo remanent magnetization (TRMs)) since intrusion of the igneous body and that the intrusion retains a primary TRM.

In situ paleomagnetic data from the Cripple Creek diatreme complex show concentrated populations of both normal and reverse polarity site mean directions [Figure 7a]. Mean normal (D = 355.6°, I = 62.0°, N = 28, α_{95} = 5.8°, k = 23.0) and reverse (D = 168.3°, I = -64.9°, N = 31, α_{95} = 4.5°, k = 33.7) polarity populations, for all accepted sites, have inclinations that are steeper than the expected mid-Tertiary field direction for this location. The reference direction we employed, D = 352.7°, I = 56.4°, is based on the average of three estimates of the mid-Tertiary paleomagnetic poles for North America [*Van der Voo*, 1993; *Diehl et al.*, 1988; *Mankinen et al.*, 1987]. A paleomagnetic reversal test [*McFadden and McElhinny*, 1990] conducted on the total number of accepted normal and reverse polarity sites yields a critical angle for the two polarities of 7.0°, with the observed angle between the mean of the normal and reverse polarity sets as 4.4°, thereby indicating a positive reversal test. The probability of exceeding this angle is between 90 and 95 percent. In the second test, we

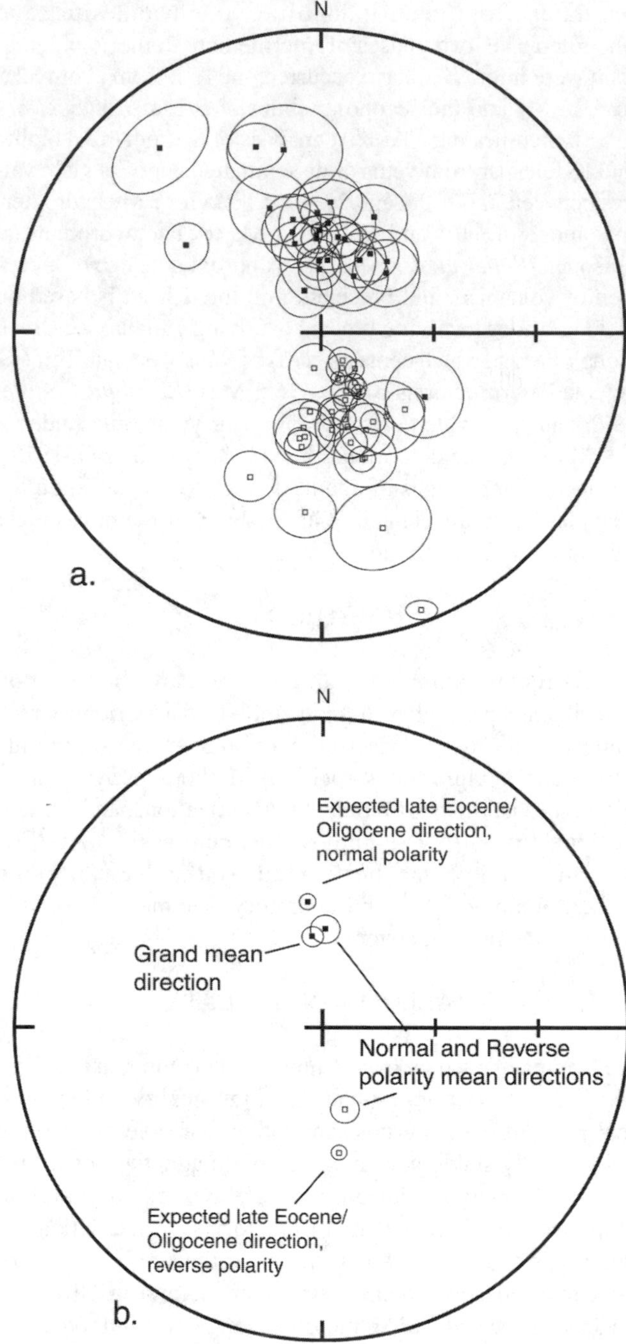

Figure 7. Equal area projection showing estimated site mean directions of magnetizations from rocks sampled at the Cripple Creek diatreme complex. Closed (open) symbols refer to lower (upper) hemisphere projections. (a) all acceptable site means (b) calculated grand mean directions for both normal and reverse polarities and the overall grand mean direction calculated from all accepted sites. Expected late Eocene/Oligocene directions derived from paleomagnetic pole positions of Van der Voo [1993], Diehl et al. [1988] and Mankinen et al. [1987].

excluded seven sites (CC9, 21, 25, 33, 64, 76, 79 [Table 1; see CDROM in back cover sleeve], considered as outliers, from the group mean populations based on the observation that their site-mean directions lay greater than two angular standard deviations from the initially estimated mean, a common practice in paleomagnetic data analysis. The revised normal polarity mean is D = 1.1°, I = 63.9° (N = 24, α_{95} = 4.2°, k = 50.3) and reverse polarity mean is D = 165.4°, I = -67.4° (N = 28, α_{95} = 3.7°, k = 61.1) [Figure 7b]. This test gives a critical angle for the two populations of 5.6°, with the observed angle between the mean of the normal and reverse polarity sets as 7.3°, indicating a slightly negative reversal test.

A grand-mean direction was determined for our accepted results from the Cripple Creek complex by inverting all reverse polarity data through the origin to yield a common polarity. This yielded a north-seeking declination and moderate to steep positive inclination result of D = 353.1°, I = 65.8° (N = 52, α_{95} = 2.9°, k = 47.6). The virtual geomagnetic pole (VGP) angular standard deviations for the two results (N= 59 and N = 52 sites) are 15.4° and 11.7°, respectively.

To evaluate the potential effects of hydrothermal processes related to the diatreme and host rocks with a well-characterized magnetization signature we sampled one roadcut exposure of ca. 1.08 Ga Pikes Peak Granite, about 0.3 km north-northeast of the CC diatreme complex [Site CC 113, Figure 8]. On a regional scale, the Pikes Peak granite has been demonstrated to have a ca. 1.11 Ga normal polarity magnetization of west-northwest and moderate to shallow positive inclination [*Spall*, 1970]. Unlike exposures of the Pikes Peak granite farther from the complex, granite at this exposure is more heavily altered as indicated by biotite and amphibole replaced by clay minerals and the pervasive epidote alteration of feldspars. In a combination of AF followed by thermal demagnetization, many samples reveal the isolation of a magnetization of north-northwest declination and shallow to moderate positive inclination [Figure 8], which we interpret to be a primary TRM. Upon further thermal demagnetization, many of these samples reveal a magnetization of south declination and moderate to shallow negative inclination [Figure 8]. This result is similar to that reported by Geissman and Harlan [2002], who interpreted this magnetization, obtained at scattered localities throughout the Pikes Peak Granite, to be secondary and of late Paleozoic age. Other samples, confined to the eastern end of the outcrop, are more strongly magnetized (NRM intensities between 50 and 100 mA/m) and yield only a north-declination, moderate positive inclination magnetization [Figure 8].

Rock Magnetic Experiments

Rock magnetic experiments were conducted on specimens from sites 28, 29, 35, 40, 47, 57, 65, 74, 84, and 89, chosen as representative of the most frequently sampled rock types in the district. Tests included isothermal remanent magnetization (IRM) acquisition and backfield demagnetization of saturation isothermal remanent magnetization (SIRM), and a comparison of AF demagnetization results on the natural remanent magnetization (NRM), anhysteretic remanent magnetization (ARM), and SIRM [Figure 5c, see CD ROM in back cover sleeve]. IRM acquisition and backfield IRM demagnetization results suggest that magnetite is the dominant primary magnetic phase in most rocks sampled [Figure 5b, see CD Rom in back cover sleeve]. AF demagnetization of NRM, ARM and SIRM [*Lowrie and Fuller*, 1971; *Johnson et al.*, 1975] shows that pseudo-single-domain (CC29, CC35, CC74, CC84) and multidomain magnetite (CC40, CC65, CC89), and hematite (CC47) are the principal magnetic carriers of rocks examined [Figure 5c, see CD ROM in back cover sleeve]. The domain states of two of the ten samples analyzed (CC28, CC57) could not be resolved and gave intermediate results [Figure 5c, see CD Rom in back cover sleeve].

GEOCHRONOLOGIC DATA

Incremental heating age spectrum analyses were performed on eight mineral separates and groundmass concentrates from the Cripple Creek district [Figure 9]. The age spectra reveal a wide range in their level of complexity that reflects both a difficulty in obtaining pristine mineral separates and in the complex hydrothermal history of the district. In some cases the age spectra are too complex to assign an unambiguous apparent age or an age precise enough to contribute to the timing of events in the area. Isochron analysis was conducted for each sample, but generally did not contribute linear arrays that significantly improved the interpretation of the age spectrum. Complex age spectra may result from alteration of the primary K-bearing phases, inclusions within grains, nonatmospheric trapped argon components, recoil of ^{39}Ar during irradiation, and/or thermal effects since emplacement of the igneous body. Refer to Appendix 1 (see CD ROM in back cover sleeve) for all ^{40}Ar/^{39}Ar data.

Sample CC74 K-feldspar from Tphk, has a complex age spectrum with the first 5 percent of ^{39}Ar released yielding aberrantly old apparent ages followed by a drop to a minimum of about 31 Ma within the initial 10 percent of ^{39}Ar release [Figure 9a]. The apparent ages increase to nearly 35 Ma before decreasing to about 32.5 Ma during the final heating step. This style of decreasing spectrum for the high-temperature steps has been documented by Foster et al. [1990] and has been shown to be caused by excess ^{40}Ar in large diffusion domains. Isochron analysis reveals a linear array for the final four steps that yield an apparent age of 31.59 ± 0.32 Ma and a ^{40}Ar/^{36}Ar$_i$ value of 308.9 ± 1.2 [Figure 10]. This isochron age

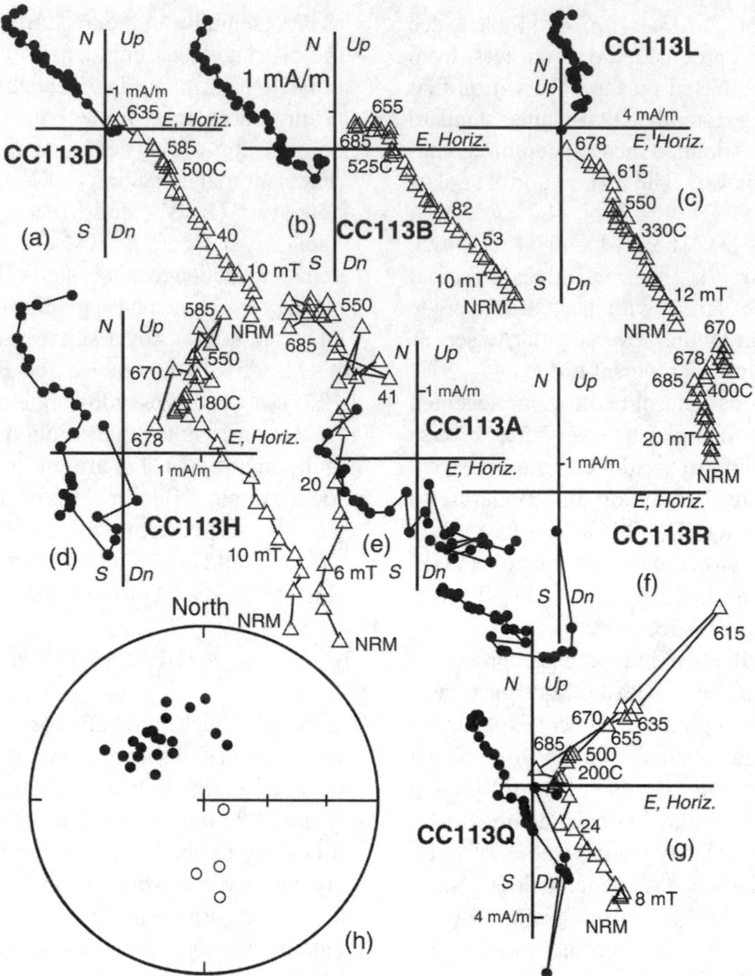

Figure 8. Paleomagnetic data from site CC113, in late Mesoproterozoic Pikes Peak Granite, exposed on Highway 67, 0.3 km north of the diatreme. (a–g) Examples of progressive demagnetization behavior, showing the endpoint of the magnetization vector measured after each demagnetization step projected onto the horizontal plane (solid circles) and the true vertical plane (open triangles). Peak alternating field (in mT) or thermal (in degrees Celsius) demagnetization steps identified for selected projections onto the true vertical plane. For several samples, thermal demagnetization above about 500°C resulted in the complete decrepitation of the specimen. (a,b) Isolation of a northwest-declination, moderate positive inclination magnetization interpreted to be a primary, thermoremanent magnetization characteristic of the Pikes Peak Granite. (c) Isolation of a north-declination, moderate positive inclination magnetization in AF and thermal demagnetization; the magnetization is inferred to be of Cenozoic age (d) Isolation of a north-declination, steep positive inclination magnetization in AF demagnetization, followed by the partial isolation of a south-declination, negative inclination remanence; both magnetizations are inferred to be of Cenozoic age (e) Isolation of a north-declination, steep positive inclination magnetization, then a west-northwest-declination steep positive inclination magnetization, followed by the partial isolation of a south-declination, moderate negative inclination magnetization; the west-nothwest declination magnetization is inferred to be a primary remanence, the reverse polarity magnetization remaining is considered to be of Cenozoic age. (f) Isolation of a west-northwest declination, steep positive inclination magetization with a remaining, and partially isolated south-declination, moderate negative inclination magnetization; the first removed magnetization is interpreted to be a primary remanence, the reverse polarity one is considered to be of Cenozoic age. (g) Isolation of a north-northwest declination, moderate positive inclination magnetization, then one of north declination and shallower positive inclination, then one of south-declination and moderate negative inclination; the first magnetization removed may be a composite of a primary Pikes Peak Granite remanence and a younger normal polarity remanence and the reverse polarity remanence is inferred to be of Cenozoic age. (h) equal area projection showing directions of magnetizations isolated at site CC113, in altered Pikes Peak Granite. Closed (open) symbols refer to lower (upper) hemisphere projections.

is our preferred closure age for CC74 K-feldspar. CC58 amphibole from Tphh provides a well-defined spectrum [Figure 9b] and has a plateau (MSWD = 0.34) of 31.12 ± 0.04 Ma for nearly 95 percent of the total ^{39}Ar released. Three samples from phonotephrite (Tpt) were analyzed [Figure 9c, 9d, 9e]. K-feldspar and groundmass separates (CC40a, CC40b respectively) from Bull Cliff and CC73 amphibole from the Isabella dike. CC40a K-feldspar [Figure 9c] has a spectrum with the initial 10 percent of ^{39}Ar released providing old apparent ages followed by a relatively flat segment at about 30 Ma, that steps up slightly in apparent age to another flat segment at about 31 Ma, over the rest of ^{39}Ar released. The high initial apparent ages are common for K-feldspar and suggest excess argon contamination. The age gradient is interpreted to reflect minor diffusive argon loss with the plateau segment defined by steps J-M of 31.15 ± 0.11 Ma recording the preferred age of the sample. CC40b groundmass concentrate [Figure 9c] yields a somewhat hump-shaped age spectrum with the initial and final steps providing younger apparent ages than most of the spectrum. The K/Ca ratio steps up during the initial heating steps and decreases slightly during the final heating steps. Isochron analysis for several combinations of chosen steps yields poor regression statistics and therefore does not support a simple excess component to explain the age spectrum. We suggest that the complexity is mainly the result of ^{39}Ar recoil and calculate an integrated age for steps C I of 31.17 ± 0.1 Ma. The initial step B is left out of this calculation, based on the interpretation that the sample has experienced some minor argon loss, however incorporation would not significantly affect the final age assignment. CC73 amphibole [Figure 9e] yields a complicated hump shaped age spectrum with an initial young apparent age step followed by a relatively flat intermediate section and finally an abrupt decrease in apparent age for the final step. The K/Ca spectrum shows 2.5 orders of magnitude variation with initial values much higher than expected for pure amphibole. The complexity of this spectrum is probably related to complex mineralogy (confirmed by petrographic inspection) of this sample and consequently we cannot assign a precise age to this sample. If the age spectrum complexity is solely caused by recoil artifacts, the integrated age would provide a reasonable age for this sample. However, the integrated age of 27.8 ± 0.3 Ma is inconsistent with known geologic relations of well-dated samples and therefore is not considered a viable age for this sample. Isochron analysis also does not aid in the interpretation of this sample. CC57 phlogopite from Tl (Pinto Dike) yields an overall readily interpretable age spectrum [Figure 9f]. Most of the age spectrum is represented by relatively consistent apparent ages that yield a weighted mean age of 30.41 ± 0.21 Ma. The MSWD value for these steps is 8.98 and indicates scatter above that predicted by analytical error alone and this may be caused by minor recoil related to high Ca contaminants.

The geologically late igneous activity is represented by xenolith samples transported by the Cresson lamprophyre Pipe. K-feldspar was separated from two of the xenoliths, CC70 [Figure 9g] and CC70b [Figure 9h]. Their age spectra are similar as both samples reveal age gradients ranging between about 25 and 29 Ma. The age gradients can be explained in one of two end-member ways. Either the xenolith-derived K-feldspars were completely degassed of radiogenic argon during pipe emplacement, so that the age gradient records post emplacement argon loss, or the xenoliths were partially degassed so that the younger apparent ages record the time of pipe emplacement and the gradient reveals the degree of resetting. For reasons discussed in the following section, we suggest the older part of the spectra records the best estimate for the emplacement age of the Cresson Pipe. Sample CC70 K-feldspar [Figure 9g] yields an age spectra with increasing apparent ages during progressive heating increments. The first 10 percent of ^{39}Ar released is represented by variations in apparent age (steps A–G), followed by a progressive increases in apparent age from 25.52 to 28.95 Ma. K/Ca ratios initially increase, level off, and then decrease during the final steps. The spectrum climbs to a maximum age of 29.0 Ma and is similar to the age assigned to sample CC70b. The final 25 percent of CC70b age spectrum has a weighted mean age of 28.38 ± 0.21 Ma [Figure 9h].

DISCUSSION

Paleomagnetism

Evaluation of post-latest Cretaceous to early Tertiary (Laramide) deformation in the Front Range of Colorado has been considered in the context of extension during the overall development of the Rio Grande rift and the possibility of regional, post-Laramide uplift and attending internal deformation of much of the foreland east and northeast of the Colorado Plateau [*Epis and Chapin* 1975; *Oppenheimer and Geissman* 1988; *Steidtmann and Middleton* 1991; *Geissman et al.*, 1992; *Yin and Ingersoll*, 1997; *Wawrzyniec et al.*, 2002]. The grand mean magnetization direction, based on the dual polarity data set, (D = 354.4°, I = 67.3°, N = 48, α_{95} = 2.9°, k = 52.1) obtained for the Cripple Creek intrusive complex has an inclination that is about 10 degrees steeper than an expected field direction (D = 352.7°, I = 56.4°) for mid-Tertiary time and a declination that is statistically indistinguishable from that predicted. Based on the data available, the discordance in inclination is applicable to the entire district. The grand-mean direction is not biased by one or a select group of intrusions that have considerably steeper inclinations than other intrusions

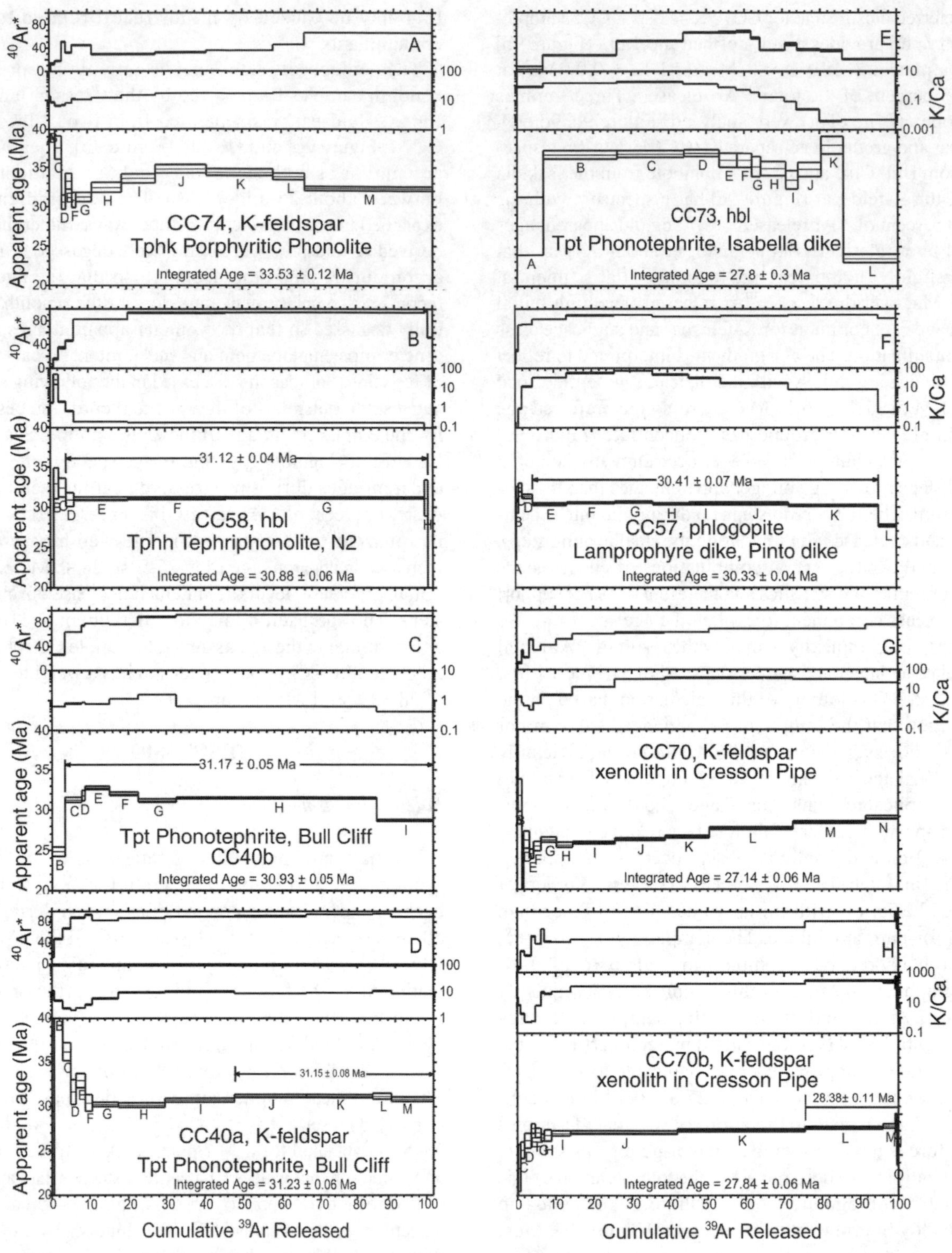

Figure 9. $^{40}Ar/^{39}Ar$ age spectra diagrams from incremental heating of samples in this contribution. Numbers below spectra are incremental-heating temperatures in degrees Celsius. Numerical errors are reported at 1σ and age spectrum steps are drawn at 2σ.

or a result of an acquired characteristic magnetization at a different period of time. The angular standard deviations (ASD) (scatter) of VGP's for the N=59 and N= 52 site mean data sets (15.4° and 11.7°, respectively) are lower than that predicted by McElhinny and McFadden's [1997] Model G, based on paleosecular variation data from lavas younger than about 5 myr. We do not find the relatively low ASD values surprising, in that it is entirely likely that the overall magnetization acquisition process in the Cripple Creek rocks differs from that on intermittently erupted lavas, and that the ensemble of data from the Cripple Creek district has smoothed paleosecular variation over the time period of magnetization acquisition. Given the dual polarity nature of the data set and the fact that the geomagnetic field was sampled over several million years during the evolution of the Cripple Creek complex, we feel that the overall ensemble of data adequately reflects an Oligocene geomagnetic field for the area. We next turn to logical ways in which to explain the observed discrepancy in inclination, including invoking district-wide tilting of about 10° in a north-side-down fashion, involving Precambrian host rocks as well, about a horizontal or sub-horizontal axis. This hypothesis is explored in greater detail below.

Overall, the paleomagnetic data contribute to our understanding of the emplacement, thermal, and alteration history of the Cripple Creek intrusive complex, in particular using polarity data and high-precision geochronologic information to correlate specific events with the Geomagnetic Polarity Time Scale (GPTS) for the mid-Tertiary [e.g., *Cande and Kent*, 1995]. Rock types whose magnetizations at specific sites are dominated by hematite, and hence required thermal demagnetization to fully isolate a well-defined and well-grouped magnetization, have both normal and reverse polarity magnetizations. We interpret this observation as indicating that hematite-formation, in conjunction with one or more phases of alteration, spanned at least one polarity reversal, and that this process is not associated with any specific, short-lived geomagnetic event.

Results from six contact tests, where the host is of one magnetic polarity and the intrusion is of opposite polarity, indicate that magnetizations in the host rocks, at distances greater than about half the width of the cross-cutting intrusion were not reset during or after emplacement of the cross-cutting intrusion. We interpret these results to indicate that, overall, the thermal history of the deposit was relatively simple in that cooling of specific intrusions was relatively fast, and that subsequent magmatism only locally modified the thermal structure of the complex. These determinations are in line with previously reported fluid inclusion data of about 250°C for peak hydrothermal conditions [*Thompson, et al.*, 1985; *Thompson*, 1996]. Admittedly, the results of these contact tests do not allow us to assess if all pre-existing magnetizations in host rocks, many of which are hydrothermally altered, are primary thermoremanent magnetizations acquired during the initial cooling of the host igneous body or formation of breccia unit, (in the case of the Cripple Creek Breccia), or secondary magnetizations acquired as a result of pervasive hydrothermal activity before emplacement of younger dikes sampled for contact tests. Specific rock types sampled in both altered and unaltered states yield consistent magnetization directions, which suggests that moderate hydrothermal activity did not considerably affect the primary NRM direction in most rocks sampled. Breccia tests conducted on the Cresson Pipe show that most clasts in the diatreme incorporated into the intrusion contain randomly-directed magnetizations. Magnetization directions at the clast level, if more than one sample was drilled per clast, are uniform. We interpret this result to indicate that breccia clasts sampled have not been magnetically reset. On the other hand, lamprophyric magma of the Cresson Pipe was fully capable of resetting crustal xenoliths or enclaves. For example, a felsic xenolith or enclave, about 25 cm in diameter, site CC70 (N = 9 samples , 4 in xenolith, 5 in Cresson Pipe (and Table 1; see CDROM in back cover sleeve) yields a magnetization that is identical to that of the host lamprophyre.

Geochronology

The new geochronologic data, combined with results of previously reported geochronologic investigations, suggest two principal episodes of igneous activity for the district. $^{40}Ar/^{39}Ar$ analysis of these rocks indicates a progression from

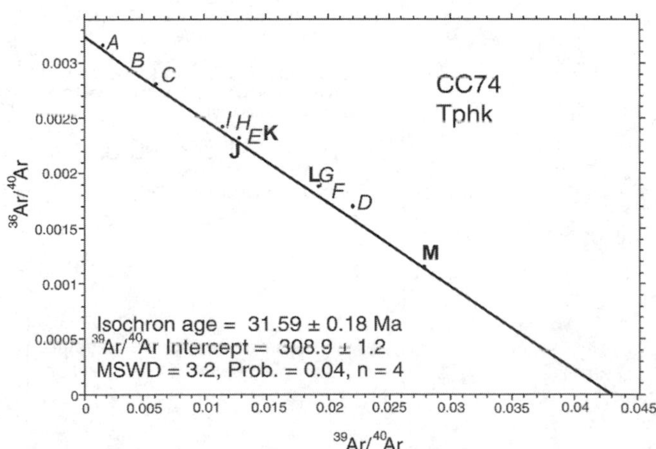

Figure 10. Inverse isochron diagram for CC74 (Tphk) plotting $^{36}Ar/^{40}Ar$ vs $^{39}Ar/^{40}Ar$ for J-M heating increments (bold type). Age is proportional to the x axis intercept and the y axis intercept is the initial argon component released.

felsic (phonolite) to intermediate and mafic (tephriphonolite and phonotephrite) magmatism to finally relatively low-volume ultramafic (lamprophyre) intrusive activity. Most of the age data are consistent with known field and cross cutting relations. Field observations show that the Pinto dike (CC57, 30.41 ± 0.21 Ma) cuts the Isabella dike (CC73), which cuts the N2 dike (CC58, 31.12 ± 0.04 Ma), all of which intrude into the Phonolite of Altman (CC74, 31.59 ± 0.32 Ma). The Bull Cliff phonotephrite (CC40, 31.15 ± 0.11 Ma) is considered to have intruded synchronously with the phonotephrite Isabella dike (CC73). The age determination of CC40 lies between the age determinations of CC57 and CC58 and is thus consistent with the known geologic relations. The emplacement age of the Cresson Pipe (CC70, 70b) remains difficult to establish. We suggest that older dates in the spectrum at ~ 28.5 Ma represent the age of the igneous body and that the age gradients measured for the K-feldspars [Figure 9g, h] reflect argon loss due to post-emplacement alteration. This interpretation is in

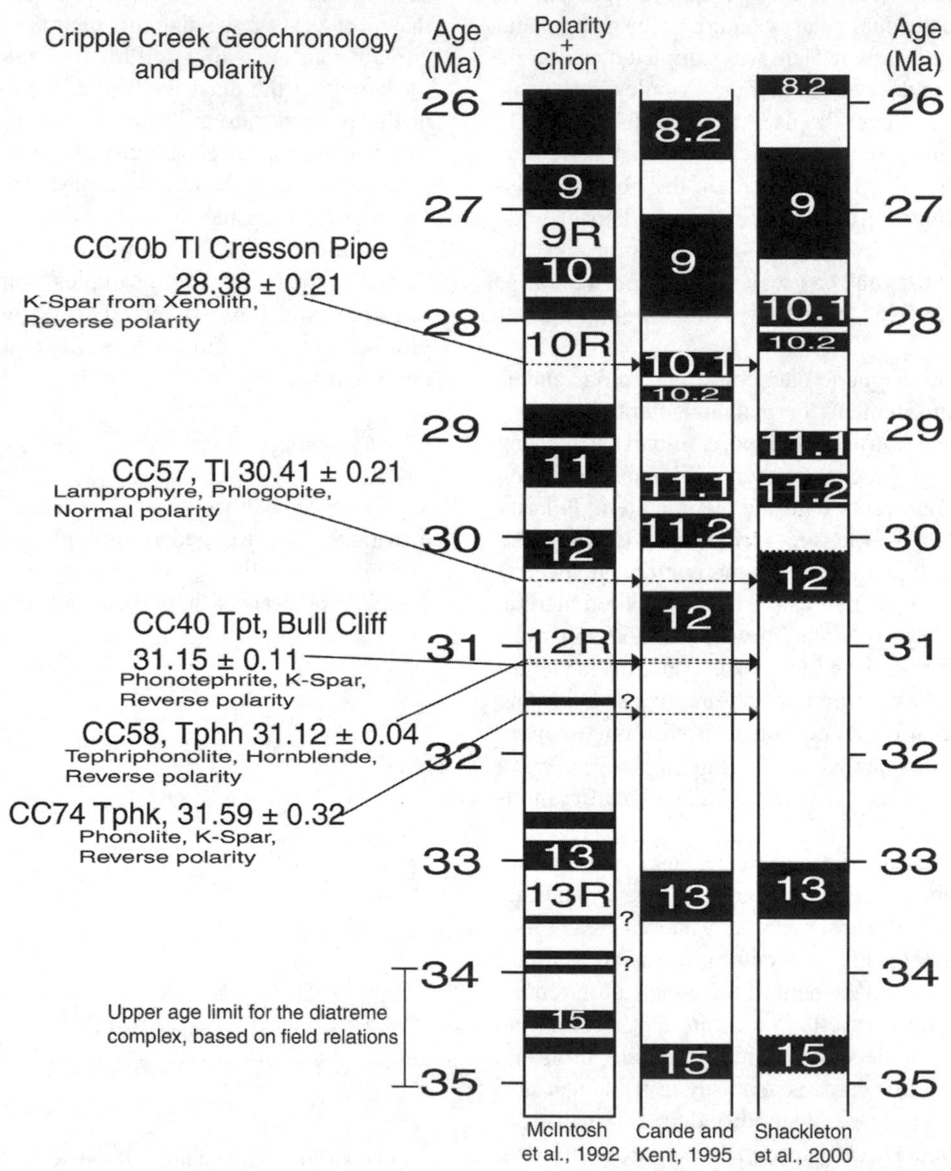

Figure 11. Oligocene geomagnetic polarity time scale, modified from McIntosh et al. [1992], Cande and Kent [1995], and Shackleton et al. [1999]. $^{40}Ar/^{39}Ar$ age data from determinations reported here combined with paleomagnetic polarity data. Age in millions of years, filled blocks represent normal polarity and open boxes represent reverse polarity.

part based on the late stage hydrothermal alteration events (28.8 and 28.2 Ma) [*Kelley et al.*, 1998] that may establish the Cresson Pipe to be no younger than this.

Six of the $^{40}Ar/^{39}Ar$ age determinations are from specific sites sampled for paleomagnetism [Figure 3a and 3b, see CD ROM in back cover sleeve]. Of these six sites, five are of reverse polarity and only one (CC57) is of normal polarity [Table 2, see CD Rom in back cover sleeve]. Based on estimates of the Oligocene geomagnetic polarity time scale [e.g., *McIntosh et al.*, 1992; *Cande and Kent*, 1995; *Shackleton et al.*, 1999] [Figure 11], several magnetic polarity reversals occurred during the time period of magmatism and the inferred time of magnetization acquisition. If magnetization acquisition was essentially synchronous with initial cooling of the igneous body and timing of closure of the K-Ar systems in specific minerals studied for geochronology, the magnetic polarity of the sampled rocks should be consistent with the polarity chrons of the geomagnetic polarity time scale on the basis of the ages assigned by $^{40}Ar/^{39}Ar$ analysis. All of our $^{40}Ar/^{39}Ar$ age data, within error, fall within the times spanned by chrons 10R or 12R and are consistent with the paleomagnetic data (polarity) of the respective rock unit [Figure 11]. In particular, we note that the data show a better correspondence with the revised polarity time scale of Shackleton et al. [1999].

Tectonic Implications

The ensemble of paleomagnetic data from the Cripple Creek diatreme complex can be most readily explained as reflecting a slight, north-side down tilting of the complex and at least some of the surrounding Precambrian host rocks. The inferred deformation affecting the Cripple Creek area is clearly not representative of all of the Front Range of Colorado, as lower Eocene strata, about 23 km northwest of Cripple Creek near Florissant, are flat-lying. To the northeast, however, the well-developed late Eocene surface on the Rampart Range north of Pikes Peak [*Epis et al.*, 1976] has a gentle north dip. Known fault and shear zones in and surrounding the Cripple Creek area [Figure 12] include the Oil Creek fault and the Ute Pass fault to the northeast and the Four Mile Creek fault to the southwest, all of which strike north-northwest. South and southeast of the diatreme, the Nipple, Skagway, and Adelaide faults strike northeast. About 24 km south of the Cripple Creek complex at the south end of the south plunging Cripple Creek Arch, are the northwest-verging Gnat Hollow anticline-syncline and associated northwest-verging thrust fault with about 500 m of displacement [*Chase et al.*, 1993]. Unlike other structures in the area, the Gnat Hollow fault displaces Paleozoic rocks. Most of the fault zone, however, is localized in the Precambrian basement and slickenlines are preserved in the cover rocks exposed in the footwall of the syncline. From slickenline orientations Chase et al. [1993] inferred a horizontal σ_1 axis slightly west of north. This σ_1 orientation is similar to σ_1 axes inferred by Erslev [2001] in north central New Mexico on the east side of the Rio Grande rift recording faulting after 25 Ma. Kinematic analysis suggested that σ_1 (and thus inferred maximum shortening direction) rotated in a counterclockwise sense from a roughly east-west orientation during the early stages of Laramide style deformation to a more north-south orientation during the mid-Tertiary [*Erslev*, 2001]. In contrast, other workers [e.g., *Yin and Ingersoll*, 1997] require no such rotation of greatest principal stress direction to explain Laramide-related features in the foreland on the margin of the Colorado Plateau.

Given the structural setting of the diatreme complex, it is possible that the north-side down tilting could have been a result of the overall compressional tectonic regime during the time of formation. The diatreme is located at the intersection of a grid of steeply-dipping northwest trending oblique dextral and northeast trending oblique sinistral faults and the axis of the Cripple Creek arch [Figure 12]. The intersection of these two fault systems is roughly in the diatreme center. Small fault population data from open pit exposures in the diatreme indicate that older diatreme rocks were shortened (early thrust fault population) along structures trending roughly east-west. These, mostly low angle, faults that cut diatreme rocks are consistent in orientation with pre-existing structural fabrics within the surrounding Proterozoic host rocks and may reflect reactivation along a preferred and long lived orientation of weakness. Steep northwest and northeast structures crosscut the low angle structures and are the dominant orientations of joints, dikes, faults, and veins that affect rocks of the complex. Cross-cutting relations indicate that these systems were active during and after diatreme formation and that the northeast and northwest systems were active simultaneously.

The structural record of the overall transition from northeast-directed Laramide shortening in the latest Cretaceous to early Tertiary to Rio Grande extension since the mid-Tertiary is poorly understood in the Front Range of Colorado. Aspects of the kinematics and associated tectonic events along the eastern margin of the Colorado Plateau during and since Laramide deformation have been the subject of some debate [*Yin and Ingersoll*, 1997; *Bird*, 1998; *Erslev*, 2001; *Wawrzyniec et al.*, 2002]. Recently, Wawrzyniec et al. [2002] proposed a general model, based on both fault kinematic and paleomagnetic data that involved dominantly northeast-directed transpression during Laramide deformation and northwest-directed transtension associated with Rio Grande rift development. This model was proposed to account for, depending on fault orientation, both dextral and sinistral shear components of deformation observed in New Mexico and south-central Colorado. Farther west, on the eastern margin of the Basin and Range province, Anderson et al. [1993] proposed a component of north-south shortening associated with Miocene and younger east-west extension in the Virgin River depression area based on brittle fault kinematics.

On the basis of the geochronologic data for the Cripple Creek Complex, any deformation affecting the complex and immediately surrounding rocks must be post main-stage Laramide deformation and either pre-Rio Grande rifting (during the early stages of diatreme formation) or syn-Rio Grande deformation. We note that all mean directions, representing a 3 m.y. long period of magmatism and alteration, from 31.6 Ma to 28.4 Ma, yield consistent mean inclinations that are steeper than expected mid-Tertiary reference values. Furthermore, many minor structures within the complex deform all rock types present in the complex [*Koschmann*, 1947 and present studies] and thus were clearly active after diatreme formation, or post 28.4 Ma. There is no correlation between

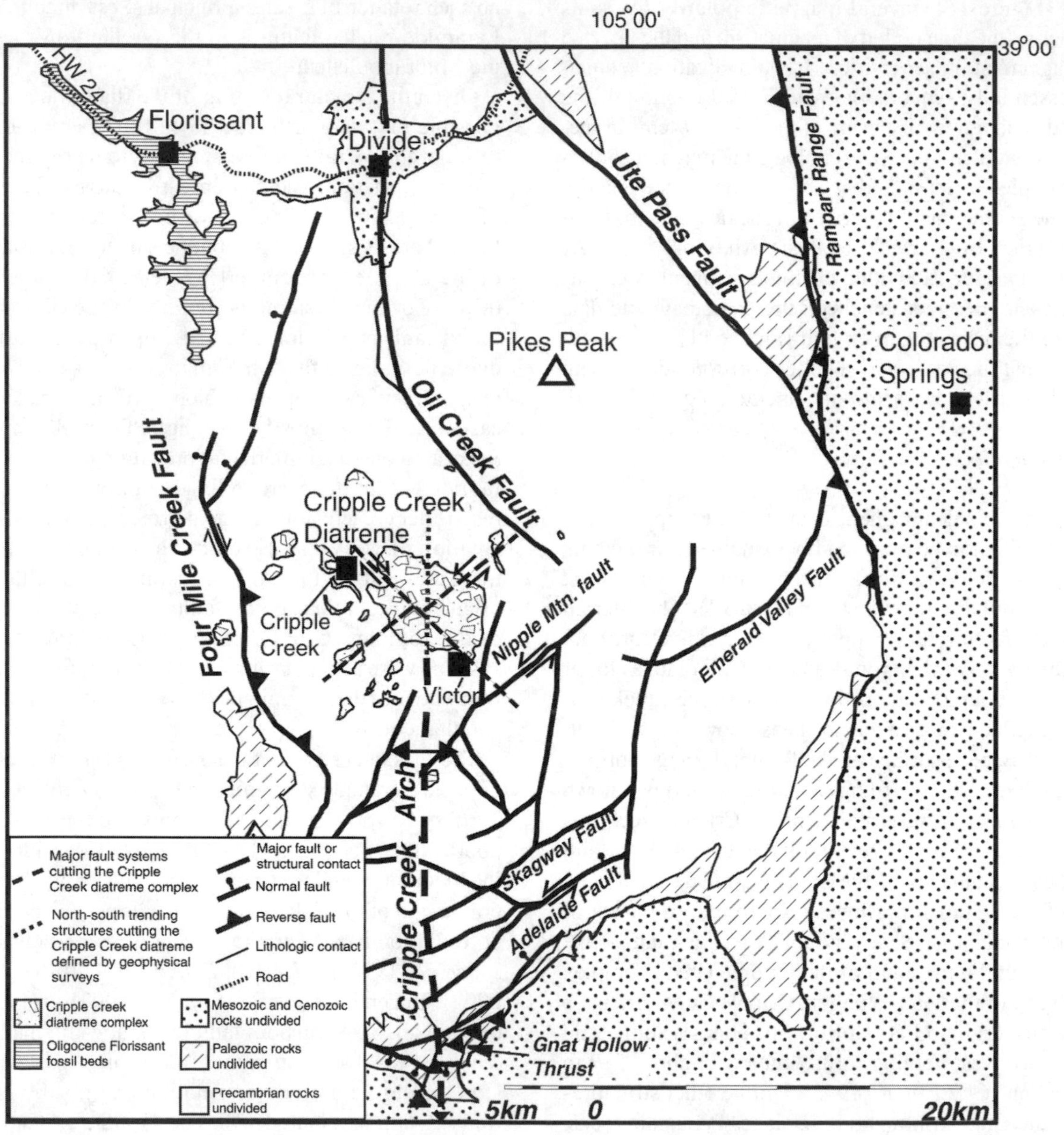

Figure 12. Simplified geologic map of the southern Front Range in the Cripple Creek area. Modified from Chase et al. [1993] and Scott et al.[1978]. Shown are major structures and geologic features in and surrounding the Cripple Creek diatreme complex.

the magnitude of inclination of the magnetization characteristic of specific intrusions and the relative age of the rocks; specifically, older rocks do not have inclinations that are steeper than those of younger rocks. We interpret the paleomagnetic data to indicate that deformation, inferred on the basis of the inclination discordancy, must have occurred after about 28 Ma, following diatreme formation. We suggest that the principal component of this deformation was a north-side down tilting of about 10 degrees about a roughly east-west axis. Structures responsible for this deformation could reflect transtensional tectonism related to Rio Grande rift development. Local structural features in the Precambrian rocks in the Cripple Creek area, with preexisting weaknesses developed during the Proterozoic and younger time, probably were critical in controlling the orientation of tilting in this area of the Front Range of Colorado. Notably, the strain field in the southern Front Range of Colorado during diatreme formation must have contained some component of north-south shortening. This is based on the early-phase low angle fault populations that offset rocks in the diatreme complex. The deformation experienced by the Cripple Creek diatreme complex and surrounding rocks since the Oligocene may not fully reflect the pattern of deformation across the entire Front Range of Colorado. Whether the tilting of the diatreme and surrounding rocks is a result of regional shortening or younger extension during this time period is, admittedly, not clear. Fault populations within the diatreme do, however, primarily reflect a compressional stress regime. Initial diatreme formation may record a short-lived period of north-south shortening during the transition from Laramide shortening to subsequent extension in the southern Front Range of Colorado.

CONCLUSIONS

Based on the $^{40}Ar/^{39}Ar$ age determinations, two episodes of igneous activity, which span a range of about 3 m.y., at the Cripple Creek complex are distinguished. Early igneous activity is indicated by $^{40}Ar/^{39}Ar$ age determinations of 31.59 ± 0.32 Ma from fresh porphyritic phonolite. Further activity is marked by intrusion of the N2 tephriphonolite dike at 31.12 ± 0.04 Ma, phonotephrite at 31.15 ± 0.11 Ma, and lamprophyre dikes at 30.41 ± 0.21 Ma. This period is followed by a hiatus of about 2 m.y. The Cresson lamprophyre pipe was emplaced at 28.38 ± 0.21 Ma and represents one of the latest intrusive events in the district. Intrusive rocks evolved from felsic to ultramafic composition during the 3 m.y. of igneous activity. $^{40}Ar/^{39}Ar$ age determinations associated with paleomagnetic data show that multiple polarity reversals took place during the formation of the diatreme complex and that, overall, the data are consistent with geomagnetic polarity time scales for the Oligocene. The paleomagnetic data are characterized by an essentially dual-polarity, well-grouped population of estimated site means, defined by nearly 60 accepted sites, the grand mean of which is some ten degrees steeper in inclination than expected mid-Tertiary values. Such an inclination discrepancy is difficult to reconcile as an artifact of the approach taken in paleomagnetic investigations (e.g., incomplete removal of secondary, superimposed magnetizations that would contaminate the preexisting signal). We interpret the inclination discrepancy to reflect a very modest, north-side down tilting of the diatreme complex and, at least locally, Proterozoic host rocks, after the complete emplacement of the diatreme complex. Such tilting could have been accommodated in a short period of either north-south shortening during the transition between Laramide-style deformation or west to northwest-directed extension during the development of the Rio Grand rift in mid-Oligocene and younger time.

Acknowledgments. We would like to thank Jon Hagstrum, Steve Harlan, and Paul Layer for insightful reviews of the manuscript. We gratefully acknowledge the Cripple Creek and Victor Gold Mining Company for access to the mine and to the Cripple Creek and Victor Gold Mining Company Mine Geology Department for insightful discussion and for field assistance. We also thank Bill McIntosh, Lisa Peters, and Rich Esser for mineral preparation and assistance in argon analysis. This study was not funded by any external granting agency. Rampe received an RPT grant from the University of New Mexico and support from the Alumni Scholarship Fund of the Department of Earth and Planetary Sciences, University of New Mexico, to aid in this research. Funds from the operating budget for the UNM Paleomagnetism Laboratory also supported this research.

REFERENCES

Anderson, R.E., and T.P. Barnhard, Aspects of three-dimensional strain at the margin of the extensional orogen, Virgin River depression area, Nevada, Utah, and Arizona. *Geol. Soc. Am. Bull., 105*, 1019–1052, 1993.

Bird, P., Kinematic history of the Laramide orogeny in Latitudes 35° –49°N, western United States, *Tectonics 17*, 780–801, 1998.

Cande, S. C., and D.V. Kent, Revised calibration of the geomagnetic polarity timescale for the Late Cretaceous and Cenozoic, *J. Geophys. Res. 100*, 6093–6095, 1995.

Chapin, S.E., and S.M. Cather, Eocene tectonics and sedimentation in the Colorado Plateau-Rocky Mountain area, *Rocky Mountain Association of Geologists*, 33–56, 1983.

Chase, R.B., C.J. Schmidt, and P.W. Genovese, Influence of Precambrian rock compositions and fabrics on the development of Rocky Mountain foreland folds, in *Laramide basement deformation in the Rocky Mountain foreland of the western United States,* edited by Schmidt, C.J., R.B. Chase, and E.A. Erslev, *Geol. Soc. of Am. Spec. Pap., 280*, 45–72, 1993.

Cross, W., and R.A.F. Penrose, Jr., Geology and mining industries of the Cripple Creek district, Colorado. 16th annual report, *U.S. Geol. Surv.*, pt. IIB, 1–109, 1895.

Dalrymple, G.B., Critical tables for conversion of K-Ar ages from old to new constants, *Isochron/West, 58*, 22–24, 1992.

De Voto, R., Mississippian and Pennsylvanian stratigraphy and history of

Colorado. in *Colorado Geology*, edited Kent, H.C., and K.W. Porter, Rocky Mountain Ass. of Geologists, 57–102, 1980.

Dickinson, W. R., M.A. Klute, M.J. Hayes, S.U. Janecke, E.R. Lundin, M.A. McKittrick, and M.D. Olivares, Paleogeographic and paleotectonic setting of Laramide sedimentary basins in the central Rocky Mountain region. *Geol. Soc. of Am. Bull., 100*, 1023–1039, 1988.

Deino, A. and R. Potts, Single-crystal $^{40}Ar/^{39}Ar$ dating of the Olorgesailie Formation, Southern Kenya Rift, *J. Geophys. Res., 95*, 8453–8470, 1990.

Diehl, J., K.M. McClannahan, and T.J. Bornhorst, Paleomagnetic results from the Mogollon-Datil volcanic field, southwestern New Mexico, and a refined mid-Tertiary reference pole for North America, *J. Geophys. Res. 93,* 4869–4879, 1988.

Epis, R.C., and C.E. Chapin, Geomorphic and tectonic implications of the post-Laramide, late Eocene erosion surface in the southern Rocky Mountains. *Geol. Soc. Am. Memoir, 144*, 45–74, 1975.

Epis, R. C., G. R. Scott, R. B. Taylor, and W. N. Sharp, Petrologic, tectonic, and geomorphic features of central Colorado, in *Studies in Colorado Field Geology*, edited Epis, R. C., and R.J. Weimer, *Professional Contributions of Colorado School of Mines*, 301–322, 1976.

Erslev, E.A., Multistage, multidirectional Tertiary shortening and compression in north-central New Mexico, *Geol. Soc. Am. Bull. 113,* 63–74, 2001.

Everitt, C. W. F., and J.A. Clegg, A field test of paleomagnetic stability, *Geophys. Jour. Royal Astronomical Society*, *6*, 312–319, 1962.

Foster, D.A., T.M. Harrison, P. Copeland, and M.T. Heizler, Effects of excess argon within diffusion domains on K-feldspars age spectra. *Geochem. Cosmochim. Acta., 54,* 1699–1708, 1990.

Geissman, J. W., and S.S. Harlan, Late Paleozoic remagnetization of Precambrian crystalline rocks along the Precambrian/ Carbonifeous nonconformity, Rocky Mountains: A relationship among deformation, remagnetization, and fluid migration, *Earth and Planet. Sci. Lett., 203*, 905–924, 2002.

Geissman, J. W., L.W. Snee, G.W. Graaskamp, R.B. Carten, and E.P. Geraghty, Deformation and age of the Red Mountain intrusive system (Urad-Henderson molybdenum deposits), Colorado: Evidence from paleomagnetic and Ar/Ar data, *Geol. Soc. Am. Bull., 104*, 1031–1047, 1992.

Hutchinson, R.M., and C.E. Hedge, Depth-zone emplacement and geochronology of Precambrian plutons, central Colorado Front Range, *Geol. Soc.* Am. Spec. Pap., 115, 424–425, 1968.

Ingersoll, R. V., Structural and stratigraphic evolution of the Rio Grande rift, northern New Mexico and southern Colorado, *International Geology Review, 43*, 876–891, 2001.

Ingersoll, R.V., W. Cavazza, W.S. Baldridge, and M. Shafiqullah, Cenozoic sedimentation and paleotectonics of north-central New Mexico: Implications for initiation and evolution of the Rio Grande rift, *Geol. Soc. Am. Bull., 102*, 1280–1296, 1990.

Irving, E., Paleomagnetism and Its Applications to Geological and Geophysical Problems, Wiley-Interscience, New York, 399 pp., 1964.

Johnson, H. P., W. Lowrie, and D.V. Kent, Stability of anhysteretic remanent magnetization in fine and coarse magnetite and maghemite particles, *Geophys. J. of the Royal Astronomical Soc., 41*, 1–10, 1975.

Kelley, K.D., S. B. Rombeger, D. W. Beaty, J. A. Pontius, L. W. Snee, H. J. Stein, and T. B. Thompson, Geochemical and geochronological constraints on the genesis of Au-Te deposits at Cripple Creek, Colorado, *Economic Geology, 93*, 981–1012, 1998.

Kelley, S.A., and C.E. Chapin, Denudational history and internal structure of the Front Range and Wet Mountains, Colorado, based on apatite-fission-track thermochronology, *New Mexico Bureau of Mines and Mineral Resources, Chapin Volume*, in press, 2002.

Kirschvink, J. L., The least squares line and plane and the analysis of paleomagnetic data. *Geophys. Jour. Royal Astronomical Soc., 62*, 699–718, 1980.

Koschmann, A.H., Structural control of the gold deposits of the Cripple Creek district, Colorado, *U.S. Geol. Surv. Bull., 955-B*, 19–58, 1949.

Lindgren, W., and Ransome, F.L., Geology and gold deposits of the Cripple Creek district, Colorado, *U.S. Geol. Surv. Prof. Pap., 54*, 516 pp., 1906.

Lipman, P.W., Volcano-tectonic setting of the Tertiary ore deposits, southern Rocky Mountains, *Arizona Geol. Soc. Digest, 14*, 199–213, 1981.

Lipman, P.W., B.R. Doe, C.E. Hedge, and T.A. Steven, Petrologic evolution of the San Juan volcanic field, southwestern Colorado: Pb and Sr isotope evidence, *Geol. Soc. Am. Bull., 89*, 59–82, 1978.

Loughlin, G.F., and A.H. Koschmann, Geology and ore deposits of the Cripple Creek district, Colorado, *Colorado Scientific Soc. Proceedings, 13,* 217–435, 1935.

Lowrie, W., and M. Fuller, On the alternating field demagentization characteristics of multidomain thermoremanent magnetization in magnetite, *J. Geophys. Res., 76*, 6339–6349, 1971.

Mahon, K.I., The New "York" regression: Application of an improved statistical method to geochemistry, *International Geology Review, 38,* 293–303, 1996.

Maughan, E.K., Permian and Lower Triassic geology of Colorado, in *Colorado Geology*, edited Kent, H.C., and K.W. Porter, Rocky Mountain Assoc. of Geologists, 103–110, 1980.

Mankinen, E., E. Larson, C.S. Gromme, M. Prevot, and R.S. Coe, The Steen Mountain (Oregon) geomagnetic polarity transition, 3. Its regional significance, *J. Geophys. Res., 92*, 8057–8076, 1987.

McDowell, F.W., K-Ar ages of igneous rocks form the western United States, *Isochron West, 2*, 16pp., 1971.

McElhinny, M., and P. McFadden, Palaeosecular variation over the past 5 Myr based on a new generalized database, *Geophys. J. International, 131*, 240–252, 1997.

McFadden, P. L., and M.W. McElhinny, Classification of the reversal test in palaeomagnetism, *Geophys. J. International, 103*, 725–729, 1990.

McIntosh, W.C., J.W. Geissman, C.E. Chapin, M.J. Kunk, and C.D. Henry, Calibration of the latest Eocene-Oligocene geomagnetic polarity time scale using $^{40}Ar/^{39}Ar$ dated ignimbrites, *Geology, 20*, 459–463, 1992.

Oppenheimer, W.L., and J.W. Geissman, Paleomagnetic data bearing on Laramide and post-Laramide deformation and magmatism in the northern Mosquito Range between Fremont and Hoosier Passes, central Colorado, *Colorado School of Mines Quarterly, 83*, 33–50, 1988.

Pontius, J.A., Gold deposits of the Cripple Creek mining district, Colorado, *Soc. of Economic Geologists Guidebook Series, 26,* 29–37, 1996.

Ross, R.J., and O. Tweto, Lower Paleozoic sediments and tectonics in Colorado, in *Colorado Geology*, edited Kent, H.C., and K.W. Porter, Rocky Mountain Assoc. of Geologists, 47–56, 1980.

Roy, J. L., and J.K. Parker, The magnetization process of certain red beds: vector analysis of chemical and thermal results, *Canadian J. Earth Sci., 11,* 437–471, 1974.

Samson S.D., and E.C. Alexander, Jr., Calibration of the interlaboratory $^{40}Ar/^{39}Ar$ dating standard, MMhb-1, *Chem. Geol. Isot. Geosci., 66*, 27–34, 1987.

Scott, G.R., R.B. Taylor, R.C. Epis, and R.A. Wobus, Geologic map of the Pueblo 1° x 2° quadrangle, south-central Colorado, *U.S. Geol. Surv., Map I-1022*, scale 1:250,000, 1978.

Shackleton, N.J., Crowhurst, S.J., Weedon, G., and Laskar, L., Astronomical calibration of Oligocene-Miocene time, *Geophys. Jour. Royal Astronomical Soc., 357*, 1909–1927, 1999.

Spall, H. C., Paleomagnetism of the Pikes Peak granite, Colorado, *Geophys. Jour. Royal Astronomical Soc., 21*, 427–440, 1970.

Steidtmann and Middleton, Fault chronology and uplift history of the southern Wind River Range, Wyoming; implications for Laramide and post-Laramide deformation in the Rocky Mountain foreland, *Geol. Soc. Am. Bull.*, 103, 472–485, 1991.

Steiger, R.H., and E. Jäger, Subcommission of geochronology: Convention on the use of decay constants in geo- and cosmochronology. Earth and Planet. Sci. Lett., 36, 359–362, 1977.

Taylor, J.R., *An introduction to error analysis*, University Science Books, Mill Valley, Calif., 1982.

Thompson, T.B., Mineral deposits of the Cripple Creek district, Colorado, *Mining Engineering, 44*, 135–138, 1992.

Thompson, T.B., Fluid evolution of the Cripple Creek hydrothermal system, Colorado, *Soc. of Economic Geologists Guidebook Series, 26*, 45–54, 1996.

Thompson, T.B., A.D. Trippel, and P.C. Dwelley, Mineralized veins and breccias of the Cripple Creek district, Colorado, *Economic Geology*, *80*, 1669–1688, 1985.

Tweto, O., Laramide (Late Cretaceous – early Tertiary) orogeny in the southern Rocky Mountains, In Curtis, B., ed., Cenozoic history of the southern Rocky Mountains: *Geol. Soc. Am. Mem., 144,* 1–44, 1975.

Tweto, O., Tectonic history of Colorado, in *Colorado Geology*, edited Kent, H.C., and K.W. Porter, Rocky Mountain Assoc. of Geologists, 5–9, 1980.

Unruh, D.M., L.W. Snee, E.E. Foord, and W.B. Simmons, Age and cooling history of the Pikes Peak Batholith and associated pegmatites, *Geol. Soc. Am. Abstr. Programs 27*, *6*, 468, 1995.

Van der Voo, R., *Paleomagnetism of the Atlantic, Tethys, and Iapetus Oceans*, Cambridge, Cambridge University Press, 411 pp., 1993.

Wallace, A.R., Regional geologic and tectonic setting of the central Colorado mineral belt: Paleozoic stratigraphy, tectonism, thermal history, and basin evolution of central Colorado, *Economic Geology Monograph, 7*, 19–28, 1990.

Watson, G. S., A test for randomness of directions, *Geophys. Jour. Royal Astronomical Soc.*, 7, 160–161, 1956.

Wawryzniec, T. F., J.W. Geissman, M.A. Melker, and M. Hubbard, Dextral shear along the eastern margin of the Colorado Plateau: A kinematic link between the Laramide orogeny and Rio Grande rifting (ca. 75 to 13 Ma), *J. Geol.*, *110*, 305–324, 2002.

Wobus, R.A., R.C. Epis, and G.R. Scott, Reconnaissance geologic map of the Cripple Creek-Pikes Peak area, Teller, Fremont, and El Paso Counties, Colorado, *U.S. Geol. Surv. Miscellaneous field studies Map*, MF-805, 1976.

Yin, A., and R. V. Ingersoll, A model for evolution of Laramide axial basins in the southern Rocky Mountains, U.S.A., *International Geol. Rev., 39*, 1113–1123, 1997.

York, D., Least squares fitting of a straight line with correlated errors, *Earth and Planet. Sci. Lett., 5*, 320–324, 1969.

Zijderveld, J. D. A., A.C. Demagnetization of rocks: Analysis of results, in *Methods in Palaeomagnetism*, edited Collinson, D. W., K.M. Creer, and S.K. Runcorn, Amsterdam, Elsevier, 254–286, 1967.

John W. Geissman, Department of Earth and Planetary Sciences, University of New Mexico, Albuquerque, NM, 87131.

Matthew T. Heizler, New Mexico Bureau of Mines and Mineral Resources, New Mexico Institute of Mining and Technology, 801 Leroy, Socorro, NM, 87801.

Marc D. Melker, Cripple Creek and Victor Gold Mining Company, P.O. Box 191, Victor, CO, 80860.

Jason S. Rampe, Newmont Mining Corp., Eastern Nevada Operations, P.O. Box 669, Carlin, NV, 89822.

Isostatic Constraints on Lithospheric Thermal Evolution: Application to the Proterozoic Orogen of the Southwestern United States

R. M. Flowers, L. H. Royden, and S. A. Bowring

Department of Earth, Atmospheric and Planetary Sciences, Massachusetts Institute of Technology, Cambridge, Massachusetts

The long-term stability of cratonic regions can be used to place first-order constraints on the thermal structure and evolution of continental lithosphere. The lithosphere retains a record of its net isostatic change, because net heating yields uplift and erosion, while net cooling yields subsidence and sedimentation. We reconstruct initial lithospheric thermal profiles compatible with isostatic stability requirements, and use a finite-difference model to compute geotherm evolution from initial conditions to steady-state. We apply this method to the Proterozoic (1.8–1.0 Ga) orogenic belt of the southwestern United States. The apparent protracted cooling histories (~ 1°C/m.y.) of the Hualapai (0.3–0.4 GPa) and Big Bug (0.3 GPa) blocks contrast with rapid cooling (25–100°C/m.y.) in the Ash Creek block (0.1–0.2 GPa). Compilation of heat production data yields values of 3.12–5.20 $\mu W/m^3$, 1.46–3.46 $\mu W/m^3$ and 0.70–1.27 $\mu W/m^3$ for the Hualapai, Big Bug and Ash Creek blocks at 1.7 Ga, respectively. Our thermal analysis indicates that hot steady-state geotherms due to high heat production in the Hualapai and Big Bug blocks are consistent with cooling at higher temperatures, while lower heat production in the Ash Creek block can explain its more rapid cooling. This study highlights the importance of heat production for laterally variable thermal regimes in heterogeneous orogenic belts, and emphasizes that domainal heat production differences must be considered when interpreting regional cooling histories. The integration of thermal records with exhumation information, heat production and heat flow data can place important constraints on feasible, isostatically consistent models for lithospheric thermal evolution.

INTRODUCTION

A detailed understanding of the thermal structure and evolution of continental lithosphere is crucial for understanding the lithospheric response to tectonic processes, because of the temperature dependency of deformation, metamorphism and melting, and a wide range of physical properties. For this reason, much attention has focused on the computation of steady-state continental geotherms using measurements of surface heat flow and models for heat production and thermal conductivity [e.g. *Chapman and Pollack,* 1977; *Chapman,* 1986; *Furlong and Chapman*, 1987; *Rudnick et al.*, 1998]. Due to uncertainty in thermal properties with increasing depth, detailed knowledge of steady-state temperature profiles in the lithosphere today is difficult to obtain. Understanding the thermal stabilization of lithosphere during evolution to the steady-state geotherm following deformation and magmatism is even more challenging.

Recent quantitative thermal models have demonstrated the important role of heat-producing elements (HPEs;

The Rocky Mountain Region: An Evolving Lithosphere
Geophysical Monograph Series 154

10.1029/154GM10

U,Th,K) for the temperature distribution of both active orogenic belts and stable cratonic lithosphere. Anomalously high temperatures within collisional orogens such as the Himalayas can be explained by basal accretion and surface erosion of HPE-enriched crust [*Royden,* 1993; *Huerta et al.,* 1996, 1999]. High-temperature, low-pressure metamorphism in cratonic crust, as documented in several Australian Proterozoic orogens, can be the result of elevated crustal heat production rather than a transient thermal event [*Sandiford and Hand*, 1998]. The dependence of thermal profiles and lithospheric rheology on the magnitude and distribution of crustal heat production indicates that upward HPE magmatic redistribution may be important for the ultimate stabilization of continental crust [*Morgan,* 1985; *Sandiford et al.,* 2002, *Sandiford and McLaren*, 2002].

Although these previous studies have recognized the important role of heat production for continental thermal regimes, there has been limited integration of PTt data for specific lithospheric sections with models for the time-dependent change of the geotherm from peak orogenesis to steady-state. The sensitivity of crustal thermal evolution to radiogenic heat production variation indicates that powerful insight can be gained into lithospheric thermal histories and heat production distribution by reconstructing cooling records from > 1000°C to < 300°C using modern high-precision U/Pb and $^{40}Ar/^{39}Ar$ thermochronologic techniques. The integration of metamorphic petrology with these high-resolution cooling histories can yield critical constraints on the thermal evolution of the entire lithospheric column by exploiting the PTt record in exposed upper and middle crustal rocks, and lower crustal and lithospheric mantle xenoliths.

The long-term stability of cratons and their insulation from plate margin processes has commonly resulted in their preservation of ancient cooling paths that are undisrupted by recent tectonic activity. Thus, cratons are an ideal location to study the thermal record of the transformation from a hot, active collisional orogen into cold, rheologically strong lithosphere. The extreme isostatic stability of cratons also imposes fundamental isostatic constraints on lithospheric evolution that have largely been neglected in previous modeling efforts. We outline an approach that integrates PTt data, radiogenic heat production measurements, heat flow data and erosion or sedimentation observations for a specific lithospheric section to better understand the thermal evolution of the entire lithospheric column. We consider the variable effects of heat production on thermal and isostatic records during lithospheric stabilization, apply this method to the Proterozoic (1.8–1.0 Ga) orogenic belt of the southwestern United States, and provide a framework with which to understand the cooling records of this region.

ISOSTATIC CONSTRAINTS ON GEOTHERM EVOLUTION FROM CRATONS

The long-term stability of cratons is attributed to thick conductive roots of ancient lithospheric mantle that shield the crust from convecting asthenosphere and protect it from deformation during craton margin collision [e.g. *Jordan*, 1978]. This lithospheric mantle is depleted in basaltic components, characterized by high seismic velocities, and buoyant relative to asthenosphere even when "cold". A wealth of information from seismic [e.g. *Dueker et al.,* 2001; *James et al.*, 2001; *van der Lee*, 2002] and mantle xenolith [e.g. *Boyd et al.*, 1985; *Rudnick and Nyblade*, 1999] studies indicates that this lithospheric "keel" extends to depths of 200–300 km. Much effort has been directed at determining the age of mantle roots, initially using Sm-Nd and Rb-Sr isotopes [e.g. *Richardson et al.,* 1984; 1993], and more recently using the Re-Os isotope system [e.g. *Pearson*, 1995, 1999; *Irvine et al*., 2001; *Shirey et al.,* 2001; 2002]. Re-Os mantle depletion dates for lithospheric peridotites from Archean cratons around the globe (Kaapvaal, Siberia, Wyoming, Tanzania) coincide with major crust-forming events, and generally lack any clear variation of age with lithospheric depth. These data suggest that > 150 km of lithosphere may have formed relatively rapidly and remained coupled to the overlying cratonic crust since its formation. A variety of models including collisional orogenesis, subduction and imbricate stacking of oceanic lithosphere, and growth by plumes have been used to explain the growth and stabilization of cratonic lithosphere [e.g. *Jordan,* 1978; *deWit et al*., 1992; *Herzberg,* 1993].

The isostatic stability of cratons during geotherm evolution, indicated by their often shallow to moderate levels of exposure (0–15 km), has important implications for the initial temperature and thickness of the lithospheric mantle root. The evolution of the lithospheric density structure parallels the time-dependent change of the geotherm, because a rock's density is largely a function of its temperature. Thus, coupling exhumation constraints with cooling histories can yield insight into the thermal and isostatic evolution of lithospheric sections.

THERMAL MODEL

The thermal lithosphere is defined as the earth's outer layer in which heat transfer is dominated by conduction [*Morgan, 1980*]. The evolution of thermal profiles is controlled by heat input from the underlying asthenosphere, the quantity and distribution of heat-producing elements within the lithosphere, and the lithospheric thermal conductivity structure. Following orogenesis, if the lithosphere remains unperturbed by subsequent tectonic events, the thermal profile will decay to steady-state. In this study we assume that the dominant control on density change during geotherm evolution is thermal expansion or contraction. We do not incorporate effects associated with phase changes or subsequent tectonic events, and assume the primary mechanism of heat transfer is conduction. Although advection and melt extraction likely occurred during lithospheric growth, we treat the initial lithospheric thermal profile as the peak geotherm immediately following crustal assembly in order to simplify our assessment of the broad controls of lithospheric thermal evolution.

We model the final steady-state thermal profile with radiogenic heat production A in a layer of thickness x_f in the upper crust:

$$T = T_s + \frac{T_m x}{L} - \frac{A}{k}\left[\frac{x^2}{2} + x\left(\frac{x_f^2}{2L} - x_f\right)\right] \quad \text{for } x > x_f \qquad (1)$$

$$T = T_s + \frac{T_m x}{L} - \frac{A}{k}\left[\frac{(x_f^2)(x-L)}{2L}\right] \quad \text{for } x < x_f \qquad (2)$$

with lithospheric thickness L, temperature at the base of the lithosphere T_m of 1300°C, surface temperature T_s of 10°C, and thermal conductivity k of 2.5 W/m°C. The first-order observation of decreasing heat production with depth has long been known from heat production distributions in exposed crustal cross-sections, low concentration of HPEs in lower crustal xenoliths relative to that contained in the upper crust, and constraints imposed by the inferred contribution of mantle and upper crustal sources to surface heat flow measurements [e.g. *Chapman and Pollack,* 1977; *Rudnick and Fountain,* 1995]. Consistent with these observations, for simplicity the geotherms in this study contain an upper crustal layer of constant HPE concentration and a lower crust that lacks heat production. Figure 1A shows sample final steady-state lithospheric geotherms, calculated from equations (1)–(2), for a range of radiogenic heat production values.

We model the initial lithospheric thermal structure as a lithospheric mantle of uniform temperature T_L emplaced beneath a crust of thickness c (Figure 1B). Although the initial lithospheric mantle temperature almost certainly was not constant with depth, this value can be thought of as the mean temperature of some depth dependent thermal gradient or the mean temperature of a lesser thickness of lithospheric mantle averaged with asthenosphere. The initial and final lithospheric thicknesses are assumed to be the same for the purposes of the isostatic calculation, because the thermal evolution of the entire final lithospheric thickness will contribute to its net isostatic change. The initial crustal temperatures are calculated as a steady-state thermal profile in equilibrium with the initial mean temperature of the lithospheric mantle. The initial thickness of the radiogenic layer differs from the final thickness by the amount of erosion or sedimentation during geotherm evolution.

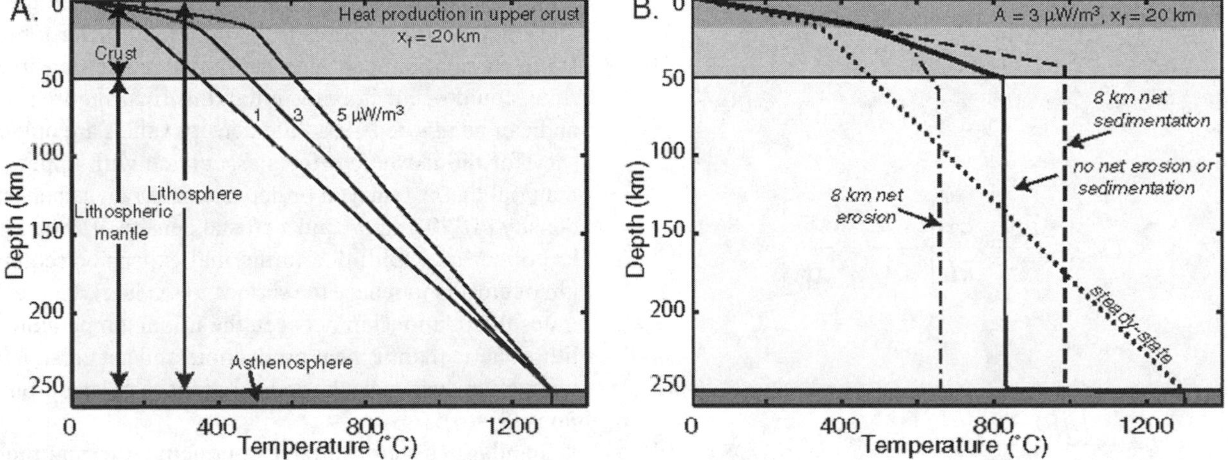

Figure 1. A) Steady-state geotherms for different radiogenic heat production concentrations (A) (values in μW/m³ associated with curves) in the upper 20 km of crust (x_f), a lithospheric thickness of 250 km, and asthenospheric temperature of 1300°C. B) Final steady-state geotherm with heat production of 3 μW/m³ in the upper 20 km of crust, with associated initial thermal profiles for 8 km of net erosion, no net erosion or sedimentation, and 8 km of net sedimentation. For each initial profile, a crustal thickness of 50 km is in equilibrium with a lithospheric mantle of uniform temperature at its base.

Knowledge of the final steady-state conductive profile and the initial lithospheric thermal structure permits calculation of the total temperature change during geotherm evolution. The mean temperature of the initial geotherm following assembly and the mean temperature of the final steady-state geotherm must be the same for no uplift or subsidence to occur. A net geotherm temperature difference between the initial and final thermal profiles induces a corresponding density change that leads to uplift or subsidence. The surface is maintained at sea level through erosion of the uplifted crust or addition of sediments to fill in the space opened by subsidence. Thus, a net heating of the lithosphere causes thermal expansion, density decrease, uplift and associated erosion at the earth's surface, while a net cooling of the lithosphere causes thermal contraction, density increase, subsidence and associated sedimentation. These erosional or sedimentation processes can further alter the lithospheric thermal structure. It is important to note that in this model surface uplift always induces erosion and subsidence always induces sedimentation, so that we commonly quantify the isostatic response as a magnitude of erosion or sedimentation. We recognize that such a causal relationship is not always true in reality, but consider this a reasonable assumption for this model.

Solutions for the mean temperature of the initial lithospheric geotherm $<T_0>$ and the mean temperature of the final lithospheric geotherm $<T_f>$ are computed by integrating equations (1) and (2), and the corresponding magnitude of erosion or sedimentation d required to maintain the surface at sea level is estimated by considering the relative densities of upper crust and asthenosphere:

$$<T_0>=T_L\left[1-\frac{(d+c)}{2L}\right]+T_s\frac{(d+c)}{L}(d+x_f)^2\left[\frac{(d+c)}{4}-\frac{(d+x_f)}{6}\right] \quad (3)$$

$$<T_f>=\frac{T_m}{2}+\frac{A}{kL}\left[\frac{x_f^2L}{4}-\frac{x_f^3}{6}\right] \quad (4)$$

$$d=\frac{L\beta\rho_m}{(\rho_m-\rho_c)}(<T_f>-<T_o>) \quad (5)$$

where β is the coefficient of thermal expansion for the lithosphere, ρ_m is the asthenospheric density, and ρ_c is the crustal density. We estimate a value of 3.3E^{-5}°C^{-1} for the coefficient

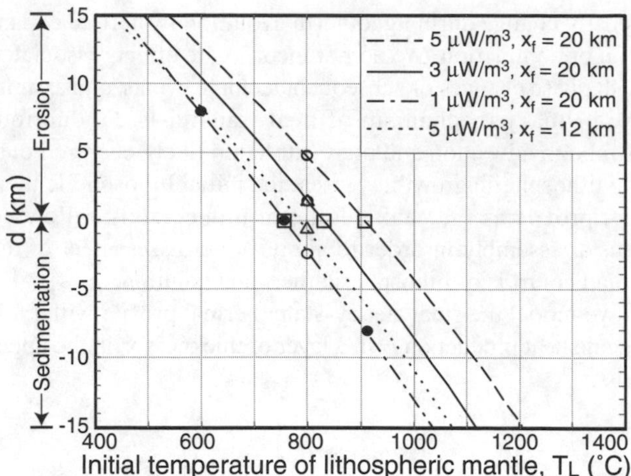

Figure 2. Net erosion or sedimentation (d) related to isostatic adjustment during geotherm evolution vs. initial temperature of emplaced lithospheric mantle (T_L). The solid, dashed, and dashed-dot curves are computed for different radiogenic heat production concentrations and a final 20 km thickness of radiogenic material. The dotted line is computed for an HPE concentration of 5 μW/m³ and a final 12 km thickness of radiogenic material. For the 1 μW/m³ curve, 8 km of net erosion, no net erosion or sedimentation, and 8 km of net sedimentation correspond to initial lithospheric mantle temperatures of 600°C, 760°C, and 915°C, and initial thicknesses of radiogenic material of 28 km, 20 km, and 12 km, respectively (solid circles). See text for additional explanation.

of thermal expansion by considering standard values for subsidence of oceanic crust as a function of distance from mid-ocean spreading ridges, and assume this value is constant with depth. Because the density change between the initial and final thermal profiles is the important factor for the isostatic calculation, complex depth-dependent density variations could occur in both initial and final thermal profiles and can be ignored. Absolute density values are only necessary for the isostatic effects associated with upper crustal material that is removed or deposited. For an asthenospheric density of 3300 kg/m³ and a crustal density of 2800 kg/m³, 1 km of surface uplift due to thermal expansion requires 6.6 km of erosion to reduce the surface to sea level. Figure 2 illustrates the relationship between the initial temperature of the lithospheric mantle, heat production, and net erosion or sedimentation during geotherm evolution, calculated from equations (3)–(5).

Standard finite difference conductive thermal modeling was exploited to constrain the time-dependent change of the geotherm from initial to steady-state conditions (example in Figure 3A). We use 2 km node spacing, linearly interpolate to obtain a smooth variation at the upper boundary position during erosion or sedimentation, and maintain a constant tem-

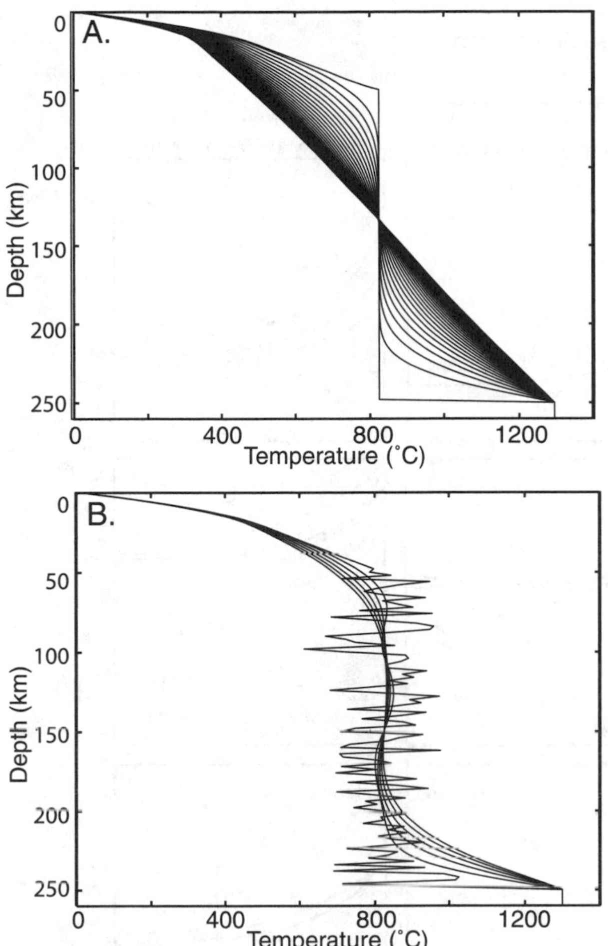

Figure 3. A) Temperature-depth-time profile showing the thermal evolution of the geotherm from initial to steady-state conditions. The geotherm is plotted in 5 m.y. increments for 150 m.y. This model uses a heat production of 3 μW/m^3, 20 km initial and final thicknesses of radiogenic material, and an initial lithospheric mantle temperature of 830°C. There is no net uplift or subsidence during geotherm evolution. B) Sensitivity test illustrating that random temperature fluctuations about a mean value decay rapidly, on the order of 10 m.y. The geotherm is plotted in 2 m.y. increments for 10 m.y. The initial conditions are the same as that in A) but the initial thermal profile of the lithospheric mantle has a standard deviation of 100°C about a mean temperature of 830°C.

perature of 1300°C at the base of the lithosphere. For simplicity, the material that is removed or deposited during erosion or sedimentation has the same HPE concentration as the upper crust. The total isostatic change calculated numerically agrees well with that determined using the analytical solution of equation (5). The numerical model's sensitivity to the simplified assumption of a uniform initial lithospheric mantle temperature is tested by examining an initial lithospheric mantle thermal structure that randomly varies about a mean temperature (Figure 3B). These initial temperature fluctuations rapidly decay within ~ 10 m.y., and suggest that a mean initial lithospheric mantle temperature is a reasonable initial thermal condition.

CONTROLS ON LITHOSPHERIC THERMAL EVOLUTION

The magnitude and distribution of heat production and the initial mean temperature of the lithospheric mantle are the primary controls on lithospheric thermal and isostatic evolution. The important influence of heat production on crustal cooling histories has been previously documented [e.g. *Morgan,* 1980; *Chapman,* 1986; *Furlong and Chapman*, 1987; *McLaren et al.,* 1999], and our discussion of heat production magnitude and distribution largely represents a review. However, these concepts have less commonly been applied to specific geologic examples, and this review is necessary to clarify our application in the following section. In addition, our analysis evaluates the isostatic consequences of heat production variation for lithospheric evolution, an aspect that has not been fully appreciated in previous studies.

We demonstrate the role of initial lithospheric temperatures, HPE concentration and HPE distribution by contrasting four final steady-state geotherms in Figure 4. For each steady-state thermal profile we evaluate cooling histories for three different initial thermal structures that undergo 8 km of net erosion, no net erosion or sedimentation, and 8 km of net sedimentation during geotherm evolution. (see CDROM in back cover sleeve) lists the parameters used in the models. We assume a total lithospheric thickness of 250 km, because seismic constraints indicate that cratonic lithosphere is typically 200–300 km thick [*James et al.,* 2001; *van der Lee,* 2002]. Thicker lithosphere decreases the mean temperature of the steady-state geotherm, because of the reduced contribution from mantle heat flow, with the reverse true for thinner lithosphere. We use our numerical model to compute the thermal evolution of rocks beginning at 16 km and 30 km lithospheric depths to explore the effects of the critical parameters on crustal cooling histories. These examples are intended only to demonstrate the basic controls on geotherm evolution, and do not necessarily represent viable geologic scenarios.

Initial Lithospheric Thermal Profile

The initial lithospheric mantle temperature exerts significant control on lithospheric thermal and isostatic histories, because it largely determines the initial mean lithospheric temperature. The cooler this initial temperature, the greater the density decrease, uplift and associated erosion during geotherm

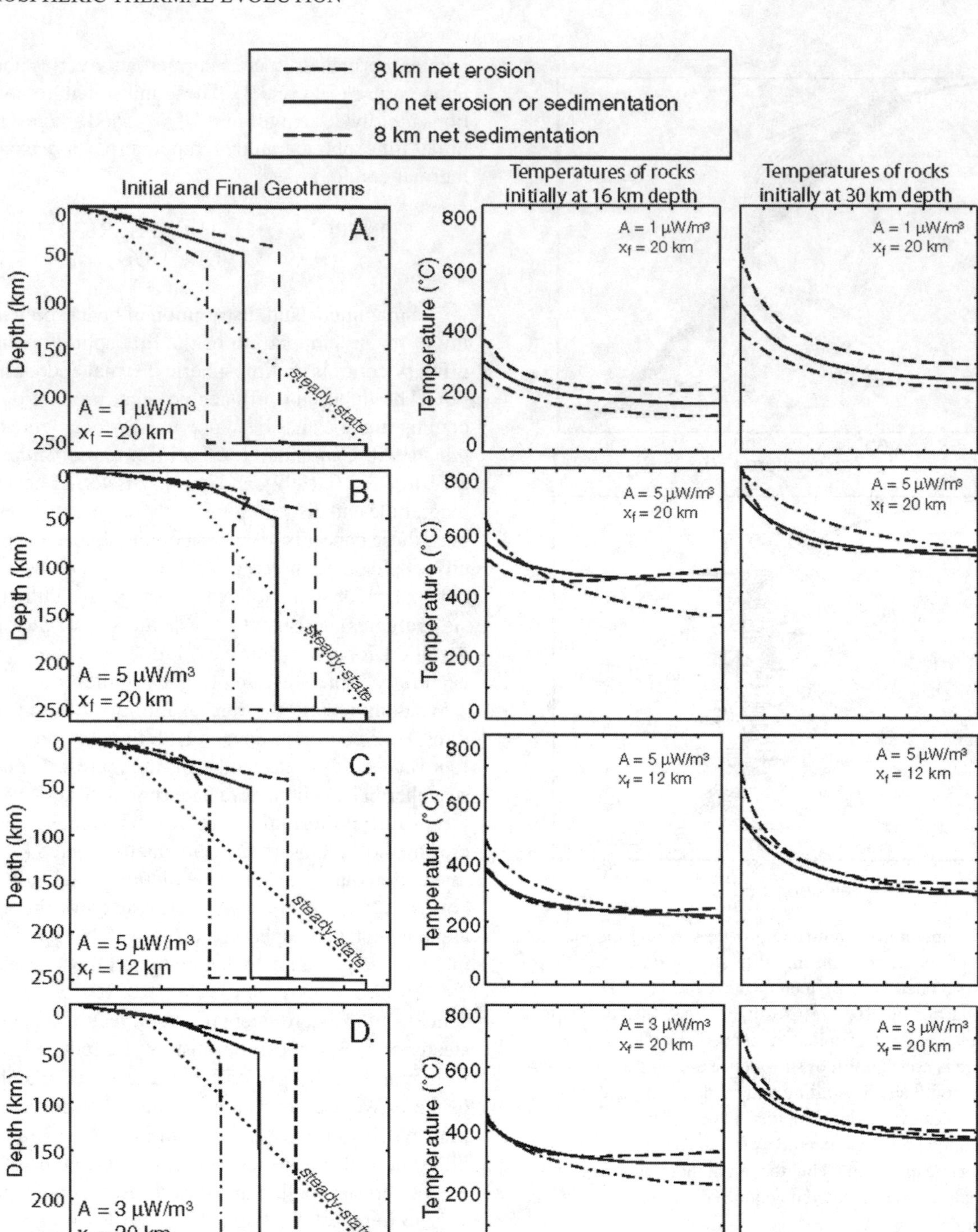

Figure 4. A) The first panel shows the final steady-state geotherm for a radiogenic heat production of 1 $\mu W/m^3$ and 20 km final thickness of the radiogenic material. Associated initial lithospheric thermal profiles for 8 km of net erosion, no net erosion or sedimentation, and 8 km of net sedimentation are also shown. Temperatures during geotherm evolution for rocks that began at 16 and 20 km crustal depths are depicted in the adjacent panels. The line styles are the same as the geotherms. B), C) and D) are similar plots, but for different steady-state geotherms (Table 1; see CDROM in back cover sleeve for a complete list of parameters used in these models). See text for additional explanation.

evolution to a given steady-state thermal profile. Figure 4A illustrates initial thermal structures that correspond to different magnitudes of erosion and subsidence during evolution to the same final steady-state geotherm with heat production of 1 $\mu W/m^3$ in the upper 20 km of crust. Initial lithospheric mantle temperatures are depicted in Figure 2 (solid circles). Thus, for the cases of 8 km of net erosion, no net erosion or sedimentation, and 8 km of net sedimentation, initial thicknesses of radiogenic material are 28 km, 20 km, and 12 km, and initial lithospheric mantle temperatures are 600°C, 760°C, and 915°C, respectively (Figure 4A).

The associated graphs of Figure 4A illustrate the thermal evolution of rocks that began at 16 km and 30 km crustal depths (different final geotherms are illustrated in Figures 4B, 4C and 4D but the same long-term patterns described below are true for all erosion, subsidence, and constant lithospheric thickness cases). An initial phase of rapid cooling can be attributed to the large temperature difference between the initial and final geotherms, such that the greater this temperature disparity at a given crustal depth, the faster the cooling. This initial temperature is dependent on our selection of initial conditions, but is insignificant for the long-term thermal history. In the uplift scenario, the crustal rocks will undergo extended cooling as they are exhumed 8 km closer to the surface. Conversely, in the subsidence case, the rocks at all crustal depths will experience long-term heating as they are transported 8 km deeper in the crust during sediment accumulation. The patterns of extended cooling or heating illustrated in Figure 4 will continue until the total magnitude of erosion or sedimentation is achieved, requiring 350–400 m.y. In contrast, crustal rocks in a lithosphere undergoing no erosion or subsidence will attain near steady-state conditions in less than 100 m.y.

Radiogenic Heat Production Concentration

Increasing the radiogenic heat production dramatically increases the mean temperature of the steady-state geotherm, as illustrated by contrasting heat productions of 1 $\mu W/m^3$ and 5 $\mu W/m^3$ for a final radiogenic layer thickness of 20 km in Figures 4A and 4B. Thus, for the case of no net uplift or subsidence, steady-state temperatures at 16 km crustal depths are ~ 150°C and ~ 450°C, respectively. In these models, higher heat production induces more rapid cooling during erosion and more rapid heating during subsidence, because the higher HPE concentration in the eroded or deposited material has a greater effect on the geotherm.

Variation in heat production induces differences in isostatic response for a given initial lithospheric mantle temperature. Higher heat production increases the mean temperature of the final steady-state geotherm, and greater net uplift and erosion occurs during geotherm relaxation. For an initial lithospheric mantle temperature of 800°C and a 20 km radiogenic layer thickness, heat productions of 1 $\mu W/m^3$, 3 $\mu W/m^3$, and 5 $\mu W/m^3$ induce 2 km of subsidence, 1.5 km of uplift, and 4.5 km of uplift, respectively (Figure 2, open circles). Alternatively, a 20 km thick HPE layer of 1 ?W/m^3, 3 ?W/m^{3}and 5 $\mu W/m^3$ correspond to initial lithospheric mantle temperatures of 760°C, 830°C and 905°C, respectively, for no net uplift or subsidence during geotherm evolution (Figure 2, open squares).

Radiogenic Heat Production Distribution

Changing the distribution of heat production in the crust can dramatically modify the lithospheric thermal profile. Distributing the same magnitude of HPEs over a greater depth increases the mean temperature of the steady-state geotherm, and maintains a higher geothermal gradient to greater depth (contrast steady-state geotherms in Figures 4C and 4D, where 60 mW of radiogenic heat is distributed over 12 km and 20 km, respectively). A shallower distribution of HPEs results in less uplift for a given initial lithospheric mantle temperature. Thus, for an initial lithospheric mantle temperature of 800°C, there is ~ 2.5 km difference in isostatic response between a 12 km thick layer of 5 $\mu W/m^3$ material and a 20 km thick layer of 3 $\mu W/m^3$ material (Figure 2, open triangles). In these models, distributing the same heat production magnitude over a greater depth slows the long-term cooling and heating rates during erosion and sedimentation, because the HPE concentration of the eroded or accumulated material is reduced and has a lesser affect on the geotherm.

APPLICATION TO THE PROTEROZOIC (1.8–1.0 GA) OROGENIC BELT OF THE SOUTHWESTERN UNITED STATES

The large control exerted by the initial lithospheric thermal structure on thermal evolution suggests that it may be possible to explain some of the diverse cooling histories preserved in different lithospheric sections solely by variations in the basic thermal parameters. We consider this hypothesis by using our model to evaluate distinct cooling records preserved in the Proterozoic (1.8–1.0 Ga) orogen of the southwestern United States. The viability of thermal evolution models that satisfy isostatic stability constraints can be evaluated by comparing model predictions of cooling histories with those constrained by detailed thermochronologic work within the geologic exposure. It is impossible to develop unique models for lithospheric thermal evolution because of the difficulty of reconstructing detailed modern geotherms using poorly constrained parameters. However, in this section we demonstrate that our approach can provide important constraints on the thermal evolution of the entire lithospheric column by using a small number of critical data on cooling histories, heat production, heat flow, and exhumation. In addition, because the effects of subsequent tectonic or magmatic events are not included in the model, geotherm evolution from the initial profile provides a useful baseline against which to consider subsequent perturbations. These testable models can be further refined with additional information on cooling records and thermal parameters.

Geologic Data

Thermochronologic data. The 1300 km wide Proterozoic basement of the southwestern United States has received much attention because it offers the opportunity to study lithospheric growth and stabilization processes [*Bowring and Karlstrom,* 1990; *Karlstrom et al.,* 2002]. The orogen is laterally segmented into distinct lithotectonic blocks with different histories, levels of exposure, and peak metamorphic conditions [*Karlstrom and Bowring,* 1988] (Figure 5). Geochronologic data indicates early assembly at 1.78 – 1.6 Ga, with subsequent magmatic activity at ~ 1.4 Ga and ~ 1.1 Ga.

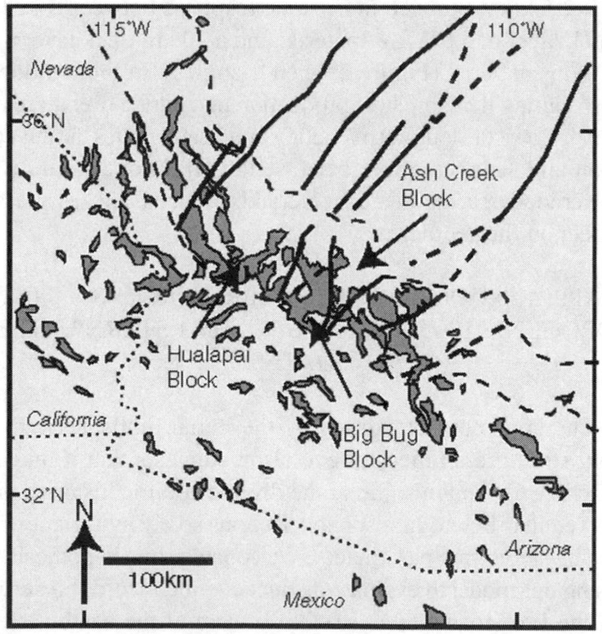

Figure 5. Geologic map of the Proterozoic orogen of the southwestern United States with exposed basement rocks in dark gray. The fault-bounded amphibolite facies Hualapai block (0.3–0.4 GPa, 500–600°C), the greenschist to amphibolite facies Big Bug block (0.3 GPa, 430–600°C), and the greenschist facies Ash Creek block (0.1–0.2 GPa) are shown. Dashed lines correspond to the boundaries between the Colorado Plateau to the northeast, the Basin and Range to the southwest, and the intervening transition zone. Dotted lines correspond to state boundaries. After *Bowring and Karlstrom,* 1990.

Detailed thermochronologic studies in the disparate crustal blocks reveal distinct cooling records following lithospheric assembly. Apparent slow cooling is recorded in both the Hualapai and Big Bug blocks. In the upper amphibolite facies Hualapai block (500–600°C, 0.3–0.4 GPa), thermochronology documents protracted cooling rates of ~ 1°C/m.y. based on U-Pb dates for metamorphic zircon (1.68 Ga), titanite (1.66 Ga), and apatite (1.52 Ga), that record cooling from 500–600°C to 400°C [*Bowring and Karlstrom,* 1990; *Chamberlain and Bowring,* 1990]. $^{40}Ar/^{39}Ar$ hornblende (1.62 Ga) and muscovite (1.1 Ma) dates [*Nyman et al.,* 1996; *Bowring and Hodges,* unpublished data], as well as thermobarometric constraints [*Williams and Karlstrom,* 1996], are also consistent with slow isobaric cooling. Within the nearby Big Bug block, metamorphic grade increases from greenschist facies (430°C, 3 kb) in the north to amphibolite facies (500–600°C, 3 kb) in the south [*Williams,* 1991]. Protracted cooling is recorded in the Horse Mountain monzogranite in the southern part of the block, with U-Pb dates for zircon (1.68 Ga), titanite (1.64 Ga), and apatite (1.47 Ga). The nearby Crazy Basin quartz monzonite records more moderate cooling rates with overlapping zircon and titanite dates at 1.7 Ga, and apatite at 1.57 Ga [*Chamberlain and Bowring,* 2000]. $^{40}Ar/^{39}Ar$ gradients in micas suggest that cooling in both of these intrusions continued at < 1°C/m.y. until ~ 1 Ga [*Hodges et al.,* 1994; *Hodges and Bowring,* 1995].

The uniformly greenschist facies Ash Creek block bounds the Big Bug block on the east, and is characterized by a markedly different cooling history. The 2 km wide Shylock shear zone, with a record of west-side up and sinistral movement, separates the two domains [*Darrach et al.,* 1991]. A quartz diorite within the Ash Creek block Cherry Creek batholith records rapid cooling following intrusion at 1.72 Ga, such that the U-Pb dates for zircon, titanite, and apatite overlap [*Chamberlain and Bowring,* 2000]. $^{40}Ar/^{39}Ar$ dates for hornblende (1.71 Ga) and biotite (1.69 to 1.62 Ga) [*Dalrymple and Lanphere,* 1971] in the Ash Creek block also indicate cooling to < 400°C within a few million years.

The distinct thermal histories preserved in the different crustal domains of this region have posed a problem for understanding lithospheric stabilization processes in the southwestern United States. In particular, the evidence for apparent protracted cooling at relatively high crustal temperatures in the Hualapai and Big Bug blocks has been enigmatic. The inferred geothermal gradient based on the thermochronologic data is at least a factor of two higher than that typically assumed for continental crust at steady-state, leading to the suggestion that hundreds of millions of years are required for the thermal stabilization of lithosphere. However, high-temperature, low-pressure metamorphism is typically attributed to transient thermal events such as plutonism [e.g. *Barton and Hanson,* 1989; *Bodorkos et al.,* 2002; *Bedard,* 2003], and the long timescales for cooling suggested by this data require another mechanism to maintain anomalously steep geotherms for hundreds of millions of years. By applying the approach outlined in this study, we develop models to test whether observed HPE variations can explain the thermal records of the crustal blocks in this region.

Heat production and heat flow data. Geochemical data, direct heat production measurements, and heat flow information are used to evaluate mean 1.7 Ga heat production in the Hualapai, Big Bug and Ash Creek blocks. Heat production decreases over time as HPEs decay, so that it is necessary to correct these values to their original concentrations during lithospheric assembly and stabilization. Refer to Table 2 for a summary of the heat production data discussed below, and for a more complete breakdown of the data by rock type.

U, Th, and K data [*Baedecker et al.,* 1998; *Dewitt written comm.* 2003] for exposed rocks of the Hualapai block are used to compute mean 1.7 Ga heat production values of 3.90 $\mu W/m^3$ for felsic intrusives (N = 196; 2.69 $\mu W/m^3$ today), 3.29 $\mu W/m^3$ for felsic volcanics (N = 97; 2.24 $\mu W/m^3$ today), and 3.12 $\mu W/m^3$ for all rocks, including mafic intrusives and mafic volcanics (N = 380; 2.14 $\mu W/m^3$ today). Direct heat production measurements reported for Hualapai rocks yield an average value of 5.2 $\mu W/m^3$ at 1.7 Ga (N = 8; 3.5 $\mu W/m^3$ today) [*Sass et al., 1994*]. Our HPE compilation and calculation of approximate 1.7 Ga heat production values cannot distinguish between the dominant suite of 1.7 Ga assembly-related plutons and later *ca.* 1.4 Ga magmatic bodies. However, because the 1.4 Ga granites were likely derived from within the crust, the 1.4 Ga granites sample HPEs that were present (but deeper) within the 1.7 Ga crustal column. Therefore, we believe using all exposed Proterozoic granites is the best method to estimate 1.7 Ga HPE concentration within the Hualapai crust. This is not an issue in the Big Bug and Ash Creek blocks because 1.4 Ga granites are not recognized in these domains.

Heat production data for the Big Bug block show a variation across the domain, correlating with rock type and metamorphic grade. Greenschist facies rocks in the northern part of the block are dominated by lower heat production metavolcanics. The metamorphic grade increases to amphibolite facies in the south where granitic intrusives are abundant. U, Th and K data [*Baedecker et al.,* 1998; *Dewitt written comm.* 2003] for all exposed rocks in the northern vs. southern parts of the block yield mean 1.7 Ga heat production values of 1.46 $\mu W/m^3$ (N = 53; 0.96 $\mu W/m^3$ today) and 2.51 $\mu W/m^3$ (N = 53; 1.66 $\mu W/m^3$ today), respectively. U, Th and K data for the slowly cooled Horse Mountain and Crazy Basin intrusions yield a mean 1.7 Ga value of 3.46 $\mu W/m^3$ (N = 11; 2.34 $\mu W/m^3$ today) [*Dewitt written comm.* 2003]. It is noteworthy that heat production values for a given rock type are relatively constant throughout the domain, but the differing proportions of exposed lithologies control the mean values for the northern and southern regions (Table 2). For example, mean 1.7 Ga heat production values in the northern vs. southern parts of the block for felsic intrusives are 3.11 $\mu W/m^3$ (N = 13; 2.03 $\mu W/m^3$ today) and 3.06 $\mu W/m^3$ (N = 38; 2.03 $\mu W/m^3$ today), respectively. It is possible that the metavolcanic-dominated greenschist facies rocks in the northern part of the block correspond to somewhat shallower crustal levels and may be underlain by a higher proportion of granitic intrusives, more comparable to the amphibolite facies southern part of the block.

Geochemical data compiled for the Ash Creek block reveal distinctly lower mean HPE concentrations than the Hualapai and Big Bug blocks. U, Th and K data from exposed rocks across the block are used to calculate a mean heat production value of 1.27 $\mu W/m^3$ at 1.7 Ga (N = 79; 0.83 $\mu W/m^3$ today) [*Baedecker et al.,* 1998; *Dewitt written comm.* 2003]. This difference is not due solely to rock type contrasts. The Cherry Creek batholith with the documented rapid cooling history yields a mean 1.7 Ga heat production of 1.08 $\mu W/m^3$ (N = 9; 0.67 $\mu W/m^3$) [*Dewitt written comm.* 2003]. Felsic intrusives across the block yield a mean 1.7 Ga value of 1.45 $\mu W/m^3$ (N = 31; 0.92 $\mu W/m^3$), contrasting with mean values of 3.12 $\mu W/m^3$ (N = 196) and 3.09 $\mu W/m^3$ (N = 106) for granitic intrusives in the Hualapai and Big Bug blocks, respectively. Direct heat production measurements indicate a 1.7 Ga value of 0.7 $\mu W/m^3$ (N = 2; 0.45 $\mu W/m^3$ today) [*Sass et al.,* 1994].

Modern mean surface heat flow data of 97 mW/m^2 for the Hualapai block (N = 8) [*Sass et al.,* 1994] and 62 mW/m^2 for the Ash Creek block (N = 2) [*Sass et al.,* 1994] are compatible with the observed heat production estimates from surface exposures. No heat flow data is reported for the Big Bug block. Applying a mantle heat flow contribution of 40 mW/m^2 (elevated above normal cratonic values to account for recent Basin and Range activity), distributing the Hualapai block modern heat production of 2.1–2.7 $\mu W/m^3$ over 21–27 km, and distributing Ash Creek block modern heat production of ~0.9 $\mu W/m^3$ over ~25 km can reasonably account for the documented variation in surface heat flow.

Thermal Analysis

Thermal stabilization. We integrate the heat production and heat flow information with the thermochronologic data to develop a suite of models that is consistent with the distinct cooling histories of the Hualapai, Big Bug and Ash Creek blocks. Seismic data suggest that the lithosphere of the southwestern United States is greater than 200 km thick [*van der Lee and Nolet,* 1997; *Dueker et al.,* 2001], and thus we apply a lithospheric thickness of 250 km for most models.

In order to explain apparent protracted cooling in the Hualapai and Big Bug blocks, we consider thermal models in which the geotherm is sufficiently hot such that steady-state temperatures at 12 km depth are always > 400°C. Recorded pressures are typically associated with peak temperatures, such that the 0.3–0.4 GPa pressures recorded in the two blocks are likely associated with orogenesis. Therefore, we assess models involving erosion or minimal isostatic change because these rocks will undergo their thermal peak early in the history.

The Hualapai and Big Bug blocks display similar heat production values in the regions of slow cooling at elevated temperatures. Therefore, we develop feasible thermal models using the Hualapai data, and consider them equally plausible for the Big Bug block. We use radiogenic heat production values of 3 $\mu W/m^3$, 4 $\mu W/m^3$, and 5 $\mu W/m^3$ to approximate the range of HPE concentration estimates for the Hualapai block and compute minimum thicknesses for the heat-producing layer of approximately 32 km, 26 km, and 22 km, respectively. These radiogenic layer thicknesses cause apparent slow cooling at 12 km crustal depths by maintaining steady-state temperatures > 400°C (Figure 6). These models require initial lithospheric mantle temperatures of 1000°C, 965°C and 940°C, respectively, for a lithosphere that undergoes no uplift or subsidence. Obviously, greater thicknesses of radiogenic material for models involving no isostatic change will result in higher steady-state temperatures. Thinner lithosphere would require lesser thicknesses of radiogenic material to attain the same elevated geotherm, with the opposite true for thicker lithosphere. For example, 200 km and 300 km lithospheric thicknesses require a 31 km and 33 km thick layer of 3 $\mu W/m^3$ heat production, respectively, to attain the same elevated geotherm at 12 km depth. Cases involving erosion during geotherm evolution will require greater thicknesses of radiogenic material and lower initial lithospheric mantle temperatures, because of cooling during exhumation of rocks as they move closer to the surface. Our analysis indicates that the 12 km of exhumation required to expose the current level of the Hualapai block did not occur during initial thermal stabilization, but rather during a later tectonic event. Temperatures > 400°C could not have been maintained in rocks that began at 12 km depth for a protracted time period if they were transported to the surface during the several hundred million years for thermal stabilization (Figure 6).

The integrated cooling rate of 25–100°C/m.y. of a magmatic body intruded into shallow crustal levels (0.1–0.2 GPa) of the Ash Creek block contrasts markedly with slow cooling in the Hualapai and Big Bug blocks. The low-grade char-

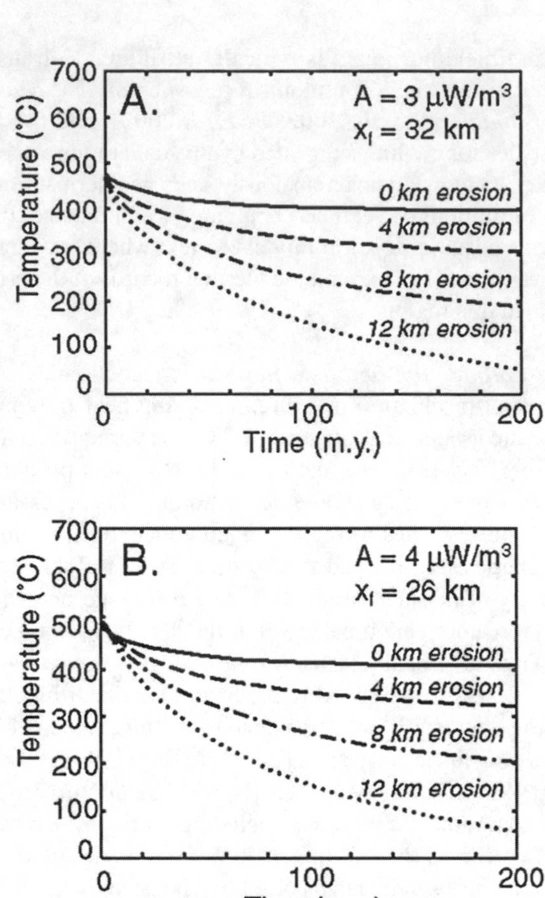

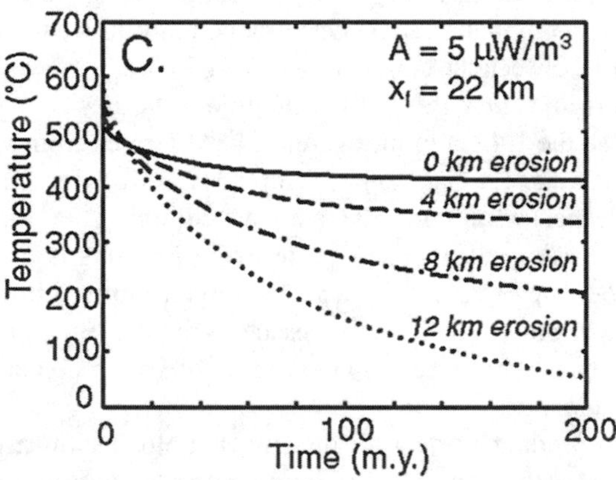

Figure 6. Thermal evolution of rocks initially at 12 km crustal depth. A) illustrates the minimum thickness of 3 $\mu W/m^3$ radiogenic material such that the model involving approximately no net uplift or subsidence maintains a steady-state crustal temperature greater than 400°C at 12 km crustal depth (solid line). The associated curves are for 4 km, 8 km, and 12 km of erosion. B) and C) are similar figures, but for different radiogenic heat production concentrations (Table 1; see CDROM in back cover sleeve for a complete list of parameters used in these models).

acter of the exposures suggests minimal isostatic change during geotherm evolution, or subsidence with erosion during a later tectonic event. The Ash Creek block does not provide tight constraints on thermal parameters, because in the shallow levels of exposure a large range of heat production magnitudes and distributions would be compatible with the rapid cooling history. The low steady-state temperatures in the country rock (always < 200°C) would cause rapid cooling of a pluton to below 400°C within a few million years of crystallization.

It would be ideal to compare the heat production and cooling histories of the different crustal blocks at the same depth, but we would predict that heat production variation should induce different isostatic responses during geotherm evolution such that they would not ultimately be exposed at the same crustal level. Thus, if our model is correct, a direct comparison is not possible using surface exposures. For example, using a 26 km thick layer of 1.3 μW/m^3 radiogenic material (approximately the 1.7 Ga Ash Creek block heat production estimate) and applying initial lithospheric mantle temperatures of 940–1000°C would induce 7–10 km of subsidence while causing rapid cooling at 6 km crustal depth (Figure 7). In contrast, this 940–1000°C temperature range in the Hualapai block induces little or no isostatic change while maintaining a sufficiently hot geotherm to satisfy its cooling history. This contrast in isostatic response is comparable to geologic exposure levels that differ by 6–10 km between the two blocks. This model requires joint exhumation by a later tectonic event.

Lateral thermal conduction across blocks characterized by distinctly different thermal regimes will be most significant for the cooling histories of rocks near domain margins. However, this is a secondary effect that we do not consider for this first order evaluation of heat production and cooling history variability between provinces. The lack of detailed thermochronologically-determined cooling histories near block boundaries currently makes this effect impossible to test. In the future, additional work to reconstruct detailed thermal records along domain margins, in tandem with two-dimensional thermal modeling, can be used to more fully evaluate the role of lateral thermal gradients.

HPE redistribution by ca. 1.4 Ga magmatism. Intrusion of anorogenic granites at *ca.* 1.4 Ga represents a major event in the southwestern United States. Major 1.42–1.40 Ga granitic plutons intruded the Hualapai block, but not the Big Bug or Ash Creek blocks. Based on the timescales for lithospheric thermal stabilization (300–400 m.y.), it is likely that near steady-state conditions had been attained in the southwestern United States prior to disturbance at *ca.* 1.4 Ga. $^{40}Ar/^{39}Ar$ data are consistent with continued slow cooling in the Hualapai block post-1.4 Ga [*Nyman et al.*, 1996; *Bowring and Hodges*, unpublished data]. Following a probable transient temperature spike due to emplacement of batholith-sized bodies of granites, the $^{40}Ar/^{39}Ar$ data suggest that the heat production distribution was appropriate to continue maintaining an elevated steady-state geotherm.

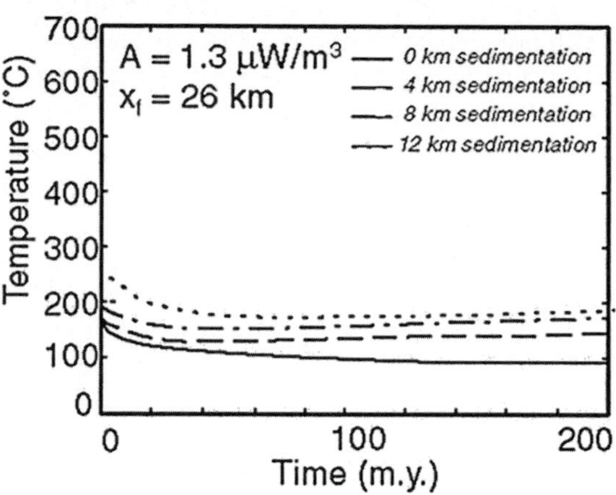

Figure 7. Thermal evolution of rocks initially at 6 km crustal depth and undergoing different magnitudes of subsidence. In all cases the temperatures are significantly below 400°C. For a 26 km thickness of 1.3 μW/m^3 radiogenic material and initial lithospheric mantle temperature of 965°C, there is approximately 8 km of net sedimentation (dash-dot curve). Note that this contrasts with Figure 6B, where a 26 km thickness of 4 μW/m^3 radiogenic material and initial lithospheric mantle temperature of 965°C induces no net uplift or subsidence. (Table 1; see CDROM in back cover sleeve) for a complete list of parameters used in these models.

We evaluate the first order impact of 1.42–1.40 Ga magmatism on the Hualapai block geotherm by considering the effects of HPE redistribution by the granites. U, Th, and K data reported for eighteen 1.4 Ga plutons across the southwestern United States [*Anderson and Bender,* 1989] yield a mean heat production value of 5.44 μW/m^3 at 1.4 Ga (4.16 μW/m^3 today). Data reported for two granites in the Hualapai block are characterized by 1.4 Ga heat production values of 5.35 μW/m^3 and 10.40 μW/m^3 (3.99 μW/m^3 and 7.86 μW/m^3, today). The highly radiogenic character of the granites suggests that this magmatism not only perturbed the geotherm on the short-term, but also had long term consequences for the steady-state structure of the Hualapai block. Derivation of the 1.4 Ga granites primarily from within the lithosphere [e.g. *Anderson,* 1983] would have redistributed HPEs to shallower lithospheric levels and caused the long-term post-1.4 Ga development of a cooler steady-state geotherm due to transport of HPEs closer to the surface (contrast Figure 4C and 4D). Thus, the documentation by $^{40}Ar/^{39}Ar$ data of a hot post-

1.4 Ga geotherm in the Hualapai block supports the U-Pb thermochronologic record of pre-1.4 Ga hot steady-state temperatures when the geotherm was even higher due to distribution of HPEs to greater depths.

APPLICATION TO OTHER GEOLOGICAL DOMAINS

"Hot" Steady-State Geotherms

Our thermal analysis suggests that the cooling histories recorded in the distinct Hualapai, Big Bug and Ash Creek blocks of the Proterozoic orogen of the southwestern United States can be explained solely by variations in heat production concentration and distribution. Analogous to the Hualapai and Big Bug blocks, it is possible that apparent slow cooling in other mid-crustal rocks can be attributed to high heat production that induces elevated steady-state geothermal gradients. For example, $^{40}Ar/^{39}Ar$ data document slow cooling in 0.3–0.4 GPa rocks in the Black Hills (1.5–1.3 Ga) [*Holm et al.,* 1997]. We compiled U, Th and K data for a variety of rocks from this area, and computed heat production values at 1.5 Ga of 2.99–5.10 $\mu W/m^3$ for three granites (1.96–4.01 $\mu W/m^3$ today) [*Gosselin et al.,* 1990; *Duke et al.,* 1992], 2.11 $\mu W/m^3$ for a trondhjemite (1.42 $\mu W/m^3$ today) [*Gosselin et al.,* 1990], 3.32 $\mu W/m^3$ for schist (2.36 $\mu W/m^3$ today) [*Gosselin et al.,* 1988], 3.60 $\mu W/m^3$ for metapelites and 3.08 for greywackes (2.56 $\mu W/m^3$ and 2.28 $\mu W/m^3$ today, respectively) [*Nabelek et al.,* 1998]. These HPE concentrations for rocks of the Black Hills are in the same range as those of the Hualapai and Big Bug blocks, and suggest that an elevated geotherm due to high heat production may also explain the maintenance of high temperatures in these 0.3–0.4 GPa rocks.

Similarly, apparent protracted cooling in a domain characterized by exceptional heat production (~ 16 $\mu W/m^3$) in the Mount Painter region of southern Australia has been attributed to elevated thermal gradients due to HPEs [*McLaren et al.*, 2002]. Highly radiogenic granites (4–10 $\mu W/m^3$ at 1.55 Ga) in the Mount Isa inlier in northern Australia have been identified as responsible for low pressure-high temperature metamorphism in the region [*McLaren et al.,* 1999]. In addition, $^{40}Ar/^{39}Ar$ dates from 1.4 – 1.0 Ga in northern New Mexico have been interpreted to represent slow cooling [*Thompson et al.,* 1996; *Karlstrom et al.,* 1997], and our analysis would predict that high heat production may also explain the thermal history of these rocks.

Heat Production Variation

Our interpretation of a sharp domainal difference in heat production between the Hualapai, Big Bug and Ash Creek blocks, consistent with both their heat flow data and cooling histories, is compatible with similar heat production variations reported for other regions. Most heat flow diversity in cratons is attributed to differences in heat production. Compilations of heat flow data around the globe document large amplitude short wavelength heat flow variations that typically coincide with boundaries between distinctive geological domains and are attributed to significant differences in bulk crustal heat production [*Jaupart and Mareschal,* 1999]. Within the Proterozoic Trans-Hudson orogen of central Canada, heat flow varies over short distances (< 50 km) across different belts of distinct compositions [*Rolandone et al.,* 2002]. A sharp increase in heat flow occurs between the Archean Kaapavaal craton and Proterozoic Natal belt of South Africa, and is attributed primarily to an increase in upper crustal heat production at the boundary [*Jones,* 1992]. Heat flow differences between the eastern and western regions of the Dharwar craton in southern India [*Gupta et al.,* 1991; *Roy and Rao,* 2000] and between the eastern and western Abitibi belt of the Superior province of Canada [*Guillou et al.,* 1994] have both been related to differences in radiogenic heat production due to varying proportions of granitic gneisses and supracrustal greenstone packages. An awareness of such heat production variability is necessary when interpreting thermochronologic records across heterogeneous packages of rocks in orogenic belts.

SUMMARY

The differences in cooling histories preserved in several crustal blocks in the Proterozoic orogen of the southwestern United States can be explained by observed variations in radiogenic heat production. High 1.7 Ga heat production for the amphibolite facies Hualapai block (3.12–5.2 $\mu W/m^3$) is consistent with the extended maintenance of high crustal temperatures in rocks at 12 km crustal depths. In the Big Bug block, heat production increases from lower values in greenschist facies metavolcanics in the north to higher values in amphibolite facies granites in the south. Apparent slow cooling recorded in the southern granites correlates with relatively high 1.7 Ga heat production values of 3.06–3.46 $\mu W/m^3$. These apparent slow cooling records and high heat production data contrast with the rapid cooling history of the Ash Creek block and associated lower 1.7 Ga mean heat production values of 1.45 $\mu W/m^3$ for felsic intrusives and 0.70 – 1.27 $\mu W/m^3$ for rocks across the domain.

High lateral variability in thermal regimes is commonly reflected in heat flow differences across geological domain boundaries in orogenic belts, and is typically correlated with heat production differences. This variability in crustal heat production concentration and distribution, and the sensitivity of crustal cooling records to this parameter, indicates that it is important to evaluate heat production data when interpreting crustal cooling records. Thus, combining high-resolution thermal histories with heat production information, heat flow data, and exhumation constraints can allow us to more fully understand the thermal structure and evolution of the lithosphere.

Acknowledgements. We thank Ed Dewitt for generously sharing his extensive geochemical dataset for the southwestern United States and thus allowing us to more fully evaluate heat production variability across this region. Critical reviews by Dennis Harry, Matt Heizler and an anonymous reviewer improved this manuscript. Constructive review by Kip Hodges and a particularly thorough reading by Carolyn Ruppel are greatly appreciated. We also thank Matt Heizler and Karl Karlstrom for helpful discussion about the thermal records and geology of the southwestern United States. This work was supported by a NSF graduate fellowship to Rebecca Flowers and by NSF Continental Dynamics (CD-ROM) grants EAR-0208215 and EAR-0003551.

REFERENCES

Anderson, J.L., Proterozoic anorogenic granite plutonism of North America, *Geol. Soc. Am. Mem, 161,* 133–154, 1983.

Anderson, J.L. and E.E. Bender, Nature and origin of Proterozoic A-type granitic magmatism in the southwestern United States of America, *Lithos, 23,* 19–52, 1989.

Baedecker, P.A., J.N. Grossman, and K.P. Buttleman, National geochemical data base: PLUTO geochemical data base for the United States: *U.S. Geological Survey digital data series, DDS-47*, 1998.

Barton, M.D. and R.B. Hanson, Magmatism and the development of low-pressure metamorphic belts: Implications from the western United States and thermal modeling, *Geol. Soc. Am. Bull., 101*, 1051–1065.

Bedard, J.H., Evidence for regional-scale, pluton-driven, high-grade metamorphism in the Archean Minto Block, Northern Superior Province, Canada, *J. of Geol, 111, 183–205,* 2003.

Bodorkos, S., M. Sandiford, N.H.S. Oliver, and P.A. Cawood, High-T, low-P metamorphism in the Paleoproterozoic Halls Creek Orogen, northern Australia: the middle crustal response to a mantle-related transient thermal pulse, *J. Met. Geol., 20,* 217–237, 2002.

Bowring, S.A. and K.E. Karlstrom, Growth, stabilization, and reactivation of Proterozoic lithosphere in the southwestern United States, *Geology, 18,* 1203–1206, 1990.

Boyd, J.R., J.J. Gurney, and S.H. Richardson, Evidence for a 150–200 km thick Archean lithosphere from diamond inclusion thermobarometry, *Nature, 315,* 387–389, 1985.

Chamberlain, K.R. and S.A. Bowring, Proterozoic geochronologic and isotopic boundary in northwest Arizona, *J. Geol., 98*, 399–416, 1990.

Chamberlain, K.R. and S.A. Bowring, Apatite-feldspar U-Pb thermochronometer: a reliable, mid-range (~450 °C), diffusion-controlled system, *Chem. Geol., 172,* 73–200, 2000.

Chapman, D.S., Thermal gradients in the continental crust, in *The Nature of the Lower Continental Crust*, edited by J.B. Dawson, D.A. Carswell, J. Hall and K.H. Wedepohl, *Geol. Soc. Spec. Pub., 24*, Oxford University Press, London, 63–70, 1986.

Chapman, D.S. and H.N. Pollack, Regional geotherms and lithospheric thickness, *Geology, 5,* 265–268, 1977.

Dalrymple, G.B. and M.A. Lanphere, $^{40}Ar/^{39}Ar$ technique of K-Ar dating; a comparison with the conventional technique, Earth Planet. Sci. Lett., 12, 300–308, 1971.

Darrach, M.E., K.E. Karlstrom, D.M. Argenbright, and M.L. Williams, Progressive deformation in the early Proterozoic Shylock shear zone, central Arizona, in *Proterozoic Geology and Ore Deposits of Arizona,* edited by K.E. Karlstrom, *Arizona Geological Society Digest, 19, Arizona Geological Society,* Tucson, AZ, 11–26, 1991.

deWit, M.J., C. Roering, R.J. Hart, R.A. Armstrong, C.E.J. de Ronde, R.W.E. Green, M. Tredoux, E. Peberdy, and R.A. Hart, Formation of and Archean continent, *Nature, 357,* 553–562, 1992.

Dueker, K., H. Yuan, and B. Zurek, Thick-structure Proterozoic lithosphere of the Rocky Mountain region, *GSA Today, 11,* 4–9, 2001.

Duke, E.F., J.J. Papike, and J.C. Laul, Geochemistry of a boron-rich peraluminous granite pluton: The Calamity Peak layered granite-pegmatite complex, Black Hills, South Dakota, *Can. Min., 30,* 811–833, 1992.

Furlong, K.P. and D.S. Chapman, Thermal state of the lithosphere, *Rev. of Geophys., 25,* 1255–1264, 1987.

Gosselin, D.C., J.J. Papike, C.K. Shearer, Z.E. Peterman, and J.C. Laul, Geochemistry and origin of Archean granites from the Black Hills, South Dakota, *Can. J. Earth Sci., 27,* 57–71, 1990.

Gosselin, D.C., J.J. Papike, R.E. Zartman, Z.E. Peterman, and J.C. Laul, Archean rocks of the Black Hills, South Dakota: Reworked basement from the southern extension of the Trans-Hudson orogen, *Geol. Soc. Am. Bulletin*, 100, 1244–1259, 1988.

Guillou, L., J.C. Mareschal, C. Jaupart, C. Gariepy, G. Bienfait, and R. Lapointe, Heat flow, gravity and structure of the Abitibi belt, Superior Province, Canada: Implications for mantle heat flow, *Earth Planet. Sci. Lett., 122,* 102–123, 1994.

Gupta, M.L., A. Sundar, and S.R. Sharma, Heat flow and heat generation in the Archaean Dharwar cratons and implications for the southern Indian Shield geotherm and lithospheric thickness, *Tectonophysics, 194,* 107–122, 1991.

Herzberg, C.T., Lithospheric peridotites of the Kaapvaal craton, *Earth Planet. Sci. Lett., 120,* 13–29, 1993.

Hodges, K.V. and S.A. Bowring, S.A, $^{40}Ar/^{39}Ar$ thermochronology of isotopically zoned micas: Insights from the southwestern USA Proterozoic orogen, *Geochim. Cosmochim. Acta, 59,* 3205–3220, 1995.

Hodges, K.V., W.E. Hames, and S.A. Bowring, $^{40}Ar/^{39}Ar$ age gradients in micas from a high-temperature-low-pressure metamorphic terrain: Evidence for very slow cooling and implications for the interpretation of age spectra, *Geology, 22,* 55–58, 1994.

Holm, D.K., P.S. Dahl, and D.R. Lux, $^{40}Ar/^{39}Ar$ evidence for Middle Proterozoic (1300–1500 Ma) slow cooling of the southern Black Hills, South Dakota, mid-continent, North America: Implications for Early Proterozoic P-T evolution and posttectonic magmatism, *Tectonics, 16*, 609–622, 1997.

Huerta, A.D. and L.H. Royden, The effects of accretion, erosion and radiogenic heat on the metamorphic evolution of collisional orogens, *J. Met. Geol., 17*, 349–366, 1999.

Huerta, A.D., L.H. Royden, and K.V. Hodges, The interdependence of deformational and thermal processes in mountain belts, *Science, 273,* 637–639, 1996.

Irvine, G.J., D.G. Pearson, and R.W. Carlson, Lithospheric mantle evolution of the Kaapvaal Craton: A Re-Os isotope study of peridotite xenoliths from Lesotho kimberlites, *Geophys. Res. Lett., 28,* 2505–2508, 2001.

James, D.E., M.J. Fouch, J.C. VanDecar, S. van der Lee, and Kaapvaal Seismic Group, Tectospheric structure beneath southern Africa: *Geophys. Res. Lett., 28*, 2485–2488, 2001.

Jaupart, C. and J.C. Mareschal, The thermal structure and thickness of continental roots, *Lithos, 48,* 93–114, 1999.

Jones, M.Q.W., Heat flow anomaly in Lesotho: Implications for the southern boundary of the Kaapvaal craton, *Geophys. Res. Lett., 19,* 2031–2034, 1992.

Jordan, T.H., Composition and development of continental tectosphere, *Nature, 274,* 745–750, 1978.

Karlstrom, K.E. and S.A. Bowring, Early Proterozoic assembly of tectonostratigraphic terranes in southwestern North America, *J. Geol., 96*, 561–576, 1988.

Karlstrom, K.E., S.A. Bowring, K.R. Chamberlain, K.G. Dueker, T. Eshete, E.A. Erslev, G.L. Farmer, M. Heizler, E.D. Humphreys, R.A. Johnson, G.R. Keller, S.A. Kelley, A. Levander, M.B. Magnani, J.P. Matzel, A.M. McCoy, K.C. Miller, E.A. Morozova, F.J. Pazzaglia, C. Prodehl, H,-M.

Pumper, C.A. Shaw, A.F. Sheeham, E. Shoshitaishvili, S.B. Smithson, C.M. Snelson, L.M. Stevens, A.R. Tyson, and M.L. Williams, Structure and evolution of the lithosphere beneath the Rocky Mountains: Initial results, *GSA Today, 12,* 4–9, 2002.

Karlstrom, K.E., R.D. Dallmeyer, and J.A. Grambling, $^{40}Ar/^{39}Ar$ evidence for 1.4 Ga regional metamorphism in New Mexico: Implications for thermal evolution of lithosphere in southwestern USA, *J. of Geol., 105,* 205–223, 1997.

McLaren, S., M. Sandiford, and M. Hand, High radiogenic heat-producing granites and metamorphism – An example from the western Mount Isa inlier, Australia, *Geology, 27,* 679–682, 1999.

McLaren, S., W.J. Dunlap, M. Sandiford, and I. McDougall, Thermochronology of high heat-producing crust at Mount Painter, South Australia: Implications for tectonic reactivation of continental interiors, *Tectonics, 21,* 2:1–18, 2002.

Morgan, P., The thermal structure and thermal evolution of the continental lithosphere, *Phys. Chem. Earth, 15,* 107–193, 1980.

Morgan, P., Crustal radiogenic heat production and the selective survival of ancient continental crust, *J. Geophys. Res., 90,* C561–C570, 1985.

Nabelek, P.I. and C.D. Bartlett, Petrologic and geochemical links between the post-collisional Proterozoic Harney Peak leucogranite, South Dakota, USA, and its source rocks, *Lithos, 45,* 71–85, 1998.

Nyman, M.W., K.E. Karlstrom, M.L. Williams, and M.T. Heizler, Metamorphic and geochronologic evidence for ca. 500 m.y. of midcrustal residence of Proterozoic rocks, Hualapai Mountains, NW Arizona, Geological Society of America, 28th Annual Meeting 28, *Geological Society of America, Boulder, CO, 375,* 1996.

Pearson, D.G., Evolution of cratonic lithospheric mantle: an isotopic perspective, in *Mantle Petrology: Field Observations and High Pressure Experimentation: A Tribute to Francis R. Boyd,* edited by Y. Fei, C.M. Bertka and B.O. Mysen, *Geochem. Soc. Spec. Publ. No 8,* 57–78, 1995.

Pearson, D.G., The age of continental roots, *Lithos, 48,* 171–194, 1999.

Richardson, S.H., J.J. Gurney, A.J. Erlank, and J.W. Harris, Origin of diamonds in old enriched mantle, *Nature, 310,* 198–202, 1984.

Richardson, S.H., J.W. Harris, and J.J. Gurney, Three generations of diamonds from old continental mantle, *Nature, 366,* 256–258, 1993.

Rolandone, C., C. Jaupart, J.C. Mareschal, C. Gariepy, G. Bienfair, C. Carbonne, and R. Lapointe, Surface heat flow, crustal temperatures and mantle heat flow in the Proterozoic Trans-Hudson Orogen, Canadian Shield, *J. Geophys. Res., 197(B12),* ETG7 1–19, 2002.

Roy, S. and R.U.M. Rao, Heat flow in the Indian shield, *J. Geophys. Res., 105(B11),* 25,587–25,604, 2000.

Royden, L.H., The steady state thermal structure of eroding orogenic belts and accretionary prisms, *J. Geophys. Res., 98(B3),* 4487–4507, 1993.

Rudnick, R.L. and D.M. Fountain, Nature and composition of the continental crust: A lower crustal perspective, *Rev. in Geophys., 33,* 267–309, 1995.

Rudnick, R.L., W.F. McDonough, and R.J. O'Connell, Thermal structure, thickness and composition of the continental lithosphere, *Chem. Geol., 145,* 395–411, 1998.

Rudnick, R.L. and A.A. Nyblade, The thickness and heat production of Archean lithosphere: Constraints from xenolith thermobarometry and surface heat flow, in *Mantle Petrology: Field Observations and High Pressure Experimentation: A Tribute to Francis R. Boyd,* edited by Y. Fei, C.M. Bertka and B.O. Mysen, *Geochem. Soc. Spec. Publ. No 8,* 3–12. 1999.

Sandiford, M. and M. Hand, Australian Proterozoic high-temperature, low-pressure metamorphism in the conductive limit, in *What Drives Metamorphism and Metamorphic Reactions?*, edited by P.J. Treloar and P.J. O'Brien, *Geol. Soc. of London, Spec. Pub., 138,* 109–120, 1998.

Sandiford, M. and S. McLaren, Tectonic feedback and the ordering of heat-producing elements within the continental lithosphere, *Earth Planet. Sci. Lett, 204,* 133–150, 2002.

Sandiford, M., S. McLaren, and N. Neumann, Long-term thermal consequences of the redistribution of heat-producing elements associated with large-scale granitic complexes, *J. Metamorphic Geol., 20,* 87–98, 2002.

Sass, J.H., A.H. Lachenbruch, S.P. Galanis Jr., P. Morgan, S.S. Priest, T.H. Moses Jr., and R.J. Munroe, Thermal regime of the southern Basin and Range Province: 1. Heat flow data from Arizona and the Mojave Desert of California and Nevada, *J. Geophys. Res., 99,* 22,093–22,119, 1994.

Shirey, S.B., R.W. Carlson, S.H. Richardson, A. Menzies, J.J. Gurney, D.G. Pearson, J.W. Harris, and U. Weichert, Archean emplacement of eclogitic components into the lithospheric mantle during formation of the Kaapvaal Craton, *Geophys. Res. Lett., 28,* 2509–2512, 2001.

Shirey, S.B., Harris, J.W., Richardson, M.J., James, D.E., Cartigny, P., Denies, P., and Viljoen, F., Diamond genesis, seismic structure, and evolution of the Kaapvaal-Zimbabwe craton: *Science, 297, 1683–1686,* 2002.

Thompson, A.G., J.A. Grambling, K.E. Karlstrom, and R.D. Dallmeyer, Mesoproterozoic metamorphism and $^{40}Ar/^{39}Ar$ thermal history of 1.4 Ga Priest Pluton, Manzano Mountains, New Mexico, *J. of Geol., 104,* 583–598, 1996.

van der Lee, S., High-resolution estimates of lithospheric thickness from Missouri to Massachusetts, USA, *Earth Planet. Sci. Lett, 203,* 15–23, 2002.

van der Lee, S. and G. Nolet, Upper mantle S velocity structure of North America, *J. Geophys. Res., 102,* 22,815–22,838, 1997.

Williams, M.L., Overview of Proterozoic metamorphism in Arizona, in *Proterozoic Geology and Ore Deposits of Arizona,* edited by K.E. Karlstrom, *Arizona Geological Society Digest, 19, Arizona Geological Society,* Tucson, AZ, 11–26, 1991.

Williams, M.L. and K.E. Karlstrom, Looping P-T paths and high-T, low-P middle crustal metamorphism: Proterozoic evolution of the southwestern United States, *Geology, 24,* 1119–1122, 1996.

S.A. Bowring, Department of Earth, Atmospheric and Planetary Sciences, Massachusetts Institute of Technology, 77 Massachusetts Ave., 54-1114, Cambridge, MA 02139.

R.M. Flowers, Department of Earth, Atmospheric and Planetary Sciences, Massachusetts Institute of Technology, 77 Massachusetts Ave., 54-1114, Cambridge, MA 02139 (email: rflowers@mit.edu).

L.H. Royden, Department of Earth, Atmospheric and Planetary Sciences, Massachusetts Institute of Technology, 77 Massachusetts Ave., 54-1114, Cambridge, MA 02139,

Contrasting Lower Crustal Evolution Across an Archean-Proterozoic Suture: Physical, Chemical and Geochronologic Studies of Lower Crustal Xenoliths in Southern Wyoming and Northern Colorado

G. Lang Farmer[1], Samuel A. Bowring[2], Michael L. Williams[3],
Nikolas I. Christensen[4], Jennifer P. Matzel[2] and Liane Stevens[3]

Crustal xenoliths from Quaternary ultrapotassic volcanic rocks at Leucite Hills in southern Wyoming and Devonian State Line kimberlite diatremes in northern Colorado have been examined to compare and contrast Archean and Paleoproterozoic lower continental crust across the Proterozoic Cheyenne Belt. The Leucite Hills xenoliths are dominantly mafic (3.5 to 13.5 wt % MgO), one or two-pyroxene (garnet absent) hornblende granulites. Peak metamorphic conditions (1.1 to 1.3 GPa and ca. 800°C) and the high xenoliths densities (2.7 to 3.1 g/cm^3) reveal that the xenoliths are lower crustal. U-Pb zircon and Nd model ages both indicate that the mafic crust is Late Archean (~2.6 Ga). The State Line xenoliths are Paleoproterozoic, two pyroxene garnet granulites with major element compositions and peak metamorphic conditions (1.1 – 1.2 GPa, 625 – 830°C) similar to the Leucite Hills xenoliths, but with higher Vp and Vs (6.6 to 7.2 km/s and 3.7 to 3.9 km/s vs. 6.2 to 6.9 km/s and 3.5 to 3.9 km/s) due to presence of garnet. Some State Line xenoliths also have low La/Yb_N (0.4 to 0.9) and likely represent restite remaining after partial crustal melting at ~1.4 Ga. These observations reveal that mafic lower crust north of the Cheyenne Belt has remained hydrous and largely unperturbed thermally since the Late Archean while Paleoproterozoic, less hydrous, lower crust in northern Colorado underwent heating and partial melting after its formation. We speculate that the Archean lower crust was shielded from the thermal events that extensively affected the Paleoproterozoic crust by thick, buoyant Archean mantle lithosphere.

[1]Department of Geological Sciences and CIRES, University of Colorado, Boulder, Colorado
[2]Department of Earth, Atmospheric and Planetary Sciences, Massachusetts Institute of Technology, Cambridge, Massachusetts
[3]Department of Geosciences, University of Massachusetts, Amherst, Massachusetts
[4]Department of Geology and Geophysics, University of Wisconsin, Madison, Wisconsin

The Rocky Mountain Region: An Evolving Lithosphere
Geophysical Monograph Series 154

10.1029/154GM11

INTRODUCTION

Physical and chemical studies of lower crustal xenoliths have become a standard part of integrated geologic, geochemical and seismic studies of the deep continental lithosphere [*Ducea and Saleeby*, 1996; *Carlson et al.*, 2000; *Corriveau and Morin*, 2000; *Holtta et al.*, 2000; *Schmitz and Bowring*, 2003]. In this paper we present the initial results of combined petrologic, geochemical, geochronologic and physical studies of the lower continental crust across the Cheyenne Belt in southern Wyoming, a primary target of the CD-ROM seismic experiments. One objective of the report is to summarize the physical properties of the lower crust beneath this region in order to aid in matching seismic velocities for the deep crust determined from seismic reflection and refraction studies to specific crustal lithologies. Of particular interest is to establish which lithologies in the lower crust could correspond to the moderately high velocity (7.xx km/sec) lower crustal layer identified beneath at least a portion of the study area [*Karlstrom et al.*, 2002; *Snelson et al.,* this volume]. However, the xenoliths also afford an opportunity to investigate the long-term geologic evolution of the deep crust across a major suture between Archean and Proterozoic continental lithosphere. As a result, our studies focus on determining if major differences in the chemical compositions of the lower crust exist across this boundary, and, if so, whether these compositional differences relate to the original processes of crust formation in the Archean and Proterozoic, or reflect later modifications of the lower crust. For example, have episodes of mafic underplating affected either the Proterozoic or Archean lithosphere?

PREVIOUS WORK

Previous petrologic and geochemical studies of mafic and ultramafic, upper mantle and lower crustal xenoliths present in the Rocky Mountain region have been reviewed elsewhere [*Eggler et al.*, 1987*a*; *Lester and Farmer*, 1998; *Selverstone et al.*, 1999] and so only a brief overview of the work pertaining to our studies of mafic xenoliths in the vicinity of the Cheyenne Belt is given here. Directly north of the Cheyenne Belt, the primary occurrences of crustal xenoliths are in the 3.0 to 0.89 Ma lamproitic lavas and volcanic necks that comprise the Leucite Hills volcanic field in south-central Wyoming (Figure 1) [*Lange et al.*, 2000]. The lower crustal xenoliths include mafic granulites, granite, anorthosite, amphibolite, and dunite, as well as unusual pyroxene-plagioclase-mica bearing nodules [*Kay et al.*, 1978; *Speer*, 1985]. None of these xenoliths had been characterized in any detail prior to this study, although the granulites xenoliths were previously known to be mafic to intermediate in composition and moderately light rare earth element (LREE) enriched with a mineral assemblage consisting of plagioclase -orthopyroxene -clinopyroxene +/- hornblende +/-biotite (garnet absent) [*Kay et al.*, 1978].

Directly south of the Cheyenne Belt, mantle and crustal xenoliths are found in kimberlitic dikes and diatremes of the Iron Mountain (Wyoming) and State Line kimberlite districts (Figure 1). Kimberlite emplacement in the State Line district

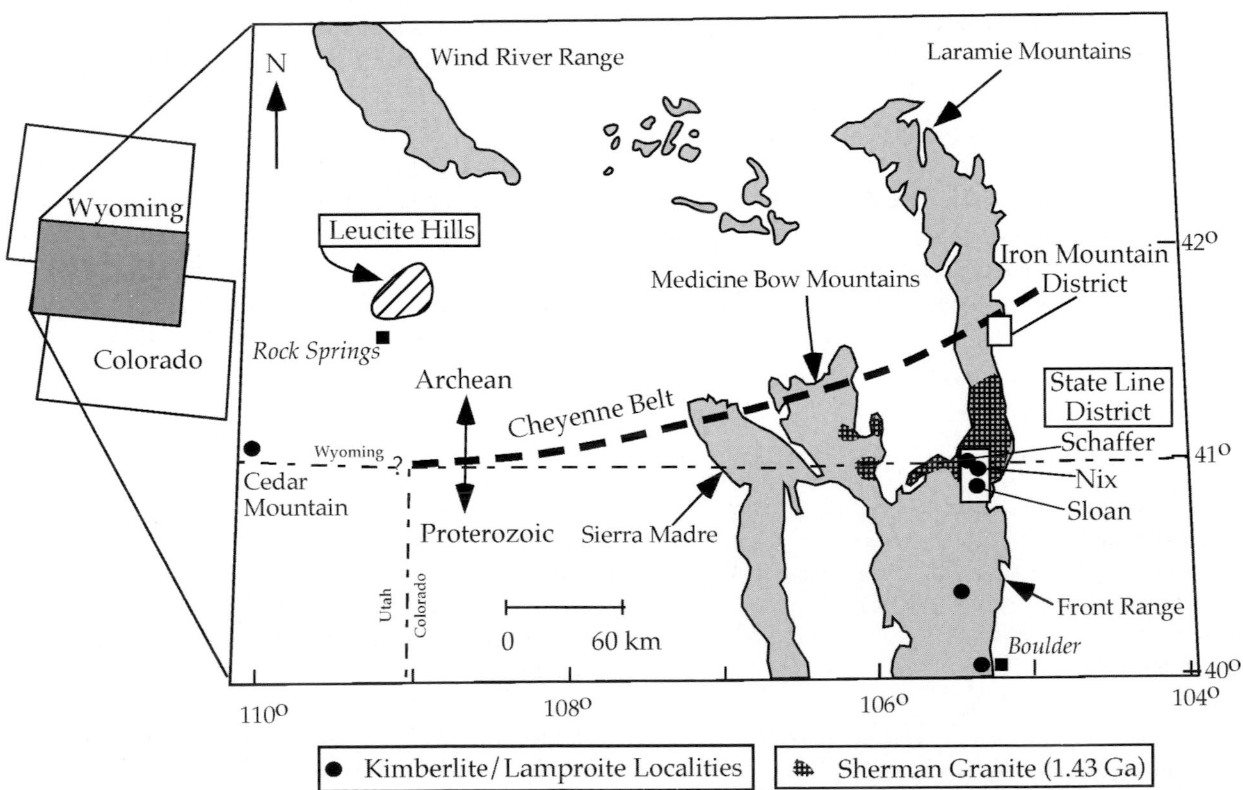

Figure 1. Location map for State Line diatremes and Leucite Hills. Modified from *Hausel* [1998].

apparently took place in two discrete events, in the Neoproterozoic and Devonian [*Lester et al.*, 2001], but only the xenoliths from the Devonian Sloan 2 and Nix pipes (Figure 1) lack extensive serpentinization, silicification, and/or carbonatization [*Bradley*, 1985; *Eggler et al.*, 1987*b*]. Lower crustal xenoliths in the State Line diatremes are primarily mafic granulites, along with less abundant anorthosite, gabbronorite, and monzogabbro [*Bradley and McCallum*, 1984; *Bradley*, 1985; *Stevens*, 2002]. The mafic granulites consist of variable proportions of plagioclase, clinopyroxene, and orthopyroxene +/- garnet and amphibole, with minor quartz (Qtz), alkali feldspar, rutile, ilmenite, zircon and apatite. Based on their trachybasalt bulk compositions, and moderate LREE enrichment, *Bradley* [1985] interpreted the mafic granulites as the crystallized equivalent of tholeiitic basaltic magmas that underplated the continental crust in this region at some point between the time of crust formation (~1.7 Ga) and Devonian kimberlite emplacement.

Overall, the available data suggest that the lowermost crust beneath both the Archean and Paleoproterozoic sides of the Cheyenne Belt is of mafic composition. However, no previous attempts have been made to define why differences in metamorphic mineral assemblages in the lower crust exist across the Cheyenne Belt (garnet absent north of Cheyenne Belt but present to the south), to determine the age and physical properties of the lower crust, itself, or to use major, trace element, and isotopic data to fully test the hypotheses that the mafic lower crust may represent the products of underplating of the continental crust.

SAMPLES/ANALYTICAL PROCEDURES

We obtained petrologic, mineral composition, isotopic, U-Pb geochronologic, and physical property data from xenoliths on both sides of the Cheyenne Belt. North of the Cheyenne Belt, a new collection of mafic to intermediate composition xenoliths was obtained from Hatcher Mesa, a 0.96 Ma ultrapotassic lava flow remnant at Leucite Hills [*Lange et al.*, 2000]. Eleven such xenoliths were chosen for detailed study (Table 1; see CDROM in back cover sleeve).

Twelve mafic xenoliths from the State Line diatremes were obtained from the collection of Dr. Malcolm McCallum, including examples of both granulites and eclogites from the Nix, Shaeffer and Sloan diatremes (Figure 1; Table 1; see CDROM in back cover sleeve). *Bradley* [1985] previously undertook petrographic and geochemical studies of the Sloan 2 (SD2) and Nix xenoliths. We obtained major and trace element (mineral and whole rock) compositions from these xenoliths (Table 1, 2, 5, 6; see CDROM in back cover sleeve), as well as Nd and Sr isotopic compositions (Table 3; see CDROM in back cover sleeve) and zircon U-Pb ages (Table 4; see CDROM in back cover sleeve). We also report zircon U-Pb ages from several granitic and amphibolite xenoliths obtained from the Sloan diatreme (SD3 xenoliths; Table 4a; see CDROM in back cover sleeve). These xenoliths were studied in order to provide a better assessment of the distribution of ages throughout the crustal column in the State Line area. We presume that these xenoliths were derived from shallower levels in the crust than the State Line mafic granulites, but they were not further characterized either chemically or petrographically.

Major and Trace Element Concentrations and Nd and Sr Isotopic Data

Major and trace element compositions of xenoliths from both localities were obtained from unleached whole rock powders by ICP-MS (Table 1, 2; see CDROM in back cover sleeve). Whole rock Nd and Sr isotopic measurements were obtained at the University of Colorado on both unleached and 1.5 N HCl leached powders. Leaching was undertaken in order to remove secondary components introduced to the xenoliths during interaction with their host kimberlites [*Rudnick*, 1992]. Chemical separations and isotopic measurements were undertaken at the University of Colorado, Boulder, using techniques described elsewhere [*Farmer et al.*, 1991]. Neodymium and strontium isotopic data were also obtained from unleached powders of clinopyroxene and plagioclase obtained through physical microdrilling (using a Merchantek® Micromilling system) for two State Line xenoliths (NX4 LC-1, SD2LC-71), and from one clinopyroxene mineral separate (from NX4-LC-1). The clinopyroxene separate was leached in dilute HCl and HF using the technique of *Zindler and Jagoutz* [1988].

Mineral Chemistry (EMP)

Electron microprobe analyses were carried out on polished thin sections at the University of Massachusetts on a Cameca SX50 electron microprobe (Table 5, 6; see CDROM in back cover sleeve). Analyses were done at 15kV and 15 nA sample current using natural silicate standards. Wavelength dispersive high-resolution compositional maps were made of individual metamorphic minerals (Grt, Cpx, Opx, Hbl, Pl; abbreviations after *Kretz*, [1983]) and of matrix domains in order to evaluate compositional zoning. Quantitative point and traverse locations were selected from compositional maps in such a way as to characteristic petrologically significant compositions (i.e. core, near-rim, rim etc.). Analyses were reduced first using the online PAP routine [*Pouchou and Pichoir*, 1985] and then refined using the Datcon software [*M.L. Williams*, unpublished]. Details of specific analyses and justification for selections of relevant compositions are presented below.

U-Pb Zircon Age Determinations

Zircon was isolated from xenoliths by standard crushing, heavy liquid, and magnetic separation techniques, and separated into different populations based on crystal morphology, color, and grain size. The zircon grains were air-abraded with pyrite after the method of *Krogh* [1982] and acid rinsed in warm 3M HNO_3 for 12 hours, followed by ultrasonication. Zircon was loaded into Teflon FEP microcapsules and washed again in 3M HNO_3 at 50°C for 2–4 hours, followed by rinsing with several capsule volumes of water. Samples were spiked with a mixed ^{205}Pb-^{233}U-^{235}U tracer and dissolved in 29M HF at 220°C for 48–120 hours followed by conversion to 6M HCl at 180°C for 18–24 hours. Pb and U were separated from mineral solutions using miniaturized HCl anion exchange chromatography procedures modified after *Krogh* [1973].

Lead and uranium were analyzed on the MIT VG Sector 54 thermal ionization multicollector mass spectrometer at M.I.T. Lead was loaded on single Re filaments with a dilute silica gel-0.1M H_3PO_4 emitter solution. Radiogenic Pb was measured either dynamically with four high-mass Faraday cups and an axial ion-counting Daly detector, peak-switching ^{205}Pb into the axial position to obtain an internal Daly-Faraday gain calibration, or by peak-switching all ion beams into the Daly detector for very small amounts of lead. An ion beam of $>0.1x10^{-13}$ A was maintained for ^{207}Pb during data acquisition. Uranium was loaded on single Re filaments either with colloidal graphite and measured as metal ions, or with silica gel and measured as UO_2^+ ions by one of two methods: in static mode on three Faraday cups for $^{238}U^+$ ($^{238}U^{16}O_2^+$) ion-beam intensities of $>0.5x10^{-13}$ A, or by peak switching all ion beams into the Daly detector for smaller amounts of uranium. Details of fractionation and blank corrections are given in Table 4 (see CDROM in back cover sleeve). Ages with propagated uncertainties were calculated using the methodology of Ludwig [1980].

Physical Property Measurements

Seventeen crustal xenoliths from the Leucite Hills and State Line xenolith suites were selected for seismic velocity measurements (Table 7, 8; see CDROM in back cover sleeve). Cylindrical cores taken from each sample were trimmed, weighed, and measured to calculate bulk densities (Table 7; see CDROM in back cover sleeve) and jacketed with copper foil for velocity measurements as a function of pressure. Two samples from Leucite Hills (LHHE-26 and LHHW-1) were large enough to obtain two perpendicular cores to investigate anisotropy. The travel times of compressional (Vp) and shear (Vs) waves were measured using a pulse transmission technique [*Christensen*, 1985], and velocities were calculated from these times and the sample lengths. Natural resonant frequencies of the transducers were 1 MHz. Measurements were performed at elevated hydrostatic confining pressure to 1GPa (equivalent to depths of ~35 km).

RESULTS

Petrography/Phase Relationships

Lower crustal xenoliths from the State Line Kimberlite District are typically fine-grained, mafic granulites [*Bradley and McCallum*, 1984; *Bradley*, 1985; *Stevens*, 2002]. The xenoliths contain various combinations of clinopyroxene, orthopyroxene, garnet, and hornblende with minor amounts of K-feldspar, ilmenite, rutile, zircon, and barite. Most samples lack a strong foliation or lineation, but several exhibit a weak preferred orientation of pyroxene and hornblende. Three general assemblages have been sampled: Cpx-Opx-Pl-Qtz; Cpx-Opx-Grt-Pl-Qtz; Cpx-Grt-Pl-Qtz. The two-pyroxene granulites tend to be equigranular with smooth, curving grain boundaries that commonly meet in 120° triple junctions (Plate 1). Garnet-bearing granulites have more complex textures, many preserving reaction and corona relationships (petrographic descriptions of three of these xenoliths are given in the Appendix). Most State Line xenoliths contain textural evidence for early Pl-Opx-Il assemblages overprinted by later Grt-Cpx-Pl assemblages (Figure 2). At least some ductile deformation occurred during or after the development of the later assemblage as suggested by the weak alignment of product Cpx and Pl.

Assemblages, textures, and mineral compositions of the State Line xenoliths are consistent with the production of garnet and clinopyroxene at the expense of early orthopyroxene and plagioclase. This model reaction can be written as:

$$\text{Opx} + \text{Pl1} = \text{Grt} + \text{Cpx} + \text{Pl2} \quad (1)$$

where Pl1 and Pl2 are a more calcic and a more sodic plagioclase, respectively. This reaction is similar to the reaction that defines the boundary between the medium- and high-pressure granulite fields [*Green and Ringwood*, 1967].

Garnet coronae are locally developed between ilmenite and plagioclase, indicating a reaction such as:

$$\text{Qtz} + \text{Pl} + \text{Il} = \text{Grt} + \text{Rt} \quad (2)$$

studied experimentally by *Bohlen and Liotta* [1986].

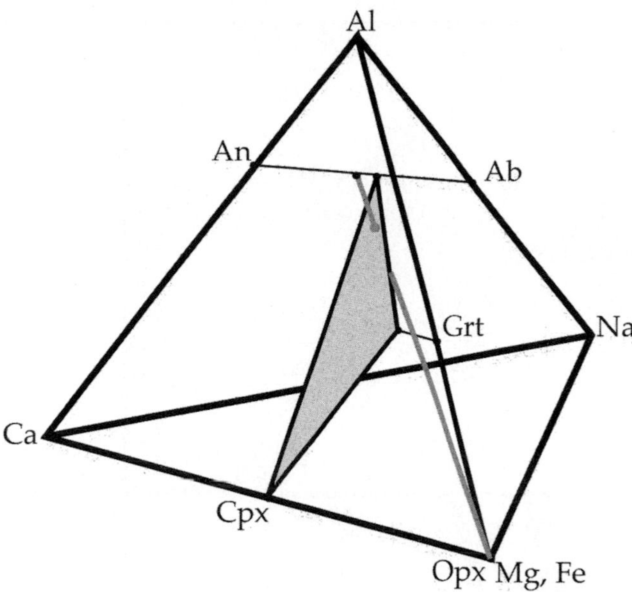

Figure 2. Phase diagram illustrating assemblages found in lower crustal xenoliths from the State Line Kimberlite District. The shaded plane represents Grt-Cpx-Pl assemblage found in sample SD2-LC74. Line shows relict Opx-Pl assemblage in samples SD2-LC120 and NX4-LC1.

Crustal xenoliths from Leucite Hills include a variety of fine-grained, intermediate and mafic metamorphic rocks, dominated by one- or two-pyroxene hornblende granulites (Plate 2). Assemblages include hornblende (40–60 modal%), biotite (20%), plagioclase (10–20%), clinopyroxene (10–20%)± orthopyroxene (10–20%) with minor apatite and ilmenite and trace amounts of barite, monazite, rutile, and zircon (Plate 2a). Elongate grains of hornblende and biotite up to 3 mm in length form a strong foliation or gneissic texture in most samples. Pyroxene and plagioclase (approximately 1 mm in diameter) are generally equant. To date, garnet has not been recognized in any of the xenolith samples from the Leucite Hills area.

Hornblende in Leucite Hills xenoliths is light brown in plane light and includes very fine opaque minerals giving it a dusty appearance (Plate 2b). A similar fine opaque dust is concentrated along hornblende grain boundaries (Plate 2b). Locally the opaque dust is so abundant within and around hornblende as to render it completely opaque. Biotite books, up to 1 mm in length, are typically associated with hornblende, and together, they define the foliation or gneissic layering. Green pleochroic clinopyroxene is slightly more abundant than the light pink to light green pleochroic orthopyroxene. Both phases contain abundant dusty inclusions similar to those in hornblende. Clinopyroxene and orthopyroxene are typically equant with smooth, irregular grain boundaries. Locally, clinopyroxene also occurs as a fine-grained foam in association with hornblende grains (Plate 2c). Several xenoliths contain skeletal orthopyroxene located in isolated pods (Plate 2e) that may be the sites of former larger orthopyroxene crystals. Plagioclase occurs as equant to slightly elongate grains, up to 2 mm in length and can locally contain fine opaque dust similar to that in hornblende. Plagioclase is most abundant as a matrix phase, but is also found included within hornblende, biotite, and clinopyroxene.

Major/Trace Element Chemistry

Whole rock MgO contents in the State Line mafic xenoliths vary from about 3.5 wt % to 13.5 wt % (Figure 3; see also Bradley, 1985). In general, Al_2O_3 and Na_2O decrease, and CaO increases, with increasing wt. % MgO (Figure 3). In contrast, Fe_2O_{3t}, K_2O and TiO_2 do not correlate with MgO. On a total alkali vs. silica plot, the xenoliths plot as basanites, basalts, and trachybasalts (Figure 4). The majority of the xenoliths are hypersthene normative (Figure 5) and subalkaline (tholeiitic) (Figure 6), in composition, as previously noted by *Bradley* [1985]. However, it is important to recognize that due to the potential loss of alkali elements during granulite metamorphism and the possible addition of these elements from the host kimberlite (see below) the bulk compositions of the State Line xenoliths are unlikely to represent protolith compositions.

Most of the State Line xenoliths are moderately LREE enriched, and have only minor Eu anomalies, if any (Figure 7a). But three of the samples (SD2-LC75, -LC71 and SH13-E18) have significantly lower La/Yb_N (0.4 to 0.9 vs. 3.0 to 14.6 (Table 2b; see CDROM in back cover sleeve) as a result of higher HREE abundances and lower LREE abundances than the other mafic xenoliths, and so have been grouped together separately (the "low La/Yb_N" group xenoliths; Figure 7b). All of the State Line xenoliths are enriched in Pb, Rb, Ba, and K, relative to other elements with similar incompatibilities (Figure 8), but do not show well-developed relative depletions in high field strength elements such as Nb and Ta.

The bulk compositions of the Leucite Hills xenoliths are similar to those of the mafic xenoliths in the State Line district (Figure 3), although the Leucite Hills xenoliths tend to have higher total alkali contents and correspond to basanites, trachybasalts, and trachyandesites (Figure 4). In addition, a large proportion of the Leucite Hills xenoliths are Si-undersaturated, containing normative nepheline rather than hypersthene (Figure 5). The Leucite Hills xenoliths are uniformly LREE enriched and, unlike their State Line counterparts, show moderate Eu anomalies [Eu/Eu*=0.4 to 1.0; (Table 2b; see CDROM in back cover sleeve) and an overall humped REE pattern due to relative enrichments in the middle REE

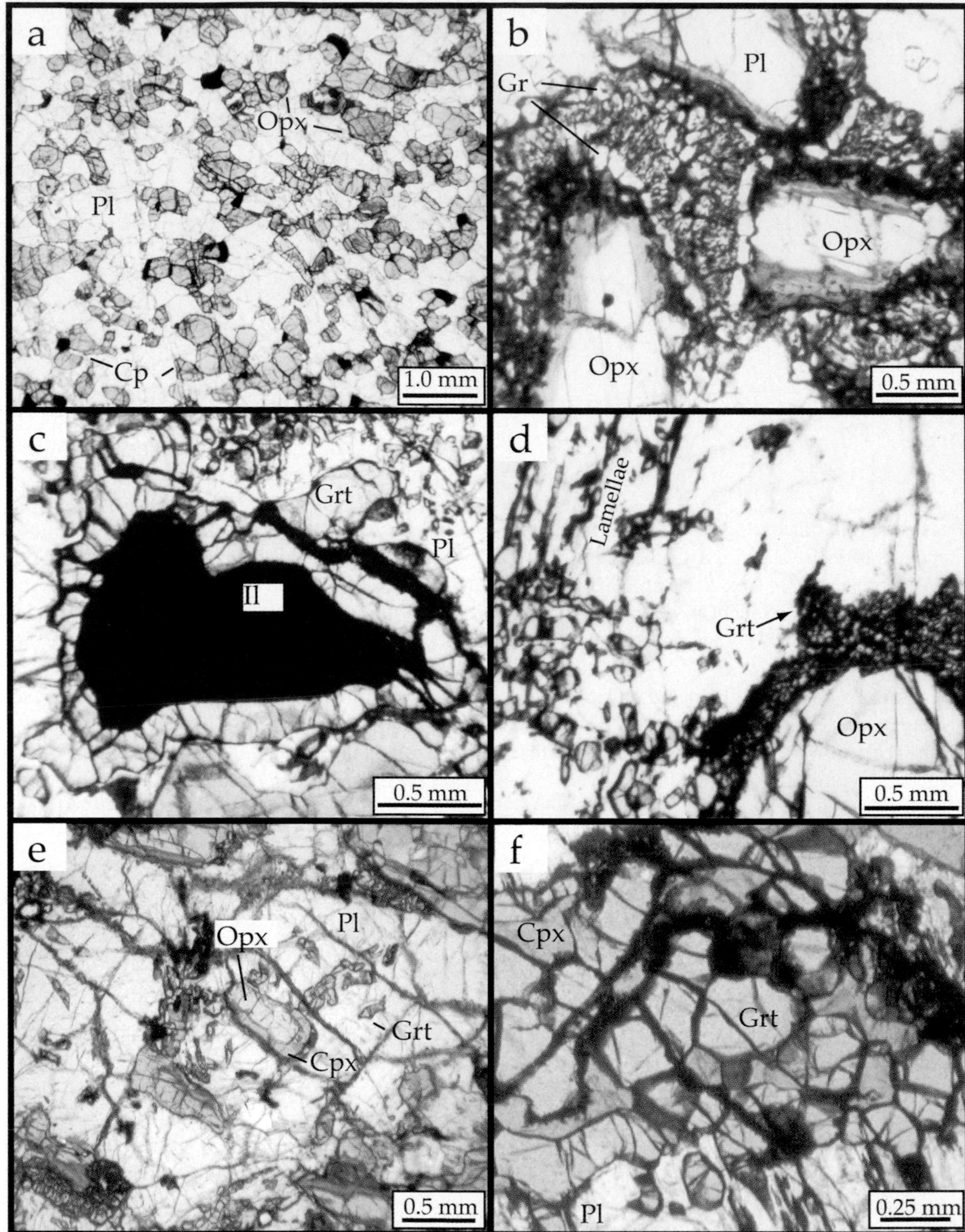

Plate 1. Photomicrographs illustrating textures in crustal xenoliths from the State Line Kimberlite District. a) Sample SD2-LC118. Opx-Cpx-Pl granulite with "typical" granulite texture: equant grains with smooth, curving grain boundaries that meet in 120° triple junctions. b) Sample SD2-LC120. "Wormy" garnet coronae between orthopyroxene and plagioclase. c) Sample SD2-LC74. Garnet coronae between ilmenite and plagioclase. d) Sample SD2-LC74. Fine-grained garnet foam around orthopyroxene and garnet lamellae aligned along twin planes in plagioclase. e) Sample SD2-LC120. Clinopyroxene coronae between orthopyroxene and plagioclase. f) Sample SD2-LC74. Grt-Cpx symplectite (see text for discussion).

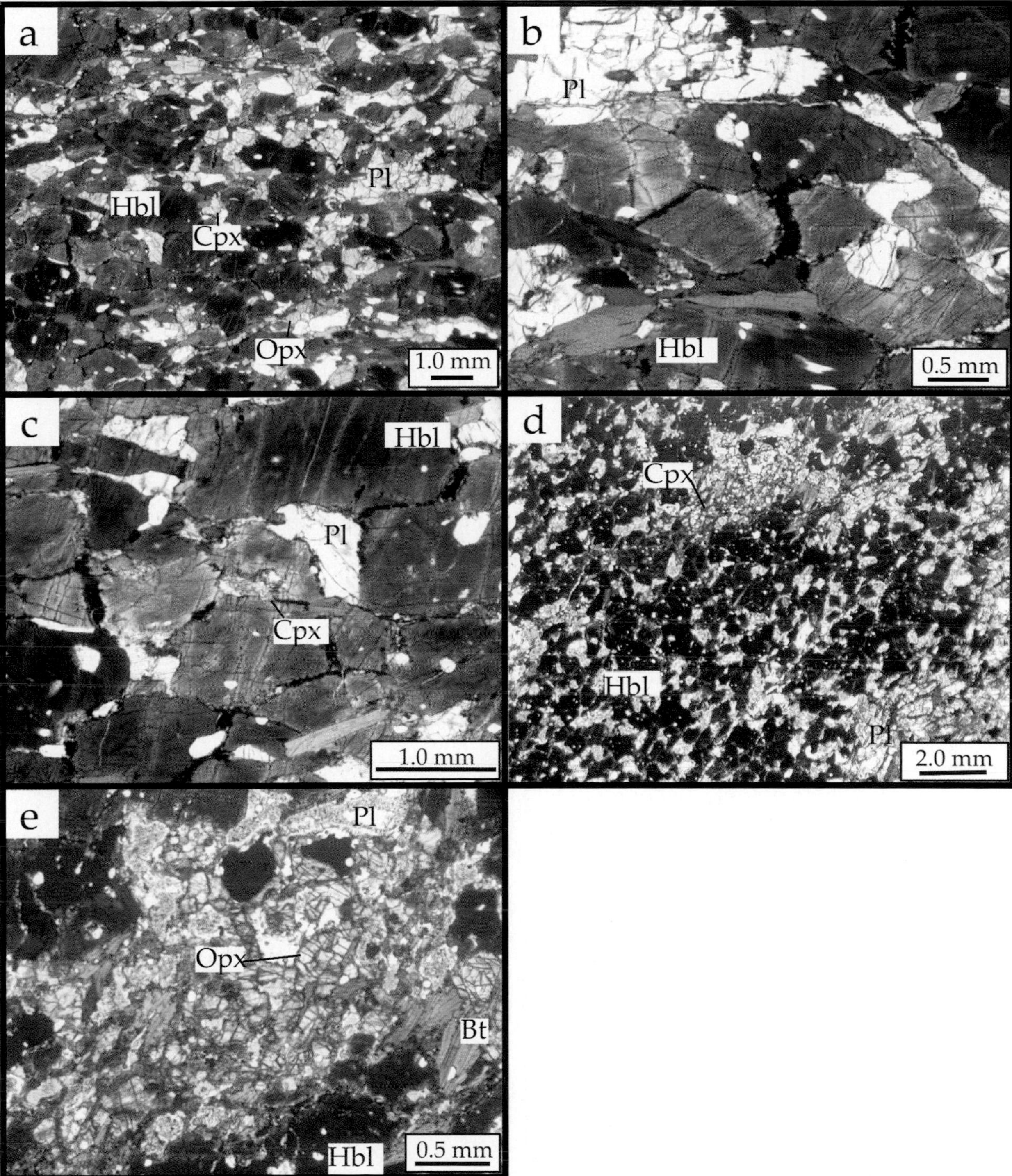

Plate 2. Photomicrographs illustrating assemblages and textures in crustal xenoliths from Leucite Hills. a) Sample LHHE-4. Hbl-Cpx-Opx-Pl granulite. Note dimensional preferred orientation of amphibole and biotite that forms foliation. b) Sample LHHE-4. Hornblende with dusty inclusions and concentrations of opaque minerals along grain boundaries. c) Sample LHHE-4. Fine-grained clinopyroxene (center of photomicrograph) surrounded by hornblende. d) Sample LHHE-26. Hbl-Cpx-Opx-Bt granulite. Foliation formed by hornblende and biotite. Note high concentration of dusty inclusions in hornblende. e) Hornblende-free "pod" of orthopyroxene within hornblende-rich matrix.

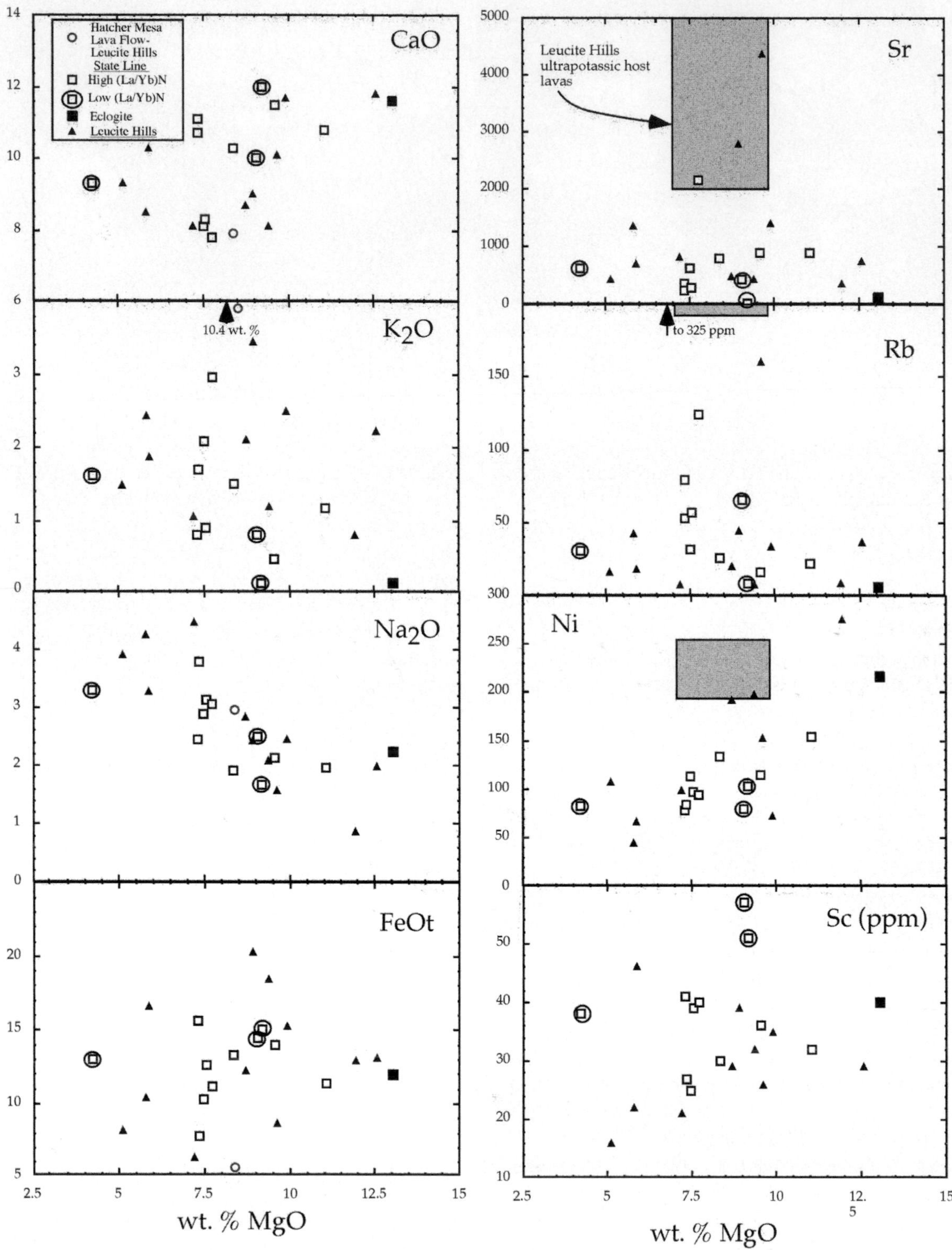

Figure 3. Wt. % MgO vs. various wt. % oxide and trace element abundances (in ppm) for State Line and Leucite Hills xenoliths. Chemical data for Leucite Hills ultrapotassic lavas from *Vollmer et al.* [1984] and *Lange et al.* [2000].

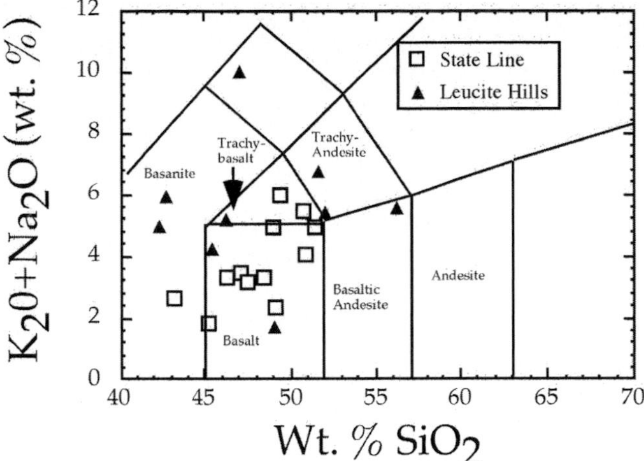

Figure 4. Total alkali vs. silica plot for State Line and Leucite Hills xenoliths.

(Figure 7c)]. All of the Leucite Hills xenoliths also have high LILE (large ion lithophile element)/HFSE (high field strength element) ratios, including the development of pronounced relative depletions in Nb and Ta (Figure 8).

Mineral Compositions

Representative mineral compositions for both State Line and Leucite Hills xenoliths are presented in Tables 5 and 6, respectively (see CDROM in back cover sleeve). In the State Line xenoliths, garnet is generally Alm_{46-55},Pyr_{23-37},Grs_{13-23}. Small garnet crystals are generally unzoned such as those in NX4-LC1. Larger crystals display a very subtle trend toward increasing X_{Mg} and decreasing X_{Ca} from core to rim (Plate 3). Rims display decreasing X_{Mg} and increasing X_{Ca}, interpreted to represent diffusional re-equilibration (probably with pyroxene). Clinopyroxene is 61–76 percent diopside with significant (near omphacitic) jadeite component. Cpx is generally unzoned except for a narrow increase in X_{Mg} near rims, interpreted to be a result of diffusional exchange with garnet. Orthopyroxene (enstatite) is 71 – 72 percent enstatite, 27 – 29 percent ferrosilite and unzoned. Plagioclase generally shows unzoned cores and decreasing X_{An} (increasing X_{Ab}) near rims (Plate 3a). The unzoned cores are interpreted to represent the early opx-bearing assemblages; the increase in albite content is interpreted to reflect the growth of garnet (and clinopyroxene) in these samples.

Hornblende from Leucite Hills generally is characterized by heterogeneous but increasing Mg ratios from inner core (ca. 0.66) to outer core (ca. 0.79), and by broad rims with nearly constant Mg-number. The broad rims compositions are interpreted to reflect near-peak metamorphic conditions. Clinopyroxene (diopside) is approximately $Di_{76-79}Hd_{21-23}$, and displays little compositional zoning, although some crystals show a subtle increase in X_{Mg} from core to rim. Orthopyroxene (enstatite) is approximately En_{65}, Fs_{35}. The Mg ratio varies between 0.63 and 0.65 with a very slight increase in both Mg and Ca from core to rim. Plagioclase is Ab_{62-65}, An_{35-40}, Or_{2-5}. Zoning profiles are generally flat across the cores of grains, but include a small, sharp increase in K at the rims.

Thermobarometry

Peak metamorphic conditions for the State Line xenolith samples were estimated based on garnet-pyroxene-plagioclase equilibria. Compositional analyses were chosen to best represent peak metamorphic conditions (Figure 9). Garnet compositions were taken at the point of maximum X_{Mg} just inward from rim reversals. Core compositions were generally used for Opx and Cpx, but because the amount of zoning is extremely small, all compositions give essentially the same results. Plagioclase compositions were generally taken at maximum X_{Ab} (near rim). Because diffusion is extremely slow in plagioclase, rim compositions are interpreted to represent peak conditions. Equilibria were characterized using the TWQ software (v. 2.02) [*Berman*, 1991; *Berman*, 1992] and database BA95 [*Berman and Aranovich*, 1996]. In general, independent reactions were chosen to correspond to the Grt-Pyx exchange thermometers and the Grt-Cpx (or Opx)-Plg barometers. Samples SD2-LC120 and NX4-LC1 (both from high La/Yb_N group) yield temperatures on the order of 700–800°C and pressure estimates near 1.2 GPa (Figure 9). Sample SD2-LC74, a low La/Yb_N sample, yields a perhaps unreasonably high temperature and pressure (>900°C; 1.5 GPa). However, if temperatures are assumed to be on the order of those for the other two samples, the pressure would be approximately 1.4 GPa.

The lack of garnet in the Leucite Hills xenoliths precludes essentially all high-confidence thermobarometers, which typ-

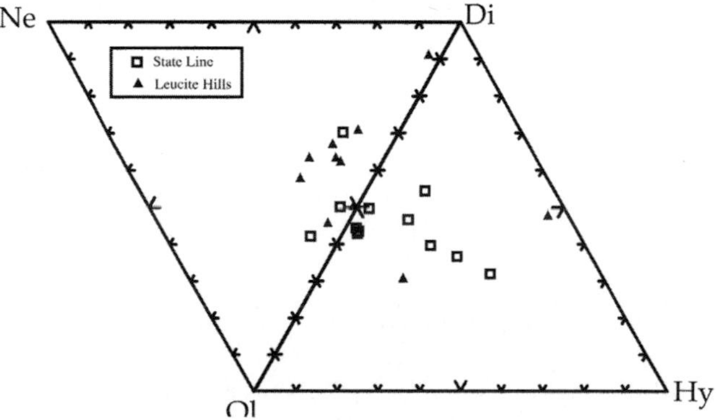

Figure 5. CIPW normative mineralogy of State Line and Leucite Hills xenoliths. Normative mineralogy calculated assuming Fe^{3+}/(total iron)=0.1.

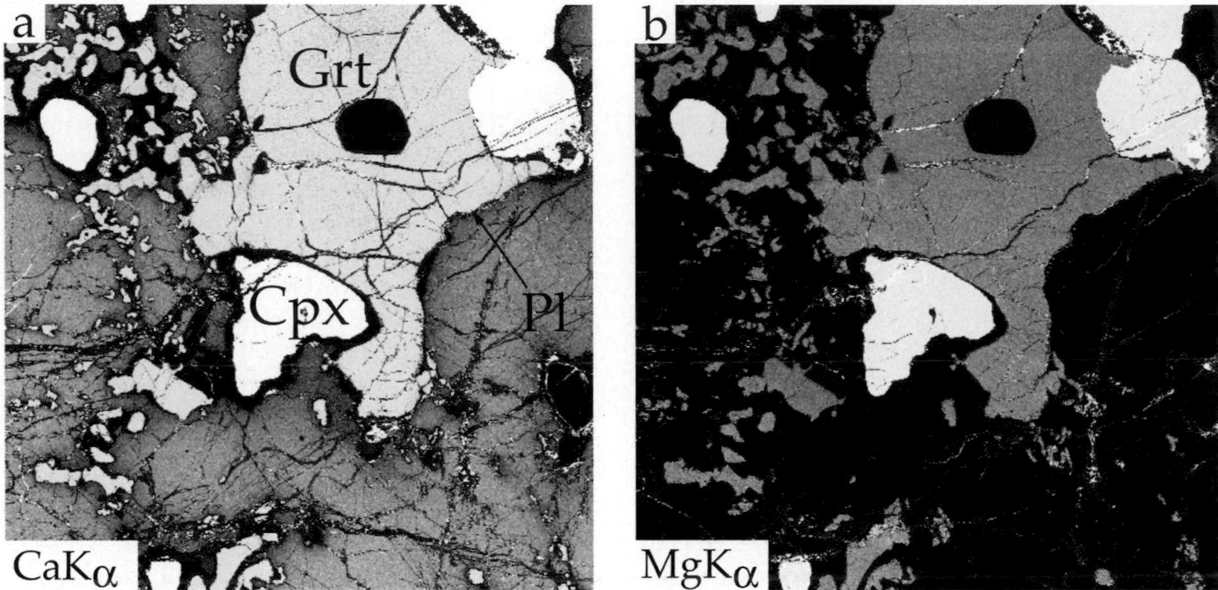

Plate 3. Compositional maps from sample SD2- LC74, State Line District. a) Ca K_α map. Note decreasing Ca at plagioclase grain boundaries (in direction shown by arrow). b) Mg K_α map. Garnet is essentially unzoned. X_{Mg} increases slightly from core to near rim and then decreases at the rim.

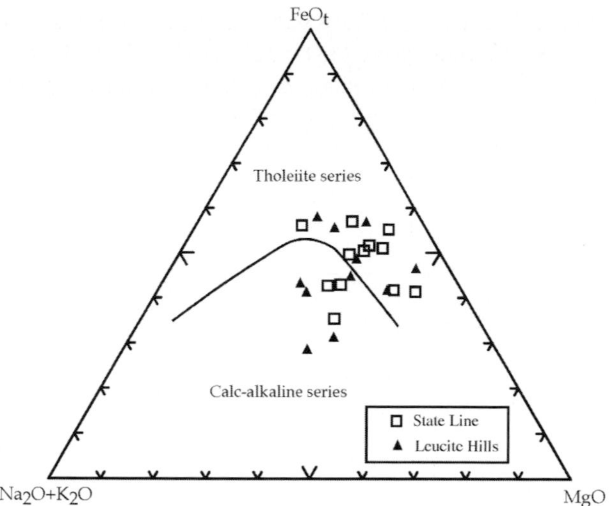

Figure 6. AFM diagram for State Line and Leucite Hills xenoliths.

ically involve Grt-Pyx or Grt-Hbl equilibria. As a result, we have attempted to constrain the general range of metamorphic conditions using Hbl-Pyx-Pl-Qtz equilibria. *Mader et al.* [1994] suggested two barometers involving Hbl-Cpx-Qtz-Pl assemblages and the thermodynamic database "JUN92" [*Berman*, 1991; *Berman*, 1992]. For three Leucite Hill xenoliths, the reaction:

$$Ts + 2Di + 2Qtz = Tr + 2An \tag{3}$$

yields extremely consistent results (Figure 10) The alternative reaction, involving pargasite and albite, yields consistent but much more scattered results that we interpret to reflect some late diffusion of Na and perhaps uncertainties associated with thermodynamic properties of Na-components in each mineral phase. Temperatures were constrained using the Hbl-Opx exchange thermometer with the thermodynamic database "JUN92" [*Berman*, 1991; *Berman*, 1992] and with two-pyroxene thermometry using the BA95 database [*Berman and Aranovich*, 1996]. Because pressure estimates are strongly dependent on temperature, temperatures were also evaluated using the Hbl-Pl thermometer of *Holland and Blundy* [1994]. Estimated conditions for three Leucite Hills xenoliths (LHHE-4, LHHE-26, and LHHE-18) are very consistent (1.1–1.3 GPa, ca. 800° C (Figure 10). Even with a relatively large uncertainty, these estimates indicate that the hornblende-bearing Leucite Hills xenoliths were derived from the lower crust.

U-Pb Geochronology

Xenoliths from the Leucite Hills, Wyoming generally contain a relatively uniform population of equant to slightly elongate and multifaceted zircon with few inclusions. U-Pb analyses of zircon from these samples are typically concordant or near concordant with $^{207}Pb/^{206}Pb$ dates of c. 2563–2703 Ma with one exception at 1067 Ga (Table 4b; see CDROM in back cover sleeve, Figure 11). Cathodoluminescence images of zircon from representative xenoliths show cores with distinct oscillatory zoning interpreted to indicate growth of zircon from a melt (Figure 12). The ca. 2.6–2.7 Ga ages obtained from the xenoliths correspond closely to ages of crust formation determined from surface geology at 2.55–2.8 Ga [*Frost et al.*, 1998]. A single titanite analysis of 2610 Ma is ~40 Ma younger than zircon from the same sample (Table 4b; see CDROM in back cover sleeve). This analysis is interpreted as a cooling date (closure temperature of ~650°C) [*Cherniak*, 1993] and indicates the lower crust in this region cooled below 650 °C shortly after assembly and has not been significantly perturbed by any younger thermal events. The one relatively young zircon analysis (1067 Ma) likely reflects a low-temperature, fluid mediated episode of zircon growth [*Ayers et al.*, 2003].

In the State Line district, the SD3 xenoliths contain zircon of a wide range of crystal sizes and shapes, color, clarity and degree of metamictization. These zircon yield a wide range of dates. The dominant population yields $^{207}Pb/^{206}Pb$ dates of 1.73–1.6 Ga (Fig 14, Table 4a; see CDROM in back cover sleeve), which are similar to crystallization ages of exposed rocks in Colorado. In addition some xenoliths (SD3-12, SD3-46, and SD3-2) yield a few grains that are ca 1.4 Ga, and are apparently related to the regional magmatic event that is well documented in exposed rocks [*Anderson and Cullers*, 1999; *Frost et al.*, 1999; *Frost et al.*, 2001]. Subordinate populations of variably discordant Archean (ca. 2.6–3.2 Ga) zircon were recovered from xenoliths with dominantly Paleoproterozoic populations (SD3-8, -12, -24, -25). It is noteworthy that the inherited Archean zircons are contained exclusively in the amphibolitic xenoliths suggesting crustal contamination of primary mafic magmas with older crust. The final population of zircon yields distinctly younger dates including a concordant analysis that is 419 Ma, a 9% discordant analysis with a Pb-Pb date of 453 Ma, and a 2% discordant analysis with a Pb-Pb date of 544 Ma. These young dates suggest growth of zircon just prior to and perhaps synchronous with kimberlite emplacement. Some of the very small, highly discordant analyses could also represent mixtures of ca 400–500 Ma zircon and Paleoproterozoic grains.

Zircon occurs as rare grains in the three mafic two-pyroxene, garnet granulites for which zircon U-Pb ages were obtained (Table 4a; see CDROM in back cover sleeve) and typically forms small (<40 μm) equant grains or larger (~100–150 μm) irregular fragments. In sample NX4-LC1 zircon are characterized by multifaceted, stubby prisms averaging about 60 to 70 μm in length. The zircon grains are generally colorless and inclusion-free, but a small percentage contains cracks. Seven zircon grains were analyzed from this sample and yield a diverse range of Pb-Pb ages. The two oldest grains are approximately 1.72 –1.7 Ga, two have intermediate ages of ca 1.62 and 1.54 Ga and three are distinctly younger (1.34, 1.37, 1.38 Ga). We regard this as evidence of multiple periods of zircon growth in a ca 1.72Ga, or older, protolith. In sample SD2-LC44, four zircons have Pb-Pb ages from 1.71 to 1.74 Ga, but one grain is Archean in age (3.1 Ga). Zircon from sample SD2-LC75 form two different morphological populations. One population consists of colorless to light brown fragments with irregular surfaces. The fragments range in size from 150

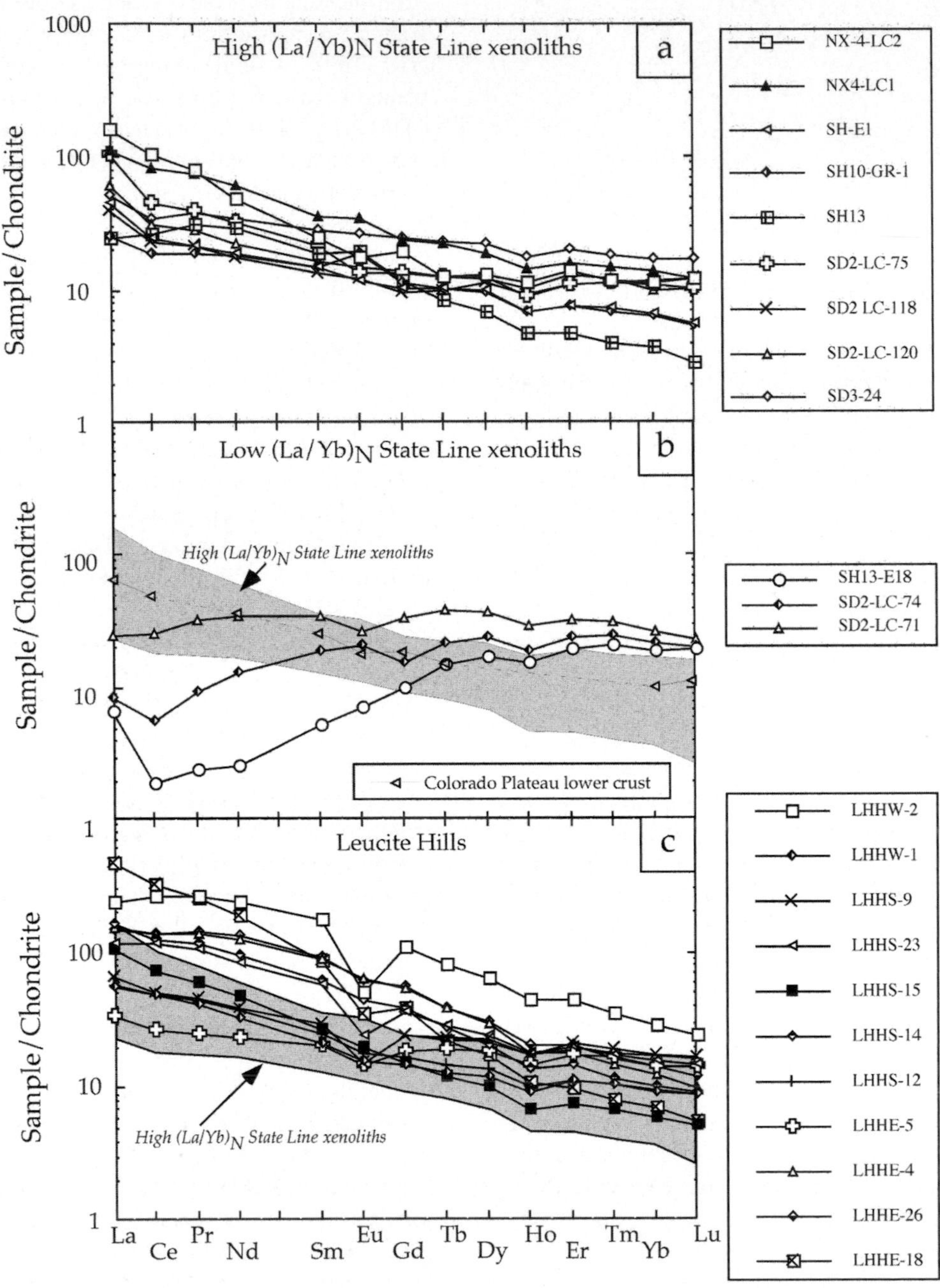

Figure 7. a) Chondrite normalized REE patterns for "high" La/Yb_N State Line xenoliths. b) Chondrite normalized REE patterns for "low" La/Yb_N State Line xenoliths. REE pattern for Colorado Plateau lower crust from *Condie and Selverstone* [1999]. c) Chondrite normalized REE patterns for Leucite Hills xenoliths.

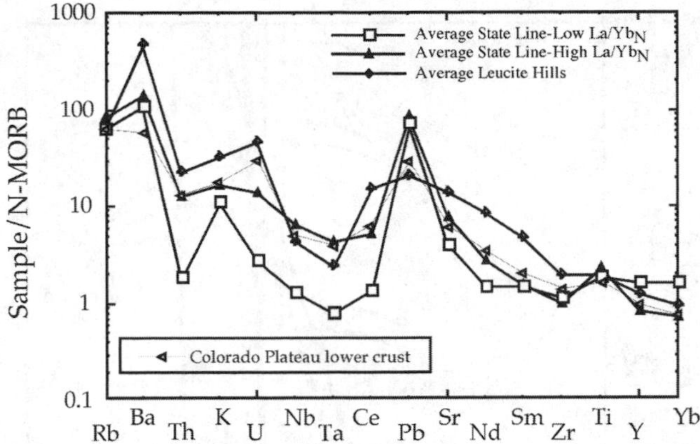

Figure 8. Average normalized trace element plot for State Line and Leucite Hills xenoliths. Normalization values from *Sun and McDonough* [1989] and *McCulloch and Gamble* [1991].

to 250 μm in length, and commonly contain cracks. The other population consists of colorless stubby prisms averaging about 60 μm in length. Both populations are relatively inclusion-free. The $^{207}Pb/^{206}Pb$ dates for both zircon populations range from 1.64 to 1.74 Ga.

In order to gain insight into the origin of the different zircon populations, several zircon grains from NX4-LC1 and SD2-LC44 were imaged using cathodoluminescence techniques. These zircon show complex growth zonation patterns (Figure 12), however, they can be distinguished based on those with sector zoning indicative of metamorphic growth and those with predominantly oscillatory zoning indicative of igneous growth. Using the catholuminescence images as a guide, we plucked zircon from our imaged mounts and analyzed individual grains. Zircons with sector zoning from NX4-LC1 yielded the $^{207}Pb/^{206}Pb$ dates of 1342.7±2.0 Ma and 1379.3±1.8 Ma (Figure 11, Table 4a; see CDROM in back cover sleeve). A single zircon grain from SD2-LC44 with an oscillatory-zoned core and bright overgrowth yielded the $^{207}Pb/^{206}Pb$ date of 1720.3±3.0 Ma (Figure 11, Table 4a; see CDROM in back cover sleeve). These analyses are interpreted to indicate that the lower crust in this region was formed during the Paleoproterozoic and that new zircon growth occurred at 1.34–1.38 Ga.

Nd and Sr Isotopes

Unleached whole rocks from the State Line diatremes have a range of measured ε_{Nd} (+16.5 to –12) and measured $^{87}Sr/^{86}Sr$ (0.7054 to 0.70962; Table 3; see CDROM in back cover sleeve, Figures 13–15). Variations in both measured and initial (calculated at 377 Ma) Sr and Nd isotopic compositions are uncorrelated (Figure 13). The low La/Yb$_N$ xenoliths and the one eclogite sample have the highest measured ε_{Nd} values (+3.4 to +16.5), while the high La/Yb$_N$ samples consistently have lower ε_{Nd} (-11.9 to 0.4). For the one sample (NX4-LC1) for which both leached and unleached samples were analyzed, the leached sample had considerably higher $^{147}Sm/^{144}Nd$ (0.1784 vs. 0.1085) and higher measured ε_{Nd} values (-6.1 vs. –10.4; Figure 14). In contrast, the $^{87}Rb/^{86}Sr$ and Sr isotopic compositions of the leached and unleached NX4-LC1 samples are similar (0.1091 vs. 0.947 and 0.70596 vs. 0.70632, respectively; Figure 16). The other leached whole rock samples analyzed have Nd and Sr isotopic compositions similar to the leached xenoliths, but tend to have higher $^{147}Sm/^{144}Nd$ (Figure 14).

The microdrilled plagioclase from SD2LC-71, one of the low La/Yb$_N$ group samples, has a low $^{147}Sm/^{144}Nd$ (0.0867) and low ε_{Nd} (-15.9) (Figure 14). The microdrilled clinopyroxene from this sample, however, has a higher $^{147}Sm/^{144}Nd$ (0.164) and higher ε_{Nd} value (-0.3). The plagioclase and clinopyroxene mineral samples also have different $^{87}Rb/^{86}Sr$ and $^{87}Sr/^{86}Sr$ ratios (0.34 vs. 1.7 and 0.70679 vs. 0.71214, respectively; Figure 16). The clinopyroxene powder obtained

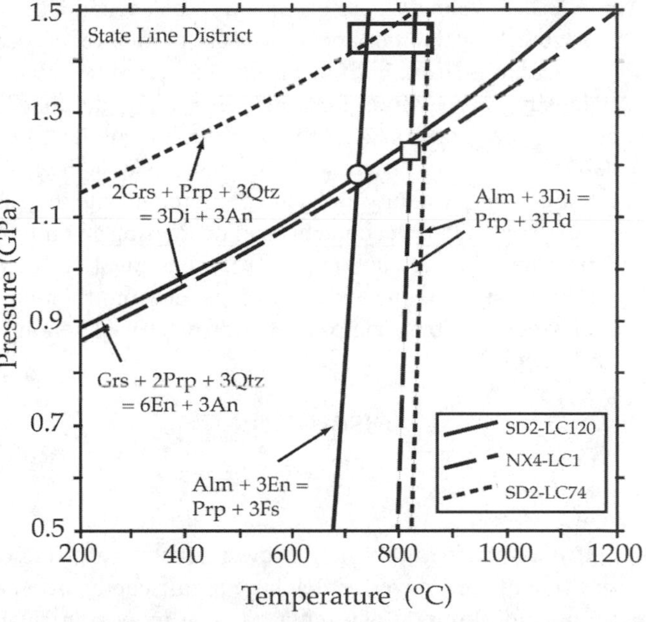

Figure 9. Pressure and temperature estimates for Grt-Pyx-Pl-Qtz-bearing xenoliths from the State Line Kimberlite District. Calculations were made using the TWQ software [*Berman*, 1991; *Berman*, 1992] and BA95 database [*Berman and Aranovich*, 1996]. Reactions were selected to correspond to the Grt-Pyx (Cpx or Opx) thermometer and Grt-Pyx-Pl-Qtz barometer [*Williams et al.*, 2000]. Note that the Grt-Opx barometer essentially corresponds to Reaction 1 (see text). The circle and square mark the intersections of barometer and thermometer reactions for SD2-LC120 and NX4-LC-1, respectively.

from microdrilling of NX4-LC1, a high La/Yb_N sample, has a similar $^{147}Sm/^{144}Nd$ to the clinopyroxene sample from SD2-LC-71, but significantly lower ε_{Nd} (-10.3), and a $^{87}Rb/^{86}Sr$ and $^{87}Sr/^{86}Sr$ of 0.3738 and 0.70922, respectively. Whole rock and minerals from SD2 LC-71 plot along an ~ 300 Ma reference Sr isochron (Figure 16), which approximates the emplacement age of the host kimberlite, but align along a significantly older (~1.0 Ga) Sm-Nd reference isochron (Figure 14).

The Leucite Hills xenolith samples, all unleached, generally have lower measured ε_{Nd} (-28.5 to -7.2) relative to the State Line xenoliths, but similar measured $^{87}Sr/^{86}Sr$ ratios (0.7036 to 0.7078; Figures 13, 14, 16). However, as with the State Line xenoliths, the Nd and Sr isotopic variations in these rocks are uncorrelated (Figure 13).

Physical Properties

Velocities at selected pressures from 20 to 1000 MPa are given in Tables 7 and 8 (see CDROM in back cover sleeve), as well as bulk densities at atmospheric pressure. The samples are arranged in order of increasing density. The two perpendicular cores from samples LHHE-26 and LHHW-1 are labeled A and B. The velocities are estimated to be accurate to 1%.

In Figure 17, velocities at 600 MPa are plotted versus density for samples from both localities. The two low velocity and density samples from Leucite Hills are granites that are probably from upper crustal rocks. Velocities of the higher density rocks are consistent with metamorphism in the lower crust. Increasing velocity and density correlate with increasing pyroxene and garnet contents and decreasing amounts of secondary alteration. The samples from the State Line locality contain significant amounts of garnet and thus have higher velocities than many of the lower crustal xenoliths from Leucite Hills.

DISCUSSION

Metamorphic History

Our data confirm that mafic lower crustal xenoliths on either side of the Cheyenne Belt have significantly different metamorphic mineral assemblages. The most significant difference is that the mafic xenoliths north of the Cheyenne Belt at Leucite Hills contain abundant hydrous mineral phases and lack garnet, while the mafic granulite xenoliths from the State Line district mainly contain anhydrous phases and abundant garnet. These mineralogical differences are unlikely to be the result of significant differences in major element composition of the mafic lower crustal rocks, as compositions of the two xenolith suites are very similar (Figures 3–6). In addition, even though the P, T constraints from

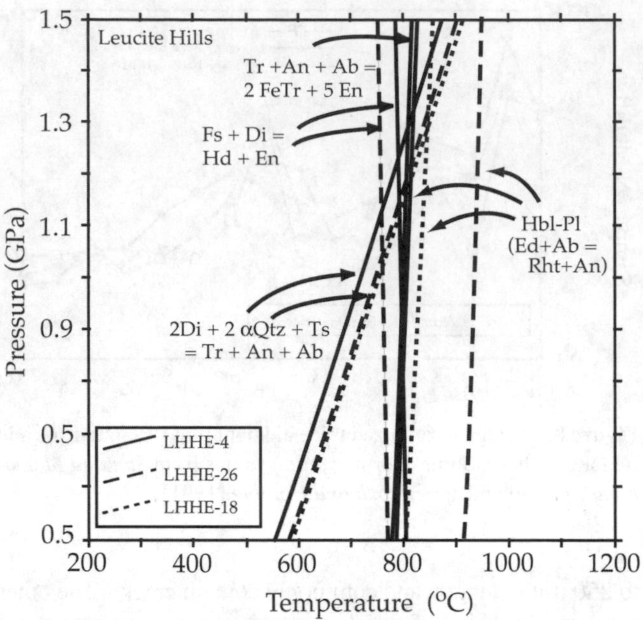

Figure 10. Pressure and temperature estimates for three Cpx-Opx-Hbl-Qtz-bearing xenoliths from the Leucite Hills. Calculations were made using the TWQ software and JUN92 database of [*Berman*, 1991; *Berman*, 1992]. Because no garnet is present in these rocks, estimates are based on Hbl-Pl-2 Pyx equilibria. Barometers are those recommended by *Mader et al.* [1994]. Thermometers are Hbl-Opx (using TWQ software and JU92 database of *Berman* [1991,1992]) and Hbl-Pl [*Holland and Blundy*, 1994]. Rht=richterite.

Leucite Hills may be somewhat less robust, our estimates suggest that both suites of xenoliths equilibrated at similar pressures, ca. 1.2–1.4 GPa. As a result, there is little evidence that the State Line xenoliths were uniformly derived from significantly deeper and/or colder portions of the crust (i.e. that greater pressures (depths) or lower temperatures are responsible for the presence of garnet in the State Line district). The main mineralogical differences between the two sets of xenoliths is more likely related to hydration, with the xenoliths indicating a significantly more hydrous lower crust for the southern Wyoming craton than for the Paleoproterozoic crust beneath the State Line District.

Nevertheless, the State Line xenoliths still record a more complex P-T-t history than their counterparts from the Leucite Hills. All of the mafic granulites studied from the State Line district contain evidence for the development of garnet and clinopyroxene from earlier orthopyroxene and plagioclase. The product assemblages typically occur as coronae or symplectites. Such symplectites and coronae are commonly interpreted to have formed during metamorphic reactions under conditions of limited deformational recrystallization and limited diffusion [*Vernon*, 1976; *Carlson*, 2002]. The growth of garnet along plagioclase twin planes and lamellae may also be indicative of garnet growth under conditions of minimal dynamic recrystallization.

The textures and phase relationships are consistent with the reaction from medium- to high-pressures granulites (Reaction 1), investigated by *Green and Ringwood* [1967] (Figure 9). Although referred to as "high-pressure granulites", the Grt-Cpx-Pl product assemblage may be produced via an increase in pressure or a decrease in temperature. In the case of State Line xenoliths, the garnet coronae and symplectites, with their euhedral and delicate crystal habits and fine grain sizes, suggestive of growth

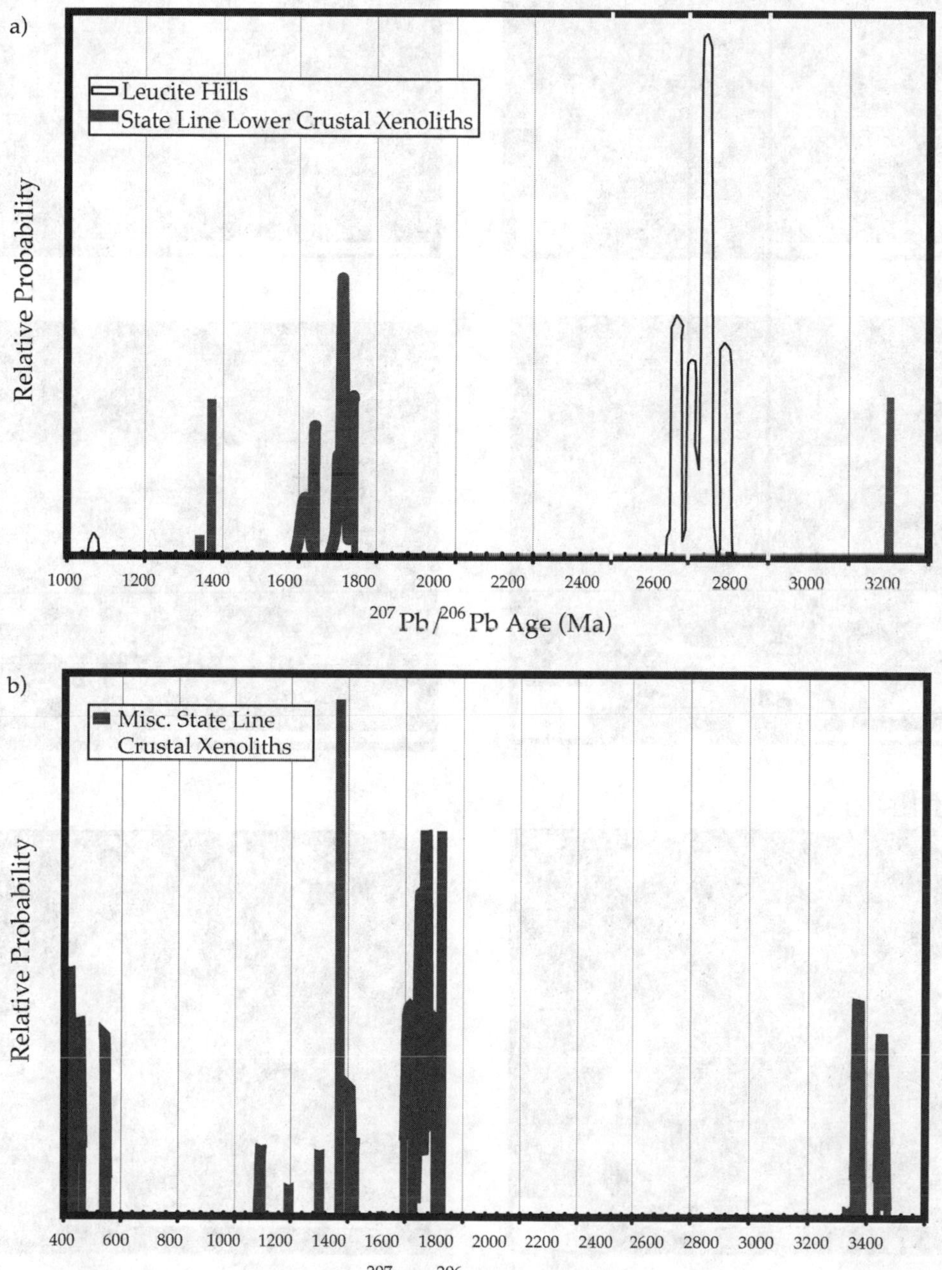

Figure 11. Probability density functions of zircon $^{207}Pb/^{206}Pb$ ages showing the distribution of zircon ages from the State Line District and Leucite Hills xenoliths. Probability density functions are constructed by summing the probability distributions of each datum with normally distributed errors.

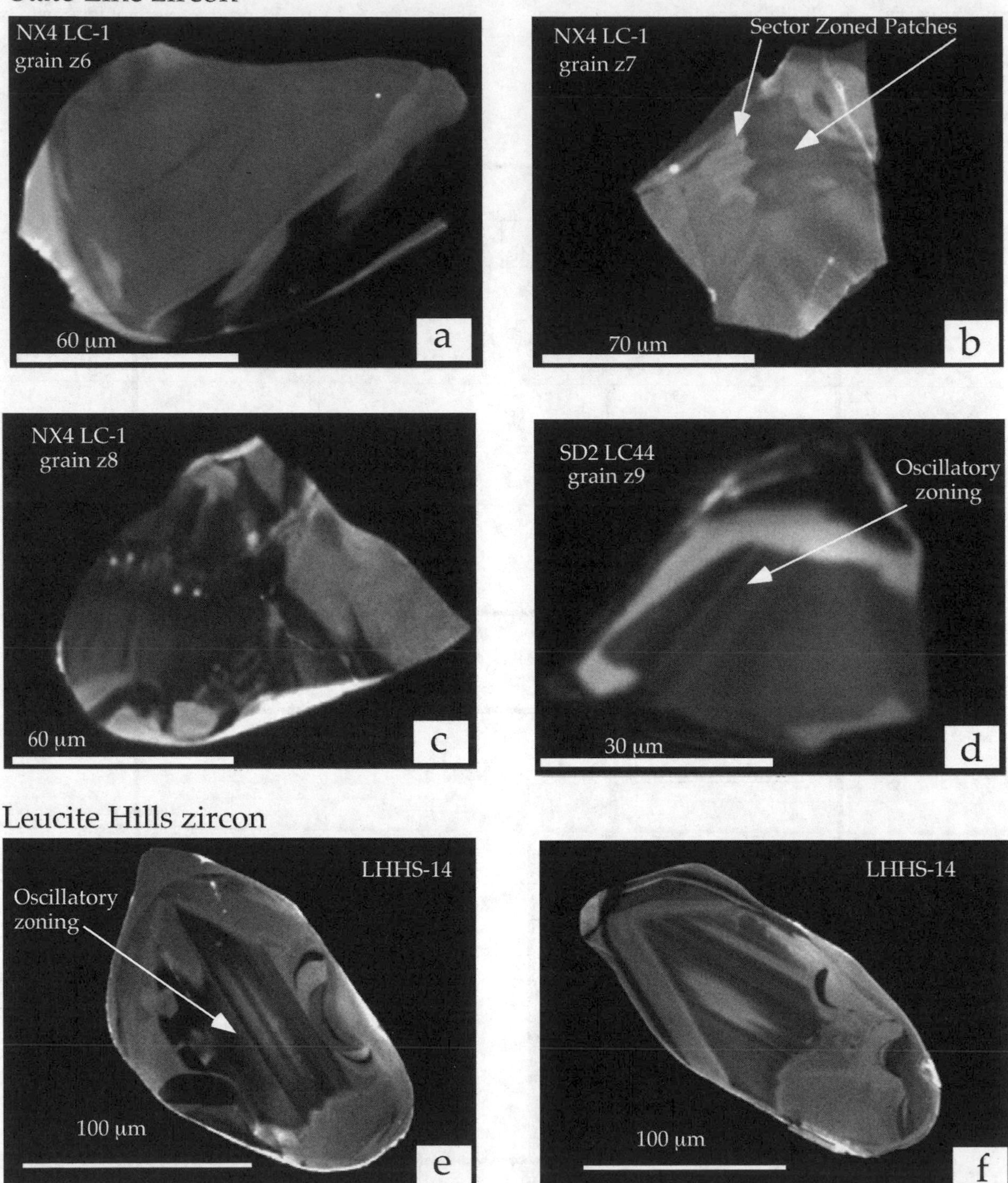

Figure 12. Cathodoluminescence images of representative zircon from the State Line diatremes (a-d) and the Leucite Hills (e-f). Grains depicted in (a) through (d) were removed from the grain mount after imaging for U-Pb isotopic analyses. Grain (b) contains sector-zoned patches and yields a Mesoproterozoic Pb-Pb age. Grain (d) contains an oscillatory-zoned core and yields a Paleoproterozoic Pb-Pb age. Zircon from the Leucite Hills contain oscillatory-zoned cores with thin overgrowths.

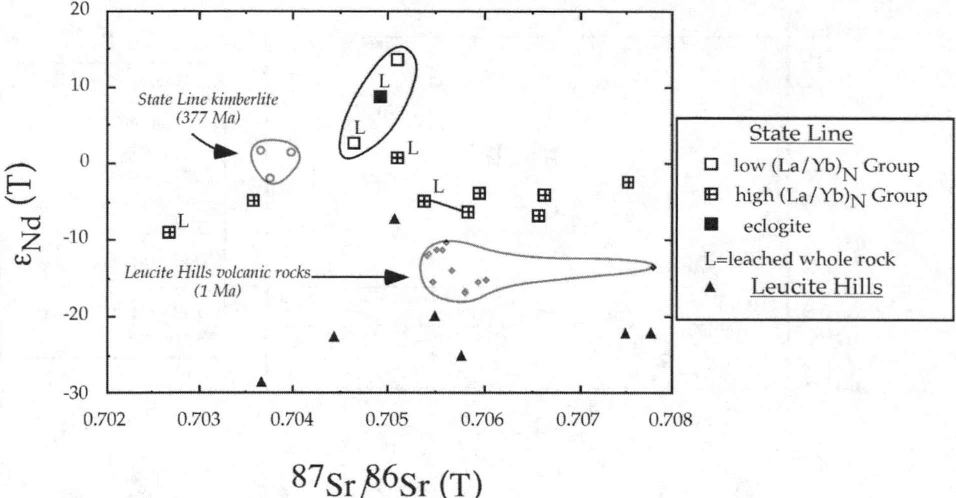

Figure 13. Whole rock ε_{Nd} (T) vs. $^{87}Sr/^{86}Sr(T)$ for unleached and acid-leached (L) State Line and Leucite Hills xenoliths. State Line xenoliths initial values calculated at 377 Ma, the presumed age of emplacement of Sloan and Nix kimberlite diatremes. Initial Sr and Nd isotopic compositions for State Line kimberlites from *Alibert and Albarede* [1988]. Solid line connects the isotopic compositions of unleached and leached whole rock material from NX4 LC-1. Initial isotopic compositions for Leucite Hills xenoliths calculated at 1 Ma. Isotopic compositions from Leucite Hills volcanic rocks from *Vollmer et al.* [1984].

under limited deformation and diffusion, are interpreted to be most consistent with garnet growth during (static) cooling. Although garnet is expected to grow at the expense of pyroxene and plagioclase with simple cooling into the high-P granulite field, some workers may be uncomfortable with significant garnet growth along a cooling path. An alternative history might involve cooling into the garnet granulite stability field and subsequent heating events within this field associated with garnet growth [*Williams et al.*, 2000]. If so, it is possible that some reheating events were significantly younger than the crystallization age of the mafic magma from which the State Line xenolith originally formed. Reheating events may have occurred during arc accretion events at 1.75–1.70 Ga, or at ca 1.37 Ga, as argued in a subsequent section. This might help to explain the range in zircon dates obtained from the xenoliths. However, given the small number of zircon grains and the small size of xenoliths it is not possible to relate the different ages of zircon to distinct xenolith mineral assemblages. As a final note, isobaric heating and reheating events would not be unexpected in these Proterozoic rocks. Surface exposures in many localities in the southwest preserve nearly constant pressures for much of the Mesoproterozoic [*Williams and Karlstrom*, 1996]. That is, there is little evidence for significant changes in crustal thickness (particularly since ~1.7 Ga) that might lead to significant pressure changes in the deep crust from which the xenoliths were derived.

Nature of Lower Crustal Protoliths

While the State Line xenoliths demonstrate the existence of mafic lower crust south of the Cheyenne Belt (at least during the Paleozoic), it is more difficult to assess whether these rocks represent cumulates, restites, or mafic magma that solidified *in situ* in the lower crust, or to determine the tectonic setting in which the mafic magmatism originally developed. A major difficulty in using the chemical data from the xenoliths to assess these issues is the likely possibility that the whole rock chemical compositions of the State Line xenoliths have been affected by contamination from the host kimberlite and/or by granulite facies metamorphism, as observed for kimberlite-hosted mafic xenoliths worldwide [*Rudnick*, 1992]. For example, *Bradley and McCallum* [1984] suggested that Sr, Rb, and K have been introduced into the xenoliths along grain boundaries and fractures. In fact, some compositional maps of State Line xenoliths show K-feldspar rims and veinlets and also barite grains on some grain boundaries. Some of these rims and grains seem to be recrystallized into the xenolith grain mosaic suggesting that the K and Ba were introduced during or before the final metamorphism but some may have been introduced during interaction with kimberlitic fluids. Our data further reveal that both high and low La/Yb_N have spikes in the abundances of Ba and Pb. Both elements could have been introduced by kimberlitic fluids, given the high contents of both elements in the host kimberlites

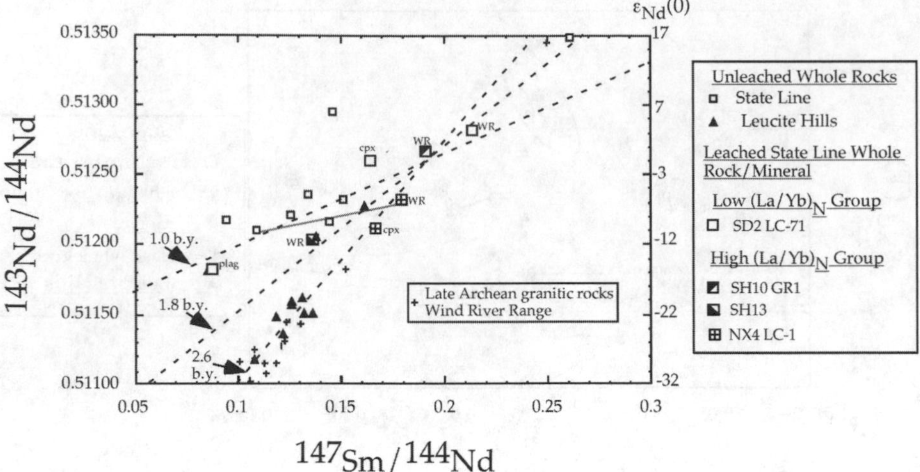

Figure 14. Measured $^{143}Nd/^{144}Nd$ vs. $^{147}Sm/^{144}Nd$ for minerals and leached and unleached whole rock xenoliths from State Line, and unleached whole rock xenoliths from Leucite Hills. Dashed lines are references isochrons for ages denoted. Data from Wind River Range granitic rocks from *Frost et al.* [1998].

[*Alibert and Albarede*, 1988]. In addition, comparison of the REE abundances of leached and unleached whole rock splits of xenolith NX4LC-1 reveal that a significant fraction of the LREE in this xenolith may reside in secondary, metasomatic phases, as evidenced by reduction in the whole rock Sm/Nd ratio after HCl leaching (Tables 2b, 3; see CDROM in back cover sleeve). These observations are significant, first because high Pb contents in basalts are generally attributed to subduction processes and could lead to the conclusion that the protoliths of the State Line xenoliths were arc-related magmas. Secondly, *Bradley* [1985] interpreted the State Line xenoliths as representing liquid compositions, in part on the basis of their mildly LREE-enriched compositions, which our data indicate could instead be the result of LREE introduction during metasomatism.

Despite the introduction of various elements during metasomatism, the abundances of several elements do co vary regularly with MgO. In all of the State Line xenoliths, Na, which is likely to be incompatible in a mafic magma undergoing crystal fractionation, increases in abundance, while the abundances of compatible elements Cr, Sc and Ni decrease, with decreasing MgO contents. Even the leached whole rock for NX4-LC1 has relatively high LREE abundances compared to other mafic xenolith suites interpreted as representing cumulate material [*Rudnick et al.*, 1986]. Therefore, we agree with *Bradley and McCallum* [1984], that the elemental variations observed between most of the granulitic State Line xenoliths, particularly for the high La/Yb_N group, are those expected for mafic igneous rocks that represent the products of the differentiation of mafic magmas involving the crystal fractionation of both olivine and pyroxene. The lack of prominent Eu anomalies in these rocks, either positive or negative and regardless of modal plagioclase contents, illustrates that plagioclase was not a fractionating phase in these magmas.

An interesting question in this regard is whether these mafic magmas were arc related, or more akin to intraplate magmas similar to modern day ocean-island or ocean plateau basalts [*Condie*, 1997]. Typically, arc and intraplate basalts are distinguished on the basis of the ratios of LILE/HFSE, the former having lower ratios, and relatively high abundances of elements (e.g. Pb) considered to be mobilized in aqueous fluids generated during subduction of oceanic lithosphere [*Pearce and Peate*, 1995]. Distinguishing these two basalt "types" is difficult in granulite grade lower crustal xenoliths, due to loss of LILE, Th, and U during high-grade metamorphism. In the State Line diatremes, the issue is further complicated by the apparent reintroduction of Rb, Ba, Pb and LREE during metasomatism that affected the xenoliths after the time of peak metamorphic conditions. As a result we are not confident that the magmas parental to the State Line xenoliths can be unambiguously fingerprinted. We do note that the absolute and relative abundance of many trace elements in the high La/Yb_N xenoliths, including the lack of prominent Eu anomalies and relatively low Nb and Tb abundances, are very similar to those calculated for Paleoproterozoic mafic lower crust beneath the southern Colorado Plateau [*Selverstone et al.*, 1999] (Figures 7b, 8). The latter has been interpreted as representing the products of arc-related magmatism, and so at least the high La/Yb_N mafic lower crust beneath

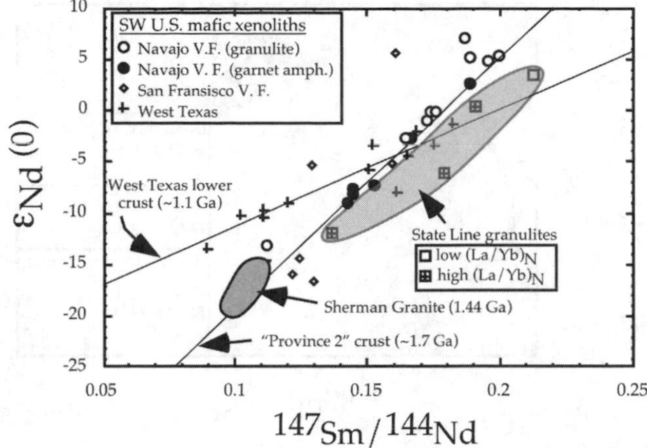

Figure 15. Measured ε_{Nd} vs. $^{147}Sm/^{144}Nd$ for leached State Line granulite xenoliths relative to lower crustal xenoliths from other localities in southwestern U.S. and from ~1.4 Ga Sherman granite (Figure 1). Data from *Wendlandt et al.* [1993], *Chen and Arculus* [1995], *Cameron and Ward* [1998], and *Frost et al.* [1999].

the northern Front Range in Colorado can be interpreted in the same fashion.

The low La/Yb_N xenoliths from the State Line diatremes, however, have much lower LREE, U, Th, Nb, Ta and Sr abundances, and higher HREE contents, at any given wt. % MgO than their higher La/Yb_N counterparts. Furthermore, the clinopyroxene in these xenoliths has extremely "humped" REE patterns (*Ridley and Farmer*, unpublished data), which could be the result of partial melting of a garnet-bearing crustal protolith [*Zou and Reid*, 2001]. We suggest that the low La/Yb_N xenoliths represent the mafic residua of partial melting of a garnet bearing, lower crustal source rock. The timing of this partial melting event, and the nature of the source rock, are both discussed further below.

The Leucite Hills xenoliths also show good correlations between major element contents, with compatible elements concentrations (Sc, Ni) generally decreasing and incompatible element concentrations (Na) increasing with decreasing wt. % MgO. The scatter in K, Rb, Sr concentrations could be the result of interaction with their ultrapotassic host magmas (Figure 3). In contrast to the State Line xenoliths, those from Leucite Hills all have well-developed negative Eu anomalies, and consistently high LILE/HFSE ratios. The negative Eu anomalies require fractionation of plagioclase from the basaltic magmas parental to the Leucite Hills xenoliths and so suggest that magmatic differentiation occurred at depths of less than ~40 km, depths at which plagioclase is an important liquidus phase in basaltic magmas [*Herzberg and O'Hara*, 1998]. The high LILE/HSFE ratios indicate that the magmas parental to the Leucite Hills xenoliths were likely produced in an arc environment.

Contrasting Age and Evolution of Lower Crust Across Cheyenne Belt

The Leucite Hills zircon form a relatively simple age population centered at ~2.6 Ga (Figure 11), similar to the ages of Archean rocks exposed in Wyoming [*Frost et al.*, 1998]. In addition, the Nd isotopic compositions of the majority of the Leucite Hills xenoliths plot on a ~2.6 Ga reference Sm-Nd isochron (Figure 14) and have late Archean Nd model ages (Table 3; see CDROM in back cover sleeve). Finally, the Nd isotopic composition of the majority of the Leucite Hills xenoliths overlap those of Late Archean granitic rocks exposed in the Wind River Range directly to the north of the Leucite Hills, themselves (Figures 1, 14). These observations suggest that the lower crust in this region formed in the Late Archean from arc related magmas during the original stabilization of this segment of continental crust and does not represent the product of underplating of the Archean crust by younger mafic magmatism. The relatively simple nature of the U-Pb zircon systematics, including the lack of younger overgrowths, makes it unlikely that these rocks have seen any significant younger metamorphic overprinting. Overall, then, we suggest that the hydrous nature of this crust was inherited from Late Archean arc magmatism, and that physical conditions in the lower crust where these xenoliths resided until the ~1 Ma Leucite Hills magmatism were never sufficient to result in breakdown of the originally formed hornblende.

South of the Cheyenne Belt, the complex U-Pb zircon ages from the State Line xenoliths are best interpreted as requiring a ~1720 Ma age for the original arc crust, with subsequent periods of zircon growth, particularly at 1.34–1.38 Ga. In contrast, *Hill and Bickford* [2001] have suggested that continental crust in central Colorado, and perhaps northern Colorado as well, was originally comprised of ~1.87 Ga crust, similar in age to rocks of the Trans Hudson Province (THP) in the north central U.S. and adjacent portions of Canada. We emphasize that none of the zircon yet analyzed from any State Line xenoliths have $^{207}Pb/^{206}Pb$ ages in the 1.8 Ga to 2.4 Ga range found for inherited components in Paleoproterozoic igneous rocks in central Colorado (Figure 11). However, because the Archean zircons (~3.2 Ga) found in several State Line xenoliths occur in igneous rocks, they must be viewed as inherited components, despite the fact that we have not documented Archean zircon with Proterozoic overgrowths. We consider this observation as clear evidence of Archean crust and/or sedimentary detritus south of the Cheyenne Belt and view the lack of any inherited zircons in the 1.8 Ga to 2.4 Ga age range in the State Line xenoliths as evidence that this segment of Colorado Precambrian crust is not a reworked portion of the THP. In light of our results, the THP cannot be regarded as the original starting material for Paleoproterozoic crust throughout Colorado.

Also supporting a Paleoproterozoic age for mafic lower crust beneath northern Colorado is the fact that the State Line xenoliths, whether low or high La/Yb_N, have chemical and whole rock Nd isotopic compositions similar to those of garnet granulites entrained in Tertiary igneous rocks from the Four Corners area in the SW U.S. (Figure 15) [*Wendlandt et al.*, 1993]. The latter xenoliths are interpreted to represent samples of ~1.7 Ga, arc-related lower crust. Mafic xenoliths from the San Francisco Volcanic Field in northern Arizona, a region underlain by Paleoproterozoic crust, also have Nd isotopic characteristic similar to those of both the Four Corners and State Line xenoliths (Figure 15) [*Chen and Arculus*, 1995]. The only mafic xenolith populations from Precambrian crustal regions with distinctly different Nd isotopic compositions are those from the West Texas volcanic field [*Cameron and Ward*, 1998]. These xenoliths are interpreted to have been derived from Grenville age (~1.1 Ga) lower crust, and the whole rock analyses define a Sm-Nd array with distinctly shallower slope than that defined by all of the mafic xenoliths derived from mafic lower crust underlying Paleoproterozoic crust (Figure 15). We consider this observation as strong evidence that the mafic lower crust sampled by the State Line xenoliths beneath the northern Front Range is Paleoproterozoic in age and does not represent mafic magmas underplated beneath this region during the Mesoproterozoic or at any subsequent time.

The low La/Yb_N xenoliths, however, do provide evidence that portions of the Paleoproterozoic mafic lower crust underwent anatexis subsequent to the time of crust formation. Unfortunately, we were unable to obtain zircon from any of the low La/Yb_N xenoliths so we have no direct, U-Pb based, estimate of when this anatexis may have occurred. Our Sm-Nd data also provide few constraints on the anatectic, or original crystallization, ages of the mafic lower crust beneath northern Colorado, primarily because the majority of our State Line samples are unleached whole rocks that are likely to contain secondary LREE introduced after the time of last isotopic equilibration for these rocks. Even our mineral isotopic data are insufficient to assess whether the age of anatexis for the low La/Yb_N xenoliths can be determined in this fashion. The plagioclase, clinopyroxene and leached whole rock samples from SD2 LC-71, for example, scatter about a ~300 m.y. Rb-Sr reference isochron (Figure16), the approximate emplacement age of the host kimberlites. This suggests that the crust sampled by this xenolith was at temperatures greater than the closure temperature for Sr in either plagioclase or clinopyroxene at the time of extraction by the kimberlite magma (~650°C and ~1,000°C, respectively, depending on cooling rate; [*Sneeringer et al.*, 1984; *Jenkin et al.*, 1995]. However, the Sm-Nd analyses from these same minerals define a crude linear array that parallels a ~1.0 Ga reference isochron (Figure 14), despite the similar closure temperatures for Sr and the LREE, at least in clinopyroxene [*Sneeringer et al.*, 1984]. The discrepancy between the Sr and Nd internal "errorchrons" in this sample could be the result of the introduction of high Rb/Sr secondary components into the xenolith during metasomatism, given that the plagioclase and clinopyroxene were unleached samples obtained through microdrilling and may have retained these secondary components. Alternatively, this discrepancy could have significance regarding the actual rates of diffusion of these elements in the lower crust, and may also yield information regarding the timing of the partial melting event that produced the unique chemical characteristics of the low La/Yb_N xenoliths. However, lacking sufficient mineral isotopic data from these rocks for these purposes, we can only state that the Nd errorchron from SD2-LC-71 suggests that resetting of the internal Nd isotopic systematics of this sample, presumably associated with anatexis, may have occurred at some point in the Mesoproterozoic. Whether or not the crustal heating event that led to lower crustal anatexis was also responsible for the episode of renewed zircon growth at 1.34–1.38 Ga recorded in the high La/Yb_N xenoliths remains a matter of speculation.

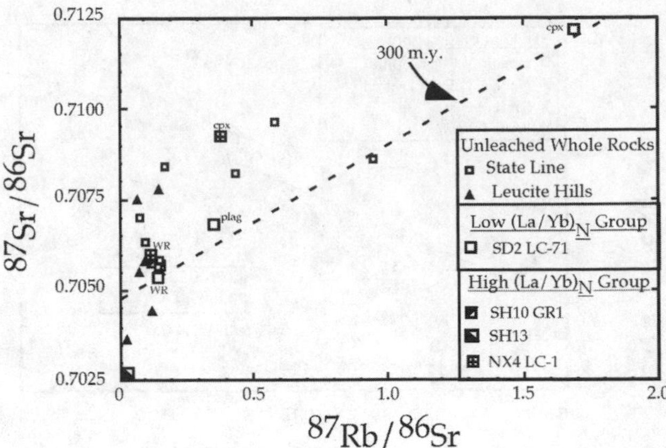

Figure 16. Measured $^{87}Sr/^{86}Sr$ vs. $^{87}Rb/^{86}Sr$ for minerals and leached and unleached whole rock xenoliths from State Line, and unleached whole rock xenoliths from Leucite Hills.

Although we do not know the exact age of lower crustal anatexis, it is interesting to note that the State Line diatremes occur in a portion of the northern Colorado underlain by the ~1.43 Ga, A-type, Sherman Batholith (Figure 1) [*Frost et al.*, 1999]. The magmas parental to the batholith have been interpreted as having been produced through anatexis of a low fO_2, tholeiitic composition, mafic crustal rock [*Frost et al.*, 1999]. The facts that the State Line mafic xenoliths are relatively anhydrous, and have subalkaline compositions, indicate that Paleoproterozoic mafic lower crust beneath

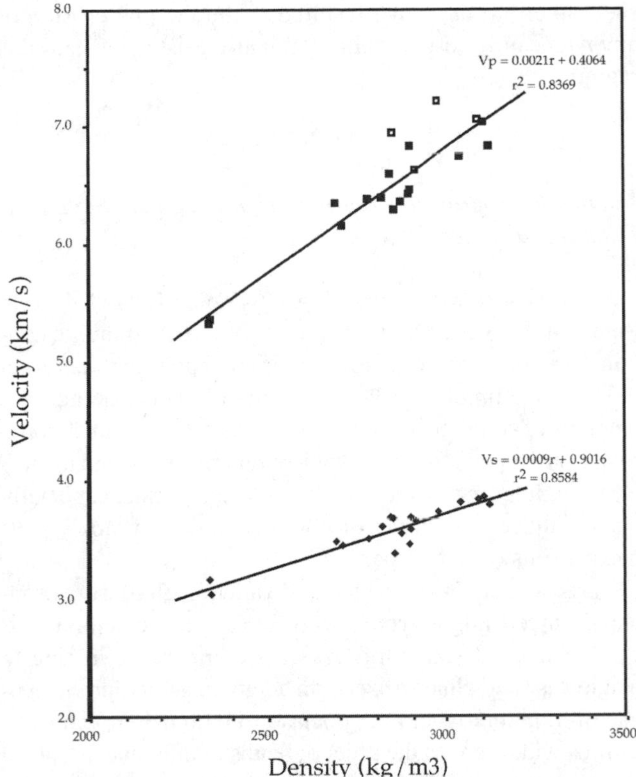

Figure 17. Velocity versus density at 600 MPa for Leucite Hills (closed symbols) and State Line (open symbols) xenoliths.

northern Colorado represents a viable source for the 1.4 Ga magmatism. Such a scenario is consistent with the similarity between the initial whole rock Nd isotopic compositions for the Sherman granite (ε_{Nd} =-1.2 to 1.8; [*Frost et al.*, 1999]) and the Nd isotopic compositions calculated for the leached State Line xenoliths at 1.4 Ga (ε_{Nd}=-2.8 to 1.5). However, it is important to emphasize that although the low La/Yb_N State Line xenoliths could represent restites related to the production of the 1.4 Ga magmatism in this area, our data provide few clues as to what triggered lower crustal melting. Injection of mafic magmas into the northern crust at ~1.4 Ga seems a likely mechanism for providing the heat necessary for anatexis, but we have no direct evidence that such an event occurred.

Significance of Physical Property Measurements

The two low velocity and density samples from Leucite Hills are granites (LHHS-6, LHHE-1), which likely originated at upper crustal levels. Velocities of the higher density rocks are consistent with a lower crustal origin. Increasing velocity and density correlate positively with increasing pyroxene and garnet content and negatively with the degree of secondary alteration. The samples from the State Line locality contain significant amounts of garnet and thus have higher velocities than many of the lower crustal xenoliths from Leucite Hills.

The primary controls on crustal compressional and shear wave velocities are mineral composition, preferred mineral orientation, temperature and confining pressure. These factors must all be considered in using laboratory data to infer crustal composition from seismic velocities. Temperature effects on seismic velocity have been documented by *Christensen* [1979]. As temperature increases velocity decreases, typically 0.1 to 0.3 km/s for most rock types at lower crustal depths. Anisotropy effects may be assessed by measuring physical properties on mutually perpendicular cores: however, because most of the lower crustal samples collected in this study appear to have relatively weak anisotropy, we invoke a seismically isotropic or weakly anisotropic lower crust beneath the Cheyenne Belt. The velocity measurements on the two Leucite Hills granulites (LHHW-1 and LHHE-26) from which we were able to obtain measurements parallel and perpendicular to layering show 1.6% and 4.6% compressional wave velocity anisotropy, respectively at 600 MPa with low velocities normal to layering.

As discussed earlier, the lower crustal xenoliths from the Leucite Hills and State Line regions differ significantly in mineralogy, suggesting that the Archean lower crust may differ from Proterozoic lower crust in seismic properties. At lower crustal pressures compressional wave velocities for Leucite Hills granulite facies rocks generally fall between 6.2 and 6.9 km/s, whereas velocities for State Line rocks vary from 6.5 to 7.1 km/s. Figure 17 compares both compressional and shear wave velocities at 600 MPa for rocks from both regions. The higher velocities for lower crustal rocks from the State Line locality originate from the presence of high velocity garnet and the anhydrous to weakly anhydrous nature of the rocks. In contrast, the Leucite Hills mafic granulites are garnet-free and contain abundant amphibole. This suggests that the mafic rocks of the Proterozoic lower crust of this region can be distinguished seismically from the garnet-free, hydrous granulites of the Archean.

Based on this discussion, it is likely that the 7.xx layer identified seismically beneath northern Colorado corresponds to garnet-bearing mafic crust, which implies that such crust remains an important component of the lower crust beneath this region [*Karlstrom et al.*, 2002]. The lack of a 7.xx layer beneath southern Wyoming [*Karlstrom et al.*, 2002] further suggests that garnet-bearing mafic lithologies do not exist beneath this region, which is consistent with the lack of garnet in the Leucite Hills xenolith population.

CONCLUSIONS

Although our data reveal that mafic lower crust is present both north and south of the Cheyenne Belt, the lower crust to the north is hydrous, garnet-free, and Late Archean in age, while to the south, the lower crust is Paleoproterozoic, generally anhydrous, and garnet bearing. Based on physical property measurements and the available seismic data, we suggest that observed variations in P-wave velocities observed in the lower crust across the Cheyenne Belt are controlled by this dichotomy in lower crustal mineralogy, with garnet-bearing lithologies south of the Cheyenne Belt producing the moderately high (~7.0 km/s) P-wave velocities observed in this region.

Our data also indicate that the lower crust beneath the southern Wyoming Craton has not been significantly heated above its present-day temperature since crust formation in the Late Archean. In contrast, the lower crust beneath northern Colorado has been affected thermally by events occurring after crust formation in the Paleoproterozoic, and after accretion of this crust against the southern margin of the Wyoming craton, most prominently manifested by metamorphic zircon growth at 1.34–1.37 Ga. However, none of the State Line xenoliths are clearly identifiable as the products of magmatic underplating of the lower continental crust in this region.

The fact that fundamental differences exist in the metamorphic history and the degree of hydration of lower crust across the Cheyenne Belt is a significant observation as it suggests that the Archean lower crust was not influenced by thermal events that thoroughly affected the Paleoproterozoic lower crust immediately to the south. This observation could bear on the long-standing debate as to the relative roles of deep lithospheric mantle roots and crustal composition in producing the relatively low surface heat flows observed in Archean cratons [*Nyblade and Pollack*, 1993; *Lenardic*, 1997]. It is possible, for example, that the lower crust beneath the southern Wyoming Craton, but not the Paleoproterozoic lower crust to the south, has been thermally insulated by thick, buoyant Archean lithospheric mantle and it is the presence of this lithosphere that accounts for the unique metamorphic history of the Archean lower crust. The issue is not straightforward, however, given that previous workers have suggested that Archean lithosphere actually extended somewhat south of the Cheyenne Belt through at least the Devonian, based on the chemical compositions of mantle xenoliths, and the occurrence of diamonds, in the State Line kimberlites [*Eggler et al.*, 1987*b*]. But it remains for future work to reconcile those lines of evidence favoring the existence of Archean mantle lithosphere immediately south of the Cheyenne Belt with our new data demonstrating extensive Proterozoic (?) reheating of the lower crust beneath the State Line area, an event that would seem likely to have also resulted in the thermal erosion of any Archean lithosphere that originally existed beneath this region.

APPENDIX

Detailed Petrographic Descriptions of Individual State Line District Mafic Xenoliths

SD2-LC74 (low La/Yb_N *group).* Sample SD2-LC74, from Pipe 2 of the Sloan Diatremes is a Cpx-Grt-Pl granulite containing plagioclase (~35%), clinopyroxene (~30%), garnet (~30%), and ilmenite (~3%) with minor K-feldspar and trace amounts of rutile, barite, and zircon. Plagioclase and clinopyroxene are equant to somewhat irregular in shape and display weak undulatory extinction. Plagioclase grains, especially where adjacent to garnet or clinopyroxene commonly have narrow rims of K-feldspar.

Garnet occurs in several textural varieties, all of which contain exsolved, aligned rutile needles. The most common variety consists of subhedral crystals comparable in size to plagioclase and clinopyroxene. In addition, garnet forms coronae on ilmenite within plagioclase-rich domains (Plate 1c). Corona widths are on the order of tenths of millimeters and are inversely proportional to ilmenite grain size. Garnet also occurs in symplectitic intergrowths with plagioclase or clinopyroxene (Plate 1d,f). Locally, symplectites are so fine grained as to constitute a fine foam-like mass (Plate 1d). Finally, small garnet grains and lamellae locally occur within plagioclase (Plate 1d). The included garnet is generally preferentially aligned along plagioclase twin planes, and unlike garnet-bearing symplectite, is generally not intergrown with other Fe-Mg-Ca phases.

SD2-LC120 (High La/Yb_N *group).* Sample SD2-LC120 from Pipe 2 of the Sloan Diatreme is a Cpx-Opx-Grt granulite containing plagioclase (35%), orthopyroxene (30%), garnet (25%), clinopyroxene (10%), with minor ilmenite and K-feldspar, and trace rutile and zircon. In general blocky, subhedral orthopyroxene grains occur in an equigranular mosaic of plagioclase with interstitial clinopyroxene and ilmenite. Garnet occurs as narrow coronae between orthopyroxene and plagioclase (Plate 1b) and as symplectitic intergrowths with fine plagioclase and clinopyroxene. Clinopyroxene occurs as isolated interstitial grains and as thin coronae between orthopyroxene and plagioclase (Plate 1e).

NX4-LC1 (High La/Yb_N *group).* Sample NX4-LC1 from Pipe 4 of the Nix Pipe, is a Cpx-Opx-Grt granulite containing plagioclase, garnet, orthopyroxene, clinopyroxene with minor ilmenite. The sample has a weak foliation defined by the alignment of slightly elongate orthopyroxene, clinopyroxene and ilmenite. Garnet occurs as coronae between orthopyroxene or ilmenite and plagioclase. Coronae are between 0.1 and 0.5 mm thick, and make up roughly 25–30% of the sample. The coronae grains have euhedral boundaries where in contact with plagioclase and commonly contain fine inclusions of plagioclase and clinopyroxene. Clinopyroxene typically occurs near or within garnet. Locally clinopyroxene occurs as rims on irregular, "relict" orthopyroxene.

Acknowledgments. Supported by National Science Foundation grants EAR-9614410 and EAR-0003747 (to Farmer), EAR-0003580 (to Christensen), EAR-9614727 and EAR-0003459 (to Williams), and EAR-0003551 (to Bowring). We are indebted to Dr. Malcolm McCallum (Emeritus, Colorado State University) for allowing us access to his extensive collection of xenoliths from the State Line district. This paper was greatly improved through reviews by Jane Selverstone, Nancy McMillan, and Doug Smith.

REFERENCES

Alibert, C., and F. Albarede, Relationships between mineralogical, chemical, and isotopic properties of some North American kimberlites, *J. Geophys. Res.*, *93*, 7643–7671, 1988.

Anderson, J.L., and R.L. Cullers, Paleo- and Mesoproterozoic granite plutonism of Colorado and Wyoming, *Rocky Mtn. Geol.*, *34*, 149–164, 1999.

Ayers, J.C., K. de la Cruz, C. Miller, and O. Switzer, Experimental study of zircon coarsening in quartzite | or -H (sub 2) O at 1.0 GPa and 1000 degrees C, with implications for geochronological studies of high-grade metamorphism, *Amer. Mineral.*, *88*, 365–376, 2003.

Berman, R.G., Thermobarometry using multi-equilibrium calculations; a new technique, with petrological applications, *Can. Mineral.*, *29*, 833–855, 1991.

Berman, R.G., *Thermobarometry with estimation of equilibrium state (TWEEQU): an IBM-compatible software package*, 1992.

Berman, R.G., and L.Y. Aranovich, Optimized standard state and solution properties of minerals; I, Model calibration for olivine, orthopyroxene, cordierite, garnet, and ilmenite in the system FeO-MgO-CaO-Al_2O_3 -TiO_2 -SiO_2, *Contrib. Mineral. Petrol.*, *126*, 1–24, 1996.

Bohlen, S.R., and J.J. Liotta, Barometer for garnet amphibolites and garnet granulites, *J. Petrol.*, *27*, 1025–1034, 1986.

Bradley, S.D., Granulite facies and related xenoliths from Colorado-Wyoming kimberlite, MSc. thesis, Colorado State University, 1985.

Bradley, S.D., and M.E. McCallum, Granulite facies and related xenoliths from Colorado-Wyoming kimberlite, in *The Mantle and Crust-Mantle Relationships*, pp. 205–217, 1984.

Cameron, K.L., and R.L. Ward, Xenoliths of Grenvillian granulite basement constrain models for the origin of voluminous Tertiary rhyolites, Davis Mountains, west Texas, *Geology*, *26*, 1087–1090, 1998.

Carlson, R.W., F.R. Boyd, S.B. Shirey, P.E. Janney, T.L. Grove, S.A. Bowring, M.D. Schmitz, J.C. Dann, D.R. Bell, J.J. Gurney, S.H. Richardson, M. Tredoux, A.H. Menzies, D.G. Pearson, R.J. Hart, A.H. Wilson, and D. Moser, Continental growth, preservation, and modification in southern Africa, *GSA Today*, *10*, 1–7, 2000.

Carlson, W.D., Scales of disequilibrium and rates of equilibration during metamorphism, *Amer. Mineral.*, *87*, 185–204, 2002.

Chen, W., and R.J. Arculus, Geochemical and isotopic characteristics of lower crustal xenoliths, San Francisco volcanic field, Arizona, U.S.A., *Lithos*, *36*, 203–225, 1995.

Cherniak, D.J., Lead diffusion in titanite and preliminary results on the effects of radiation damage on Pb transport, *Chem. Geol.*, *110*, 177–194, 1993.

Christensen, N.I., Compressional wave velocities in rocks at high temperatures and pressures, critical thermal gradients, and low velocity zones, *J. Geophys. Res.*, *84*, 6849–6857, 1979.

Christensen, N.I., Measurement of dynamic properties of rock at elevated pressures and temperatures, in *Measurement of Rock Properties at Elevated Pressure and Temperature: A Symposium*, edited by J. Pincus, and E.R. Hoskins, American Society of Testing and Materials, Philidelphia, 1985.

Condie, K.C., Contrasting sources for upper and lower continental crust: the greenstone connection, *J. Geol.*, *105*, 729–736, 1997.

Condie, K.C., and J. Selverstone, The crust of the Colorado Plateau; new views of an old arc, *J. Geol.*, *107*, 387–397, 1999.

Corriveau, L., and D. Morin, Modeling 3D architecture of western Grenville from surface geology, xenoliths, styles of magma emplacement, and Lithoprobe reflectors, *Can. J. Earth. Sci.*, *37*, 235–251, 2000.

Ducea, M.N., and J.B. Saleeby, Buoyancy sources for a large, unrooted mountain range, the Sierra Nevada, California: Evidence from xenolith thermobarometry, *J. Geophys. Res.*, *101*, 8229–8244, 1996.

Eggler, D.H., M.E. McCallum, and M.B. Kirkley, Kimberlite-transported nodules from Colorado-Wyoming; A record of enrichment of shallow portions of an infertile lithosphere, in *Mantle metasomatism and alkaline magmatism*, edited by E. Mullen-Morris, and J. Dill-Pasteris, pp. 77–90, Boulder, Colorado, 1987a.

Eggler, D.H., J.K. Meen, F. Welt, F.O. Dudas, K.P. Furlong, M.E. McCallum, and R.W. Carlson, Tectonomagmatism of the Wyoming province, *Col. School Mines Quart.*, *82*, 25–40, 1987b.

Farmer, G.L., D.E. Broxton, R.G. Warren, and W. Pickthorn, Nd, Sr, and O isotopic variations in metaluminous ash-flow tuffs and related volcanic rocks at the Timber Mountain/Oasis Valley Caldera Complex, SW Nevada: implications for the origin and evolution of large-volume silicic magma bodies, *Contrib. Mineral. Petrol.*, *109*, 53–68, 1991.

Farmer, G.L., and D.J. DePaolo, Origin of Mesozoic and Tertiary granite in the western United States and implications for pre-Mesozoic crustal structure 1. Nd and Sr isotopic studies in the geocline of the northern Great Basin, *J. Geophys. Res.*, *88*, 3379–3401, 1983.

Frost, C.D., J.M. Bell, B.R. Frost, and K.R. Chamberlain, Crustal growth by magmatic underplating: Isotopic evidence from the northern Sherman batholith, *Geology*, *29*, 515–518, 2001.

Frost, C.D., B.R. Frost, K.R. Chamberlain, and B.R. Edwards, Petrogenesis of the 1.43 Ga Sherman Batholith, SE Wyoming: a reduced, rapakivi-type anorogenic granite, *J. Petrol.*, *40*, 1771–1802, 1999.

Frost, C.D., B.R. Frost, K.R. Chamberlain, and T.P. Hulsebosch, The late Archean history of the Wyoming Province as recorded by granitic magmatism in the Wind River Range, Wyoming, *Precamb. Res.*, *89*, 145–173, 1998.

Green, D.H., and A.E. Ringwood, An experimental investigation of the gabbro to eclogite transformation and its petrologic applications, *Geochem. Cosmochem. Acta*, *31*, 767–833, 1967.

Hausel, W.D., *Diamonds and Mantle Source Rocks in the Wyoming Craton with a Discussion of Other U.S. Occurrences*, Wyoming State Geological Survey Report of Investigations, 93 pp., 1998.

Herzberg, C., and M.J. O'Hara, Phase equilibrium constraints on the origin of basalts, picrites, and komatiites, *Earth Sci. Rev.*, *44*, 39–79, 1998.

Hill, B.M., and M.E. Bickford, Paleoproterozoic rocks of central Colorado: Accreted arcs or extended older crust, *Geol. Soc. Amer. Bull.*, *29*, 1015–1018, 2001.

Holland, T.J.B., and J.D. Blundy, Non-ideal interactions in calcic amphiboles and their bearing on amphibole plagioclase thermometry, *Contrib. Mineral. Petrol.*, 433–447, 1994.

Holtta, P., I. Huhma, I. Manttari, P. Peltonen, and J. Juhanoja, Petrology and geochemistry of mafic granulite xenoliths from the Lahtojoki kimberlites pipe, eastern Finland, *Lithos*, *51*, 109–133, 2000.

Jaffey, A.H., K.F. Flynn, L.E. Glendenin, W.C. Bentley, and A.M. Essling, Precision measurement of half-lives and specific activities of ^{235}U and ^{238}U, *Phys. Rev. C*, *4*, 1889–1906, 1971.

Jenkin, R.T., G. Rogers, A.E. Fallick, and C.M. Farrow, Rb-Sr closure temperatures in bi-mineralic rocks; a mode effect and test for different diffusion models, *Chem. Geol.*, *122*, 227–240, 1995.

Karlstrom, K.E., S.A. Bowring, K.R. Chamberlain, K.G. Dueker, T. Eshete, E.A. Erslev, G.L. Farmer, M. Heizler, E.D. Humphreys, R.A. Johnson, G.R. Keller, S.A. Kelley, A. Levander, M.B. Magnani, J.P. Matzel, A.M. McCoy, K.C. Miller, E.A. Morozova, F.J. Pazzaglia, C. Prodehl, H.M.

Rumpel, C.A. Shaw, A.F. Sheehan, E. Shoshitaishvili, S.B. Smithson, C.M. Snelson, L.M. Stevens, A.R. Tyson, and M.L. Williams, Structure and evolution of the lithosphere beneath the Rocky Mountains; initial results from the CD-ROM experiment, *GSA Today*, *12*, 4–10, 2002.

Kay, S.M., R.W. Kay, J. Hangas, and T. Snedden, Crustal xenoliths from potassic lavas, Leucite Hills, Wyoming, *Geol. Soc. Am. Abstr. Prog.*, *10*, 432, 1978.

Krogh, T.E., A low-contamination method for hydrothermal decomposition of zircon and extraction of U and Pb for isotopic age determinations, *Geochem. Cosmochim. Acta*, *37*, 485–494, 1973.

Krogh, T.E., Improved accuracy of U-Pb zircon ages by the creation of more concordant systems using an air abrasion technique, *Geochem. Cosmochim. Acta*, *46*, 637–649, 1982.

Kretz, R. 1983. Symbols for rock-forming minerals. *American Mineralogist* **68**, 277–279.

Lange, R.A., I.S.E. Carmichael, and C.M. Hall, 40Ar/39Ar chronology of the Leucite Hills, Wyoming: eruption rates, erosion rates, and an evolving temperature structure of the underlying mantle, *Earth. Planet. Sci. Lett.*, *174*, 329–340, 2000.

Lenardic, A., On the heat flow variation from Archean cratons to Proterozoic mobile belts, *J. Geophys. Res.*, *102*, 709–721, 1997.

Lester, A., and G.L. Farmer, Lower crustal and upper mantle xenoliths along the Cheyenne Belt and vicinity, *Rocky Mtn. Geol.*, *33*, 293–304, 1998.

Lester, A.P., E.E. Larson, G.L. Farmer, C.R. Stern, and J.A. Funk, Neoproterozoic kimberlite emplacement in the Front Range, Colorado, *Rocky Mtn. Geol.*, *36*, 1–12, 2001.

Ludwig, K.R., Calculation of uncertainties of U-Pb isotope data, *Earth Planet. Sci. Lett.*, *46*, 212–220, 1980.

Mader, U.K., J.A. Percival, and R.G. Berman, Thermobarometry of garnet-clinopyroxene-hornblende granulites from the Kapuskasing structural zone, *Can. J. Earth. Sci.*, *31*, 1134–1145, 1994.

McCulloch, M.T., and J.A. Gamble, Geochemical and geodynamical constraints on subduction zone magmatism, *Earth Planet. Sci. Lett.*, *102*, 358-374, 1991.

Nyblade, A.A., and H.N. Pollack, A global analysis of heat flow from Precambrian terrains; implications for the thermal structure of Archean and Proterozoic lithosphere, *J. Geophys. Res.*, *98*, 12,207–12,218, 1993.

Pearce, J.A., and D.W. Peate, Tectonic implications of the composition of volcanic arc magmas, *Annu. Rev. Eatht Planet. Sci.*, *23*, 252–285, 1995.

Pouchou, J.L., and F. Pichoir, "PAP" phi-rho-Z procedure for improved quantitative microanalysis, in *Microbeam Analysis*, edited by J.L. Armstrong, pp. 104–106, San Fransisco Press Inc., San Fransisco, 1985.

Rudnick, R.L., Xenoliths-samples of the lower continental crust, in *Continental Lower Crust*, edited by D.M. Fountain, R. Arculus, and R.W. Kay, pp. 269–316, Elsevier, Amsterdam, 1992.

Rudnick, R.L., W.F. McDonough, M.T. McCulloch, and S.R. Taylor, Lower crustal xenoliths from Queensland, Australia: evidence for deep crustal assimilation and fractionation of continental basalts, *Geochem. Cosmochim. Acta*, *50*, 1099–1115, 1986.

Schmitz, M.D., and S.A. Bowring, Ultrahigh-temperature metamorphism in the lower crust during Neoarchean Ventersdorp rifting and magmatism, Kaapvaal Craton, Southern Africa, *Geol. Soc. Amer. Bull.*, *115*, 533–548, 2003.

Selverstone, J., A. Pun, and K.C. Condie, Xenolithic evidence for Proterozoic crustal evolution beneath the Colorado Plateau, *Geol. Soc. Amer. Bull.*, *111*, 590–606, 1999.

Sneeringer, M., S.R. Hart, and N. Schimizu, Strontium and samarium diffusion in diopside, *Geochem. Cosmochem. Acta*, *48*, 1589–1608, 1984.

Speer, J.T., Xenoliths of the Leucite Hills volcanic rocks, Sweetwater County, Wyoming, MSc. thesis, University of Wyoming, 1985.

Stacey, J.S., and J.D. Kramers, Approximation of terrestrial lead isotope evolution by a two-stage model, *Earth Planet. Sci. Lett.*, *26*, 207–221, 1975.

Stevens, L., Petrology of mafic granulite xenoliths across the Cheyenne Belt, Colorado and Wyoming, and implications for the nature of the lower crust, MSc. thesis, University of Massachusetts, Amherst, 2002.

Sun, S.-s., and W.F. McDonough, Chemical and isotopic systematics of oceanic basalts: implications for mantle composition and processes, in *Magmatism in the Ocean Basins*, edited by A.D. Saunders, and M.J. Norry, pp. 313–345, 1989.

Vernon, R.H., *Metamorphic Processes*, George Allen and Unwin, London, 1976.

Vollmer, R., P. Ogden, J.-G. Schilling, R.H. Kingsley, and D.G. Waggoner, Nd and Sr isotopes in ultrapotassic volcanic rocks from the Leucite Hills, Wyoming, *Contrib. Mineral. Petrol.*, *87*, 359–368, 1984.

Wendlandt, E., D.J. DePaolo, and W.S. Baldridge, Nd and Sr chronostratigraphy of Colorado Plateau lithosphere; implications for magmatic and tectonic underplating of the continental crust, *Earth. Planet. Sci. Lett.*, *116*, 23–43, 1993.

Williams, M.L., and K.E. Karlstrom, Looping P-T paths and high-T, low-P middle crustal metamorphism: Proterozoic evolution of the southwestern United States, *Geology*, *24*, 1119–1122, 1996.

Williams, M.L., E.A. Melis, C.F. Kopf, and S. Hanmer, Microstructural tectonometamorphic processes and the development of gneissic layering; a mechanism for metamorphic segregation, *J. Metamorph. Geol.*, *18*, 41–57, 2000.

Zindler, A., and A. Jagoutz, Mantle cryptology, *Geochem. Cosmochem. Acta*, *52*, 319–333, 1988.

Zou, H., and M.R. Reid, Quantitative modeling of trace element fractionation during incongruent dynamic melting, *Geochem. Cosmochim. Acta*, *65*, 153–162, 2001.

S. A. Bowring, Department of Earth, Atmospheric and Planetary Sciences, Massachusetts Institute of Technology, 77 Massachusetts Ave., Cambridge, MA 02129-4307, sbowring@MIT.EDU (Bowring).

N. I. Christensen, Department of Geology and Geophysics, 156 Weeks Hall, 1215 West Dayton St., Madison, WI 53706, chris@geology.wisc.edu.

G. L. Farmer, Department of Geological Sciences and CIRES, Campus Box 399, University of Colorado, Boulder, CO 80309, farmer@cires.colorado.edu.

J. P. Matzel, Atmospheric and Planetary Sciences, Massachusetts Institute of Technology, 77 Massachusetts Ave., Cambridge, MA 02129-4307, sbowring@MIT.EDU (Bowring).

L. Stevens, and M. L. Williams, Department of Geosciences, University of Massachusetts, Amherst MA 01003, mlw@geo.umass.edu (Williams).

$^{40}Ar/^{39}Ar$ Thermochronologic Record of 1.45–1.35 Ga Intracontinental Tectonism in the Southern Rocky Mountains: Interplay of Conductive and Advective Heating with Intracontinental Deformation

Colin A. Shaw[1]

Department of Earth and Planetary Sciences, University of New Mexico, Albuquerque, New Mexico

Matthew T. Heizler

New Mexico Bureau of Geology and Mineral Resources, Socorro, New Mexico

Karl E. Karlstrom

Department of Earth and Planetary Sciences, University of New Mexico, Albuquerque, New Mexico

New and compiled $^{40}Ar/^{39}Ar$ data from the southern Rocky Mountains and SW U.S. provide significant constraints on the Mesoproterozoic thermal evolution of the region. Our main conclusions are that the Proterozoic provinces of the southwest were affected by a regional thermal episode 1.45–1.35 Ga coeval with intracontinental tectonism and plutonism and that the highest temperatures were recorded in southern Colorado and northern New Mexico where there are few ~1.4 Ga plutons. In northern and central Colorado biotite and muscovite dates of ~1.4 Ga and discordant hornblende dates of 1.7–1.4 Ga are consistent with variable ~1.4 Ga temperatures of ~300–500°C with narrow (<10 km) thermal aureoles around 1.4 Ga plutons and abrupt temperature changes across mylonite zones. In contrast, compiled data from the Archean Wyoming craton preserve dates up to 3.2 Ga indicating that regional temperatures at the level of exposure did not exceed ~300°C during the Proterozoic. Hornblende data from southern Colorado and northern New Mexico yield a narrow range of both hornblende and mica dates ~1.45–1.35 Ga suggesting mid-crustal temperatures >500°C. This area is characterized by a paucity of 1.4 Ga plutons, more penetrative ~1.4 Ga ductile deformation, and higher grade metamorphism. We postulate that this region may represent a deeper level of exposure (15–20 km) beneath a mid-crustal layer with a high concentration of ~1.4 Ga plutons emplaced near the brittle-ductile transition (~10–15 km).

The Rocky Mountain Region: An Evolving Lithosphere
Geophysical Monograph Series 154

10.1029/154GM12

[1]Present address: Department of Geology, University of Wisconsin, Eau Claire, Wisconsin

INTRODUCTION

The lithosphere of the southwestern United States comprises a mosaic of arc terranes that were assembled to the southern margin of the Archean Wyoming craton between 1.8 and 1.6 Ga [Figure 1; *Karlstrom and Bowring, 1993*]. This 1200-km-wide swath of juvenile lithosphere was profoundly modified by a widespread 1.45–1.35 Ga (~ 1.4 Ga) tectonic episode involving A-type granitic magmatism, deformation, and metamorphism [e.g. *Silver et al.*, 1977; *Anderson*, 1983; *Anderson and Morrison*, 1992; *Bickford and Anderson*, 1993; *Nyman et al.*, 1994; *Karlstrom and Humphreys*, 1998; *Shaw et al.,* 1999; *Williams et al.*, 1999]. Laramide basement uplifts expose Proterozoic mid-crustal rocks throughout the region [*Williams and Karlstrom,* 1996]. $^{40}Ar/^{39}Ar$ thermochronologic data provide important constraints on the thermal evolution of the middle crust and provide insight into processes of continental growth, stabilization, and intracontinental deformation that are hidden in younger, less deeply exhumed orogenic belts.

In this paper we argue that $^{40}Ar/^{39}Ar$ and K/Ar mineral dates from throughout the Proterozoic provinces of the Southwest record a widespread thermal episode coeval with ~1.4 Ga plutonism and deformation [*Shaw et al.*, 1999]. The overwhelming majority of mica $^{40}Ar/^{39}Ar$ dates from Colorado and New Mexico are less than 1.45 Ga. Muscovite and biotite ages from many localities are nearly concordant. Hornblende data are more dispersed between 1.7 and 1.38 Ga and many yield highly complex intermediate ages. These observations are best explained by variable degrees of thermally enhanced diffusive argon loss and/or recrystallization of mica and hornblende at ~1.4 Ga. Thus, argon dates from minerals with different closure or recrystallization temperatures can be used to provide information about peak ~1.4 Ga temperatures in Proterozoic mid-crustal rocks that are now exposed on the surface. Spatial trends in the data suggest that temperature variations due to inherited lithospheric structure, proximity to plutons, and depth of exposure influenced ~1.4 Ga peak temperatures, and thus the degree of argon loss and/or recrystallization.

METHODS AND ASSUMPTIONS

We employ a statistical approach to distill regional trends from a large thermochronologic dataset that reflects variations in mineral composition, grain size, local or transient thermal irregularities, depth of exposure, and other factors. Several simplifications are necessary to facilitate this analysis. (1) For our compilations, we assign a single plateau or preferred date to $^{40}Ar/^{39}Ar$ age spectra that is calculated from the mean of individual steps weighted by the inverse variance. This approach provides a broad view of regional trends at the expense of detailed information on cooling histories from individual samples. New data are summarized in Appendix 1 (see CDROM

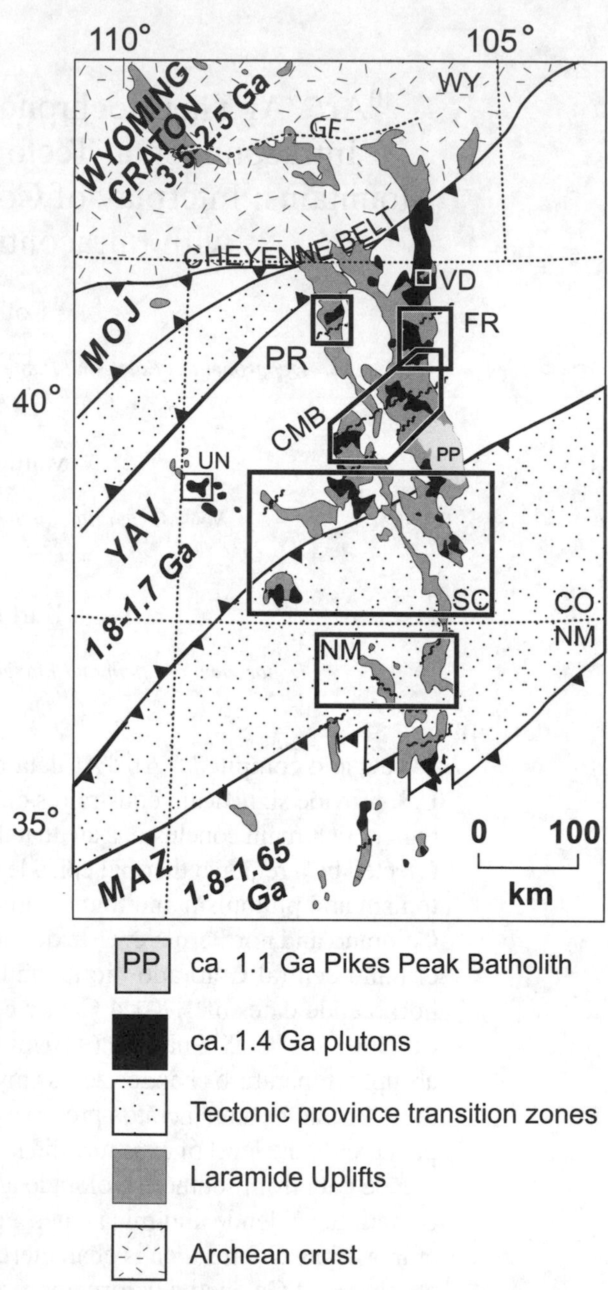

Figure 1. Index map showing generalized Precambrian geology of the southern Rocky Mountains, with crustal provinces and ages from *Karlstrom and Bowring,* 1988 (MOJ: Mojave, YAV: Yavapai, MAZ: Mazatzal). Stipple indicates transition zones between provinces. Study areas discussed in text are indicated: PR: Park Range (Figure 3), FR: Front Range (Figure 4), VD: Virginia Dale (no detailed map), CMB: Colorado mineral belt (Figure 5), SC: South-Central Colorado Transect (Figure 6), UN Uncompahgre (Figure 6, inset).

in back cover sleeve) and detailed incremental release spectra are presented in Appendix 2 (see CDROM). (2) We use both K-Ar mineral dates and $^{40}Ar/^{39}Ar$ dates from the literature as a supplement to the new $^{40}Ar/^{39}Ar$ dates (Appendixes 3 and 4; see CDROM). This is justified by the comparison presented in Figure 2 that shows broad correlation between total gas dates (roughly equivalent to K/Ar dates) and plateau dates from studies where both are available. (3) We use the widely cited nominal closure temperature (T_c) ranges [*McDougall and Harrison*, 1999]. Assuming slow cooling, our assigned closure temperatures are: hornblende, T_c = 525° C; muscovite, T_c = 350° C; biotite, T_c = 300° C.

Many hornblende samples yield highly complex spectra with ages between 1.65 and 1.45 Ga that do not correspond to any recognized regional tectonic episode. We interpret these spectra as geologically meaningless 'mixed' ages. produced by episodic partial loss of argon and/or recrystallization of hornblende during a ~1.4 Ga thermal episode. An alternate interpretation is that these ages record blocking of argon diffusion during very slow cooling after ~1.75–1.65 Ga tectonism [e.g. *Hodges et al.,* 1994; *Chamberlain and Bowring,* 2001]. We examine mechanisms that might account for partial resetting of argon thermochronometers below and discuss evidence supporting a ~1.4 Ga thermal event in a separate section.

One mechanism for producing disturbed spectra is homogenization during laboratory heating of partial argon loss profiles that are created in nature by thermally enhanced diffusion [*Harrison*, 1981; *Baldwin et al.,* 1990; *Kelley and Turner,* 1991]. In rare cases well-behaved diffusion profiles are exhibited by amphiboles that have undergone partial argon loss [e.g. *Harrison and McDougall,* 1980b], but more commonly spectra do not reflect simple diffusive degassing of the sample during the experiment. The lack of simple diffusion profiles in amphibole spectra that record partial argon loss is related to instability of the amphibole in the ultra high-vacuum, high-temperature argon extraction furnace [cf. *Lee et al.,* 1991]. Despite the lack of clear diffusion spectra in incremental heating experiments, disturbed hornblende spectra can give a qualitative indication of partial argon loss that reflects diffusion profiles within crystals [*Kelley and Turner,* 1991].

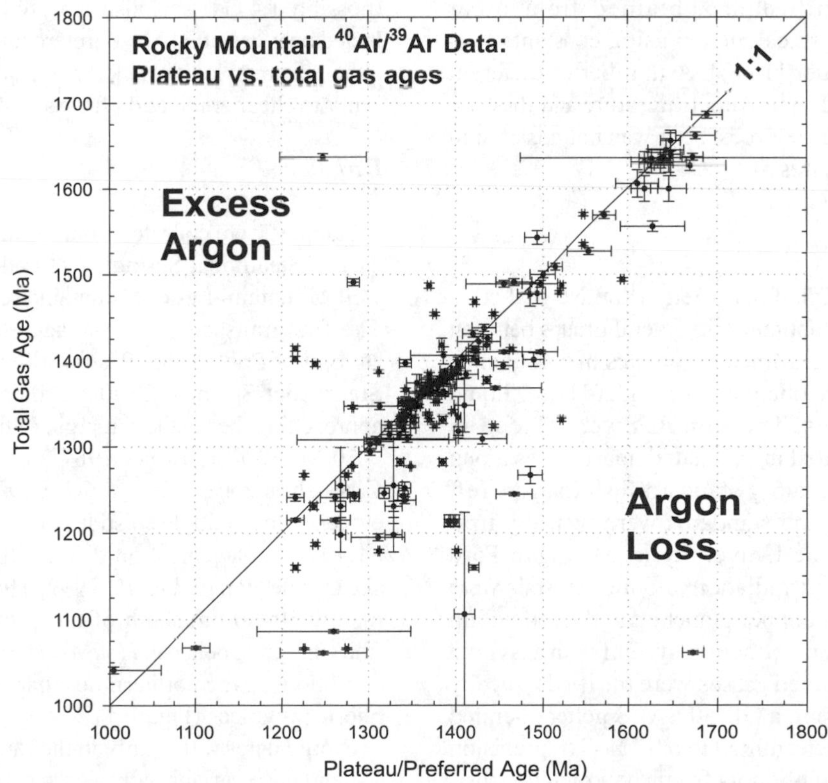

Figure 2. Plot of 260 $^{40}Ar/^{39}Ar$ plateau dates vs. $^{40}Ar/^{39}Ar$ total gas dates from this study and recently published studies [*Harlan et al.*, 1994; *Holm et al.*, 1997; *Karlstrom et al.*, 1997a; *Marcoline et al.*, 1999]. Although there is considerable scatter, broad equivalence in ages suggests that K/Ar ages (equivalent to total gas ages) and $^{40}Ar/^{39}Ar$ plateau ages should yield similar first-order information about temperature conditions. Most data lie below the 1:1 line showing that, on average, argon loss dominates over excess argon among the samples plotted.

Another mechanism to produce mixed ages is to partially recrystallize existing mineral grains or to grow rims or overgrowths on older cores. For instance, *Marcoline et al.* [1999] inferred that ~1.7 Ga actinolite had been partially to completely replaced by hornblende during 1.4 Ga regional amphibolite facies metamorphism in central New Mexico producing a range of geologically meaningless intermediate ages. Because the closure temperature for argon diffusion in hornblende [~500–550°C, *McDougall and Harrison,* 1999] and the lower temperature stability of hornblende [~550° C, *Spear and Gilbert,* 1982] are similar, temperatures inferred from complex hornblende spectra are fairly insensitive to the mechanism of argon loss.

For all new data we use a newly proposed age for Fish Canyon Tuff (FCT) sanidine standard and total ^{40}K decay constant. Over the past several years it has become increasingly obvious that, for K-Ar ages to be consistent with U/Pb dates, there needs to be a change in one or both of the standard values chosen for the FCT age and total decay constant [*Renne et al.*, 1998; *Min et al.*, 2000; *Schmitz and Bowring*, 2001; *Kwon et al.*, 2002]. We have applied the recently suggested values by Kwon et al. [2002] that yield apparent ages likely to be in better agreement with dates obtained from other geochronometers than those calculated using constants suggested by Steiger and Jäger [1977]. A number of different decay constants were used in the older literature and they are not always identified in the sources, so it was not possible to recalculate all compiled dates.

Analytical Methods

Samples were prepared and analyzed at the New Mexico Geochronology Research Laboratory in several groups between August 1994 and June 2002. Mineral separates of hornblende, muscovite, and biotite were obtained by standard heavy liquid and magnetic techniques. The samples were placed in machined Al discs and sealed in evacuated quartz tubes along with monitors of interlaboratory standard Fish Canyon Tuff (28.27 Ma). The samples and standards were typically irradiated for 100 hours at the University of Michigan Ford research reactor. Following irradiation, monitor crystals were placed in holes drilled in a copper planchett and fused within an ultra-high vacuum argon extraction system with a Synrad CO_2 continuous laser. Evolved gasses were purified typically for two to four minutes using a GP-50 SAES getter operated at 450°C. J-factors were determined to a 0.1% (1σ) precision by analyzing single crystal aliquots from each monitor position. The samples were step-heated in a double vacuum Mo resistance furnace or with a defocused laser beam. Evolved gas was cleaned during heating with a SEAS GP-50 getter operated at ~450° C and for an additional 3–5 minutes in a second stage containing two GP-50 getters (450° C, 20° C). Argon isotopic compositions were determined with an MAP 215–50 mass spectrometer in electron multiplier mode with a typical sensitivity of 2x10^{-16} moles/pA. Extraction system and mass spectrometer blanks and backgrounds were measured repeatedly throughout the analyses and typically were under 200, 1.0, 0.5, 0.5, and 1.3 x 10^{-18} moles at masses 40, 39, 38, 37, and 36, respectively. Corrections for interfering nuclear reactions were determined using K-glass and CaF_2. The total decay constant (5.476 x 10^{-10} /y) was taken from *Kwon et al.,* [2002] and relative isotopic abundances from *Steiger and Jäger* [1977]. Uncertainties are reported at the 1σ level and include uncertainty in J-value precision.

NEW ^{40}AR/^{39}AR RESULTS FROM COLORADO AND NEW MEXICO

Data presented in this section include analyses from a variety of rock types collected in several mountain ranges in Colorado and northern New Mexico (Figures 1, 3). Samples include metasedimentary and metaigneous rocks of Paleoproterozoic age. Some pegmatite and amphibolite dikes of uncertain age (possibly 1.4 Ga) were also analyzed. Many samples were collected near or within Mesoproterozoic shear zones and plutons although no mylonitic rocks were analyzed. The effects of proximity to shear zones and plutons will be discussed below.

Park Range

Samples were collected from two transects in the Park Range east of Steamboat Springs, Colorado and also from the Farwell Mountain–Lester Mountain area to the north (Figure 3). The first transect is an E-W age-elevation profile from the summit of Rabbit Ears Pass (2800 m) to the range front near Steamboat Springs (2200m), the second is a N-S traverse approaching the southern margin of the Mt. Ethel Pluton [1439 ± 50 Ma, Rb-Sr; *Tweto*, 1987] across the Fish Creek–Soda Creek shear zone [FCSC, *Snyder*, 1980]. The shear zone shows evidence for multiple episodes of deformation including some ~1.4 Ga displacement on discrete mylonite zones within the shear zone [*Barinek et al.,* 1999] The study area lies approximately 50–80 km south of the Archean–Proterozoic suture [the Cheyenne belt, e.g. *Houston et al.*, 1989] so the data reflect conditions in the northernmost part of the Proterozoic lithospheric provinces (Figure 1).

Biotite dates fall mainly in the range 1350–1460 Ma (Plate 1a) and have variably complex spectra (Appendix 2). There is no apparent trend in the biotite dates with respect to elevation or distance from the Mt. Ethel pluton (Figure 3). The variation in individual biotite ages likely represents local temperature variations as well as a range in the effective closure temperature of individual biotite samples due to differences in com-

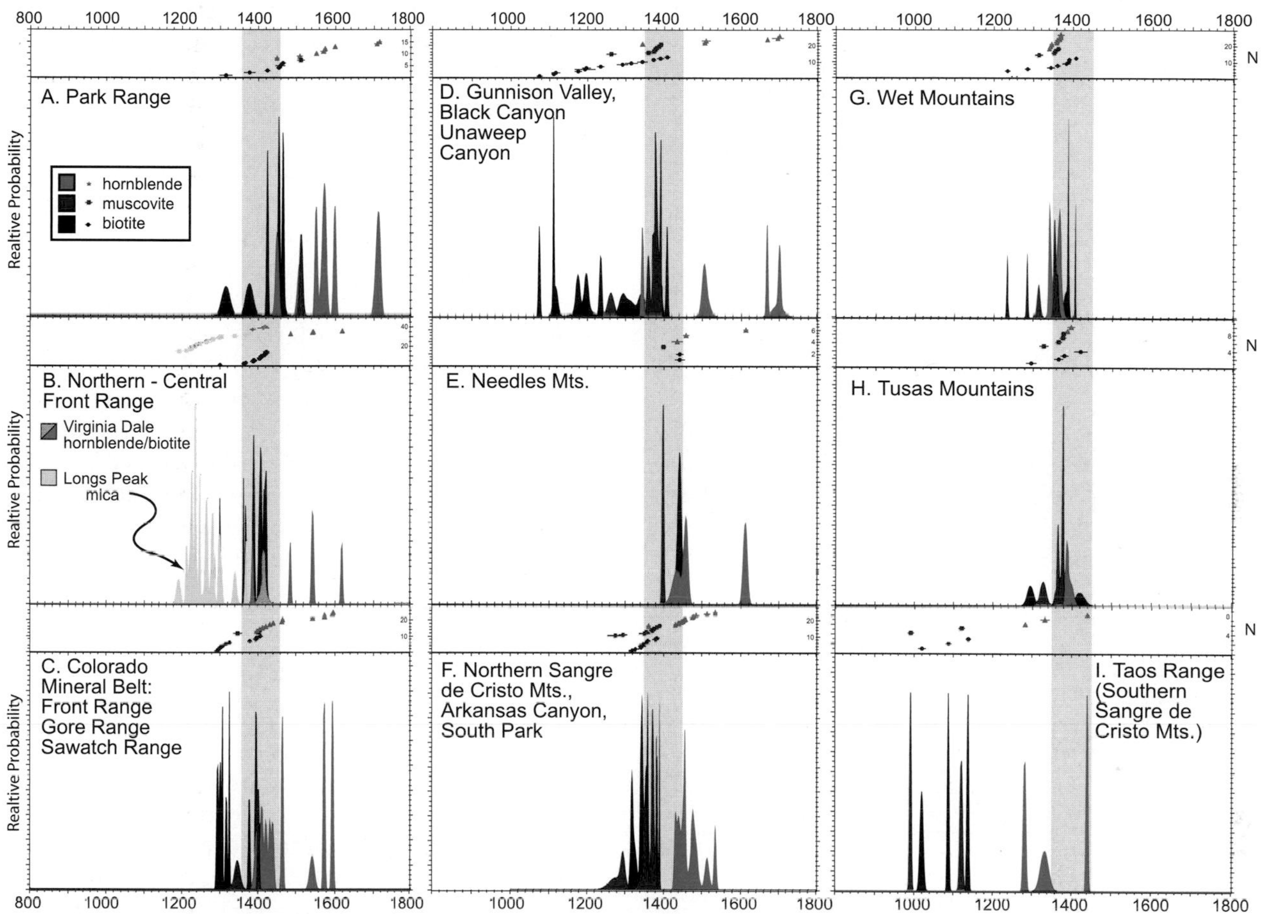

Plate 1. Relative probability diagrams (histograms) summarizing data from Proterozoic provinces of the southern Rocky Mountains (Colorado and Northern New Mexico). Data are grouped by mountain range or region (see Figure 1 for locations). Gray bar represents approximate time span of ca. 1.40 Ga tectonism (1.45–1.35 Ga). Data from northern, central, and southwestern Colorado (Park Range, northern Front Range, Colorado mineral belt, Gunnison Valley, Arkansas Valley, Needles Mountains) are characterized by mica dates of ~ 1.35–1.45 Ga hornblende ages 1.70–1.45 Ga. Data from the Wet Mountains, and northern New Mexico are characterized by hornblende and mica ages of < 1.45 Ga indicating partial to complete resetting by higher ~1.4 Ga temperatures. Cumulative probability plots provide a convenient visual summary of our results. However, it should be noted that these plots are models of the data distribution based on assumptions of the form of probability distributions. Reduced data tables, $^{40}Ar/^{39}Ar$ incremental release spectra, and a summary table of plateau and preferred ages are presented in Appendixes 1 and 2.

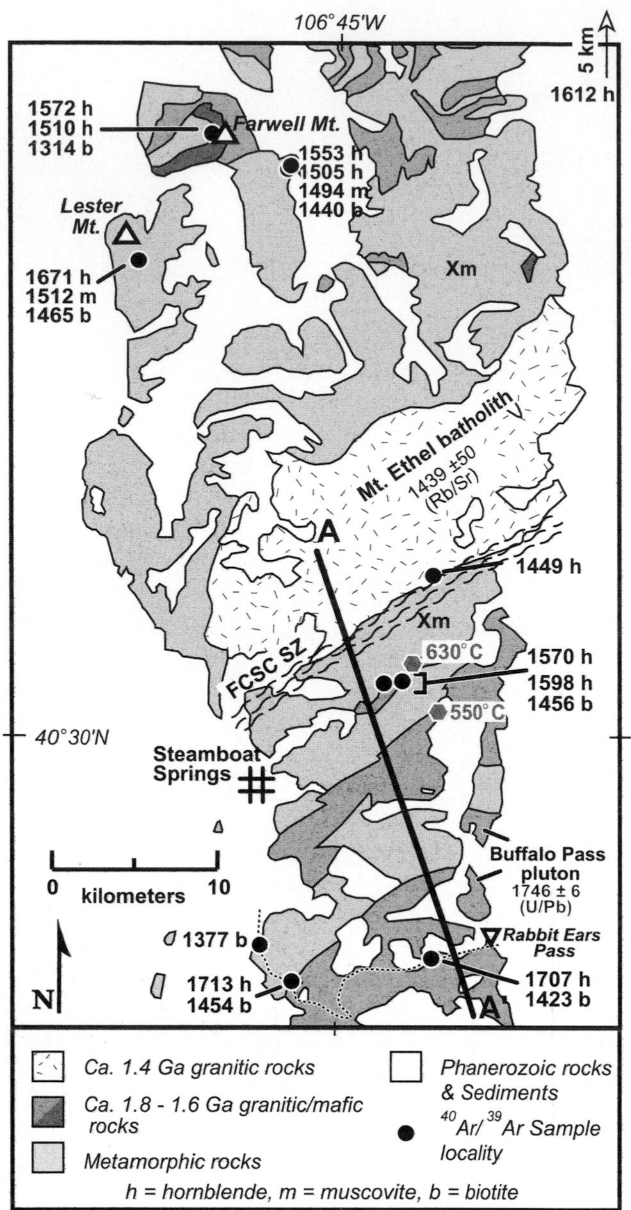

Figure 3. A) Generalized geologic map of the central Park Range study area, North Central Colorado [*Snyder*, 1980; *Tweto*, 1989; *Green*, 1992] showing ca. 1.7 Ga Buffalo Pass Pluton, Paleoproterozoic schist and gneiss, ~1440 Ma (Rb-Sr) Mt. Ethel Batholith [*Reed et al.*, 1993], and Fish Creek-Soda Creek shear zone (FCSZ).

position and grain size [*McDougall and Harrison*, 1999]. The fact that biotite samples far from the pluton margin record the same range of cooling dates as samples adjacent to the pluton is interpreted to indicate that peak temperatures exceeded biotite closure temperatures (~300°C) throughout the study area and cooled through about 300°C between about 1450 and 1350 Ma. Thus, we interpret biotite closure temperature (~300°) as a lower limit on regional mid-crustal temperatures in the Park Range at ~1.4 Ga.

In contrast to the biotite data, hornblende dates range between 1.45 and 1.71 Ga increasing systematically with distance from the pluton (Figure 3). Dates from the contact of the Mt. Ethel batholith are concordant with the age of the pluton. Older ages prevail farther from the pluton. About 7 km south of the pluton (within the FCSC shear zone) hornblende gives 1570–1600 Ma dates. At distances >15 km from the pluton hornblende dates range from 1500–1713 Ma (Figure 3). The preservation of Paleoproterozoic dates far from the pluton and shear zone suggest that regional temperatures did not exceed hornblende closure (<550°C) at 1.4 Ga. Adjacent to the pluton hornblende ages were likely reset by diffusive argon loss or by recrystallization at temperatures >500° C. Argon systematics of samples from within the shear zone may have been reset by any combination of thermal, fluid, or mechanical processes [e.g. *Cosca et al.,* 1992].

Metamorphic data are consistent with temperature estimates based on $^{40}Ar/^{39}Ar$ data. *Barinek et al.* [1999] calculated metamorphic temperatures of ~540°C within the FCSC Shear zone increasing to ~630°C at ~400 MPa toward the margin of the Mt. Ethel batholith from Fe-Mg equilibria in post-tectonic garnet cores and matrix biotite (Figure 3). Garnet rims and matrix biotite from outside the shear zone yield temperatures of ~550°C.

Northern Front Range

Data from the northern Front Range includes recalculated $^{40}Ar/^{39}Ar$ dates first presented by *Shaw et al.* [1999] from the Big Thompson Canyon west of Loveland, Colorado and new data from Longs Peak and from upper Fall River canyon in Rocky Mountain National Park (Figure 4). In addition, three dates were obtained from Virginia Dale, approximately 5 km south of the Wyoming-Colorado border (Figure 1).

Like those in the Park Range, hornblende ages from the central Front Range [*Shaw et al.*, 1999] are highly variable with ages ranging from about 1.40 Ga to 1.62 Ga (Plate 1b). *Shaw et al.* [1999] interpreted the wide range of hornblende ages to reflect varying degrees of argon loss and/or recrystallization during a prograde metamorphic event around 1.4 Ga [*Selverstone et al.*, 1997]. Peak temperatures during this metamorphism were in the staurolite and andalusite stability fields in semi-pelitic rocks [>500°C; *Shaw et al.*, 1999].

In some cases different hornblende samples from a single locality yield $^{40}Ar/^{39}Ar$ dates that differ by as much as ~130 m.y. Samples with younger ages typically have actinolitic cores [*Shaw et al.,* 1999]. The younger ages could be caused by lower average closure temperatures for multiphase crystals or by mixing of argon retained in Paleoproterozoic actino-

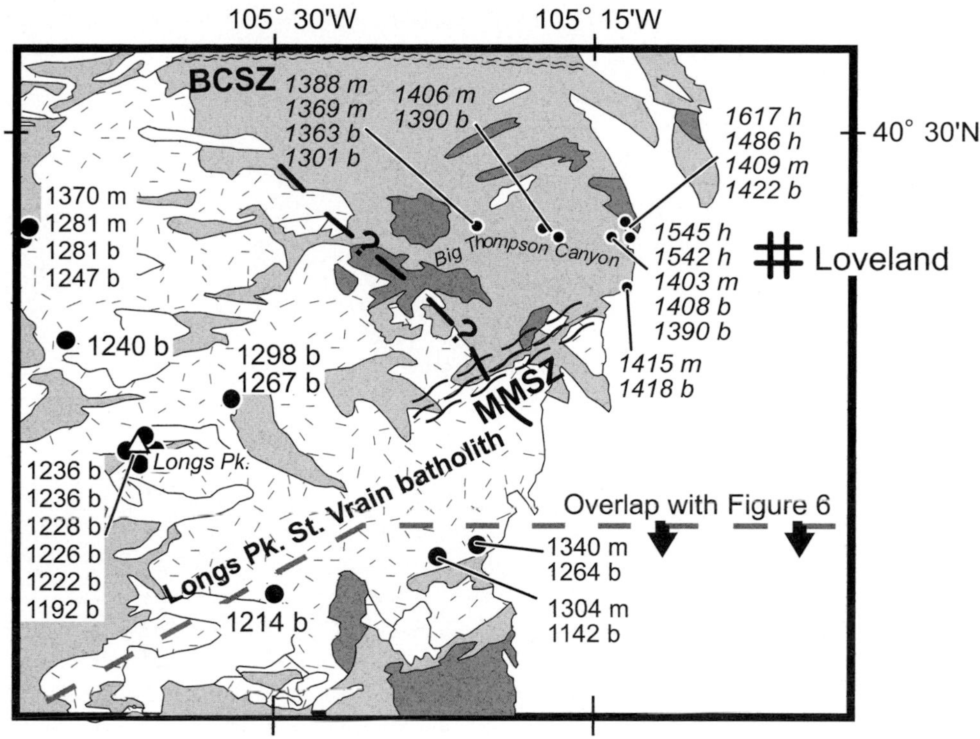

Figure 4. Generalized geologic map of the northern Front Range study area, Colorado [*Tweto*, 1989; *Green*. 1992] showing sample locations and preferred 40Ar/39Ar ages [*Shaw et al.*, 1999]. Longs Peak - St. Vrain Batholith ~1420 Ma, [*Peterman et al.*, 1968]. Shear Zones with documented ca. 1.4 Ga movement are shown [*Cavosie and Selverstone*, 2003; *McCoy et al.*, this volume; *Selverstone et al.*, 1997]: BCSZ: Buckhorn Creek shear zone. ISRSZ: Idaho Springs-Ralston Shear Zone, MMSZ Moose Mt. Shear Zone, See Figure 3 for map symbol definitions.

litic cores and recrystallized aluminum-rich rims [e.g. *Marcoline et al.,* 1999]. Discordant hornblende dates from other study areas may have a similar cause.

Mica from near the eastern edge of the Precambrian exposure in Big Thompson Canyon yields dates between about 1.39 and 1.42 Ga but ages generally decrease toward the west and south where biotite dates as young as about 1.20 Ga are found within the Longs Peak–St. Vrain batholith (Figure 4). Although no major mapped structures separate the two groups of samples and paleomagnetic evidence shows that post 1.4 Ga tilting of crystalline rocks in the range was modest [*Kellog*, 1973] it is possible that the Longs Peak data represent a somewhat deeper crustal level. Alternatively, the young dates could record anomalously slow cooling due to radiogenic heat production within the batholith [e.g. *McLaren et al.,* 1999].

Colorado Mineral Belt: Sawatch Range, Gore Range, and Central Front Range

The Colorado mineral belt (Figure 1) is a northeast-trending concentration of Laramide intrusions and mineral deposits that follows a system of Precambrian shear zones [*Tweto and Sims*, 1963] that were active at ~1.7 and ~1.4 Ga [*Shaw et al.*, 2001; *McCoy et al.* this volume]. Samples were collected from areas near the Homestake shear zone (HSZ) in the Sawatch and Gore ranges and the Idaho Springs–Ralston (ISR) shear zone in the Front Range (Figure 5).

The range of hornblende and biotite dates in the Colorado mineral belt is similar to the range of dates in the Park and northern Front Range (Plate 1c). Hornblende dates far from Mesoproterozoic plutons range from about 1600 to 1440 Ma. Also, as in the Park Range, the thermal effects of plutons are manifested by complete resetting of hornblende dates near the Mt. Evans batholith [1442 ± 2 Ma, U-Pb; *Reed and Snee,* 1991] and Silver Plume batholith [1422 ± 3 Ma, *Tweto,* 1987] (Figure 5).

A transect along the eastern margin of the northern Sawatch Range crosses the Homestake shear zone (Figure 5). Hornblende ages range from 1650 to 1400 Ma and, with one exception, mica ages are tightly clustered around 1400 Ma. Despite ~1.4 Ga dip-slip motion on mylonite zones within the HSZ [*Shaw et al.,* 2001] there is no clear contrast between $^{40}Ar/^{39}Ar$ dates across the shear zone although the oldest hornblende dates and youngest mica dates occur close to the shear zone.

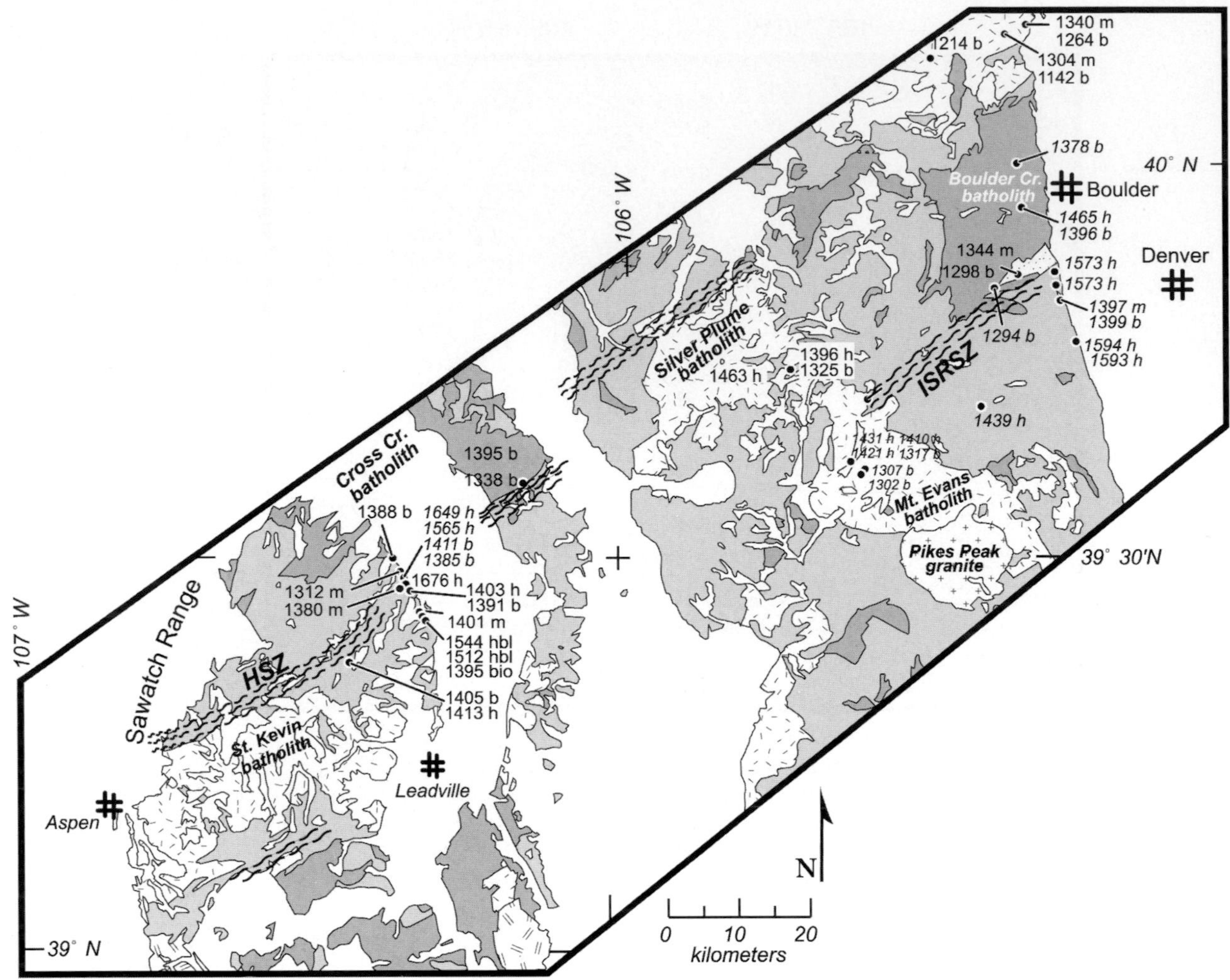

Figure 5. Generalized geologic map of the Colorado mineral belt study area including the northern Sawatch Range, southern Gore Range, and central Front Range [*Tweto*, 1989; *Green*, 1992]. Pikes Peak granite ~1.1 Ga, Mt. Evans batholith [1442 Ma, *Aleinikoff et al.*, 1990], Boulder Creek Batholith [1714 Ma, *Premo and Fanning*, 1997], St. Kevin Batholith, 1400 ±50 Ma [*Pearson et al.*, 1966]. Shear Zones with documented ca. 1.4 Ga movement are shown [HSZ, Homestake shear zone, ISRSZ, Idaho Springs Ralston Shear zone, *McCoy et al.*, this volume; *Shaw et al.*, 2001]. See Figure 3 for map symbol definitions.

Southern Colorado: Unaweep Canyon, Black Canyon, Gunnison Valley, Needle Mountains, Northern Sangre de Cristo Range, and Wet Mountains

Several Ranges in southern Colorado and northern New Mexico including the Wet Mountains, Taos Range, and Tusas Range yield no $^{40}Ar/^{39}Ar$ hornblende dates >1.45 Ga (Plate 1, Appendix 1). The contrast between these areas and the areas in northern and central Colorado discussed above is evident in a 250-km-long WNW–ESE transect across central Colorado from the Black Canyon of the Gunnison to the southern Wet Mountains (Figure 6). Hornblende dates from the first 180 km of this transect (Black Canyon to Arkansas River gorge; Figures 3d, 3e, 3f) range from 1400–1720 Ma like other hornblende dates from farther north in Colorado (e.g. Park and central Front Range). However, hornblende dates from the Wet Mountains south of the Arkansas River are restricted to the interval 1390–1350 Ma (Plate 1g).

Mica dates in the Black Canyon and Gunnison valley range between 1.08 Ga and 1.40 Ga–anomalously young compared with other areas in the southern Rocky Mountains. This may reflect a different post-1.4-Ga cooling history related to Neoproterozoic deformation along the northwest trending Uncompahgre uplift or to slower exhumation and cooling.

Two biotite samples from the Needle Mountains yield ages of ~1440 Ma (Figures 3e, 7) and a single fine-grained muscovite from a low temperature shear zone yields a plateau age of 1398 Ma. Hornblende dates from samples located outside the areole of the 1.4 Ga Electra Lake gabbro are complex, but parts of their age spectra retain 1.6 to 1.7 Ga apparent ages (Appendix 2). Samples within and near the gabbro yield nearly concordant hornblende and biotite dates between 1.43 and 1.46 Ga.

All the hornblende dates from the Wet Mountains are younger than 1.4 Ga and are interpreted to record post-1.4 Ga cooling through ~500°C. In some cases hornblende dates are older than mica ages from the same locality. Because most of our samples from the southern Wet Mountains were collected within 10 km of the margin of the 1.36 Ga San Isabel pluton [*Reed et al.*, 1993] we cannot rule out local effects of the pluton as the cause of the <1.4 Ga dates in this area.

The San Isabel pluton was probably emplaced at a deeper level than most 1.4 Ga plutons in Colorado. Based on Al in hornblende geobarometry the San Isabel batholith intruded at 15–20 km depth [500–700 MPa; *Cullers et al.*, 1992], somewhat deeper than the Oak Creek pluton in the northern Wet Mountains [10–15 km or 300–400 MPa; *Cullers et al.*, 1993] and significantly deeper than the Longs Peak–St. Vrain batholith in the central Front Range [7–12 km or 200–300 MPa; *Anderson and Thomas,* 1985]. The penetrative ductile ~1.4 Ga deformation that characterizes the Wet Mountains south of the Arkansas River [*Siddoway et al.*, 2000] also differs from the focused deformation on discrete mylonite zones that is typical of northern and central Colorado [e.g. *Shaw et al.,* 2001; *McCoy et al.,* this volume].

Based on thermochronologic and pressure data and on the contrast in structural style cited above, we infer that the younger ages recorded by rocks in the Wet Mountains are, at least in part, a result of a deeper exposure level. This deeper exposure is not a function of differential Phanerozoic uplift because the depth below the sub-Paleozoic unconformity does not differ substantially.

Northern New Mexico: Tusas and Taos Ranges

All argon dates from the Tusas and Taos ranges of northern New Mexico are less than about 1.45 Ga (Figures 3h, 3i, 8). Hornblende data from the Taos Range are quite young compared to other hornblendes from the southwestern United States. Samples from the lowest elevations yield apparent

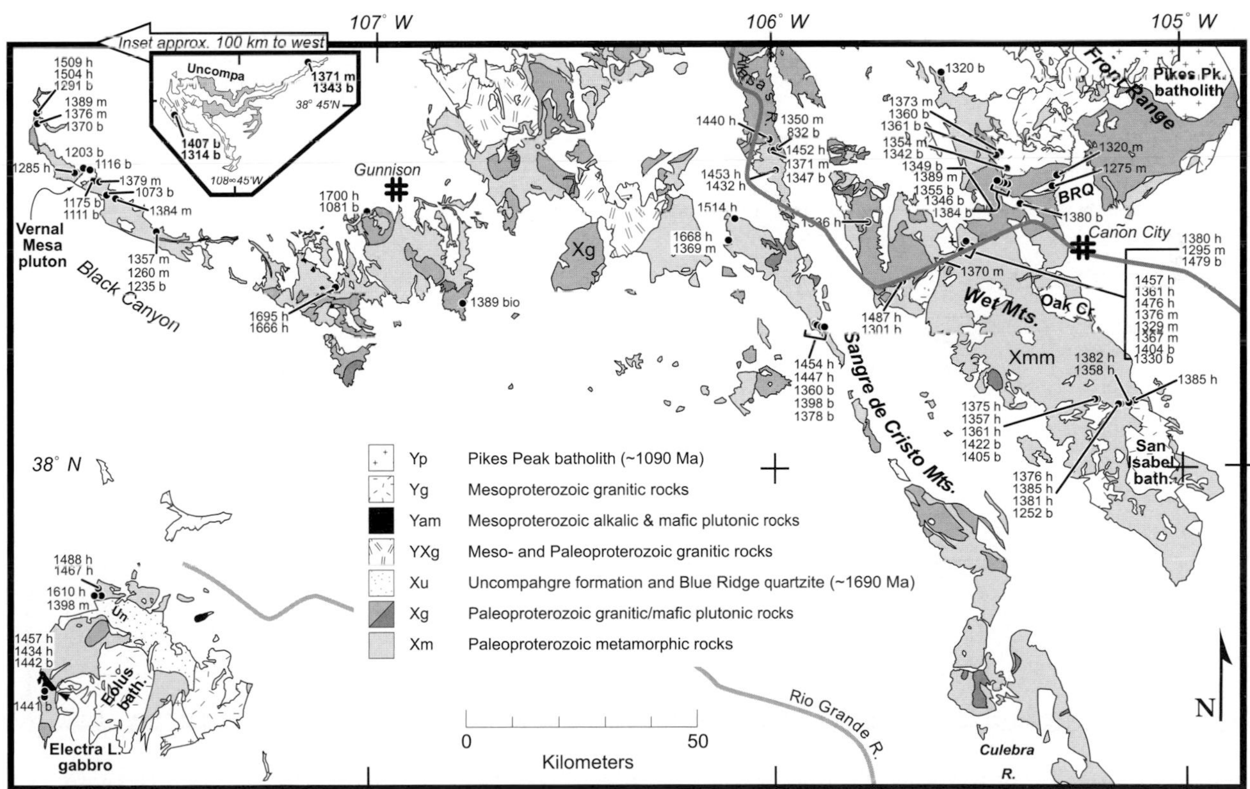

Figure 6. A) Generalized geologic map of south-central Colorado [*Tweto*, 1989; *Green*, 1992] with sample localities and preferred $^{40}Ar/^{39}Ar$ dates. Hornblende dates decrease systematically from the Black Canyon of the Gunnison area in the northwest to the Wet Mountains in the southeast.

ages between about 0.60 and 1.00 Ga and higher elevation samples approach 1.42 Ga (albeit these older ones are very complex). Unlike many hornblende samples, the young samples here have simple K/Ca spectra (Appendix 2) and therefore we do not attribute their young age to alteration. These hornblende spectra may be recording partial argon loss associated with Mid-Tertiary igneous activity in the area that is supported by a completely degassed (21 Ma) biotite (Appendix 2). Spectra from hornblende samples with older dates are also complex (Appendix 2) and do not unambiguously constrain the high temperature cooling history of this region. However, many hornblendes from northern New Mexico reported by Karlstrom et al. [1997] have fairly well-behaved age spectra consistent with complete resetting at 1.4 Ga. Several of these samples are from well outside the Questa Caldera and are therefore not suspected of being affected by mid-Tertiary heating. The complexity of hornblende spectra from the Tusas Mountains and Cerro Colorado is interpreted to reflect alteration and some of these dates may be unreliable.

A single published hornblende date from the Cimarron River tectonic unit of 1692 ± 3 Ma [*Grambling and Dallmeyer*, 1993] suggests that at least this one block did not reach 500° C (Figure 7) at 1.4 Ga. However, hornblendes collected less than 3 km away, across the transpressive Fowler Pass shear zone [*Carrick and Andronicos*, 2001] in the Eagle's Nest tectonic unit, record dates of about 1400 Ma. In addition, metamorphic monazite from this tectonic block has an age of 1430 Ma [*Grambling et al.*, 1992]. The striking difference in hornblende dates across the Fowler Pass shear zone has been interpreted to reflect motion on this structure during or subsequent to metamorphism.

Mica from northern New Mexico yields a wide range of apparent ages < 1.45 Ga. Pristine coarse-grained muscovite from pegmatites in the Tusas and Taos Range yields flat spectra with plateau ages of about 1375 Ma (Appendixes 1, 2). Finer-grained muscovite and biotite samples yield more disturbed age spectra with younger apparent ages between 975 and ~1300 Ma. In slight contrast, all of the micas from Cerro Colorado in the extreme southern Tusas are between 1.34 and 1.36 Ga. These new data are consistent with previous mica results from northern New Mexico presented by Karlstrom et al. [1997]. South of the study area in the southern Santa

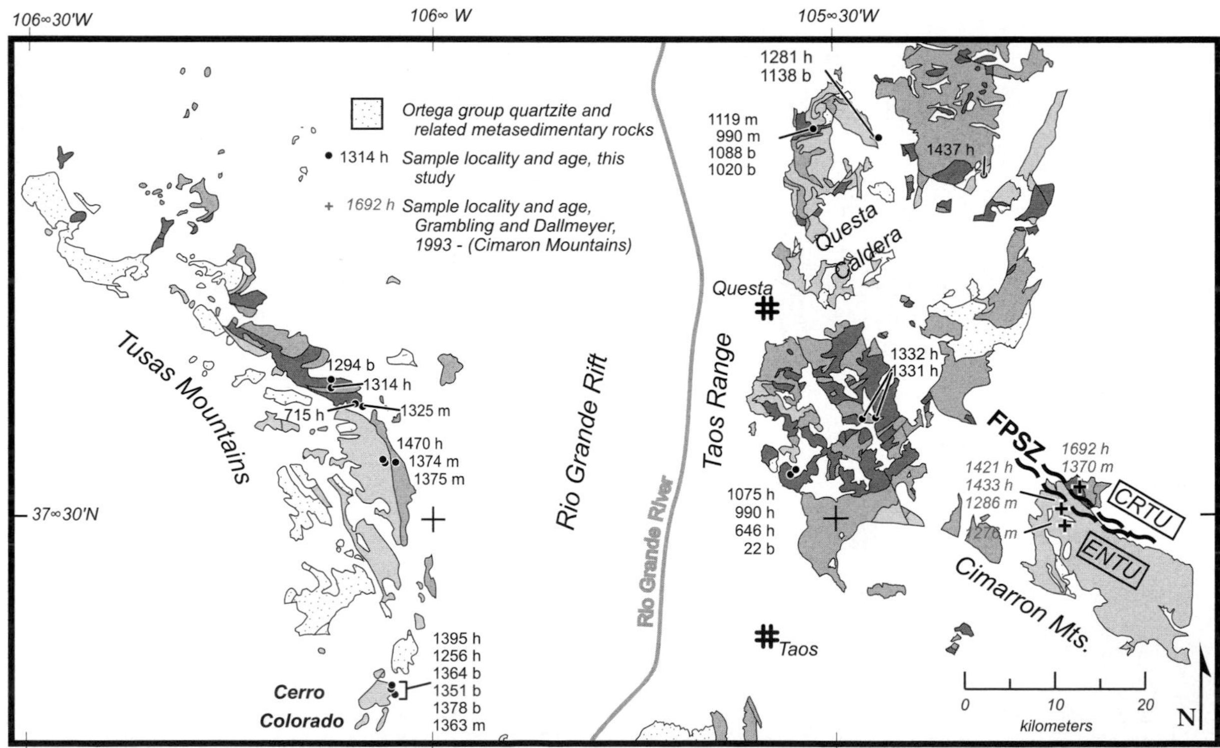

Figure 7. Generalized geologic map of the Tusas and Taos ranges of northern New Mexico with sample localities and preferred $^{40}Ar/^{39}Ar$ dates [*Green and Jones*, 1997; *Anderson and Jones,* 1994]. Data from Grambling and Dallmeyer [1993] shows sharp contrast in $^{40}Ar/^{39}Ar$ hornblende dates across the Fowler Pass Shear Zone (FPSZ) in the Cimarron Mountains. Hornblende dates >1650 Ma are preserved in the Cimarron River tectonic unit (CRTU) to the northeast of the shear zone. In the Eagle's Nest tectonic unit (ENTU) hornblende dates are < 1450 Ma. See Figure 3 for additional map symbol definitions.

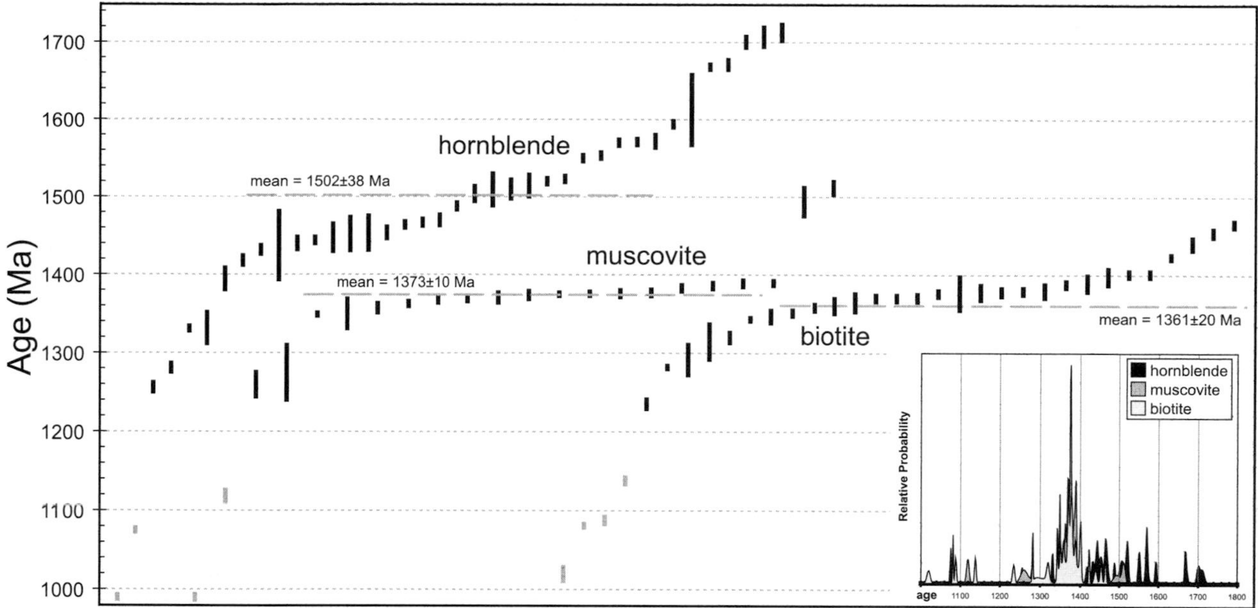

Figure 8: $^{40}Ar/^{39}Ar$ hornblende (n = 38), muscovite (n = 22), and biotite (n = 33) dates from samples collected more than 5 km from Mesoproterozoic pluton margins and more than 2 km from the center lines of shear zones with ~1.4 Ga movement. Vertical bars represent analytical uncertainty (2σ). Data arranged from youngest to oldest.

Fe range, Sandia, and Manzano Mountains hornblende dates range between ~1.4 and ~1.7 Ga similar to central Colorado [*Karlstrom et al.*, 1997a; *Karlstrom et al.*, in press, 2003].

We interpret the predominance of hornblende dates <1.4 Ga to record cooling from high temperatures (>500°C). This conclusion is independently supported by evidence for ~1.4 Ga amphibolite facies (>500°C, 4 kbar) metamorphism and penetrative deformation throughout the region [*Grambling*, 1988; *Grambling et al.*, 1989], including amphibolite-facies metamorphism in the Tusas Range [*Williams et al.*, 1999], near granulite-facies metamorphism in the Taos and Rincon ranges [1421 Ma, U-Pb metamorphic monazite dates; *Pedrick*, 1995; *Read et al.*, 1999], and upper amphibolite-facies metamorphism in the Picuris Range, [~1.4 Ga, U-Pb garnet and staurolite; *Wingsted et al.*, 1996].

EVIDENCE FOR A REGIONAL THERMAL EPISODE AT ~1.4 GA

The overwhelming majority of mica $^{40}Ar/^{39}Ar$ ages from the southern Rocky Mountains are less than 1.45 Ga whereas hornblende ages range mainly 1.8–1.3 Ga (Plate 1, Appendix 1). This suggests that most of the mid-crustal rocks sampled for this study cooled through argon closure temperatures for mica ~1.4 Ga. However, many samples were collected near ~1.4 Ga plutons and/or shear zones for the purpose of investigating local thermal effects. In order to test for a regional thermal event it is necessary to remove these samples from consideration and focus on data from samples that represent ambient temperatures far from pluton margins and shear zones.

Thermal modeling and empirical studies suggest that for plutons similar in size and depth of emplacement to ~1.4 Ga plutons of the southern Rockies thermal aureoles probably extend no more than several kilometers from the pluton margin [e.g. *Harrison and McDougall,* 1980a; *Peacock,* 1989; *Wohletz*, 2002]. Because of the possibility of convective heat transport, fluid alteration, argon loss due to deformation, and other factors we have filtered out all data points that are within 5 km of exposed plutons and within 2 km of the center line of shear zones with documented or suspected ~1.4 Ga deformation. The remaining subset of the $^{40}Ar/^{39}Ar$ data [this paper and *Shaw et al.,* 1999] should provide a representative sample reflecting regional thermal conditions at the level of exposure. The filtered data subset comprises 38 hornblende, 22 muscovite, and 33 biotite samples and includes data from the Park Range, Front Range, Gunnison River Valley, Arkansas Gorge, northern Sangre de Cristo Range, Taos Range, and Tusas Range (Figure 1).

In the filtered data subset, as in the entire dataset, hornblende dates range mainly between ~1.7 and 1.4 Ga (Figure 8). Many hornblende samples yield highly complex age spectra (Appendix 2). Several hornblende dates <1.4 Ga occur in the Taos and Tusas Ranges (Figure 7). A number of mica dates <1.35 Ga occur throughout the region, but only six mica dates are >1.4 Ga and all of these are from the Park Range (Figure 3).

We interpret the relatively narrow range (1.40–1.35 Ga) of mica dates in the filtered dataset as recording cooling through

350–300°C after ~1.4 Ga throughout Colorado and northern New Mexico. With the exception of a few outlying data points muscovite and biotite ages are fairly tightly clustered and approximately normally distributed about mean dates of 1373 ±10 and 1361 ±20 Ma (2σ), respectively. We interpret the narrow variation in ages to reflect limited random variation in several factors that affect $^{40}Ar/^{39}Ar$ ages including temperature, cooling rate, mineral composition, and grain size. Because the filtered dataset does not include samples from near any identifiable ~1.4 Ga heat sources or shear zones we infer that the data reflect first order trends in the regional thermal evolution of the middle crust.

Regional cooling through mica closure temperatures at ~1.4 Ga would be consistent with either cooling following prolonged residence at temperatures above mica closure or with cooling following a regional thermal event. In Arizona *Hodges et al.* [1994] and *Chamberlain and Bowring* [2001] argued that discordance between several thermochronometers recorded protracted cooling from high temperature. Specifically the Big Bug block remained above 425 °C for approximately 250 Ma following the relatively shallow (~10 km) emplacement of the ca. 1700 Ma Crazy Basin quartz monzonite [cf. *Hodges et al* 1994; *Hames and Bowring,* 1994]. To explain the unusually steep and persistent geothermal gradients that this interpretation implies *Flowers et al.* [this volume] invoked anomalously high heat generation values.

In contrast, we argue that evidence for ~1.4 Ga metamorphism in the Park Range [*Barinek,* 1999], northern and central Front Range [*Selverstone et al.*, 1997; *Shaw et al.*, 1999], Wet Mountains [*Siddoway et al.,* 2000], northern New Mexico [*Grambling*, 1988; *Grambling et al.*, 1989; *Williams et al.*, 1999; *Pedrick*, 1995; *Read et al.*, 1999; *Wingsted et al.*, 1996] and Arizona [*Williams*, 1991] is most consistent with a regionally extensive thermal episode.

REGIONAL ~1.4 GA TEMPERATURES

If $^{40}Ar/^{39}Ar$ ages in the southwestern U.S. primarily reflect cooling following a ~1.4 Ga thermal event then first-order variations in $^{40}Ar/^{39}Ar$ and K-Ar dates can be used to broadly estimate peak temperatures during the episode. In order to map regional ~ 1.4 Ga temperatures for exposed basement rocks, we use $^{40}Ar/^{39}Ar$ thermochronologic data from minerals with different closure temperatures to bracket maximum temperatures (T_{max}) for exposed basement rocks. For example, where all of the K-bearing minerals used in this study (hornblende, muscovite, and biotite) retain Paleoproterozoic dates (> 1.6 Ga) we infer that these rocks have not been heated above about 300° C since 1.6 Ga. Where hornblende retains Paleoproterozoic dates, and biotite and muscovite yield a range of mixed dates (1.6–1.4 Ma), we infer that T_{max} was between about 300° C and 500° C at 1.4 Ga; in other words, hot enough to partially reset mica to varying degrees, but below the closure temperature of hornblende. Where hornblende yields variable dates (1.6–1.4 Ga) consistent with partial resetting and muscovite and biotite record complete resetting at ca. 1.4 Ga, we infer that 350°C < T_{max} < ~500° C. Where $^{40}Ar/^{39}Ar$ ages of all K-bearing minerals, including hornblende, are reset to ca. 1.4 Ga, we infer that T_{max} > 500°C. Where hornblende consistently records ca. 1.4 Ga dates and U-Pb monazite and zircon dates on metamorphic minerals are also ca. 1.4 Ga, we infer T_{max} > 600° C. In several areas our thermochronologic estimates of T_{max} are independently verified by thermobarometric data [e.g. *Williams and Karlstrom*, 1996; *Barinek et al.*, 1999].

We recognize that there are complexities including short wavelength irregularities in the temperature field, variations in mineral composition and diffusion parameters, different cooling rates, variable argon retentivities for minerals or for domains within minerals, and other factors that could affect argon loss [e.g. *Dodson*, 1973; *McDougall and Harrison*, 1999]. However, the systematic effects of local and regional temperature differences are much greater and more consistent than the randomly varying effects of variable closure temperature. The large quantity of data makes it likely that average closure temperatures of the population of samples in this study is similar to nominal closure temperatures. Consideration of generalized trends in the extensive regional data set smoothes local irregularities and allows us to construct a first-order quantitative estimate of the regional metamorphic field gradient.

Compiled Thermochronologic Data

Over 650 $^{40}Ar/^{39}Ar$ and K/Ar dates compiled from the literature extend our analysis of ~ 1.4 Ga peak mid-crustal temperatures beyond the Southern Rocky Mountain region (Appendixes, 3, 4). In particular, data from Wyoming [e.g. *Giletti*, 1968; *Hills and Armstrong*, 1974], New Mexico [e.g. *Karlstrom et al.*, 1997a; *Marcoline et al.*, 1999; *Karlstrom et al.*, in press, 2003], and Arizona [e.g. *Shafiqullah et al.*, 1980] add important information to the regional thermal picture. Contrasts in the distribution of $^{40}Ar/^{39}Ar$ and K-Ar dates from different lithospheric provinces are evident in the regional cumulative probability diagrams in Figure 9.

Central Wyoming, Western North Dakota, and Southwestern Montana (Archean Craton, Great Falls Tectonic Zone, and Trans Hudson Orogen)

Thermochronologic and geologic evidence from Wyoming show that the interior of the Archean craton was not strongly affected by 1.4 Ga tectonism at the present level of exposure

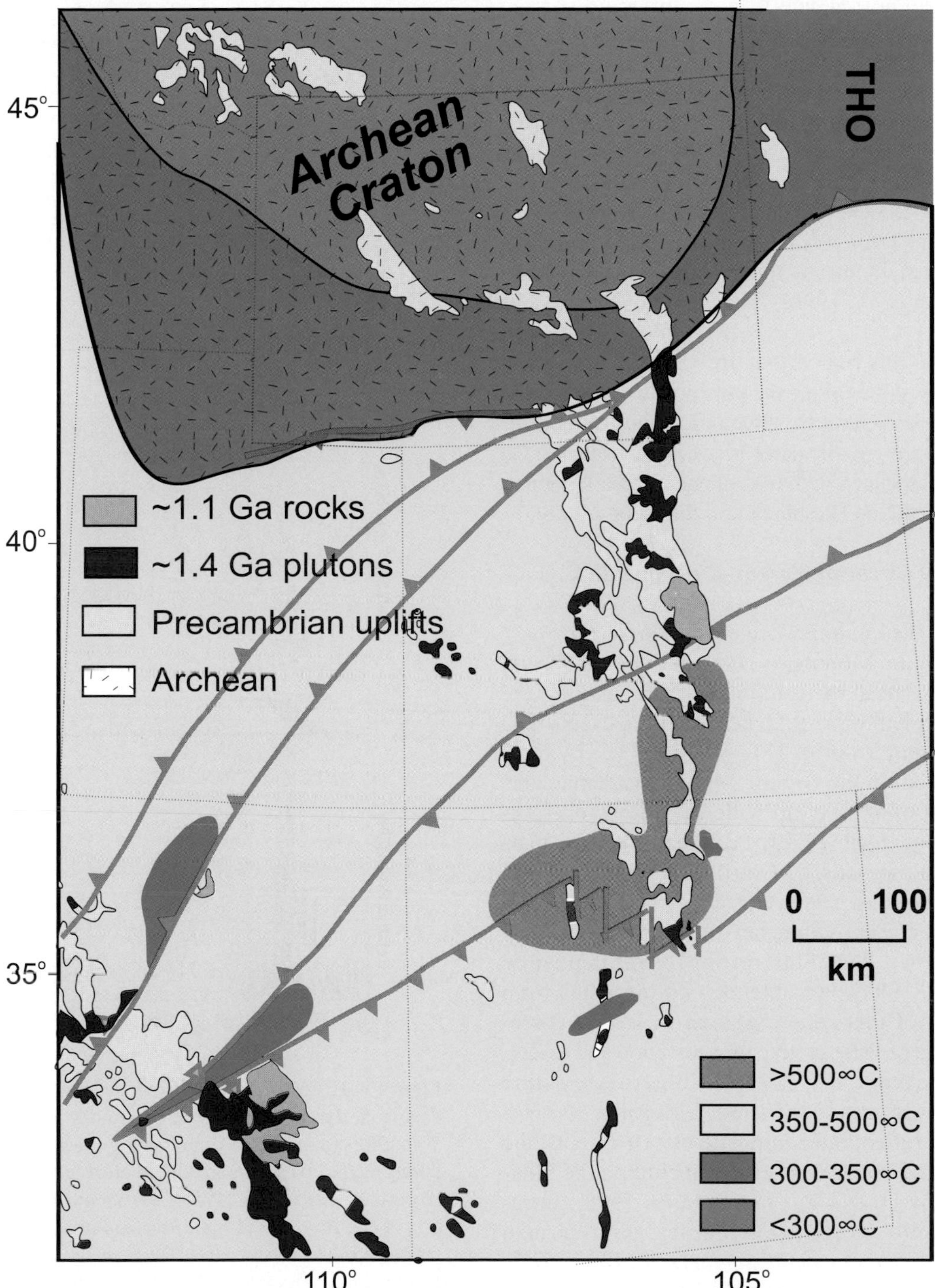

Plate 2. Map showing first-order estimate of regional ca. 1.4 Ga temperature inferred from new and compiled thermochronologic data. Precambrian outcrop shown in gray. Core of Wyoming craton remained below about 300°C (biotite retained Archean and Paleoproterozoic ages). Attenuated margin reaches temperatures of 300–350°C (micas partially reset), Proterozoic provinces reach average temperatures of 350–550°C (mica reset, hornblende partially to completely reset). High-T zone in southern Colorado - northern New Mexico reaches temperatures above 500–550°C. Sample locations shown as small dots. Data compiled from National Geochronologic Database (NGDB), New Mexico Geochronologic Database (NMGDB) and published papers (Appendix 4).

(Plate 2). Mica and hornblende from within the Wyoming craton largely record Archean and Paleoproterozoic dates (Figure 9a) [*Giletti*, 1968; *Peterman and Hildreth*, 1978]. The fact that biotite and muscovite were not reset during the 1.4 Ga thermal event indicates that temperatures at exposed crustal levels probably did not exceed ~300° C in the interior of the craton. Previous studies have concluded that regional metamorphism predates 2.5 Ga in the core of the craton and ~1.8 Ga near its margins [*Houston and others*, 1993]. Although ~1.4 Ga mafic dikes do intrude the interior of the Wyoming craton [e.g. *Harlan et al.*, 1996], it is devoid of voluminous ~1.4 Ga plutons.

Most dates from this region that are <2.0 Ga fall within Proterozoic tectonic zones along the northwestern and eastern margins of the craton (Great Falls tectonic zone, Trans Hudson orogen). Paleoproterozoic dates between ~1.8 and 1.6 Ga record tectonism associated with the collision of the Wyoming craton with the Medicine Hat block and Superior craton.

Southern Wyoming (Archean Margin–Cheyenne Belt)

K/Ar data from the Laramie Range [*Hills and Armstrong*, 1974] and the Granite Mountains [*Peterman and Hildreth*, 1978] show that Archean rocks of the cratonic margin were more strongly affected by post-Archean thermal events than rocks of the cratonic interior (Figures 10b, 12). A thermochronologic front in the Granite Mountains demarcates the northern limit of a ~200 km wide transition zone corresponds with the attenuated Early Proterozoic passive margin intruded by voluminous 2.1 Ga rift-related mafic dikes [*Premo and Van Schmus,* 1989; *Cox et al.*, 2000] and, presumably, the region where sediments of the Proterozoic passive margin (Snowy Pass Supergroup) were deposited [*Karlstrom et al.*, 1983]. Hornblende cooling dates from the northern part of this zone range up to 1.8 Ga corresponding to the age of Paleoproterozoic accretion and assembly along the Cheyenne belt Archean/Proterozoic suture (Figure 9b). Younger ages to the south [*Peterman and Hildreth,* 1978] may reflect the composite effects of episodic loss during Proterozoic tectonic events including the Medicine Bow orogeny [1.78 Ga; *Chamberlain*, 1998], intrusion of the Laramie anorthosite complex and Sherman Granite [~1.43 Ga; *Scoates and Chamberlain*, 1995, 2003; *Frost et al.*, 1999], intrusion of mantle-derived mafic dikes at ~1.5 Ga [*Chamberlain et al.,* 2003], and ~1.4 Ga tectonism.

Although the role of the ~1.4 Ga thermal episode in resetting argon systematics is uncertain, thermochronologic data can constrain the maximum temperatures that could have been attained. Minimal resetting of hornblende dates suggests that temperatures did not exceed about 500° C in the attenuated

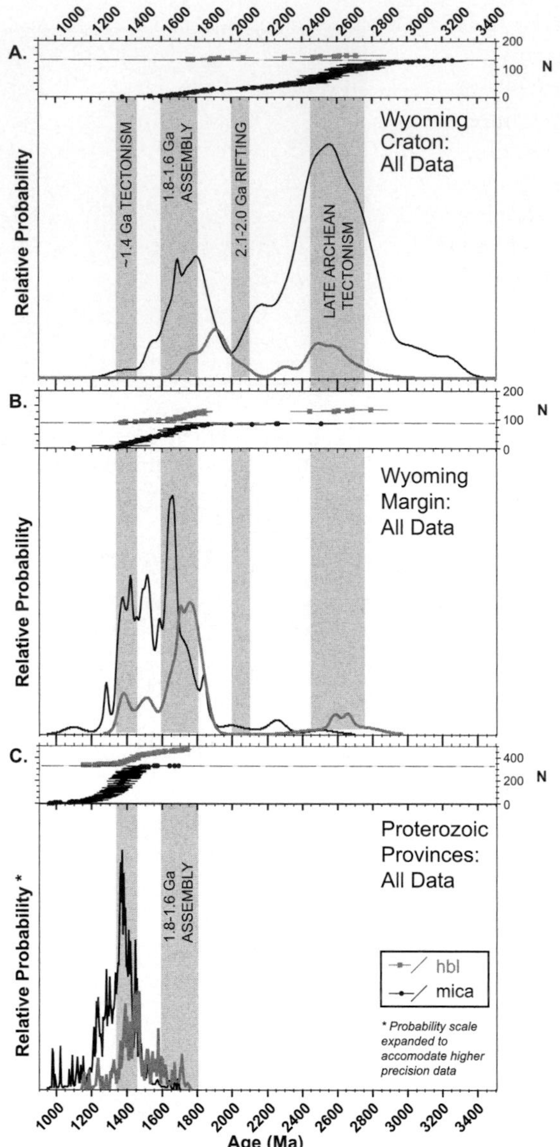

Figure 9. Histograms summarizing compiled $^{40}Ar/^{39}Ar$ and K-Ar data from the central and southern Rocky Mountains. (A) Archean Wyoming craton, (B) Wyoming margin where crust was attenuated during ~2.1–2.0 Ga rifting, (C) Proterozoic provinces. Data in (A) & (B) from: *Aldrich et al.* [1958]; *Bassett and Giletti* [1963]; *Bayley et al.* [1973]; *Brookins* [1968]; *Burwash et al.* [1962]; *Butler* [1966]; *Crittenden and Sorensen* [1980]; *Gast et al.* [1958]; *Giletti* [1966]; *Giletti* [1971]; *Goldich et al.* [1966]; *Hanson and Gast* [1967]; *Hayden and Wehrenberg* [1960]; *Heimlich and Armstrong* [1972]; *Heimlich and Banks* [1968]; *Hills and Armstrong* [1974]; *Holm et al.* [1997]; *Jahn* [1967]; *Kistler et al.* [1969]; *Laughlin* [1969]; *Marvin and Dobson* [1979]; *Peterman* [1981]; *Peterman and Hedge* [1964]; *Peterman and Hildreth* [1978]; *Reed and Zartman* [1973]; *Reid et al.* [1975]; *Schwartzman* [1966]; *Schwartzman and Giletti* [1968]; *Whelan* [1970]. See Appendix 4 for full references. Data in (C) from all references in Appendix 4.

southern margin of the Wyoming craton at the level of exposure.

Proterozoic crust in Colorado apparently reached higher temperatures (350–550°C) than the Archean margin. The marked contrast in the thermochronologic record of the Archean craton, the attenuated Archean cratonic margin, and the accreted Proterozoic terranes suggest that lithospheric architecture exerted a strong control on the thermal response to the ~ 1.4 Ga episode [cf. *Chamberlain et al.,* 2003].

Arizona and New Mexico (Proterozoic Provinces)

Thermochronologic data from Yavapai province rocks of central and northern Arizona are characterized by sharp contrasts in cooling ages in different tectonic blocks (Plate 2). In the Grand Canyon ~1.4 Ga mica ages in the east are juxtaposed across the 96-mile shear zone with rocks preserving older ages [*Karlstrom et al.*, 1997b]. The Ash Creek block preserves 1.6 Ga hornblende ages whereas the Crazy Basin has micas with rims as young as 1270 Ma consistent with very slow cooling [*Hodges et al.*, 1994]. Thermochronologic data from Arizona are consistent with contrasting 1.4 Ga temperatures superimposed on different post-1.6 Ga cooling histories within km-scale blocks. We interpret this to indicate post-1.4 Ga movement on shear zones such as the Shylock and 96-mile zone [e.g. *Karlstrom et al.*, 1997b].

The Mazatzal block (sensu stricto) preserves some of the lowest grade rocks in the Southwest (sub-greenschist grade) that were not appreciably heated at 1.4 Ga [*Karlstrom and Bowring*, 1993]. However, in Mazatzal province rocks south of the Slate Creek shear zone, inferred ~1.4 Ga temperatures increase abruptly to 550° C, with two aluminum silicates present [*Williams*, 1991]. In southern Arizona, as in southern New Mexico, background 1.4 Ga temperatures were probably near 350° C, with hotter conditions near 1.4 Ga plutons [*Williams*, 1991].

In westernmost Arizona and southern California (Mojave province), 1.4 Ga metamorphism also seems to be low grade as it did not reset argon ages. The Lower Granite Gorge of the Grand Canyon has 1.4 Ga K-feldspar, as do rocks just beneath the Cambrian unconformity at Gold Butte. Likewise, rocks of the Death Valley region were less than 300° C at 1.4 Ga as shown by preserved >1.6 Ga K-Ar mica dates.

DISCUSSION

The contrast between the lithospheric architecture of the Archean craton and that of the accreted Proterozoic arc terranes

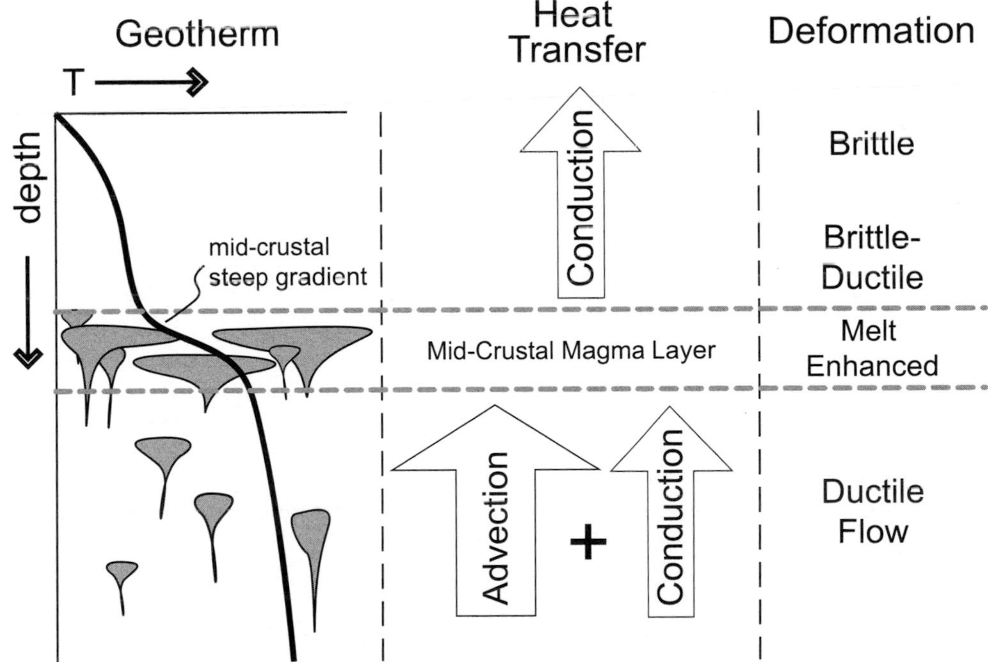

Figure 10. Cartoon illustrating how a magma-rich layer could produce steep geothermal gradients in the middle crust. Advective heat transport below the level of magma emplacement is more efficient than conductive transport above. This forces steep geothermal gradients to form in the middle crust separating low-temperature upper crust from higher temperature lower crust. Deformation below the magma layer is by penetrative ductile flow; deformation above the layer is focused in narrow shear zones.

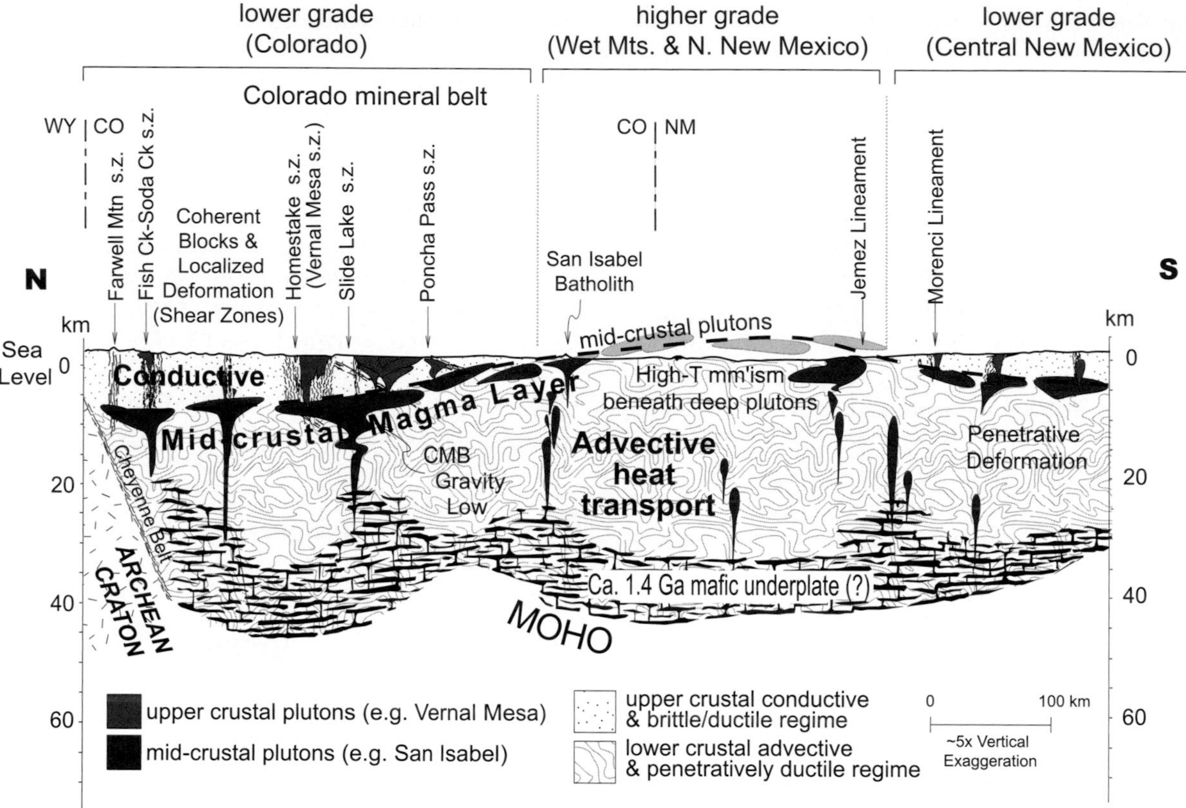

Plate 3. Cartoon illustrating our preferred model for higher temperatures, deeper pluton emplacement depths, and more penetrative deformation in southern Colorado and northern New Mexico. Plutons pool near the brittle ductile transition due to a combination of neutral buoyancy at this level and rheologic impediments to further ascent. Plutons rise into the upper part of crust at shear zones (especially extensional bends, e.g. St. Kevin/ Homestake). The transition from magmatic advective heat transport to conductive transport in the results in steep geothermal gradients in the middle crust. Few plutons occur in the ductile flow domain because there is little impedance to ascent. Deeper exposure levels in southern Colorado and northern New Mexico expose the hotter advective domain.

appears to have exerted a strong control on ca. 1.4 Ga mid-crustal temperatures. The lack of pervasive resetting of argon systematics in rocks of the Wyoming craton could indicate that the Wyoming lithospheric mantle was thick, cold, and relatively infertile by ca. 1.4 Ga [*Karlstrom and Humphreys*, 1998] explaining both the lower average temperatures revealed by thermochronologic data and the lack of ca. 1.4 Ga plutons within the craton. South of the geochronologic front in the Granite Mountains [Figure 1; *Peterman and Hildreth,* 1978; *Chamberlain et al.,* 2003], the Archean lithospheric mantle that had been thinned and modified by 2.1–2.0 Ga rifting [*Karlstrom and Houston*, 1984; *Premo and Van Schmus,* 1989] may have been less resistant to heating [*Hills and Armstrong*, 1974; *Peterman and Hildreth*, 1978] and perhaps more conducive to magma generation and ascent (although the Sherman batholith and Laramie anorthosite complex, which intrude the Cheyenne belt, are the only large ~1.4 Ga plutons of the craton margin). It is also possible that some of the <1.8 Ga ages in this domain may record partial resetting during the ~1.4 Ga themal episode. *Chamberlain et al.* [2003] postulate a similar mechanism to account for the younger ages in the area of ~2.1 Ga extension, but attribute argon loss to a ~1.5 Ga mantle event.

The geochronologic front could also delineate a region of deeper crustal exposure along the southern margin of the Wyoming province [*Peterman and Hildreth*, 1978]. One possible mechanism to produce contrasting exposure levels would be differential uplift during Mesoproterozoic tectonism. However, this is difficult to verify because there are few constraints on the Mesoproterozoic paleodepth of rocks in southern Wyoming.

The Proterozoic provinces of the Southwest display broadly similar thermochronologic signatures suggesting that the accreted terranes experienced similar thermal histories. However, important trends in the data are evident. Ambient temperatures throughout much of the Yavapai province in Colorado and Arizona probably averaged more than 300°C at ca 1.4 Ga. Paleodepths are well constrained in a few key locations around pluton aureoles and in areas of higher regional temperatures such as northern New Mexico. Published estimates mainly fall in the range of 7–15 km [200–500 MPa; *Williams*, 1991; *Williams and Karlstrom*, 1996; *Selverstone et al.*, 1997; *Barinek et al.*, 1999] suggesting that regional background temperatures were somewhat elevated (i.e. >30°C/km) with respect to average geothermal gradients for the Proterozoic.

A region extending from the central Wet Mountains in Colorado into northern New Mexico is characterized by a paucity of hornblende ages >1.45 Ga (Figures 7, 8). We suggest that this region represents a high-T domain with inferred temperatures >500°C (Plate 2). Surprisingly, this region is also distinguished by a paucity of ~1.4 Ga plutons. Deeper pluton emplacement depths in the Wet Mountains [inferred from hornblende geobarometry, *Cullers et al.*, 1992; *Cullers et al.*, 1993], the change in structural style of 1.4 Ga deformation [*Siddoway et al.*, 2000], and higher metamorphic pressures in northern New Mexico [*Williams et al.*, 1999] suggest that high temperatures in this region may reflect a somewhat deeper level of exposure.

A few isolated blocks within the higher temperature domain including the Cimarron River tectonic unit (Figure 7) [*Grambling and Dallmayer*, 1993], and the Zuni Mountains [*Karlstrom et al.*, in press, 2003], appear to have remained relatively cool during the Mesoproterozoic. Abrupt changes in cooling ages across shear zones suggest that steep geothermal gradients may have prevailed in the middle crust (now exposed at the surface) such that relatively small dip-slip displacements juxtaposed blocks from different crustal depths with contrasting thermal histories. Such juxtaposition of cool and hot blocks is not evident in ~1.4 Ga shear zones in central Colorado (e.g. Homestake and Idaho Springs–Ralston, Figure 5).

We propose a model for the thermal structure of the crust at ~1.4 Ga that explains the apparent steep mid-crustal thermal gradients and the curious paucity of ~1.4 Ga plutons intruding the high grade rocks of northern New Mexico and southern Colorado. In our model large volumes of A-type granitic magma were emplaced near the brittle-ductile transition where they would have encountered a rheologic barrier to further ascent [e.g. *Brown et al.*, 1998]. The transition from relatively efficient advective plus conductive heat transport in the ductile lower crust to relatively inefficient conductive transport in the brittle upper crust resulted in steep geothermal gradients in the middle crust (Plate 3). Transient thermal anomalies around plutons continually altered the complex and dynamic interface between lower crustal and upper crustal thermal regimes. Completely reset hornblende ages in the neighborhood of plutons are consistent with narrow thermal aureoles. Movement on shear zones during and after the ~1.4 Ga episode [*Grambling*, 1993; *Carrick and Andronicos*, 2001; *Shaw et al.*, 2001] further modified the interface by juxtaposing hotter and colder crustal blocks by dip-slip or oblique displacement.

We speculate that the high-T domain in northern New Mexico and southern Colorado represents a window eroded through the fossil thermal interface that corresponded to the magma-rich mid-crustal layer (Figure 10). The high grade rocks beneath the magma layer are exposed in a broad upwarp of the pluton-rich layer and paleo-isotherms; most ~1.4 Ga plutons have been eroded from this area. Ascending granitic magmas advected heat into the lower crust but were only emplaced upon reaching the brittle-ductile transition where magmas ponded and crystallized. The resulting mid-crustal magmatic layer blanketed the deeper crust and maintained higher temperatures as long as plutonism continued. This residence at high-T conditions completely reset argon systematics. Elsewhere in the Southwest, where nearly 20% of exposed Precambrian rocks are ~1.4 Ga granitoids [*Anderson and Morrison*, 1992], the current level of basement exposure lies near, or somewhat above the magma-rich layer (Plate 2). In several areas including the Park Range and Sawatch Range in Colorado ~1.4 Ga plutons may fill pull-apart structures in transpressional shear systems that facilitated magma transport to shallower levels [Figures 4, 6; *McCoy et al.*, this volume; *Shaw et al.*, 2001]. Above the mid-crustal magma layer rocks would have been subjected to a transient thermal pulse that reset argons systematics in mica and partially reset argon systematics in hornblende [*Shaw et al.*, 1999].

Contrasting styles of ~1.4 Ga deformation in Colorado and northern New Mexico also seem to reflect contrasting temperature regimes. Deformation within the Colorado mineral belt in central Colorado (Figures 1, 6)–presumably above the mid-crustal magma layer–was characterized by intense, localized shearing on discrete greenschist facies mylonite zones [e.g. *McCoy et al.*, this volume]. Pseudotachylyte associated with these ductile shear zones suggests that deformation occurred near the brittle-ductile rheologic transition [*Shaw et al.*, 2001]. In the Wet mountains and Tusas range–below the mid-crustal magma layer–more penetrative deformation involved large scale folding and fabric transposition [*Williams et al.*, 1999; *Siddoway et al.*, 2000]. Emplacement of magmas at the brittle-ductile transition would have modified the rheologic boundary by temporarily raising mid-crustal temperatures and by adding strong granitic plutons to the crust resulting in a complex interplay between deformation and plutonism.

According to petrogenetic models, the voluminous A-type granites of the Southwest imply a large mafic complement in the form of a mafic restite, cumulate, or underplate [*Vander Auwera et al.*, 2003; *Klimm et al.*, 2003; *Frost et al.*, 1999, 2001; *Anderson and Thomas*, 1985; *Anderson and Morrison*, 1992]. Seismic evidence from the CD-ROM experiment [*Keller and CD-ROM Group*, this volume; *Levander et al.*, this volume] revealed a 5–10 km-thick layer near the base of the crust with high seismic velocities that could, in part, be a basaltic underplate/restite layer. However, xenolith studies [*Selverstone et al.*, 1999; *Farmer et al.*, this volume] suggest that 1.4 Ga rocks may comprise only a relatively small proportion of a composite lower crustal high velocity layer. Modeling the distribution of ~1.4 plutons as a relatively thin layer restricted to the mid-crust results in more tenable estimates of the volume of mafic complement that would be required in the lower crust.

Structural data supporting regional contraction or transpression [*Nyman et al.*, 1994; *Kirby et al.*, 1995; *Gonzales et al.*, 1996; *Nyman and Karlstrom*, 1997; *McCoy et al.*, 2000; *Shaw et al.*, 2001], geochemical evidence for juvenile additions of crust along the southeastern margin of Laurentia [e.g. *Nelson and DePaolo*, 1985; *Bowring et al.*, 1992], and evidence for convergence along the southeastern margin of Laurentia [*Mosher*, 1998; *Rivers and Corrigan*, 2000; *Karlstrom et al.*, 2001] suggest that ca. 1.4 Ga episode in the continental interior occurred in a convergent or transpressional setting. Based on the large volume and regional extent of the proposed ~1.4 Ga magma layer we propose that this interface could be a deeply exhumed analog to melt- or fluid-rich layers that have been inferred in modern intracontinental orogenic settings [e.g. *Nelson and Project_INDEPTH*, 1996; *Babeyko et al.*, 2002; *Brasse et al.*, 2002]. This interpretation suggests a model wherein 1.4 Ga tectonism, plutonism, and metamorphism occurred beneath an intracontinental orogenic plateau comparable to modern Tibet or the Andean Altiplano-Puna plateau. One problem with this model is the lack of A-type igneous chemistry in these modern analogs.

Modern orogens and orogenic plateaus frequently undergo synchronous shortening and extension both parallel and perpendicular to the orogen [e.g. *Metzger et al.*, 1991; *Burchfiel et al.,* 1992; *Hurtado et al.,* 2001]. We speculate that topographically driven syncontractional extension at ~1.4 Ga could reconcile structural evidence for SE-NW shortening with the geochemistry of the plutons. In this model relatively shallow mantle melting required to produce the tholeiitic parental melts inferred from Sm-Nd signatures of some plutons [*Frost et al.*, 1999] could be related to asthenospheric upwelling driven by thermal or convective removal of an orogenically thickened continental lithosphere [e.g. *Houseman et al.*, 1981; *Collins*, 1994; *Inger*, 1994]. Upwelling asthenosphere could also account for heating of the middle-crust and regional low-P series metamorphism [e.g. *Loosveld and Etheridge*, 1990] and, possibly, the partial removal of any mafic underplate related to ~1.4 Ga magmatism.

CONCLUSIONS

New $^{40}Ar/^{39}Ar$ data from the Rocky Mountains and published data from throughout the southwestern United States constrains the maximum temperatures associated with ~1.4 Ga tectonism at the level of exposure. Important features of the regional temperature field include: (1) the middle crust of Proterozoic provinces south of the Cheyenne belt was pervasively heated to ~300°–500°C, (2) the highest mid-crustal temperatures occurred in northern New Mexico where exposed 1.4 Ga plutons are scarce, (3) temperatures at exposed crustal levels of the Archean craton remained below the closure temperature of biotite (~300°C) except in a 200 km-wide domain of attenuated lithosphere at the margin of the craton where temperatures increased southward to 300–450°C. The contrast in the ca. 1.4 Ga thermal history of Archean and Proterozoic lithosphere suggests a strong lithospheric control on the thermal response of these two provinces. Argon data reveal a thermal maximum in northern New Mexico that may correspond to a somewhat deeper mid-crustal exposure level. Steep thermal gradients are implied by the juxtaposition of blocks with strongly contrasting thermal histories across narrow dip-slip shear zones.

Based on our image of ca. 1.4 Ga mid-crustal temperatures, we suggest that current exposure in Proterozoic lithospheric provinces broadly coincides with a ca. 1.4 Ga mid-crustal level of granite emplacement where regional heating enhanced by advection in the lower crust gave way to conductive heat transfer in the upper crust. Where rocks from below this layer are exposed, argon systematics have been completely reset in all minerals and deformation is dominated by penetrative ductile flow. Where rocks from above the mid-crustal magma layer are exposed, hornblende partially retained accumulated argon and deformation is focused into narrow greenschist-facies mylonite zones.

Acknowledgements. Sample preparation and analysis were carried out at the New Mexico Geochronology Research Laboratory (NMGRL) at the New Mexico Institute of Mining and Technology. Lisa Peters, Richard Esser, Vladimir Ispolatov and numerous students provided help in sample preparation and analysis. Funding was provided by the CD-ROM project (NSF Continental Dynamics program grant EAR 9614787, Karlstrom and Heizler; EAR 0003500, Karlstrom), the Caswell-Silver foundation (Kelly Silver Graduate Fellowship, Shaw), and the University of New Mexico (Research, Project, and Travel Grant, Shaw). We thank Jane Selverstone, John Geissman, and Mousumi Roy for discussions and reviews of an earlier version of this paper. We thank Dan Holm and Kevin Chamberlain for critical and constructive formal reviews of the manuscript.

REFERENCES

Anderson, J.L., Proterozoic anorogenic granite plutonism of North America, in *Proterozoic Geology; Selected Papers From An International Proterozoic Symposium*, edited by J. L.G. Medaris, C.W. Byers, D.M. Mickelson, and W.C. Shanks, pp. 133–154, Geol. Soc. Am., Boulder, 1983.

Anderson, J.L., and J. Morrison, The role of anorogenic granites in the Proterozoic crustal development of North America, in *Proterozoic Crustal Evolution*, edited by K.C. Condie, pp. 263–299, Elsevier, Amsterdam, 1992.

Anderson, J.L., and W.M. Thomas, Proterozoic anorogenic two-mica granites: Silver Plume and St. Vrain batholiths of Colorado, *Geology*, *13*, 177–180, 1985.

Anderson, O. J., and G. E. Jones, Geologic Map of New Mexico: *New Mexico Bureau of Mines and Mineral Resources Open-file Report 408-A and B*, 1:500,000, 1994.

Babeyko, A.Y., S.V. Sobolev, R.B. Trumbull, O. Oncken, and L.L. Lavier, Numerical models of crustal scale convection and partial melting beneath the Altiplano-Puna Plateau, *Earth Planet. Sci. Lett.*, *199*, 373–388, 2002.

Baldwin, S.L., T.M. Harrison, and J.D. Fitz Gerald, Diffusion of ^{40}Ar in hornblende, *Contrib. Mineral. Petrol. 105,* 691–703, 1990.

Barinek, M.F., C.T. Foster, and P.P. Chaplinsky, Metamorphism and deformation near the ~1.4 Ga Mount Ethel pluton, Park Range, Colorado, *Rocky Mt. Geol*, *34*, 21–35, 1999.

Bickford, M.E., and J.L. Anderson, Middle Proterozoic Magmatism, in *Precambrian: Conterminous U.S.*, edited by J.C. Reed, Jr., M.E. Bickford, R.S. Houston, P.K. Link, D.W. Rankin, P.K. Sims, and W.R. Van Schmus, 281–292, Geol. Soc. Am., Boulder, 1993.

Bowring, S.A., T.B. Housh, W.R. Van Schmus, and F.A. Podosek, A major Nd isotopic boundary along the southern margin of Laurentia, *Eos, Trans., AGU, suppl.*, *73* (14), 333, 1992.

Brasse, H., P. Lezaeta, V. Rath, K. Schwalenberg, W. Soyer, and V. Haak, The Bolivian Altiplano conductivity anomaly, *J.of Geophys. Res., B, Sol. Earth and Planets*, *107*, (5), 15 pp., 2002.

Brown, L.D., A.R. Ross, and K.D. Nelson, Deep seismic bright spots, magmatism and the origin of the Tibet Plateau, *Abstr. Prog. - Geol. Soc. Am.*, *30* (7), 296, 1998.

Burchfiel, B.C., Z. Chen, K.V. Hodges, Y. Liu, L.H. Royden, C. Deng, and J. Xu, *The South Tibetan detachment system, Himalayan Orogen: extension contemporaneous with and parallel to shortening in a collisional mountain belt*, Geological Society of America (GSA), Boulder, 1992.

Carrick, T.L., and C.L. Andronicos, A study of the Fowler Pass shear zone in the Proterozoic basement of the Cimarron Mountains, northern New Mexico, *Abstr. Prog. - Geol. Soc. Am.*, *33* (5), 24, 2001.

Chamberlain, K.R., Medicine Bow Orogeny; timing of deformation and model of crustal structure produced during continent-arc collision, ca. 1.78 Ga, southeastern Wyoming, *Rocky Mountain Geology*, *33* (2), 259–277, 1998.

Chamberlain, K.R., and S.A. Bowring, Apatite-feldspar U-Pb thermochronometer; a reliable, mid-range (approximately 450° C), diffusion-controlled system, *Chem. Geol.*, *172* (1–2), 173–200, 2001.

Chamberlain, K.R., C.D. Frost, and B.R. Frost, Early Archean to Mesoproterozoic evolution of the Wyoming Province: Archean origins to modern lithospheric architecture, *Can. J. Earth Sci., 40,* 1357–1374.

Collins, W.J., Upper- and middle-crustal response to delamination; an example from the Lachlan fold belt, eastern Australia, *Geology*, *22* (2), 143–146, 1994.

Cosca, M.A., J.C. Hunziker, S. Huon, and H. Masson, Radiometric age constraints on mineral growth, metamorphism and tectonism of the Gummfluh Klippe, Brianconnais domain of the Prealpes, Switzerland, *Contributions to Mineralogy and Petrology*, *112* (4), 439–449, 1992.

Cox, D.M., C.D. Frost, and K.R. Chamberlain, 2.01 Kennedy dike swarm, southeastern Wyoming: Record of a rifted margin along the southern Wyoming province, *Rocky Mt. Geol*, *35* (1), 7–30, 2000.

Cullers, R.L., T.J. Griffin, M.E. Bickford, and J.L. Anderson, Origin and chemical evolution of the 1360 Ma San Isabel batholith, Wet Mountains, Colorado: a mid-crustal granite of anorogenic affinities, *Geol. Soc. Am. Bulletin*, *104*, 316–328, 1992.

Cullers, R.L., J. Stone, J.L. Anderson, N. Sassarini, and M.E. Bickford, Petrogenesis of Mesoproterozoic Oak Creek and West McCoy Gulch plutons, Colorado: an example of cumulate unmixing of a mid-crustal, two-mica granite of anorogenic affinity, *Precamb. Res.*, *62*, 139–169, 1993.

Dodson, M.H., Closure temperature in cooling geochronological and petrological systems, *Contrib. Mineral. petrol.*, *40*, 259–274, 1973.

Farmer, L., S.A. Bowring, N.I. Christiansen, M.L. Williams, J. Matzel, and L. Stevens, Contrasting lower crustal evolution across an Archean-Proterozoic suture: Physical, chemical and geochronologic studies of lower crustal xenoliths in southern Wyoming and northern Colorado, in *Lithospheric Structure and Evolution of the Rocky Mountain Region*, edited by K.E. Karlstrom, and G.R. Keller, AGU, Washington, this volume, 2003.

Frost, C.D., J.M. Bell, B.R. Frost, and K.R. Chamberlain, Crustal growth by magmatic underplating; isotopic evidence from the northern Sherman Batholith, *Geology*, *29* (6), 515–518, 2001.

Frost, C.D., B.R. Frost, K.R. Chamberlain, and B.R. Edwards, Petrogenesis of the 1.43 Ga Sherman Batholith, SE Wyoming, USA; a reduced, rapakivi-type anorogenic granite, *J. Petrol.*, *40* (12), 1771–1802, 1999.

Giletti, B.J., Isotopic geochronology of Montana and Wyoming, in *Radiometric Dating for Geologists*, edited by E.I. Hamilton, and R.M. Farquhar, pp. 111–146, Interscience, New York, 1968.

Gonzales, D.A., K.E. Karlstrom, and G.S. Siek, Syncontractional crustal anatexis and deformation during emplacement of ~1435 Ma plutons, western Needle Mountains, Colorado, *J. Geol.*, *104* (2), 215–223, 1996.

Grambling, J.A., A summary of Proterozoic metamorphism in northern New Mexico: The regional development of 520° C, 4 kbar rocks, in *Metamor-*

phism and CrustalEvolution of the Western United States, edited by W.G. Ernst, pp. 447–465, Prentice Hall, Englewood Cliffs, 1988.

Grambling, J.A., S.A. Bowring, and R.D. Dallmayer, Middle Proterozoic cooling ages in the Cimarron Mountains, northern New Mexico: U-Pb and $^{40}Ar/^{39}Ar$ constraints, *Geol. Soc. Am. - Abstr. Prog.*, *24*, 92, 1992.

Grambling, J.A., and R.D. Dallmayer, Tectonic evolution of Proterozoic rocks in the Cimarron Mountains, northern New Mexico, *J. Metamorph. Geol.*, *11*, 739–755, 1993.

Grambling, J.A., M. Williams, R. Smith, and C. Mawer, The role of crustal extension in the metamorphism of Proterozoic rocks in northern New Mexico, in *Proterozoic geology of the Southern Rocky Mountains*, edited by J.A. Grambling, and B.A. Tewksbury, pp. 87-110, Geol. Soc. Am., Boulder, 1989.

Green, G.N., The Digital Geologic Map of Colorado in ARC/INFO Format, *USGS Open File Report 92–0507*, 1:500,000, 1992.

Green, G.N., and G.E. Jones, The Digital Geologic Map Of New Mexico In Arc/Info Format, *USGS Open-File Report 97–52*, 1:500,000, 1997.

Hames, W.E., and S.A. Bowring, An empirical evaluation of the argon diffusion geometry in muscovite, *Earth and Planetary Science Letters*, *124* (1–4), 161–169, 1994.

Harlan, S.S., J.W. Geissman, L.W. Snee, and R.L. Reynolds, late Cretaceous remagnitization of Proterozoic mafic dikes, southern Highland Mountains, southwestern Montana: A paleomagnetic and $^{40}Ar/^{39}Ar$ study, *Geol. Soc. Am. Bulletin*, *108* (6), 653–668, 1996.

Harlan, S.S., L.W. Snee, J.W. Geissman, and A.J. Brearly, Paleomagnetism of the Middle Proterozoic Laramie anorthosite complex and Sherman Granite, southern Laramie Range, Wyoming and Colorado, *J. Geophys. Res.*, *99*, 997–10,020, 1994.

Harrison, T.M., Diffusion of ^{40}Ar in hornblende. *Contrib. Mineral. Petrol. 78*, 324–331, 1981.

Harrison, T.M., and I. McDougall, Investigations of an intrusive contact, Northwest Nelson, New Zealand; I, Thermal, chronological and isotopic constraints, *Geochimica et Cosmochimica Acta*, *44* (12), 2005–2020, 1980a.

Harrison, T.M., and I. McDougall, Investigations of an intrusive contact, Northwest Nelson, New Zealand; II, Diffusion of radiogenic and excess ^{40}Ar in hornblende revealed by $^{40}Ar/^{39}Ar$ age spectrum analysis, *Geochimica et Cosmochimica Acta*, *44* (12), 2005–2020, 1980b.

Heizler, M.T., Slow-cooling or reheating: Can SW USA thermochronological data be reconciled?, *Geol. Soc. Am. - Abstr. Prog.*, *34* (6), 180, 2002.

Hills, F.A., and R.L. Armstrong, Geochronology of Precambrian rocks in the Laramie Range and implications for the tectonic framework of southern Wyoming, *Precamb. Res.*, *1*, 213–225, 1974.

Hodges, K.V., W.E. Hames, and S.A. Bowring, $^{40}Ar/^{39}Ar$ age gradients in micas from a high-temperature - low pressure metamorphic terrain: Evidence for very slow cooling and implications for the interpretation of age spectra, *Geology*, *22*, 55–58, 1994.

Holm, D.K., P. Dahl, S., and D.R. Lux, $^{40}Ar/^{39}Ar$ evidence for middle Proterozoic (1300–1500 Ma) slow cooling of the southern Black Hills, South Dakota, midcontinent, North America: Implications for Early Proterozoic P-T evolution, *Tectonics*, *16* (4), 609–622, 1997.

Houseman, G.A., D.P. McKenzie, and P. Molnar, Convective instability of a thickened boundary layer and its relevance for the thermal evolution of continental convergent belts, *J.of Geophys. Res.*, *86* (7), 6115–6132, 1981.

Houston, R.S., E.M. Duebendorfer, K.E. Karlstrom, and W.R. Premo, A review of the geology and structure of the Cheyenne Belt and Proterozoic rocks of southern Wyoming, in *Proterozoic geology of the Southern Rocky Mountains*, edited by J.A. Grambling, and B.J. Tewksbury, pp. 1–12, Geol. Soc. Am., Boulder, 1989.

Houston, R.S., and others, The Wyoming province, in *Precambrian: Conterminous U.S.*, edited by J.C. Reed, Jr., M.E. Bickford, R.S. Houston, P.K. Link, D.W. Rankin, P.K. Sims, and W.R. Van Schmus, pp. 121–170, Geol. Soc. Am., Boulder, 1993.

Hurtado, J.M., Jr., K.V. Hodges, and K. Whipple, Neotectonics of the Thakkhola Graben and implications for Recent activity on the South Tibetan fault system in the central Nepal Himalaya, *Geological Society of America Bulletin*, *113* (2), 222–240, 2001.

Inger, S., Magmagenesis associated with extension in orogenic belts: examples from the Himalaya and Tibet, *Tectonophysics*, *238*, 183–197, 1994.

Karlstrom, K.E., K.-I. Åhall, S.S. Harlan, M.L. Williams, J. McLelland, and J.W. Geissman, Long-lived (1.8–1.0 Ga) convergent orogen in southern Laurentia, its extensions to Australia and Baltica, and implications for refining Rodinia, *Precamb. Res.*, *111* (1–4), 5–30, 2001.

Karlstrom, K.E., J.M. Amato, M.L. Williams, M.T. Heizler, C.A. Shaw, and A.S. Read, Proterozoic tectonic evolution of the New Mexico region: a synthesis, in *Geology of New Mexico*, New Mexico Geological Society, Socorro, in press, 2003.

Karlstrom, K.E., and S.A. Bowring, Proterozoic orogenic history of Arizona, in *Precambrian: Conterminous U.S.*, edited by W.R. Van Schmus, pp. 171–334, Geol. Soc. Am., Boulder, 1993.

Karlstrom, K.E., R.D. Dallmeyer, and J.A. Grambling, $^{40}Ar/^{39}Ar$ evidence for 1.4 Ga regional metamorphism in New Mexico: implications for thermal evolution of lithosphere in the southwestern USA, *J. Geol.*, *105*, 205–223, 1997a.

Karlstrom, K.E., A.J. Flurkey, and R.S. Houston, Stratigraphy and depositional setting of the Proterozoic Snowy Pass Supergroup, southeastern Wyoming; record of an early Proterozoic Atlantic-type cratonic margin, *Geol. Soc. Am. Bulletin*, *94* (11), 1257–1274, 1983.

Karlstrom, K.E., M.T. Heizler, and M.L. Williams, $^{40}Ar/^{39}Ar$ muscovite thermochronology within the upper granite gorge of the Grand Canyon, *Eos, Transactions, AGU*, *78* (46, Suppl.), 784, 1997b.

Karlstrom, K.E., and R.S. Houston, The Cheyenne belt: analysis of a Proterozoic suture in southern Wyoming, *Precamb. Res.*, *25*, 415–446, 1984.

Karlstrom, K.E., and E.D. Humphreys, Persistent influence of Proterozoic accretionary boundaries in the tectonic evolution of southwestern North America: Interaction of cratonic grain and mantle modification events, *Rocky Mt. Geol*, *33* (2), 161–179, 1998.

Keller, G.R., and CD-ROM Working Group, Contrasting lower crustal evolution across an Archean-Proterozoic suture: Physical, chemical and geochronologic studies of lower crustal xenoliths in southern Wyoming and northern Colorado, in *Lithospheric Structure and Evolution of the Rocky Mountain Region*, edited by K.E. Karlstrom, and G.R. Keller, AGU, Washington, this volume.

Kelley, S.P., and G. Turner, Laser probe ^{40}Ar - ^{39}Ar measurements of loss profiles within individual hornblende grains from the Giants Range Granite, northern Minnesota, U.S.A., *Earth Planet. Sci. Lett.*, *107*, 634–648, 1991.

Kellog, K.S., A paleomagnetic study of various Precambrian Rocks in the northeastern Colorado Front Range and its bearing on Front Range rotation, Ph.D. thesis, University of Colorado, Boulder, 177 pp., 1973.

Kirby, E., K.E. Karlstrom, C.L. Andronicos, and R.D. Dallmeyer, Tectonic setting of the Sandia Pluton; an orogenic 1.4 Ga granite in New Mexico, *Tectonics*, *14* (1), 185–201, 1995.

Klimm, K., F. Holtz, W. Johannes, and P.L. King, Fractionation of metaluminous A-type granites; an experimental study of the Wangrah Suite, Lachlan fold belt, Australia, *Precamb. Res.*, *124* (2–4), 327–341, 2003.

Kwon, J., K. Min, P.J. Bickel, and P.R. Renne, Statistical methods for jointly estimating the decay constant of ^{40} K and the age of a dating standard, *Mathematical Geol.*, *34* (4), 457–474, 2002.

Levander, A.R., C. Zelt, and B. Magnani, Crust and upper mantle velocity structure of the southern Rocky Mountains from the Jemez lineament to the Cheyenne belt, in *Lithospheric Structure and Evolution of the Rocky Mountain Region*, edited by K.E. Karlstrom, and G.R. Keller, AGU, Washington, this volume.

J.K.W. Lee, T. C. Onstott, K. V. Cashman, R. J. Cumbest and D. Johnson. 1991: Incremental heating of hornblende in vacuo: Implications for $^{40}Ar/^{39}Ar$ geochronology and the interpretation of thermal histories. *Geology*, *19*, No. 9, pp. 872–876

Lister, G.S., and S.L. Baldwin, Modeling the effect of arbitrary P-T-t histories on argon diffusion in minerals using the MacArgon program for the Apple Macintosh, *Tectonophysics*, *253*, 83–109, 1996.

Loosveld, R.J.H., and M.A. Etheridge, A model for low-pressure facies metamorphism during crustal thickening, *J. Metamorph. Geol.*, *8*, 257–267, 1990.

Marcoline, J., M. Heizler, L.B. Goodwin, S. Ralser, and J. Clark, Thermal, structural, and petrologic evidence for 1.4 Ga metamorphism and deformation in central New Mexico, *Rocky Mt. Geol*, *34*, 93–119, 1999.

McCoy, A.M., K.E. Karlstrom, M.L. Williams, and C.A. Shaw, Proterozoic ancestry of the Colorado mineral belt: ca. 1.4 Ga shear zone system in central Colorado, in *Lithospheric Structure and Evolution of the Rocky Mountain Region*, edited by K.E. Karlstrom, and G.R. Keller, AGU, Washington, this volume.

McCoy, A.M., C.A. Shaw, K.E. Karlstrom, M.L. Williams, and M. Jercinovic, Geometry, kinematics, and protracted deformation along the Colorado Mineral Belt mylonite network, *Geol. Soc. Am. Abstracts with Program*, *32*, 230–231, 2000.

McDougall, I., and T.M. Harrison, *Geochronology and Thermochronology by the $^{40}Ar/^{39}Ar$ Method*, 212 pp., Oxford University Press, Oxford, 1999.

McLaren, S., M. Sandiford, and M. Hand, High radiogenic heat-producing granites and metamorphism; an example from the western Mount Isa Inlier, Australia, *Geology*, *27* (8), 679–682, 1999.

Mezger, K., B.A. van der Pluijm, E.J. Essene, and A.N. Halliday, Synorogenic collapse; a perspective from the middle crust, the Proterozoic Grenville Orogen, *Science*, *254* (5032), 695–698, 1991.

Min, K., R. Mundil, P.R. Renne, and K.R. Ludwig, A test for systematic errors in $^{40}Ar/^{39}Ar$ geochronology through comparison with U/Pb analysis of a 1.1-Ga rhyolite, *Geochim. Cosmochim. Acta*, *64*, 73–98, 2000.

Mosher, S., Tectonic evolution of the southern Laurentian Grenville orogenic belt, *Geol. Soc. Am. Bulletin*, *110* (11), 1357–1375, 1998.

Nelson, B., and D.J. DePaolo, Rapid production of continental crust 1.7 to 1.9 b.y. ago: Nd isotopic evidence from the basement of the North American mid-continent, *Geol. Soc. Am. Bulletin*, *96*, 746–754, 1985.

Nelson, K.D., and Project INDEPTH, Partially molten middle crust beneath southern Tibet: Synthesis of Project INDEPTH Results, *Science*, *274*, 1684–1688, 1996.

Nyman, M.W., and K.E. Karlstrom, Pluton emplacement processes and tectonic setting of the 1.42 Ga Signal Batholith, SW USA; important role of crustal anisotropy during regional shortening, *Precamb. Res.*, *82* (3–4), 237–263, 1997.

Nyman, M.W., K.E. Karlstrom, E. Kirby, and C.M. Graubard, Mesoproterozoic contractional orogeny in western North America: Evidence from ca. 1.4 Ga plutons, *Geology*, *22*, 901–904, 1994.

Peacock, S.M., Thermal modeling of metamorphic pressure-temperature-time paths, a forward approach, in Spear, F.S and Peacock, S.M., *Metamorphic Pressure-Temperature-Time Paths. Short course in Geology*. Am. Geophys. Union, Washington, 57–102, 1989.

Pearson, R.C., C.E. Hedge, H.H. Thomas, and T.W. Stearn, Geochronology of the St. Kevin Granite and neighboring Precambrian rocks, northern Sawatch Range, Colorado, *Geol. Soc. Am. Bulletin*, *77*, 1109–1120, 1966.

Pedrick, J., *Polyphase tectonometamorphic history of the Taos Range, northern New Mexico*, Ph.D. thesis, University of New Mexico, Albuquerque, 146 pp., 1995.

Peterman, Z.E., and R.A. Hildreth, *Reconnaissance geology and geochronology of the granite mountains, Wyoming, U.S. Geol. Surv. Prof Paper 1055*, 22 pp., 1978.

Premo, W.R., and C.M. Fanning, SHRIMP U-Pb zircon ages for Big Creek Gneiss, Wyoming and Boulder Creek batholith, Colorado; implications for timing of Paleoproterozoic accretion of the northern Colorado Province, *Rocky Mountain Geology*, *35* (1), 31–59, 2000.

Premo, W.R. and W.R. Van Schmus, Zircon geochronology of Precambrian rocks in southeastern Wyoming and northern Colorado, in *Proterozoic geology of the Southern Rocky Mountains*, edited by J.A. Grambling, and B.A. Tewksbury, pp. 13–33, Geol. Soc. Am., Boulder, 1989.

Read, A.S., K.E. Karlstrom, J.A. Grambling, S.A. Bowring, M. Heizler, and C. Daniel, A middle-crustal cross section from the Rincon Range, northern New Mexico: Evidence for 1.68 Ga pluton-influenced tectonism and 1.4 Ga regional metamorphism, *Rocky Mt. Geol*, *34* (1), 67–91, 1999.

Reed, J.C., Jr., M.E. Bickford, and O. Tweto, Proterozoic accretionary terranes of Colorado and southern Wyoming, in *Precambrian: Conterminous U.S.*, edited by J.C. Reed, Jr., M.E. Bickford, R.S. Houston, P.K. Link, D.W. Rankin, P.K. Sims, and W.R. Van Schmus, pp. 110–121, Geol. Soc. Am., Boulder, 1993.

Reed, J.C., Jr., and L.W. Snee, 1.4-Ga Deformational and Thermal Events in the Central Front Range, Colorado., *Geol. Soc. Am. Abstr. Prog.*, *23* (4), 58, 1991.

Renne, P.R., C.C. Swisher, A.L. Deino, D.B. Karner, T.L. Owens, and D.J. DePaolo, Intercalibration of standards, absolute ages and uncertainties in $^{40}Ar/^{39}Ar$ dating, *Chem. Geol.*, *145*, 117–152, 1998.

Rivers, T., and D. Corrigan, Convergent margin on southeastern Laurentia during the Mesoproterozoic: Tectonic implications, *Can.n J. Earth Sci.*, *37*, 1–25, 2000.

Schmitz, M.D., and S.A. Bowring, U-Pb zircon and titanite systematics of the Fish Canyon Tuff; an assessment of high-precision U-Pb geochronology and its application to young volcanic rocks, *Geochimica et Cosmochimica Acta*, *65* (15), 2571–2587, 2001.

Scoates, J.S. and Chamberlain, K.R., 1995, Baddeleyite (ZrO2) and zircon (ZrSiO4) from anorthositic rocks of the Laramie Anorthosite complex, Wyoming: Petrologic consequences and U-Pb ages: American Mineralogist, v. 80, p. 1319–1329.

Scoates, J.S. and Chamberlain, K.R., 2003, Geochronologic, geochemical and isotopic constraints on the origin of monzonitic and related rocks in the Laramie anorthosite complex, Wyoming, USA. Precambrian Research, v. 124, p. 269–304.

Selverstone, J., M. Hodgins, C. Shaw, J.N. Aleinikoff, and C.M. Fanning, Proterozoic tectonics of the northern Colorado Front Range, in *Geologic history of the Colorado Front Range*, edited by D.W. Bolyard, and S.A. Sonnenberg, pp. 9–18, Rocky Mountain Association of Geologists, Denver, CO, 1997.

Selverstone, J., A. Pun, and K.C. Condie, Xenolithic evidence for Proterozoic crustal evolution beneath the Colorado Plateau, *Geol. Soc. Am. Bull.*, *111*, 590–606, 1999.

Shafiqullah, M., P.E. Damon, D.J. Lynch, S.J. Reynolds, W.A. Rehrig, and R.H. Raymond, K-Ar Geochronology and Geologic History Of Southwestern Arizona and Adjacent Areas, *Ariz. Geol. Soc. Digest*, *12*, 201–260, 1980.

Shaw, C.A., and K.E. Karlstrom, The Yavapai-Mazatzal Boundary in the southern Rocky Mountains, *Rocky Mt. Geol*, *34* (1), 37–52, 1999.

Shaw, C.A., K.E. Karlstrom, M.L. Williams, M.J. Jercinovic, and A.M. McCoy, Electron-microprobe monazite dating of ca. 1.71–1.63 Ga and ca. 1.45–1.38 Ga deformation in the Homestake shear zone, Colorado; origin and early evolution of a persistent intracontinental tectonic zone, *Geology*, *29* (8), 739–742, 2001.

Shaw, C.A., L. Snee, W., J. Selverstone, and J.C. Reed, Jr., $^{40}Ar/^{39}Ar$ thermochronology of Mesoproterozoic metamorphism in the Colorado Front Range, *J. Geol.*, *107* (1), 49–67, 1999.

Siddoway, C.S., R.M. Givot, C.D. Bodle, and M.T. Heizler, Dynamic vs. Anorogenic setting for Mesoproterozoic plutonism in the Wet Mountains, Colorado: Does the interpretation depend on level of exposure?, *Rocky Mt. Geol*, *35*, 91–111, 2000.

Silver, L.T., M.E. Bickford, W.R. Van Schmus, J.L. Anderson, T.H. Anderson, and L.G. Medaris, Jr., The 1.4–1.5 b.y. transcontinental anorogenic plutonic perforation of North America, *Geol. Soc. Am. Abstr. Prog.*, *9* (7), 1176–1177, 1977.

Snyder, G.L., Geologic map of the central part of the northern Park Range, Jackson, and Routt Counties, Colorado, *Miscellaneous Investigations Series I–1112*, 1:48,000, 2 sheets, 1980.

Spear, F.S., and M.C. Gilbert, Experimental studies of amphibole stability; amphiboles in metamorphic rock compositions. *Reviews in Mineralogy, vol. 9B*, 268–278, 1982.

Steiger, R.H., and E. Jäger, Sub-commission on geochronology: Convention on the use of decay constants in geo- and cosmochronology, *Earth Planet. Sci. Lett.*, *36* (3), 359–362, 1977.

Turner, G., The distribution of potassium and argon in chondrites, *International Series of Monographs on Earth Sciences*, *30*, 387–398, 1968.

Tweto, O.L., *Rock units of the Precambrian basement in Colorado, USGS Professional Paper 1321–A*, 54 pp., U.S. Geological Survey, 1987

Tweto, O.L., Geologic Map of Colorado, State Geologic Map, 1:500,000, 2 sheets, 1989.

Tweto, O.L., and P.K. Sims, Precambrian ancestry of the Colorado mineral belt, *Geol. Soc. Am. Bull.*, *74*, 991–1014, 1963.

Vander Auwera, J., M. Bogaerts, J.-P. Liegeois, D. Demaiffe, E. Wilmart, O. Bolle, and J.C. Duchesne, Derivation of the 1.0–0.9 Ga ferro-potassic A-type granitoids of southern Norway by extreme differentiation from basic magmas, *Precamb. Res.*, *124* (2–4), 107–148, 2003.

Williams, M.L., Overview of Proterozoic metamorphism in Arizona, *Ariz. Geol. Soc. Digest*, *19*, 11–26, 1991.

Williams, M.L., and K.E. Karlstrom, Looping P-T paths and high-T, low-P middle crustal metamorphism; Proterozoic evolution of the Southwestern United States, *Geology*, *24* (12), 1119–1122, 1996.

Williams, M.L., K.E. Karlstrom, A. Lanzirotti, A.S. Reed, J.L. Bishop, C.E. Lombardi, and M.B. Wingsted, New Mexico middle crustal cross sections: 1.65 Ga macroscopic geometry, 1.4 Ga thermal structure and continued problems in understanding crustal evolution, *Rocky Mt. Geol*, *34* (1), 53–66, 1999.

Wingsted, M., M.L. Williams, and A. Lanzirotti, Timing constraints on porphyroblast growth and fabric development in the southern Picuris Range, north central New Mexico; implications for the protracted tectonic/thermal evolution of the middle crust, *Geol. Soc. Am. Abstr. Prog.*, *28*, 495, 1996.

Wohletz, K.H., Heat, *computer program*, University of California, Los Alamos National Laboratory, Los Alamos, NM, 2002.

Characterization and Age of the Mesoproterozoic Debaca Sequence in the Tucumcari Basin, New Mexico

Jose F.A. Amarante[1], Shari A. Kelley[1], Matthew T. Heizler[2], Melanie A. Barnes[3], Kate C. Miller[4], and Elizabeth Y. Anthony[4]

Petrographic, geochronologic, and well log data from two deep oil wells (Mescalero 1 and State Mr. Jones 1) in the Tucumcari Basin, New Mexico are used to characterize a thick vertical section of the Mesoproterozoic Debaca sequence and portions of the underlying crystalline basement. This information is coupled with industry seismic data to constrain the geometry of the northernmost extent of the Mesoproterozoic Debaca basin. The Debaca sequence, a weakly metamorphosed sedimentary-volcanic package, is composed of volcaniclastic sandstone, tuffaceous sandstone, rhyolite, quartz-rich dolostone, dolomitic quartzite, sandstone, and arkose. The sequence rests disconformably on deeply weathered quartz syenite of the underlying Mesoproterozoic Panhandle Igneous Complex. Numerous sills and dikes of gabbro cut the Debaca sequence, imparting contact metamorphism to the adjacent rock units. U-Pb SHRIMP and ^{40}Ar/^{39}Ar ages for samples from the Mescalero 1 well indicate that Debaca sequence was formed ca. 1105–1332 Ma. Zircon from a quartz diorite (gradationally below the syenite) that is part of the crystalline basement beneath the Debaca sequence yields a SHRIMP U-Pb age of 1332 ± 18 Ma. Eight detrital grains of zircon from the arkose, the basal unit of the Debaca sequence, yield SHRIMP U-Pb zircon ages ranging from 1308 ± 52 to 1708 ± 14 Ma. ^{40}Ar/^{39}Ar ages on hornblende and biotite indicate a mean emplacement age of 1105±3 Ma for the gabbro. Three pulses of magmatism, one bimodal episode at 1.33 Ga, one felsic episode at ~1.26 Ga, and one mafic episode at 1.09 Ga, have been identified in Mescalero 1.

1. INTRODUCTION

Exposed Mesoproterozoic sedimentary successions ranging in age from 1.24 to 1.26 Ga that rest upon older crystalline basement in Arizona (Apache and Unkar Groups) and west Texas (Thunderbird Group, Allamoore Formation) offer important insights into this sparsely represented time in Earth's history in the southwestern United States [*Wrucke*, 1989; *Elston and McKee*, 1982; *Timmons et al.*, 2003; *Pittenger et al.*, 1994; *Bickford et al.*, 2000]. Most notably, these rocks preserve evidence of continental to marginal marine deposition in basins that formed in response to deformation preceding the NW-directed, ~1.1 Ma Grenville collisional event [*Timmons et al.*, 2003]. Mesoproterozoic sedimentary and volcanic rocks (Debaca sequence) that appear to be stratigraphically equivalent to the exposed sections are extensively preserved in the subsurface of the Texas panhandle and in eastern New Mexico (Figure 1). The upper part of the Debaca sequence has been penetrated in numerous oil wells in Texas and New Mexico, so that the aerial distribution of the unit is reasonably

[1]Department of Earth and Environmental Science, New Mexico Institute of Mining and Technology, Socorro, New Mexico
[2]New Mexico Bureau of Geology and Mineral Resources, Socorro, New Mexico
[3]Geosciences, Texas Tech University, Lubbock, Texas
[4]Department of Geological Sciences, The University of Texas at El Paso, El Paso, Texas

The Rocky Mountain Region: An Evolving Lithosphere
Geophysical Monograph Series 154

10.1029/154GM13

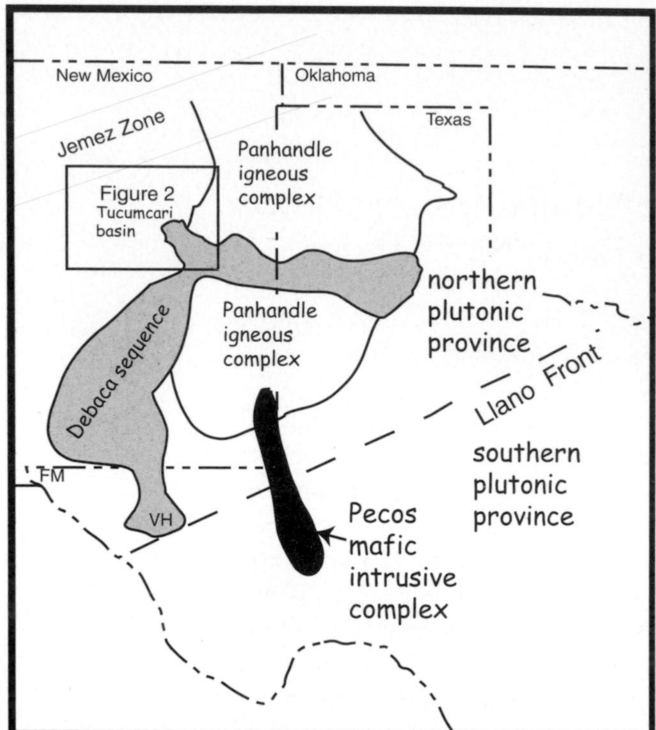

Figure 1. Generalized map showing the location of the Tucumcari Basin with respect to the Proterozoic basement sequences: plutonic province, the Panhandle igneous complex, the Pecos mafic intrusive complex, the Llano front and the Debaca sequence. Modified from *M.A. Barnes et al.* [1999].

well know, but the vertical succession is poorly constrained [*Flawn*, 1956; *Muehlberger et al*., 1967; *Denison and Hetherington*, 1969]. Here we present lithologic and geochronologic data from two deep oil wells that penetrate the entire Mesoproterozoic Debaca sequence and portions of the underlying juvenile Mesoproterozoic crystalline basement of southern Laurentia. We couple this information with industry seismic data to crudely constrain the geometry of the northernmost extent of the Mesoproterozoic Debaca basin.

In 1996, Labrador Oil Company drilled two deep wells, Mescalero 1 and State Mr. Jones 1, in the Tucumcari Basin in Guadalupe County, east-central New Mexico (Figure 2). The Tucumcari Basin, a series of deep sub-basins bound by northeast-, northwest- and east-trending normal faults, developed during early Pennsylvanian to early Permian time in response to Ancestral Rocky Mountain deformation [*Broadhead and King*, 1988; *Broadhead*, 2001]. Mescalero 1, located in T. 6 N, R. 22 E, section 1, penetrated ~1768 meters of Mesozoic and Paleozoic sediments and ~2652 meters of basement rocks, totaling 4420 meters. State Mr. Jones 1, located in T. 7 N, R. 23 E, sec 32, went through ~1900 m of Mesozoic to Paleozoic rocks, and ~1750 m of basement. The drill cuttings for Mescalero 1 and geophysical logs for both wells are on file at the New Mexico Bureau of Geology and Mineral Resources in Socorro, New Mexico. Our analysis of the well cuttings from Labrador Oil Mescalero 1, which is located about 15 km from an industry seismic profile, was undertaken to provide a basis for interpreting reflectors on industry seismic lines in the Tucumcari Basin and the southern end of the CD-ROM NM-1 seismic line [*Eshete,* 2001]. In the course of our investigation [*Amarante*, 2001], we determined that Mescalero 1 had intersected a complete section of Mesoproterozoic Debaca sequence. Furthermore, industry seismic reflection data show the geometry of the associated Mesoproterozoic basin. This particular well provides important new information about the Mesoproterozoic history of east-central New Mexico and offers an excellent opportunity to compare and contrast geologic and geochronologic data from Mesoproterozoic successions across eastern New Mexico, Arizona (Apache and Unkar Groups) and west Texas (Thunderbird Group, Allamoore Formation). The three goals of this project are: (1) petrologic and geochemical characterization of the basement rocks encountered in Mescalero 1; (2) establishing the age of the sedimentary sequence in the well and evaluating its relationship with presumed correlative units (e.g., Castner Marble and Allamore Fm.); and (3) determining the geophysical log signatures of the sequence so that the rocks can be tied to reflectors on seismic lines.

2. BACKGROUND

Flawn [1956], upon examining well cuttings and core from Precambrian basement terranes in the subsurface of Texas and southeastern New Mexico, defined the Swisher Gabbroic Terrane in the Texas panhandle and eastern New Mexico. This terrane is composed of carbonates and arkosic siltstone interbedded with diabase and cut by gabbroic intrusions. Although contact metamorphism adjacent to mafic intrusions is locally important, these rocks do not appear to have experienced significant regional metamorphism. *Flawn* [1956] noted that the Swisher Terrane overlies the Panhandle Volcanic Terrane, a sequence of rhyolite tuffs and granite intrusions. *Flawn* [1956] also recognized that the Precambrian sedimentary rocks in the Franklin Mountains of west Texas share similarities with rocks encountered in the subsurface of southeastern New Mexico. He called this sequence of rocks in southeastern New Mexico the "Metasedimentary and Metavolcanic Terrane." Subsequent work by *Foster and Stipp* [1961], *Muehlberger et al.* [1966; 1967], *Denison and Hetherington* [1969], and *Denison et al.* [1984] added lithologic and new geochronologic data to establish age relationships and further refine the identity of the subsurface terranes. *Muehlberger et al.* [1967] identified the De Baca Terrane based on well samples from De Baca County, New Mexico. This terrane is composed of weakly metamorphosed quartzite, siltstone and sandy carbonates with diabase and gabbro intrusions. These authors note that the lithologies in the De Baca Terrane are quite similar to those in the Swisher Terrane and that the two units occupy the same stratigraphic position, although the amount of diabase is greater in the latter. *Denison et al.* [1984] combined the two names into the Debaca-Swisher terrane and *Barnes et al.* [2002] proposed the name the Debaca sequence for the entire area (Figure 1).

Figure 1 shows the location of the 1.26 Ga Mesoproterozoic Debaca sequence with respect to other Proterozoic basement features, including the older, underlying plutonic and Panhandle igneous complexes and the younger Pecos mafic intrusive complex of *M.A. Barnes et al.* [1999, 2002]. The **plutonic province** is made up of deformed granitic gneiss, diorite, and granite. The term plutonic province was defined by *M.A. Barnes et al.* [1999, 2002] to encompass previously designated Proterozoic "terranes" such as the Chaves granite terrane of *Muehlberger et al.* [1967], and the Red River mobile belt and Fisher metasedimentary terrane of *Flawn* [1956], and the Llano province of *Denison et al.* (1984). The **Panhandle igneous complex** is composed of undeformed 1.34 – 1.37 Ga granite and rhyolite [M.A. *Barnes et al.*, 2002]. The **Debaca sequence** consists of weakly metamorphosed sedimentary and volcanic rock units that were deposited in a shallow basin on the Panhandle igneous complex and older basement and later intruded by gabbro. Volcanic ashes from outcrops of presumed Debaca sequence in the Franklin Mountains (*Pittenger et al.*, 1994) and Van Horn area (*Bickford et al.*, 2000) yield an age of approximately 1.26 Ga for this unit. The **Pecos mafic complex** is a 1.1 Ga layered intrusion that appears to be related to extension during the 1.1 Ga Grenville Orogeny [*C.G. Barnes et al.*, 1999]. The Llano front, which is defined by a gravity and magnetic anomaly [*Mosher*, 1998], separates basement affected by Grenville contractional deformation from older basement.

3. METHODS

3.1. Petrography

The entire set of cuttings from Mescalero 1 was examined using a binocular microscope and a total of twenty-eight thin sections were prepared from representative sections. Great care was used in identifying and removing cuttings that had caved from uphole. Geophysical logs for the well proved very

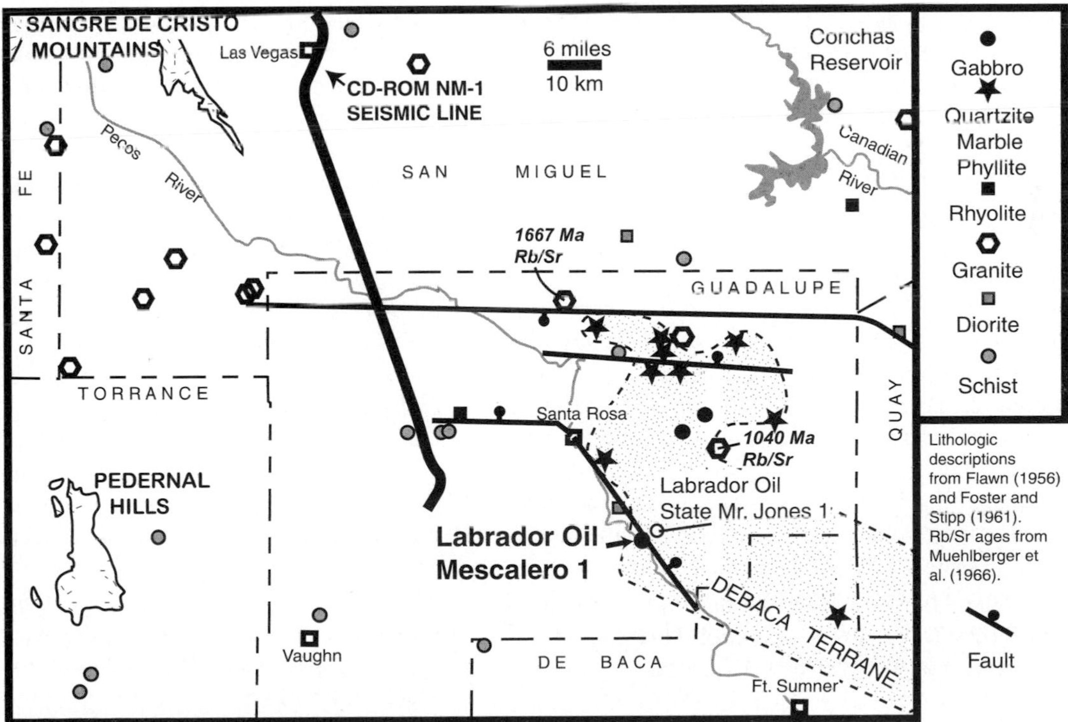

Figure 2. Map showing the location of Labrador Oil Mescalero 1 and State Mr. Jones 1, as well as other wells penetrating basement drilled in the Tucumcari Basin region. The identification of Debaca sequence, especially in wells drilled after 1960, is based on macroscopic examination of cuttings. Pennsylvanian faults from *Broadhead* [2001].

useful in lithological identification and recognizing lithologic breaks. Some minerals in the thin sections were examined in greater detail using the Cameca SX-100 electron microprobe at the New Mexico Bureau of Geology and Mineral Resources.

3.2. Geochronologic Analysis

$^{40}Ar/^{39}Ar$ ages on hornblende and biotite from gabbroic samples and U-Pb SHRIMP zircon ages from arkose and quartz diorite were determined. Biotite and hornblende are common in the gabbroic bodies near the top of the section, but datable minerals were not found in the gabbroic units lower in the section. Granite and rhyolite were also evaluated as potential candidates for geochronologic analyses, but insufficient datable material was recovered from these rocks. Samples for U-Pb geochronology were milled and a heavy mineral concentrate was obtained using a Wilfley table, a Frantz magnetic separator, and heavy liquids at New Mexico Tech. $^{40}Ar/^{39}Ar$ ages were determined at the New Mexico Geochronology Research Laboratory following the methods of *McIntosh et al.* [2002] (Table 1; see CDROM in back cover sleeve). The analytical data are presented in Table 2; see CDROM in back cover sleeve. All ages are calculated using a revised decay constant for ^{40}K of 5.476×10^{-10} /year [*Kwon et al.*, 2002]. The decay constant has not been formally evaluated like those presented in *Steiger and Jäger* [1977], however, it does represent a value that is based on comparison with U-Pb analyses and is thought to better reflect the true decay constant.

SHRIMP U-Pb zircon analyses were performed in the geochronology laboratory of Australia National University using the methods of *Williams* [1998]. The results are reported in *M. A. Barnes* [2001] and M.A. *Barnes et al.* [2002].

3.3. Geochemistry

Seven representative samples from selected rock units were identified for major and trace-element chemical analyses. Approximately 25 grams of chips were collected for each sample. Each sample represents a composite of approximately 30-m of the drill hole. Care was taken to avoid exotic material (uphole contamination) in these samples. Samples were analyzed for ten major element oxides and eleven trace elements (Table 3; see CDROM in back cover sleeve) by inductively-coupled plasma atomic emission spectroscopy at Texas Tech University. Details of the method and analytical uncertainties are given in M.A. *Barnes* [2001].

3.4. Well Log Analysis

A complete suite of modern geophysical logs was run in both Labrador wells, including gamma ray, neutron porosity, density porosity, resistivity, and sonic logs. This suite of logs, coupled with the well cuttings from Mescalero 1, provides a rare opportunity for detailed characterization of the rocks. The observations made for this well can be applied to similar rocks, well log signatures, and seismic reflections elsewhere in the region. The sonic, density, and gamma logs for both wells were digitized and the sonic log from State Mr. Jones 1 was used to produce a synthetic seismogram because it is closer to the seismic line. The digitized logs were correlated with the petrographic analyses and average values of sonic velocity, density, and gamma response were determined for each rock type. The average values were plotted on cross-plots to determine which physical properties were most diagnostic for determining lithology.

4. RESULTS

4.1. Petrography

Petrographic study of twenty-eight thin sections of cuttings from Mescalero 1 led to identification of four intrusive rock types (gabbro, quartz syenite grading to diorite, and granite), and a weakly metamorphosed sedimentary-volcanic sequence, with localized intense contact metamorphism (hornfels and marble) adjacent to the intrusive bodies (Plate 1a). The sedimentary-volcanic sequence is composed of volcaniclastic sandstone, tuffaceous sandstone, rhyolite, quartz-rich dolostone, dolomitic quartzite, sandstone, and arkose (Plate 1a). This sedimentary-volcanic package matches the description of the Debaca sequence in the subsurface of the west Texas Panhandle and southeastern New Mexico [*Denison and Hetherington*, 1969]. The following is a brief report of the major rock units encountered downhole in the well, keyed to Plate 1a. Complete descriptions and photomicrographs can be found in *Amarante* [2001].

1) The **gabbro** is a gray, medium to coarse-grained crystalline rock, with a cumulate to sub-ophitic texture. Plagioclase feldspar, pyroxene, olivine, and amphibole are the main mineral phases. In general, the plagioclase is tabular and displays a sub-parallel orientation. Biotite, magnetite, apatite, and titanite are accessory minerals. Chlorite, hematite, epidote, and calcite are present as alteration minerals. The two largest gabbroic bodies are at depths of 1759–1859 m and 1905–2155 m. A number of small dikes are present in the lower part of the well. The gabbro appears to be the youngest rock unit within the basement, because it caused metamorphism in other Proterozoic rock units.

2) The **hornfels** has a spotted texture developed by reddish-brown biotite, indicating contact metamorphism. The groundmass is cryptocrystalline and dominated by fine-grained microlites of quartz, muscovite, sericite, carbonate, and lesser

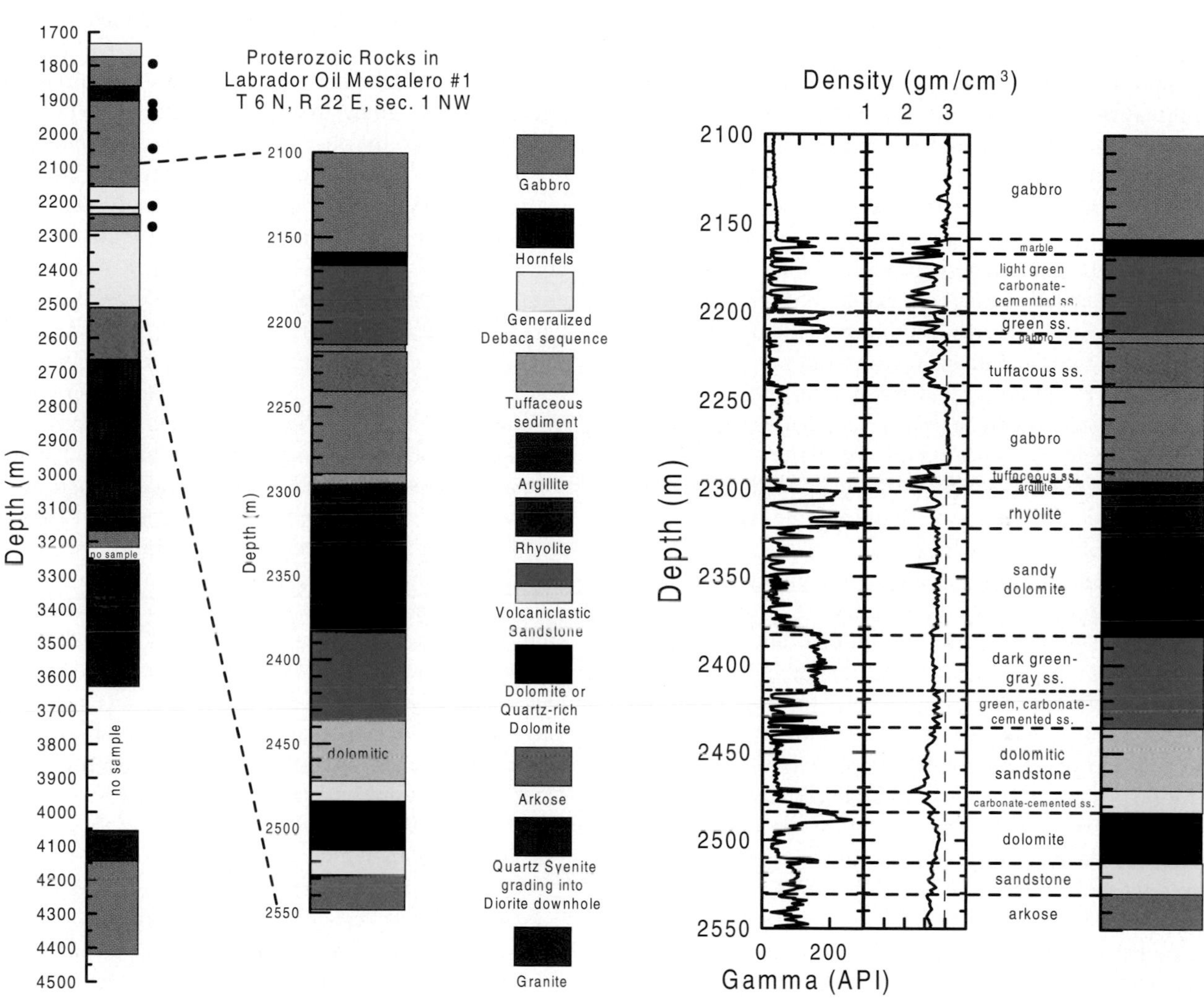

Plate 1 a. Stratigraphic column of the Mesoproterozoic rocks in Mescalero **b.** Relationship between lithology, gamma ray activity and density in Mescalero 1.

amounts of chlorite and pyrite. The hornfels is approximately 46 m thick and is located between two bodies of gabbro close to the top of the Proterozoic section (Plate 1a).

3) **Sedimentary-volcanic sequence** (2155–2664 m):

A) The **volcaniclastic sandstone** is green to gray and is fine to medium grained. The sandstone contains angular to subangular grains of quartz, feldspar, and lithic fragments. The lithic fragments are from at least three different sources, plutonic, sedimentary, and volcanic rocks, and are within a fine- to medium-grained matrix. Fragments of intrusive rocks are composed mostly of potassium feldspar, plagioclase and quartz. The potassium feldspar commonly exhibits a well-developed "tartan" twinning. The presence of plagioclase may indicate that the sediment was relatively proximal to its igneous source, and that little chemical weathering occurred. Fragments of volcanic rocks are generally strongly altered to white phyllosilicates (muscovite, clay), kaolinite, and carbonate. Fragments of quartz containing abundant fluid inclusions are common. The matrix is dominated by microlites of feldspar, quartz, mafic minerals, clay and carbonates, which occur as irregular patches. Accessory minerals include apatite, zircon, and pyrite.

B) The **tuffaceous sandstone** is fine-grained, light green to white, and is poorly reactive with cold dilute hydrochloric acid. Quartz, feldspar, and fragments of devitrified glass-shards dominate the rock. Carbonate, zeolite, and randomly oriented muscovite are also observed.

C) The **rhyolite** is primarily aphanitic, although portions of the unit are porphyritic. The rock is light green, dark green, to brown in color. The rock locally shows flow structure and contains remnants of phenocrysts such as biotite, hornblende, titanite, and feldspar that are totally replaced by one or more of the following minerals: chlorite, carbonate, and muscovite. Pyrite and magnetite are disseminated in the groundmass.

D) The **quartz-rich dolostone** is fine-to coarse-grained, white to light gray or dark gray in color, and has a clastic texture. Quartz, dolomite, and calcite dominate this rock type. Quartz makes up about 50 percent of the rock by volume; dolomite and calcite make up the remaining 50 percent. The rock effervesces selectively with cold dilute hydrochloric acid.

E) The **dolomitic quartzite** is fine-to coarse-grained, and is light gray, white, or black in color. It has a banded appearance, dominated by quartz, dolomite, and calcite. Quartz grains are commonly smoky and occur in dark-colored bands intercalated with light colored bands, which are composed of dolomite and calcite. Dolomite crystals are well developed, filling pore spaces between quartz grains, which suggests that dolomite formed paragenetically later than quartz. Calcite is finely crystalline and coats quartz grains.

F) The **arkose** is fine- to medium-grained and pink to orange in color. The rock is made up quartz, potassium feldspar, plagioclase, and lithic fragments of quartz syenite, and chert. The matrix is predominantly composed of smaller grains of quartz and feldspar cemented by calcite and iron oxide. Quartz fragments rich in fluid inclusions are abundant in this rock; however, fragments of inclusion-free quartz are also present, suggesting at least two sources for quartz fragments. Carbonate and white phyllosilicate (muscovite, clay) locally replace plagioclase. The arkose contains abundant fragments of the underlying quartz syenite at the base of the section.

4) The **quartz syenite to diorite** (2665–3739 m) is medium to coarsely crystalline, and is pink, light orange to light gray. Microscopically, the quartz syenite shows a hypidiomorphic granular texture that is locally granophyric or myrmekitic, with intergrowths of potassium feldspar and quartz. The rock is made up of potassium feldspar, plagioclase and lesser amounts of biotite, hornblende and quartz. Accessory minerals are magnetite, amphibole, biotite, ilmenite, apatite, and zircon. Alteration minerals are sericite, muscovite, and chlorite. This rock exhibits variation in composition, passing from quartz syenite to diorite downhole. Near its top, the quartz syenite exhibits a well-developed weathering profile (red due to oxidation of iron-bearing minerals), indicating a nonconformable contact between the quartz syenite and overlying arkose. As will be discussed below, this zone is a strong reflector on a seismic line in the Tucumcari Basin.

5) The **granite** (3740 to 4100 m) has a porphyritic texture with large phenocrysts of plagioclase (1.2–2.0 mm), potassium feldspar, and quartz in a matrix composed mostly of quartz, feldspar, and mafic minerals (hornblende and biotite), in decreasing order. Accessory minerals are magnetite, apatite, and ilmenite. Feldspar is partially replaced by white phyllosilicates. The age relationship between the granite and the other basement units cannot be determined.

4.2. Geochronology

Results of the $^{40}Ar/^{39}Ar$ geochronologic analyses are presented in Figure 3. The dated samples come from four distinct gabbroic bodies near the top of the section (Plate 1a; black circles). Biotite was collected from well depths ranging from 1917 to 2292 m, whereas three hornblende separates were obtained from depths of 1798, 2051–54, 2222 m. The age spectra are variable, however all samples yield apparent ages of ~1105 Ma (Figure 3). Most of the biotite age spectra have a characteristic shape with initial steps yielding young apparent ages followed by a steep rise to ages between 1115–1135 Ma and then an overall decrease to a relatively flat section (Figure 3). The initial complexity could be caused by slight excess argon contamination and/or ^{39}Ar recoil artifacts [cf. *McDougall and Harrison*, 1999]. The hornblende samples are also variable. The spectrum for hornblende at 1978 m is very complex (Figure 3a) and no plateau age is assigned. The

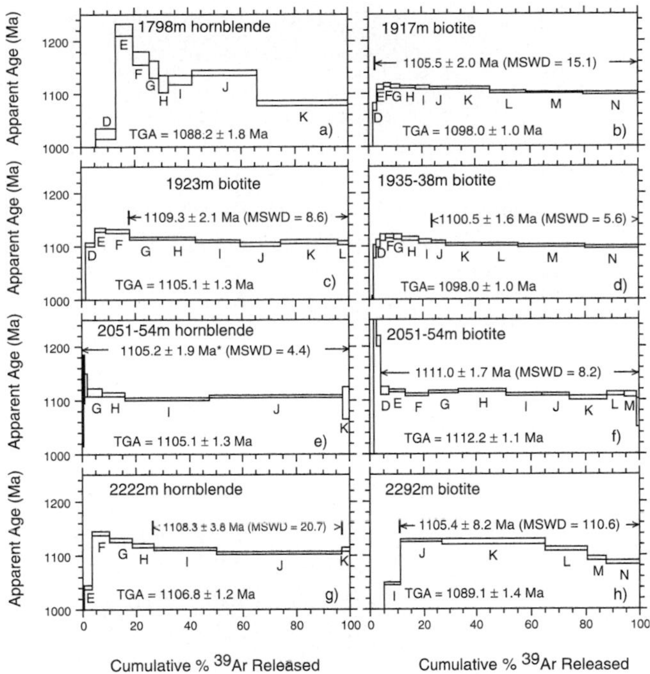

Figure 3. $^{40}Ar/^{39}Ar$ age spectra for gabbros in Mescalero 1.

hornblende age spectrum from the 2222–25 m interval is saddle-shaped, and two steps comprising about 74% of the ^{39}Ar released yield a poorly defined apparent age of 1108 ± 4 Ma (Figure 3g). In contrast, hornblende from the 2051–54 m interval yields a flat age spectrum with a precise plateau age of 1105.2± 1.9 Ma (Figure 3e). This latter hornblende also has a cogenetic biotite pair that gives an analytically indistinguishable plateau age of 1110.0 ± 1.7 Ma (Figure 3f). The seven plateau ages pass the Chi-square test with an MSWD of 1.6 and indicate that within analytical uncertainty all of the samples are indistinguishable in age. The seven ages represent a single population with a combined weighted mean emplacement age of 1106 ± 3.4 Ma. As noted previously, these gabbros appear to be the youngest units in the Proterozoic section because they intrude the other Proterozoic rock units.

Zircon from a quartz diorite (sample 3140 m) that is part of the crystalline basement beneath the arkosic sequence yields a SHRIMP U-Pb age of 1332 ± 18 Ma [*M.A. Barnes*, 2001; *M.A. Barnes et al.*, 2002]. This sample is mafic (Table 3; see CDROM in back cover sleeve), and thus our initial question was whether this sample is associated with the gabbroic sills and dikes that intrude the granites and syenites of the crystalline basement or whether it is a mafic phase of the syenitic intrusive rocks. Both petrographic observation and the well log signature suggest that the quartz diorite is gradational with the syenite, and thus our interpretation is that the date obtained represents the age of syenitic magmatism. This date of 1332 Ma is very similar to the youngest date (1339 (29 Ma) obtained for the Panhandle igneous complex in west Texas [*M.A. Barnes et al.*, 2002]. The sample dated at 1339 Ma is from a late-phase syenitic intrusion that crosscuts 1360 to 1380 Ma granites and quartz monzonites.

Geochronologic analysis of eight detrital grains of zircon indicates that the arkose, the basal unit of the Debaca sequence, yields SHRIMP U-Pb zircon ages ranging from 1308 ± 52 to 1708 ± 14 Ma [*M.A. Barnes et al.*, 2001]. These data indicate that the sediment was both locally (from the underlying quartz syenite) and distally derived (from older basement exposures such as those in the Pedernal Hills or the Sangre de Cristo Mountains, Figure 2). Paleoproterozoic source regions have also been documented for the Mesoproterozoic Lanoria Quartzite of the Franklin Mountains, where *Patchett and Ruiz* [1989] use Nd data to document significant input of Paleoproterozoic material into the Lanoria Quartzite.

In summary, the geochronology for the Mescalero 1 well documents a pulse of magmatism at 1330 Ma, forming the crystalline basement upon which the Debaca sequence lies, and a pulse of magmatism at 1105 Ma, forming a series of dike and sills that cut the Debaca sequence. In the interval between these pulses, an episode of rhyolitic volcanism occurred during deposition of the metasedimentary section. The similarity of the stratigraphy of the sedimentary package in Mescalero 1 compared to that exposed in the Franklin and Van Horn mountains can be used to argue that this package is correlative with the west Texas Debaca sequences. If this is the case, the age of the Debaca sequence in Mescalero 1 is likely ~ 1260 Ma [*Bickford et al.*, 2000]. Finally, ages of detrital zircon in the basal arkose in the Debaca sequence require both Paleoproterozoic and Mesoproterozoic sources.

4.3. Geochemistry

The igneous rocks penetrated by Mescalero 1 fall into three groups: (1) ~1330 Ma syenitic and dioritic rocks that are part of the basement complex, (2) a rhyolitic sample from the Debaca sequence (presumably circa 1260 Ma, and (3) ~1105 Ma gabbroic intrusives in the Debaca metasedimentary rocks (Table 3; see CDROM in back cover sleeve). The dioritic and quartz syenitic rocks are compositionally distinct: the SiO_2 content in the former unit is around 50 wt.%, whereas in the latter, SiO_2 is around 64 wt.% (Table 3; see CDROM in back cover sleeve). The dioritic rocks tend to cluster tightly in terms of trace element composition (Table 3; see CDROM in back cover sleeve), but the quartz syenite samples show wide ranges in trace element composition, particularly Rb, Sr, Y, and V. This variability may be a function of lithologic heterogeneity.

In the geochemical classification of *Frost et al.* [2001], the syenitic rocks plot as magnesian, alkali-calcic to alkalic and the Debaca rhyolite as magnesian, calc-alkalic to claciccalcic (Figure 4). Looking specifically at the three samples of the quartz syenite, we see that two of the samples (2664 and 2682 m) are borderline alkalic while the third (2987 m) is alkali-calcic (Table 3; see CDROM in back cover sleeve, Figure 4). Previous studies of the ca. 1.4 Ga granites of the area have established that outcrop samples from northern New Mexico are alkali-calcic (e.g., the Priest and Sandia plutons; Figure 4) [*Thompson and Barnes*, 1999]. In contrast, the basement core samples from west Texas are alkalic to alkali-calcic (Figure 4) [*M.A. Barnes et al.*, 1999, 2002]. Given that two of the three samples from the Mescalero 1 well plot on the compositional boundary between the two groups, it is not possible to say at this time to which affinity the Mescalero samples belong. Figure 4 also shows the compositional fields for circa 1.1 Ga alkalic granites of the Pikes Peak batholith [*Smith et al.*, 1997] and the Red Bluff granitic suite (RBGS) [*Shannon et al.*, 1997]. These younger units are characterized by their ferroan compositions, unlike nearly all of the older felsic rocks of the basement.

Two groups of mafic rocks were analyzed: 1) gabbroic samples at depths of 1862, 2063–87, and 2134–58 m, which intrude the Debaca sequence and are dated at 1105 Ma, and 2) dioritic samples from depths of 3140–70, 3170–3201, and 3210–22 m, from the underlying ca. 1340 Ma basement. Both groups are classified as alkaline on the basis of total alkalies vs. silica, but there are substantial differences in the geochemistry of the two mafic rock types that indicate they are distinct groups (Figure 5). The 1105 Ma gabbros have, for instance, lower values of SiO_2, Ba, Zr, and Y and higher concentrations of Al_2O_3, Fe_2O_3, MgO, CaO, Sr, and Cr than the 1330 Ma dioritic samples (Table 3; see CDROM in back cover sleeve).

The compositions of mafic rocks from the Mescalero 1 well are compared to mafic noritic rocks from the Pecos mafic intrusive complex [*Kargi and Barnes*, 1995], mafic dikes in the Red Bluff granite suite [*Shannon et al.*, 1997; *C.G. Barnes et al.*, 1999], and mafic samples from the northern Panhandle igneous complex of New Mexico and Texas [*M.A. Barnes et al.*, 2002] in Figure 5. Such comparisons must take into account the fact that nearly all analyzed samples of the Pecos mafic intrusive complex are cumulates, so that their compositions do not represent melt compositions. We further note that Pecos mafic intrusive complex cumulates are rich in orthopyroxene, a mineral that is absent in all of the other mafic rocks. The Mg/(Mg+Fe) values of the Mescalero rocks (~0.45) are consistent with emplacement of evolved basaltic magmas, i.e., magmas that had undergone fractionation prior to emplacement. Further, in keeping with their alkaline nature, these rocks have Zr and TiO_2 contents that are identical to those of the Red Bluff granite suite mafic dikes and alkaline samples from the Panhandle igneous complex, but distinct from the tholeiitic rocks of the Pecos mafic intrusive complex (Figure 5).

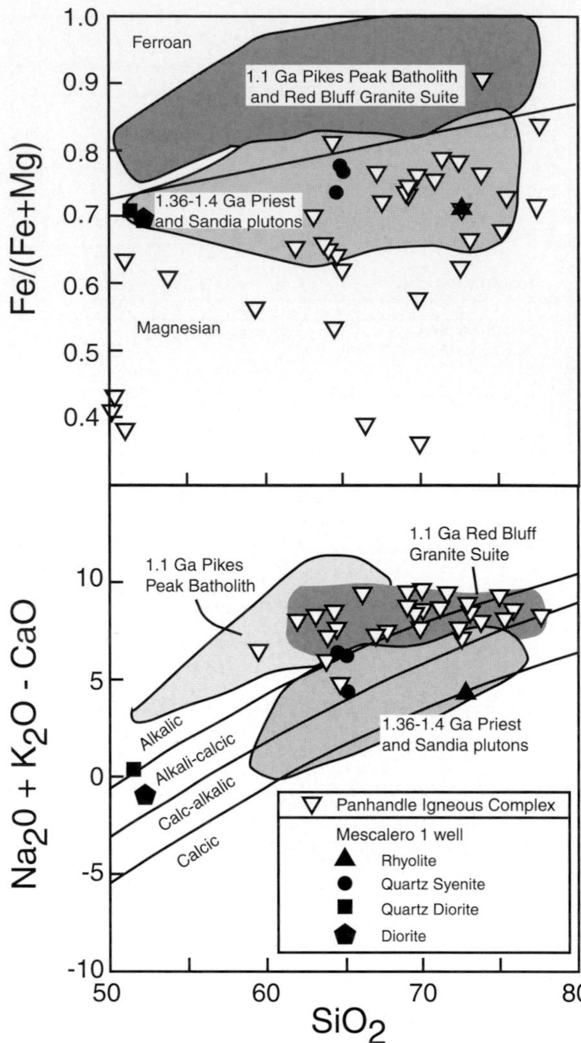

Figure 4. Compositional plots for Panhandle intrusive complex rocks and the Debaca sequence rhyolite from Mescalero 1 using the classification scheme of *Frost et al.* [2001]. Data from Panhandle intrusive complex rocks in the Texas panhandle [*Barnes*, 2001] and fields for the Pikes Peak batholith, the Red bluff granite, and other 1.35 to 1.40 Ga intrusives are shown for comparison.

The mafic rocks throughout the region show variable Nb content, which is significant because Nb has been used to interpret the tectonic setting of the rocks as either arc/back-arc or OIB-like. Following *Norman et al.* [1987] and *C.G. Barnes et al.* [1999], *M.A. Barnes et al.* [2002] recognized that the mafic rocks can be either high Nb (≥ 12 ppm Nb) or low Nb (< 12 ppm Nb). Many of the Red Bluff granite suite mafic dikes are low Nb, while, for instance, the 1105 Ma dikes from

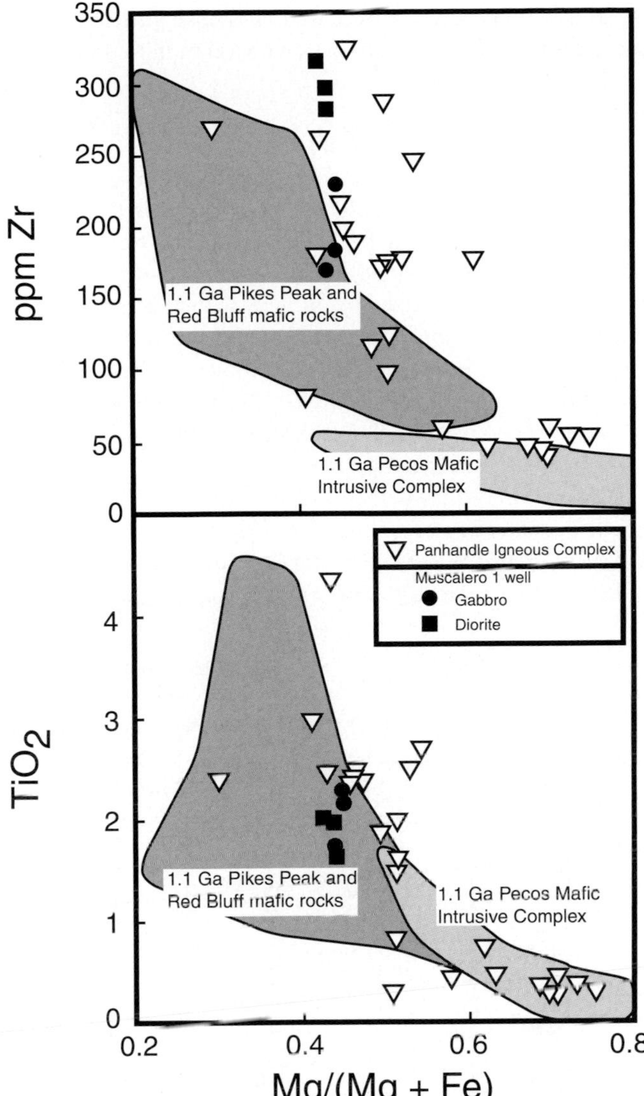

Figure 5. Compositional plots comparing 1.1 and 1.3 Ga mafic rocks from Mescalero 1 with the Pecos Mafic Intrusive Complex, and mafic rocks associated with the Red Bluff and Pikes Peak intrusives. Data from the Panhandle intrusive complex in Texas are also shown.

the Mescalero 1 well are high Nb. Previous authors have interpreted these signatures differently, with *Norman et al.* [1987] attributing a back-arc setting to the Red Bluff granite suite rocks, while *C.G. Barnes et al.* [1999] argued that the low Nb signature is inherited from the source. Their reasoning was based on the extensional, A-type characteristics of the felsic magmatism in the Franklin Mountains at 1.1 Ga. *C.G. Barnes et al.* [1999] also showed that the mantle source regions for circa 1.1 Ga magmatism were heterogeneous on a regional scale. Thus, one would not expect uniform compositions over regions as large as eastern New Mexico and western Texas.

Finally, little can be said concerning the analyzed rhyolitic sample from the Debaca sequence, except that it can be classified as magnesian and alkali-calcic and is apparently more silica-rich and less alkalic than the ca. 1.34 Ga felsic magmatism (Figure 4). Additional analyses will be necessary to determine its petrologic affinities.

4.4. Well Log Analysis

The total gamma ray and bulk density logs for the main portion of the Debaca sequence are shown in Plate 1b and a cross-plot of these physical parameters keyed to lithology is shown in Figure 6. The geophysical logs for the entire well, including the igneous rocks of the underlying Panhandle complex, are depicted in Figure 7. Each lithology encountered in the well has a distinctive log signature. Some of the rock units have very consistent log characteristics. For example, gabbro invariably has low gamma ray values (14–46 API) and high values of density (~3.0 gm/cm^3). Dolostone has a bulk density of 2.7 – 2.8 gm/cm^3 and low gamma values of 40 – 80 API. Argillite have low values of density and gamma (Plate 1b), and syenitic to dioritic basement rocks have a density of 2.7 gm/cm^3 and gamma value of 116 API (Figure 6 and 7). In contrast, the volcaniclastic sediment shows highly variable values of gamma ray activity and density. Part of this variability (between 2170 to 2200 m) can be attributed to wash out (excessive borehole diameter) during drilling. The remainder of the variability of physical properties is related to variations in the bulk composition and cementation of the sandstone. For instance, the tuffaceous sandstones have a low bulk density and surprisingly low gamma activity (15 API, Plate 1b). The green sandstones at ~2205 m and ~2400 m that are cemented by quartz have a significantly higher gamma activity (132–140 API) than those cemented by carbonate at ~2170 m and ~2430 m (37–68 API). The carbonate portions of the Debaca sequence, the dolostone and sandy dolostone, also vary considerably in composition. Some sections have log characteristics indicative of nearly pure dolomite (~2350 m and ~2500 m), while other sections appear to contain a substantial amount of sand (~2450 m and ~2490 m). The lowest sandy dolostone, based on the gamma log (Plate 1b), incorporated a high amount of uranium, giving rise to smoky quartz in this interval. Thin sections from this interval indicate brecciation of the dolomite.

The detailed analysis of the cuttings and logs from Mescalero 1 allows lithologic correlation to neighboring wells. Labrador Oil State Mr. Jones 1 is located ~5 km to the northeast of Mescalero 1. The cuttings for the State Mr. Jones 1 well unfortunately were destroyed, but a complete suite of geophysical logs is available. The gamma and density logs from State Mr. Jones 1 are compared to those from Mescalero 1 in Figure 7. The

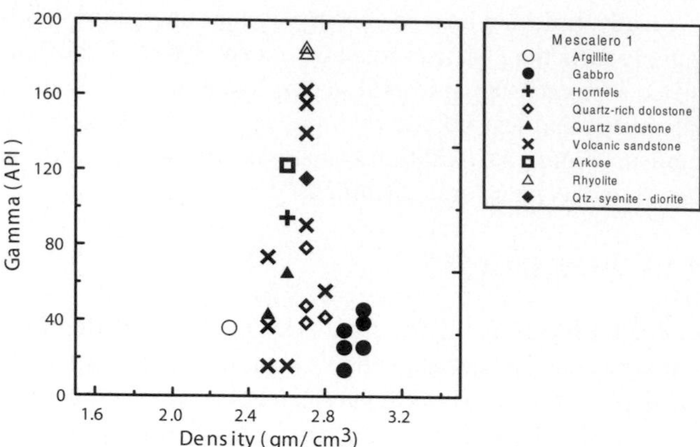

Figure 6. Cross-plot diagram showing the relationship between gamma ray activity and density for different rock types in Mescalero 1.

gabbros encountered in State Jones Mr. 1 are easily recognizable. A significant Pennsylvanian fault separates Mescalero 1 and State Mr. Jones 1 [*Broadhead*, 2001], dropping the State Mr. Jones 1 section down 160 m to the east (Figure 2).

The upper 400 m of the Proterozoic section in State Mr. Jones 1 is composed of gabbro; the hornfels and uppermost sedimentary package found in Mescalero 1 are not readily apparent on the log for State Mr. Jones 1. The distinctive gamma ray and density pattern of the upper Debaca, middle gabbro, and middle Debaca sequence in Mescalero 1 is also present in State Mr. Jones 1 (Figure 7). This package, particularly the gabbro, is about the same thickness in both wells. The arkose (or lower Debaca sequence) appears to thin to the east between the two wells.

The relative stratigraphic position and fairly constant thickness of the gabbros within the Debaca sequence suggest that these bodies may be sills (Figure 7). The gabbros located deeper in the drillholes within the Panhandle igneous complex vary in thickness and relative position with respect to the nonconformity, indicating that these bodies may be dikes (Figure 7). The nature of the Panhandle igneous complex basement beneath the Debaca sequence in State Mr. Jones 1 appears be different than that in Mescalero 1. The quartz syenite in Mescalero 1 has an average gamma value of 116 API and density of 2.7 gm/cm^3. In contrast, the basement beneath the Debaca sequence in State Mr. Jones 1 has an average value of ~80 API and an average density of 2.6 gm/cm^3. Note that the density in State Mr. Jones 1 tends to increase downhole between the nonconformity and the top of the gabbro at ~3000 m and again below the gabbro to 3225 m. The gamma activity trends parallel those in density, but the values decrease. The downhole increase in density and decrease in gamma activity is also recorded in the quartz syenite to quartz diorite to diorite pluton in Mescalero 1. These patterns reflect increasing mafic composition downward in these intrusions.

4.5. Seismic Analysis

The well log data can be correlated to seismic reflection data on the basis of a synthetic seismogram generated from the sonic and density logs in the State Mr. Jones 1 well. The synthetic seismogram (Figure 8) shows that the top of Precambrian basement is characterized by an increase in acoustic impedance. The underlying Debaca sequence is quite reflective as would be expected from layers of sedimentary rocks. Below the base of the Debaca sequence, reflections in the synthetic seismogram come from the gabbroic sills inferred to intrude the Panhandle Igneous Complex.

Correlation of the synthetic seismogram to a N-S trending seismic reflection profile (Figure 8) that lies approximately 15 km northeast of the State Mr. Jones 1 well yields new information on the geometry of the Debaca sequence and the gab-

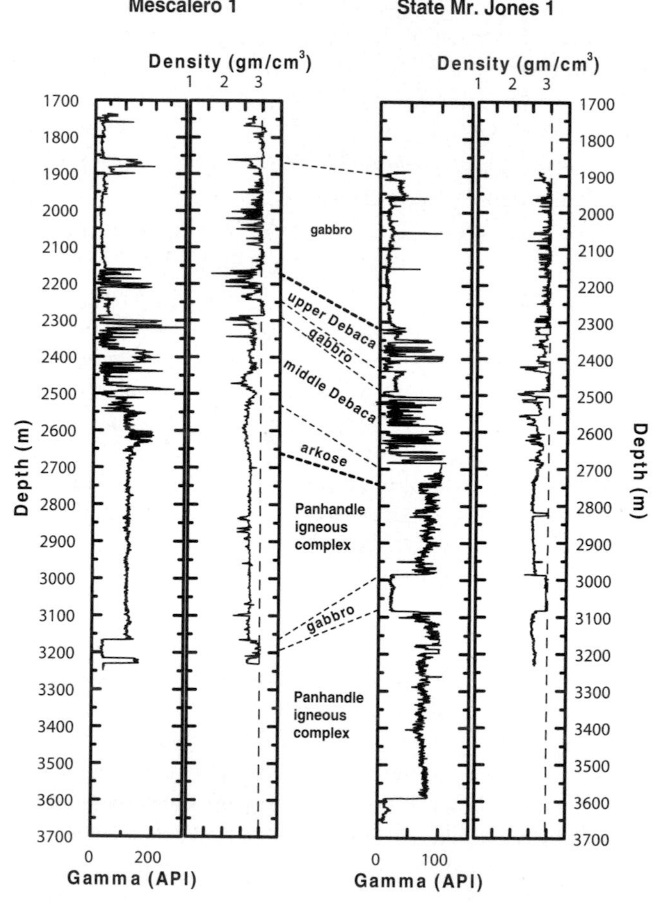

Figure 7. Correlation of gamma and density logs for Mescalero 1 and State Mr. Jones 1.

broic sills. The top of Precambrian basement is characterized by a strong, continuous reflection that is offset in places by Phanerozoic faults. A second south-dipping reflection 0.1 to 0.3 s below is interpreted as the unconformity at the base of the Debaca sequence similar to that seen in the Mescalero 1 well. We interpret the intervening package of reflectors as the Debaca sequence. The profile shows that the Debaca sequence both onlaps the unconformity and thins to the north. The Debaca sequence appears to pinch out toward the north end of the line. The location of the pinch out is consistent with the northern terminus of the sequence shown on Figure 2.

The seismic data below the unconformity is largely unreflective except for a bright reflector that dips to the south between 1.5 and 2.1 s on the south half of the line. We interpret this feature to be a mafic sill that intrudes the unreflective Panhandle igneous complex. This interpretation is supported by the log data from the State Mr. Jones 1 well, which show a number of high density, high velocity layers within the Panhandle igneous complex. Industry seismic data throughout the Tucumcari basin exhibit similar reflections. [*Eshete,* 2001]. On the basis of the well data presented here, we suggest that most of these reflections represent gabbroic sills that were intruded ca. 1105 Ma.

There are two major implications of these results for the interpretation of the CD-ROM NM line 1. First, the subcrop map (Figure 2) and the industry seismic data suggest that it is unlikely the Debaca sequence is a candidate for shallow reflections on that line because the CD-ROM lies 40 km west of the closest Debaca subcrop. Wells in the vicinity of the CD-ROM NM-1 line penetrate only metamorphic and igneous basement (Figure 2). Second, the data presented here do suggest that gabbroic sills of ca. 1.1 Ga are common in the subsurface of this region and thus that it may be more reasonable to interpret many of the bright sub-horizontal reflectors (*Eshete et al.*, this volume) as sills of this age.

5. DISCUSSION

5.1. Summary of Lithology, Age, and Basal Relationships

Detailed petrologic and geochronological studies of basement rocks penetrated by the well Mescalero 1 indicate that the sedimentary and volcanic rocks in this drillhole most likely correlate with the Debaca sequence of west Texas and southeast New Mexico. The Debaca sequence in this well consists of arkose, sandstone, dolomitic sandstone, quartz-rich dolostone, volcaniclastic sediment, rhyolite, tuffaceous sediment, dolomitic marble, and hornfels. Several bodies of gabbro, both sills and dikes, intrude the sequence. As a result of the intrusion of gabbro, some carbonate-rich rocks were converted to marble and some volcanics were metamorphosed to hornfels. The U-Pb SHRIMP ages determined for eight detrital grains of zircon (from 1708 ± 14 to 1308 ± 52 Ma) (M.A. Barnes, 2001) in the arkose suggest that the Debaca sequence was accumulated during a period in which regional erosion prevailed, resulting in deposition of zircons from

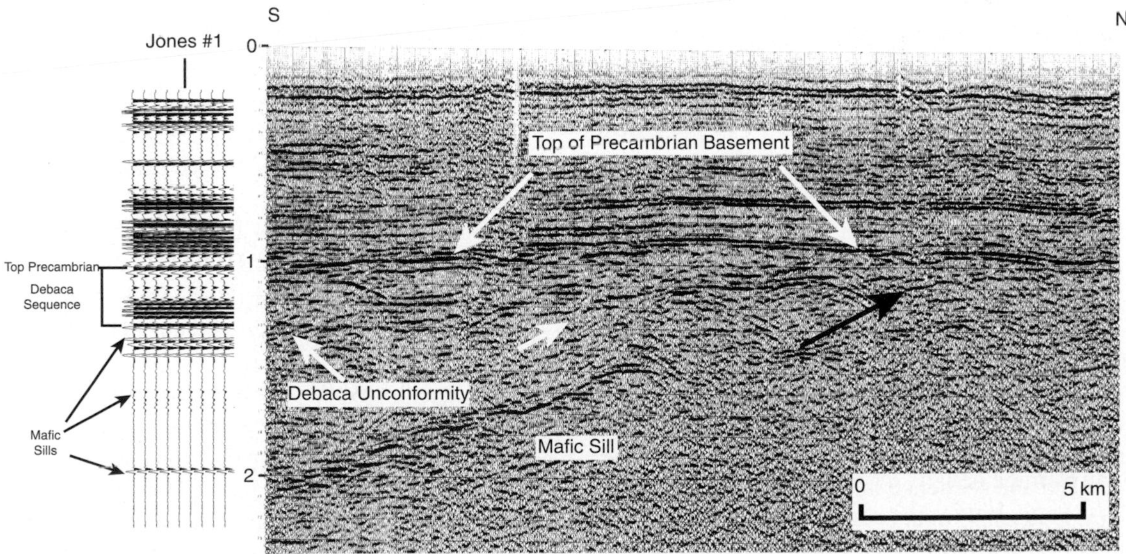

Figure 8. Correlation of seismic reflection data in the Tucumcari Basin with a synthetic seismogram from the State Mr. Jones 1 well. Debaca sequence reflectors appear to onlap an unconformity and pinch out to the north. A bright reflection between 1.5 and 2.1 s is interpreted to be a mafic sill. Seismic data provided by Seismic Exchange Inc.; interpretation is that of the authors. Vertical exaggeration is 2:1 assuming a velocity of 5 km/s.

source rocks of different ages. The U-Pb zircon age determined for the diorite (1332 ± 18 Ma), which underlies the Debaca sequence, is coincident with the youngest age determined for detrital grains of zircon from the arkose, suggesting that the underlying basement is one of the sources for the arkose. This interpretation is supported by petrographic analysis that identified a weathering profile in the quartz syenite below the arkose, and the presence of abundant fragments of quartz syenite in the arkose, especially in the lower portion of this unit near the contact with quartz syenite. This erosional contact shows up clearly on seismic reflection lines across the Tucumcari Basin (Figure 8). The Debaca sequence pinches out to the north. The association of immature volcaniclastic sandstones, carbonates, and felsic volcanics in upper part of the Debaca sequence suggest deposition in a shallow-water, extensional basin.

5.2. Correlation With Other Mesoproterozoic Rocks in West Texas, New Mexico, and Arizona

The rocks of the Debaca sequence in Mescalero 1 have lithologic and age affinities with surface exposures in the Franklin Mountains and the Van Horn area of west Texas (Table 4; see CDROM in back cover sleeve, Figure 9), as well as small outcrops in the Sacramento Mountains in south-central New Mexico. The Franklin Mountains contain a metasedimentary sequence of stromatolitic Castner Marble, which is overlain by Mundy Breccia, Lanoria Quartzite, and the Coronado Hills Conglomerate of the Thunderbird Group. A U-Pb zircon age for a volcanic ash layer within the Castner Marble yielded an age of 1260 ±20 Ma [*Pittenger et al.*, 1994; *Bickford et al.*, 2000]. Mapping of the basaltic Mundy Breccia has confirmed that it is syndepositional with the underlying Castner Marble [*Ballard*, 1997]. The similarity in age of these units is significant in that it establishes a bimodal character to magmatism at Debaca time. The contact between the Mundy Breccia and the Lanoria Quartzite is an erosional unconformity [*Ballard*, 1997, *Seeley*, 1999], as is the contact between the Lanoria Quartzite and the Coronado Hills Conglomerate [*Seeley*, 1999]. Detailed mapping of both contacts has led to the interpretation that the hiatus represented by the upper erosional unconformity is substantially greater than for the lower contact. Based of this interpretation, the Lanoria Quartzite may be more closely related to the Debaca sequence.

Rocks of the Debaca sequence in Mescalero 1 may correlate with Allamore and Tumbledown Formations in the Van Horn, Texas area [*Bickford et al.*, 2000]. The Allamore Formation is dominated by carbonates with interbedded basalt and felsic tuff, while the Tumbledown Formation is made up of basalt flows with volcanic sandstone and felsic tuffs. U-Pb ages of rhyolites, tuffs, and ashes in the Allamore and

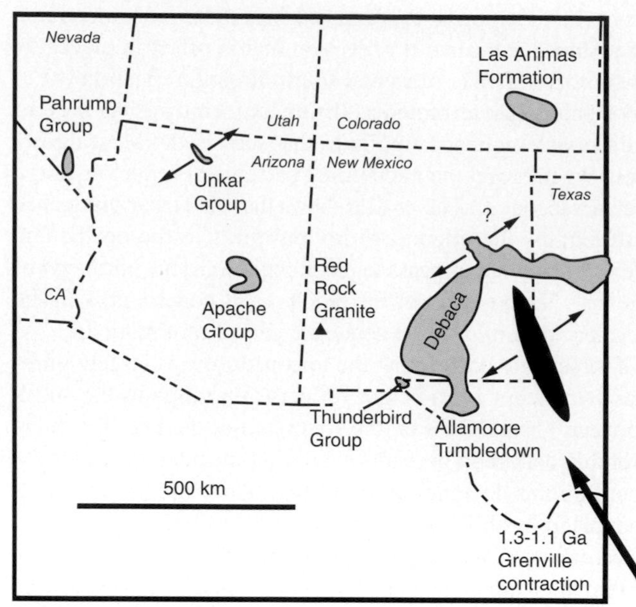

Figure 9. Distribution of 1.1–1.3 Ga sedimentary rocks in the southwestern U.S. shown in gray. Triangle is Red Rock granite in the Burro Mountains. Black oval is the Pecos Mafic Intrusive Complex (PMIC). Possible intracontinental extension directions and Grenville contraction direction from *Timmons et al.* [2001] and *Mosher* [1998]. Figure modified from *Timmons et al.*, 2001. Pahrump Group [*Prave*, 1999] and Las Animas Formation [*Tweto*, 1983] shown for reference.

Tumbledown formations are 1243 to 1256 Ma [*Bickford et al.*, 2000]. A small outcrop of basement rocks including quartzite, siltstone, and shale cut by diabase dikes at the western base of the Sacramento Mountains is assumed to be Debaca sequence [*Denison and Hetherington*, 1969], however, this assumption needs to be proved by geochronological study. *Denison and Hetherington* [1969] note that north of an E-W line through the center of Otero County, New Mexico, carbonates in the Debaca sequence are rare and the section becomes more argillacous and arkosic in New Mexico; however, our results suggest that this generalization is not totally true. The Debaca sequence north of this line is not as strongly metamorphosed as it is to the south [*Denison and Hetherington*, 1969].

The age the gabbro in Mescalero 1 overlaps with the ages of other late Mesoproterozoic igneous rocks in west Texas. For example, the Pecos mafic igneous complex (PMIC) in Texas has a U-Pb zircon age of 1163 ±4 Ma, as reported by *Keller et al.* [1989]. PMIC magmatism was approximately coeval with felsic magmatism in the Franklin Mountain of west Texas and the Sacramento Mountains of New Mexico, where *Shannon et al.* [1997] report a U-Pb zircon age of 1120 ±35 Ma for the Red Bluff Granite Suite, and *Bickford et al.* [2000] report a U-Pb zircon age of 1111 ±20 Ma for rhyolite from the upper part of the Thunderbird Group. In addition, Mescalero 1 contains 1.33 Ga syenitic and dioritic rocks that can be correlated temporally with rocks of similar composition in the Texas panhandle [*Barnes et al.,* 2002], These correlations are important because these relationships establish three intervals of bimodal magmatism (1.33 Ga, 1.26 Ga, and 1.1 Ga) during Mesoproterozoic time in New Mexico and west Texas.

Mesoproterozoic exposures in Arizona (Figure 9) also have affinities with the section represented in the Mescalero 1 well. The Apache Group in Arizona has U-Pb zircon, K-Ar and Rb-Sr isochron ages that are similar to those in Mescalero 1 (Table 4; see CDROM in back cover sleeve). *Wrucke* [1989] published a U-Pb zircon ages of 1150 ±30 Ma and 1100 ±15 Ma for diabase in the Little Dragon Mountains, southern Arizona. A K-Ar age on biotite for the Sierra Ancha sill indicates an age of 1140 ±40 Ma. U-Pb zircon on an ash bed within the Apache Group indicate that unit was deposited between 1300 and 1250 Ma ± 10 Ma [George Gehrels, 2000, personal communication].

The ages of Cardenas basalt and basaltic sills that overlie and intrude the basal portion of Unkar Group located in the Grand Canyon region of Arizona are also ca. 1100 Ma. *McKee and Noble* [1976] reported an Rb/Sr date of 1090±70 Ma for the Cardenas lava, which was followed by an estimate of 1070±30 Ma by *Elston and McKee* [1982]. *Larson et al.* [1994] obtained four new Rb/Sr data points and combined their results with those of *Elston and McKee* [1982] to establish an-eruption age of 1103±66 Ma. This geochronologic information, together with the field evidence reported by *Timmons et al.* [2001], indicate that the Unkar Group was deposited in an extensional basin created by a NW trending fault system related to the 1.3 to 1.1 Ga Grenville collision.

Two principal observations concerning the stratigraphic packages summarized in Table 4 (see CDROM in back cover sleeve) arise from this study. First, all of these sequences have two characteristic lithologic assemblages, an older ~1250 Ma sandstone-shale-carbonate-felsic volcanic assemblage and a younger ~1100 Ma gabbro-basalt-diabase (dikes and sills) assemblage (Table 4; see CDROM in back cover sleeve). Second, the lithologic assemblages are essentially time-equivalent (with some data limitations). The 1250 Ma time interval was dominated by sedimentation and felsic volcanism (rhyolite and rhyolitic ash). During this period, carbonate deposition includes the Bass Limestone in the Unkar Group, the Mescal Limestone in the Apache Group, the Castner Marble in the Franklin Mountains, and carbonate units in the Debaca sequence. For the first three, a shallow marine paleodepositional environment has been proposed based on the presence of stromatolites (Table 4; see CDROM in back cover sleeve). Stromatolites have not been identified in the Debaca sequence due to the fact that well cuttings do not readily preserve such structures. However, based on the similarities between the Debaca sequence and other areas discussed, we propose that the Debaca sequence in Mescalero 1 may also have been deposited in shallow marine conditions. The 1100 Ma event was dominated by mafic igneous activity that resulted in emplacement of gabbro and diabase in all sequences examined in Table 4 ; see CDROM in back cover sleeve.

Based on the lithologic and timing similarities we propose that all of these sequences were deposited in one or more large, northwest trending basins that extended from Arizona through New Mexico and west Texas. The NW-trend of the northern Debaca Basin penetrated by Mescalero 1 roughly parallels the trend of other Mesoproterozoic basins in the southwestern U.S. (Figure 9), and appears to be related to extension associated with Grenville contraction. Volcanism throughout the region resulted in the deposition of rhyolite and rhyolitic ash in intimate association with the carbonate units. The magmatic roots of a possible source of some of these rhyolitic ashes lies in the Burro Mountains of southwestern New Mexico, where Rämo et al. [2003] have recently obtained an age of 1250 Ma on the Red Rock granite (Figure 9). To test this hypothesis, more investigation is needed. Detailed stable isotope studies are recommended to determine possible chemical similarities in the carbonate units. Similarly, trace element geochemistry is recommended to determine the composition and possible source of the rhyolitic ash beds and to facilitate correlation of these beds between sequences.

6. CONCLUSIONS

Four major intrusive rock types (gabbro, quartz syenite grading to diorite, granite) and a sedimentary-volcanic sequence are present in Mescalero 1. The sedimentary-volcanic sequence fits descriptions of the Debaca sequence and the quartz syenite, diorite, and granite belong to the underlying Panhandle igneous complex discussed by previous workers in eastern New Mexico and west Texas. Seven gabbroic samples yield analytically indistinguishable plateau ages with a combined weighted mean emplacement age of 1105±3 Ma. Because the gabbros intrude the other Proterozoic rock units, they appear to be the youngest units within the basement; therefore, the entire sequence of basement rock types cut by Mescalero 1 is older than 1.09 Ga. The diorite, which underlies the Debaca sequence, yields a SHRIMP U-Pb zircon age of 1332 ±18 Ma. These ages are similar to those found for correlative strata in the Franklin Mountains and the Van Horn area. The age and geochemical data from Mescalero 1, when correlated regionally to Mesoproterozoic surface exposures in the southwestern United States and to well data from the Texas panhandle, can be used to identify three pulses of volcanism, one bimodal episode at 1.33 Ga, one felsic episode at ~1.26 Ma, and one mafic episode at 1.09 Ga.

Detailed analysis of the cuttings in Mescalero 1 indicates that the basal arkose of the Debaca sequence rests on deeply weathered quartz syenite of the underlying Panhandle igneous complex. This erosional contact shows up clearly on a seismic reflection line across the Tucumcari Basin. The seismic line shows that the Mesoproterozic Debaca basin thins to the north. This interpretation agrees with our preliminary observations that only 10 to 30 m of Debaca sequence composed mainly of white carbonate-cemented sandstone is preserved in wells to the north of the Labrador wells, where the Debaca sequence sits on rhyolite or granite (see Figure 2). In addition, the cuttings and well log analyses in Mescalero 1 can be tied to the seismic reflection line to document the geometry of gabbroic bodies within the basement (Figure 8).

Each of the major rock units has a characteristic signature on the geophysical logs. For example, gabbro has a density of ~3.0 gm/cm^3 and low total gamma activity. In contrast, the quartz syenite has a density of ~ 2.70 gm/cm^3 and a total gamma ray of ~ 115 GAPI. Analysis of natural gamma-ray spectral logs indicates that the volcaniclastic sediment generally has high uranium values. This type of analysis is useful for correlating logs drilled into Proterozoic basement, as illustrated in Figure 7, and has the potential to help with our continuing efforts to map out the geometry of the Mesoproterozoic Debaca basin. The geometry of the basin may hold the key to understanding the tectonic development of the basin and depositional setting of the sedimentary package.

Acknowledgments. Continental Dynamics NSF grant EAR-9614787 (CD-ROM) and a grant from the Roswell Geological Society funded this project. Mike Williams, Kent Nielsen, and Tim Denison provided invaluable reviews of the manuscript. Discussions with Ron Broadhead, Brian Brister, Cal Barnes, and George Asquith were very helpful. We are particularly grateful to Ron Broadhead for pointing out the significance of these cuttings and to Cal Barnes, who helped with this study in many ways. Cal Barnes first recognized that the Mescalero well contained the Debaca sequence and his contributions to the geochemistry section of the manuscript strengthened it considerably. Partial funding was received from Texas Advanced Research Project grants to K. Miller (Project Number 003661-0056-1999) and C. Barnes. Collecting in the Franklin Mountains State Park (TX) was done under permit 34-95. Seismic data provided by Seismic Exchange Inc.

REFERENCES

Amarante, J.F.A., Characterization of the basement rocks in the Mescalero 1 well, Guadalupe County, New Mexico, M.S. thesis, 73 pp, New Mexico Institute of Mining and Technology, Soccoro, NM, 2001.

Anderson, J. L., Proterozoic anorogenic granite plutonism of North America, in *Proterozoic Geology: Selected Papers from an International Proterozoic Symposium*, edited by L. G. Medaris, Jr., C. W. Byers, D. M. Milkelson, W. C. Shanks, Geol. Soc. Amer. Mem. 161, 133–154, 1983.

Ballard, J.L.W., The depositional history of the Mundy Breccia and the lowermost member of the Lanoria Formation, M.S. thesis, 180 pp., University of Texas at El Paso, 1997.

Barnes, M.A., The petrology and tectonics of the Mesoproterozoic margin of Southern Laurentia, Ph.D. thesis, 190 pp., Texas Tech University, Lubbock, TX, 2001

Barnes, C.G., W. M. Shannon, and H. Kargi, Diverse Mesoproterozoic basaltic magmatism in west Texas, *Rocky Mountain Geology*, *34*, 263–273,1999.

Barnes, C.G., B.R. Burton, T.C. Burling, J.E. Wright, and H.R. Karlsson, Petrology and geochemistry of the late Eocene Harrison Pass pluton, Ruby Mountains core complex, northeastern Nevada, *Journal of Petrology*, *42*, 901–929, 2001.

Barnes, M. A., C. R. Rohs, E.Y. Anthony, W.R. Van Schmus, and R.E. Denison, 1999, Isotopic and elemental chemistry of subsurface Precambrian igneous rocks, west Texas and eastern New Mexico: *Rocky Mountain Geology*, *34*, 245–262.

Barnes, M.A., Anthony, E.Y., Williams, I., and Asquith, G.B., Architecture of a 1.38–1.34 Ga granite-rhyolite complex as revealed by geochronology and isotopic and elemental geochemistry of subsurface samples from west Texas, USA, *Precambrian Research*, *119*, 9–43, 2002.

Bickford, M. E., K. Soegaard, K.C. Nielsen, and J.M. McLelland, Geology and geochronology of Grenville-age rocks in the Van Horn and Franklin Mountains area, west Texas: Implications for the tectonic evolution of Laurentia during the Grenville, *Geol. Soc. of Amer. Bull.*, *112*, 1134–1148, 2000.

Broadhead, R.F., New Mexico elevator basins 1 – Petroleum systems studied in the southern ancestral Rocky Mountains, *Oil and Gas Journal*, 32–38, 2001.

Broadhead, R. F., and W.E. King, Petroleum Geology of Pennsylvanian and Lower Permian strata, Tucumcari Basin, east-central New Mexico, *NM Bur. of Mines and Mineral Resources, Bulletin 119*, 77pp., 1988.

Denison, R. E., and E.A. Hetherington, Jr., Basement rocks in far west Texas and south-central New Mexico, in *Border stratigraphy symposium*: edited by F.E. Kottlowski and D.V. Lemone, New Mexico Bureau of Mines and Mineral Resources Circular 104, 1–16, 1969.

Denison, R. E., E.G. Lidiak, M.E. Bickford, and E.B. Kisvarsanyi, Geology and geochronology of Precambrian rocks in the central interior region of the United States, *U.S. Geological Survey Professional Paper 1241-C*, 11–14, 1984.

Elston, D. P. and E.H. McKee, Age and correlation of the late Proterozoic Grand Canyon disturbance, northern Arizona, *Geol. Soc. of Amer. Bull.*, *93*, 681–699, 1982.

Eshete, T.G., Structure of northeastern New Mexico from deep seismic reflection profiles: Implications for the Proterozoic structure of southwestern North America, Ph.D. dissertation, 133 pp., University of Texas, El Paso, 2001.

Flawn, P.T., Basement rocks of Texas and southeast New Mexico, *Bureau of Economic Geology Publication 5605*, 261 pp., 1956.

Foster, R. W., and T.F., Stipp, Preliminary geologic and relief map of the Precambrian rocks of New Mexico: *New Mexico Bureau of Mines and Mineral Resources Circular 57*, 37 pp, 1961.

Frost, B.R., C.G. Barnes, W.J. Collins, R.J. Arculus, D.J. Ellis, and C.D. Frost, 2001, A geochemical classification for granitic rocks, *Journal of Petrology*, *42*, 2033–2048, 2001.

Heizler, M.T., W.C. McIntosh, L. Peters, and R.P. Esser, Operating procedures and age calculations for the New Mexico Geochronology Research Laboratory, Open-File Report xx, p. xx, 2002.

Kargi, H. and C. Barnes, A Grenville-age layered intrusion in the subsurface of west Texas: Petrology, petrography, and possible tectonic setting, *Can. J. Earth Sci.*, *32*, 2159–2166, 1995.

Keller, G. R., J.M. Hills, M.R. Baker, and E.T. Wallin, Geophysical and geochronological constraints on the extent and age of mafic intrusions in the basement of west Texas and eastern New Mexico, *Geology*, *11*, 1049–1052, 1989.

Kwon, J., K. Min, P. Bickel, and P.R. Renne, Statistical methods for jointly estimating decay constant of ^{40}K and age of a dating standard, *Mathematical Geology*, *34* (4), 457–474, 2002.

Larson, E. E., P.E. Patterson, and F.E. Mutschler, 1994, Lithology, chemistry, age, and origin of the Proterozoic Cardenas Basalt, Grand Canyon, Arizona, *Precambrian Research*, *65*, 255–276.

McDougall, I., and T.M. Harrison, *Geochronology and thermochronology by the* $^{40}Ar/^{39}Ar$ *method*, Oxford Press, New York, 269 pp., 1999.

McKee, E.H., and D.C. Noble, Age of the Cardenas Lavas, Grand Canyon, Arizona, *Geol. Soc. Am. Bull.*, *87*, 188–1190, 1976.

McIntosh, W.C., M. Heizler, L. Peters, and R. Esser, $^{40}Ar/^{39}Ar$ geochronology at the New Mexico Bureau of Geology and Mineral Resources, *New Mexico Bureau of Geology and Mineral Resources Open File Report OF-AR-1*, 10 pp., 2003.

Mosher, S., 1998,Tectonic evolution of the southern Laurentian Grenville orogenic belt, *Geol. Soc. Amer. Bull.*, *110*, 1357–1375, 1998.

Muehlberger, W. R., C.E. Hedge, R.E. Denison, and R.F. Marvin, Geochronology of the mid-continent regions, United States: Part 3. Southern area, *J. Geophys. Res.*, *71*, 5409–5426, 1966.

Muehlberger, W. R., R.E. Denison, and E.G. Lidiak, Basement rocks in continental interior of United States, *AAPG Bull.*, *51*, 2351–2380, 1967.

Norman, D. I., K.C. Condie, and R.W. Smoth, Geochemical and Sr and Nd isotopic constraints on the origin of late Proterozoic volcanics and associated tin-bearing granites from the Franklin Mountains, west Texas, *Can. J. Earth Sci.*, 24, 830–839, 1987.

Patchett, P.J., and J. Ruiz, Nd isotopes and the origin of Grenville-age rocks in Texas: implications for Proterozoic evolution of the United States mid-continent region, *J. Geol.*, *97*, 685–695, 1989.

Pittenger, M. A., K.M. Marsaglia, and M.E. Bickford, 1994, Depositional history of the Middle Proterozoic Castner Marble and basal Mundy Breccia, Franklin Mountains, west Texas, *Journal of Sedimentary Research*, *64*, 282–297, 1994.

Prave, A.R., Two diamictites, two cap carbonates, two del^{13}C excursions, two rifts: the Neoproterozoic Kingston Peak Formation, Death Valley, California, *Geology*, *27*, 339–342, 1999.

Rämo, O.T., V.T. McLemore, M.A. Hamilton, P.J. Kosunen, M. Heizler, M., I. Haapala, Intermittent 1630–1220 Ma magmatism in central Mazatzal Province: New geochronologic piercing points and some tectonic implications, *Geology*, *31*, 335–338.

Seeley, J.M., Studies of Proterozoic tectonic evolution of the southwestern United States, Ph.D. dissertation, 312 pp., University of Texas at El Paso, 1999.

Shannon, W., C. Barnes, and M. Bickford, M., Grenville magmatism in west Texas: Petrology and geochemistry of Red Bluff granitic suite, *Journal of Petrology*, 38, 1279–1305, 1997.

Smith, D.R., C. Barnes, W. Shannon, R. Roback, and E. James, Mid-Proterozoic granitic magmatism in Texas, *Precambrian Research*, *85*, 53–79, 1997.

Steiger, R.H., and E. Jager, Subcommission on geochronology: convention on the use of decay conatants in geo-and cosmochronology, *Earth and Planetary Science Letters*, *36*, 359–362, 1977.

Timmons, J. M., K.E Karlstrom, C.M Dehler, J.W. Geissman, and M.T. Heizler, Proterozoic multistage (~1.1 and ~0.8 Ga) extension in the Grand Canyon Supergroup and establishment of northwest and north-south tectonic grains in the southwestern United States, *Geol. Soc. Amer. Bull.*, *113*, 163–180, 2001.

Timmons, J. M., K.E. Karlstrom, M.T. Heizler, S. A. Bowring, G. E. Gehrels, and L. Crossey, Tectonic inferences from the ca. 1254–1100 Ma Unkar Group and Nankoweap Formation, Grand Canyon: Intracratonic deformation and basin formation during protracted Greenville orogenesis, 2003 (in press).

Thompson, A.G., and G.G. Barnes, 1.4-Ga peraluminous granites in central New Mexico: Petrology and geochemistry of the Priest pluton, *Rocky Mountain Geology*, *34*, 223–243, 1999.

Tweto, Ogden, Las Animas Formation (upper Precambrian) in the subsurface of southeastern Colorado, *U.S. Geological Survey Bulletin 1529-G*, 22 p., 1983.

Williams, I.S., U-Th-Pb geochronology by ion microprobe, *Applications of Microanalytical Techniques to Understanding Mineralizing Processes*, edited by M.A. McKibben, W.C. Shanks, III, and W.I. Ridley, Review on Economic Geology, 7, 1–35, 1998.

Wrucke, C. T., The middle Proterozoic Apache Group, Troy Quartzite, and associated diabase of Arizona, *Geologic Evolution of Arizona*, edited by J.P. Jenney and S.J. Reynolds, Geological Society Digest, 17, 239–258, 1989.

Jose F.A. Amarante and Shari A. Kelley, Department of Earth and Environmental Science, New Mexico Institute of Mining and Technology, Socorro, NM, 87801

Elizabeth Y. Anthony and Kate C. Miller, Department of Geological Sciences, The University of Texas at El Paso, El Paso, TX 79968

Melanie A. Barnes, Geosciences, Texas Tech University, Lubbock, TX 79409

Matthew T. Heizler, New Mexico Bureau of Geology and Mineral Resources, Socorro, NM, 87801

Background and Overview of Previous Controlled Source Seismic Studies

Claus Prodehl[1], Roy A. Johnson[2], G. Randy Keller[3], Catherine M. Snelson[4], and Hanna-Maria Rumpel[1]

Controlled source seismic techniques offer useful tools to study the structure of the earth's crust and upper mantle in considerable detail. The seismic reflection technique uses source–receiver offsets that are relatively small in comparison to the intended maximum depth of penetration, and the goal is to form an image of the subsurface structure through sophisticated data processing. The seismic refraction technique uses source–receiver offsets that are large in comparison to the intended maximum depth of penetration. Sophisticated interpretation methods have been developed to derive velocity models from these data. Reflection and refraction data are traditionally gathered along profiles and produce 2-D results, but new instrumentation is making surveys with an element of 3-D coverage possible at a crustal scale. From the standpoint of seismic refraction and reflection surveys, coverage for the Southern Rocky Mountains and surrounding areas before the 1999 CD-ROM project was relatively poor. Reflection surveys targeting the deep crust were particularly sparse. Of the pre-1999 refraction data, several profiles are at least partly unreversed, most have only a few widely spaced shotpoints, and for most profiles, the interval between recording stations is typically >10 km. The thickest crust in the region (~50 km) does not correlate directly with the highest topography. The Southern Rocky Mountains are bisected by the Rio Grande rift whose crust thins from north to south and is at least 5 km thinner than that of adjacent regions. There is also evidence for crustal thinning across southern Wyoming.

[1] Geophysikalisches Institut, University of Karlsruhe, Karlsruhe, Germany
[2] Department of Geosciences, University of Arizona, Tucson, Arizona
[3] Department of Geological Sciences, University of Texas at El Paso, El Paso, Texas
[4] Department of Geoscience, University of Nevada Las Vegas, Las Vegas, Nevada

The Rocky Mountain Region: An Evolving Lithosphere
Geophysical Monograph Series 154

10.1029/154GM14

1. INTRODUCTION

The CD-ROM project offered a unique opportunity to delineate detailed crustal and upper mantle structure along a transect of substantial length (~1000 km, Figure 1) by applying a variety of seismic techniques including deep seismic reflection and refraction profiling using controlled seismic sources, as well as passive-source teleseismic tomography and receiver-function studies. In addition, there have been several recent seismic studies in the region besides CD-ROM. Understanding the results of such a variety of techniques requires at least a basic appreciation of each of them. The purpose of this paper is to briefly overview the aspects of controlled-source

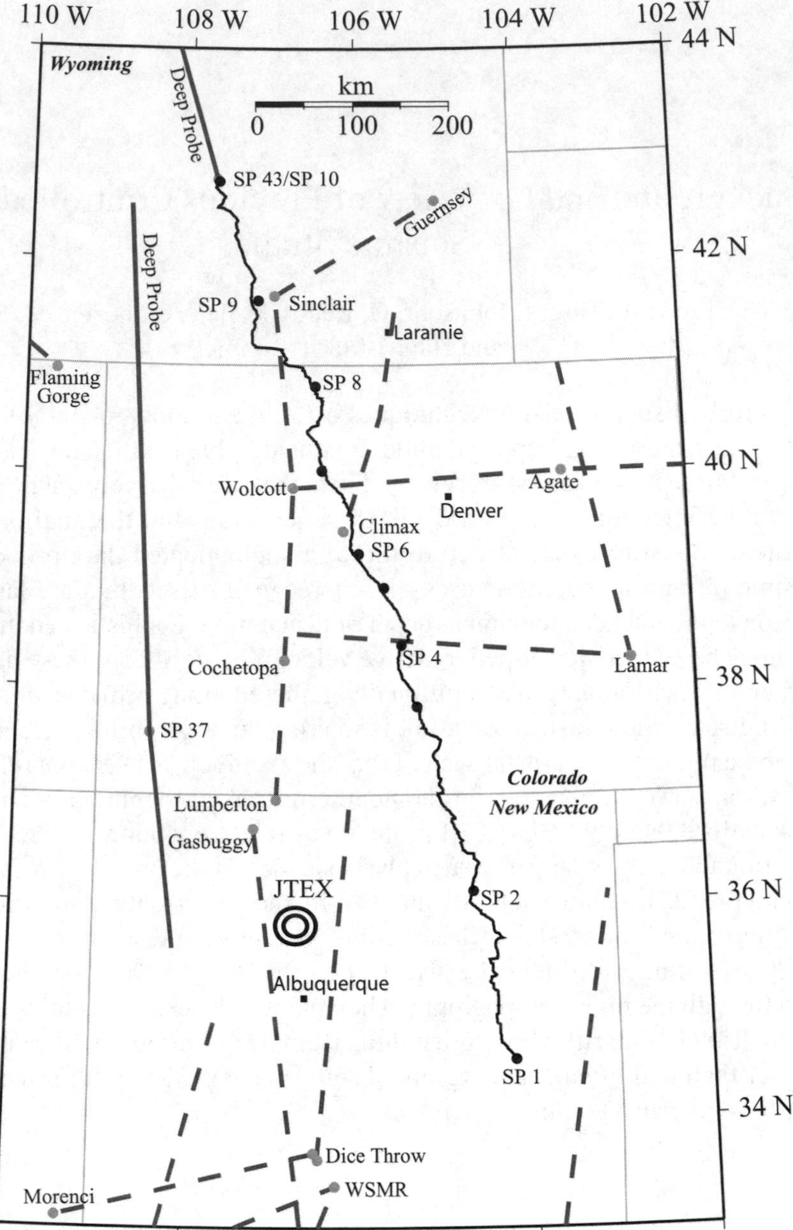

Figure 1. Index map of seismic studies in the Rocky Mountain region prior to CD-ROM in 1999. Solid circles are shotpoints used in these studies; stars denote major cities. Detailed information about these profiles can be found in Table 1 (see CDROM in back cover sleeve).

seismic techniques needed to understand their strengths and weaknesses and to summarize the results of studies of this type in the Southern Rocky Mountain region prior to CD-ROM (Figure 1). Our goal is also to complement the overview of passive seismic techniques provided by *Sheehan et al.* [this volume]. In general, seismic techniques offer the highest resolution geophysical methods to study the earth's crust and uppermost mantle. Detailed overviews of the various seismic methods employed to study lithospheric structure, their advantages and their limitations, have been published by *Braile et al.* [1995] and *Mooney* [1989]. Modern seismic studies using controlled energy sources such as underwater and borehole explosions, Vibroseis, and commercial quarry blasts are capable of excellent resolution to depths of several tens of kilometers (the crust and uppermost mantle). In order to obtain optimal resolution, detailed controlled-source studies require hundreds of instruments and a coordinated program of reflection (near-vertical-angle) and refraction (wide-angle) recording. Passive-source techniques provide valuable, but less detailed, constraints on lithospheric structure, and the deeper we look into the mantle the more we must depend on these techniques. Controlled-source and passive-source techniques are highly complementary and can each provide information throughout the lithosphere. A quantitative integration of the full spectrum of these techniques would be a very powerful tool but has yet to be achieved.

2. THE REFRACTION (WIDE-ANGLE REFLECTION) PROFILING TECHNIQUE

As classically practiced, seismic refraction profiling has provided much of the data available on the crust, such as thickness, gross velocity structure, and uppermost mantle P-wave velocities (Pn). However, modern instrumentation provides the capability to resolve much more detailed information about crustal structure. Crustal-scale seismic refraction data are usually gathered along linear profiles several hundred kilometers in length. Up to the late 1980s, they were exclusively recorded in analog form. Since the mid-70s, major improvements have emerged for the analysis of crustal seismic refraction data because of rapidly increasing computing and graphics capabilities. In particular, the capability to digitize the analog traces enabled further processing of the digital data and the application of numerical modeling algorithms for two and three-dimensional structures. Nowadays, all data is acquired in digital form.

In the seismic refraction technique, the distance from the source to the most distant receiver is several times the maximum depth of penetration intended, while in reflection seismic surveys the depth of reflectors may be much larger than the extent of a recording array. Elementary texts on seismic exploration leave the impression that only critically refracted waves that travel along interfaces (i.e., head waves) are interpreted in seismic refraction studies. However when employed in crustal structure investigations, the term refraction survey is misleading because many reflected phases are also interpreted. As these surveys are directed toward the interpretation of waves that intersect interfaces at angles greater than a few degrees, they are often termed wide-angle reflection experiments. Also, the so-called refracted waves are usually diving waves that do not travel strictly along a "refraction" horizon, but with increasing distance, penetrate or dive into the medium below the corresponding velocity boundary. This is particularly true if the boundary is not a sharp discontinuity, but rather a thin transition zone where the velocity does not jump discontinuously from the upper to the lower layer. In this case, the velocity gradient changes rapidly from a smaller to a larger value causing an increase of velocity over a limited depth range.

Wide-angle surveys yield data that are not straightforward to interpret. The strength of such surveys is their ability to resolve in situ velocities, but ambiguities may arise when complex structures are encountered. The first step in interpreting wide-angle seismic data is correlating the observed phases, and this subjective task may be difficult beyond the interpretation of first arrivals only. However, a thorough investigation of the earth's crust and upper mantle requires the analysis and correlation of both first and later arrivals from seismogram to seismogram along a profile (Figs. 2 and 3). Digital data processing helps with this procedure by allowing for improvement of the original data and their readability. Such processing was cumbersome or impossible to apply to analog data. These processes include frequency and velocity filtering, slant stacking of traces, and various modifications to the display of traces such as automatic gain control and variable-area shading.

To facilitate such a correlation of first and secondary arrivals, all seismograms along a profile are usually arranged into a record section according to their distance from the shot-point using reduced travel times (Figure 2). The reduced travel time Tr is defined as the observed travel time (T) minus distance (D) divided by the reduction velocity (Vr): Tr = T - D/Vr. For investigation of compressional (P) waves propagating in the crust, Vr = 6 km/s is a suitable and widely used reduction velocity, while Vr = 8 km/s is normally used for P-wave investigations of the uppermost mantle. As examples of such record sections, a set of unpublished wide-angle seismic profiles are shown in Figures 2 and 3. These data were collected in 1965 along a profile that approximately follows the meridian 107° W from southern Colorado to southern Wyoming, and were interpreted as part of a study by *Prodehl and Pakiser* [1980]. For comparison, *Rumpel et al.* [this volume], *Snelson et al.* [this volume], and *Levander et al.* [this volume] present much denser seismic refraction P-data obtained during the CD-ROM seismic survey.

After the display and correlation of the recorded seismic waves, the next step is the inversion of the observed time-distance data into velocity-depth functions, V(z). The simplest case occurs when the correlated travel time curves are straight-line segments. Formulas have been developed to calculate depth and dip of corresponding layers under the assumption that the earth's crust consists of layers with constant velocities separated by discontinuities at which the velocity increases discontinuously from V(i) to V(i+1) [e.g., *Bullen and Bolt*, 1985; *Steinhart and Meyer*, 1961]. Curved segments of correlated travel time curves can occur in first as well as in

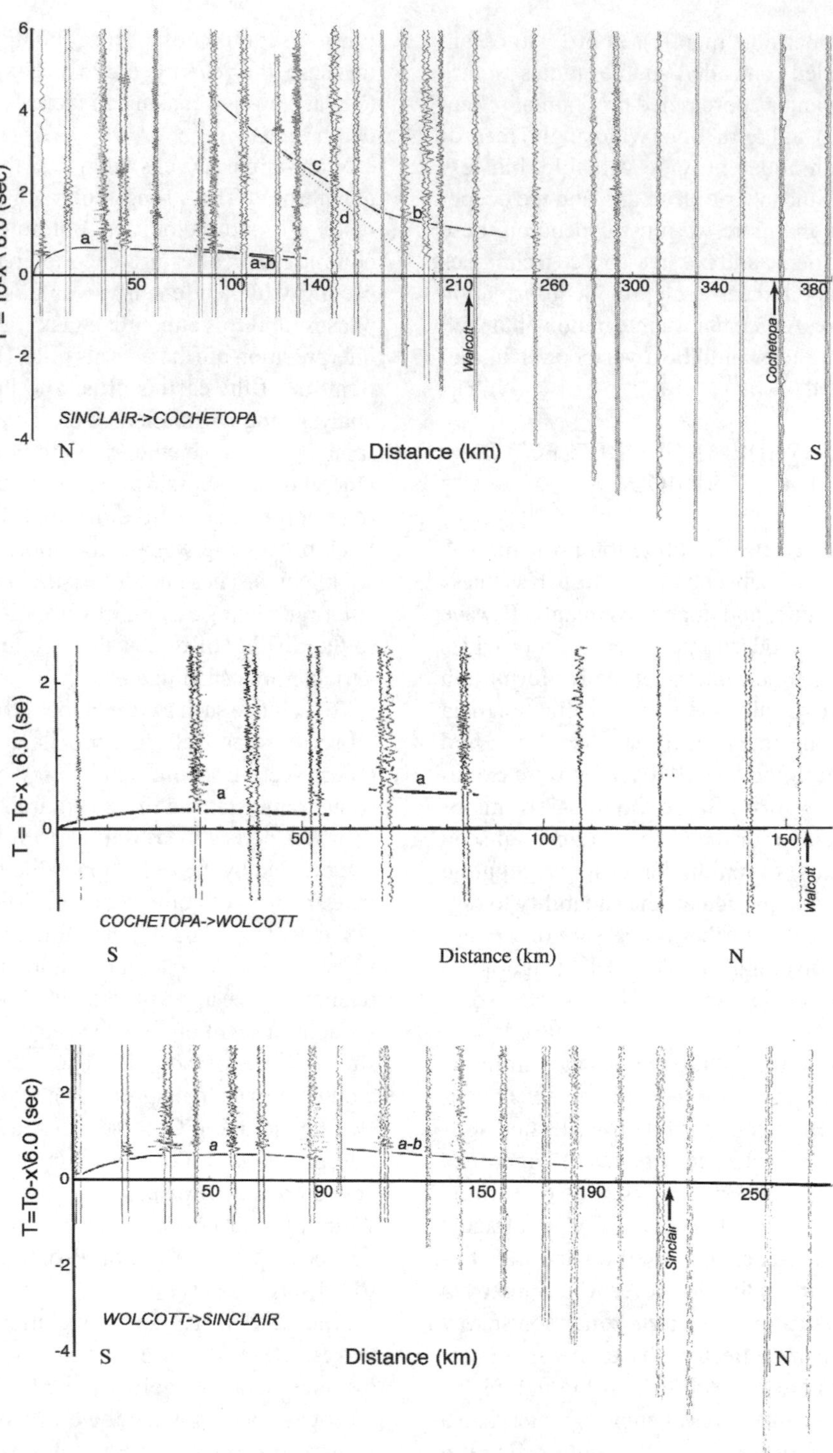

Figure 2. Previously unpublished refraction record sections along a N-S line parallel to the axis of the southern Rocky Mountains extending from Sinclair, Wyoming, to Lumberton, New Mexico. These data were recorded in 1965 [*Prodehl and Pakiser*, 1980]. Shotpoint locations are shown in Figure 1. The phases identified are labeled as: a - refraction from the upper crust (Pg); a-b - reflection from the upper crust; b – reflection from the middle crust (PiP); c – reflection from the Moho (PmP); d – refraction from the upper mantle (Pn).

later arrivals. Curved segments of first-arrival phases may indicate a vertical velocity gradient in the corresponding layer. When later arrivals are fitted by hyperbolic travel time curves, they are interpreted as reflections from first-order discontinuities or from thin transition zones with strong velocity gradients. A general method for inversion of travel time data into depth values is the Wiechert-Herglotz method [e.g., *Giese*, 1976]. This method can be applied to determine the complete velocity-depth function if the travel-time curve, T(D), forms a continuous function including cusps, and if the derivative, dD/dT, increases continuously along the travel time curve.

However for many seismic refraction observations, the conditions required by the Wiechert-Herglotz method are not fulfilled. *Fuchs and Landisman* [1966] were among the first to use a less restricted method to calculate the velocity-depth model for a set of observed travel time data. They started with an initial model, which may be obtained from intercept times and apparent velocities for each travel time branch. Velocities and depths were then varied in a trial-and-error procedure until a reasonable fit between observed and calculated travel times is obtained. *Giese* [1976] discusses several other approximation formulas developed by various authors and describes a special approximation method to invert travel time data directly into velocity-depth values. This method was applied by *Prodehl* [1979] to a large number of seismic refraction profiles throughout the western United States. Each of these methods allows for inclusion of velocity inversions, but generally assumes homogeneity in the horizontal direction.

Synthetic seismogram techniques using ray theoretical and reflectivity methods have been developed to include the analysis of observed amplitudes and waveforms and thus check and refine velocity models obtained by travel time analysis only [e.g., *Fuchs and Mueller*, 1971; *Kind*, 1976; *Cerveny et al.*, 1977; *McMechan and Mooney*, 1980; *Kennett*, 1983; *Sandmeier and Wenzel*, 1986]. All these methods assume an elastic half space consisting of homogeneous, horizontal and isotropic layers. In the reflectivity method [*Fuchs*, 1968], the complete wave field is computed, including all shear waves, converted phases (P to S or vice versa), multiple phases, tunnel waves, etc. For a system of homogeneous and parallel layers, elementary seismograms can be computed exactly from ray theory [*Mueller*, 1970]. The reflectivity method at present, however, can only be applied to one-dimensional velocity-depth distributions, which means that the method cannot be used directly if laterally varying structures have to be modeled. In practice, two-dimensional velocity distributions can be simulated by successive application of one-dimensional solutions at different points along a profile.

The need for modeling complex structures in which the geometry is two-dimensional (and ultimately three-dimensional) and velocities vary laterally within layers, has driven a large amount of research on ray theory. In the ray theoretical approach, the wave field is separated into elementary waves corresponding to groups of kinematically analogous waves and the eikonal equation (whose solutions are approximate solutions to the wave equation) is solved. In practice, this approach is limited to modeling body waves (P-waves and S-waves) and examples of ray tracing results can be seen in *Levander et al.* [this volume], *Snelson et al.* [this volume], and *Rumpel et al.* [this volume]. Numerical tests have shown that an approximate ray method [*Cerveny et al.*, 1977] yields useful synthetic seismograms. With certain modifications, this technique yields satisfactory results even near points of reversal or near critical distances. Using ray theory, computer methods have been developed that calculate travel times and amplitudes for rays traveling through laterally inhomogeneous structures [e.g., *Cerveny and Psencik*, 1983]. A combination of 1-D and 2-D modeling and interpretation procedures is usually most effective. The ray method has been further developed [e.g., *Cerveny*, 1985], and efforts have been made to include the effects of anisotropy and inhomogeneity on the propagation of high-frequency elastic waves in the ray-tracing method [e.g., *Gajewski and Psencik*, 1987]. A graphically sophisticated and widely used ray tracing package (MACRAY) was developed by *Luetgert* [1992] and allows interactive modeling.

Some authors have studied interpretation methods that were previously only applied to seismic reflection data and have applied them to densely recorded refraction data. For example, *McMechan et al.* [1982], *Milkereit et al.* [1985] and *McMechan and Fuis* [1987] applied normal movement (NMO) velocity analysis and common-midpoint (CMP) stacking methods.

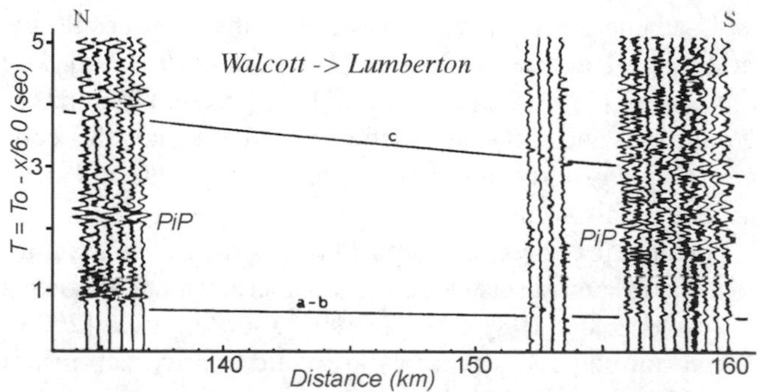

Figure 3. Previously unpublished detailed refraction record section along a profile extending from Walcott, Colorado toward Lumberton, New Mexico. These data were recorded in 1965 [*Prodehl and Pakiser*, 1980]. Shotpoint locations are shown in Figure 1. The phases identified are labeled as: a-b - reflection from the upper crust; PiP – reflection from the middle crust; c – reflection from the Moho (PmP).

In order to speed the tedious process of deriving 2-D velocity models from the many sources and receivers that characterize modern refraction surveys and provide estimates on resolution, *Zelt and Smith* [1992] developed a travel time inversion method that simultaneously models reflection and refraction arrivals from a number of sources to multiple receivers. Numerous trial and error iterative forward modeling steps are thus replaced by an effective minimization of travel time residuals by the least-squares method. In addition, the inversion method provides an estimate of resolution and uncertainty of the individual model parameters. A good starting model, which may be obtained by a careful application of the methods discussed above, is recommended in order to obtain a rapid and successfully converging inversion. This approach has been applied in many studies around the world including the CD-ROM [*Levander et al.*, this volume; *Rumpel et al.*, this volume] and DEEP PROBE (Figure 1) data [*Gorman et al.*, 2002].

For a 3-D-approach, *Hole* [1992] describes a non-linear high-resolution tomographic technique that exclusively uses first-arrival picks. In addition to the case of truly 3-D geometries [e.g., *Guterch et al.*, 2001; *Sroda et al.*, 2002], this approach is useful if first-arrival picks are available in large numbers and if the topography of the study area results in a crooked-line geometry of a profile. The *Hole* [1992] code uses a finite-difference approximation to the eikonal equation to calculate travel times [*Vidale*, 1988, 1990], and the model space consists of a 3-D velocity model defined on a uniformly spaced grid. Ray paths are back-projected through the array of calculated travel times to obtain the travel time at any given receiver for a source in the model space. *Snelson et al.* [this volume] applied this method successfully to the 1999 CD-ROM data and updated the 3-D first arrival model by forward modeling of later arrivals (prominent reflections) using an enhancement of the Hole code that provides for the modeling of floating reflectors [*Hole and Zelt*, 1995]. *Zelt and Barton* [1998] and *Zelt et al.* [1999] present a similar 3-D approach and show successful applications, and this code was applied to derive the tomographic model shown in *Levander et al.* [this volume].

In summary, there are important and restrictive assumptions made in most seismic refraction interpretations as pointed out by *Mooney* [1989], *Zelt* [1999], and *Zelt et al.* [2003]. Planar and layered structures are valid approximations, at least for a starting model. Lateral velocity variations are usually smaller than vertical ones and often occur on the same distance scale as the shotpoint spacing. Principal crustal phases are usually correctly identified and have not been confused with multiples, phase conversions, and noise. Detailed comparisons of interpretations of identical data sets have established that a final velocity model is more a function of the particular phase identification than of the interpretation method [e.g., *Ansorge et al.*, 1982; *Finlayson and Ansorge,* 1984, *Mooney and Prodehl*, 1984]. The lower crust is a region where ambiguity in how to correlate phases is common, and differing interpretations of lower crustal phases for the CD-ROM data are presented in *Levander et al.*, [this volume] and *Snelson et al.* [this volume].

In addition to analyzing information such as ray coverage, the fit between observed and calculated travel times and amplitudes, and resolution estimates provided by inversion algorithms [e.g., *Gorman et al.*, 2002], recent studies have used a combination of tomographic and ray tracing approaches to access the uncertainty in velocity models [e.g., *Fuis et al.*, 2003]. *Zelt* [1999] and *Zelt et al.* [2003] have formalized this approach, and it is applied to the CD-ROM data by *Levander et al.* [this volume]. Given the increasing gap in signal quality, the number of sources, and the number of receivers between modern surveys and ones done prior to the emergence of instrumental resources provided through PASSCAL (~1990), it is hard to generalize about resolution of key elements of crustal velocity models. However, the uncertainties in the interpretation of most modern refraction data may be estimated as follows: 1) the velocity of the upper crust typically has an error of ~3 percent; 2) the determination of the Moho depth and the upper mantle velocity for reversed profiles that are several hundred kilometers long has an error of 3–5 percent; and 3) deep crustal velocities often have an error of ~10 percent [*Mooney*, 1989]. Since the lower crust turned out to be a major feature of interest in the CD-ROM project, it is important to note that better errors (~5%) for lower crustal layers will result if they are thick enough to produce clear first arrivals.

3. THE SEISMIC REFLECTION TECHNIQUE

The reflection technique, as practiced by industry primarily in the search for hydrocarbons, can produce very detailed images of the uppermost crust by recording nearly vertically traveling waves as they reflect off discontinuities in the earth. Fundamental aspects of this technique are reviewed in many books, such as *Sheriff and Geldart* [1995] and *Yilmaz* [2001]. The goal of seismic reflection studies is to produce a clear image of the subsurface so its interpretation will be nearly self-evident. In addition, the velocities of intervals in the image can be estimated and subtleties in the data, such as amplitude variations with distance from the source, can reveal details about the composition and physical properties of rocks in the region imaged. Extensive petroleum exploration in the Rocky Mountain region has resulted in the recording of many 1000's of kilometers of seismic reflection profiles and a small but very useful fraction of these data have been published [e.g., *Cline and Keller*, this volume and *Quezada et al.*, this volume] providing significant insight about specific geologic features. In spite of the considerable cost of data acquisition, deep seismic reflection profiling of the entire crust emerged in the 1970's because geoscientists interested in deep earth structure and the processes at work in the earth recognized the value of reflection data early in the evolution of the method. With modification of industry data acquisition parameters and the use of larger sources, organizations such as the Consortium for Continental Reflection Profiling (COCORP), which was funded by the National Science Foundation, initiated large-scale crustal profiling efforts in the US. Other national efforts for deep seismic reflection profiling were established soon after the formation of COCORP. These included the British Institutions Reflection Profiling Syndicate (BIRPS), the Deutsches Kontinentales Reflexionsseimisches Programm (DEKORP), and LITHOPROBE, a Canadian effort that featured transects that were highly integrated with co-located geological and complementary geophysical studies and served as a model for the CD-ROM effort.

Compared to the refraction method, the reflection technique is configured to record seismic waves (usually P-waves) at receivers that are closely spaced (< 50 m) and whose distance from the source is relatively small (0 - 15 km). *Morozova et al.* [this volume] and *Magnani et al.* [this volume, a, b] discuss recording parameters and interpretations of the CD-ROM profiles in Wyoming and New Mexico respectively. For effective analysis of crustal velocities, the maximum desirable source-receiver offset is on the order of the depth to which the image is intended to extend. Reflections from deeper crustal or even upper mantle levels commonly are recorded and imaged well using shorter source-receiver offsets, but consequently, reliable velocity information for deep reflections generally is lacking. Since the 60's, common-midpoint (CMP, or sometimes CDP for common depth-point) profiling techniques [*Sheriff and Geldardt*, 1995] have been used with extraordinary success to improve S/N in reflection images. In this fundamental technique, many repetitive, closely spaced sources are employed and many 100's of groups of receivers are deployed. This multiplicity allows traces to be added to enhance the usually weak reflected signals and to determine the velocities that are needed to form an image via CMP processing. An individual seismogram on a typical seismic reflection record section is in fact the sum of many 1000's of individual seismic pulses that have traveled through the earth. The massive processing effort required to create seismic reflection images far exceeds that required in other seismic methods. In fact, the high level of computations required has driven many important advances in computer technology over the decades.

Ultimately, the goal is to recover the reflectivity of the earth and use it to solve for the velocity structure usually employing the vertical-incidence reflection coefficient for each interface:

$$R = (V_2 \rho_2 - V_1 \rho_1) / (V_2 \rho_2 + V_1 \rho_1), \text{ where}$$

V_1 is the velocity above the reflector,
V_2 is the velocity below the reflector,
ρ_1 is the density above the reflector, and
ρ_2 is the density below the reflector.

Reflection coefficients are measures of the changes in physical properties of juxtaposed rocks; large reflection coefficients indicate large differences in rock velocities and/or densities and generally result in strong reflections on seismic profiles. Thus, along with reflection geometries, reflection amplitudes are important indicators of geological relationships.

It is important to note that reflection and refraction techniques generally measure different aspects of the often highly complex velocity structure of the crust. Refraction techniques, employing very large source–receiver offsets rely on critically refracted arrivals whose measured velocities are determined by relatively horizontal travel at the deepest point of the energy's penetration. In simple horizontally layered sequences, the measured velocities are predominantly those of the upper portions of higher velocity layers in which waves travel horizontally. In contrast, reflection techniques maintain relatively small source–receiver offsets (compared to the depths of subsurface structures) and velocities are determined from changes in arrival times at increasing offsets for waves following roughly vertical travel paths. Despite the differences between the natures of the velocities measured using refraction and reflection techniques, reliable comparisons between, say, the depth to the Moho based on refraction and carefully designed reflection methods show considerable agreement.

After appropriate processing, reflection times represent the time that seismic energy takes to travel from the surface to a reflector and back (i.e., two-way time) that is a consequence of the depth to the reflector and the average velocity between the reflector and the surface. Velocities determined from shot records or CMP gathers during processing remove the component related to simple average velocity in order to resolve a component that depends on a weighted average velocity known as stacking or RMS (Root-Mean-Square) velocity. The weighting comes from the fact that waves will spend more time (relatively) in thick layers than in thin layers, which in turn affects the total travel time. Stacking and RMS velocities, though technically distinct, are very similar (and indeed are identical for horizontal, homogeneous layers). Most importantly, stacking or RMS velocities can be used to measure the velocity of the material in the interval between two reflectors, known as interval velocity [*Dix*, 1955], which is a major goal of the technique. Interval velocities determined from seismic reflection surveys in sedimentary basins show very good correlation with more direct measurements in boreholes. However, interval velocities are best resolved where data quality is very high and horizontal layers characterize the subsurface. For crustal-scale reflection surveys, these ideal conditions do not exist and, in general, reflection velocity resolution diminishes with depth.

A common-midpoint-stacked (CMP) seismic profile simulates what might be recorded if the experiment were conducted with a source and single receiver, both of which are located together (i.e., the offset is zero for each one-trace record), and the source and receiver are moved along the profile. This type of recording is of course not used in practice, but the coincident-source-receiver concept is useful for visualizing some of the distortions that arise in reflection profiling and that must be corrected in order to provide meaningful subsurface images. Most importantly, it's easy to envision that events recorded at the receiver in coincident source and receiver profiles result from reflections at normal incidence (perpendicular) to the reflecting boundaries. When reflectors are not horizontal, reflections will arrive at the receiver along paths that are not directly below it. However, regardless of the direction from which a reflection emanated, it will be plotted directly below the source-receiver position along the surface since that is where it was recorded. In essence, a reflection at a particular time on a single seismic trace could have come from anywhere on a surface (referred to as an aplanatic surface [*Waters*, 1987]) defined by the distance between that reflecting surface and the receiver and by the average velocity along its travel path. Because of this effect, dipping reflections on CMP profiles are not in their true positions and must be corrected by the process of migration. Unmigrated reflections appear to have less dip and are shifted in a down-dip direction laterally from the true position of the reflecting interface; migration steepens these events and shifts them to their proper up-dip spatial positions. In addition, diffracted energy from lateral discontinuities or structures with curvatures less than a Fresnel zone, which complicate unmigrated reflection images and disrupt resolution, can be collapsed back to a "point" through migration.

Accurate migration is generally possible when the velocity structure along the reflection path is known. Even approximately correct velocities often will produce migrations that improve structural resolution and, thus, migration has become a standard procedure in the processing of reflection data as computer capabilities have allowed. There are numerous approaches used to migrate seismic reflection data, many of which are detailed by *Yilmaz* [2001], including migration of time sections (time migration) and migration to form depth sections (depth migration). Both time and depth migration can be applied either before or after CMP stacking depending on the algorithms used. Unfortunately, migrations become less accurate and even may introduce artifacts for the deeper parts of crustal reflection profiles due to lack of detailed velocity information at great depth, complex reflection travel paths, and loss of data from the bottoms and sides of profiles as the migration aperture increases with depth [*Warner*, 1987]. Furthermore, two-dimensional reflection profiles can only be properly migrated if all of the reflections were generated and propagated within the plane of the profile. For deep crustal profiles, for which the targets of the imaging are usually quite complex, this probably never is the case. Reflected energy from 3-D structures can only be correctly migrated with adequate 3-D data. While its use in industry is becoming standard, large crustal-scale 3-D reflection surveys have not yet been conducted. Present practice is to use both migrated and unmigrated profiles to analyze and interpret deep crustal reflection images, and the skill and experience of the interpreter is an important factor in avoiding incorrect interpretation of artifacts or improperly migrated features.

Although seismic profiles look enticingly like geologic cross sections, the fact that they still are scaled in travel time means that they are inherently distorted, at least in a vertical direction. For example, a low-velocity part of the section might occupy just as much of the time scale as a thick, high-velocity zone. Furthermore, since velocities tend to increase with depth in the crust, deeper parts of the section are compressed in time profiles. One consequence of this effect is that a planar, dipping structure, such as a fault plane, can appear in seismic time profiles to be listric, so care in evaluating true dip relationships in time profiles is especially important. If the velocity in the subsurface is known from calculating stacking velocities, or from independent measurements such as a refraction survey, the seismic profile can be converted to depth and may, therefore, be interpreted more directly. As noted earlier, certain migration procedures do this conversion as part of the migration process.

Interpretation of crustal seismic reflection profiles ultimately requires considerable geological insight and incorporation of independent geological information to provide constraints on interpretations. To an important extent, interpretation is an interactive process between geophysical and geological interpretation and data processing [e.g., *Johnson and Smithson*, 1985] because choices made during data processing (e.g., velocity analysis, filter application, migration procedures, etc.) affect the final images that will be used for interpretation. Within the context of broad regional geologic and tectonic analysis, interpretations often focus on groups of events or individual events with particular geometries, anomalous amplitudes, juxtapositions or other characteristics that are interpreted to have geological significance. A crustal-scale seismic reflection image generally shows gross crustal structure, which then is analyzed in more specific detail. Often initially categorized by their geometric relations, seismic events frequently are analyzed quantitatively to provide additional information on factors such as reflection amplitude, true dip, spatial relationships with other structures, etc. As the papers in this volume show, such analyses can provide important insights about reflection coefficients and polarities (and thereby the nature of rocks juxtaposed in the subsurface), whether certain events are noise or artifacts, and whether certain events have spatial relations that link them to other important geological structures and to other important aspects of the data or subsurface geology. Of course, where possible, comparisons with geologic analogues or direct ties with surface exposures are used to constrain possible interpretations, and this approach is a key focus of papers in this volume by *Magnani et al.*, a, b; *Morozova et al.*, *Quezada et al.*, and *Cline and Keller*.

In recent years, considerable resources have been expended by the exploration industry on efforts to record multicomponent seismic reflection data, which allows the recording of both P-waves and S-waves. A strong motivator for this interest is the fact that compressional waves and shear waves respond differently to the presence of fluids within fractures and pore spaces in rocks. Comparisons of the compressional and shear wave fields can be used to constrain volumetric estimates and determine likely pore-fluid compositions in petroleum reservoirs. Furthermore, shear waves often can be used to image subsurface structures that are obscured to P-wave imaging such as "gas chimneys" above reservoirs, which strongly affect P-wave propagation, but which have a limited effect on S-waves. In practice, industry-related multicomponent surveys are most commonly conducted in marine environments as a means to enhance production success, and involve ocean-bottom cables that record one vertical- and two horizontal-components along with the pressure wave field (often referred to as 4-component or 4C recording).

In crustal reflection and refraction profiling, the use of multicomponent recording (generally acquiring one vertical component and one or two horizontal components) has become more common, but is more the exception rather than the rule. In these larger-scale crustal studies, the principle advantage of multicomponent recording is that better constraints on crustal compositions can be determined by evaluation of Poisson's ratios or, equivalently, Vp/Vs ratios [*Kern and Richter*, 1981; *Domenico*, 1984; *Fountain and Christensen*, 1989; *Holbrook et al.*, 1992; *Christensen*, 1996; *Satarugsa and Johnson*, 2000]. Although P-wave velocities are sensitive to rock composition (e.g., the velocity of granite generally is significantly less than that of gabbro), many very different crystalline rocks in the crust have velocities that fall within ranges that overlap [*Christensen and Fountain*, 1975; *Christensen*, 1979; *Holbrook et al.*, 1992; *Christensen and Mooney*, 1995], making interpretation of compositions based solely on P-wave velocities problematical. Because S-waves respond differently to rock compositions and fabrics, inclusion of S-wave velocities in interpretations often can be used to distinguish between crustal lithologies that would be unresolvable using P-wave velocities alone. For example, in analysis of compressional-wave reflection and refraction data beneath the Ruby Mountains in northeastern Nevada, *Satarugsa and Johnson* [1998] interpreted crustal compositions that had large possible variations; later analysis that included shear-wave data [*Satarugsa and Johnson*, 2000] constrained the crustal composition beneath the Ruby Mountains from seventeen possible rock types in the middle and lower crust to just six. As part of the CD-ROM experiment, limited 3-component data were acquired along most of the northern profile and were used by *Shoshitaishvili* [2002] to constrain near-surface lithologies beneath the surface location of the Archean-Proterozoic suture in the Sierra Madre Mountains (Cheyenne Belt).

In addition to compositional resolution, multicomponent data provide information on velocity anisotropy due to metamorphic fabrics [*Brocher and Christensen*, 1990; *McDonnough and Fountain*, 1993] and stress-aligned fractures [*Crampin*, 1985]. *Johnson and Hartman* [1991] interpreted low values of Poisson's ratio in the Colorado Plateau near Flagstaff Arizona to be related to closure of fluid-filled fractures in the upper crust, and *Satarugsa and Johnson* [2000] found that crustal anisotropy beneath the Ruby Mountains metamorphic core complex was significantly lower than found in laboratory measurements of mylonitic rock samples from the area [*McDonnough and Fountain*, 1993]. *Satarugsa and Johnson* [2000] also noted that the measurements of anisotropy from the S-wave data were consistent with a fast shear-wave orientation aligned approximately parallel to the regional maximum horizontal stress in the Nevada part of the Basin and Range Province, suggesting that bulk anisotropy in the upper crust is controlled by stress-induced fractures. Multicomponent data also provide the ability to detect seismic-event arrival azimuths to help distinguish events that arrive from below a profile from those that might arrive obliquely from the side of a profile. This is especially important to help constrain interpretations in complex structural environments.

While nearly always seen as highly desirable, multicomponent recording frequently is not feasible in crustal-scale academic surveys because appropriate equipment is not available, or because of the increase in cost that such recording often entails. Using 3-components at a single station, rather than the more common use of a single vertical component, requires three times the number of recording channels to cover the same distance. Often there is a trade-off between greater surface coverage (profile length) and the benefits of full-wavefield recording. Furthermore, extra field effort is needed to ensure correct and consistent alignment of the horizontal-component geophones. However, differences in wave-propagation mechanisms, and thus velocities, also make inclusion of shear-wave data non-trivial. Shear waves propagating through rocks with Poisson's ratios of 0.25 travel with velocities that are nearly half those of the corresponding compressional waves (Vp/Vs = 1.73). Consequently, shear waves always arrive after the compressional waves, and often are difficult to resolve in the presence of P-wave codas, scattered energy, and converted P- to S-wave reflected and refracted energy. Thus, significant care is required in the analysis of such data. Nevertheless, as larger numbers of new-generation recording systems become available (for example, through IRIS – Incorporated Research Institutions for Seismology), more surveys are taking advantage of the benefits of multi-component profiling.

4. RESOLUTION

When discussing seismic data, resolution is a major consideration. Technically, this concept is somewhat different in the refraction and reflection techniques, but as a practical matter, resolution refers to the ability of a measurement to detect the presence of some feature and the accuracy to which this feature can be located and its velocity determined. In lithospheric studies, the main parameters obtained by controlled-source seismic measurements are thickness of the crust, average velocity of the major layers that constitute the crust and uppermost mantle, geometry of structures such as faults, uplifts, and sedimentary basins, and major velocity discontinuities in the uppermost mantle. Because of uncertainties in determining crustal velocities, an inherent ambiguity exists between the arrival time of an event and its depth, and this ambiguity increases with increasing time/depth. As a result, it would be hard to argue that any existing controlled-source data set in a continental area resolves the position of the Moho to better than ± 2 km [e.g., *Zelt et al.*, 2003]. On the other hand, the depths to shallower structures are generally much better constrained. Refraction studies so far provide the best constraints on large-scale crustal and upper mantle (Pn) velocities; modern studies usually determine this later parameter to ± 0.1-0.2 km/s with shallower velocities determined with this resolution or better. In general, the depths of events in both refraction and reflection profiles may not be determined precisely, but reflection data generally show structural details on a scale much less than the depth uncertainty, and thus, the relative resolution is often on the order of 10s to 100s of meters.

Natural attenuation in the earth increases as frequency increases, and this fact is a fundamental limitation in seismology that limits resolution because high frequencies are needed to "see" small (relative to wavelength) features. For typical seismic frequencies, anelastic attenuation is an approximately linear function of frequency. In addition to this physical phenomenon, the quality of a data set is usually considered to be a function of the strength of the signal relative to noise (usually expressed as signal-to-noise ratio – S/N) and the number of seismograms recorded. The interval between recording sites is also an important factor if phases are to be correlated along a recording array in refraction studies and if high-quality images are to emerge from reflection surveys. Up to a point, the more data one has, the higher the resolution obtained will be.

The strength of a signal depends not only on the quality of the energy produced by a controlled-source signal (i.e., an underwater or a borehole explosion, an airgun, a mechanical source such as Vibroseis, or a quarry blast) but is also a function of local geology as manifested by scattering and attenuation. In reflection studies, the number of traces that share common midpoints is a major contribution to producing a high S/N. In principle, S/N increases by the square root of the number of traces added together for a single midpoint location in the processing sequence. On the other hand, increasing source energy increases S/N roughly linearly, but with the common effect that larger impulsive sources tend to produce lower dominant frequencies. Thus, there often is a trade-off between signal penetration and frequency content. Empirical experience [*Brocher and Hart*, 1991] suggests that impulsive sources (e.g., dynamite) may produce better reflection images than Vibroseis sources, but other studies indicate that Vibroseis signals can be used for wide-angle recording to offsets of 100 km or more [*Kubichek et al.*, 1984]. Most commonly, the choice of sources is dictated by cost and logistical factors. The lesson from the excellent results obtained in many of the Lithoprobe transects [e.g., *White et al.*, 1994; *Clowes*, 1998; *Cook et al.*, 1998; *Snyder*, 2002] is that a low-noise environment (away from cities, roads, and other human activity) perhaps is the most important factor in producing high S/N in crustal reflection profiles. For this reason, some surveys are recorded at night when ambient noise conditions are lowest.

Experience in the exploration industry has demonstrated clearly the efficacy of obtaining dense, 3-D subsurface data to improve resolution and enhance geological interpretations. Such efforts require very large numbers of recording channels and source points and can be extraordinarily expensive; this has limited their use, so far, mainly to petroleum development and smaller-scale petroleum exploration work, although small-scale 3-D surveys have been acquired by some academic groups. Use of similar 3-D techniques in crustal-scale reflection experiments has been limited by cost and availability of appropriate equipment, but greater numbers of newer

IRIS recording systems are making the prospect for such large-scale experiments more likely. However, the inherent loss of resolution with depth will likely require somewhat different deployment strategies and processing techniques, and will inherently result in lower resolution images of deep crustal targets than can be obtained for exploration targets in sedimentary basins.

Different interpretation techniques can often lead to models that appear to be different results, and this situation arises more as the quality and quantity of data, and thus resolution, decreases. For refraction studies in the 1960s and early 70s, most of the lines in the Southern Rocky Mountains had a station spacing of 15 km and more (Table 1; see CDROM in back cover sleeve), but this value generally decreased over the years reaching 800 m in the CD-ROM profiles. Early workers appreciated the value of closely spaced recordings and locally employed 500-m spacing [e.g., *Steeples and Miller*, 1989; *Prodehl and Pakiser*, 1980]. The CD-ROM refraction profile is a recent example of a refraction profiling experiment [*Levander et al.*, this volume; *Rumpel et al.*, this volume; *Snelson et al.*, this volume] with relatively high resolution, and the POLONAISE'97 [*Sroda et al.*, 2002] and CELEBRATION 2000 experiments [*Guterch et al.*, 2001] are examples where a significant degree of 3-D coverage was obtained.

Independent from the number of available instruments, physical laws limit the resolution of even a perfect data set. The earth strongly attenuates high frequencies; thus, signals that penetrate deeply into the earth will, by definition, be dominantly low frequency in nature. The effects of this phenomenon can be best demonstrated using reflection and refraction signals. An estimate of resolution can be obtained by applying the relationship: velocity = frequency x wavelength. A typical crustal velocity is 6 km/s, and a typical dominant frequency for a deep-crustal signal is 10–15 Hz (cycles/s). Thus, a typical dominant wavelength is on the order of 0.5 km. A common rule of thumb employed in reflection surveys is that the tops and bottoms of layers thinner than 1/4 of a wavelength cannot be resolved individually, and thus such data cannot easily detect features smaller than about 120 m in thickness. Signals that penetrate to the Moho probably have dominant frequencies of no greater than 5–15 Hz, and their limit of resolution would be about 300–100 m. These estimates probably are optimistic, especially for refraction profiles.

A bigger factor in the limitation of resolution for deep crustal reflectors is the fact that reflections result from interactions with an area on a reflecting surface rather than simply reflecting from a point. This area is known as the Fresnel zone, and its dimension is a function of signal frequency, average velocity above the reflector, and depth [*Sheriff and Geldart*, 1995]. For an average crustal velocity of 6 km/s, reflections from a depth of 20 km arise from areas of about 5 km; for reflections from 30 km, the area is about 6 km. Reflections from structures at lower crustal depths with continuous areas less than these form diffractions – broadly arcuate events that generally will produce complex, interfering "reflection" patterns. Outcrops of deep crustal rocks generally show highly complex structures, few of which could be considered to form a planar area on the scale of a reflecting Fresnel zone. Thus, resolution of individual deep crustal reflectors from most seismic profiles is very difficult. In principle, migration should collapse diffractions to a point, and thus provide lateral resolution on the order of a wavelength, but accurate migration is dependent on a precise knowledge of the crustal velocity structure, and further requires that the reflector have no dip perpendicular to the seismic profile. Migration of relatively shallow crustal data is highly effective in improving structural resolution, but migration of deep crustal data requires very large migration apertures [*Warner*, 1987], and optimal conditions for migration seldom exist for deep data. Often, a suite of migrations using different velocities and other parameters provides some improvement in deep seismic data, and such an approach is useful as an interpretive tool. Alternatively, migration of line segments can provide improvements in evaluating structural trends in very deep data. Most commonly, deep crustal reflections form interfering patterns from complex geological structures. However, large-scale features often are resolvable when regional layering or through-going deformation causes reflection and/or diffraction patterns to align. Thus, in spite of the inherent difficulties in resolving individual, very-small-scale features, reflection profiles provide remarkable resolution of larger-scale structures often of fundamental structural or tectonic significance. Many such structures are described and interpreted in the papers in this volume [e.g., *Magnani et al.* a and b; *Morozova et al.*].

5. PREVIOUS SEISMIC REFRACTION PROFILES

Surprisingly few deep seismic surveys have been undertaken in the Rocky Mountain region prior to 1999, considering the high level of geologic interest in the area. In addition, the early surveys (prior to 1980) were generally of low resolution. Thus, interpretations of them [e.g., *Toppazada and Sanford*, 1976; *Prodehl and Pakiser*, 1980; *Prodehl and Lipman*, 1989] led to simplified earth models, in particular concerning the upper crust. However, the early studies provide a general picture of crustal structure in the region, but few details have been resolved. Figure 1 shows all major seismic refraction profiles of crustal dimensions that were recorded in and around the southern Rocky Mountain region of the United States before 1999.

In the 1960's the U.S. Geological Survey recorded a large number of seismic refraction profiles in the western United

States, but only a few touched the Rocky Mountain area [e.g., *Stewart and Pakiser*, 1962; *Jackson et al.*, 1963; *Jackson and Pakiser*, 1965; *Willden*, 1965] (Fig. 1). Most of these data were reinterpreted by *Prodehl* [1970, 1979], *Prodehl and Pakiser* [1980], and *Prodehl and Lipman* [1989], but previously unpublished profiles are shown in Figures 2 and 3. The Southern Rocky Mountains and adjacent Great Plains region were the focus of two campaigns, in 1961 (two lines from Lamar, Colorado [*Jackson et al.,* 1963]) and in 1965. In addition, in 1964, two large explosions at the Climax Molybdenum Mine, Colorado were used to record a 360 km unreversed S-N line that followed the Front Range [*Jackson and Pakiser*, 1965]. The 1965 campaign was comprised of a long N-S line extending from Sinclair, Wyoming to Lumberton, New Mexico, traversing primarily the ranges in western Colorado [*Prodehl and Pakiser*, 1980]. In addition, two E-W lines extended from the Great Plains into the mountains [*Prodehl and Lipman*, 1989], a northern line in Wyoming (Guernsey to Sinclair) and a southern line in Colorado (Agate to Wolcott). The 1965 survey also included a line in the Great Plains east of Denver that extended from Agate, Colorado to Concordia, Kansas [*Steeples and Miller*, 1989]. Two lines through the Middle Rocky Mountains (American Falls, Idaho to Flaming Gorge, Utah) and through the Colorado Plateau (Hanksville, Utah to Chinle, New Mexico) were part of the U.S. Geological Survey effort in 1963 [*Prodehl*, 1979]. The American Falls-Flaming Gorge line was later supplemented by an unreversed line extending from the Bingham Canyon copper mine near Salt Lake City, Utah to the northeast [*Braile et al.*, 1974]. The Yellowstone - Snake River Plain experiment of 1978 and 1980 also extended into the Middle Rocky Mountains region in northwestern Wyoming [*Smith et al.*, 1982].

In Wyoming, controlled source seismic data on crustal structure are very sparse. A particularly interesting observation is that, according to interpretations of the older data, the crust thins rather strongly (about 10 km) from northern Colorado to southern Wyoming. Although detail is lacking, the refraction profiles (Fig. 1) extending from Wolcott, Colorado to Sinclair, Wyoming, from Sinclair, Wyoming to Guernsey, Wyoming, and northward from Climax, Colorado to Laramie, Wyoming all show evidence of this effect [*Jackson and Pakiser*, 1965; *Prodehl and Pakiser*, 1980; *Prodehl and Lipman*, 1989]. *Johnson et al.* [1984] used these data and gravity modeling to derive a crustal model of the Cheyenne belt region that shows this effect extending into central Wyoming. In westernmost Wyoming, refraction profiles show abrupt thickening from the Basin and Range province into Wyoming [*Willden*, 1965; *Prodehl*, 1979; *Braile et al.*, 1989]. The DEEP PROBE results [*Henstock et al.*, 1998; *Snelson et al.*, 1998; *Gorman et al.*, 2002] show that crust thickens in central Wyoming and remains thick (~50 km) across Montana, which is in agreement with earlier work in Montana [*McCamy and Meyer*, 1964; *Asada and Aldrich*, 1966; *Prodehl and Lipman*, 1989].

Other than DEEP PROBE, crustal refraction surveys in Montana consist of the very early investigations in 1959 [*Meyer et al.,* 1961] and a deep investigation of the LASA array area [*Warren et al.*, unpublished U. S. Geological Survey data]. These data covered the area of the Northern Rocky Mountains and were also discussed by *McCamy and Meyer* [1964], *Asada and Aldrich* [1966], and *Prodehl and Lipman* [1989].

The central Rio Grande rift area is only covered by a few unreversed lines with large station spacing [*Olsen et al.*, 1979; *Toppozada and Sanford*, 1976], and there is no refraction data in the Colorado portion of the rift. Data recorded in the southern Rio Grande rift in the 1980's are denser, but were recorded by analog systems and are only partly reversed. These experiments are summarized by *Sinno et al.* [1986], *Keller et al.* [1990] and *Schneider and Keller* [1994].

The lithospheric structure of the Colorado plateau has been the subject of much recent interest. The old refraction profile of *Rolle*r [1965] indicated a crustal thickness of ~42 km. However, subsequent surface wave [*Keller et al.*, 1979] and COCORP deep reflection profiles [*Hauser and Lundy*, 1989] indicated that it was somewhat thicker (at least 45 km). The PACE project (Pacific to Arizona Crustal Experiment) produced a data set for the southern plateau that has been interpreted in two ways. *Wolfe and Cipar* [1993] show the crust to be nearly 50 km thick and to have a fast lower crustal layer (Vp ~ 7.3 km/s) just above the Moho. However, *Parsons et al.* [1996] derived a similar model for the portion of the profile north of the Grand Canyon but found the crust to be about the same thickness as suggested by *Roller* [1965] to the south. This issue is important because of the geologic similarities between the plateau, the Southern Rocky Mountains and the adjacent Great Plains and will be addressed further in *Keller et al.* [this volume]. Northeast of Flagstaff, Arizona, *Johnson and Hartman* [1991] recorded densely-sampled P- and S-wave data from the PACE shots to develop an upper crustal model of Poisson's ratio on the Colorado Plateau that indicated shallow crustal (<4 km) values were ~0.24, which decreased between 6 and 15 km depth to ~0.21.

None of the refraction profiles mentioned above, except for the PACE project, have the close station spacing (1–3 km) and shot spacing (<50 km) that is nowadays the norm. Particularly in the Southern Rocky Mountains, maximum recording distances of 360 km were reached, but the average recording station interval was close to 16 km (Figs. 2 and 3), [*Prodehl and Pakiser*, 1980]. Exceptions are the early Great Plains lines from Lamar, Colorado, and the Agate – Concordia line, where, over offsets of several 10s of kilometers, geophone spacings of 500 m were obtained, and the recent Jemez

Tomography EXperiment (JTEX) experiment in northern New Mexico. However, the JTEX experiment focused on the upper crust of the Valles caldera rather than the determination of crustal structure in the region [*Keller et al.*, 1998]. The only recent long refraction profile recorded in Colorado and Wyoming before 1999 was the DEEP PROBE experiment whose particular purpose was to probe the upper mantle along a line traversing the western United States from Canada almost to Mexico approximately along 110°W. It was recorded with dense station spacing (1.25 km). Because of the goals of this experiment, it involved only a few large shots at very large spacing (about 750 km); one located at the U.S./Canadian border, one in central Wyoming (SP 43 in Fig.1), one in southwestern Colorado (SP 37) and one in southwestern New Mexico (SP 33) [*Henstock et al.*, 1998; *Snelson et al.*, 1998; *Gorman et al.*, 2002].

6. PREVIOUS SEISMIC REFLECTION PROFILES

A few deep seismic reflection surveys have been carried out in the Southern Rocky Mountain area. The Consortium for Continental Reflection Profiling (COCORP) recorded a series of profiles near Socorro, New Mexico [*Brown et al.*, 1980] in the central portion of the Rio Grande rift, and these results agreed very well with the refraction results of *Olsen et al.* [1979] that showed a crust that was ~35 km thick with a very strong mid-crustal reflector. COCORP also conducted two surveys in Wyoming. One consisted of a profile that crossed the Wind River uplift [*Smithson et al.*, 1979, 1980] providing an excellent image of the Wind River thrust to mid-crustal depths. In southeastern Wyoming, COCORP [*Allmendinger et al.*, 1982; *Brewer et al.*, 1982; *Johnson and Smithson*, 1985, 1986] and the University of Wyoming [*Smithson et al.*, 1977; *Gohl and Smithson*, 1994; *Speece et al.*, 1994; *Templeton and Smithson*, 1994] also conducted surveys in the Laramie region. These surveys provided interesting images of the structural fabric of the crust but only modest constraints on deep crustal structure.

Thanks to exploration for oil and gas, a number of seismic reflection profiles have been released for publication and provide important information about the upper crust. The data published prior to 1988 where summarized by *Smithson and Johnson* [1989]. The Rocky Mountain Association of Geologists and the Denver Geophysical Society published an atlas of seismic data for the Rocky Mountain region [*Gries and Dyer*, 1985]. In addition, there are a few record sections, and cross-sections derived from them, in a volume edited by *Lowell and Gries* [1983]. Finally, a volume edited by *Keller and Cather* [1994] contained a considerable number of reflection profiles in the Rio Grande rift region. In addition to the deep seismic reflection data produced by the CD-ROM project, two papers in this volume contain significant amounts of newly released seismic reflection data [*Cline and Keller*, this volume; *Quezada et al.*, this volume].

7. DISCUSSION

Our overall knowledge of crustal and uppermost mantle structure of the United States, as based on geophysical studies, was last summarized in a comprehensive volume "Geophysical Framework of the United States" [*Pakiser and Mooney*, 1989]. *Smithson and Johnson* [1989] reviewed seismic reflection studies in the Rocky Mountain region in this volume. In addition, *Prodehl and Lipman* [1989] reviewed the results of previously published interpretations of crustal profiles mentioned above and added a few reinterpretations. In particular, they summarized the main features of crustal structure in a generalized cross section traversing the Southern Rocky Mountains and Rio Grande rift at approximately 106°W and in several contour maps. The full set of these contour maps is provided on the CDROM in the back cover sleeve along with page-sized versions of all of the record sections from *Prodehl and Lipman* [1989]. *Keller et al.* [1998] updated their synthesis and added crustal thickness estimates in Colorado and New Mexico from receiver-function determinations [*Murphy* 1991, *Sheehan et al.*, 1995, *Kilbride,* 2000].

In brief, the results of seismic reflection and refraction studies of the crust and uppermost mantle conducted prior to the CD-ROM effort provide the following general view of lithospheric structure. From a physiographic point of view, the Southern Rocky Mountains occupy the crest of a broad uplift [*Eaton*, 1986] and the Rio Grande rift is an extensional feature that follows the crest of this uplift [*Cordell,* 1982]. Thus, it seems clear these features share much in terms of their tectonic evolution. In the Rio Grande rift, distinct crustal thinning (about 5 km relative to adjacent areas) has been documented from Albuquerque, New Mexico, southward. The area of thinned crust widens southwards as does the physiographic expression of the rift. Its gradual thinning from the rift shoulders to the rift valley may reflect the thermal regime that existed prior to rifting. The thickest crust in the region appears to be associated with both the Southern Rocky Mountains in Colorado and the Great Plains in Colorado and New Mexico and does not correlate directly with topography. These observations indicate that the mantle is playing a significant role in the attainment of isostatic balance in the area [e.g., *Sheehan et al.*, 1995]. In Colorado, major magmatic modification of the crust during the Phanerozoic appears to have been limited to the San Juan Mountains region in the south and the Colorado Mineral belt [*McCoy et al.*, this volume]. The San

Juan Mountains are associated with a very large gravity low [*Plouff and Pakiser*, 1972] and are traversed by the Cochetopa – Lumberton refraction profile that indicates the presence of a low-velocity anomaly [Fig. 2b – delay of phase a; *Prodehl and Lipman*, 1989]. There is evidence of the crust thinning by about 10 km northward from Colorado into Wyoming (e.g., Fig. 2a). This thinning could be a relic of the Proterozoic rifting of the southern margin of the Archean Wyoming craton.

Acknowledgments. Financial support was provided by the National Science Foundation - Continental Dynamics Program (EAR-9614269 to Keller and EAR-9614208 and EAR-0003577 to Johnson) and the Deutsche Forschungsgemeinschaft (DFG). The data compilation effort is part of the GEON (Geoscience Network) project. Instruments used in the CD-ROM field program were in part provided by the PASSCAL facility of the Incorporated Research Institutions for Seismology (IRIS) through the PASSCAL Instrument Center and the University of Texas at El Paso. The facilities of the IRIS Consortium are supported by the National Science Foundation. We thank Matthew Averill for a helpful review.

REFERENCES

Allmendinger, R. W., Brewer, J. A., Brown, L. D., Kaufman, S., Oliver, J. E., and Houston, R. S., COCORP profiling across the Rocky Mountain front in southern Wyoming; Part 2, Precambrian basement structure and its influence on Laramide deformation, *Geol. Soc. Am. Bull., 93*, 1253–1263, 1982.

Ansorge, J., Prodehl, C., and Bamford, D., Comparative interpretation of explosion seismic data, *J. Geophys., 51*, 69–84, 1982.

Asada, T., and Aldrich, L. T., Seismic observations of explosions in Montana, *in J. S. Steinhart and T. J. Smith (eds), The Earth Beneath the Continents*, American Geophysical Union, *Mono. 10*, Washington, D.C., 382–390, 1966.

Braile, L.W., Hinze, W. J., von Frese, R. R. B., and Keller, G. R., Seismic properties of the crust and uppermost mantle of the conterminous United States and adjacent Canada, in *Pakiser, L. C., and Mooney, W. D. (Editors): Geophysical framework of the continental United States*, Geol. Soc. Am., *Memoir 172*, 655–680, 1989.

Braile, L. W., Keller, G. R., Mueller, S., and Prodehl, C., Seismic techniques, in *Olsen, K. H. (Editor), Continental rifts: Evolution, structure, tectonics*, Elsevier, Amsterdam, 61–92, 1995.

Braile, L. W., Smith, R. B., Keller, G. R., Welch, R. M., and Meyer, R. P., Crustal structure across the Wasatch Front from detailed seismic refraction studies, *J. Geophys. Res., 79*, 2669–2676, 1974.

Brewer, J. A., Allmendinger, R. W., Brown, L. D., Oliver, J. E., and Kaufman, S., COCORP profiling across the Rocky Mountain front in southern Wyoming; Part 1, Laramide structure, *Geol. Soc. Am. Bull., 93*, 1242–1252, 1982.

Brocher, T. M., and Hart, P. E., Comparison of Vibroseis and explosive source methods for deep crustal seismic reflection profiling in the Basin and Range Province, *J. Geophys. Res., 96*, 18,197–18,213, 1991.

Brown, L. D., Chapin, C. E., Sanford, A. R., Kaufman, S., and Oliver, J. E., Deep structure of the Rio Grande rift from seismic reflection profiling, *J. Geophys. Res., 85*, 4773–4800, 1980.

Bullen, K. E., and Bolt, B. A., *An introduction to the theory of seismology*, Cambridge University Press, Cambridge, England, 499 p., 1985.

Cerveny, V., Ray synthetic seismograms for complex two-dimensional and three-dimensional structures, *J. Geophys., 58*, 2–26, 1985.

Cerveny, V. and Psencik, I., SEIS83 - numerical modeling of seismic wave fields in 2-D laterally varying layered structure by the ray method, in *Documentation of Earthquake Algorithms,* edited by E. R. Engdahl, *World Data Center A for Solid Earth Geophys, Boulder, Rep. SE-35,* 36–40, 1983.

Cerveny, V., Molotkov, I. A., and Psencik, I., *Ray method in seismology*, Univerzita Karlova, Praha, 214 p. 1977.

Christensen, N. I., Compressional wave velocities in rocks at high temperatures and pressure, critical thermal gradients, and crustal low-velocity zones. *J. Geophys. Res., 84*, 6849–6863, 1979.

Christensen, N. I., Pore pressure, seismic velocities and crustal structure, in *Pakiser, L.C., and Mooney, W. D. (eds.), Geophysical framework of the continental United States*, Geol. Soc. Am., Memoir 172, 783–798, 1989.

Christensen, N. I., Poisson's ratio and crustal seismology, *J. Geophys. Res., 101*, 3139–3156, 1996.

Christensen, N. I., and Fountain, D. M., Constitution of the lower continental crust based on experimental studies of seismic velocities in granulite, *Geol. Soc. Am. Bull., 86*, 227–239, 1975.

Christensen, N. I., and Mooney, W. D., Seismic velocity structure and composition of the continental crust: a global view, *J. Geophys. Res., 100*, 9761–9788, 1995.

Clowes, R. M., Cook, F. A., and Ludden, J. N., Lithoprobe leads to new perspectives on continental evolution, *GSA Today, 8*, 1–7, 1998.

Cook, F. A., van der Velden, A. J., Hall, K. W., and Roberts, B. J., Tectonic delamination and subcrustal imbrication of the Precambrian lithosphere in northwestern Canada mapped by Lithoprobe, *Geology, 26*, 839–842, 1998.

Cordell, L., Extension in the Rio Grande rift, *J. Geophys. Res., 87*, 8561–8569, 1982.

Crampin, S., Evaluation of anisotropy by shear-wave splitting, *Geophysics, 50*, 142–152, 1985.

Dix, C. H., Seismic Velocities from surface measurements, *Geophysics, 20*, 68–86, 1955.

Domenico, S. N., Rock lithology and porosity determination from shear and compression wave velocity, *Geophysics 41,* 1188–1195, 1984.

Eaton, G. P., A tectonic redefinition of the southern Rocky Mountains, *Tectonophysics, 132*, 163–193, 1986,

Finlayson, D. M., and Ansorge, J., Workshop proceedings; interpretation of seismic wave propagation in laterally heterogeneous structures, *Bureau of Mineral Resources, Geology and Geophysics Report, 258*, Canberra, Australia, 207 p., 1984.

Fountain, D. M., and Christensen, N. I., Composition of the continental crust and upper mantle: A review, in *Pakiser, L.C., and Mooney, W. D. (eds.), Geophysical framework of the continental United States*, Geol. Soc. Am., Memoir 172, 711–742, 1989.

Fuchs, K., The reflection of spherical waves from transition zones with arbitrary depth-dependent elastic moduli and density, *J. Phys. Earth, 16*, 27–41, 1968.

Fuchs, K., and M. Landisman, Detailed crustal investigation along a north-south section through the central part of western Germany, in *Steinhart, J. S., and Smith, T. J. (eds.), The earth beneath the continents*, Geophys. Mono., 10, AGU, Washington, D.C., 433–452., 1966.

Fuchs, K., and Mueller, G., Computation of synthetic seismograms with the reflectivity method and comparison with observations, *Geophys. J. R. Astr. Soc., 23*, 417–433, 1971.

Fuis, G. S., R. W. Clayton, P. M. Davis, T. Ryberg, W. J. Lutter, D. A. Okaya, E. Hauksson, C. Prodehl, J. M. Murphy, M. L. Benthien, S. A. Baher, M.D. Kohler, K. Thygesen, G. Simila, and G. R. Keller, 2003, Fault systems of the 1971 San Fernando and 1994 Northridge earthquakes, southern California: Relocated aftershocks and seismic images from LARSE II, *Geology, 31*, 171–174, 2003.

Gajewski, D., and Psencik, I., Computation of high-frequency seismic wavefields in 3-D laterally inhomogeneous anisotropic media, *Geophys. J. R. Astr. Soc., 91*, 383–411, 1987.

Giese, P., Depth calculation, in *Giese, P., Prodehl, C., and Stein, A. (eds.), Explosion seismology in central Europe - data and results*, Springer, Berlin-Heidelberg, 146–161, 1976.

Gohl, K., and Smithson, S. B., Seismic wide-angle study of accreted Proterozoic crust in SE Wyoming, *Earth and Planet. Sci. Lett., 125*, 293–305, 1994.

Gorman, A. R., Clowes, R.M. Ellis, R. M., Henstock, T. J., Spence G.D., Keller, G. R., Levander, A. R. Snelson, C. M., Burianyk, M. J. A., Kanasewich, E. R., Asudeh, I., Hajnal, Z., and Miller, K. C., Deep Probe: imaging the roots of western North America, *Canadian Journal of Earth Sciences, 39*, 375–398, 2002.

Gries, R. R., and Dyer, R. C. (Eds.), *Seismic Exploration of the Rocky Mountain Region, Rocky Mountain Association of Geologists and Denver Geophysical Society*, 300 p., Denver, Colorado, 1985.

Guterch, A., Grad, M., and Keller, G. R., Seismologists Celebrate the New Millennium with an Experiment in Central Europe, EOS, *Trans. AGU, 82*, 529, 534–535, 2001.

Hauser, E. C., and J. Lundy, COCORP deep reflections: Moho at 50 km (16 s) beneath the Colorado Plateau, *J. Geophys. Res., 94*, 7071–7081, 1989.

Henstock, T. J., Levander, A., Snelson, C. M., Keller, G. R., Miller, K. C., Harder, S. H., Gorman, A. R., Clowes, R. M., Burianyk, M. J. A., and Humphreys, E. D., Probing the Archean and Proterozoic lithosphere of western North America, *GSA Today, 8*, 1–5, 1998.

Holbrook, W. S., Mooney, W. D., and Christensen, N. I., The seismic velocity structure of the deep crustal continental crust. in *Fountain*, D. M., Arculus, R., and Kay, R. W. (Eds.), Continental Lower Crust, Elsevier, Amsterdam, 1–43, 1992.

Hole, J. A., Non-linear high resolution three-dimensional seismic travel tomography, *J. Geophys. Res., 97*, 6553–6562, 1992.

Hole, J. A., and Zelt, B. C., Three-dimensional finite-difference reflection travel times, *Geophys. J. Int., 121*, 427–434, 1995.

Jackson, W.H., and Pakiser, L.C., Seismic study of crustal structure in the southern Rocky Mountains, *U.S. Geol. Surv. Prof. Paper 525-D*, 85–92, 1965.

Jackson, W.H., Stewart, S.W., and Pakiser, L.C., Crustal structure in eastern Colorado from seismic refraction measurements, *J. Geophys. Res., 68*, 5767–5776, 1963.

Johnson, R. A., and Hartman, K. A., Upper crustal Poisson's ratios in the Colorado Plateau from multicomponent wide-angle seismic recording: in Meissner, R., et al., eds, *Continental Lithosphere: Deep Seismic Reflections*, American Geophysical Union, Geodynamics Series, v. 22, p. 323–328, 1991.

Johnson, R. A., K. E. Karlstrom, S. B. Smithson, and R. S. Houston, Gravity studies across the Cheyenne belt, a Precambrian suture in southeastern Wyoming, *J. Geodynamics, 1*, 445–472, 1984.

Johnson, R. A., and S. B. Smithson, Thrust faulting in the Laramie Mountains, Wyoming, from reanalysis of COCORP data, *Geology, 13*, 534–537, 1985.

Johnson, R. A., and S. B. Smithson, Interpretive processing of crustal seismic reflection data; examples from the Laramie Range COCORP data, in *Barazangi, M., and Brown, L. (eds.), Reflection seismology; a global perspective*, Am. Geophys. Un. Geodyn. Series, *13*, 197–208, 1986.

Keller, G. R., and Cather, S. M. (eds.), Basins of the Rio Grande rift: structure, stratigraphy, and tectonic setting, Geol. Soc. Am., Boulder, Colorado, *Spec. Paper* 291, 304 p., 1994.

Keller, G. R., L. W. Braile, and P. Morgan, Crustal structure, geophysical models, and contemporary tectonism of the Colorado Plateau, *Tectonophysics*, 61, 131–147, 1979.

Keller, G. R., P. Morgan, and W. R. Seager, Crustal structure, gravity anomalies and heat flow in the southern Rio Grande rift and their relationship to extensional tectonics, *Tectonophysics, 174*, 21–37, 1990.

Keller, G. R., C. M. Snelson, A. F. Sheehan and K. G. Dueker, Geophysical studies of the crustal structure in the Rocky Mountain Region, A Review, *Rocky Mountain Geology, 33*, 217–228, 1998.

Kennett, B. L. N., *Seismic Wave Propagation in Stratified Media*, Cambridge University Press, Cambridge, England, 341p., 1983.

Kern, H., and Richter, A., Temperature derivatives of compressional and shear wave velocities in crustal and mantle rocks at 6 Kbar confining pressure, *J. Geophys., 49*, 47–56, 1981.

Kilbride, F., Receiver function studies in the Southwestern United States, and correlation between stratigraphy and Poisson´s ratio in southwest Washington State. *PhD Dissertation, University of Texas at El Paso,* 207 p., 2000.

Kind, R., Computation of reflection coefficients for layered media, *J. Geophys., 42*, 191–200, 1976.

Kubichek, R. F., Humphreys, C., Johnson, R. A., and Smithson, S. B., Long range recording of Vibroseis data: simulation and experiment: *Geophys. Res. Lett., 11*, 809–812, 1984.

Lowell, J. D., and Gries, R. R., eds., *Rocky Mountain Foreland Basins and Uplifts*, Denver, Colorado, Rocky Mountain Association of Geologists, 392 p., 1983.

Luetgert, J. H., MacRay – interactive two-dimensional raytracing for the Macintosh, *U. S. Geol. Surv. Open-File Rept., 92–356*, 1992.

Masson, F., Jacob, B., Prodehl, C., Readman, P., Shannon, P., Schulze, A., and Enderle, U., A wide-angle seismic traverse through the Variscan of SW Ireland, *Geophys. J. Int., 134*, 689–705, 1998.

McCamy, K., and Meyer, R. P., A correlation method of apparent velocity measurement, *J. Geophys. Res., 69*, 691–699, 1964.

McDonough, D. T., and Fountain, D. M., P-wave anisotropy of mylonitic and infrastructural rocks from a Cordilleran core complex: the Ruby-East Humboldt Range, Nevada, *Phys. Earth Planet Inter. 78*, 319–336 1993.

McMechan, G. A., and Fuis, G. S., Ray equation migration of wide-angle reflections from southern Alaska, *J. Geophys. Res., 92*, 407–420, 1987.

McMechan, G.A., and Mooney, W. D., Asymptotic ray theory and synthetic seismograms for laterally varying structure; theory and application to the Imperial Valley, California, *Bull. Seis. Soc. Am., 70*, 2021–2035, 1980.

McMechan, G. A., Clayton, R., and Mooney, W. D., Application of wave field continuation to the inversion of refraction data, *J. Geophys. Res., 87*, 927–935, 1982.

Meyer, R.P., Steinhart, J.S., and Bonini, W.E., Montana 1959, in *Steinhart, J. S., and Meyer, R. P. (eds.), Explosion studies of continental structure*, Carnegie Inst. of Washington, *Publ. 622*, 305–343, 1961.

Milkereit, B., Mooney, W. D., and Kohler, W. M., Inversion of seismic refraction data in planar dipping structure, *Geophys. J. Roy. Astr. Soc., 82*, 81–103, 1985.

Mooney, W. D., Seismic methods for determining earthquake source parameters and lithospheric structure, in *Pakiser, L.C., and Mooney, W.D. (Editors), Geophysical framework of the continental United States*, Geol. Soc. Am., *Memoir 172*, 11–34, 1989.

Mooney, W. D., and Prodehl, C. (Editors), Proceedings of the 1980 workshop of the IASPEI on the seismic modeling of laterally varying structures; contributions based on data from the 1978 Saudi Arabian refraction profile, *U.S. Geol. Surv., Circ., 937*, 158 p., 1984.

Mueller, G., Exact ray theory and its application to the reflection of elastic waves from vertically inhomogeneous media, *Geophys. J. R. Astr. Soc., 21*, 261–283, 1970.

Murphy, B. P., Determination of shear wave velocity structure in the Rio Grande Rift through receiver function and surface wave analysis, *M. S. Thesis, University of Texas at El Paso*, 57 p., 1991.

Olsen, K. H., Stewart, J. N., and Keller, G. R., Crustal structure along the Rio Grande rift from seismic refraction profiles, in *Riecker, R.E. (ed.), Rio Grande rift; tectonics and magmatism,*. Am. Geophys. Un., Washington, D.C., 127–143, 1979.

Pakiser, L.C., and Mooney, W. D. (Editors), Geophysical framework of the continental United States, *Geol. Soc. Am., Memoir 172*, 826 p., 1989

Parsons, T., J. McCarthy, W. M., Kohler, C. J., Ammon, H. M. , Benz, J. A. , Hole, and E. E. Criley, Crustal structure of the Colorado Plateau, Arizona: Application of new long-offset seismic data analysis techniques, *J. Geophys. Res., 101*, 11,173 –11,194, 1996.

Plouff, D., and L. C. Pakiser, Gravity study of the San Juan Mountains, Colorado, *U.S. Geological Survey, Professional Paper 800-B*, B183–B190, 1972.

Prodehl, C., Seismic refraction study of crustal structure in the western United States, *Geol. Soc., Am. Bull., 81*, 2629–2646, 1970.

Prodehl, C., 1979. Crustal structure of the western United States - a reinterpretation of seismic refraction measurements from 1961 to 1963 in comparison with the crustal structure of central Europe. *U.S. Geol. Survey Prof. Paper 1034*, 74 pp.

Prodehl, C., and Lipman, P. W., Crustal structure of the Rocky Mountain region, in *Pakiser, L.C., and Mooney, W. D. (Editors), Geophysical framework of the continental United States*, Geol. Soc. Am., *Memoir 172*, 249–284, 1989.

Prodehl, C., and Pakiser, L. C., Crustal structure of the southern Rocky Mountains from seismic measurements, *Geol. Soc. Am. Bull., 91*, 147–155, 1980.

Roller, J. C., Crustal structure in the eastern Colorado Plateaus province from seismic-refraction measurements. *Bull. Seismol. Soc. Am., 55*, 107–119, 1965.

Sandmeier, K.-J., and Wenzel, F., Synthetic seismograms for a complex model, *Geophys. Res. Lett., 13*, 22–25, 1986.

Satarugsa, P., and Johnson, R. A., Crustal velocity structure beneath the eastern flank of the Ruby Mountains metamorphic core complex: results from normal-incidence to wide-angle seismic data, *Tectonophysics, 295*, 369–395, 1998.

Satarugsa, P., and Johnson, R. A., Constraints on crustal composition beneath a metamorphic core complex: results from 3-component wide-angle seismic data along the eastern flank of the Ruby Mountains, Nevada, *Tectonophysics, 329*, 223–250, 2000.

Schneider, R. V., and Keller, G. R., Crustal structure of the western margin of the Rio Grande rift and Mogollon-Datil volcanic field, southwestern New Mexico and southeastern Arizona, in *Keller, G. R., and Cather, S. M. (eds.), Rio Grande rift, structure, stratigraphy, and tectonic setting,* Geol. Soc. Am., Boulder, Colorado, Spec. Paper 291, 207–226, 1994.

Sheehan, A.F., Abers, G.A., Jones, C.H., and Lerner-Lam, A.L., Crustal thickness variations across the Colorado Rocky Mountains from teleseismic receiver functions, *J. Geophys. Res., 100*, 20,391–20,404, 1995.

Sheriff, R. E., and Geldart, L. P., *Exploration Seismology, 2nd Edition,* Cambridge University Press, Cambridge, England, 592 p., 1995.

Shoshitaishvili, E., Geophysical investigation of Archean and Proterozoic crustal-scale boundaries in Wyoming and Colorado with emphasis on the Cheyenne Belt, *PhD Dissertation, University of Arizona,* 338 p., 2002.

Sinno, Y. A., Daggett, P. H., Keller, G. R., Morgan, P., and Harder, S. H., 1986. Crustal structure of the southern Rio Grande rift determined from seismic refraction profiles. J. Geophys. Res., 91, 6143–6156, 1986.

Smith, R.B., Schilly, M. M., Braile, L. W., Ansorge, J., Lehmann, J. L., Baker, M. R., Prodehl, C., Healy, J. H., Mueller, St., and Greensfelder, R. W., The 1978 Yellowstone - eastern Snake River Plain seismic profiling experiment, Crustal structure of the Yellowstone region and experiment design, *J. Geophys. Res., 87*, 2583–2596, 1982.

Smithson, S. B., and Johnson, R. A., Crustal structure of the western United States based on reflection seismology, in *Pakiser, L.C., and Mooney, W. D. (eds.), Geophysical framework of the continental United States*, Geol. Soc. Am., *Memoir 172*, 577–612, 1989.

Smithson, S. B., Shive, P. N., and Brown, S. K., Seismic reflections from Precambrian crust, *Earth and Planet. Sci. Lett.*, 37, 333–338, 1977.

Smithson, S. B., Brewer, J. A., Kaufman, S., Oliver, J. E., and Hurich, C. A., Structure of the Laramide Wind River uplift, Wyoming from COCORP deep reflection data and gravity data, *J. Geophys. Res., 84*, 5955–5972, 1979.

Smithson, S. B., Brewer, J. A., Kaufman, S., Oliver, J. E., and Zawislak, R. L., Complex Archean lower crustal structure revealed by COCORP crustal reflection profiling in the Wind River Range, Wyoming, *Earth and Planet. Sci. Lett., 46*, 295–305, 1980.

Snelson, C. M., Henstock, T. J., Keller, G. R., Miller, K. C., Levander, A., Crustal Structure of the Uppermost Mantle Structure along the DEEP PROBE seismic profile, *Rocky Mountain Geology, 33*, 181–198, 1998

Snyder, D. B., Clowes, R. M., Cook, F. A., Erdmer, P., Evenchick, C. A., van der Velden, A. J., and Hall, K. W., Proterozoic prism arrests suspect terranes; insights into the ancient Cordilleran margin from seismic reflection data, *GSA Today, 12*, 4–10, 2002.

Speece, M. A., Frost, B. R., and Smithson, S. B., Precambrian basement structure and Laramide deformation revealed by seismic reflection profiling in the Laramie Mountains, Wyoming, *Tectonics, 13*, 354–366, 1994.

Sroda, P., Czuba, W., Grad, M., Gaczynski, E., Guterch, A., and POLONAISE Working Group, Three-dimensional seismic modelling of crustal structure in the TESZ region based on POLONAISE'97 data, *Tectonophysics, 360*, 169–185, 2002.

Steeples, D.W., and R. D. Miller, Kansas refraction profiles, *Kansas Geol. Surv. Bull., 226*, 129–164, 1989.

Steinhart, J. S., and Meyer, R. P., Explosion studies of continental structure, *Carnegie Inst. of Washington, Publ. 622*, 409 p., 1961.

Stewart, S.W., and Pakiser, L.C., Crustal structure in eastern New Mexico interpreted from the GNOME explosion., *Bull. Seismol. Soc. Am., 52*, 1017–1030, 1962.

Templeton, M. E., and Smithson, S. B., Seismic reflection profiling of the Cheyenne Belt Proterozoic suture in the Medicine Bow Mountains, southeastern Wyoming; a tie to geology, *Tectonics, 13*, 1231–1241, 1994.

Toppozada, T. R., and A. R. Sanford,. Crustal structure in central New Mexico interpreted from the GASBUGGY explosion, *Bull. Seismol. Soc. Am., 6*, 877–886, 1976.

Vidale, J. E., Finite-difference calculation for traveltimes, *Bull. Seis. Soc. Am., 78*, 2062–2076, 1988.

Vidale, J. E., Finite-difference calculation of travel-times in three dimensions, *Geophysics, 55*, 521–526, 1990.

Warner, M., Migration: why doesn't it work for deep continental data?, *Geophys. J. R. Astr. Soc., 89*, 21–26, 1987.

Waters, K. H., *Reflection Seismology; a Tool for Energy Resource Exploration, 3rd Edition.,* John Wiley and Sons, New York, NY, 538 p., 1987.

White, D. J., Lucas, S. B., Hajnal, Z., Green, A. G., Lewry, J. F., Weber, W., Bailes, A. H., Syme, E. C., and Ashton. K., Paleo-Proterozoic thick-skinned tectonics; Lithoprobe seismic reflection results from the eastern Trans-Hudson Orogen, *Canadian Journal of Earth Sciences, 31*, 458–469, 1994.

Willden, R., Seismic refraction measurements of crustal structure beneath American Falls reservoir, Idaho and Flaming Gorge reservoir, Utah, *U.S. Geol. Surv. Prof. Paper 525-C*, 44–50, 1965.

Wolf, L. W., and J. J. Cipar, Through thick and thin: A new model for the Colorado Plateau from seismic refraction data from Pacific to Arizona crustal experiment, *J. Geophys. Res., 98*, 19,881–19,894, 1993.

Yilmaz, O., *Seismic Data Analysis: Processing, Inversion, and Interpretation of Seismic Data*, Society of Exploration Geophysicists, Tulsa, OK, 2027 p., 2001.

Zelt, C.A., and R. B. Smith, Seismic traveltime inversion for 2-D crustal velocity structure, *Geophys. J. Int., 108,* 16–34, 1992.

Zelt, C.A. and P. J. Barton, Three-dimensional seismic refraction tomography; a comparison of two methods applied to data from the Faeroe Basin, *J. Geophys. Res., 103*, 7187–7210, 1998.

Zelt, C. A., Modelling strategies and model assessment for wide-angle seismic traveltime data, *Geophys. J. Int., 139*, 183–204, 1999.

Zelt, C. A., A. M. Hojka, E. R. Flueh, and K. D. McIntosh, 3-D simultaneous seismic refraction and reflection tomography of wide-angle data from the central Chilean margin, *Geophys. Res. Lett., 26*, 2577–2580, 1999.

Zelt, C. A., K. Sain, J. V. Naumenko, and D. S. Sawyer, Assessment of crustal velocity models using seismic refraction and reflection tomography, *Geophys. J. Int., 153*, 609–626, 2003.

Roy A. Johnson, Department of Geosciences, University of Arizona, Tucson, AZ 85721–0077, Johnson@geo.arizona.edu.

G. Randy Keller, Department of Geological Sciences, University of Texas at El Paso, 500 University Ave., El Paso, TX 79968, keller@geo.utep.edu.

Claus Prodehl and Hanna-Maria Rumpel, Geophysikalisches Institut, University of Karlsruhe, Hertzstr.16, Karlsruhe, Germany 76187, claus.prodehl@epost.de; Hanna-Maria.Rumpel@gpi.uni-karlsruhe.de.

Catherine M. Snelson, Department of Geoscience, University of Nevada Las Vegas, 4505 Maryland Parkway, MS 4010, Las Vegas, NV 89154–4010, csnelson@unlv.nevada.edu.

Inter-Wedging Nature of the Cheyenne Belt – Archean-Proterozoic Suture Defined by Seismic Reflection Data

E. A. Morozova, X. Wan, K. R. Chamberlain, S. B. Smithson

Department of Geology & Geophysics, University of Wyoming, Laramie, Wyoming

R. Johnson

Department of Geosciences, University of Arizona, Tucson, Arizona

K. E. Karlstrom

Earth & Planetary Sciences, University of New Mexico, Northrop Hall, Albuquerque, New Mexico

New seismic reflection data from the CD-ROM project (Continental Dynamics of the Rocky Mountains) show that the Archean-Proterozoic suture in Wyoming, the Cheyenne belt, consists of a crustal-scale, conjugate thrust wedge, where Archean and Proterozoic crust were thrust into each other. Moderately S-dipping reflections extend to depths of 15–18 km and surface at the mapped shear zones of the Cheyenne belt; these terminate close to a north-dipping reflection that extends to about 24 km depth and converges at the surface at the Farwell-Lester Mountain (FLM) area with several strong, south-dipping reflections. The FLM zone is marked by dismembered ophiolites suggestive of an ocean basin and may represent a back-arc basin subsequently closed by continued south-dipping subduction, resulting in a cryptic suture. Arcuate criss-crossing reflections in Archean basement to the north of the Cheyenne belt are related to folding, inferred conjugate thrusting and an antiformal duplex stack. This stack formed during Paleoproterozoic suturing in supracrustal rocks comprising at least the upper 24 km of the crust. Our analysis of wide-angle seismic data does not reveal a 7 km/s lower crustal layer that could be interpreted as underplate in southern Wyoming or northern Colorado. A suture, marked by complex interwedging of crustal blocks, was probably steepened by continued convergence to the south.

INTRODUCTION

The question of how Precambrian lithosphere formed and evolved is well posed by the rocks of the uplifts of the central Rocky Mountains, which were well studied geologically but, until recently, were poorly surveyed using modern geophysical methods. The target of our study is the Cheyenne belt (CB) suture in southeastern Wyoming, the long-lived boundary between the Archean craton to the north and accreted Proterozoic island-arc blocks to the south (Figure 1). The CB has been the focus of detailed geologic studies in the Sierra Madre, Medicine Bow, and Laramie Mountains because it is a well exposed Precambrian suture. Tectonic models have suggested that the 1.78 Ga Green Mountain arc in northern Colorado was sutured against the Archean craton along north-verging-thrust shear zones of the CB from 1.78–1.76 Ga [*Hills and*

The Rocky Mountain Region: An Evolving Lithosphere
Geophysical Monograph Series 154

10.1029/154GM15

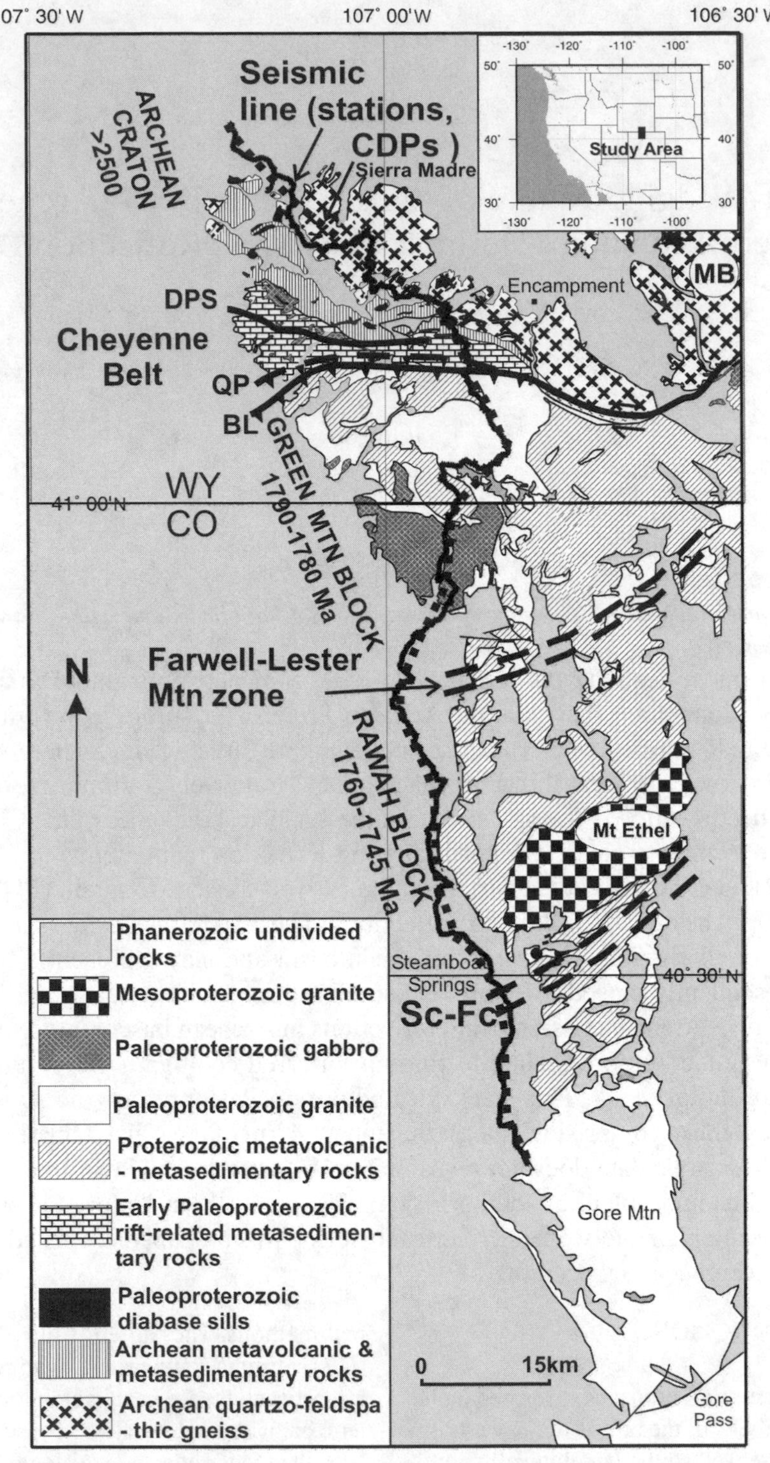

Figure 1. Geologic map of study area after Love and Christensen (1995) and Tweto (1979) shows the location of the CD-ROM seismic profile. Main geologic units are indicated. DPS- Divine Peak synclinorium; MB – Medicine Bow Mountains; QP- mylonite, Quartz Peak thrust fault; BL – Battle Lake thrust fault, cataclastic fault of the CB; Sc-Fc – Soda-creek – Fish creek shear zone.

Houston, 1979; *Houston et al.,* 1979, 1989, 1993; *Duebendorfer and Houston,* 1986, 1987, 1990; *Karlstrom et al.,* 1983; *Karlstrom and Houston,* 1984; *Karlstrom and Humphreys,* 1998; *Houston and Graff,* 1995; *Chamberlain,* 1998] (Figure 1). Recently, the CB became a major scientific target of a large multidisciplinary research project, Continental Dynamics of the Rocky Mountains (CD-ROM), which brought together geoscientists from 18 institutions. The project combines refraction, reflection and passive-source seismic data as well as surface geology and xenolith studies. The seismic line in SE Wyoming and NE Colorado is the first deep crustal reflection profile recorded on land in the U.S. in the 8 years since the last COCORP survey line.

REGIONAL GEOLOGIC SETTING

The two main areas of exposure of the CB are in the Medicine Bow Mountains and the Sierra Madre. In both ranges, Archean crystalline basement of the Wyoming craton is overlain by Archean and Early Proterozoic supracrustal rocks and intruded by 2.1– to 2.0 Ga mafic sills and dikes. The Early Proterozoic rocks and dikes record rifting and development of a passive margin, with an attenuated margin and a network of S-dipping normal faults that became inverted during suturing [*Hills and Houston,* 1979; *Karlstrom et al.,* 1983; *Karlstrom and Houston,* 1984; *Houston et al.,* 1989]. Just south of the Cheyenne belt is the Green Mountain block, an island-arc terrane accreted to the Wyoming craton at about 1.78–1.76 Ga. (Figure 1). This block contains amphibolite-grade volcanic rocks and turbidities and associated intermediate and gabbroic plutons that represent somewhat deeper levels of the island arc, all intruded by synkinematic granites.

There is a significant lateral variation in the CB between its two main exposures. In the Medicine Bow Mountains, the zone is about 10 km wide and consists of mylonitic strands within Proterozoic gneisses, plus more discrete thrusts in the miogeoclinal Snowy Pass Supergroup. Shear sense indicates northward vergence which, combined with the absence of Proterozoic arc magmatism to the north, suggests a south-dipping Proterozoic paleosubduction system for the suturing [*Hills and Houston,* 1979; *Karlstrom and Houston,* 1984; *Houston et al,* 1993; *Duebendorfer and Houston,* 1990]. Available ($(_{Nd}$ values from isotope studies in the Medicine Bow Mountains have been interpreted to suggest a relatively shallow southward dipping subduction zone because of the presence of older material incorporated into younger docked terranes to the south of the suture zone [*Ball and Farmer,* 1991].

In contrast, the CB in the Sierra Madre is narrower, and the surface expression of the CB in the Sierra Madre is marked by a shear zone that is steep and mylonitic in the east and moderately dipping and cataclastic in the west where our seismic profile crosses it. The eastern segment is correlative with the CB in the Medicine Bow Mountains. The western cataclastic segment is a moderately dipping thrust fault (50–150 m wide), which has been interpreted to have overridden, truncated, and dismembered an earlier continuous ductile shear zone [*Duebendorfer and Houston,* 1990]. Thus, the continent-island arc suture exposed on the surface in the Sierra Madre may be detached from its root by this later thrusting. ($(_{Nd}$ data from the Sierra Madre suggest mainly juvenile materials to the south implying a steeper suture zone and a more abrupt lithospheric transition [*Chamberlain,* 1998].

THE RESULTS OF PREVIOUS GEOPHYSICAL STUDIES

In the Laramie Mountains, a reflection dipping southeast at 55° was recorded by COCORP to a depth of 12 km; this was interpreted as the Archean-Proterozoic suture [*Allmendinger et al.,* 1982]. The lower crust on both sides of this reflection is non- reflective below 7s (~22 km) at pre-critical and normal-incidence geometries. No distinct Moho reflections were found by either common-depth-point [*Allmendinger et al.,* 1982; *Smithson et al.,* 1980; *Speece et al.,* 1994] or wide-angle studies [*Gohl and Smithson,* 1994]. A broad range of lower crustal reflections appears at large offsets on wide-angle data [*Gohl and Smithson,* 1994], interpreted as generated by relatively small steps in velocity. These characteristics coupled with lower crustal velocities of 6.5 to 7.5 km/s, were interpreted as evidence for lower crustal, layered mafic intrusions in an island arc setting [*Gohl and Smithson,* 1994].

A similar survey in the Medicine Bow Mountains revealed that two south-dipping reflectors extend to depths of about 14 km, at ~ 60° true dip and project to the surface at the locations of mylonitic zones that mark the Cheyenne belt [*Templeton and Smithson,* 1994]. Minor, discontinuous reflectors in the lower crust were interpreted as possible mafic underplate; no Moho was imaged at near vertical incidence [*Templeton and Smithson,* 1994].

Wide-angle data across the Cheyenne belt in southern Wyoming have been interpreted to indicate an increase in crustal thickness from 40–46 km in the southern Archean Wyoming craton to 48–50 km in the central Colorado Province [*Prodehl and Pakiser,* 1980; *Prodehl and Lipman,* 1989; *Henstock et al.,* 1998; *Snelson et al.,* 2001]. A step in Moho boundary at the CB has also been proposed based on gravity data from the Laramie Mountains, Medicine Bow Mountain, and Sierra Madre [*Johnson et al.,* 1984], and 53-km thick crust in central Colorado is supported by teleseismic receiver-function analysis [*Sheehan et al.,* 1995].

NEW SEISMIC DATA FROM THE CD-ROM PROFILE ACROSS THE CHEYENNE BELT

The CD-ROM Sierra Madre seismic profile starts in the southern Wyoming Archean craton and continues south across exposed Archean supracrustal rocks of the Sierra Madre, perpendicularly crossing the Cheyenne belt and the accreted 1.78–1.76 Ga Green Mountain block (Figure 1). The overall N-S orientation of the seismic line follows the axis of Sierra Madre, parallels the strike of Laramide uplifts, and gives a well situated cross section of the Proterozoic crustal assemblage. The seismic data were collected in the fall of 1999 with four 50,000-lb-force vibrators, 25-m station intervals, 100-m source intervals and a 4 ms sample interval into 1000 channels giving a 25-km recording spread. Data collection resulted in 2124 records with 25- to 30-s record lengths and a nominal fold of 125.

Two crustal blocks characterized by different types of reflectivity can be identified (block A and P) from the seismic section (Figure 2 and 3). The first (block A) is associated with the Archean craton in the northern part of the profile (to the north of the CB suture) and the second (block P) is to the south of the CB and is related to accreted island arc terranes of the Green Mountain block. Most continuous seismic reflections (length> 1 km) dip at > 10° (Figure 2) except for the short possible Moho reflection in the northern part of the profile and several discontinuous reflections (4, 14, 15 in the Archean and 24 in the Proterozoic part of the section).

SEISMIC REFLECTIVITY OF THE WYOMING ARCHEAH CRATON

The upper 20 km of the crust of the Archean craton (block A on Figs 2 and 3) is densely reflective with numerous north- (12 and 13 in Figure 2) and south-dipping, numbered (2, 5, 8, 9, 10, 11and 14 Figure 2), arcuate, criss-crossing events forming a complex unmigrated reflection pattern (Figure 2). Most of the dipping upper crustal reflections project to the surface and correlate well with the surface geology (Figure1, 2 and 3). The dominant feature is a crustal-scale antiform mapped by Houston and Graff (1995), which coincides with a change in dip of the surface foliation and a refolded antiformal axis just north of the Divide Creek synclinorium, indicated by reflection 5 (Figures 2 and 3). We interpret this as an antiformal duplex stack of south-verging thrusts and recumbent folds because most prominent reflections are north-dipping (numbered 5, 8, 9 and 10 on Figures 2 and 3) to a depth of 15 km. At about 18–20 km depth, the reflections become horizontal (numbered 15 on Figure 2), and a discontinuous series of south-dipping, strong reflections (numbered 16 and 17) appear in the middle and lower crust. Deeper events in this northern part of the profile include a weakly reflective Moho at about 13 seconds (40 km) and weaker north-vergent, sub-parallel, steeply dipping events between 9 and 16 s (16 and 17 on Figures 2 and 3).

SEISMIC REFLECTIVITY OF THE PROTEROZOIC GREEN MOUNTAIN BLOCK

The reflectivity changes immediately adjacent to and south of the CB. The reflections in this area form a complex pattern, dominantly south-dipping, but they are much weaker (this contrast is not related to differences in signal penetration [*Morozova et al.,* 2002], and less continuous presumably resulting from strong, steep foliation in the Proterozoic rocks. Two sub-parallel, south-dipping weak events just to the south of the CB (6 and 7 on Figures 2 and 3) dip at about 40 degrees (migrated in Figure 3) and when correlated with moderately dipping events 6a and 7a extend to depths of 15 to 18 km (Figure 3) and can be correlated directly with two major faults associated with the surface expression of the CB, the Quartzite Peak and Bottle Lake faults [*Duebendorfer and Houston,* 1990]. Weak criss-crossing events (19 and 20 on Figure 3) may truncate the S-dipping events. Another discontinuous weak reflection dips to the north to a depth of about 24 km (numbered 21 on Figures 2 and 3) and strong, more continuous reflections, (numbered 23) dip to the south to a depth of 23 km and, if correlated with the reflection numbered 23a, extended to a depth of 33 km. The above reflections together with the south-dipping reflection 22 converge to the surface in the Farwell-Lester Mountain (FLM) zone. The interpretation of the reflections alone could be questionable. Mapped discontinuities in geochronologic age and metamorphism suggest an isotopic-metamorphic break at the point where reflections approach the surface. Dismembered ophiolites, sillimanite nodules, and ore mineralization are suggestive of a former ocean basin. These combined observations indicate the presence of a cryptic suture zone and the reflector geometrics described above are consistent with such an interpretation. Note that with only either the geology or geophysics, interpretation of this zone is incomplete, but together they make sense. Because of the narrowness of the exposed Green Mountain block and the implied presence of sea floor, we interpret the FLM zone as a former back-arc basin that closed after renewed subduction. The greater abundance of south-dipping reflections in this zone suggests later continuation of south-dipping subduction.

The distinctive change in reflectivity between the Archean and Proterozoic crustal blocks coincides with a prominent, north-pointing, mid-crustal wedge that is formed by connecting the strongest reflections beneath the CB (reflections 1, 2, 3 and 4 on Figs 2 and 3). This boundary could be con-

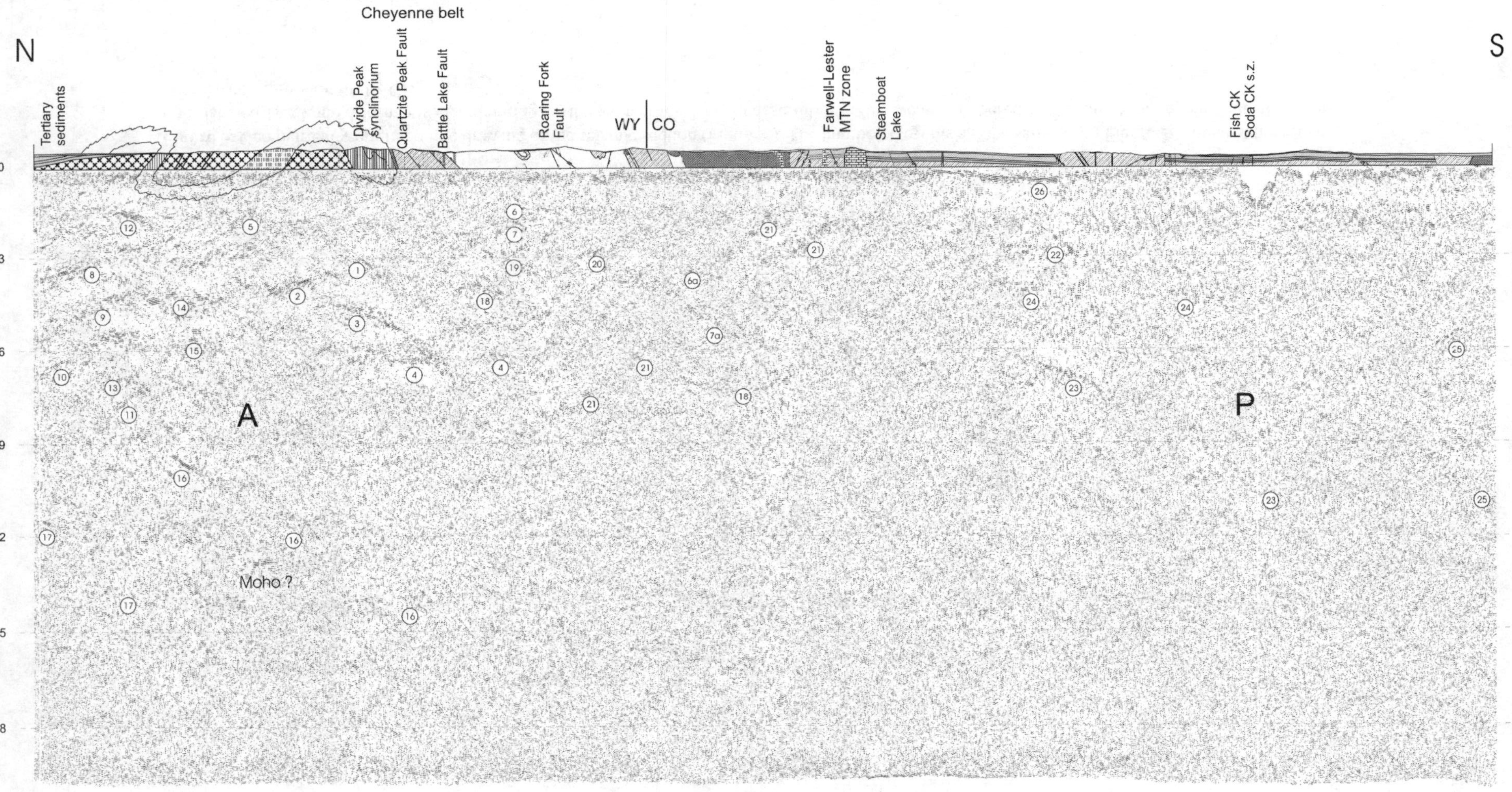

Figure 2. Stacked seismic section below geologic cross-section. No vertical exaggeration. Numbers indicate the most prominent reflections described in the text. The reflections numbered 22 and 23 dipping south are stronger than reflections 21 dipping north indicating the south-dipping cryptic suture.

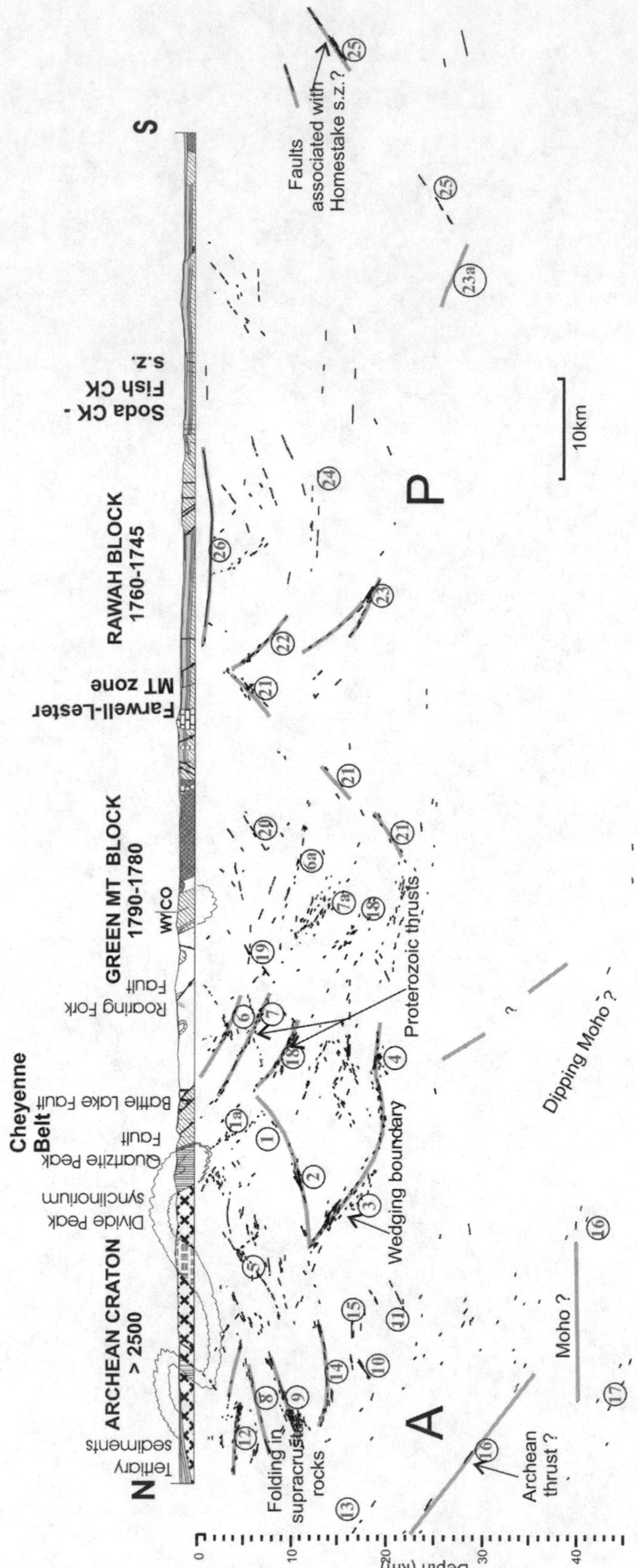

Figure 3. Geometrically migrated line drawing of the seismic section (Figure 2). The numbered events are the same as on Figure 2. Note that the reflections labeled 1, 2,3 and 4 form a wedge pointing north (see text for detailed description). The events numbered 6 and 7 are Proterozoic faults that form the surface expression of the CB.

tinued to the surface by connecting the wedge to the south-dipping reflection 1a (Figure 3), resulting in a steep zigzag boundary between two crustal blocks. We interpret this zigzag boundary as the image of the present-day CB suture

DISCUSSIONS AND CONCLUSIONS

The new reflection data offer a dramatic new view of the geometry and tectonic history of the Cheyenne belt suture. The suture itself is imaged as a complex series of wedges (Figures 2–4) resulting in an interdigitization of the Archean and Proterozoic crust, similar to "crocodiles" of Meissner [1989]. We interpret this pattern to imply that a process of crustal wedging [*Oxburgh,* 1972; *Beaumont et al.,* 1994] is responsible for the present geometry of the CB suture. A similar process has been proposed by Cook et al. (1998) and Snyder et al. (2002) for the collision between Archean and Proterozoic blocks across the Slave (northwest Canada) and Baltic margins. This crustal wedging is accomplished by a series of north- and south-vergent thrust faults in the upper 20–25 km of the crust that take up the strain at different or overlapping times. The ca. 1.78 Ga suturing of the Green Mountain block to the Archean craton was followed by 150 m.y. of shortening from 1.74 to 1.63 Ga as additional island-arc terranes were docked to the south [*Chamberlain et al.,* 1993; *Tyson et al.,* 2001]. This continued regional shortening may have produced far-field strain to steepen the CB suture. This interpretation differs from the more traditional view of the CB suture in which a thick band of sub-parallel, moderately dipping faults is conceived to mark a suture.

We interpret the FLM zone as a back-arc basin that broke up the original arc. This explains the oceanic affinities and the narrowness of the Green Mountain block, much narrower than typical arcs. The southern fragment then incorporated into the newly accreted arcs to the south and may be represented by the Salida terrane identified by Reed et al., (1987), who earlier proposed a back-arc basin in this area. This basin then closed along south-dipping subduction which is correlated with the more numerous south-dipping reflections that project to the FLM zone. Several interpretations [*Dueker,* 2001, *Tyson et al.,* 2002] proposed a flip in subduction resulting in a north-dipping subduction zone north of the CB. Such interpretations are not supported by primary features of the reflection wavefield. Their interpretation extrapolates the north-dipping FLM reflections for 40 km and continues the north-dipping subduction zone right through south-dipping reflections (16 and 17 on Figures 2 and 3) in the Archean lowermost crust. Our interpretation agrees better with the available data, and we suggest that the positive tomography anomaly, interpreted as a high-velocity mantle slab [*Dueker et al.,* 2001], is delamination of eclogite "caught in the act". The older Archean crust may have been thicker and had more time to pass through a favorable temperature-viscosity window [*Jull and Kelemen,* 2001] to promote delamination; to the north, Archean crust is even thicker [*Gorman et al.,* 2002]. Several compelling lines of evidence suggest long-term south-dipping subduction early in CB development. These observations include: the lack of Proterozoic arc magmatism north of the Cheyenne belt, south-dipping lower crustal reflections north of the CB (features 16 and 17 on Figure 2), lack of

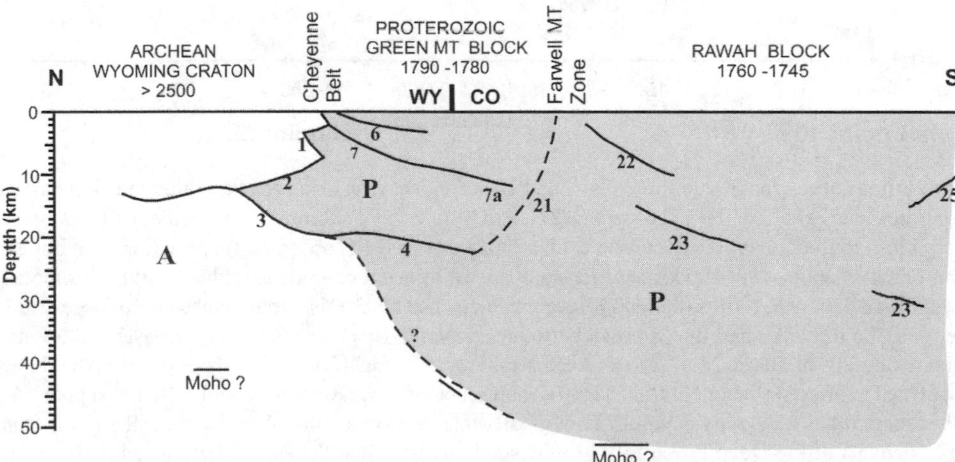

Figure 4. Interpretation of crustal structure across the CB area. Proterozoic island- arc terrane (labeled P) is wedged into the Archean Wyoming craton (labeled A). Crust thickens from 40 under the Archean Wyoming craton to ~45 km beneath the Proterozoic.

reflectivity in the lower crust south of the CB, and the north vergence of Cheyenne belt shear zones.

The Proterozoic arc crust to the south of the CB is distinctly less reflective, and similarly low reflectivity has been found for some modern arc crust [*Snyder et al.,* 1995; *Makovsky and Klemperer,* 1999]. Beneath the CB, the Moho boundary is poorly reflective at vertical incidence except for a small segment in the northern part of the section (Figure 3).

We have modeled the northern part of the wide-angle seismic data across the CB earlier presented by Snelson et al., (2001) in order to reconcile the two seismic datasets. We interpreted a short reflection in CDP data at about 42 km depth north of the CB as the Moho, and wide-angle data suggested that the Moho is deeper. The wide-angle PmP reflections, however, do not have depth points underneath the CB and don't overlap. Nor do the CDP and wide-angle profiles coincide over most of their length. In addition, a Moho uplift indicated by the curvature of the PmP arrivals (Figure 5) occurs just north of the CB and may account for part of a positive tomographic anomaly [*Dueker et al.,* 2001]. Also, as indicated by first arrivals (Figure 5) and PmP asymptotes with velocities not higher than 6.5 km/s (Figure 5), no 7 km/s layer is found in the lower crust in southern Wyoming and northernmost Colorado. Our short reflection at 13.5 s (about 42 km) in the Archean is interpreted as the Moho rather than the top of a lower crustal underplate because our analysis of the wide-angle, reflection-refraction profile (Figure 5) does not reveal the presence of a so-called 7xx layer [*Snelson et al.,* 2001] in this area.

The paucity of lower crustal reflections south of the CB (except for the ones associated with the FLM zone and Homestake area, Figure 3) is not related to the signal-to-noise ratio [*Morozova et al.,* 2002], but here is attributed to small velocity contrasts. This is similar to the corresponding Proterozoic

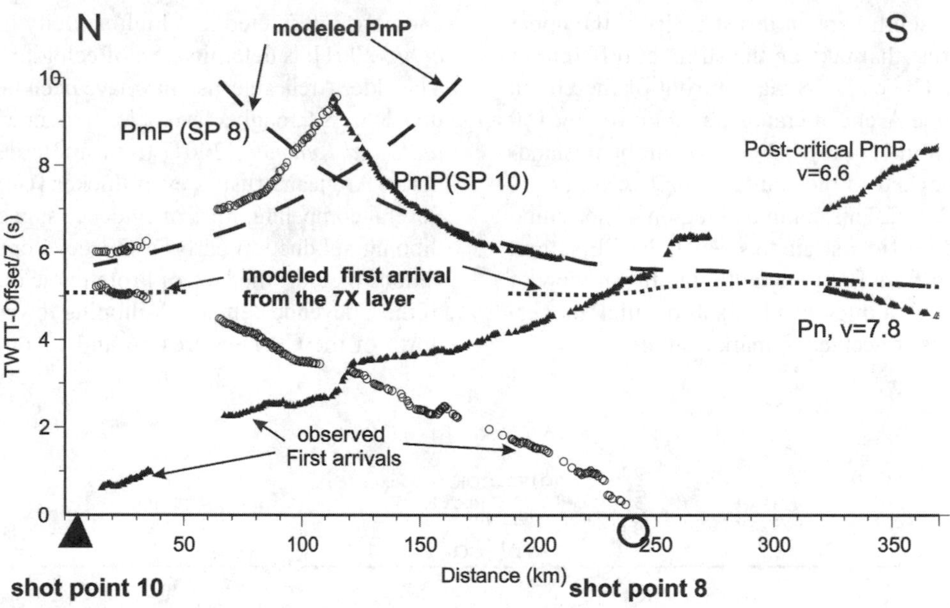

Figure 5. Travel time curves for shot points (SPs) 8 and 10 are shown with observed (triangles and circles) and calculated (dashed line) times using the RAYINVR program (Zelt and Smith, 1992). Reduction velocity is 7 km/s. We use a velocity model very close to the one used by Snelson et al., (2001) that well fits observed first arrivals from upper and middle crust up to an offset of about 180–200 km (the fit is not shown to avoid complicating the figure). This model has a high velocity lower crust of about 7.1 km/s (called 7X layer by Snelson et al., 2001). Note the absence of observed first arrivals with a velocity of about 7 km/s and thus a misfit between calculated first arrivals from a 7X layer and observed arrivals. If a layer with a velocity of about 7.1–7.2 km/s were present in this area (Snelson et al., 2001) we would have observed this as a first arrival at offsets of about 200–250 km. The presence of a high velocity layer is also not supported by the post-critical PmP event that has a velocity asymptote not greater than 6.6 km/s rather than 7 km/s. All arrivals from SP 10 are very similar to arrivals of the Deep Probe SP 43 to the south, detonated at the same location as SP 10, and that resulted in an interpretation without a high velocity lower crust just to the north of the CB suture (Snelson et al., 1998). Also note the extreme curvature of PmP reflections at near offsets from both SPs and thus a misfit between calculated and observed travel-times curves at near offsets. Uplift in Moho boundary between SPs 8 and 10 is necessary to fit the PmP arrivals at near offsets.

terrane to the northeast where no lower crustal reflections were found at vertical incidence [*Speece et al.,* 1994], but where coincident high reflectivity was found at wide angles of incidence. This may also be interpreted in terms of lenticular scatterers in the lower crust [*Levander and Holliger,* 1992].

In particular, Moho reflectivity beneath the Proterozoic Green Mountain block is very weak or not imaged at all, although an increase in average amplitude is attributed to a defuse Moho zone (Figure 3). Lack of Moho reflectivity suggests that gradient zones or layers with low contrast in acoustic impedance (or both) characterize the crust-mantle boundary beneath the Proterozoic part of the profile.

The steepening of foliation and shear zone segments, due to protracted shortening along and south of the CB suture, decreased the overall reflectivity of the arc rocks and caused further inter-wedging. The complex nature of the suture was likely conditioned by a pre-existing geometry involving listric normal faults on the rift margin.

Acknowledgments. We are grateful to R.S. Houston and K. Dueker for valuable comments and discussions of the manuscript. We thank D. B. Snyder and an anonymous reviewer for help in improving the manuscript, Dawson Geophysical company for the data acquisition, and NSF, and especially the Continental Dynamics Program (grant #9614862) that made the project possible.

REFERENCES

Allmendinger, R.W., J. A. Brewer, L. D. Brown, S. Kaufman, J. E. Oliver, and P. S. Houston, COCORP profiling across the Rocky Mountains Front in southern Wyoming, Part 2: Precambrian basement structure and its influence on Laramide deformation: *Geological Society of America Bulletin, 93*, 1253–1263, 1982.

Ball, T. T., and G. L. Farmer, Identification of 2.0 to 2.4 Ga Nd model age crustal material in the Cheyenne Belt, southeastern Wyoming: Implications for Proterozoic accretionary tectonics at the southern margin of the Wyoming craton, *Geology, 19*, 360–366, 1991.

Beaumont, C., P. Fullsack, and J. Hamilton, Styles of crustal deformation in compressional orogens caused by subduction of the underlying lithosphere, *Tectonophysics, 232*, 119–132, 1994.

Chamberlain, K. R., S. C. Patel, B. R. Frost, G. L. Snyder, Thick-skinned deformation of the Archean Wyoming Province during Proterozoic arc-continent collision, *Geology, 21*, 11, 995–998, 1993.

Chamberlain, K.R., Medicine Bow Orogeny: Timing of deformation and model of crustal structure produced during continent-arc collision, ca 1.78 Ga, southeastern Wyoming, *Rocky Mountain Geology, 33*, 259–277, 1998.

Cook, F. A., A. J. van der Velden, K. W. Hall, B. J. Roberts, Tectonic delamination and subcrustal imbrication of the Precambrian Lithosphere in northwestern Canada mapped by LITHOPROBE, *Geology 26,* 9, 839–842, 1998.

Duebendorfer, E.M., and R. S. Houston, Kinematic history of the Cheyenne belt: A Proterozoic suture in southeastern Wyoming, *Geology, 14*, 171–174, 1986.

Duebendorfer, E.M., and R. S. Houston, Proterozoic accretionary tectonics at the southern margin of the Archean Wyoming craton, *Geol. Soc., Am. Bull., 98,* 554–568, 1987.

Duebendorfer, E.M., and R. S., Houston, Structural analysis of a ductile-brittle Precambrian shear zone in the Sierra Madre, Wyoming: western extension of the Cheyenne belt, *Precambrian Research, 48,* 21–39, 1990.

Gohl, K., and S. B. Smithson, Seismic wide-angle study of accreted Proterozoic crust in southeastern Wyoming: *Earth and Planetary Science Letters, 125,* 293–305, 1994.

Gorman A. R, R. M. Clowes, R. M. Ellis, T. J. Henstock, G. D. Spence, G. R. Keller, A. Levander, C. M. Snelson, M. J. A. Burianyk, E. R. Kanasewich, I. Asudeh, Z. Hajnal, K. C. Miller, Deep Probe; imaging the roots of western North America. In: *The Lithoprobe Alberta basement transect, 39;* 3, Pages 375-398. 2002.

Henstock, T.J., A. Levander, C. M. Snelson, G. R. Keller, K. C. Miller, S. H. Harder, A. R. Gorman, R. M. Clowes, M. J. A. Burianyk, E. D. Humphreys, Probing the Archean and Proterozoic lithosphere of western North America, *GSA Today, 8,* 7, 1–5, 1998.

Hills, F.A., and R. S., Houston, Early Proterozoic tectonics of the central Rocky Mountains, North America, *Contributions to Geology, 17,* 89–109, 1979.

Houston, R.S., K. E. Karlstrom, F. A. Hills, and S. B. Smithson, The Cheyenne belt: A major Precambrian crustal boundary in the western United States, Geol. Soc. Am. Abstr. with Programs, 11, 446, 1979.

Houston, R.S., E. M. Duebendorfer, K. E. Karlstrom, and W. R. Premo, A review of the geology and structure of the Cheyenne belt and Proterozoic rocks of southern Wyoming, in Grambling, J.A., and Tewksbury B.J., ed., Proterozoic geology of the southern Rocky Mountains: *Geol. Soc. Am. Spec. Pap., 235,*1–12, 1989.

Houston, R. S., 1993, Wyoming Province, in Reed, et al., The Geology of North America: GSA, C–2, 121–170, 1993.

Houston, R. S., and P. J. Graff, Geologic map of Precambrian rocks of the Sierra Madre, Carbon County, Wyoming and Jackson and Routt Counties, Colorado: O.S. Geological Survey Map I-2452, 1995.

Johnson, R.A., K. E. Karlstrom, S. B. Smithson, and R. S. Houston, Gravity profiles across the Cheyenne belt: A Precambrian crustal suture in southern Wyoming, *Journal of Geodynamics, 1*, 445–472, 1984.

Jull M., and Kelemen P. B., On the conditions for lower crustal convective instability, *J. Geoph. Res, B, 106,* 4, 6423–6446, 2001

Karlstrom, K.E., A. J., Flurkey, and R.S Houston, Stratigraphy and depositional setting of Proterozoic metasedimentary rocks in southeastern Wyoming: record of an Early Proterozoic Atlantic-type margin, *Geological Society of America Bulletin, 97,* 257–1294, 1983.

Karlstrom, K.E., and R. S. Houston, The Cheyenne belt: analysis of a Proterozoic suture in southern Wyoming, *Precambrian Research, 25*, 415–446, 1984.

Karlstrom, K.E., and E. D. Humphreys, Persistent influence of Proterozoic accretionary boundaries in the tectonic evolution of southwestern North America: Interaction of cratonic grain and mantle modification events, *Rocky Mountain Geology, 33,* 161–179, 1998.

Levander, A. R., and K. Holliger, Small-scale heterogeneity and large-scale velocity structure of the continental crust, *J. Geoph. Res, 97,* 8797–8804, 1992

Lester A., and G. L. Farmer, 1998, Lower crustal and upper mantle xenoliths along the Cheyenne belt and vicinity, *Rocky Mountain Geology, 33,* 293–304, 1998.

Makovsky, Y., and S. L. Klemperer, Measuring the seismic properties of Tibetan bright spots, evidence for free aqueous fluids in the Tibetan middle crust, *J. Geoph. Res, B, 104,* 5, 10795–10825,1999.

Meissner, R., Rupture, creep, lamellae and crocodiles: happenings in the continental crust, *Terra Nova, 1,* 17–28, 1989.

Morozova, E. A., X. Wan, K. R. Chamberlain, S. B. Smithson, I.B. Morozov, N. Boyd, R. A. Johnson, K. E. Karlstrom, A. R. Tyson, C. T. Foster, Geometry of Proterozoic sutures in the central Rocky Mountains from seismic reflection data: Cheyenne belt and Farwell Mountain structures, *Geophys. Res. Letters*, in press, 2002.

Oxburgh, E., Flake tectonics and continental collision, *Nature, 239*, 202–204, 1972.

Pfiffner, O.A., S. Ellis, C. Beaumont, Collision tectonics in the Swiss Alps: Insight from geodynamics modeling, *Tectonics, 19,* 6, 1065–1094, 2000.

Prodehl, C., and P. W. Lipman, Crustal structure of the Rocky Mountains region, in Pakiser, L.C., and Mooney, W.R., eds, Geopysical framework of the continental United States: Boulder, Colorado, Geological Society of America Memoir 172, 249–284, 1989.

Prodehl, C., and L. C. Pakiser, Crustal structure of the Southern Rocky mountains from seismic measurements, *Geological Society of America Bulletin, 91,* 147–155, 1980.

Reed, J. C., M. E. Bickford, W. R. Premo, J. N. Aleinikoff, and J. S. Pallister, Evolution of the early Proterozoi Colorado province: constrains from U-Pb geochronology, *Geology, 15,* 861–865, 1987.

Sheehan, A.F., G. A. Abers, C. H. Jones, and A. L. Lerner-Lam, Crustal thickness variations across the Colorado Rocky Mountains from teleseismic receiver functions, *J. Geophys. Res, 100,* 20,391–20,404, 1995.

Smithson, S.B., J. A. Brewer, S. Kaufman, J. E. Oliver, and R. L. Zawislak, Complex Archean lower crustal structure reveled by COCORP crustal reflection profiling in the Wind River Range, Wyoming: *Earth and Planetary Science Letters, 46,* 295–305, 1980.

Snelson, C.M., H. Rumpel, G. K. Keller, C. Prodehl, K. C. Miller, S. H. Harder, Velocity structure along the CD-ROM'99 seismic refractionwide-angle reflection experiment, GSA Abstracts with programs, 33, 5, 2001.

Snyder, D. B., Lithospheric growth at margins of cratons, in Deep seismic probing of the continents and their margins, ed .H. Thybo, *Tectonophysics, 355,* 1–4, p. 7–22, 2002.

Snyder, D. B., H. Prasetyo, D. J. Blundell, C. J. Pigram, A. J. Barber, A. Richardson, and S. Tjokrosaproetro, A dual doubly vergent orogen in the Banda Arc continent-arc collision zone as observed on deep seismic reflection profiles, *Tectonics, 15, (1),* 34–53, 1996.

Speece, M.A., B. R. Frost, and S. B. Smithson, 1994, Precambrian basement structure and Laramide deformation revealed by seismic reflection profiling in the Laramide Mountains, Wyoming, *Tectonics, 13,* 354–366, 1994.

Templeton, M.E., and S. B. Smithson, 1994, Seismic reflection profiling of the Cheyenne belt Proterozoic suture in the Medicine Bow Mountains, southeastern Wyoming: A tie to geology, *Tectonics, 13,* 1231–1241, 1994.

Tyson,A. R, E. A. Morozova, K .E. Karlstrom, K. R. Chamberlain, S. B. Smithson, K. G. Dueker, C. T. Foster, Proterozoic Farwell Mountain-Lester Mountain suture zone, northern Colorado; subduction flip and progressive assembly of arcs *Geology (Boulder)*, 30; 10, Pages 943–946. 2002.

K. R. Chamberlain, Department of Geology and Geophysics, University of Wyoming,, Laramie, WY, 82071

R. Johnson, Department of Geosciences, University of Arizona, Tucson, AZ,

K. E. Karlstrom, Earth and Planetary Sciences, University of New Mexico, Northrop Hall, Albuquerque, NM

E. A. Morozova, Department of Geology and Geophysics, University of Wyoming,, Laramie, WY, 82071

S. B. Smithson, Department of Geology and Geophysics, University of Wyoming, Laramie, WY, 82071

X. Wan, Department of Geology and Geophysics, University of Wyoming,, Laramie, WY, 82071

Seismic Investigation of the Yavapai-Mazatzal Transition Zone and the Jemez Lineament in Northeastern New Mexico

Maria Beatrice Magnani and Alan Levander

Department of Earth Science, Rice University, Houston, Texas

Kate C. Miller and Tefera Eshete

Department of Geological Sciences, University of Texas at El Paso, El Paso, Texas

Karl E. Karlstrom

Department of Earth and Planetary Sciences, University of New Mexico, Albuquerque, New Mexico

A new seismic reflection profile of the Precambrian lithosphere under the Jemez Lineament (JL) (northeastern New Mexico, USA) shows impressive reflectivity throughout the crust. The upper crust is characterized by a 2 km thick undeformed Paleozoic and Mesozoic sedimentary sequence above the Precambrian basement. At a depth of 5–8 km, undulating reflections image a Proterozoic nappe cropping out in the nearby Rincon Range. To the south the upper crust is seismically transparent except for south dipping reflections at 2–10 km depth. The middle-lower crust, from 10–45 km depth, shows oppositely dipping reflections that converge in the deep crust (35–37 km) roughly at the center of the profile. To the north the reflectivity dips southward at 25° to a depth of 33 km before fading in the lower crust. In the southern part of the profile a crustal-scale duplex structure extends horizontally for more than 60 km. We interpret the oppositely dipping reflections as the elements of a doubly vergent suture zone that resulted from the accretion of the Mazatzal island arc to the southern margin of the Yavapai proto-craton at ~1.65–1.68 Ga. Subhorizontal high amplitude reflections at 10–15 km depth overprint all the reflections mentioned above. These reflections, the brightest in the profile, are interpreted as mafic sills. Although their age is unconstrained, we suggest that they could be either 1.1 Ga or Tertiary-aged intrusions related to the volcanic activity along the JL. We further speculate that the Proterozoic lithospheric suture provided a pathway for the basaltic magma to penetrate the crust and reach the surface.

1. INTRODUCTION

Geologic and geophysical investigations in the southwestern US suggest that Proterozoic accretionary boundaries have had a persistent influence on the tectonic and magmatic evolution of the Southern Rocky Mountain lithosphere for as long as 1.8–1.6 Ga since initial continental assembly. Here, the northeast-striking fabric established during cratonic assembly has provided, to varying degrees, a preferred path for the modification of the lithosphere during the various Phanerozoic orogenic and magmatic events that modified the western margin of the North American plate.

The Rocky Mountain Region: An Evolving Lithosphere
Geophysical Monograph Series 154

10.1029/154GM16

During cratonic assembly, tectonic elements of various scales up to lithospheric-scale were incorporated in the Laurentia super-continent and the boundaries between these elements persisted as shallowly to steeply dipping structural and chemical boundaries, some of which appear to have been reactivated during subsequent magmatism or/and tectonic deformation. Geophysical, geological and geochemical data show that these northeast-striking boundaries influenced the sedimentation throughout the Paleozoic, for example, controlling the location of mineralization and mantle-derived magmatism in the Tertiary (see Colorado Mineral Belt, Jemez Lineament as examples) [*Tweto and Sims*, 1963].

Despite the long-lived influence of some of these assembly boundaries, the details of the nature, structure, and tectonic role of these features have not been determined clearly. The southern segment of the Continental Dynamics of ROcky Mountains (CD-ROM) active source seismic experiment [*Prodhel et al.*, this volume] targeted the boundary separating the Yavapai and Mazatzal terranes, two Paleoproterozoic island arc terranes with different geochemical signatures. As discussed below, the location and nature of this boundary is still hotly debated. We have interpreted from the reflection profiles that the Yavapai-Mazatzal boundary in central New Mexico represents a doubly vergent accretionary suture and that it coincides with the modern-day Jemez Lineament, a Tertiary-Quaternary volcanic trend stretching from Arizona across New Mexico to southern Colorado. From the regional point of view the geometry of the features recognized in the seismic data provides an explanation for the elusive and diffuse character of this boundary. More generally the coincidence of the Proterozoic suture with the Tertiary volcanic centers of the Jemez Lineament suggests that collision zones play a persistent (and active) role in the evolution of the lithosphere. In the case of the Proterozoic suture studied, the weakness zone in the crust, and likely in the mantle, acted as a locus of tectonism and provided a preferred pathway for the mantle to penetrate the crust. The seismic data document a multifaceted tectonic history with crustal scale structures that have survived since continental accretion, and Tertiary igneous structures that are an expression of the processes that modify the lithosphere today.

2. THE YAVAPAI-MAZATZAL TRANSITION-ZONE

The Southern Rocky Mountains region has experienced a complex geologic history recording lithospheric assembly during Paleo-Proterozoic time (1.8–1.6 Ga), intracratonic magmatism during the Mesoproterozoic (1.44–1.35 Ga), incipient rifting (1.1–0.5 Ga), development of the ancestral Rockies during the Paleozoic (350–290 Ma) and, more recently, Laramide tectonism during the Cretaceous-Paleogene (75–45 Ma) and Tertiary extension, magmatism and uplift [see *Karlstrom and Humphreys*, 1998 for a review].

The Proterozoic continental assembly produced an area of accretion ~1200 km wide that includes major crustal provinces defined on the basis of their composition and age: the Archean Wyoming Province (with protoliths and deformation between 2.5–3.5 Ga), the Mojave Province (pre-1.8 Ga crustal material to a 1.75 Ga arc), the Yavapai Province (1.76–1.72 Ga juvenile arc crust deformed during 1.70 Ga Yavapai orogeny) and the Mazatzal province (1.7–1.6 Ga supracrustal rocks) (see Karlstrom et al., [this volume] for a review). The exposed boundaries separating these provinces have a general northeast strike but exhibit extremely different characters. Unlike the Cheyenne belt, which abruptly separates the Archean and Proterozoic crust, boundaries between the Proterozoic terranes of the Mojave and Yavapai and Yavapai and Mazatzal are diffuse and transitional zones where changes include isotopic (Pb and Nd) differences of the crust, timing of deformation (before 1.7 Ga for the Yavapai Province, between 1.66 and 1.60 Ga for the Mazatzal Province), and character of crustal xenoliths (P-T paths and rock types). The Yavapai-Mazatzal crustal boundary in the Southern Rocky Mountains is a 300 km wide region with tectonically intermixed rocks of the two provinces. The northern edge of the transition zone is defined by the northern limit of the Mazatzal-age (1.7–1.65 Ga) deformation in southern Colorado [*Shaw and Karlstrom*, 1999]. The southern margin of the transition zone coincides approximately with the Jemez Lineament in northern New Mexico.

3. THE JEMEZ LINEAMENT

The Jemez Lineament is defined on the basis of the NE-trending alignment of Tertiary-Quaternary volcanic centers [*Mayo*, 1958] of variable-width extending more than 800 km from the White Mountains-Springerville, Arizona, volcanic field at the southern margin of the Colorado Plateau, through the Jemez Mountains [*Aldrich*, 1986], to the Raton-Clayton center and into the western Great Plains of New Mexico. The volcanic rocks along the lineament range in age from 16.5 Ma to 1200 B.P. [*Gardner and Goff*, 1984] and are dominantly basaltic with a few silicic volcanoes. The alkali basalts show great petrologic variation along the lineament, attributed to distinct source zones in the mantle, different crustal contaminants and variable differentiation. Petrologic observations on the alkali basalts at the northeastern end suggest a cratonic mantle source that is distinct from the asthenospheric signature of the magmas in the southern end of the lineament [*Perry et al.*, 1987]. The volcanic activity along the Jemez lineament exhibits no time progression: volcanism began at 13.2 Ma in the middle of the lineament, at about 9.8 Ma at the southwestern end, and at 8.2 Ma at the northeastern limit.

Karlstrom and Daniel [1993] have proposed that the Jemez Lineament is a broad zone coinciding with the southern edge of the Yavapai-Mazatzal crustal boundary in northern New Mexico, as it also corresponds to a band of northeast-trending magnetic highs located south of the volcanic centers [*Zietz*, 1982] that exhibit right lateral offsets across north-striking faults.

Tomographic studies of the mantle underneath the Jemez lineament in New Mexico show a pronounced low-velocity anomaly of 1–2% within the regional 5–7% low-P-wave-velocity zone. The anomaly is 200 km long and 100 km wide and lies in the depth range of 50–160 km [*Spence and Gross*, 1990] suggesting the presence of melt in upper mantle rocks. Similar conclusions come from a *Dueker et al.*'s [2001] tomographic investigation, which also images a south-dipping low-velocity anomaly between depths of 100 and 300 km.

Tectonic activity along the Jemez Lineament is also recorded by incisions along the Canadian River in northeastern New Mexico. These geomorphic features support a Plio-Pleistocene arching and flexure along the Jemez Lineament with uplift rates of ~0.06–0.07 mm/yr [*Wisniewski and Pazzaglia*, 2002].

4. CD-ROM YAVAPAI-MAZATZAL TRANSITION-ZONE PROFILE

One of the two CDROM reflection profiles collected in 1999 [*Prodhel et al.*, this volume] was designed to investigate the crust of the Jemez Lineament and the transition zone between the Yavapai and Mazatzal Provinces. The profile extends north and south of the town of Las Vegas, New Mexico, which sits at the southern edge of the Jemez Lineament. The data were acquired with Vibroseis® sources and 1001 active recording channels on a 25 km-long seismograph array, with a 25 m group interval, and 100 m source interval. Nominally the array configuration was a 10 km/15 km split spread. Sweep frequencies ranged 8–60 Hz and sweep time was 20 s. A "listening" time of 45 s resulted in record lengths of 25 s. The profiles extend north-south for approximately 170 km, just to the east of the Front Range Fault that borders the Proterozoic outcrops of the Sangre de Cristo Mountains (Figure 1), and between the volcanic centers that define the Jemez Lineament and the band of magnetic highs. Exposures of uplifted Precambrian rocks in the Sangre de Cristo Range west of the seismic line show complex polyphase deformation and metamorphism that include several episodes of Paleoproterozoic shortening and Mesoproterozoic emplacement of plutons at middle and upper crustal depths.

4.1. Seismic Data Processing

The seismic processing sequence (Table 1) focused on improving the S/N ratio by removing organized noise. In order to account for differences in outcrops north and south of Las Vegas, we chose different filter parameters for the two segments of the southern CD-ROM profile. However the processing parameters of the two seismic segments are very similar, permitting the combination of the final migrated sections into a composite display.

The seismic data generally exhibit good S/N ratio with reflections visible at offsets up to 14 km and up to 10 s time on shot gathers. The most commonly identified noise on the data includes reverberations generated in the subsurface, surface (Rayleigh) waves, air waves, 60 Hz electrical power-line noise, and incoherent ambient or cultural noise.

After correlating the data, we resampled to 6 ms and assigned a crooked-line geometry. The common-midpoint bin size is 12.5 m. Careful trace editing and air-wave mutes preceded a frequency-wavenumber filter applied to mitigate surface wave noise. We applied prestack static elevation corrections to source and receivers using a constant replacement velocity of 3800 m/s shifting the traces to a final reference datum elevation of 2200 m.

We applied predictive deconvolution to the data in order to reduce the surface and inter-bed reverberations present in the seismic records. Trace autocorrelations indicated that the dominant period of the main reverberation was about 70 ms for the northern part of the profile (north of Las Vegas) and about 65 ms for the southern part (south of Las Vegas). Our tests showed that a prediction distance of 42 ms and 30 ms gave the best results for the northern and the southern sectors respectively, with deconvolution-operator lengths of 500 ms (north) and 200 ms (south) and a white noise level of 0.1%.

A time-variant bandpass filter (8-12-45-55 Hz at t=0-3 s, 8-12-38-42 Hz at t= 3-6 s to 8-12-32-36 Hz at t=6-20 s) attenuated power line noise and frequencies outside the bandwidth of interest. We applied an automatic gain control (AGC) function with a time gate of 1 s in order to increase the amplitude of weak reflections and to gain deeper portions of the data.

We performed velocity analysis using common velocity stacks (CVS) and common-midpoint (CMP) velocity spectrum analysis in order to determine normal move-out corrections (NMO). We used CVS to develop a preliminary stacking velocity field, which we then refined with velocity-spectrum measurements made every 600 m along the profile. Velocity estimates deteriorate below 8–10 s. Data were then NMO-corrected and stacked. On the stack section, we identified several prominent reflections then computed and applied surface-consistent residual static corrections to prestack data, followed by another velocity-spectrum analysis to further refine the stacking velocity field. The results of the processing are shown in Figure 2.

We depth migrated the stacked section to 60 km using a Kirchhoff algorithm. For depth migration we converted the final stack-

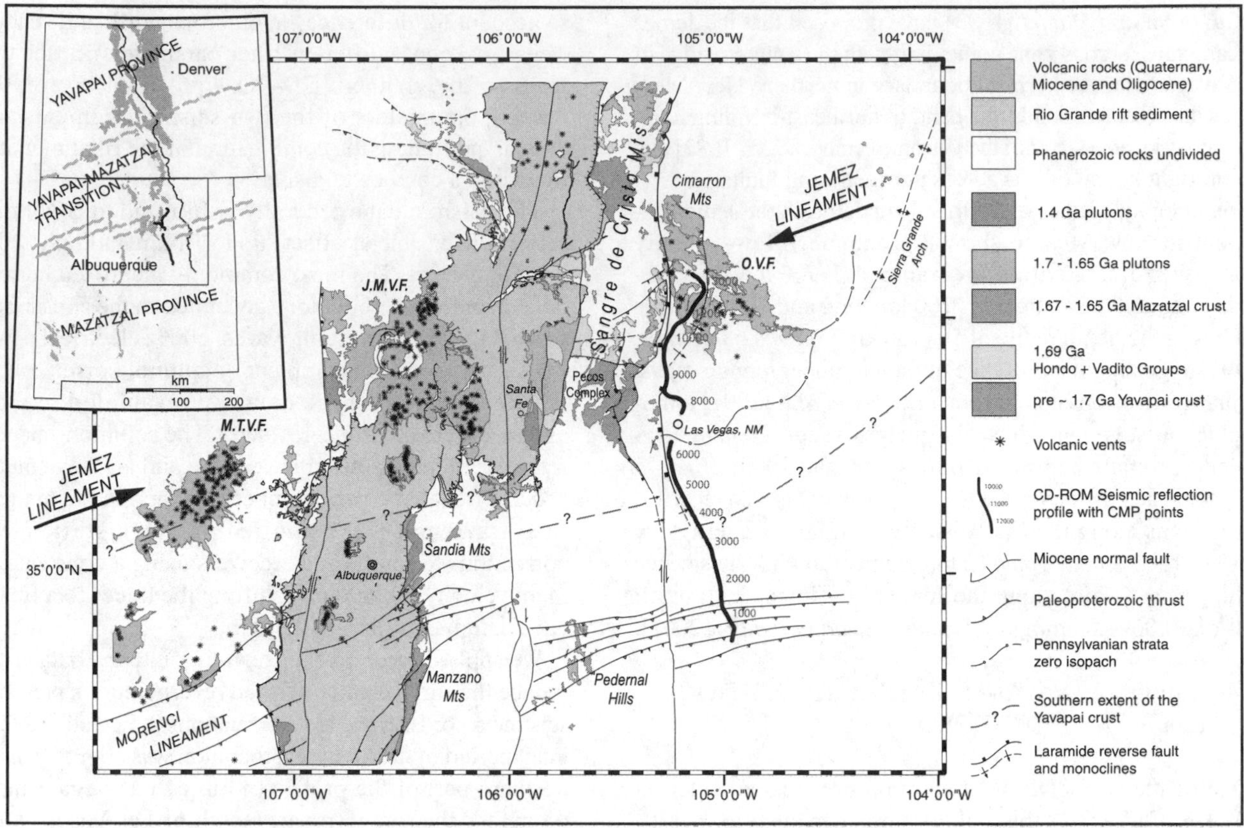

Figure 1. Location map showing the two segments of the CD-ROM seismic lines, the Jemez Lineament and the Proterozoic outcrops in the area of investigation. **M.T.V.F.-** Mount Taylor Volcanic Field, **J.M.V.F.-** Jemez Mountain Volcanic Field, **O.V.F.-** Ocate Volcanic Field.

ing velocities to interval velocities in depth. This velocity field was then smoothed and reduced in magnitude. We performed several migrations, testing different fractions (90% – 80%) of the migration velocity field. We also migrated the data with the refraction velocity model obtained from the CD-ROM seismic refraction profile recorded almost coincident with the reflection profile [*Snelson*, 2002; *Levander et al.*, this volume; *Prodhel et al.*, this volume]. Migration with the refraction velocity model led to overmigration of shallow (0–20 km depth) events while correctly migrating the deeper portion of the profile. The final velocity function resulted in the integration of both fields (velocity field determined by CMP analysis reduced by 10% from 0–35 km and refraction velocities for depths >35 km). The section was adequately migrated to 60 km, with an aperture of 20 km. For display purposes, traces were equalized along the seismic profile, amplitudes were squared and the absolute value of each sample was plotted (Figure 3).

4.2. Observations

Figure 3a is the depth-migrated profile shown to a depth of 50 km. The seismic image exhibits impressive reflectivity at upper-middle crustal depths with conflicting dips, some reflections at lower crustal depths, and a reflection-free upper mantle. This structurally complicated image is consistent with the long and complex geologic history of this area. Reflectors are identified as A-E in Figure 3b.

At shallow depth (2 km), a single high-amplitude, subhorizontal and laterally continuous event crosses the northern half of the profile (Figure 3b: Event A). Above this marker is an almost undeformed sedimentary section that gradually thins southward until it disappears south of ~CDP 8000. At the very southern end of the profile (~CDP 0–2000) the same sedimentary packages are visible again at a depth of 2 km for a distance of 25 km.

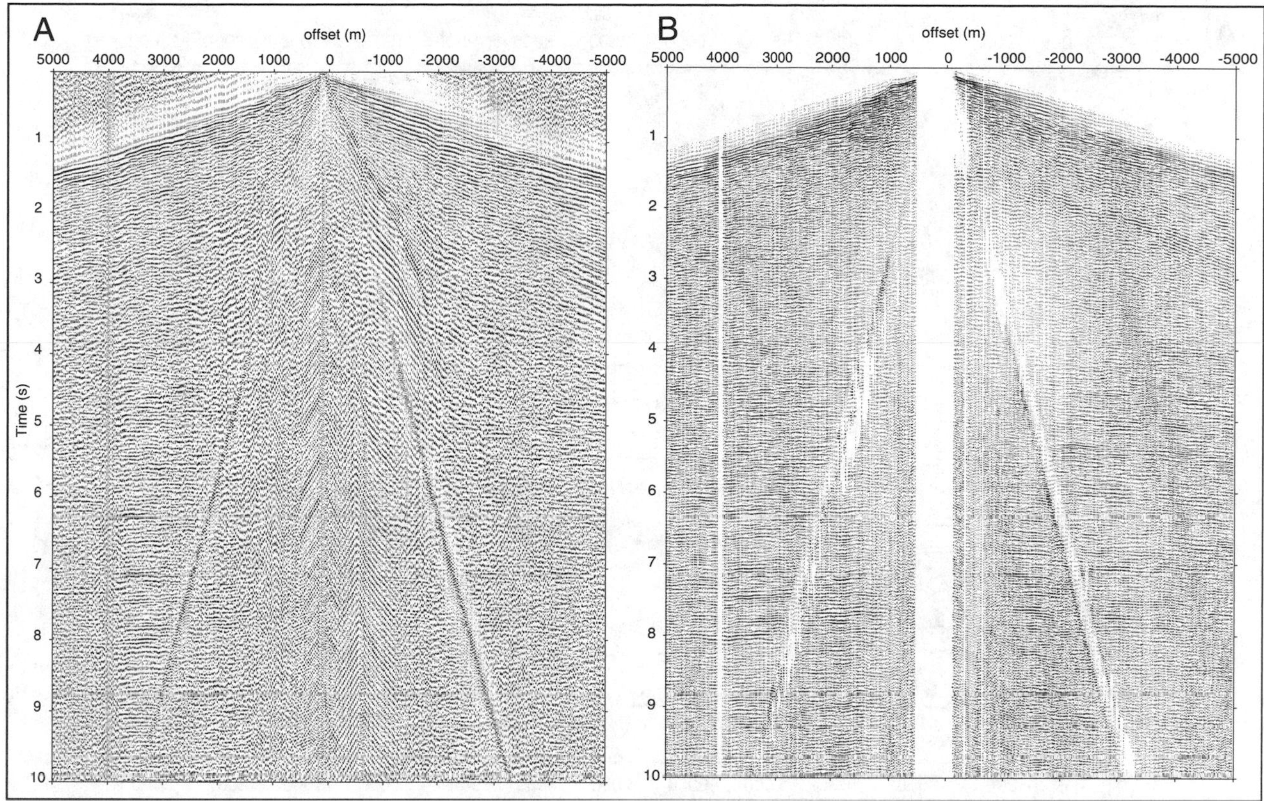

Figure 2. Comparison between the same shot gather pre a) and post b) seismic processing. Air-wave as well as low frequency surface-wave noise is removed from the processed data and reflections emerge throughout the middle and lower crust.

In the north beneath the discontinuity (A) and extending to 8 km depth are other continuous strong reflections that depict a gently to strongly folded and faulted structure (B1, CDP 9500–13500). Figure 4 shows the northern end of this feature and the overlying unconformity (A). To the south, this reflectivity dies out into a featureless region south of CMP 9500. South of Las Vegas, NM (~CDP 7000, Figure 3), the upper crust appears mostly transparent except for three south dipping reflectors at 3–10 km depth (E, Figure 3) at the southern end of the profile (CDP 0–2500).

At 8–14 km depth is a system of bright nearly continuous reflections (C, Figure 3) that form an arch extending more than 100 km with an apex beneath Las Vegas (CDP 7000). In the north these bright reflections are visible at CDP 10000–12500 at 14–25 km depth, where they overprint an 8 km-thick band of south-dipping reflectors (D1). The D1 reflectors extend from 10 km depth at the northern edge of the profile, to ~33 km depth near CDP 11000, a distance of about 34 km. No additional reflectivity is observed at lower and middle crustal depths in the north. To the south, the entire middle and lower crust is occupied by continuous north-dipping bright reflections (D2) that delineate what we interpret to be a crustal duplex. The structure has a maximum thickness of 27–30 km and extends more than 50 km (CDP 1500–6800), thinning to 10 km at about 33 km depth at CDP 6800. South of the crustal duplex, another set of north-dipping reflectors enters the profile at 25 km depth and is traceable down to a depth of 43 km at CDP 2000 (D3).

Despite careful processing and analysis (see discussion below) no clear and continuous reflections from the Moho discontinuity were detected. To the south, the reflectivity dies out uniformly at about 40–45 km whereas to the north only scattered reflectors can be traced below 33 km depth.

4.3. Interpretation

The interpretation (Figure 3b) of reflectors A and B is based on borehole data, surface outcrops and projections from geologic maps [*Baltz*, 1999]. Reflector A can be interpreted with certainty as the Great Unconformity, above which is an almost undeformed section of Phanerozoic sedimentary sequences [*Powell*, 1876]. The Great Unconformity and the overlying sedimentary rocks exhibit almost no deformation for a distance of 90 km except for the southward thinning of the Phanerozoic

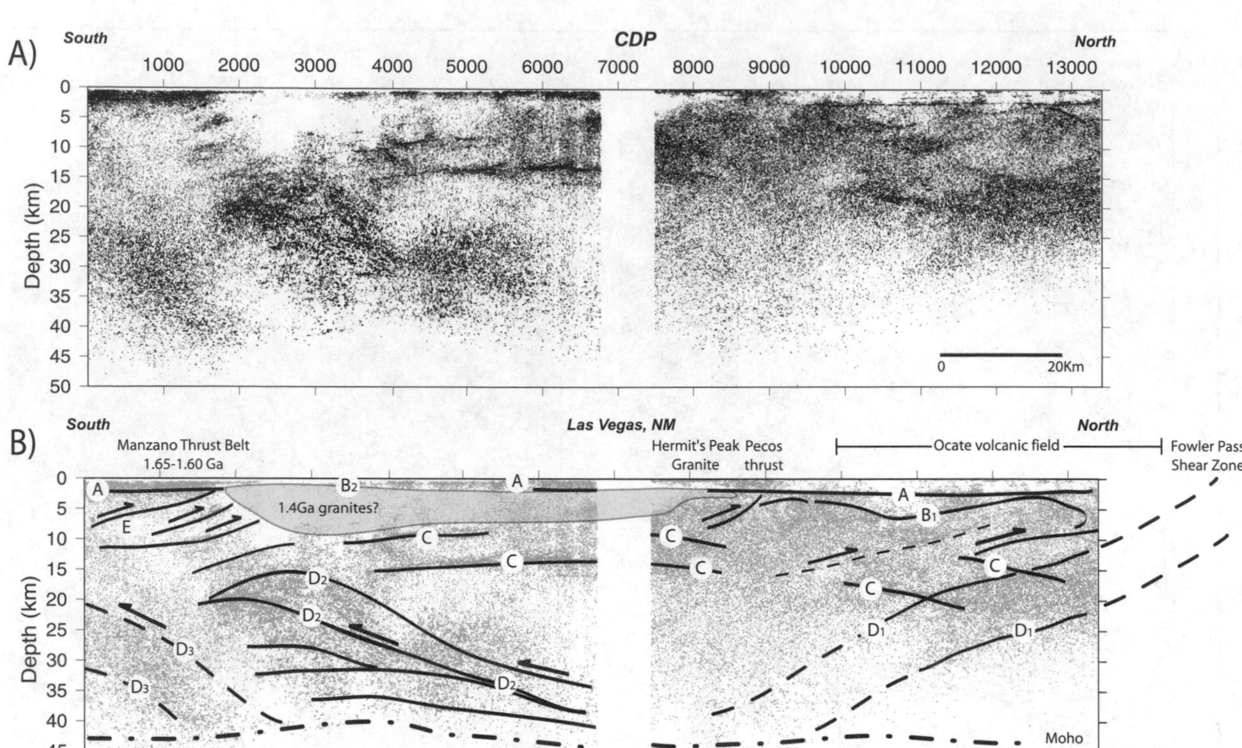

Figure 3. a) Seismic profile migrated displayed to a depth of 50 km; b) Interpretation. A) Great Unconformity; B1) Allochtonous Proterozoic supracrustal rocks; B2) top of the "granitic" crust; C) Proterozoic (1.1 Ga) or Tertiary mafic intrusions; D1, D2, D3) Proterozoic doubly vergent suture associated with the arc-continent collision during the 1.68–1.65 Ga Mazatzal orogeny; E) Manzano thrust belt (1.60 Ga).

sequences due to the uplift associated with the Sierra Grande arch. The uniformity of the Great Unconformity and overlying sediments shows that little deformation occurred this far east after the Pennsylvanian-Mississippian. The absence of shallow, well-organized reflections between CDP 1800–5000 is partially due to late Paleozoic high-angle faults that dissect the sedimentary cover and the underlying Precambrian rocks [*Broadhead and King*, 1988], obliterating the seismic signal. The base of the shallow reflectivity (B2) is interpreted as the contact between the Precambrian basement and the inferred granitic crust (see below).

We interpret the undulating horizon and overturned fold (B1), imaged just below the Great Unconformity at a depth of 3–5 km, as the subsurface expression of Proterozoic nappes that were uplifted by Laramide thrusts in the Rincon Range (southern Sangre de Cristo Mountains) to the west (Figure 1). Exposed in these ranges is the lower limb and part of the hinge of a north-facing fold-nappe that is imbricated by several south-dipping, top-to-the-north ductile thrusts, and that is cored by the 1.68 Ga Guadalupita granitic gneiss [*Riese*, 1969; *Read et al.*, 1999]. The granite, exposed at the Guadalupita and El Oro domes, intruded the supracrustal rocks and deformed synchronously with the overturned fold [*Read et al.*, 1999]. The seismic image (Figures 3 and 4) exhibits remarkable correspondence with the surface geology, showing clearly the reflections that compose the whole nappe on the northern end of the profile (CDP ~9000–13700), with the overturned limb defined by a 1200 m-thick quartzite layer (Ortega Formation), a regional stratigraphic marker in the 1.69 Ga Hondo Group. The section exposed in outcrops shows that the bedding of the recumbent fold is pervasively deformed in tight parasitic folds and suggests that the structure imaged by the seismic section is, in fact, the envelope of these folds. This horizon is cut by several Precambrian age thrust faults that sole at depths of 7–8 km. The seismically transparent crust above the Ortega Formation is likely a result of a dominant vertical foliation that overprints and obscures any previous geometry, producing a transparent or diffractive seismic response [*Levander et al.*, 1994]. Geobarometry and geothermometry data [*Read et al.*, 1999] indicate that the nappe formed at a paleodepth of

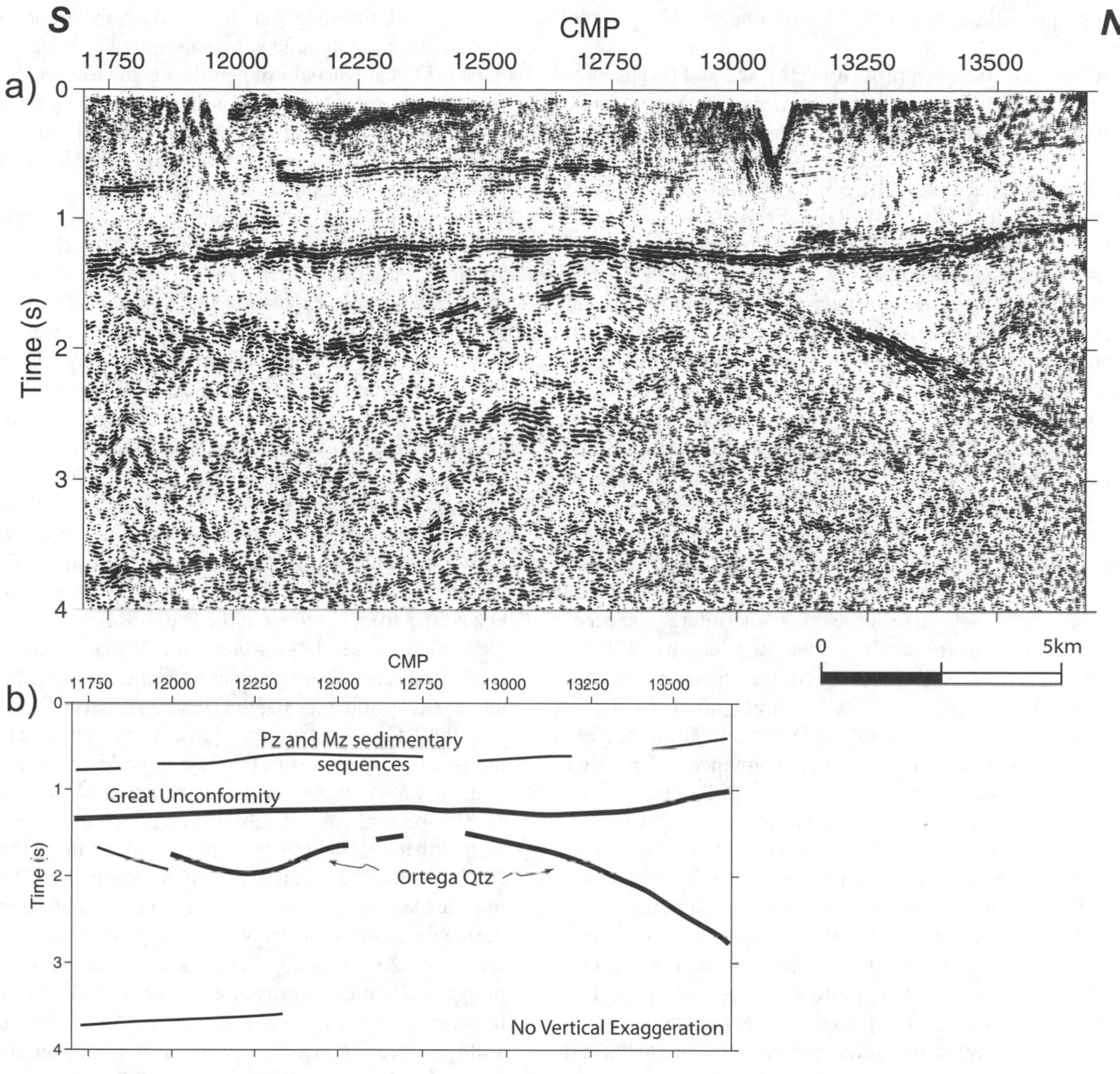

Figure 4. a) Detail of the nappe structure; b) Interpretation. The nappe is outlined by the ~1200 m-thick cross-bedded nearly pure quartzites of the Ortega Formation. The Proterozoic nappe shows a remarkable correspondence with the compressive structure outcropping at the Rincon Range, to the west. The Pennsylvanian-Mississippian Great Unconformity separates the Paleozoic and Mesozoic sedimentary rocks of the Las Vegas basin from the Proterozoic basement rocks.

12–18 km (4–6 kbar pressure and 650 °C), making it a mid-crustal, likely brittle-ductile transition zone structure.

Where the reflectivity loses coherence south of CDP 9000 (Figure 3), the south-dipping Pecos Thrust ductile shear zone crops out above the ~1.4 Ga Hermit's Peak granite [*Read et al.*, 1999]. Due to the lack of continuous Proterozoic outcrops south of Las Vegas, the extent of this magmatic body is unknown. However we suggest that the transparent upper crust imaged by the seismic profile from CDP 2000 to CDP 8500 is made up of granitic bodies that intruded the crust during several Proterozoic tectonic events. Granitic plutons, 1.4 Ga in age, are exposed in numerous locations in New Mexico and particularly in the Sandia and Manzano Mountains (Priest and Sandia plutons), 120 km west of the seismic profile. Although a 1.6 Ga age cannot be ruled out, we speculate that the granitic bodies were emplaced during the 1.4 Ga "anorogenic" magmatic event [*Karlstrom and Humphreys*, 1998].

Since none of the deep reflectors (D1, D2 and D3) reaches the surface within the length of the seismic profile, or can be unambiguously associated with any nearby outcropping structure, the interpretation of the deeper portion of the profile is based on the island arc collision model for growth of the southwest U.S. [*Hamilton*, 1981; *Bowring and Karlstrom*, 1990] and mechanical models for development of doubly-vergent compressional orogens [e.g. *Willet et al.*, 1993]. The southward-dipping reflection package (D1) and the crustal duplex structure (D2) define a doubly-vergent compressional orogen that we speculate formed during the 1.65 Ga Paleoproterozoic collision of the Mazatzal island arc with the Yavapai proto-continent, synchronous with the mid to upper crustal nappe structure (events B1) lying above its northern half. The main elements of the orogen, D1 and D2, are interpreted respectively as a north-verging, south-dipping shear zone system associated with the thrusting of Mazatzal supracrustal rocks above the Yavapai basement and as a north-dipping south-verging duplex structure that accommodates the back thrusting of the Mazatzal tectonic units. The suture zone occupies the present thickness of the crust for a length of 170 km and represents the crustal expression of the lithospheric boundary between the Yavapai and the Mazatzal provinces. Rocks above the suture are tectonically and magmatically mixed elements of both the Yavapai and Mazatzal provinces, whereas we speculate that rocks below (D1) are Yavapai basement. Such an interpretation is supported by exposures of the Fowler Pass shear zone in the Cimarron Mountains, north of the profile, which may be the projection to the surface of D1 reflectivity.

The Fowler Pass shear zone is considered a fundamental Proterozoic crustal boundary as it juxtaposes profoundly different rock types characterized by diverse deformation histories: high-grade quartzites and syenites (interpreted as derived from a supracrustal passive-margin environment) to the south, and greenschist-grade calc-alkaline arc rocks and greenstones (associated with a convergent margin environment) to the north [*Grambling et al.*, 1993]. The shear zone has a long history of reactivation (*Carrick*, 2002) and, although no constraints on activity before 1.4 Ga exist, the oldest fabric on both sides of the fault is ca. 1.7 Ga (Andronicos, pers. comm). According to surface geology, rocks above D1 are mixed metasedimentary, meta-volcanic and meta-plutonic rocks related to both the Mazatzal and the Yavapai blocks, tectonized during large-scale thrusting of an island arc onto the southern edge of the proto-North American continent. To the south, the duplex structure is probably formed by Mazatzal basement rocks thrust to the south to accommodate shortening associated with the arc-continent collision. The north-dipping reflectivity labeled D3 could represent the deeper portion of an outer back thrusting system. At a much shallower depth, the three south-dipping reflectors (E) south of CDP 2500, underlying the Great Unconformity reflector and truncated by it, are interpreted as the eastward continuation of the 1.65–1.60 Ga Manzano Thrust Belt [*Thompson et al.*, 1991], exposed at the Manzano Mountains and Pedernal Hills, 120 km and 60 km west of the seismic profile respectively.

The interpretation of the complex set of bright reflections (C on Figure 3) relies mainly on their seismic character (high amplitude, continuity, depth range, etc.) and on observations from both surface exposures and well data. The reflections stand out clearly on individual field records as well as on the depth-migrated section, with amplitudes of 9–10 dB above the background reflectivity (Figure 6). We interpret them as mafic intrusions, based also on the evidence of many other seismic profiles around the world that imaged reflectors with similar attributes [e. g. *Goodwin et al.*, 1989; *Ross and Eaton*, 1997; *Papasikas and Juhlin*, 1997]. The age of these intrusions is more difficult to constrain and we propose two alternative possibilities: Mesoproterozoic or Tertiary. Evidence for a major mafic magmatic episode in New Mexico of 1.1 Ga age comes from outcrops in the Pecos Mafic Intrusive Complex [*Keller et al.*, 1989; *Adams and Miller*, 1995] and from well data where a 1.09 Ga gabbro intruded metasedimentary and metavolcanic rocks of the Debaca sequence [*Amarante et al.*, 2000]. On the other hand, Tertiary to recent basaltic magmatism is widespread in this region and is both associated with the Rio Grande rift and with the Jemez Lineament [*Baltz*, 1999; *Green and Jones*, 1997]. Although the Rio Grande rift magmatism has not affected the area where the seismic profile is located, the northern part of the seismic profile crosses into the Quaternary Ocate volcanic field, one of the centers that define the Jemez Lineament. The composition of the exposed lavas ranges from alkali olivine basalts to dacites 8.0–0.3 my in age and indicates crustal contamination by mixing of a basaltic melt with a crustal granodiorite [*Nielsen and Dungan*, 1985] suggesting that these magmas have ponded at several crustal levels during their ascent through the crust. We speculate that they have used preexisting zones of weakness or faults of Precambrian or/and Laramide age as conduits from the mantle. These observations indicate that the complex, layered, bright reflections could represent the intrusive counterpart of the extrusive rocks associated with the Jemez Lineament. Due to the very recent age of the magmatic activity, it is also possible that the bright reflections correspond to molten magma.

Lower crustal reflectivity changes along the profile and fades out at about 40–43 km south of Las Vegas. To the north, the reflective packages terminate shallower at about 33–35 km and only intermittent reflectivity is observed in the lower crust. No reflections from the Moho are detected along the pro-

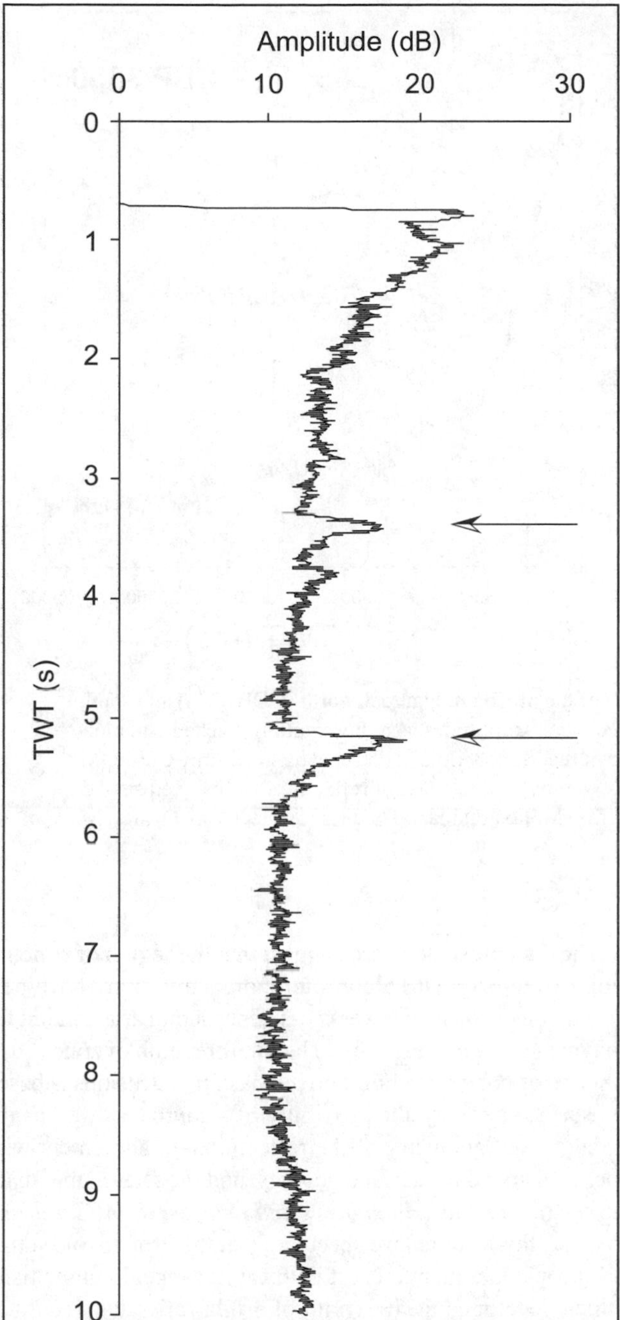

Figure 6. Amplitude plot of the bright reflections (C) roughly at CDP 5200. Amplitude decay curve is averaged over several traces of a shot gather. Arrows indicate the locations of the bright reflections where the amplitudes are 9–10 dB above the background reflectivity. The slight increase of amplitude strength at the bottom of the curve (~9 s) is due to the reflectivity of the duplex structure (reflectors labeled D2 in Figure 3).

file. CD-ROM seismic refraction data, collected parallel to the reflection profile at a distance of 50 km, show a 43–45 km thick crust [*Snelson*, 2002; *Levander et al.*, this volume], in good agreement with the observation of the termination of the lower crustal reflectivity on the reflection profile.

The absence of continuous Moho reflections was investigated through analysis of amplitude and frequency decay patterns [*Barnes*, 1994]. Analyses were performed on shot gathers with minimal processing. Decay plots of the southern and the northern segment of the profile show that amplitude values are above the noise levels down to 17 s time to the north and down to 18 s to the south, after which no further decay is observed (Figure 5 a and b). Coincident refraction data detect the Moho at about 13.5 s (TWT) along the reflection profile, the depth at which the deepest reflectivity is observed on the stacked section. Moho reflections are also visible at short offsets in the refraction data at frequencies of 1–6 Hz, well below the frequency range of the reflection data. The analysis therefore shows that signal penetration is adequate to image the Moho and upper mantle reflectivity and that the lack of strongly reflective zones or reflections at these depths is probably a genuine geologic feature.

5. DISCUSSION

Interpretation of reflector (A), the Great Unconformity, and (B1), a remarkable Precambrian nappe structure, are well correlated with local and regional geology. The Great Unconformity is a province-wide feature well documented in local stratigraphic studies [*Baltz*, 1999]. The nappe is an accretionary structure in the paleo-middle crust that we discuss further below.

The interpretation of deeper portion of the profile (i.e. reflection package (D), the Precambrian suture) is based on the accretionary model for the continents [*Hamilton*, 1981] and geodynamic models of bivergent orogens [*Willet et al.*, 1993]. The interpretation of the almost flat bright events (C) as the Proterozoic or Tertiary-modern magmatic features is based on circumstantial geologic evidence and the reflection character.

The accretionary model for the lithospheric growth of the southwestern US invokes the southward amalgamation of island-arc lithospheric fragments to the craton along a northeast-striking continental margin [*Karlstrom and Bowring*, 1990]. In this setting we interpret reflectors (D1 and D2) and the overlying nappe structure (B) as part of a lithospheric suture that resulted from the collision of the Mazatzal island arc and the Yavapai Province during the Paleoproterozoic. The north-verging structures (reflectors B and D1) and the south-verging crustal duplex (D2 reflectors) define the doubly vergent orogen that identifies at depth the northeast-trending boundary between the Yavapai and Mazatzal Provinces as defined by *Karlstrom and Bowring*, [1988]. The nappe

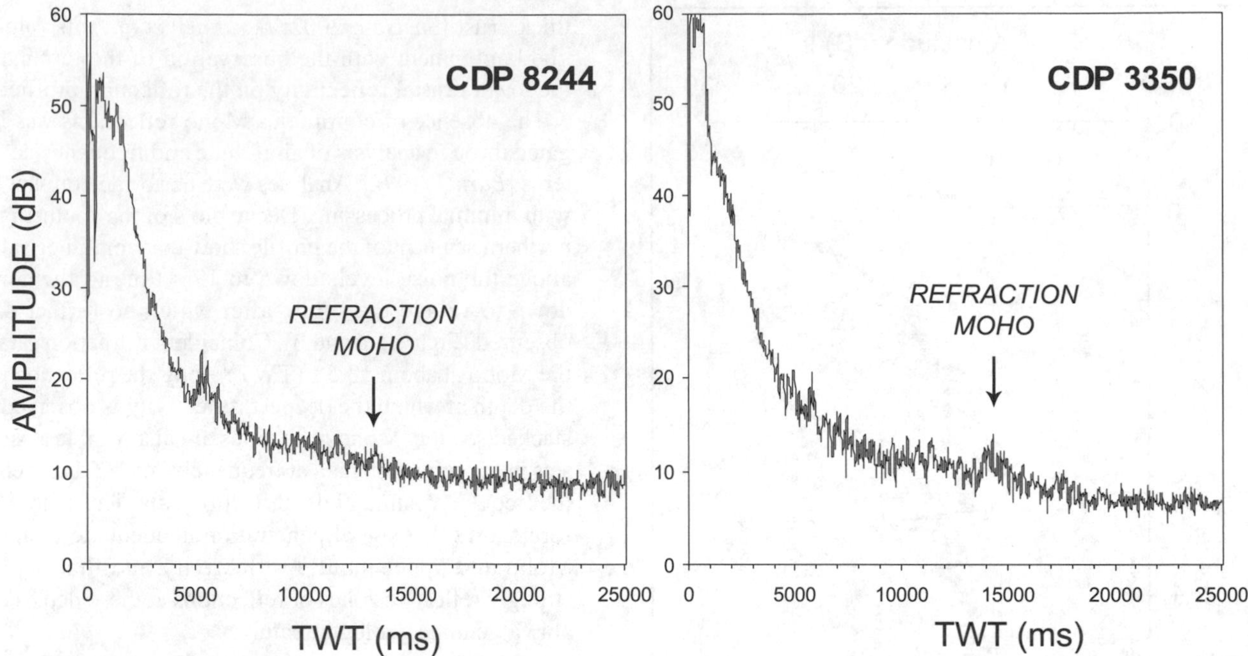

Figure 5. Amplitude-decay curves of two different CDP locations of the profile in Figure 3, north (CDP 8244) and south (CDP 3350) of Las Vegas. No processing was applied on analyzed data. Amplitudes were horizontally stacked and plotted. Amplitudes appear to flatten at about 17 s, well below the time of the Moho (~13.5 s) detected by the nearby CD-ROM refraction profile [*Levander et al.,* this volume; *Snelson et al.,* this volume]. An increase of reflectivity is observed at 14.5 s in the southern segment (CDP 3350) corresponding to the refraction Moho boundary. The bright reflections (C) are visible at ~5 s on CDP 8244.

structure, formed at paleo-midcrustal levels, deformed in the brittle-plastic region, and was probably at least partly decoupled from the deeper suture structures (D1) by a series of thoroughgoing thrust faults that deepen beneath the nappe into broader shear zones with decreasing seismic signature as the rheological regime becomes more ductile. Surface outcrops suggest that the uppermost part of the suture (above D1 and D2) is composed of intermixed metasedimentary, metavolcanic and metaplutonic rocks of both the Yavapai and Mazatzal terranes that were pervasively deformed in the ductile regime during the Mazatzal orogeny [*Conway and Silver*, 1989]. The geometry of the low angle accretionary structures that propagate throughout the crust provides an explanation for the diffuse character of the Mazatzal-Yavapai boundary.

The seismic fabric of the imaged Proterozoic suture resembles in scale and geometry that of younger doubly vergent orogens like the Alps. Similarly to the Alpine orogeny we infer that the collision between the Mazatzal arc and the Yavapai continent followed the subduction of the Yavapai passive margin and that the bivergent geometry was achieved through the continuous convergence after the collision [*Schmid et al.,* 1996].

The data presented here suggest that the region of penetration of the original Paleoproterozoic suture may have persisted as a crust-mantle weakness zone, acting as a conduit for magma ascent into the crust. The interpretation of zone (C) as a series of ponded mantle-derived basaltic intrusions is based on surface geology, the proximity of a source, as well as the strength and geometry of the reflections. In agreement with these seismic data, surface geology, and the teleseismic mantle seismic results [*Humphreys and Dueker*, 1994; *Zurek and Dueker*, this volume] we speculate that the Proterozoic suture has provided a mantle-crust pathway for recent magmatism, which developed the overprint of bright reflectivity (C).

6. CONCLUSIONS

The seismic data examined here cross the Cenozoic Jemez Lineament in northern New Mexico and show two Proterozoic-age crustal structures, forming what we interpret as part of a major bivergent accretionary suture, that have largely survived subsequent tectonism. The Proterozoic suture marks the boundary between the Yavapai crust below, and tectonically mixed Yavapai-Mazatzal crust of the transition region above.

We interpret the overprinting reflections as basaltic intrusions, spatially coincident with the extrusive rocks of the Tertiary-Quaternary Jemez Lineament volcanic centers and with the bivergent orogen, suggesting that the suture has persisted in the crust as a zone of weakness for magma penetration until the present. Teleseismic results [*Dueker et al.*, 2001] show low velocities in the upper mantle to 120 km depth suggesting that the suture persists into the mantle, providing a source for basaltic magma generation.

Acknowledgments. The authors would like to thank Dawson Geophysical for the acquisition of the data, Christopher Andronicos and the CD-ROM working group for the fruitful discussions. This research was supported by the National Science Foundation grants EAR 9614777, EAR 003539, EAR 0208020 to Rice University; EAR-9614269, EAR-0003578, and EAR-0207794 to the University of Texas at El Paso.

REFERENCES

Adams, D. C., and Miller, K. C., Evidence for late middle Proterozoic extension in the Precambrian basement beneath the Permian basin, *Tectonics*, 14, 1263–1272, 1995.

Aldrich, M.J., Jr., Tectonics of the Jemez Lineament in the Jemez Mountains and Rio Grande Rift: *J. of Geophys. Res.*, 91, 1753–1762, 1986.

Amarante, J.F.A. and Kelley, S. A., Miller, K. C., Characterization of the Proterozoic Basement in Mescalero #1, Guadalupe County, New Mexico, AAPG Rocky Mountain Section Meeting, *Am. Ass. Pet. Geol. Bull.*, 84, 1235, 2000.

Baltz, E.H., Map of the bedrock geology and cross sections of southeastern Sangre de Cristo Mountains and western part of Las Vegas Basin, in *Stratigraphic framework of upper Paleozoic rocks, southeastern Sangre de Cristo Mountains, New Mexico, with a section on speculations and implications for regional interpretation of Ancestral Rocky Mountains paleotectonics,* New Mexico Institute of Mining and Technology, Memoir 38, scale 1:125,000, 1999.

Barnes, A.E., Moho reflectivity and seismic signal penetration, *Tectonophysics*, 232, 299–307, 1994.

Bowring, S.A. and Karlstrom, K.E., Growth, stabilization, and reactivation of Proterozoic lithosphere in the Southwestern United States, *Geology*, 18, 12, 1203–1206, 1990.

Broadhead and King, Petroleum geology of Pennsylvanian and Lower Permian strata, Tucumcari Basin, east-central New Mexico, *New Mexico Bureau of Mines and Mineral resources*, Bulletin 119, 1988.

Brown, L.D., Krumhansl, P.A., Chapin, C.E., Sanford, A.R., Cook, F.A., Kaufman, S., Oliver, J.E., and Schilt, F.S., COCORP seismic reflection studies of the Rio Grande rift, in *Rio Grande rift: Tectonics and magmatism*, edited by Riecker, R.E., Washington, D.C., American Geophysical Union, 169–174, 1979.

Carrick, T.L., Proterozoic Tectonic Evolution of the Cimarron Mountains, Northern New Mexico, M.S. Thesis, University of Texas, El Paso, 76, 2002.

Christensen, N.I. and Mooney, W.D., Seismic velocity structure and composition of the continental crust: A global view, *J. of Geophys. Res.,* 100, 7, 9761–9788, 1995.

Conway, C.M. and Silver, L.T., Early Proterozoic rocks of 1710–1610Ma age in central to southwestern Arizona, in *Geologic evolution of Arizona* edited by Jennings, J. and Reynolds, S., Arizona Geological Society Digest 17, 501–540, 1989.

Dueker, K., Yuan, H. and Zurek, B., Thick-structured Proterozoic lithosphere of the Rocky Mountain region, *GSA Today*, 11, 2, 4–9, 2001.

Gardner, J.N. and Goff, F., Potassium-argon dates from the Jemez volcanic field: Implications for tectonic activity in the north-central Rio Grande rift, Field Conf. Guidebook, *New Mexico Geol. Soc.*, 35, 75–81, 1984.

Goodwin, E. B., Thompson, G., Okaya, D. A., Seismic identification of basement reflectors; the Bagdad reflection sequence in the Basin and Range Province-Colorado Plateau transition zone, Arizona, *Tectonics*, 8, 821–831, 1989.

Grambling, J.A. and Dallmeyer, R.D., Tectonic evolution of Proterozoic rocks in the Cimarron Mountains, northern New Mexico, USA, *J. of Metamorphic Geol.*, 11, 739–755, 1993.

Green, G.N. and Jones, G.E., The digital geologic map of New Mexico in ARC/INFO format, *Open File Report OF-97–52*, United States Geological Survey, 1997.

Hamilton, W., Crustal evolution by arc magmatism, *Royal Society of London Philosophical Transactions,* ser. A, 301, 279–291, 1981.

Humphreys, E. and Dueker, K., Physical state of the Western U.S. upper mantle, *J. of Geophys. Res.*, 99, 5, 9635–9650, 1994.

Jarchow, C.M., Thompson, G.A., Catchings, R.D. and Mooney, W.D., Seismic evidence for active magmatic underplating beneath the Basin and Range province, western United States, *J. of Geophys. Res.,* 98, 22095–22108, 1993.

Jones, T D. and Nur, A., The nature of seismic reflections from deep crustal fault zones, *J. of Geophys. Res.*, 89, 5, 3153–3171, 1984.

Karlstrom, K.E. and Bowring, S.A., Early Proterozoic assembly of tectonostratigraphic terranes in southwestern North America, *J. of Geology*, 96, 561–576, 1988.

Karlstrom, K.E. and Daniel, C.G., Restoration of Laramide right–lateral strike slip in northern New Mexico by using Proterozoic piercing points: Tectonic implications from the Proterozoic to the Cenozoic, *Geology*, 21, 1139–1142, 1993.

Karlstrom, K.E. and Humphreys, E.D., Persistent influence of Proterozoic accretionary boundaries in the tectonic evolution of southwestern North America: interaction of cratonic grain and mantle modification events, *Rocky Mountain Geology*, 33, 161–180, 1998.

Keller, G. R., Hills, J. M., Baker, M. R., Wallin, E. T., Geophysical and geochronological constraints on the extent and age of mafic intrusions in the basement of West Texas and eastern New Mexico, *Geology,* 17, 1049–1052, 1989.

Levander, A., Hobbs, R.W., Smith, S.K., England, R.W., Snyder, D.B. and Holliger, K., The crust as a heterogeneous "optical" medium, or "crocodiles in the mist", *Tectonophysics*, 232, 1–4, 281–297, 1994

Levander, A., Zelt, C. and Magnani, M.B., Crust and upper mantle velocity structure of the Southern Rocky Mountains from the Jemez Lineament to the Cheyenne Belt, *This Volume*.

Mayo, E.B., Lineament tectonics and some ore districts of the southwest, *Trans. Am. Inst. Min. Metall. Pet. Eng.*, 211, 1169–1157, 1958.

Nielsen, R.L. and Dungan, M.A., The petrology and geochemistry of the Ocate volcanic field, north-central New Mexico, *Geol. Soc. of Am. Bull.*, 96, 296–312, 1985.

Papasikas, N. and Juhlin, C., Interpretation of reflections from the central part of the Siljan Ring impact structure based on results from the Stenberg-1 borehole, *Tectonophysics*, 269, 237–245, 1997.

Perry, F.V., Baldridge, W.S. and De Paolo, D.J., Role of astenosphere and lithosphere in the genesis of Late Cenozoic basaltic rocks from the Rio Grande rift and adjacent regions of the southwestern United States, *J. of Geophys. Res.*, 92, 9193–9213, 1987.

Powell, J.W., Exploration of the Colorado River of the West, *Smithsonian Institution*, 291pp., 1876.

Prodhel, C., Johnson, R.A., Keller, G.R., Snelson, C. and Rumpel, H.-M., Background and overview of previous controlled source seismic studies, *This Volume*.

Read, A.S., Karlstrom, K.E., Grambling J.A., Bowring, S.A., Heizler, M. and Daniel, C., A middle-crustal cross section from the Rincon Range, northern New Mexico: Evidence for 1.68 Ga, pluton-influenced tectonism and 1.4Ga regional metamorphism, *Rocky Mountain Geology*, 34, 67–91, 1999.

Riese, R.W., Precambrian geology of the southern part of the Rincon Range [M.S. Thesis]: Socorro, New Mexico Institute of Mining and Technology, 103, 1969.

Ross, G.M and Eaton, D.W., Winagami reflection sequence: seismic evidence for post-collisional magmatism in the Proterozoic of Western Canada, *Geology*, 25, 3, 199–202, 1997.

Schmid, S.M., Pfiffner, O.A., Froitzheim, N., Schoenborn, G. and Kissling, E., Geophysical-geological transect and tectonic evolution of the Swiss-Italian Alps, *Tectonics*, 15, 1036–1064, 1996.

Snelson, C. M., Investigating seismic hazards in the western Washington and crustal growth in the Rocky Mountains, [Ph.D. Thesis], University of Texas, El Paso, 248pp., 2002.

Shaw, C. and Karlstrom, K.E., The Yavapai-Mazatzal crustal boundary in the Southern Rocky Mountains, *Rocky Mountains Geology*, 34, 37–52, 1999.

Spence, W. and Gross, R.S., A Tomographic glimpse of the upper mantle source of magmas of the Jemez Lineament, New Mexico, *J. of Geophys. Res*., 95, 10,829–10,849, 1990.

Thompson, A.G., Grambling, J.A. and Dallmeyer, R.D., Proterozoic tectonic history of the Manzano Mountains, central New Mexico, *Open Report 137,* New Mexico Bureau of Mines and Mineral Resources Bulletin, 71–77, 1991.

Tweto, O. and Sims, P.K., Precambrian ancestry of the Colorado mineral belt, *Geol. Soc. of Am. Bull.*, 74, 8, 991–1094, 1963.

Willett, S., Beaumont, C. and Fullsack, P., Mechanical model for the tectonics of doubly vergent compressional orogens, *Geology*, 21, 4, 371–374, 1993.

Wisniewski, P.A. and Pazzaglia, F., Epeirogenic controls on Canadian River incision and landscape evolution, Great Plains of northeastern New Mexico, *J. of Geology*, 110, 4, 437–456, 2002.

Zietz, I., Composite magnetic anomaly map of the United States: U.S. Geological Survey Geophysical Investigation Map GP-954-A, scale 1:2,500,000, 1982.

Zurek, B. and Dueker, K., Lithospheric stratigraphy beneath the Southern Rocky Mountains, This Volume.

Tefera Eshete, Department of Geological Sciences, University of Texas at El Paso, 500 W. University Ave, El Paso, TX, 79902.

Karl E. Karlstrom, Department of Earth and Planetary Sciences, University of New Mexico, 200 Yale Blv. NE, Albuquerque, NM 87131.

Alan Levander, Department of Earth Science, MS 126, Rice University, 6100 Main St., Houston, TX, 77005.

Maria Beatrice Magnani, Department of Earth Science, MS 126, Rice University, 6100 Main St., Houston, TX, 77005.

Kate C. Miller, Department of Geological Sciences, University of Texas at El Paso, 500 W. University Ave, El Paso, TX, 79902.

Listric Thrust Faulting in the Laramide Front of North-Central New Mexico Guided by Precambrian Basement Structures

Maria Beatrice Magnani and Alan Levander

Department of Earth Science, Rice University, Houston, Texas

Eric A. Erslev and Nicole Bolay-Koenig

Department of Geosciences, Colorado State University, Fort Collins, Colorado

Karl E. Karlstrom

Department of Earth and Planetary Sciences, University of New Mexico, Albuquerque, New Mexico

New seismic reflection images and structural analyses of the Laramide front in the Sangre de Cristo Mountains of northern New Mexico indicate the importance of listric faulting and basement weaknesses in basement-involved foreland structures. At the surface, recumbent fault-propagation folds with west-tilted backlimbs indicate that the frontal Laramide structures in this region are west-dipping reverse faults that shallow with depth. Minor fault kinematics indicates a combination of fold-perpendicular shortening and regional ENE shortening. Seismic reflection images show that these listric faults extend to depths of at least 12 km. Continued westward tilts beyond the seismic coverage suggest that these faults eventually flatten into a middle crustal detachment zone underneath the center of the range. Although this geometry could seem to be consistent with neoformed Laramide thrusts that developed in response to ENE horizontal shortening, several lines of evidence suggest this geometry was at least partially guided by pre-existing basement weaknesses. Firstly, strong but diffuse reflectivity in the middle crust about 10 km below the Laramide thrusts mimics the shape of the Laramide frontal faults. These are not likely to be a Laramide blind thrust system because no Laramide deformation is seen where they would surface to the east under the Great Plains. Secondly, N- and NNE-striking folds and faults are cut by numerous discordant structures, indicating oblique slip on pre-existing faults during transpressive deformation. We suggest that the thrusting of basement rocks of the Sangre de Cristo Range over Paleozoic sedimentary rocks was guided by pre-existing basement structures of probable Precambrian age.

The Rocky Mountain Region: An Evolving Lithosphere
Geophysical Monograph Series 154

10.1029/154GM17

1. INTRODUCTION

The Laramide orogeny is commonly associated with stresses generated by the low-angle subduction of the Farallon and Kula plates under North America [*Engebretson et al.*, 1985;

Cross, 1986; *Coney*, 1987; *Stock and Molnar*, 1988; *Bird*, 1998; and *Humphreys et al.*, 2001]. This deformation in the Rocky Mountain foreland began at about 80–75 Ma and ended in the northern and southern Cordillera at about 56 Ma and 35 Ma, respectively [*Dickinson et al.*, 1988]. Laramide contraction resulted in formation of variously oriented basement-cored arches that are bounded by narrow zones of faulting, folding, and overturning of Paleozoic and Mesozoic strata (Figure 1). Some puzzling aspects of the basement-involved thrust faults is that they strike in a wide range of directions and vary in dip from high to low angle (Figure 1).

The controls on Laramide deformation have been the subject of extensive debate [e.g. *Erslev*, 1993, 2001; *Woodward et al.*, 1997; *Bird*, 1998; and *Cather*, 1999]. Surface studies indicate abundant evidence for reactivation of earlier faults [e.g., 12 examples cited by *Brown*, 1988] suggesting that pre-existing weaknesses play an important role in localizing Laramide structures, particularly high-angle faults. Pennsyl-

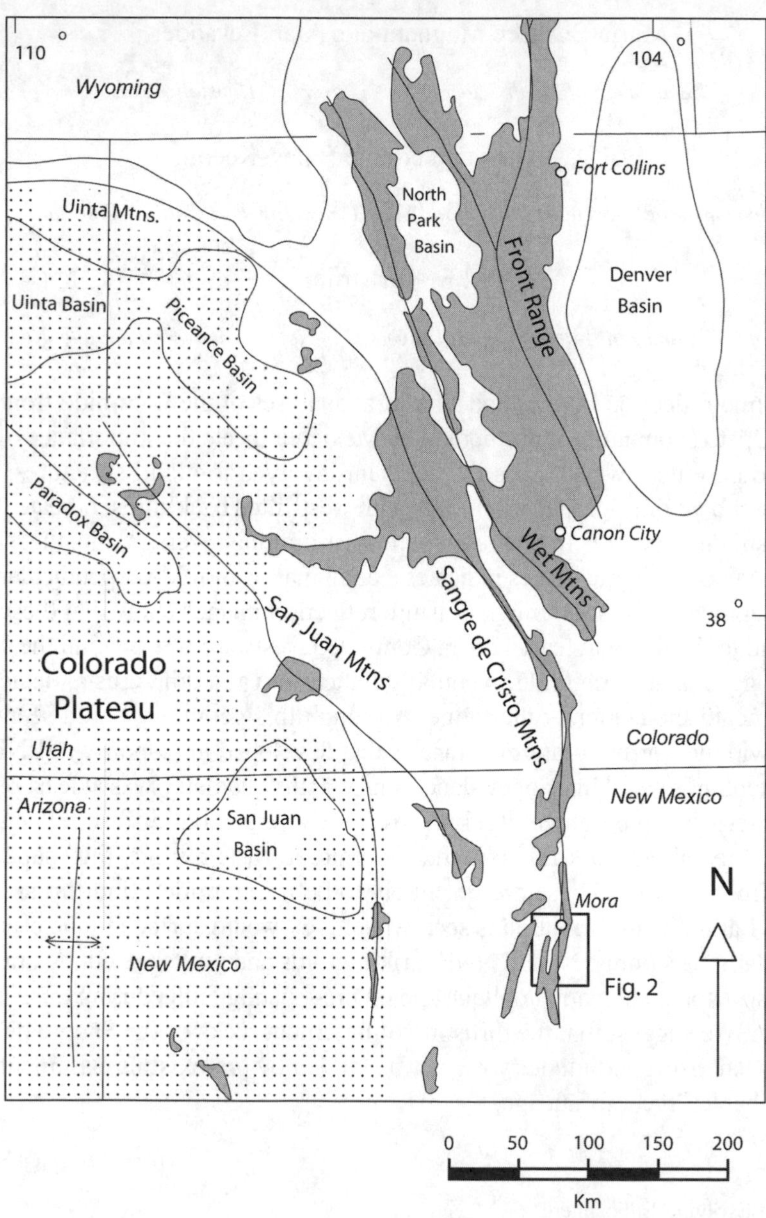

Figure 1. Simplified Laramide tectonic map of the southern Rocky Mountains showing basement exposures (gray), Laramide sedimentary basins, and the extent of the Colorado Plateau (coarse stipple).

vanian faults generated during the Ancestral Rocky Mountain orogeny were reactivated in numerous areas in the southern Rockies [e.g. *Tweto*, 1975; and *Stone*, 1986). *Marshak et al.* [2000] postulated that Ancestral Rockies and Laramide faults resulted from tectonic inversion of a Precambrian normal fault system. *Timmons et al.* [2001] showed that NW- and N-striking normal faults in basement formed at 1.1 and 0.8 Ga respectively. However, the role played by pre-existing weaknesses at deeper structural levels has not been thoroughly investigated due to the scarcity of deep crustal data.

Direct observations of deeper Laramide fault geometries have been very limited, and the presence of mid-crustal detachments where faults may merge is debated [*Erslev*, 1993; *Tikoff and Maxson*, 2001]. The overall geometry of Laramide arches suggests the presence of detachments or a fault-propagation folding mechanism at depth [*Lowell*, 1983; *Oldow*, 1989; *Erslev*, 1993]. Only at one location, the Wind River thrust, Wyoming, has the deep geometry of a Laramide fault system been imaged to middle crustal depths [*Smithson et al.*, 1978, 1979; *Lynn et al.*, 1983; *Sharry et al.*, 1986]. The most recent seismic interpretations of *Sharry et al.* [1986] show a NE-dipping fault to 22 km depth where it appears to merge with a complex set of duplexes reaching a depth of 26 km. Gravity information [*Hurich and Smithson*, 1982] suggesting no Moho offsets combined with the minimum 21 km of slip on the fault strongly indicates that the Wind River fault flattens at depth into a subhorizontal detachment. In addition, a listric fault geometry that flattens into a detachment can explain the minimal deformation and remarkably uniform tilt of the NE-dipping backlimb of the Wind River arch [*Erslev*, 1986].

This paper presents the results of one of the seismic reflection experiments of the Continental Dynamics - Rocky MOuntain project (CD-ROM) carried out in north-central New Mexico. The experiment targeted the subsurface continuation of the frontal thrust faults underlying the Sangre de Cristo mountain range (Figure 2). This location provides an ideal opportunity to test for the influence of pre-existing faults because *Baltz and Myers* [1999] documented numerous structural highs and depocenters associated with the Pennsylvanian Ancestral Rocky Mountain deformation on the flank of the Taos trough. Their mapping defines a series of back-rotated Laramide fault blocks, which suggest basement detachment at shallow levels (10–15km) using the basement block balancing approach of *Erslev* [1986].

2. GEOLOGIC SETTING

The eastern front of Laramide arches in the southern Rocky Mountains of Colorado and northern New Mexico is remarkably linear (Figure 1) and trends N-S on average. In Colorado, this deformation front is dominated by NNW-striking thrust faults and folds that step right to the north, forming a ragged, sawtooth pattern of uplifted units. These structural steps range in size from 50 km between the en echelon Sangre de Cristo, Wet Mountain, and Front Range arches to 10 km along the margin of the northeastern Front Range. South of the Sangre de Cristo arch, the Laramide deformation front makes a major step to the west where Laramide structures are heavily overprinted by the more recent Rio Grande deformation. The study area (Figure 2) is located where the Laramide front makes an arcuate curve towards the west just north of this major step defining the end of the Sangre de Cristo Mountains.

Recent minor fault measurements in Mesozoic strata along the eastern front of Laramide deformation (Figure 3b, and c; *Erslev*, 2001; *Wawrzyniec et al.*, 2002; *Erslev et al.*, in press) generally show ENE-WSW shortening by thrust and strike-slip faulting, indicating transpressive deformation between the Colorado Plateau and the interior of the North American craton. The amount of regional strike-slip displacement parallel to the N-S Laramide front is highly contested, with cumulative estimates ranging from minimal [*Woodward et al.*, 1997] to more than 100 km [*Cather*, 1999]. The lack of strike-slip fault zones cutting the continuous Permian and Mesozoic strata which wrap around the southern end of the Sangre de Cristo arch suggests that major Laramide strike-slip faults, if they exist, must be west of the arch axis.

Baltz and Myers [1999] compiled their geologic mapping of the southeastern Sangre de Cristo Mountains (Figure 2) and the stratigraphic evidence for extensive Ancestral Rocky Mountain deformation in north-central New Mexico. Within the context of the complicated history of pre-Laramide deformation, the diverse arrays of Laramide faults and folds along the southeastern flank of the Sangre de Cristo arch (Figure 2 and 4a) are not surprising. Fold axes and faults show trimodal distributions, with major NNW-SSE modes and lesser N-S and NNE-SSW modes. The close correlation of fault and fold patterns is expected because most of the folds are fault-propagation and detachment folds whose axes trend parallel to the strike of the underlying faults. Measurements of minor fault orientations taken from the Mora area (Figure. 3a and 4b) are oblique to fault strikes farther north (Figure 3b and c), suggesting anomalous ESE-WNW shortening and compression. The fault localities, indicated by Xs in Figure 2, are, however, adjacent to NNE-trending folds and may not sample the full regional shortening direction. This is further indicated by the dominance of NW-striking left-lateral faults in the Mora minor fault data (Figure 3a). Some degree of oblique slip is required by the diverse fold axis orientations and the fact that folds and faults change orientation along strike.

Along the NNE-trending fold and fault systems in the area (Figure 2), *Baltz and Myers* [1999] mapped smaller SE-NW faults with left-lateral separations offsetting the traces of the

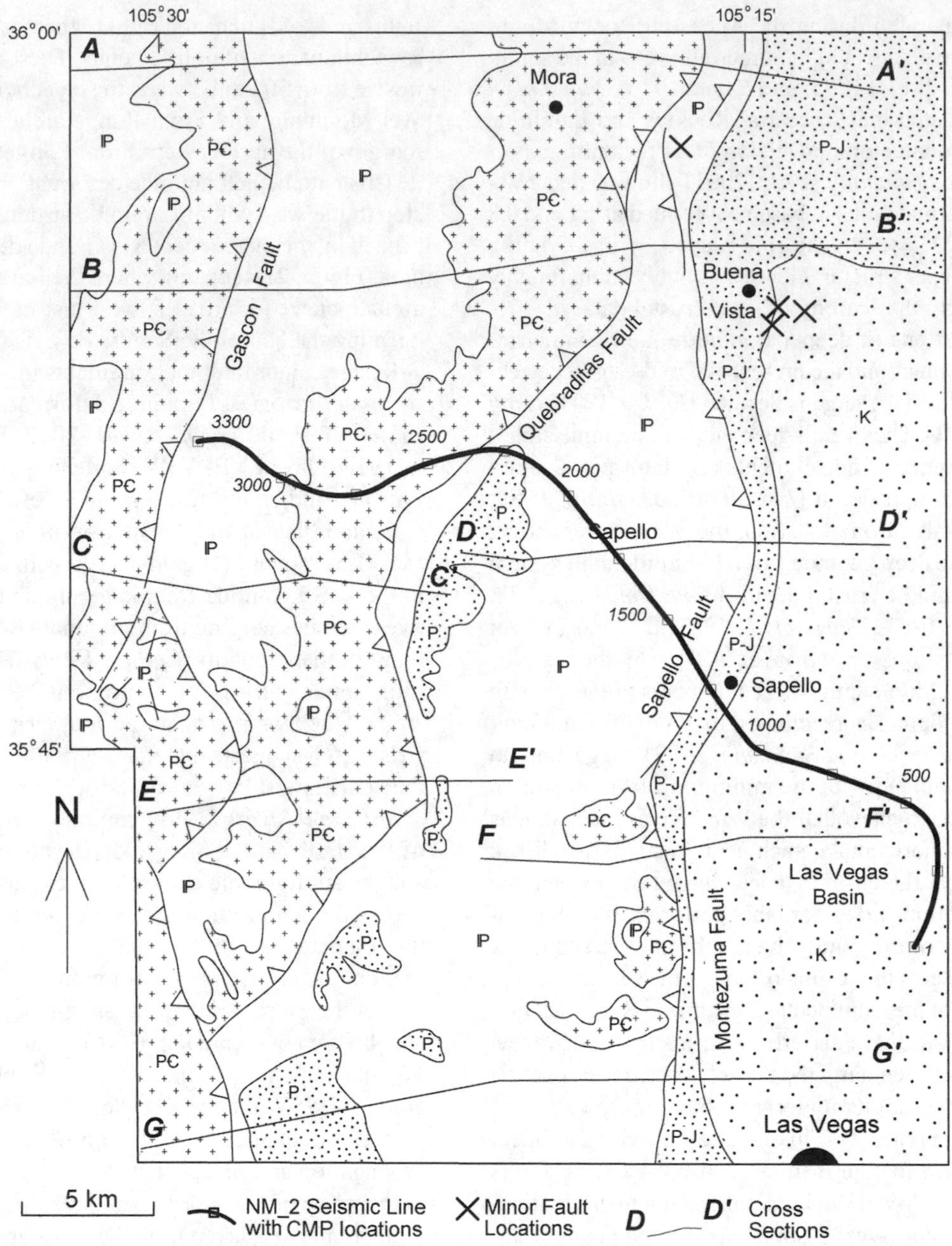

Figure 2. Simplified geologic map of the eastern flank of the Sangre de Cristo arch near Mora, New Mexico after *Baltz and Myers* [1999]. The heavy solid line shows the seismic acquisition line and the adjacent line segments show cross-sections in Figure 6. Localities sampled for minor faults are shown as Xs. Abbreviations for units are **PC** – Proterozoic crystalline basement, **IP** – Pennsylvanian and older sedimentary rocks, **P** – Permian sedimentary rocks, **P-J** – Permian to Jurassic sedimentary rocks and **K** – Cretaceous sedimentary rocks

main faults and anticlines. In comparison, such offsetting faults are rare along NNW-trending fold and fault systems. Analogous cross-cutting faults were mapped along parts of the NNE-trending East Kiabab monocline in Utah, which *Tindall and Davis* [1999] interpreted to be due to combined right-lateral and reverse slip on the underlying fault. Thus, we interpret the map pattern in the southeastern Sangre de Cristo arch as highly suggestive of ENE-shortening on NNW-striking

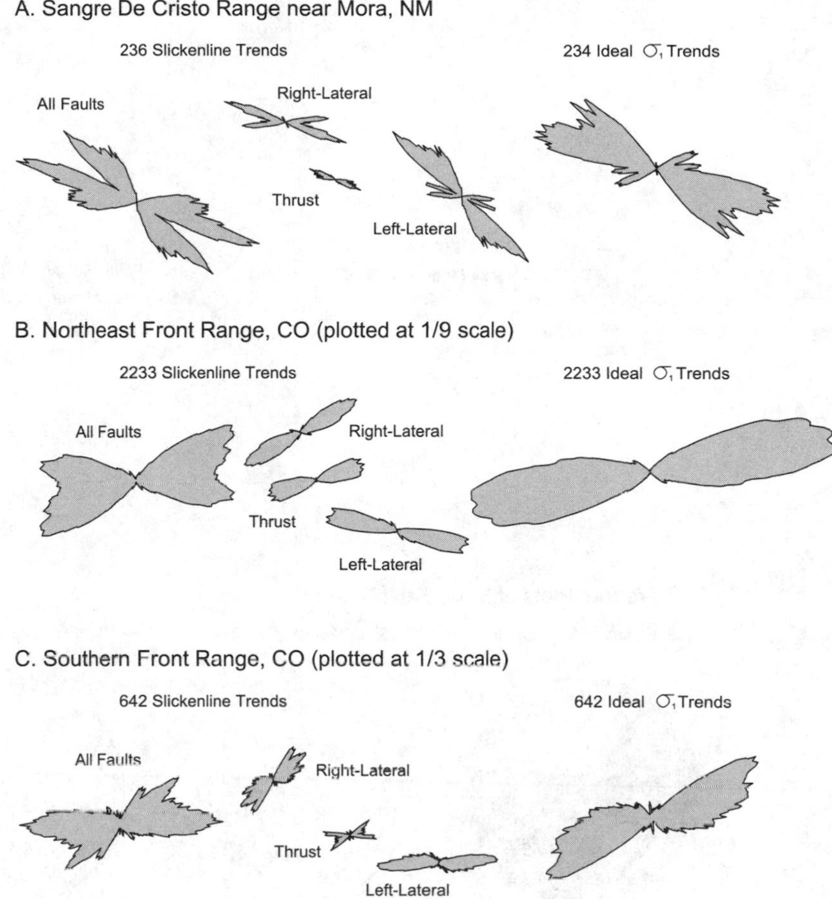

Figure 3. Minor fault measurements along the Laramide deformation front near a) Mora, New Mexico [*Bolay-Koenig and Erslev*, 1999], b) Fort Collins, CO [*Holdaway*, 1998], and c) Canon City, CO [*Jurista et al.*, 1996]. An angle of 25° between the slickenline and ideal sigma 1 directions was used to calculate ideal sigma 1 axes. Rose diagrams are smoothed using a 10° window and show trends of linear elements whose plunge is less than 45°.

thrust faults and oblique slip on NNE-striking faults, which probably have both right-lateral and thrust components to their slip.

The structural geometries defined by *Baltz and Myers* [1999] suggest listric Laramide faulting in 3D. In the area of the seismic line, minimal changes in Paleozoic stratigraphic thickness [*Baltz and Myers*, 1999] and minor post-Laramide reactivation of the faults indicate that the basement unconformity, which is well defined in the area (Figure 2), provides a good datum for evaluating the Laramide deformation. Cross-sectional profiles from *Baltz and Myers* [1999] (Figure 5) of faults and the basement unconformity show the contractional nature of the fault-propagation fold structures. The consistent backward tilts of the basement blocks (Figure 5) indicates rotational fault-bend folding [*Erslev*, 1986] on listric, or concave upward, faults at depth The fact that backlimb tilts increase from west to east in nearly every hanging wall (Figure 5) suggests that the fault curvature increases toward the surface, causing progressively more tilting of the hanging wall as the fault tip is approached.

3. SEISMIC DATA AND INTERPRETATION

One of the seismic lines acquired during the 1999 Continental Dynamics of ROcky Mountains (CD-ROM) seismic experiment was collected with the goal of investigating the geometry of the Laramide thrusts in the southern Rocky Mountain. The surface trace of the 40 km long profile (NM_2) (Figure 2) extends from the Las Vegas basin in northern New Mexico into the basement-cored arch of the Sangre de Cristo

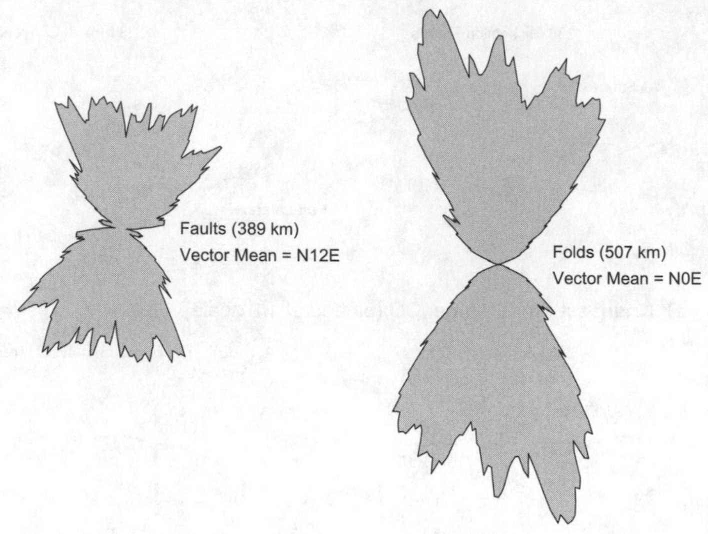

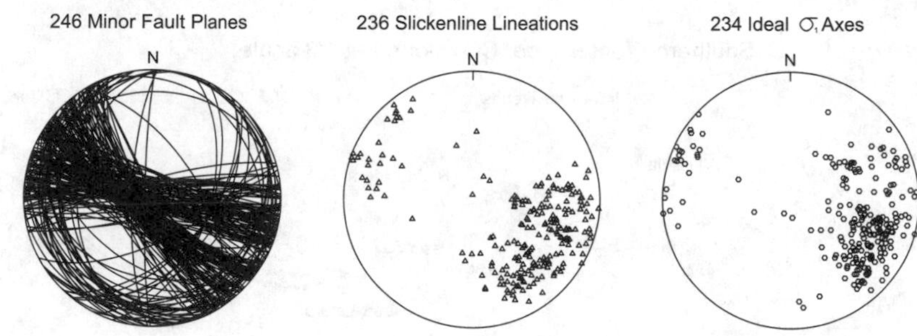

Figure 4. a) Smoothed (10° window) rose diagrams of faults and folds mapped by *Baltz and Myers* [1999]; b) Stereonet plots of fault planes, slickenlines and ideal sigma 1 directions for minor faults measured near Mora, NM.

Mountains. It crosses nearly normal to the strike of the frontal Sapello Fault near Mora and oblique to the Quebraditas Fault that thrusts Proterozoic over Paleozoic units (Figure 2 and 6). Impassable mountains prevented the extension of the profile farther to the west.

The seismic data were recorded with 4 large mass Vibroseis sources using 1001 active recording channels on a 25 km long seismograph array. The source interval and the group interval were 100 m and 25 m, respectively. Sweep time was 20 s. Listening time was 45 s, giving a record length of 25 s. The seismic data were processed using the Promax® processing software. Noisy traces were edited out of the dataset. Several filters were applied to attenuate and remove the surface waves, and amplitude corrections were performed to improve the S/N ratio. The processing sequence included predictive deconvolution, static corrections (refraction and surface consistent residual) to a 2200m datum plane, velocity analysis and stacking. The stack section was depth migrated with a variable velocity 2-D Kirchhoff algorithm using an attenuated velocity field derived from the prestacking velocity analysis. Basement velocities vary from 4700–5800 m/s and sedimentary rocks of the Las Vegas basin were migrated with a velocity ranging from 3300–4200 m/s. The data were not depth migrated using the velocity field derived from the seismic refraction experiment [*Levander et al.*, this volume] due to the different tectonic setting of the two profiles.

The seismic image NM_2 (Figure 6 and 7) shows features that correspond well with the structures mapped at the surface. Each geologic unit exhibits a distinctive seismic reflectivity: the Proterozoic rocks (CDP 2300–2800) exhibit a very diffuse reflectivity, the Paleozoic rocks (CDP 1250–2300) show layering, and Mesozoic sedimentary rocks of the Las Vegas Basin

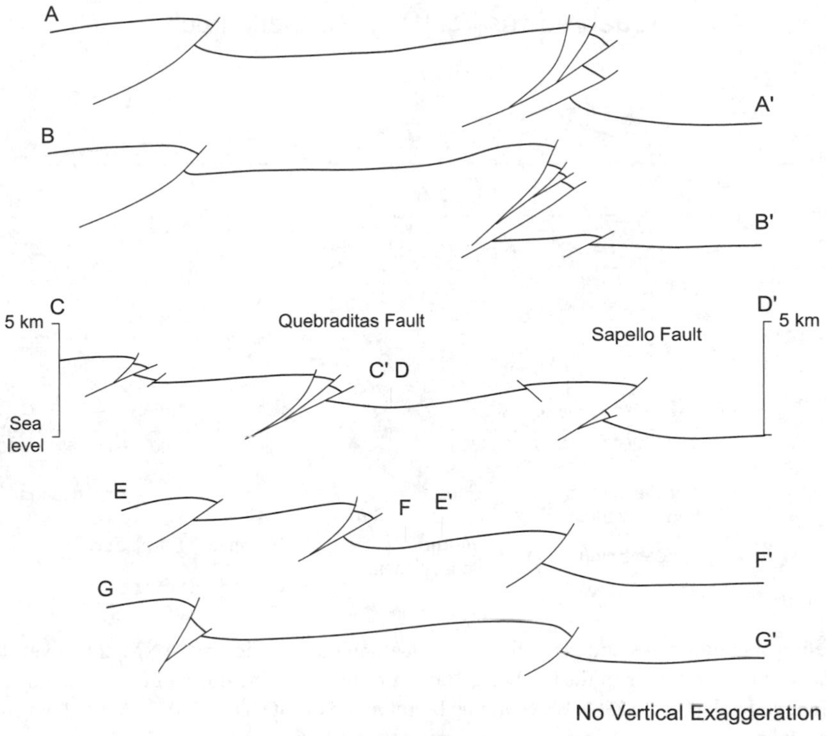

Figure 5. Top of basement and fault geometries from cross sections of *Baltz and Myers* [1999]. These unexaggerated section locations are in Figure 2. Composite cross section C-D' is highlighted because it is closest to the CD-ROM seismic line.

(CDP 1–1250) exhibit a well-preserved sedimentary stratification that depicts a 2 km deep, nearly undeformed basin that thins southeastward. A west-dipping reflector at depth of 2 km (A in Figure 8) marks the base of the Las Vegas Basin. Below, at a depth of about 5 km, a set of relatively bright, discontinuous reflectors (D_2 in Figure 8) is visible between CDP 1000–2000.

The two reflectors A and D_2 are cut by a west-dipping zone of reflectivity labeled (C) at about CDP 1500 and CDP 2000, respectively. This west-dipping structure (C) surfaces at CDP 1250 and corresponds with the Sapello fault, the frontal thrust of the Sangre de Cristo Mountains in this area. From the surface the reflectivity dips 30° west and flattens to a nearly horizontal orientation between 10–11km. Two bright, west-dipping reflections (D_1) at about 5 km depth (CDP 2300–2700) appear to coincide with the sharp bend in the line where it intersects the Quebraditas fault. Careful analysis on CDP and shot gathers ruled out the possibility of out of plane energy reflected on the fault plane and indicated that the bright reflections (D_1) represent the energy coming from an interface below the trace of the profile. The Quebraditas fault appears to correlate with weak reflections that surface at CDP 2350.

The deeper portions of the seismic profile are dominated by highly reflective sets of west-dipping reflections (B) that appear to parallel the Sapello thrust, but at 10 km deeper levels. The structures appear to be characterized by two different sets of reflections. The shallower part is composed by two sharp, high amplitude reflectors (B_1 and B_2 at 10 km and 14 km depth respectively) that converge from the eastern end of the profile to 16 km around CDP 1400. At greater depths the seismic character of the structure changes and a zone of diffuse but bright reflectivity (B) reaches the west end of the line at a depth of 23 km. Although the continuity and the character of the reflections are variable as a function of depth, we consider these zones of reflectivity to be tectonically and genetically related, as shown below. A moderately west-dipping grain is also visible in the data at this depth, and is especially noticeable between the C and B reflections, where it seems to parallel both structures. The deeper lower crust is almost featureless in the seismic profile except for few weak subhori-

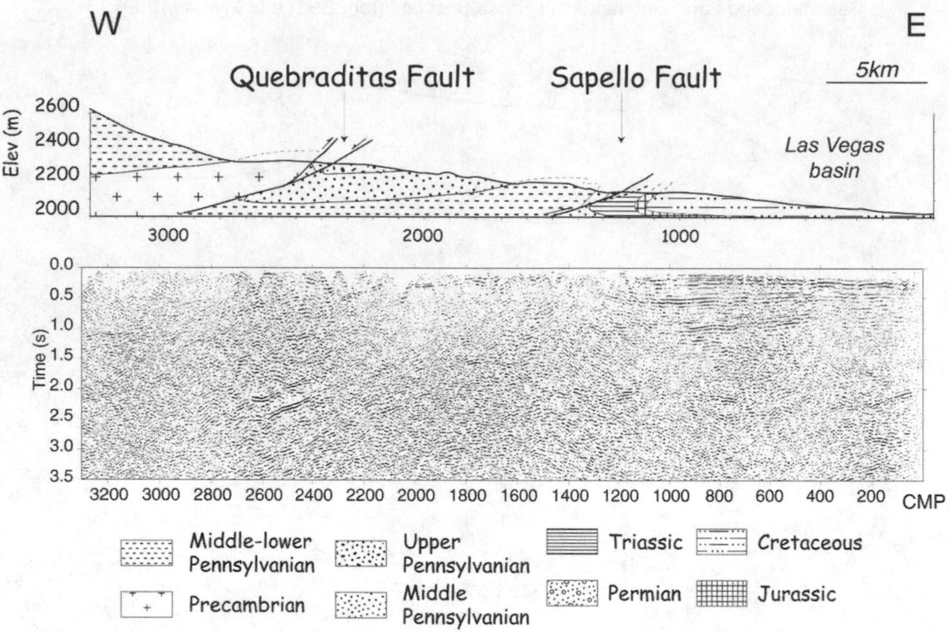

Figure 6. Cross section showing rocks and faults cropping out along the seismic profile NM_2 (above); first 3s of the stack section (below). The sedimentary rocks of the Las Vegas Basin are clearly visible down to 1.1s depicting a broad syncline cut to the west by the Sapello Fault (CDP 1250). West-dipping bright reflectors at CDP 2400–2600 (2–2.5s) are interpreted as the Proterozoic supracrustal rocks in the hangingwall of the Sapello thrust. Seismic data are plotted with no vertical exaggeration.

zontal reflections that are traceable at about 34–35 km depths, which is likely too shallow to be a Moho reflection based on the CD-ROM refraction data [*Levander et al.*, this volume].

The interpretation of the upper part of the seismic line is well constrained by surface geology and borehole data in the area [*Baltz and Meyers*, 1999]. The basement of the Las Vegas basin (A), on the easternmost end of the profile, is imaged at about 2 km depth. This depth corresponds to the depth of the Pennsylvanian "Great Unconformity" with underlying Precambrian basement [*Powell*, 1876] observed in boreholes [*Baltz and Myers,* 1999]. The basin is almost undeformed and shallows eastward. To the west, the Mesozoic sedimentary units stop at the mapped position of the Sapello Fault (CDP 1250), where they come in tectonic contact with the Paleozoic units that core the Sangre de Cristo arch. The Sapello Fault (C) is steeply dipping at the surface but quickly flattens to apparent dips of ~30°W and further shallows to a 10° apparent dip at 10 km depth (relative to 2200m average surface elevation) in the westernmost end of the profile. In spite of its weak signal, the Quebraditas fault can also be traced to a depth of 5 km.

The two bright reflectors, visible in the hanging wall of the Sapello Fault (D_1) and the highly reflective packages (D_2) beneath the Las Vegas basin likely represent the Proterozoic supracrustal rocks that make up El Oro Mountains on the southern end of the Rincon Range (Vadito, Ortega and Hondo Group) (Read et al., 1998) and that crop out immediately northwest of the profile. This interpretation is corroborated by the observations on a N-S reflection profile, acquired just east of the Sapello frontal thrust [*Magnani et al.*, this volume], that images the highly reflective, nearly pure quartzites of the Ortega Fm. folded by several south-dipping ductile shear zones at a depth of 5 km beneath the Great Unconformity.

The interpretation of the structure (B) and the above west-dipping reflectivity, imaged at middle crustal depth below the Sangre de Cristo uplift, the frontal thrust and the Las Vegas basin, is not constrained by any surface and subsurface observation and is, therefore, conjectural. At first glance this west-dipping grain looks like a master detachment surface to which the Sapello and other Laramide faults would sole to the west. However, the sedimentary rocks in the Las Vegas basin above the Great Unconformity appear to be entirely undeformed by this fault system, suggesting that the age of the system is greater than Pennsylvanian-Mississippian and that the west-dipping fabric in the middle crust is an inherited feature rather than a Laramide structure. It is possible that this zone was reactivated during the Laramide, as suggested below in the discussion of structural balancing, as a mechanism to raise the western part of the basement unconformity, but the strength

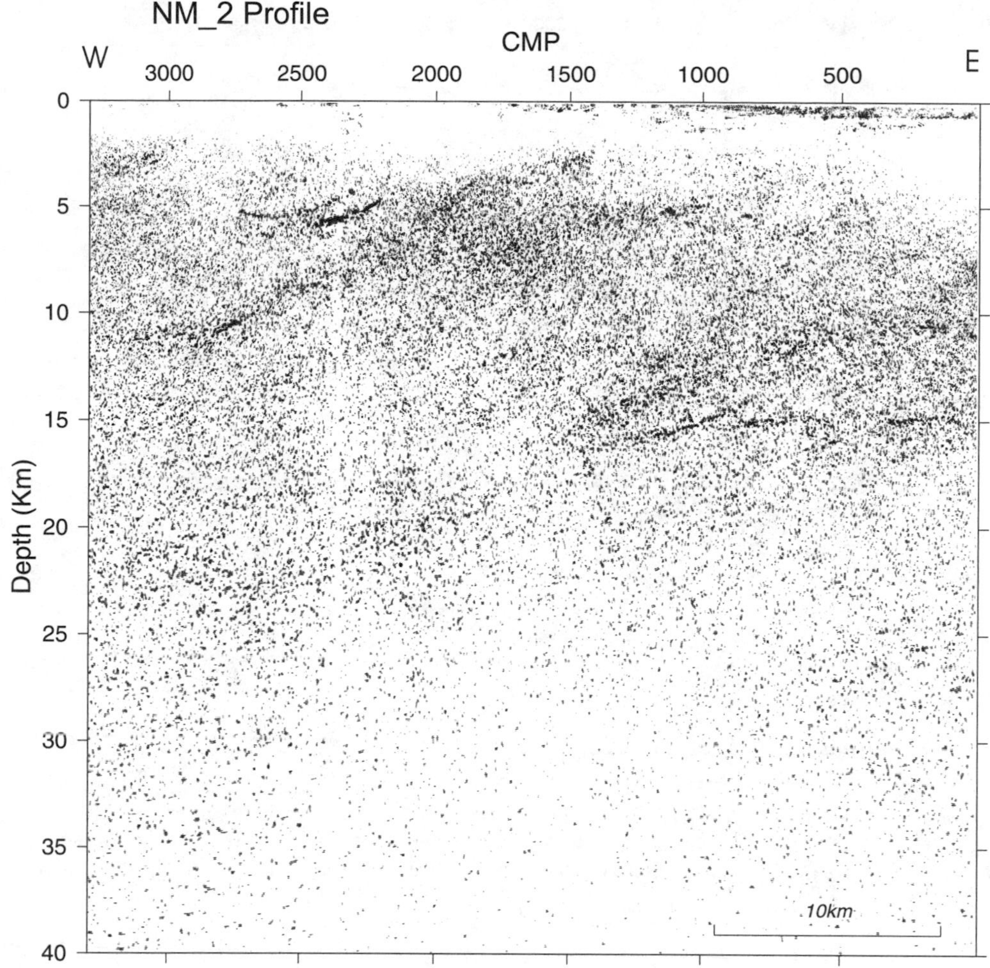

Figure 7. Energy plot of the depth-migrated reflection seismic section (see Figure 2 for location), with depth from 2200 m average surface elevation plotted versus horizontal distance.

of the reflector is inconsistent with the minimal Laramide slip actually forming the reflector itself. To the east this zone of reflectivity shallows to ~15 km depth and continues into the two sharp bright reflectors (B_1) and (B_2). The interpretation of these reflectors relies on the observations along the north-south tie profile [*Magnani et al.*, this volume] that show a system of high–amplitude, subhorizontal reflectors at 10–15 km depths that match the depth and seismic character of the two reflectors (B_1) and (B_2) observed in the NM_2 profile. The bright amplitude is indicative of a large impedance contrast that, at this depth (10–15 km) can be explained by the presence of fluids or mafic material intruded in the granitic crust. Reflectors with similar characteristics, imaged in many seismic profiles worldwide, have been interpreted as crustal magmatic intrusions [e. g. *Goodwin et al.*, 1989; *Jarchow et al.*, 1993]. Due to their bright reflectivity, the remarkable lateral continuity along the north-south profile and the evidence of repeated magmatic activity in this area throughout the geologic history [*Keller et al*, 1989; *Aldrich*, 1986], we interpret them as mafic sills of Proterozoic or Tertiary age, therefore suggesting that the reflectors (B_1) and (B_2) on the NM_2 line are also mafic sills. The age of these mafic intrusions is more difficult to constrain but because they seem to cross cut a broad zone of inferred Paleoproterozoic reflectivity in the cross line, we interpret them to be of either Neoproterozoic (1.1 Ga) or Tertiary-Quaternary age. Also, in this line of section, they seem to parallel the broad zone of reflectivity (B), suggesting that the sills were emplaced parallel to zones of weakness in the Precambrian crust, possibly exploiting pre-existing structural weaknesses.

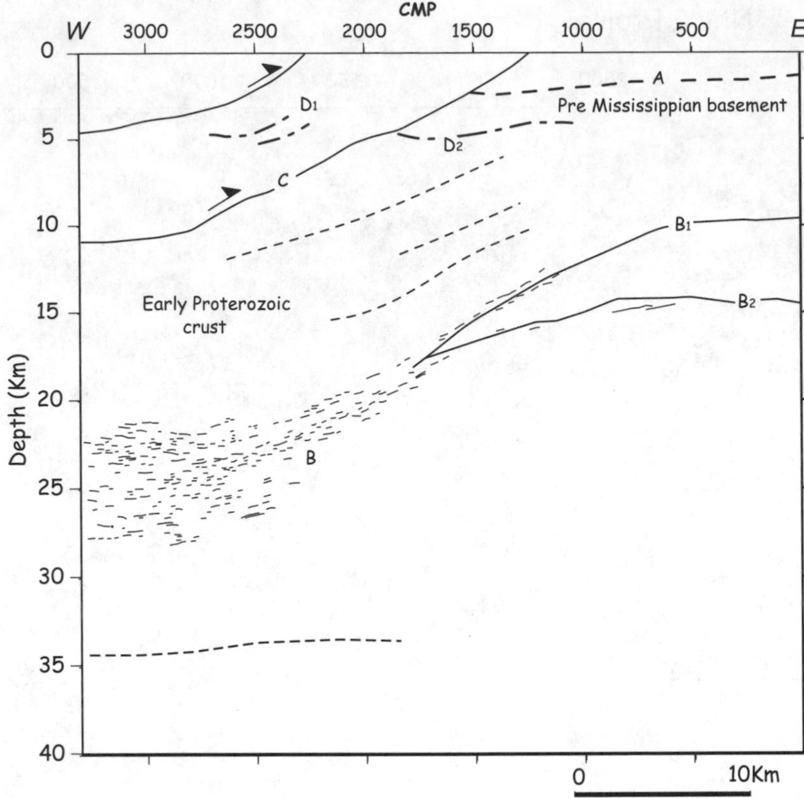

Figure 8. Geologic interpretation of the seismic data. **A** – Great Unconformity, **B'** and **B"** – Mafic sills of Proterozoic or Cenozoic age, **B** – Proterozoic (?) zone of weakness, **C** - Sapello thrust fault, **D** - supracrustal Proterozoic rocks (Hondo and Vadito Group).

4. RESTORATION OF LARAMIDE STRUCTURES

The proposed geometries of the Sapello and Quebraditas faults, as determined from surface geology [*Baltz and Myers*, 1999] and the seismic data, were tested by 2D restorations of the basement unconformity using the 2DMove program of Midland Valley Exploration. An east-west, composite cross-section (C-D' from *Baltz and Myers*, [1999]) was simplified to show minimal footwall basement folding and extended to depth by projecting the seismic reflectors interpreted as extensions of the Sapello and Quebraditas faults into the plane of the cross-section (Figure 9a). Points on the postulated fault zone reflections were projected parallel to each fault and corrected from depth relative to 2200 m (Figure 7) to depth relative to sea level (Figure 9).

Restoration of this simplified cross-section (Figure 9b) was accomplished using fault-parallel flow [*Egan et al.*, 1997], which provides a close approximation for basement deformation on arcuate faults. Fault-propagation folding of the hanging wall near the shallow fault tip area was not restored because folding in Rocky Mountain basement-involved structures is largely limited to shallow levels in the crust and does not appear to influence the deeper levels [*Erslev*, 1991]. Any simple shear components parallel to the fault zone will likewise not influence the deeper geometry of the faults and the larger-scale rotational fault-bend folding of the overlying basement blocks [*Erslev*, 1986].

The restoration in Figure 9b shows the general viability of the listric fault geometries indicated by both the basement block tilts and the deep crustal reflections. Two weaknesses in the restoration stand out, however. First, the non-planarity of the basement unconformity on the block backlimbs, which show antiformal shapes, suggests that fault curvatures are greater closer to the surface than were interpreted in this section. Secondly, restoration of the faults does not flatten the basement unconformity overall, with the western side of the section still substantially higher than the eastern side of the section. This latter problem can be explained in two ways. The excess elevations of the basement unconformity to the west can be remedied by steepening

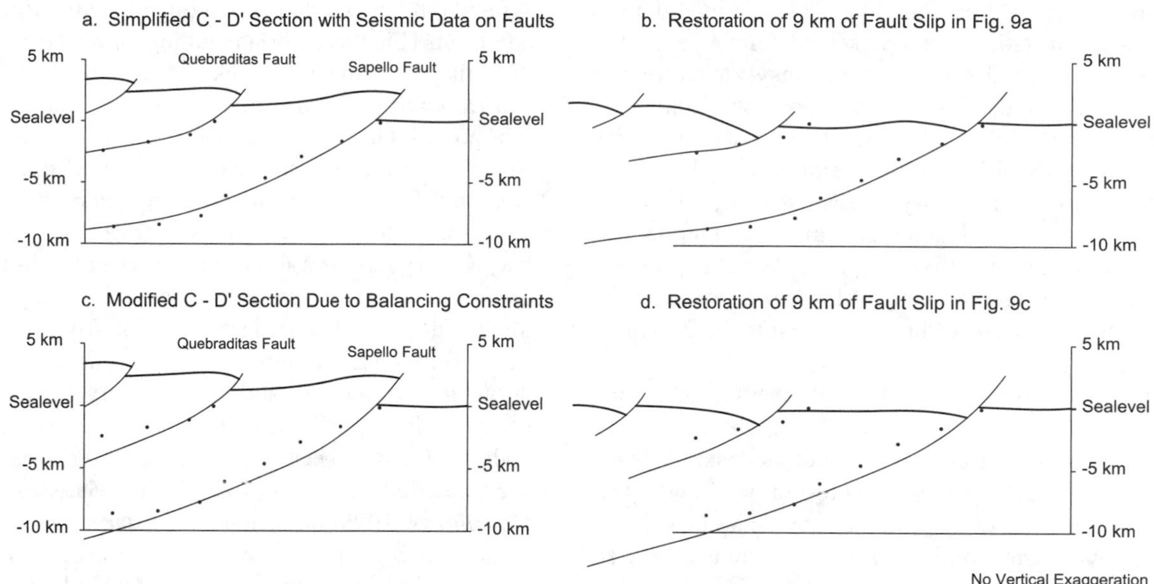

Figure 9. a) Simplified unexaggerated section C-D' with all basement folding put into the hanging wall and fault trajectories at depth projected parallel to the faults from the seismic interpretation in Figure 8; b) Restoration of simplified section C-D' using fault-parallel flow and neglecting shallow fault-propagation folding in the hanging wall; c) Simplified section C-D' modified to allow optimal restoration by fault-parallel flow; d) Improved restoration of modified simplified section C-D' using fault-parallel flow and neglecting shallow fault-propagation folding in the hanging wall.

the faults, which when restored, drop the basement unconformity to lower levels. This interpretation is not inconsistent with the seismic interpretation because an underestimation of the seismic velocity of the basement as well as edge effects during the depth migration processing could have resulted in incorrect depths to the reflectors in Figure 7. Alternatively, the excess elevations of the basement unconformity to the west could be due to blind duplexing at depth, an option that is consistent with the geometry of the deep reflections under these faults. This option is acknowledged, but was not pursued any further because deeper structural involvement was not necessary to balance the sections.

Simplified cross-section C-D' was further modified according to the above observations so that faults have increased curvature near the surface and steeper dips at depth (Figure 9c). This results in a better, although admittedly non-unique, restoration (Figure 9d), with a more uniform restored elevation of a more planar basement unconformity. This interpretation has the faults continuing to flatten to the west, and is consistent with the westerly tilt of the basement unconformity from the western edge of the cross-section all the way to the Pecos Valley, 14 km to the west.

5. DISCUSSION AND CONCLUSIONS

The Laramide faults and shear zones that bound the Proterozoic uplifts are notorious for their complicated deformation history and their geometry at depth is still hotly debated. The seismic line images listric Laramide basement thrusts that can be traced from surface outcrops to a depth of 10–15 km below sea level where they appear to sole into an upper-crustal zone of reflectivity. These thrusts appear to dip up to 50–55° in surface exposures, rapidly shallow to 30° dips in the shallowest part of the seismic section, and then more gradually flatten to 20–10° at the western end of the seismic profile.

The interpretation of the deeper structure is more speculative. The reflectors in the middle crust at 22–24 km beneath the Rockies ramp up to 14 km beneath the eastern front of the range and extend under the Great Plains. The character of the reflectivity along the structure appears to change from being a 2 km thick zone of scattered short reflectors to two narrow and sharp reflectors. Based on our interpretation of a crossing seismic line, we interpret the brightness of the shallow features as resulting from mafic sills that intruded subsequent to the development of the detachment system.

The age and origin of the pre-existing west-dipping structure along which the sills intruded are difficult to define. We suggest three alternative explanations. (1) It formed during the Laramide deformation and represents a Laramide master detachment fault. (2) It formed during Pennsylvanian Ancestral Rocky Mountain deformation and possibly reactivated during the Laramide deformation. (3) It formed prior to the Ancestral Rockies and Laramide events during the early-middle Proterozoic accretionary stages [*Bowring and Karlstrom,* 1990] and thus any Phanerozoic movement represents fault reactivation. The first two hypotheses are supported by its flat-ramp-flat geometry, parallelism to the Sapello Fault and depth (from 10–15 km to the east to 23–25 km to the west).

Model 1 might also be compatible with the middle crustal flow model of *McQuarrie and Chase* [2000], which envisions a 10 km wedge of middle crust (at about this depth) flowing eastwards from the Sevier highlands to the west, under the Colorado Plateau and Rocky Mountains and causing surface uplift of the region. However intracrustal flow from the Sevier highlands should have disrupted the seismic signature of pre-existing structures and this is in disagreement with the available observations on seismic profiles. Moreover, the absence of any Laramide uplift/anticline/dome in front of the deep structure makes the presence of a mid-crustal master detachment fault unlikely. It seems unlikely that the feather edge of the injected wedge material would have no surface expression; instead, if such features existed, they might be analogous to blind thrusts in front of thrust belts that are expressed at the surface as broad upwarps and anticlines [*Fermor and Price*, 1987].

The only mapped structure in the Great Plains comparable in scale to the imaged middle crustal seismic feature (B) is the Sierra Grande Uplift [*Woodward*, 1988]. The Sierra Grande uplift is considered Pennsylvanian in age and developed during the Ancestral Rockies deformation, although later motion may have occurred in Laramide time. Reactivation of Ancestral Rockies arches and basins during mid-Cretaceous – Eocene time has been recorded for Las Animas dome, which may represent the continuation of the Sierra Grande uplift to the north [*Volk*, 1972; *Tweto*, 1983; *Merewether*, 1987].

The third hypothesis, a possible Precambrian origin for the structure, provides an alternative explanation for the absence of a major uplift associated with the observed structure. The Rockies expose 1.4 Ga middle crustal rocks that were emplaced at 3–4kbar, whereas the same age rocks in the Great Plains are rhyolitic volcanic and sedimentary materials that have never been buried to great depths. The timing of this post- 1.4 Ga differential uplift is not well constrained. Some uplift, perhaps several km, took place in the Laramide as the Rockies went up relative to the Great Plains; some uplift, again perhaps several kilometers, took place in the Ancestral Rockies event, but some may also have taken place during 1.1 and 0.8 Ga extensional faulting along the Rocky Mountain front. This Precambrian faulting may have been the time of initiation of the N-S Rocky Mountain trend [*Karlstrom and Humphreys*, 1998]. If so, the deep reflectors may represent pre-1.4 Ga faults responsible for this uplift, consistent with the Proterozoic inheritance for mid-continent arches of Ancestral Rockies age suggested by *Marshak et al.* [2001].

We speculate that the deep crustal structures observed on our line probably represent shear zones/basement faults that formed during the Proterozoic and were reactivated in Ancestral Rocky Mountains and/or Laramide deformation. They appear to have been exploited as a minimum strength surface by mafic sills during the Precambrian and/or the Tertiary. By analogy to the north-south tie profile [*Magnani et al.*, this volume], mafic bodies could have been emplaced during the magmatism that accompanied the rifting event at 1.1 Ga [*Karlstrom and Humphreys*, 1998] or during the Tertiary magmatic activity associated with the Jemez Lineament in New Mexico and Arizona [*Aldrich*, 1986; *Gardner and Goff*, 1984; *Perry et al.*, 1987].

Acknowledgments. We thank Dawson Geophysical for the acquisition of the reflection seismic data. Funding for this research was provided by NSF Grants EAR 9614777, EAR 003539, EAR 0208020 to Rice University; and NSF Grant EAR-0003582 and a supplement of EAR-9614787 to Colorado State University. Midland Valley Exploration provided the 2DMove program for cross-section restoration.

REFERENCES

Aldrich, M.J., Jr., Tectonics of the Jemez Lineament in the Jemez Mountains and Rio Grande Rift, *J. of Geophys. Res.*, 91, 1,753–1,762, 1986

Amarante, J.F.A. and Kelley, S. A., Miller, K. C., Characterization of the Proterozoic Basement in Mescalero #1, Guadalupe County, New Mexico, AAPG Rocky Mountain Section Meeting, *Am. Assoc. Pet. Geol. Bull.*, 84, 1235, 2000.

Baltz, E.H. and Myers, D.A., Stratigraphic framework of upper Paleozoic rocks, southeastern Sangre de Cristo Mountains, New Mexico with a section on speculations and implications for regional interpretation of ancestral Rocky Mountains paleotectonics, *New Mexico Bureau of Mines and Mineral Resources Memoir,* 48, 269, 5 sheets, 1999.

Bird, P., Kinematic history of the Laramide orogeny in latitudes 35°–49° N, western United States, *Tectonics,* 17, 780–801, 1998.

Bolay-Koenig, N.V. and Erslev, E.A., Basement control of Laramide deformation in the Southern Rocky Mountains, *Geol. Soc. Am.* Ann. Meet., 31, 7, 131, 1999.

Bowring, S.A. and Karlstrom, K.E., Growth, stabilization and reactivation of Proterozoic lithosphere in the Southwestern United States, *Geology,* 18, 12, 1203–1206, 1990.

Brown, W. G., Deformation style of Laramide uplifts in the Wyoming foreland, *in Interaction of the Rocky Mountain foreland and the Cordilleran thrust belt*, edited by Schmidt, C.J. & Perry, W.J. Jr., *GSA Memoir,* 171, 53–64, 1988.

Cather, S.M., Implication of Jurassic, Cretaceous and Proterozoic piercing lines for Laramide oblique-slip faulting in New Mexico and rotation of the Colorado Plateau, *Geol. Soc. Am. Bull.,* 111, 849–868, 1999.

Coney, P.J., The regional tectonic setting and possible causes of Cenozoic extension in the North America Cordillera, in *Continental extensional tectonics,* edited by Coward, M.P., Dewey, J.F. and Hancock, P.L., *Geological Society of London Special Publication* 28, 177–186, 1987.

Dickinson, W.R., Klute, M.A., Hayes, M.J., Janecke, S.U., Lundin, E.R., McKittrick, M.A. and Olivares, M.D., Paleogeography and paleotectonic setting of Laramide sedimentary basins in the central Rocky Mountain region, *Geol. Soc. Am. Bull.,* 100, 1023–1039, 1988.

Egan, S. S., Buddin, T. S., Kane, S. J., and Williams, G. D., Three-dimensional modeling and visualization in structural geology: New techniques for the restoration and balancing of volumes, in *Proceedings of the 1996 Geoscience Information Group Conference on Geological Visualization, Electronic Geology,* V. 1, Paper 7, 67–82, 1997.

Engebretson, D.C., Cox, A. and Gordon, R.G., Relative motions between oceanic and continental plates in the Pacific Basin, *Geol. Soc. Am. Special Paper* 206, 59, 1985.

Erslev, E., Basement balancing of Rocky Mountain foreland uplifts, *Geology,* 14, 3, 259–262, 1986.

Erslev, E.A., Trishear fault-propagation folding, *Geology,* 19, 617–620, 1991.

Erslev, E., Thrusts, back-thrusts, and detachments of Rocky Mountain foreland arches, in *Laramide basement deformation in the Rocky Mountain foreland of the western United States* edited by Schmidt, C.J., *Geol. Soc. Am. Special Paper* 280, 339–358, 1993.

Erslev, E., Multistage, multidirectional Tertiary shortening and compression in north-central New Mexico, *Geol. Soc. Am. Bull.,* 113, 63–74, 2001.

Fermor, P.R. and Price R.A., Multiduplex structure along the base of the Lewis thrust sheet in the southern Canadian Rockies, *Bull. Can. Pet. Geol.,* 35, 2, 159–185, 1987.

Gardner, J.N. and Goff, F., Potassium-argon dates from the Jemez volcanic field: Implications for tectonic activity in the north-central Rio Grande rift, *Field Conf. Guidebook. N.M. Geol. Soc.,* 35, 75–81, 1984.

Goodwin, E. B., Thompson, G., Okaya, D. A., Seismic identification of basement reflectors; the Bagdad reflection sequence in the Basin and Range Province-Colorado Plateau transition zone, Arizona, *Tectonics,* 8, 821–831, 1989.

Holdaway, S.M. and Erslev, A.E., Multiple directions of folding and faulting during oblique convergence of a basement-cored arch in the Southern Rocky Mountains, *Am. Ass. Pet. Geol. Abstr.* Annual Meeting Suppl., 1998.

Hurich, C.A. and Smithson, S.B., Gravity interpretation of the southern Wind River Mountains, Wyoming, *Geophysics,* 47, 11, 1550–1561, 1982.

Humphreys, E.D, Erslev, E. and Farmer, G.L., A metasomatic origin for Laramide tectonism and magmatism, Geological Society of America, Rocky Mountain Section, 53rd annual meeting, *Abstracts with Programs – Geol. Soc. Am.,* 33, 5, 50, 2001.

Jarchow, C.M., Thompson, G.A., Catchings, R.D. and Mooney, W.D., Seismic evidence for active magmatic underplating beneath the Basin and Range province, western United States, *J. Geophys. Res.,* 98, 22095–22108, 1993.

Jurista, B.K., Fryer, S.L. and Erslev, E.A., Laramide faulting and tectonics of south-central Colorado, *Am. Ass. Pet. Geol. Bull.,* 79, 6, 920, 1995.

Karlstrom, K.E. and Humphreys, E.D., Persistent influence of Proterozoic accretionary boundaries in the tectonic evolution of southwestern North America: interaction of cratonic grain and mantle modification events, *Rocky Mountain Geology,* 33, 161–180, 1998.

Keller, G.R., Hills, J.M., Baker, M.R. and Wallin, E.T., Geophysical and geochronological constraints on the extent and age of mafic intrusions in the basement of West Texas and eastern New Mexico, *Geology,* 17, 1049–1052, 1989.

Levander, A., Zelt, C. and Magnani M.B., Crust and upper mantle velocity structure of the Southern Rocky Mountains from the Jemez Lineament to the Cheyenne Belt, this volume.

Lowell, J.D., Foreland detached deformation, *Am. Ass. Pet. Bull.,* 67, 8, 1349, 1983.

Lynn, H.B., Quam, S., Thompson, G.A., Depth migration of the COCORP Wind River, Wyoming, seismic reflection data, *Geology,* 11, 8, 462–469, 1983.

Marshak, S., Karlstrom, K. and Timmons, J.M., Inversion of Proterozoic extensional faults: An explanation for the pattern of Laramide and Ancestral Rockies intracratonic deformation, United States, *Geology,* 28, 735–738, 2000.

Magnani, M.B., Levander, A., Miller, K., Eshete, T. and Karlstrom K., Seismic investigations target the Yavapai-Mazatzal transition zone and the Jemez Lineament in northeastern New Mexico, this volume.

McQuarrie, N. and Chase, C.G., Raising the Colorado Plateau, *Geology,* 28, 1, 91–94, 2000.

Merewether, E.A., Oil and gas plays of the Las Animas arch, southeastern Colorado, *U.S. Geol. Surv. Open-File Rep. 87–450D,* 22, 1987.

Oldow, J.S., Mesozoic transpression in the Cordilleran Orogen, Abstract with Programs, *Geol. Soc. Am.,* 21, 5, 125, 1989.

Perry, F.V., Baldridge, W.S. and DePaolo, D.J., Role of astenosphere and lithosphere in the genesis of Late Cenozoic basaltic rocks from the Rio Grande rift and adjacent regions of the southwestern United States, *J. Geophys. Res.,* 92, 9193–9213, 1987.

Powell, J.W., Exploration of the Colorado River of the West, *Smithsonian Institution,* 291, 1876.

Sharry, J., Langan, R.T., Jovanovich, D.B., Jones, G.B., Hill, N.R. and Guidish, T.M., Enhanced imaging of the COCORP seismic line, Wind River Mountains, *Geodynamic Series,* 13, 223–263, 1986.

Smithson, S.B., Brewer, J.A., Kaufman, S., Olivier, J.E. and Hurich, C.A., Nature of the Wind River thrust, Wyoming, from COCORP deep reflection data and from gravity data, *Geology,* 6, 648–652, 1978.

Stock, J. and Molnar, P., Uncertainties and implications of the Late Cretaceous and Tertiary position of North America relative to the Farallon, Kula and Pacific plates, *Tectonics,* 7, 1339–1384, 1988.

Stone, D.S., Geometry and kinematics of thrust-fold structures in central Rocky Mountain foreland, *Am. Ass. Pet. Geol. Bull.,* 70, 8, 1057, 1986.

Tikoff, B. and Maxson, J., A lithospheric buckling model for the Laramide Orogeny, *J. Czech Geol. Soc.,* 45, 3–4, 269–270, 2000.

Timmons, J.M., Karlstrom, K.E., Dehler, C.M., Geissman, H.W. and Heizler, M.T., Proterozoic multistage (ca.1.1 and 0.8 Ga) extension recorded in the Grand Canyon Supergroup and establishment of northwest- and north-trending tectonic grains in the Southwestern United States, *Geol. Soc. Am. Bull.,* 113, 2, 163–180, 2001.

Tindall, S.E. and Davis, G.H., Monocline development by oblique-slip fault-propagation folding; the East Kaibab Monocline, Colorado Plateau, Utah, *J. Struct. Geol.,* 21, 10, 1303–1320, 1999.

Tweto, O., Laramide (late Cretaceous-early Tertiary) orogeny in the southern Rocky Mountains, *Geol. Soc. Am.,* Memoir n.144, Cenozoic history of the southern Rocky Mountains, 1–44, 1975.

Tweto, O., Las Animas Formation (Upper Precambrian) in the subsurface of southeastern Colorado: *U.S. Geol. Surv. Bull. 1529-G,* 14 p, 1983.

Volk, R.W., The Denver Basin and the Las Animas arch, in *Geologic atlas of the Rocky Mountain region,* edited by Mallory, W.W., Denver, Colorado, *Rocky Mountain Association of Geologists,* 281–282, 1972.

Wawrzyniec, T.F., Geissman, J.W., Melker, M.D. and Hubbard, M., Dextral shear along the eastern margin of the Colorado Plateau; a kinematic link between Laramide contraction and Rio Grande rifting (ca. 75–13 Ma), *J. Geol.,* 110, 3, 305–324, 2002.

Woodward, L.A., Tectonic map of the Rocky Mountain region of the United States, in *Sedimentary cover – North American craton* edited by Sloss, L.L.: U.S.: Boulder, Colorado, Geology of North America, *Geol. Soc. Am., D-2,* Plate 2, 1988.

Woodward, L.A., Anderson, O.J. and Lucas, S.G., Mesozoic stratigraphic constrains on Laramide right slip on the east side of the Colorado Plateau, *Geology,* 25, 843–846, 1997.

Nicole Bolay-Koenig and Eric E. Erslev, Department of Earth Resources, Colorado State University, Fort Collins, CO 80523.

Karl E. Karlstrom, Department of Earth and Planetary Sciences, University of New Mexico, 200 Yale Blv. NE, Albuquerque, NM 87131.

Alan Levander and Maria Beatrice Magnani, Department of Earth Science, MS 126, Rice University, 6100 Main St., Houston, TX 77005.

An Integrated Geophysical Study of the Southeastern Sangre de Cristo Mountains, New Mexico: Summary

Oscar Quezada[1], G. Randy Keller, and Christopher Andronicos

Department of Geological Sciences, PACES, University of Texas at El Paso, Texas

The Sangre de Cristo Mountains in northern New Mexico are the uplifted eastern flank of the Rio Grande rift and expose a number of important geologic features whose origins are of great geologic interest (Figure 1). We have investigated this area, and here we present an integrated analysis of a variety of geophysical data that features almost 100 km of newly released seismic reflection data (Figure 1). The southeastern Sangre de Cristo Mountains are the site of a pronounced and yet poorly understood gravity minimum. By integrating different geophysical and geological data, we have focused on this gravity anomaly hoping that modeling the major upper crustal structures related to it aid in unraveling the complex tectonic evolution of the southern Rocky Mountains.

The seismic reflection profiles provide many insights about the evolution of the upper crust in this region. For example, the faulting interpreted from the northernmost SC1 (Figure 2) seismic profile is radically different than the relatively unbroken base of the Paleozoic section seen further south. Due to their relatively low dip angle, these faults could be related to east-verging Laramide-age compression and were later reactivated by Tertiary extension (*Adam Read*, personal communication). Modeling of Bouguer gravity anomalies shows that the presence of a low density Guadalupita pluton (1.68 Ga) and the combined effects of adjacent Tertiary (Las Vegas basin) and Paleozoic (Rainsville trough) strata agrees best with the seismic and geologic data. The seismic reflection data rule out the presence of an extensive sub-horizontal detachment in the first 10–12 km of the crust (as had been previously suggested) and show that there is a high degree of deformation in the Precambrian basement underneath the Paleozoic strata of the Sangre de Cristo uplift. The high amplitude intra-basement reflectors interpreted in the seismic data reveal the NE–SW trending features that illustrate a section of the amalgama-

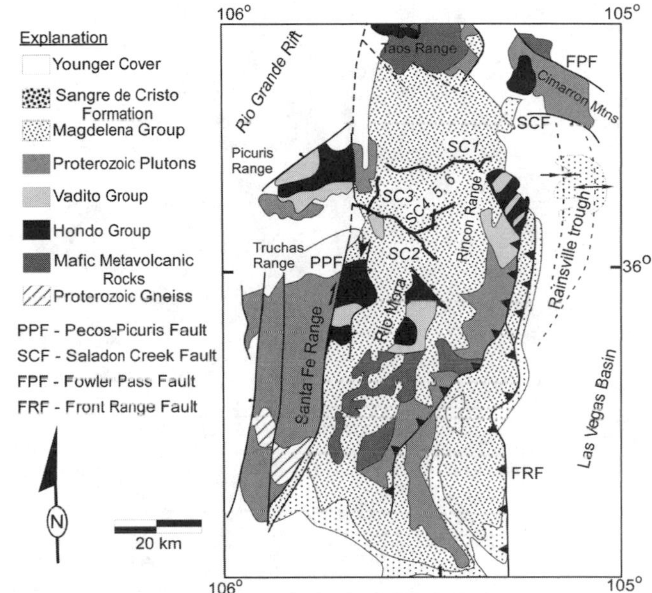

Figure 1. Tectonic index map of the Sangre de Cristo Mountains region, New Mexico. The solid lines labeled SC1–SC6 are the seismic reflection lines presented and interpreted in this paper.

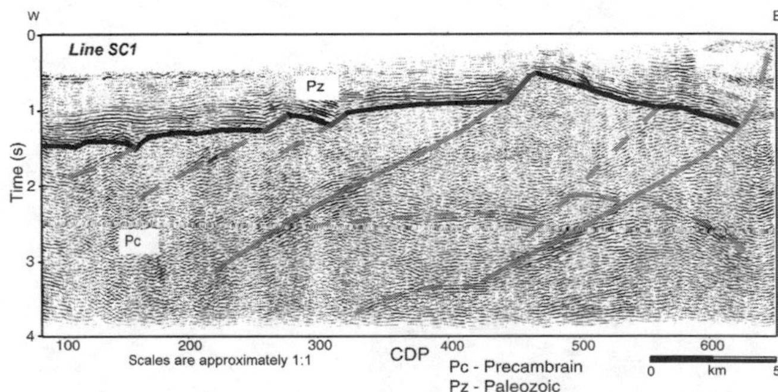

Figure 2. Interpreted version of seismic reflection line SC 1. The numbers along the horizontal axis are Common Depth Point numbers. The black, sub-horizontal line is the interpreted top of the Precambrian basement. The lighter gray lines are interpreted faults and intrabasement reflectors.

[1] Now at Anadarko, Houston, Texas

The Rocky Mountain Region: An Evolving Lithosphere
Geophysical Monograph Series 154

10.1029/154GM18

tion and subsequent deformation of the shallow crust in northern New Mexico.

The major structures interpreted in this integrated study can be classified into two distinct deformational episodes. (1) Proterozoic accretion (1.7–1.6 Ga), which we propose is represented by the imbricate-like intra-basement reflectors on the seismic reflection profiles, as well as, intrusion and 1.6 Ga metamorphism of the Guadalupita pluton in the Rincon Range; (2) Laramide-age compression with late Tertiary reactivation. Paleozoic basin and uplift deformation associated with the Ancestral Rocky Mountains was not directly observed in the seismic data. However, previous studies as well as our gravity modeling indicate that the Taos and Rainsville troughs were created during this period.

The complete version of this paper including color diagrams can be found in the CDROM in the back cover sleeve.

Christopher Andronicos and G. Randy Keller, Department of Geological Sciences, University of Texas at El Paso, 500 W. University Ave., El Paso, TX, 79968.

Oscar Quezada, Anadarko, Houston, TX.

An Integrated Geophysical Analysis of the Great Divide Basin and Adjacent Uplifts, Southwestern Wyoming: Summary

Veronica J. Cline

ChevronTexaco Energy Technology Company, Houston, Texas

G. Randy Keller

Department of Geological Sciences, University of Texas at El Paso, El Paso, Texas

Southwestern Wyoming is located at the margin of the Archean Wyoming craton but has experienced significant deformation as a result of both the Sevier and Laramide orogenies. This study focuses on the nature and extent of this deformation and its interactions with structures within the Precambrian basement. We used about 350 km of newly released industry seismic reflection data along with gravity data, satellite imagery, and drilling information in an integrated analysis focusing on the north-south trending Rock Springs uplift, the northwest-trending Wind River uplift and the west-east trending Sweetwater uplift. These features form arches that are bounded by the Green River, Wind River, Great Divide, and the Washakie basins (Fig. 1). An example of the seismic data is shown in Figure 2 displays structural complexity at the northeast boundary of the Great Divide basin involving high-angle reverse faults with northeast dips. The fault that lies roughly in the middle of the line is interpreted to be the southeastern extension of the Wind River thrust, and the fault at the northeast end of the line is interpreted to be the Mormon Trail thrust. A gravity profile was modeled as a medium to integrate all of the data. This model of the upper crust indicates the presence of inhomogeneities in the Archean basement that have not been recognized previously. The basement northeast of the Wind River

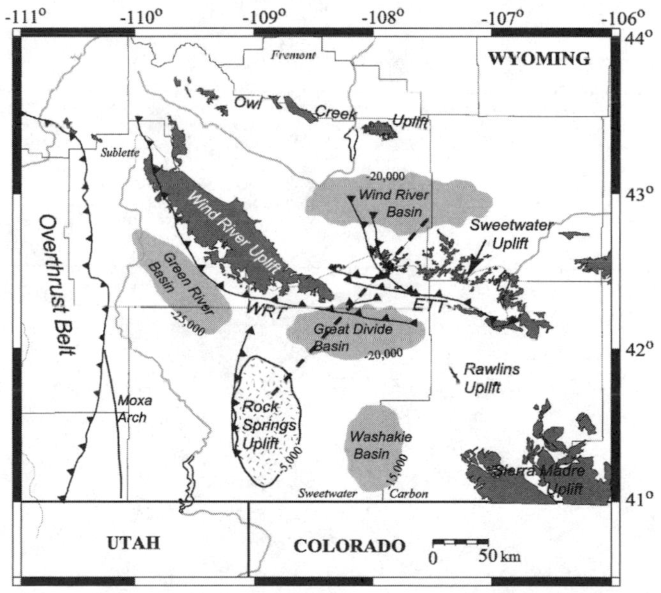

Figure 1: Index map of the southwestern Wyoming region. The outlines of Precambrian outcrops (dark gray) and uplifts (stippled) and thrust faults (lines with teeth) were digitized from the literature. The deepest portions of the basins were also digitized from the literature and are shown as light gray areas enclosed by a contour line that represents depth (feet) below sea level. WRT–Wind River thrust; ETT-Emigrant Trail thrust. The dashed line is the location of the integrated earth model constructed in this study. The seismic reflection profile shown in Figure 2 covers most of the southwest portion of this model.

The Rocky Mountain Region: An Evolving Lithosphere
Geophysical Monograph Series 154

10.1029/154GM19

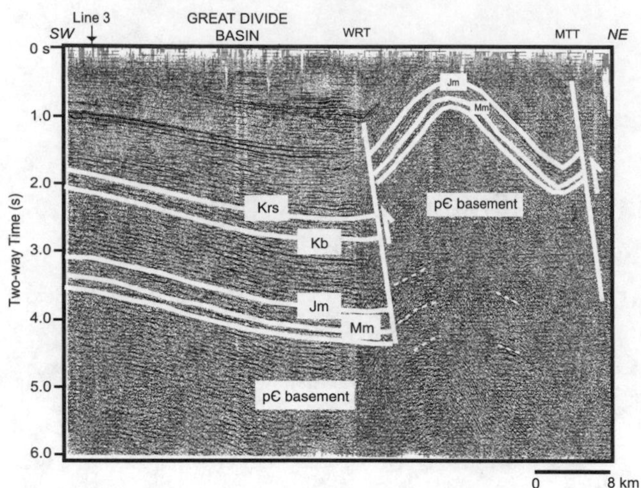

Figure 2: Interpreted version of seismic Line 4b that shows the southern extension of the Wind River thrust. Krs-Rock Springs Formation (Upper Cretaceous); Kb-Baxter (Cody) Shale (Upper Cretaceous); Jm–Morrison Formation (Jurassic); Mn–Madison Formation (Mississippian). WRT–Wind River thrust; MTT–Mormon Trail thrust. This record section extends to a depth of about 10 km. Vertical exaggeration is ~3.4.

thrust contains considerable reflectivity indicating folding or fabric that either reflects or controls Laramide structures. The interweaving of reflectors in one line resemble imbricate structures shown in the CD-ROM Cheyenne belt deep reflection profile and could be related to an ancient structural boundary within the basement. Our analysis shows that the multiple thrusts bounding the Sweetwater uplift occur near major inhomogeneities in the Precambrian basement. Spatial relations we observe are consistent with the hypothesis that anastomosing arches characterize Laramide foreland deformation because the large positive gravity anomalies associated with the Wind River and Sweetwater uplifts smoothly merge.

The complete version of this paper including color diagrams can be found in the CDROM in the back cover sleeve.

Veronica J. Cline, ChevronTexaco Energy Technology Co., 4800Fournace Pl., Bellaire, TX 77401

G. Randy Keller, Department of Geological Sciences, University of Texas at El Paso, El Paso, TX 79968

Results of the CD-ROM Project Seismic Refraction/Wide-Angle Reflection Experiment: The Upper and Middle Crust

Hanna-Maria Rumpel and Claus Prodehl

Geophysical Institute, Karlsruhe University, Germany

Catherine M. Snelson

Department of Geoscience, University of Nevada, Las Vegas

G. Randy Keller

Department of Geological Sciences, University of Texas at El Paso, El Paso

During the field experiments of the Continental Dynamics Rocky Mountains Project (CD-ROM) in 1999, a 950 km long refraction / wide-angle reflection seismic profile was acquired in the southern Rocky Mountains. This 950 km long seismic line extended from northern New Mexico to central Wyoming. The sedimentary thickness along the line varies between <100 m near the Arkansas River, Colorado and > 3 km in the North Park basin, Colorado. The Precambrian basement in New Mexico is ~2 km deep. In Colorado and Wyoming, portions of the line are located on basement exposures intervened by basins up to 2.5 km in depth. The velocities at the top of the upper crust vary between 5.75 km/s and 6.15 km/s and increase gradually to (v_{max} = 6.35 km/s) at 25 km depth. The data suggest that the uppermost Precambrian basement consists of late Proterozoic volcanic and sedimentary rocks or felsic intrusions. The northern Wet Mountains are underlain by high velocities in the upper crust, and Cambrian mafic rocks crop out in these mountains. The Colorado Mineral belt correlates with a broad zone of relatively low velocities in the Precambrian crust indicating the presence of felsic intrusions. The average thickness of the upper crust is 20 km in the Great Plains and 25 km in the Rocky Mountains region. Strong lateral velocity variations occur in the upper crust. Major vertical velocity steps are prominent at the upper to middle crust. The middle crust velocities increase from 6.60 to 6.75 km/s and its average thickness is ~10 km.

1. INTRODUCTION

The lithosphere of the southern Rocky Mountain region formed in Precambrian time, as discussed by *Karlstrom et al.* [this volume]. Its complex tectonic history has lead finally to the present-day configuration which we see geophysically. Previous studies of crustal structure based on active source studies [*Prodehl et al.*, this volume; *Keller et al.*, 1998], and passive source studies [*Sheehan*, this volume] provide a valuable regional framework and foundation for our more detailed work. The main aim of the interdisciplinary CD-ROM project is to study the Proterozoic assembly of southwestern North America, which is a challenge because of overprinting by

The Rocky Mountain Region: An Evolving Lithosphere
Geophysical Monograph Series 154

10.1029/154GM20

Phanerozoic orogeny processes such as the Ancestral Rocky Mountain orogeny [*Kluth and Coney*, 1981], the Laramide orogeny, and the formation of the Cenozoic Rio Grande rift [*Keller and Baldridge*, 1999].

A major element of the CD-ROM seismic effort was a refraction / wide-angle reflection profile (Figure 1). For the sake of brevity, we refer to this result as a refraction profile, but the reader should keep in mind that many wide-angle reflections were recorded and modeled. This profile was designed to provide a crustal and uppermost mantle velocity model that would be both a useful result in its own right and a source of velocity control for the passive seismic and deep seismic reflection groups. With a total length of ~950 km, the refraction profile parallels and connects the shorter and dis-

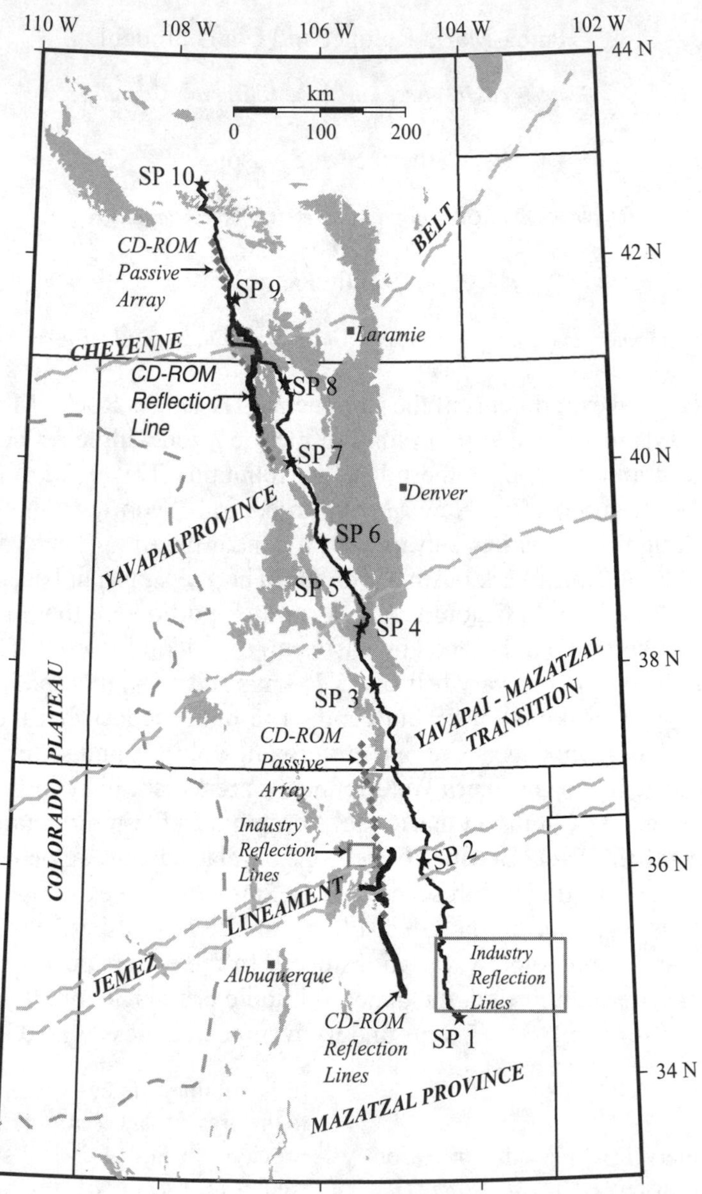

Figure 1. Index map showing Precambrian basement exposures and tectonic provinces for the Rocky Mountain region. Black stars are shotpoint locations for the CDROM refraction / wide-angle reflection profile, the thin black line represents the seismic stations of the profile. The thick black lines are the reflection profiles, the gray diamonds are the passive array. Areas of industry reflection lines are marked by quadrangles. The shot points are Fort Sumner (1), Wagon Mound (2), Gardner (3), Canon City (4), Guffey (5), Fairplay (6), Kremmling (7), Walden (8), Rawlins (19) and Day Loma (10).

continuous seismic reflection lines [*Morozova et al.*, this volume] and teleseismic profiles [*Zurek and Dueker, Yuan and Dueker*, this volume] which were directed perpendicular to the strike of the Proterozoic provinces.

In New Mexico, the refraction profile is located on the Great Plains (Figure 1, shot point (SP) 1–2), and it enters the Rocky Mountains south of the Wet Mountains around Gardner, CO (SP3). From this point, it extends through South Park (SP 4–6) and North Park (SP 7–8) that are located between the Front and Park Range. In Wyoming, it crosses the Sierra Madre continuing north to the Rawlins uplift (SP 9) and terminates in the Gas Hills region (SP 10). More than 1200 stations were deployed along this 950 km long transect. The receiver spacing was 800 m and the shot point spacing averaged 100 km (less in Colorado and more to the north and south). The charge sizes varied between 900 kg for the shot points in the center of the profile to 4.5 t at the northern end (Day Loma, Wyoming; SP 10).

We chose to interpret the data from the refraction profile from two different perspectives and thus employed two different modeling and interpretation scenarios. The first scenario resulted from the fact that the CD-ROM was not funded to acquire passive seismic and deep seismic reflection data in central Colorado, and thus, the refraction experiment was configured to collect more detailed data in this region. This extra effort required additional seismic sources and seismic recorders and was made possible by a grant from the German Science Foundation. In this paper, we

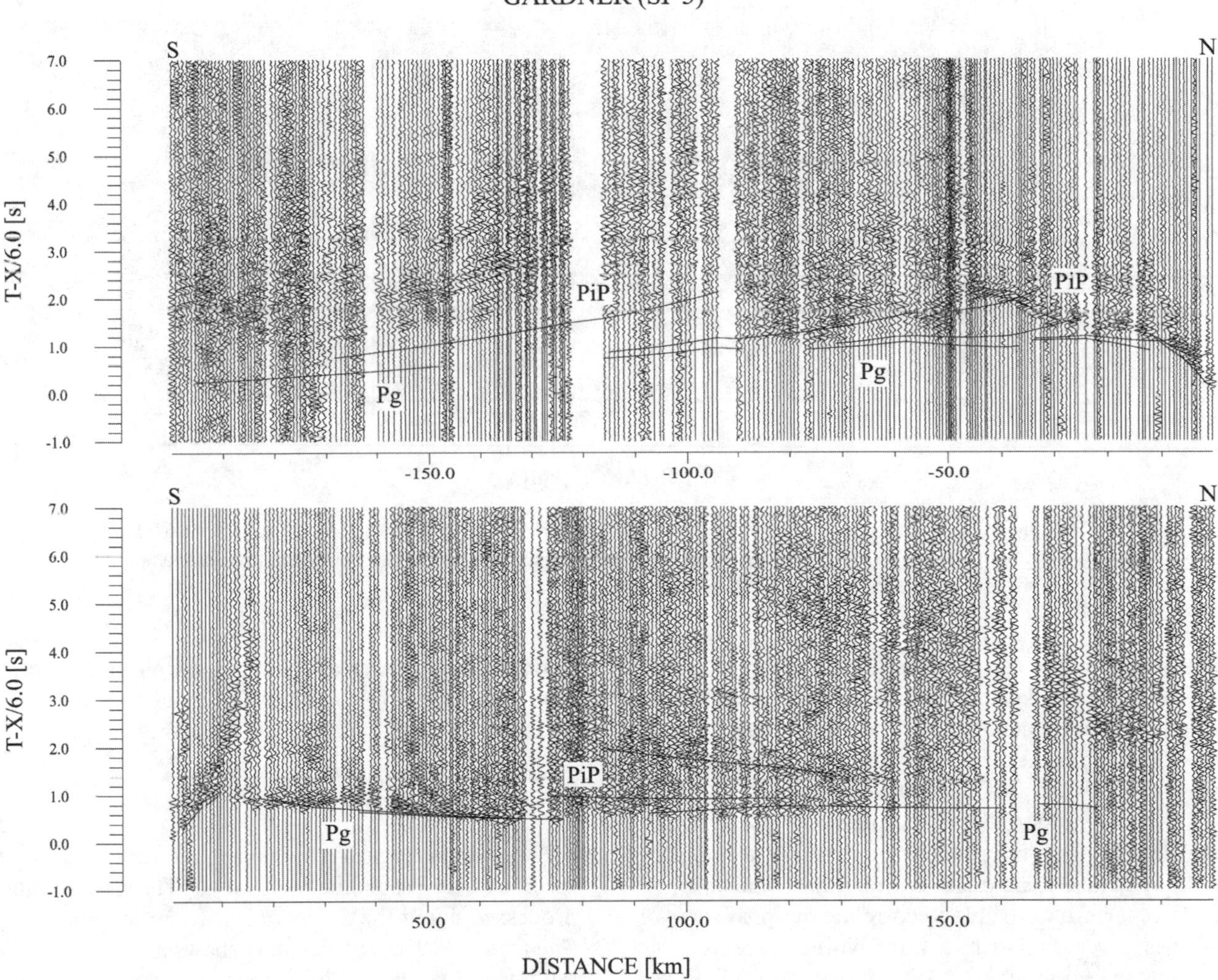

Figure 2. Seismic record section for south and north of Gardner (SP 3). Travel time reduction velocity is 6.0 km/s. Theoretical travel time curves calculated from the shot point to the receiver locations are overlayed. Phases seen on the section are first arrivals (Pg) and intracrustal reflections (PiP).

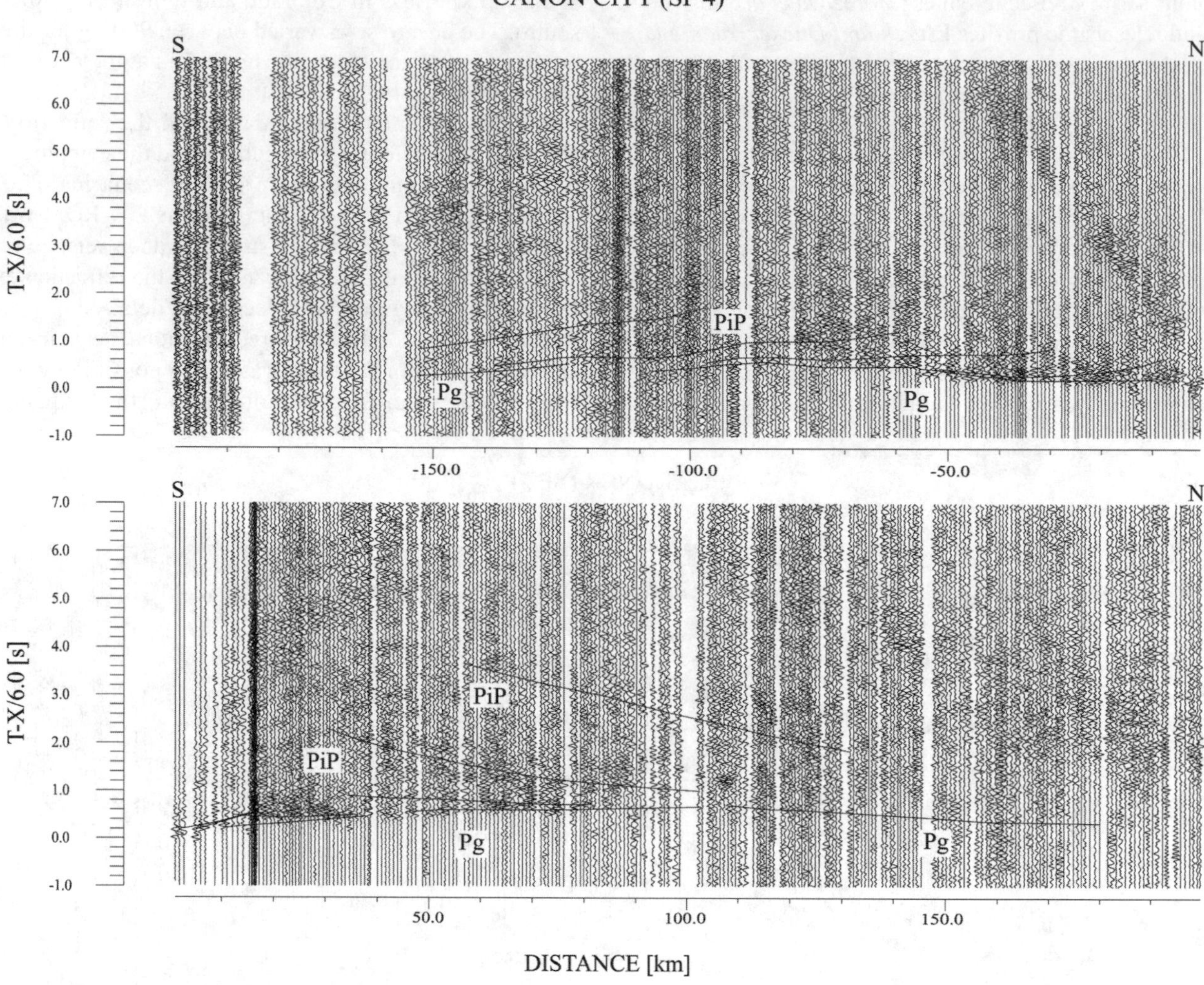

Figure 3. Seismic record section for south and north of Canon City (SP 4). Travel time reduction velocity is 6.0 km/s. Theoretical travel time curves calculated from the shot point to the receiver locations are overlayed. Phases seen on the section are first arrivals (Pg) and intracrustal reflections (PiP).

emphasize the results of our attempt to extract as much detail as possible from the refraction data. Because of the design of the experiment and the nature of the data, we will, to some extent focus on the upper and middle crust and the central portion of the profile. Our approach in this paper involved picking numerous crustal phases, most of which were reflections. This approach introduces an element of subjectivity because some phase correlations could be questioned. However, we were able to correlate our phase arrival picks across several record sections while honoring constraints such as reciprocity. In spite of our effort to extract as much detail as possible, the model that we derived undoubtedly portrays a structure that is considerably smoother than reality. The second scenario took a more regional view of our data set and emphasized the lowermost crust and upper mantle. The results of this effort are described by *Snelson et al.* [this volume].

2. ANALYSIS

The seismic refraction / wide-angle reflection data set was recorded with a sampling rate of 4 ms. The seismic data were processed using the *SeismicHandler* processing software [*Stammler*, 1993]. The raw data shows a low signal to noise ratio which can be much improved by applying a filter. The main frequency content is below 30 Hz. Noise peaks can be found at 60 Hz (powerline), 90 Hz (instrument noise) and 10 Hz (either subtone of the 60 Hz peak or cultural noise like a

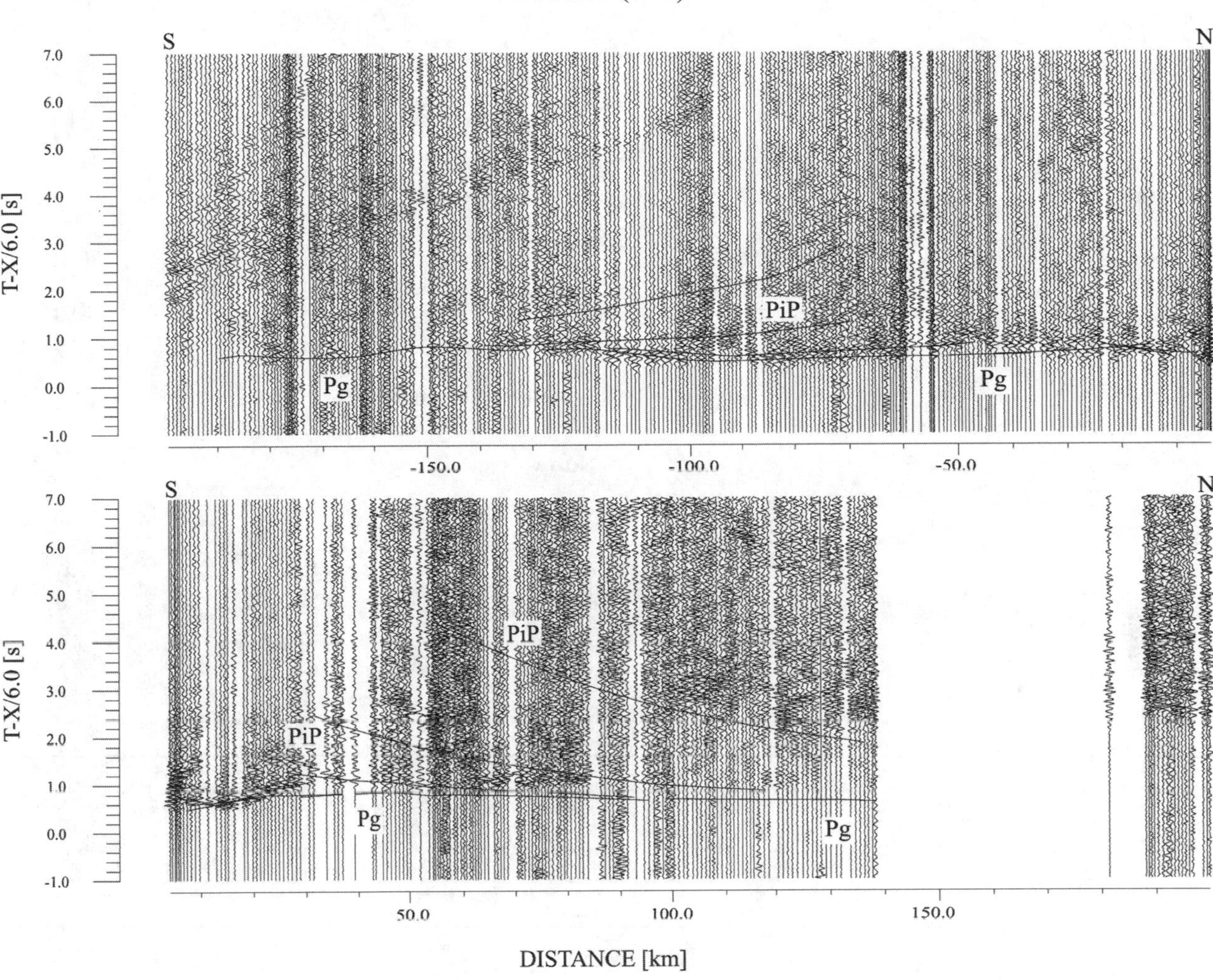

Figure 4. Seismic record section for south and north of Guffey (SP 5). Travel time reduction velocity is 6.0 km/s. Theoretical travel time curves calculated from the shot point to receiver locations are overlayed. Phases seen on the section are first arrivals (Pg) and intracrustal reflections (PiP).

water pump). After some experimentation, we found that a 4 – 15 Hz bandpass filter along with a very narrow bandstop at 10 Hz produced the best results. Data examples can be found in Figure 2–6 for the shot points in Colorado. The sections for the shot points in New Mexico and Wyoming can be found in *Snelson et al.* [this volume].

In a further step, the arrival times for the different phases in the P-wave sections were picked. The following phases were correlated: First arrivals, which represent the diving waves through the crust (pg) and intracrustal reflection (PiP and PIP for the deepest one); reflection from the crust-mantle boundary (PmP) and the refracted wave through the upper mantle (Pn). Since this paper focuses on the upper to middle crust PIP, PmP and Pn are not discussed here. The overall data quality is good. The first arrivals are very clear and can be followed for a long distance (200 km or more). The PiP reflections are weaker but can be found throughout a large portion of the data.

To ensure a pick consistency for the weaker phases we compared every picked phase with their counterpart of the reversed shot. Our confidence in our phase correlations is reflected in our estimates of the uncertainties for the phase picks. We used ±0.05 s for the Pg and (0.1 s for the later arrivals (PiP).

Following the phase identification 1D modeling was carried out for each shot point and the results were used to create a 2D model. Additionally, the thickness of the sedimentary cover known from other geophysically and geological data was integrated. The initial 2D model was then improved by ray tracing. We used the RAYINVR code by *Zelt and Smith* [1992] and

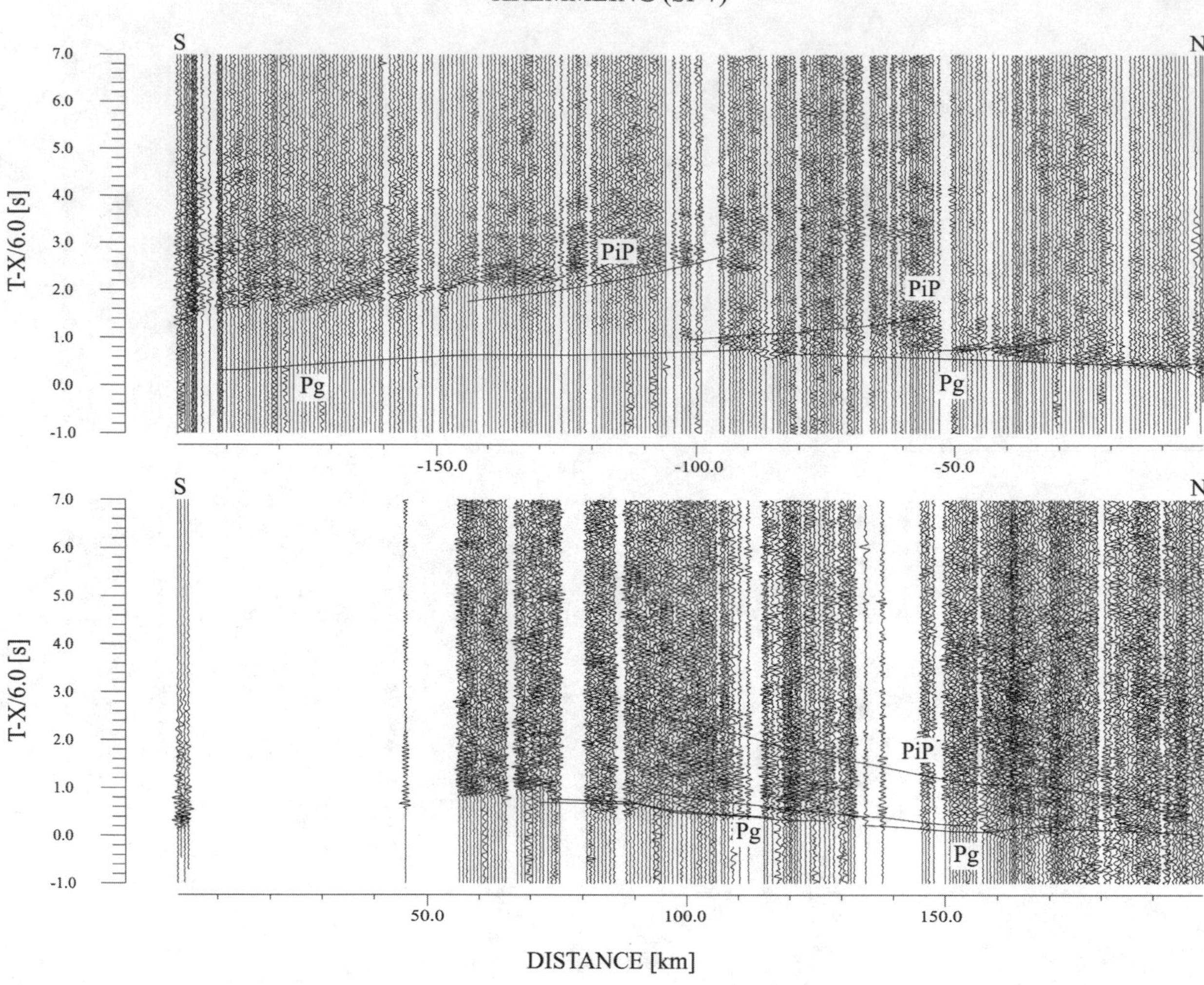

Figure 5. Seismic record section for south and north of Kremmling (SP 7). Travel time reduction velocity is 6.0 km/s. Theoretical travel time curves calculated from the shot point to receiver locations are overlayed. Phases seen on the section are first arrivals (Pg) and intracrustal reflections (PiP).

modeled layer by layer from top to bottom in a forward routine and verified the results by inversion. The inversion was carried out applying the DMPLSTSQR code by *Zelt and Smith* [1992], which uses the method of damped least-squares. For that the partial derivates of travel times with respect to model parameters and the travel time residuals were calculated. Careful modeling taking into account the data variations along the profile with topography and geology, the ray coverage for different geologic features from various angles, shot and reverse shot constellations led to a final P-wave model, which explains more than 90% of our travel time data. The fit of the theoretical travel times is shown in Figures 2–6. Following our phase uncertainties, the adjustment for the upper and middle crust is good. We achieved a final P-wave velocity model that explains our data and will be discussed in detail in the following paragraph. The introduced model uncertainties were ±1.0 km for z-nodes defining the interfaces and ±1.0 km/s for the velocity nodes. The used data points were 2765 from 3003 picked arrival time (92%). The travel time residual T_{RMS} is 0.18 s and the X^2-parameter, which defines the over or under parametrization of the model is $X^2 = 4.67$ ($0 < X^2 < 12$ is recommended) [*Zelt and Smith*, 1992].

3. P-WAVE VELOCITY MODEL

The P-wave velocity model was set up to be 953 km long and 60 km deep. For this paper we zoomed in on the upper / middle crust (Plate 1). It nominally consists of 9 layers where

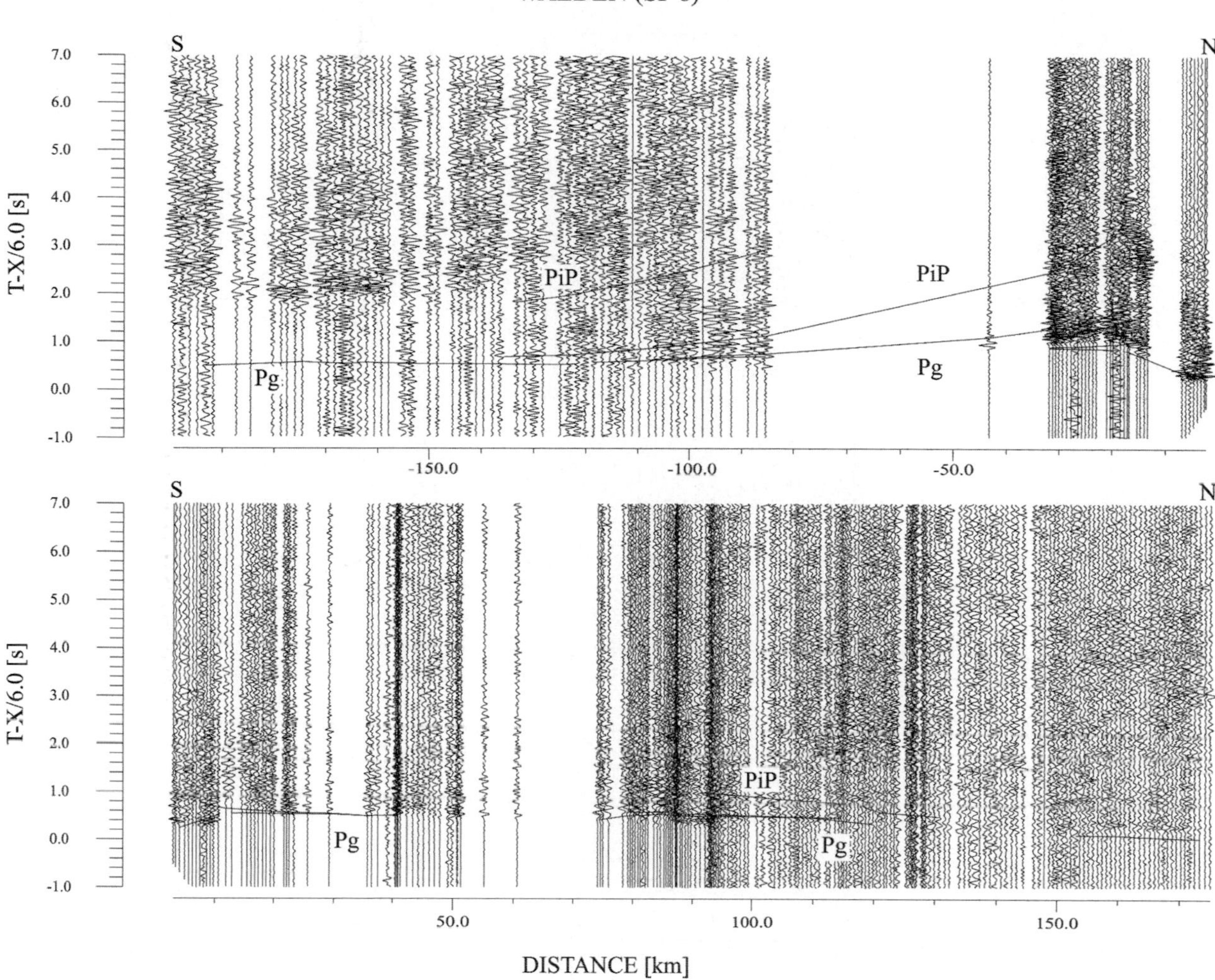

Figure 6. Seismic record section for south and north of Walden (SP 8). Travel time reduction velocity is 6.0 km/s. Theoretical travel time curves calculated from the shot point to receiver locations are overlayed. Phases seen on the section are first arrivals (Pg) and intracrustal reflections (PiP).

sedimentary layers at the top locally vanish along Precambrian basement out-crops. The elevation differences (1.30–3.50 km above sea level) were also taken into account by an additional 4 km at the top of the model space. In the following the abbreviation *mkm* refers to *m*odel *k*ilo*m*eter and describes the distance from the southern end of the velocity model. For example Wagon Mound (SP 2) is located at mkm 173.

To answer the question of how thick the sedimentary cover along the profile is, we first have to define the top of the basement. One possibility is to define the depth at which the velocity reaches the typical values of 6.0 km/s as seismic basement. However, we have both the geologic control and motivation to map the basement surface as carefully as possible. Published contour maps of the depth to the Precambrian basement [*Mallory*, 1972; *Suleiman and Keller*, 1985; *Blackstone*, 1993] indicate that it occurs at a shallower depth than the 6.0 km/s isovelocity in our models (Plates 1 and 2). We therefore defined the top of the Precambrian basement along most of the model to occur where strong velocity contrasts occur above velocity values of 5.50 – 6.00 km/s.

In New Mexico the maximum thickness of the sedimentary cover reaches ~2 km and it is ~3 km in Colorado and Wyoming (Plate 2). There is at least a thin veneer of sediments along most of the profile, but exceptions are the Wet Mountains in Colorado and the Sierra Madre in Wyoming (Precambrian basement out-crops). Large lateral velocity gradients are present in the sediments, detailed geologic interpretations are presented in the discussion below.

For the purpose of our analysis, we will define the upper crust to consist of Precambrian basement material whose

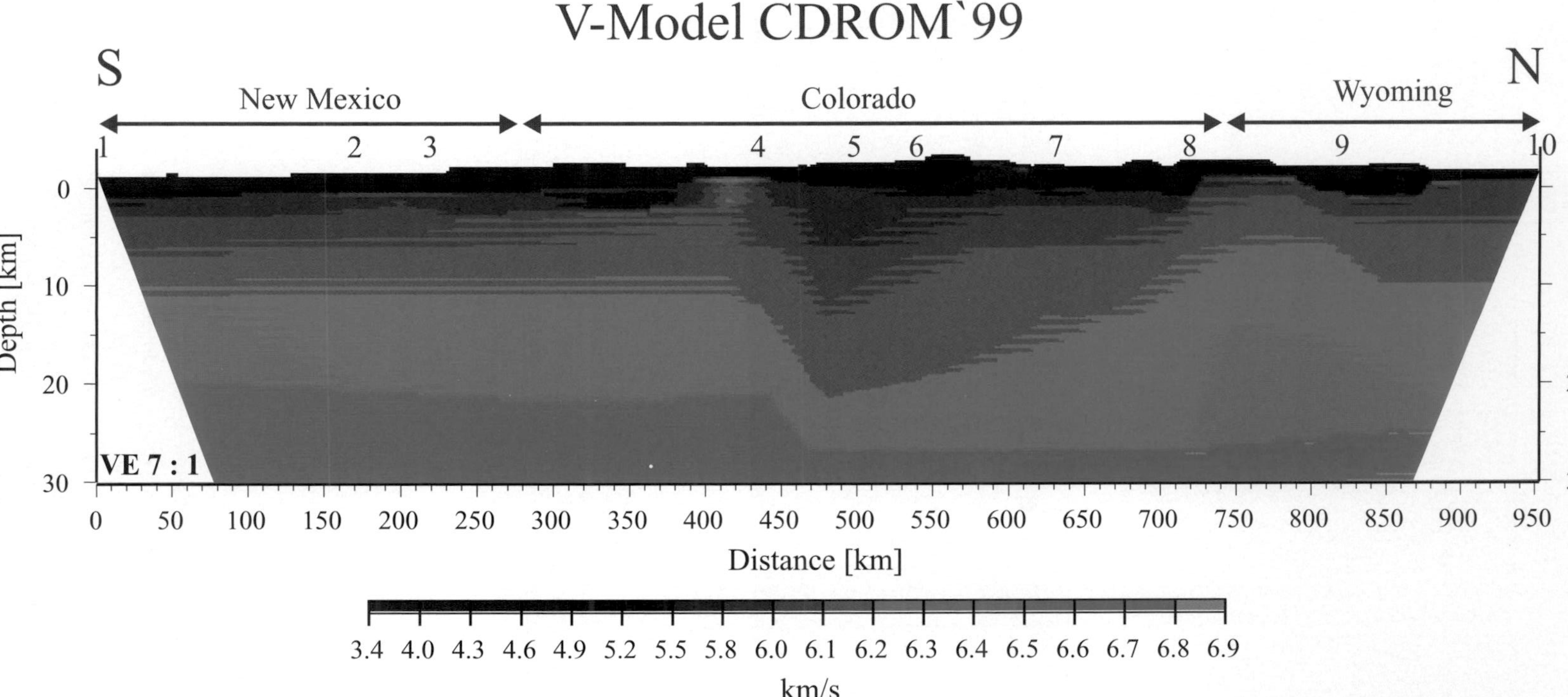

Plate 1. Two-dimensional P-velocity model for the upper 30 km of the CDROM reflection / wide-angle reflection profile. The velocities range from 3.4 km/s in the sediments (bluish) to 6.9 km/s in the middle crust (greenish). Shot Points are marked by numbers at the top of the model as well as the U.S. states crossed by the profile.

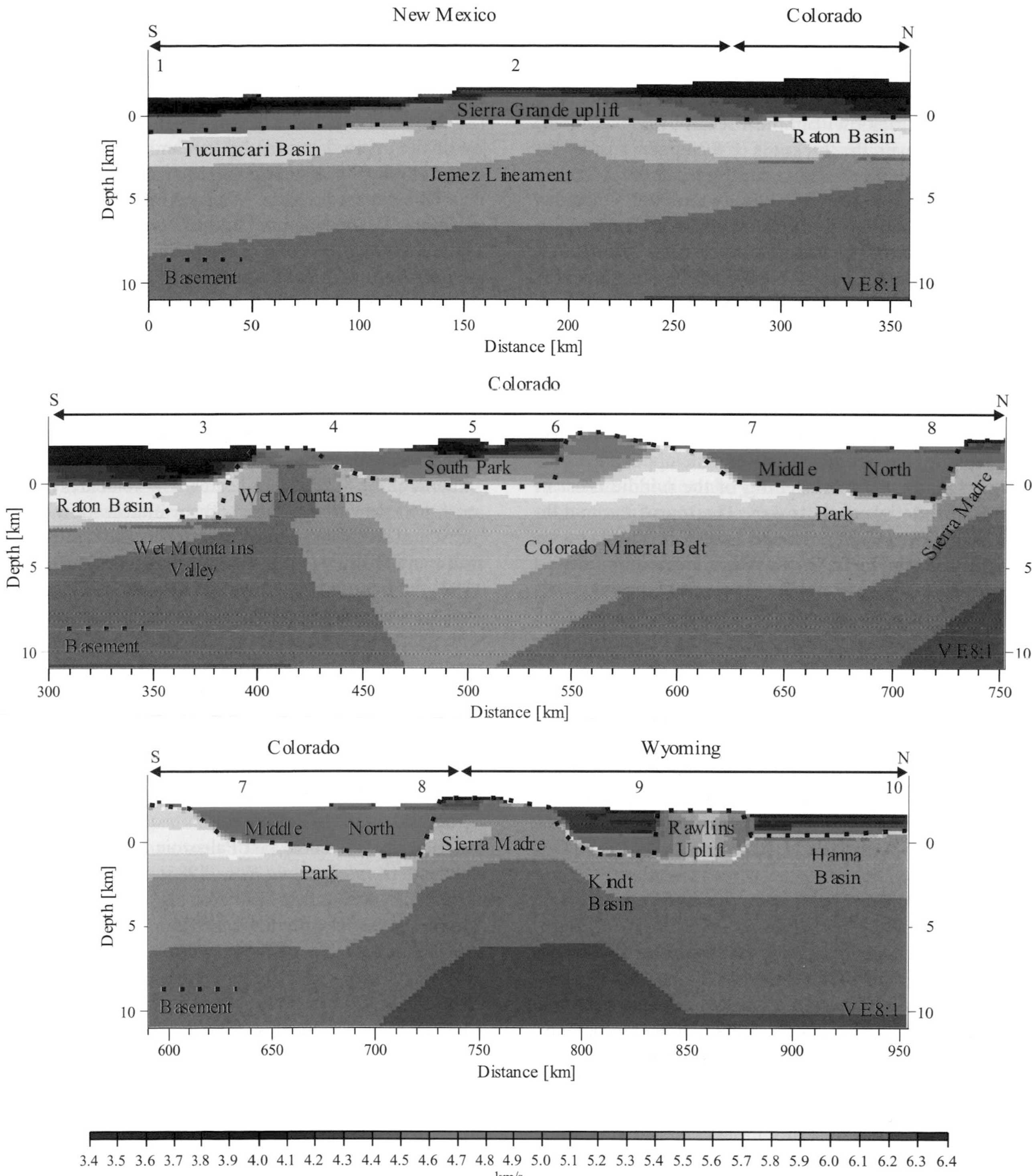

Plate 2. P-wave velocity model for the upper 15km along the CD-ROM profile. The three segments are New Mexico, Colorado and Wyoming from top to bottom. The numbers at the top of each part mark the shot points. The dotted line follows the Precambrian basement. The sediments show blue to green colors. The crystalline crust is in yellow to red colors.

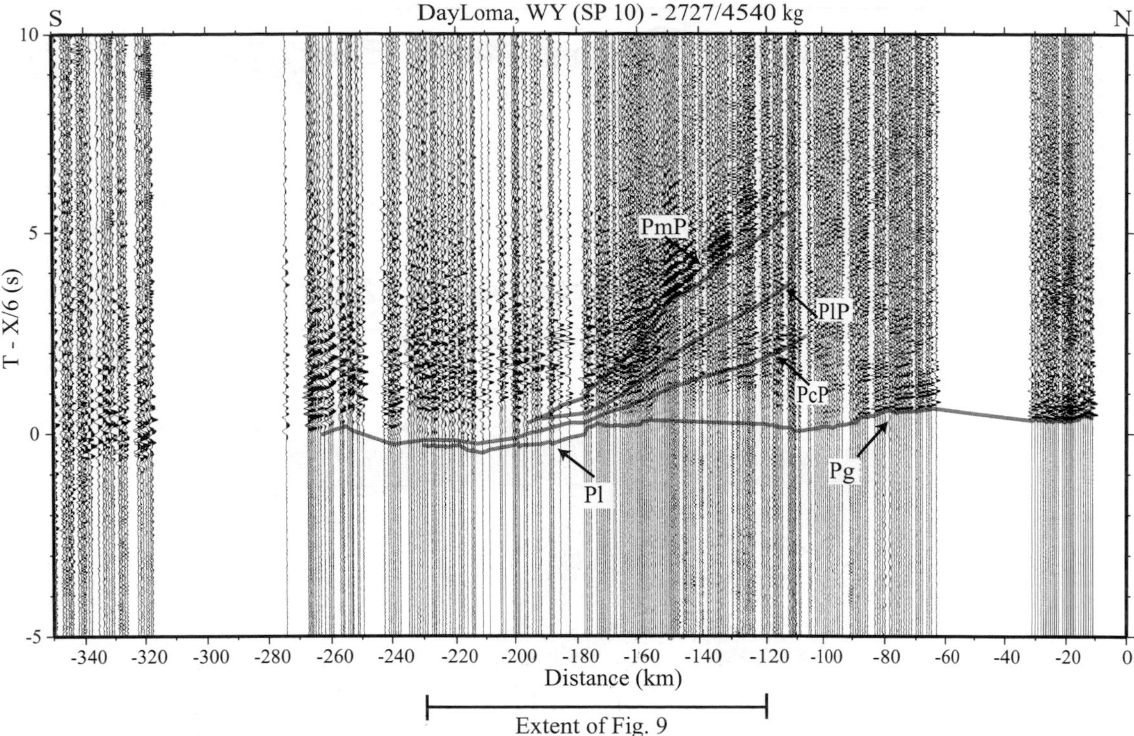

Figure 5. Seismic record section for Day Loma, WY (SP 10). Bar at base of figure indicates the lateral extent of Figure 9 relative to this figure.

down to and up from a surface placed within the model that was derived from the tomography [*Hole and Zelt,* 1995]. The layer depth and geometry is then adjusted until the observed and calculated traveltimes fit well and the RMS is ~ 0.200 s. This procedure was used to determine the location of the top of the mid-crustal interface (PcP), the top of the lowermost crustal layer (PlP), and the top of the Moho (PmP). During this process, we observed that the depth to these interfaces constrained by reflections is typically shallower than the depth one would have picked based on the tomographic velocity model alone (Plate 1b). This reflects the tendency of tomography to produce continuous velocity-depth relationships and suggests that the velocity in the crust may be somewhat slower at shallower depths than is required by the tomographic results.

The addition of reflecting interfaces in the model provided additional constraints on crustal structure along the profile. The resulting final model (Plate 1c) shows that the mid-crustal interface is at a depth of about 25 km at the southern end of the model and increases in depth to about 30 km before rising to about 25 km at the northern end of the profile. The top of the lowermost crustal interface is at about 35 km depth at the southern end of the profile and deepens under central Colorado to about 45 km before thinning at the northern end of the profile to about 40 km. The Moho depth at the southern end of the profile is about 45 km. The Moho deepens to about 55 km under central Colorado and rises at the northern end of the profile to about 45 km.

4.3 Reflectivity Modeling

The CD-ROM data are characterized by large amplitude arrivals associated primarily with post-critical reflections from deep interfaces on most of the record sections (e.g., Figures 2 – 9). At large offsets, the move-out of the arrival that was generally the most energetic phase was relatively low (~6 km s^{-1}). Because this apparent velocity is approximately the average velocity of the column of material above the reflector and because there were arrivals that appeared to originate from deeper interfaces, we suspected that this wide-angle reflection was not associated with the Moho, but originated in the mid-crust (PcP). To confirm that this key interpretation was valid, synthetic seismograms were created to better model amplitudes seen within the seismic record sections [*Fuchs and Müller,* 1971; *Sandmeier,* 1990]. This technique is limited to 1-D models of the earth, but calculates full wave field synthetic waveforms for a user-defined set of offsets and time interval.

The Ft. Sumner (SP 1) record section was chosen as the best candidate for amplitude modeling, because the data quality was high and multiple phases from the deep crust were present (Figure 2). The 1-D reflectivity model is based on the final tomographic model (Plate 1b; Figure 10 inset). In order to create the large amplitude arrival that has a slow apparent velocity at post-critical offsets, the 1-D model includes a sharp boundary at ~22 km depth, which is correlative to the classic mid-crustal Conrad discontinuity (Figure 10a). In order to reproduce the PlP phase another relatively sharp boundary is necessary at ~35 km depth (Figure 10a). The high-velocity lowermost crust is represented by a zone of gradational velocity increase (7.0 to 7.4 km s^{-1}) down to a depth of 42 km, where a transition to mantle velocities of about 7.8 km s^{-1} occurs.

The synthetic record section produced from this 1-D model (Figure 10) shows waveforms that are similar to the original data. At pre-critical offsets (~100 km), the relative amplitudes of the PcP, PlP, and PmP phases are comparable to the observed data, as is the moveout of the reflections (Figure 10). Both the Pl and PlP phases from the lowermost crust match those in the original data. At post-critical offsets (>220 km), PcP has an apparent velocity of about 6.0 km s^{-1}, as in the observed data. PmP and PlP merge into a complex waveform whose relative amplitude is similar to the observed data near the critical point for PmP, but at greater offsets, the amplitude of this phase is larger in the synthetic seismograms than in the observed data. We could not simultaneously match the observed amplitudes of the pre-critical and post critical amplitudes of the PlP/PmP reflections, which suggest complexity that is beyond the 1-D approximation of the reflectivity method. In general, the observed data (Figures 2 – 10) include a considerable amount of reflectivity that follows the PmP and PcP phases at pre-critical offsets and the PcP phase at post-critical offsets. We consistently picked the beginning of this reflectivity as the arrival time of the phase, but this reflectivity is an indication of complex layering that is not 1-D in nature.

The same pattern of deep reflectors (PcP, PlP, and PmP) is evident along the profile as far north as southernmost Wyoming. However, the confidence with which we could identify all three deep reflections decreased at the northern end of the profile. Our approach of identifying these phases on each record section represents a bias towards making consistent phase correlations along the profile. Because of the integrated nature of our modeling approach, this bias does not greatly alter the overall crustal structure. However, in the following analysis, the reflecting interfaces appear more continuous than they probably are in the earth. This approach to phase correlations is different than that of *Levander et al.* [this volume] and explains some of the resultant differences in velocity models. However, comparison of the various velocity models

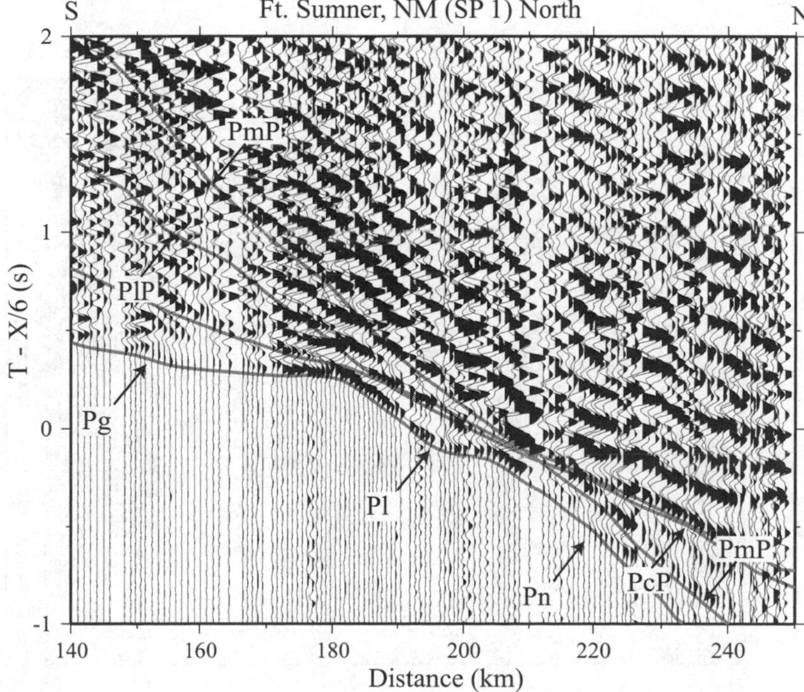

Figure 6. Close-up of the seismic record section from Figure 2 for the shot point at Fort Sumner, NM (SP 1). Notice the clear first arrivals that we interpret as the Pl phase.

shows that the tectonic implications of these differences are not significant [*Keller et al.*, this volume].

5. MEASURES OF RESOLUTION

A sense for the model resolution can be gained by evaluating the traveltime fits, ray coverage, and RMS error. Unfortunately with the tomographic inversion, a resolution matrix is not created because the technique is non-linear [*Hole,* 1992]. We have also used the uncertainties obtained from RAYINVR by *Rumpel et al.* [this volume] as a guide for discussing the relative uncertainties in our model.

5.1. Ray Coverage

The ray coverage or hit count represents the number of rays that pass through a particular cell. The velocity estimate for a given cell may be considered more reliable for a cell with higher hits counts. Considering the modest number of shots in the experiment, the ray coverage from first arrivals is adequate except for the deep portion of the northern third of the profile (Figure 11a). The inclusion of the reflected phases greatly increases the ray coverage in that part of the model in particular. Using the velocity model produced by *Rumpel et al.* [this volume], the ray coverage for the reflected arrivals as output from RAYINVR is shown in Figure 11b. Figure 11c shows the bounce points for the reflections as well as some of the refracted arrivals, but it is apparent that the main control for the lower portion of the model is the reflected arrivals.

5.2. Estimated Resolution

Based on forward and inversion modeling in collaboration with *Rumpel et al.* [this volume] and the tomographic inversion of the first arrivals and reflections, we believe that the resolution of the depth for the deep interfaces is +/- 2 km if the velocity structure and phase identification is assumed to be completely accurate. However, if one considers the velocity uncertainty, then the uncertainty for deeper interfaces could be as much as +/- 3 km. The estimated uncertainty related to the upper crustal velocity field is +/- 0.1 km s^{-1} at a depth of about 25 km and then increases to about +/- 0.2 km s^{-1} in the lower crust and upper mantle. Figure 11c shows the locations where rays crossed the various interfaces within the model space. These reflection points show where each interface is being sampled in the model space. As expected, the resolution is best for the central portion of the model.

6. INTEGRATION WITH GRAVITY DATA

The Bouguer gravity map of the southern Rocky Mountains (Plate 2, see CDROM in back cover sleeve) is characterized by a long wavelength, ca. 150 mGal gravity low

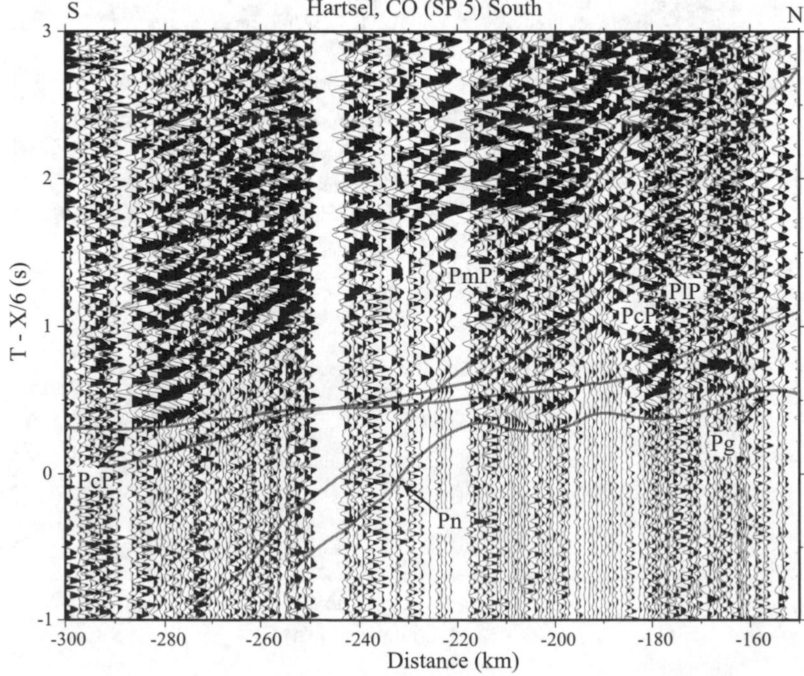

Figure 7. Close-up of the seismic record section for Hartsel, CO (SP 5) to the south. Notice the high amplitudes of the PcP phase.

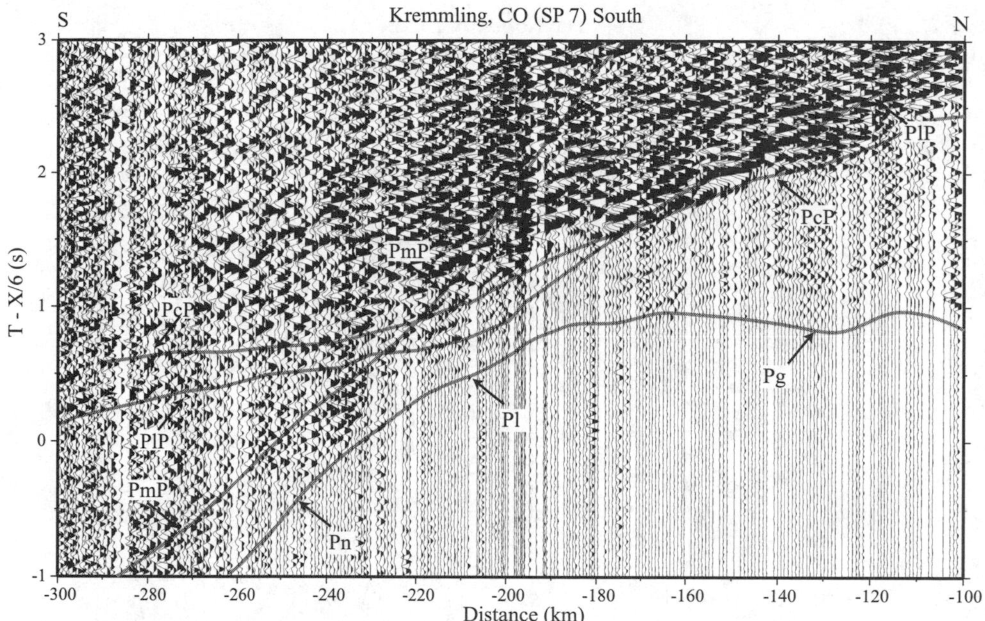

Figure 8. Close-up of the seismic record section for Kremmling, CO (SP 7) to the south. Notice the high amplitudes of the PcP phase. Also there are clear first arrivals that we interpret as Pl.

centered on the highest topography in Colorado. Definition of the source of this anomaly is controversial and is critical to understanding the mechanism of isostatic compensation of the topography. Elements of this anomaly have been variously interpreted as resulting from a shallow crustal source [*Isaacson and Smithson,* 1976; *McCoy et al.,* this volume], a combination of lateral density variations in the crust and Moho relief [*Li et al.,* 2002] and significant density variations in the mantle [*Sheehan et al,* 1995]. The new controls on crustal velocity and thickness as well as upper mantle velocity presented here, together with the Deep Probe results [*Gorman et al.,* 2002] allow us to further investigate the source of this anomaly and address its implications for isostatic compensation.

6.1. Density Modeling

In our gravity modeling, we used an implementation of the 2.5D algorithm of *Cady* [1980]. We employed an iterative forward modeling approach that incorporated all available geological and geophysical constraints in order to match the observed Bouguer anomaly values. The initial gravity model was derived from the final CD-ROM tomographic velocity model and our Deep Probe velocity model [*Snelson et al.,* 1998]. Initial density values were calculated from experimentally determined velocity/density relationships [*e.g., Christensen and Mooney,* 1995]. To guide modeling of the upper crust, a geologic cross-section along the profile was constructed using geologic maps and compiled drill hole data in the region [*Suleiman and Keller,* 1985; *Blackstone,* 1993; *Snelson et al.,* 1998; *Treviño and Keller,* this volume]. Additional constraints were provided by various seismic reflection and refraction profiles [*Behrendt et al.,* 1969; *Wellborn,* 1977; *Applegate and Rose,* 1985; *Beggs,* 1985; *Gries and Dyer,* 1985; *Kaplan and Skeen,* 1985; *Lange and Wellborn,* 1985; *Prodehl et al.,* this volume]. Features discernible in the detailed upper crustal velocity model of *Rumpel et al.* [this volume] correlate very well with surface geology and were used as a constraint in the density modeling and taken as an additional guide to the location of major upper crustal features important to the gravity model (Figure 12).

The final gravity model (Figure 12) reveals a number of new aspects of the lithospheric structure of the Rocky Mountains. As expected, known geologic features of the uppermost crust contribute significantly to short-wavelength gravity anomalies. Four upper crustal density anomalies play a significant role in fitting the intermediate wavelength (~300 km) features in New Mexico and Colorado. The occurrence of these features is evident in the detailed velocity modeling of *Rumpel et al.* [this volume] and is consistent with other geologic evidence for both low and high-density intrusions within the crystalline crust. In particular, to match the steep gravity gradients on either side of the gravity low in the central portion of the profile, an upper crustal body with density 2600 kg m^{-3} must be surrounded by two high-

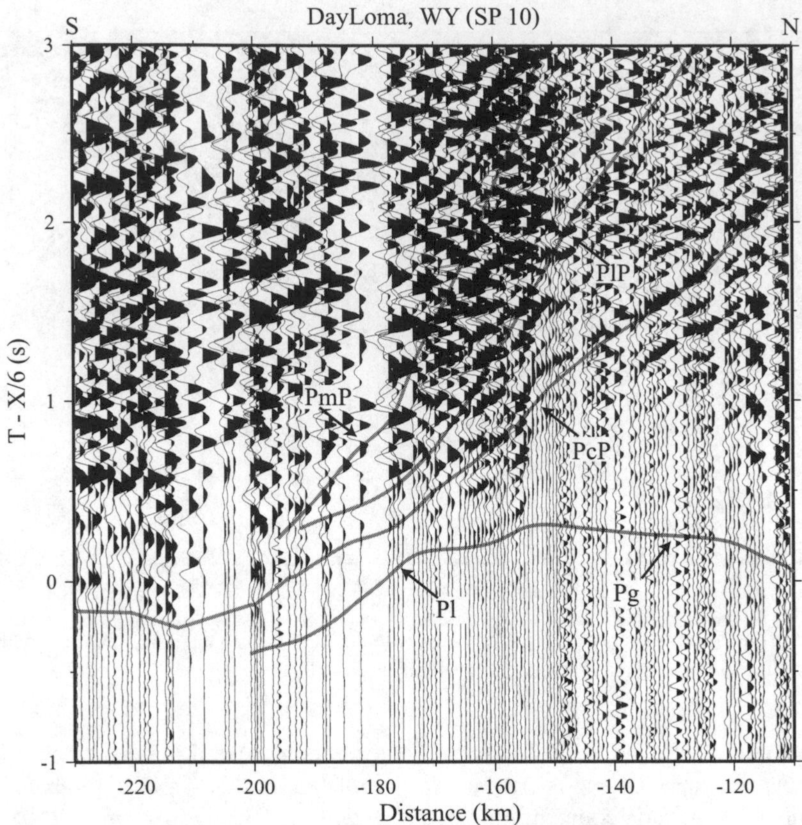

Figure 9. Close-up of the seismic record section from Figure 5 for Day Loma, WY (SP 10). Notice the high amplitudes of the PcP phase.

density bodies of 2900 kg m^{-3} with similar depth extent to 10 km. When these bodies are removed from the model (Figure 12a), a much broader, lower-amplitude (ca. 80 mGal) gravity low remains suggesting a deeper feature is controlling the gravity signature.

The large gravity low at a distance of ~650 km in the model is part of a prominent northeast-trending anomaly that correlates with the Colorado mineral belt (Figure 12) [*McCoy et al.,* this volume]. A number of Laramide felsic intrusions crop out along this belt, which also has the same approximate trend as Precambrian terrane boundaries. No younger features are present that could explain this long-wavelength anomaly, and thus, it is attributed to a series of large felsic bodies, probably intrusions, in the upper crust.

The Sierra Grande uplift region (northeastern New Mexico) is along the northwest extension of the Wichita-Amarillo uplift [*Suleiman and Keller,* 1985], which is part of the Southern Oklahoma aulacogen. Particularly in Oklahoma, mafic intrusions of Cambrian age are well documented under the Wichita uplift [e.g., *Keller and Baldridge*, 1995]. In addition, the 1.1 Ga Pecos mafic igneous complex extends northward from Texas and southeastern New Mexico, where it is well documented in drill holes and seismic reflection data [*Adams and Miller,* 1995]. The gravity modeling of *Suleiman and Keller* [1985] showed that the basement relief in northeastern New Mexico is insufficient to produce the observed gravity high that crosses the Sierra Grande uplift region, and thus they include mafic material in the upper crust of their crustal models. Thus, the southern mafic body in Figure 12d could either be due to Cambrian or late Proterozoic magmatism.

The Wet Mountains area is associated with a gravity high and the dense upper crustal body placed in the model is consistent with the deep origin of the Precambrian rocks exposed in the mountains [*C. Andronicos,* personal communication]. If the highly metamorphosed rocks at the surface are from depths of 25 km or more, then rocks that were once lower crustal are present in the upper crust below them. The origin of the gravity high associated with the Sierra Madre/Park Range is less clear. The simple body used in the model does not fit some short wavelength features in the observed data, and it is centered just south of the Cheyenne belt. The exposed geology in this area is multifaceted and includes large mafic and felsic plutons [*e.g., Foster et al.,* 1999]. In addition, the presence of the Cheyenne belt and several major shear zones probably produce local anomalies that are beyond the scale of this study.

With the upper crustal density structure well constrained by known geology and velocity modeling, a southward decrease in mantle density is required to match the observed data. This density decrease is completely consistent with the southward decrease in mantle velocity observed in data from the Deep Probe experiment [*Henstock et al.,* 1998; *Snelson et al.,* 1998]. When the modeled gravity is recalculated using a single density for the mantle (Figure 12b) the overall shape of the calculated curve remains similar to the observed curve, but the calculated values are less than the observed values by >50 mGal on the north end of the model. Thus, the long wavelength gravity low that extends from about 200 to 1000 km in the model clearly correlates with the both the shape of the Moho and the topography, but lateral density variations in the upper mantle (Figure 12c) are needed to produce a fit on the north end.

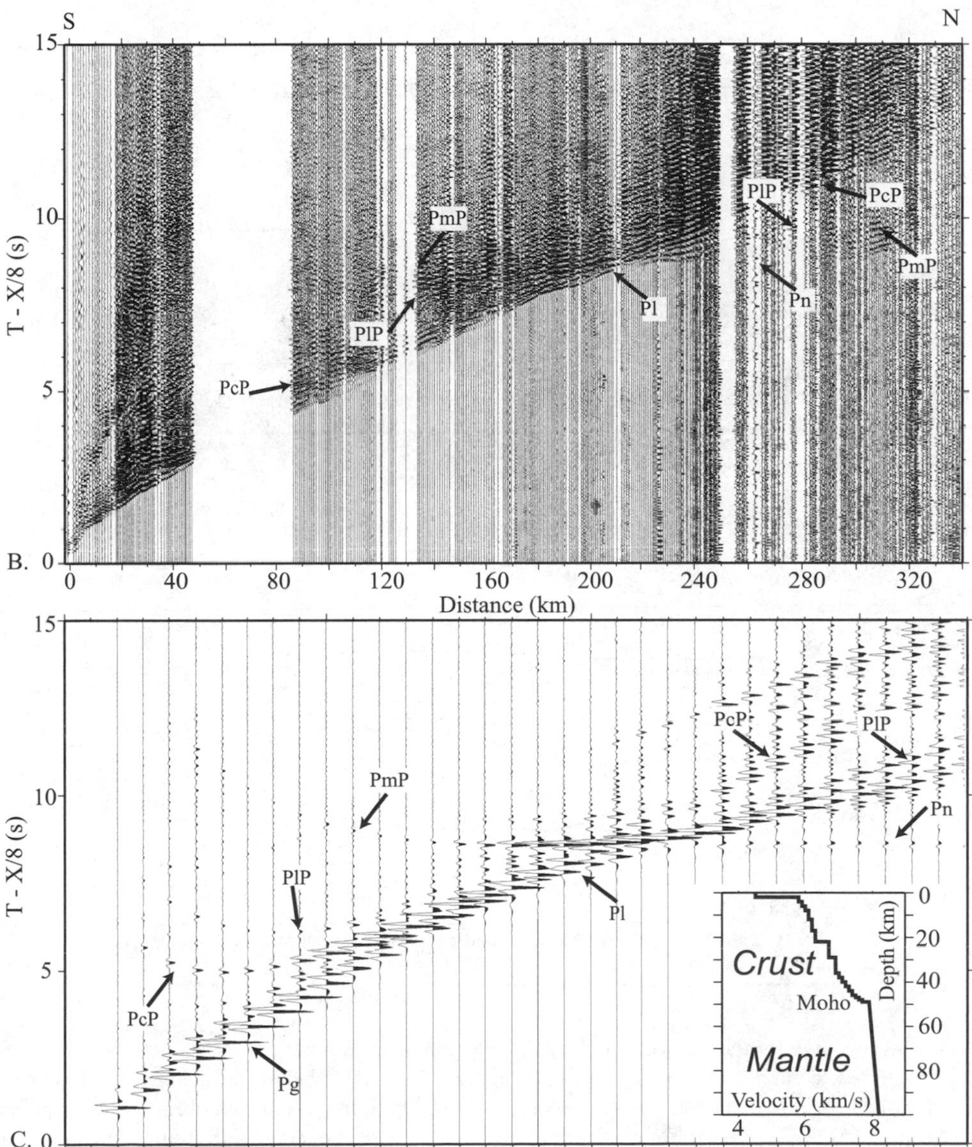

Figure 10. (a) The 1-D velocity-depth model used to produce the synthetic record section in (b). (b) Seismic record section for Fort Sumner, NM (SP1) reduced at 8 km s^{-1}. (c) The synthetic seismic record; notice the high amplitudes of the reflected phases, which is similar to (b).

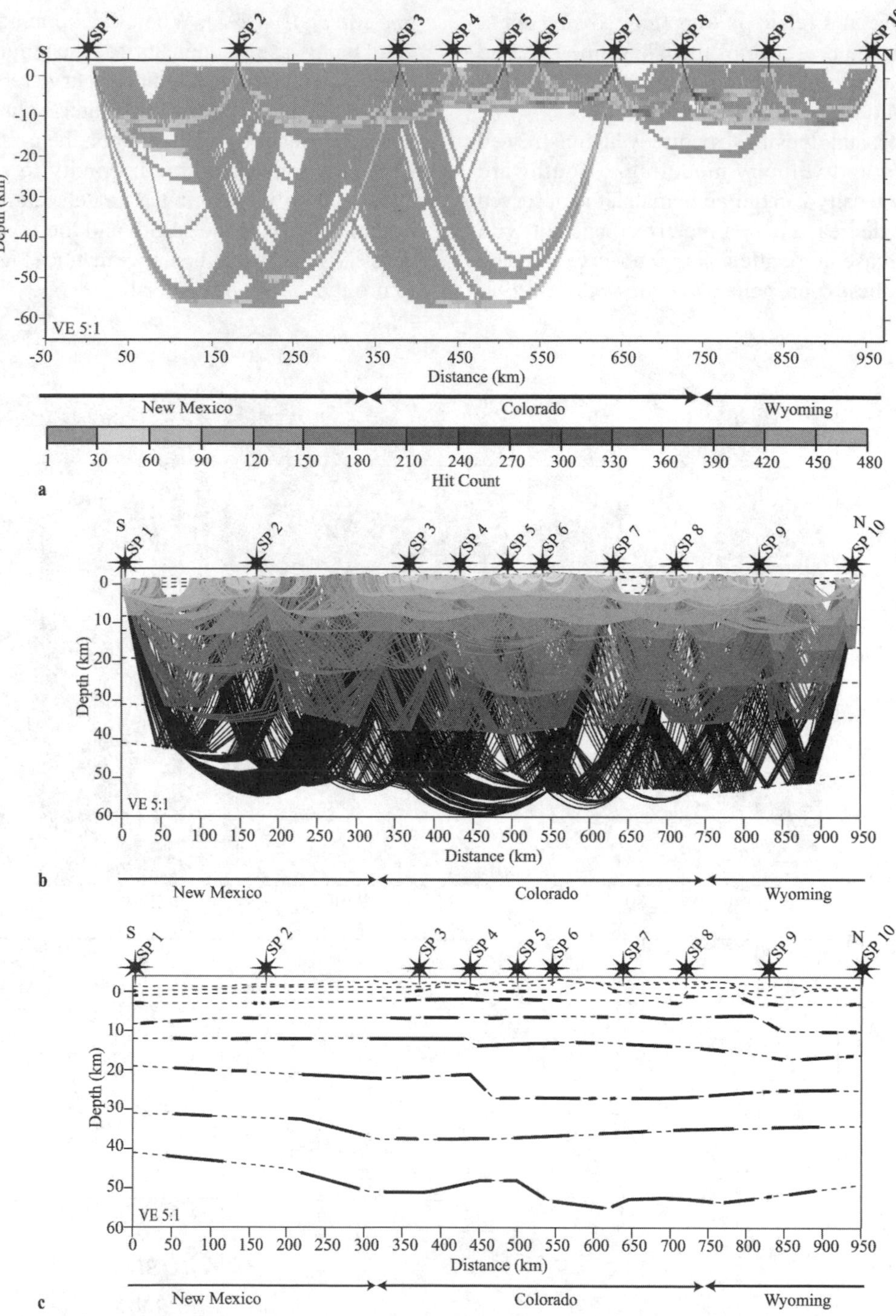

Figure 11. (a) Ray coverage for the 3-D velocity model. Hit count is the number of rays that encountered each cell. High hits are dark gray to black and low hits are lighter shades of gray. The 3-D model has been compressed to two dimensions for display. Thus, the hit count in a given cell represents a sum of hits from cells in the y-direction. Shot points are noted at the top of the plot. (b) Ray coverage from travel time modeling by *Rumpel et al.* [this volume]. The northern portion of model shows only reflections and no refractions as indicated by the lack of diving waves. Light gray rays are upper crustal, medium gray is middle crustal, and dark gray to black are lower crustal and mantle reflections. (c) Reflection points from travel time modeling by *Rumpel et al.* [this volume]. The model is well resolved from the wide-angle reflections. Dark black lines are the bounce points for the model.

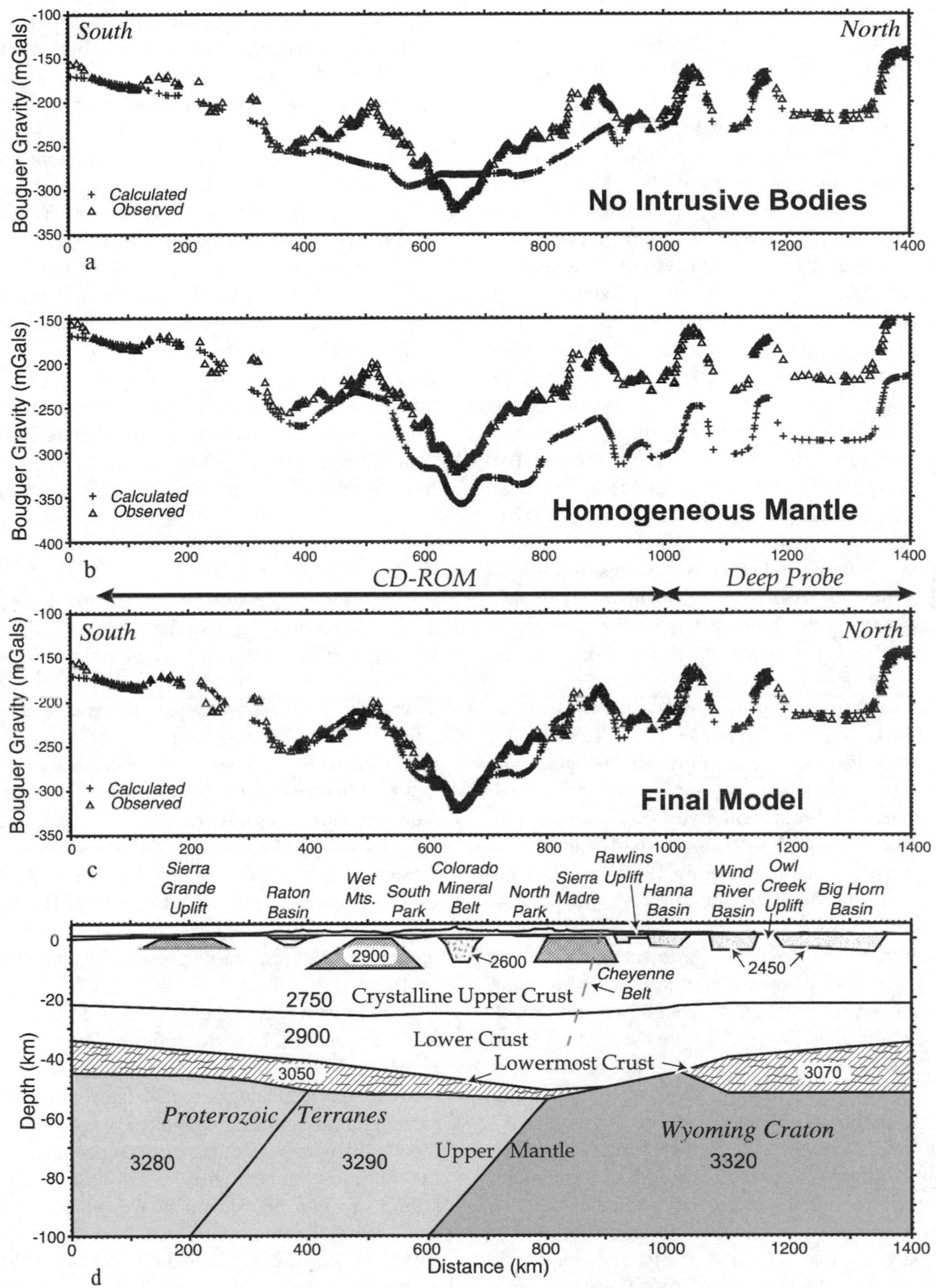

Figure 12. (a) 2.5-D gravity values along the CD-ROM transect showing the effects of removing the intrusive bodies in the upper crust in (d); (b) 2.5-D gravity values along the CD-ROM transect showing the effects of making the mantle in (d) homogeneous; (c) 2.5-D gravity values along the CD-ROM transect showing fit of values calculated from the final model (d) to the observations; and (d) final 2.5-D density model along the CD-ROM transect. The density values are in kg m^{-3}.

7. DISCUSSION

7.1. Mid-Crustal Velocity Boundary

Below the laterally heterogeneous velocity field of the upper crust [*Rumpel et al.,* this volume], the middle crust is characterized by a well-defined boundary at 20 to 25 km depth, which is marked by a change in velocity from 6.4 to 6.6 km s^{-1}. This interface location is defined by prominent PcP reflected energy found on all the shot records. The energy from the PcP is characterized by a long coda that may result from multiple reflections within the upper crust (Figures 2 and 10). Our modeling suggests that the main reflecting boundary dips gently northward beneath the Great Plains to reach a maximum depth of ca. 26 km, as the profile crosses the Rocky Mountain Front near model coordinate 500 km (Plate 1). It then rises gently northward to ca. 23 km depth at the north end of the model. Whereas reflections from this boundary are prominent in the wide-angle data presented here, the near-vertical incidence data collected as part of the CD-ROM effort exhibit no corresponding sub-horizontal reflectivity at this level [*Magnani et al.,* this volume a, b; *Morozova et al.,* this volume]. This observation suggests that the boundary is characterized by a velocity gradient that it is transparent to the shorter wavelength content of the near-vertical incidence data.

More recent results from inferred exposures of the Conrad discontinuity [*Salisbury and Fountain,* 1994; *Lana et al.,* 2003] and from xenoliths [*Sachs and Hansteen,* 2000] indicate that the Conrad is both a metamorphic and compositional boundary. Comparison of representative 1-D velocity-depth functions from the CD-ROM velocity model to laboratory measurements of rock velocities (Figure 13) shows that the mid-crustal velocity discontinuity along the CD-ROM profile also marks the transition from model velocities appropriate for lower-grade rocks with more felsic compositions to velocities appropriate for higher grade rocks with more mafic compositions. Refraction results from the central United States [*Braile,* 1989] and earlier results from the Rocky Mountains [*Prodehl and Lipman,* 1989] map a similar velocity discontinuity near 20 km depth, although the Deep Probe results from western Colorado and New Mexico do not require its presence [*Snelson et al.,* 1998] (Figure 13). Thus, most of the evidence suggests that this boundary is a widespread feature of the crust in the Great Plains and Rocky Mountains.

We interpret the observation that the mid-crustal boundary extends from the Great Plains through the Rocky Mountains as evidence that it is a Paleproterozoic/Mesoproterozoic-age boundary that has remained essentially undisturbed by Phanerozoic tectono-magmatic events such as the late Paleozoic Ancestral Rockies event and the Laramide orogeny. Reflectivity interpreted as the signature of the Proterozoic orogens in the near-vertical incidence data [*Magnani et al.,* this volume b; *Morozova et al.,* this volume] crosscuts the discontinuity suggesting that the formation of this boundary must have been contemporaneous with the last Proterozoic orogenic pulse. Otherwise, this reflectivity should have been more profoundly disrupted by subsequent differentiation processes.

It is likely that subsequent magmatic events may have served to enhance this boundary. Models for stabilization and evolution of continental crust [*e.g., Bohlen and Mezger,* 1989; *Nelson,* 1991; *Keller et al.,* this volume] commonly invoke episodic injection of the crust by mafic magmas as a mechanism for thickening the crust and pushing the lower crust towards a more mafic composition. The 1.4 Ga "anorogenic" magmatic event that affected both the Great Plains and the Rocky Mountains, as well as voluminous Tertiary magmatism in the southern Rockies, are at least two candidates for times when this velocity boundary may have been enhanced.

7.2. High Velocity Lower Crust

The lowermost crust of our velocity model is characterized by a ca. 10 km thick layer with a velocity of ca. 7.2 km s^{-1}. This feature is defined from wide-angle reflections (PlP) (Figures 2 – 9) and a number of shot records (Figures 2 – 3, 5 – 6, 8 – 9) that we interpret to contain a refracted arrival from the lower crust. A single continuous layer is the simplest way to model this feature; however, the data allow for the possibility that the thickness of this layer may vary significantly beneath New Mexico and southern Colorado, as is the case in the velocity model of *Levander et al.* [this volume]. The model presented in this paper can be considered a thick end member for the high-velocity lower crust compared to the lower thicknesses derived by *Levander et al.* [this volume]. Furthermore, the data do not require its presence at the north end of the profile.

Laboratory measurements (Figure 13) [*e.g., Christensen and Mooney,* 1995] and theoretical calculations [*Furlong and Fountain,* 1986] show that velocities ca. 7.2 km s^{-1} are appropriate for mafic lithologies including gabbroic rocks and mafic garnet granulite. A high-velocity lower-crustal layer is common to many regions of the central and western U. S. including the mid-continent [*Braile,* 1989], the Colorado Plateau [*Wolf and Cipar,* 1993], and the Deep Probe profile [*Snelson et al.,* 1998]. Such a layer is also observed in other regions of the world, with the Baltic shield [*e.g., EUROBRIDGE Seismic Working Group,* 1999] being a notable example. These features are most commonly interpreted as mafic magmatic underplates emplaced during a major melting event, and are consistent with observed high velocities and magmatic history of the region.

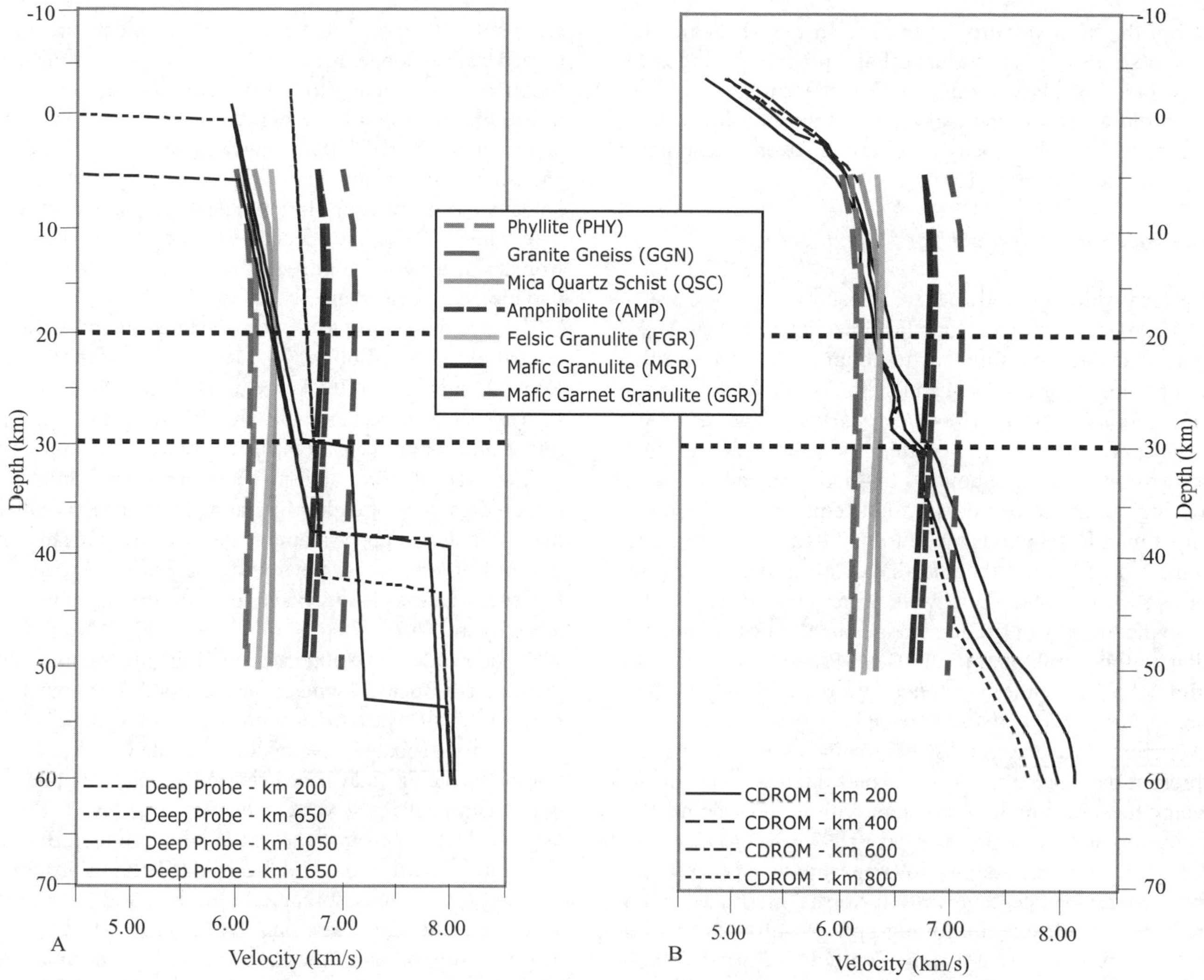

Figure 13. Comparison of laboratory measurements of selected rocks types [*Christensen and Mooney,* 1995] to 1-D velocity-depth functions extracted from the (a) Deep Probe and (b) CD-ROM velocity models. Horizontal dashed lines at 20 km and 30 km depth indicate possible range for the Conrad discontinuity. Velocity is in km s-1.

We also interpret the high-velocity layer on the CD-ROM profile as a magmatic underplate. It probably is in part a relic of the initial formation of the crust, but based on its continuity across the Great Plains-Rocky Mountain tectonic boundary, we propose that a significant portion of it was likely emplaced during the voluminous ca. 1.4 Ga magmatic event [*Anderson,* 1989]. This event profoundly modified the crust of the Great Plains to form the southern Granite-Rhyolite province and also led to the intrusion of many plutons along Proterozoic shear zones in the Yavapai and Mazatzal provinces [*Karlstrom and Humphreys,* 1998]. The high-velocity layer beneath the Baltic Shield is also associated with magmatism that is chemically similar to, but, at ca. 1.5 Ga, somewhat older [*e.g., Gaál and Gorbatschev,* 1987]. Subsequent magmatic episodes, including events at ca. 1.1 Ga, 500 Ma, and in Tertiary time may also have contributed material to this layer in the southern Rocky Mountains [*Keller et al.,* this volume].

The unusually thick (~20 km) high-velocity (~7.2 km s^{-1}) lower-crustal layer observed along the Deep Probe transect (Figure 13) [*Henstock et al.,* 1998; *Snelson et al.,* 1998; *Gorman et al.,* 2002] likely represents a mafic underplate that is Archean or Paleoproterozoic in age rather than Mesoproterozoic and younger. This layer extends for 700 km from central Wyoming into southern Alberta, where it terminates at the north side of the Medicine Hat block [*Gorman et al.,* 2002; *Clowes et al.,* 2002]. These regions lack evidence for ca. 1.4 Ga magmatism. However, in Wyoming and Montana, *Chamberlain et al.* [2003] have mapped at least five thermal events of Archean age that could represent the signature of underplating events. By contrast, xenoliths interpreted as originating from the high-velocity layer beneath the Medicine Hat Block in southern Alberta have Paleoproterozoic ages of 1814–1745 Ma [*Davis et al.,* 1995]. In addition, the layer beneath these Archean crustal blocks is not physically continuous with the high-velocity lower-crustal layer present in the CD-ROM model (Figure 13).

7.3. Crustal Thickness and Upper Mantle Velocity

Our final velocity model (Plate 1c) shows that the crust is 40 to 45 km thick crust beneath the Great Plains of New Mexico and that the crust thickens to about 50 km beneath the high topography of southern and central Colorado. Previous models [*Johnson et al.,* 1984; *Prodehl and Lipman,* 1989; *Snelson et al.,* 1998; *Prodehl et al.,* this volume] suggest that the crust thins again to about 40 to 45 km beneath southern Wyoming, although our data do not require such thinning.

Upper mantle velocities range from 7.8 to 7.9 km s^{-1} along the entire profile, and the analysis of *Levander et al.* [this volume] shows a distinct southward decrease of velocity southward to the vicinity of the Jemez lineament. These values are similar to those obtained in the Deep Probe velocity model south of the Cheyenne belt [*Henstock et al.,* 1998] and the earlier results of *Prodehl and Lipman* [1989]. These values are low compared to averages for stable continents and suggest that the present-day mantle is warm and buoyant. Low upper mantle velocities are broadly consistent with large-scale mantle velocity models [*e.g., Grand et al.,* 1997; *van der Lee and Nolet,* 1997] that map slow, hot upper mantle beneath the uplifted orogenic plateau of western North America, and with more detailed teleseismic tomography results [*Zurek and Dueker,* this volume] that map generally low P-wave velocities in the upper 200 km of the mantle in this region. Asthenospheric upwelling beneath the modern Rio Grande rift likely plays an important role in reducing upper mantle velocities along the CD-ROM profile.

7.4. Implications for Isostatic Compensation of Topography

By constraining crustal thickness in Colorado through analysis of receiver functions, *Sheehan et al.* [1995] were able demonstrate that a significant portion of the support for the excess topography must come from the mantle. However, they were unable to document the role of crustal density variations in detail because the receiver function method does not produce information on lateral velocity variations in the crust, nor can it estimate Moho depth as accurately as refraction data because of the long wavelengths involved. *Li et al.* [2002] used shear wave velocity structure based on measurements of Rayleigh wave dispersion concluded that the density anomaly most responsible for the high topography must reside in the crust rather than the mantle. Surface wave data are characterized by even longer wavelengths than those used in receiver functions, thus leading to lower resolution results. Furthermore, the inversion for a match to the Bouguer anomaly employed by *Li et al.* [2002] may have underestimated density variation in the mantle due to the sensitivity of shear wave velocity to the presence of melt. The density model presented here (Figure 12) has the advantage of having been constructed from a shorter wavelength seismic source sensitive to lateral variations in crustal velocity.

The modeling serves to separate the three main elements that contribute to low Bouguer gravity values in the southern Rocky Mountains: (1) thickened crust beneath the high topography; (2) a decrease in mantle density beneath the high topography; and (3) low density intrusive bodies in the upper crust. Thickened crust contributes to a long wavelength gravity low centered on the Colorado Mineral belt (Figure 12a), whereas the difference in upper mantle density contributes to a mass deficit in the southern portion of the profile (Figure 12b). Upper crustal low-density bodies increase the magnitude of the low (Figure 12c).

We conclude that compensation of the high topography in the southern Rocky Mountains results from a compound interplay of density variations within the lithosphere, most of which have probably developed since late Cretaceous time. Crustal thickening may have occurred during Laramide shortening as a result of lower crustal flow [*e.g., Bird,* 1998] or large-scale detachments [*e.g., Erslev,* this volume]. Buoyancy in the upper crust would have been added during Tertiary-age magmatism in the Colorado mineral belt and the San Juan volcanic field [*e.g., Stein and Crock,* 1990; *Decker,* 1995]. Thermal support from the mantle may have been added in Cenozoic time, possibly as a result of removal of the mantle lithosphere [*e.g., Decker,* 1995; *Humphreys et al.,* 2001].

8. CONCLUSIONS

The CD-ROM refraction profile crosses a major modern physiographic province and a number of major Precambrian tectonic boundaries, yet displays remarkably simple velocity structure. We interpret this as evidence that the crustal architecture developed primarily during Paleo- and Mesoproterozoic tectonism and that subsequent Phanerozoic events caused only minor modification or enhancement of the existing structure. We propose that a two-layer crust developed during accretion of Yavapai and Mazatzal crust from 1.76 to 1.6 Ga. Differentiation of the crust into an upper felsic layer and a lower more mafic layer probably occurred primarily through melting of the lower crust to produce widespread felsic plutons at ca. 1.6 Ga that now crop out at the surface. Later magmatic episodes, particularly at 1.4 Ga and in Tertiary time may have enhanced this basic structure through melting of lower crustal rocks, emplacement of felsic rocks in the upper crust, and segregation of mafic components to the lower crust. Underplating of basaltic melts to the base of the crust at 1.4 Ga regionally, and perhaps at 1.1 Ga and 500 Ma locally caused formation of a third, high-velocity layer at the base of the crust, as well as crustal thickening. At a crustal scale, the only evidence for Laramide shortening is a modest deepening of the mid-crustal discontinuity beneath the high topography and a corresponding thickening of the crust.

The control on lithospheric architecture provided by the refraction velocity model permit a new assessment of the factors that contribute to the isostatic compensation of high topography in the southern Rocky Mountains. A density model derived from the velocity model shows that a multifaceted interplay of density variations within the lithosphere is required to explain the Bouguer gravity low associated with the range in central and southern Colorado. These density variations include crustal thickening, a transition from high density to low-density mantle, and low-density material in the upper crust. All of these features likely developed since late Cretaceous time as a result of Laramide-age shortening and magmatism. This is yet another demonstration that, in general, a number of mechanisms are commonly at work to accomplish compensation of high topography on the continents.

Acknowledgements. We wish to thank the CD-ROM Working group for making this project come to life. Thanks are due to the many volunteers who helped deploy the instruments. Thanks are also due to the UTEP and IRIS/PASSCAL instrument centers for the use of the Texans and RefTeks and personnel during the experiment. In addition, we thank the BLM, Microgeophysics, Buckley Powder, Crass Drilling, and Young's Drilling for various services and support. We would also like to acknowledge the landowners who allowed access to their property, including Hage & Webb Land and Cattle, Fallon Ranch, Pettigrew Ranch, Carlos Pena, Vermejo Park Ranch, Tercio Ranch, Fred Harte, Stella Wheatley, the Brand's Ranch, the Boettcher Ranch, Kim & Felix Najera, and Robert Wrigley. We wish to thank Robert Schneider and Lorraine Wolf for their helpful reviews. Database and computational support was provided by the Pan American Center for Earth and Environmental Studies, which is supported by NASA. Funding was provided by the National Science Foundation - Continental Dynamics Program (EAR-9614269) and Deutsche Forschungsgemeinschaft (DFG).

REFERENCES

Adams, D. C., and G. R. Keller, Possible extension of the Midcontinent rift in west Texas and eastern New Mexico, *Can. J. Earth Sci.*, 31, 709–720, 1994.

Adams, D. C., and K. C. Miller, Evidence for late middle Proterozoic extension in the Precambrian basement beneath the Permian basin, *Tectonics*, 14, 1263–1272, 1995.

Anderson, J. L., Proterozoic anorogenic granites of the southwestern U. S., in *Arizona Geological Digest*, 19, 211–238, 1989.

Applegate, J. K., and P. R. Rose, Structure of the Raton basin from a regional seismic line, in *Seismic Exploration of the Rocky Mountains*, edited by R. R. Gries and R. C. Dyer, Rocky Mountain Association of Geologists and Denver Geophysical Society, pp. 259–265, Denver, Colorado, 1985.

Beggs, H. G., Interpretation of geophysical data from the central South Park Basin, Colorado, in *Exploration Frontiers of the Central and Southern Rockies*, edited by H. K. Veal, Rocky Mountain Association of Geologists, pp. 67–76, Denver, Colorado, 1977.

Behrendt, J. C., P. Popenoe, R. E. Mattick, A geophysical study of North Park and the surrounding ranges, Colorado, *Geol. Soc. Am. Bull.*, 80, 1523–1538, 1969.

Bird, P., Kinematic history of the Laramide orogeny in latitudes 35(– 49(N, western United States, *Tectonics*, 17, 780–801, 1998.

Blackstone, D. L., Jr., Precambrian basement map of Wyoming; outcrop and structural configuration, in *Laramide basement deformation in the Rocky Mountain foreland of the Western United States*, edited by C. J. Schmidt, R. B. Chase, E. A. Erslev, Geological Society of America, Special paper *280*, 335–337, Geological Society of America, Boulder, Colo., 1993.

Bohlen, S. R., and K. Mezger, Origin of granulite terranes and the formation of the lowermost continental crust, *Science*, 244, 326–329, 1989.

Braile, L. W., Crustal structure of the continental interior, in *Geophysical Framework of the continental United States*, edited by L. C. Pakiser and W. D. Mooney, *Geological Society of America, Memoir 172*, 285–315, Geological Society of America, Boulder, Colo., 1989.

Cady, J. W., Calculation of gravity and magnetic anomalies of finite-length right polygonal prisms, *Geophysics*, 45, 1507–1512, 1980.

Chamberlain, K. R., C. D. Frost, and B. R. Frost, Early Archean to Mesoproterozoic evolution of the Wyoming Province: Archean origins to modern lithospheric architecture, *Can. J. Earth Sci.*, 40, 1357–1374, 2003.

Christensen, N. I., and W. D. Mooney, Seismic velocity structure and composition of the continental crust: A global review, *J. Geophys. Res.*, 100, 9761–9788, 1995.

Clowes, R. M., M. J. A. Burianyk, A. R. Gorman, and E. R. Kanasewich, Crustal velocity structure from SAREX, the southern Alberta Refraction Experiment, *Can. J. Earth Sci.*, 39, 351–373, 2002.

Davis, W. J., R. Berman, and B. Kjarsgaard, U-Pb geochronology and isotopic studies of crustal xenoliths from the Archean Medicine Hat Block, northern Montana and southern Alberta: Paleoproterozoic reworking of the Archean crust, in *1995 Alberta Basement Transect Workshop*, compiled by G. R. Ross, Lithoprobe Secretariate, The University of British Columbia, Lithoprobe report 47, 329–334, 1995.

Decker, E. R., Thermal regimes of the Southern Rocky Mountains and Wyoming Basin in Colorado and Wyoming in the United States, *Tectonophysics*, 244, 85–106, 1995.

Dueker, K., H. Yuan, and B. Zurek, Thick Proterozoic lithosphere of the Rocky Mountain region, *GSA Today*, 11, 4–9, 2001.

Erslev, E. A., 2D Laramide geometries and kinematics in the Rocky Mountains of the western United States, this volume.

EUROBRIDGE Seismic Working Group, Seismic velocity structure across the Fennoscandia-Sarmatia suture of the East European Craton beneath the EUROBRIDGE profile through Lithuania and Belarus, *Tectonophysics*, 314, 193–217, 1999.

Foster, C. T., M. K. Reagan, S. G. Kennedy, G. A. Smith, C. A. White, J. E. Eiler, and J. R. Rougvie, Insights into the Proterozoic geology of the Park Range, Colorado, *Rocky Mt. Geol., 34*, 7–20, 1999.

Fuchs, K., and G. Müller, Computation of synthetic seismograms with the reflectivity method and comparison with observations, *Geophys. J. R. Astr. Soc*, 23, 417–433, 1971.

Furlong, K. P. and D. M. Fountain, Continental crustal underplating; thermal considerations and seismic-petrologic consequences, *J. Geophys. Res.*, 91, 8285–8294, 1986.

Gaál, G., and R. Gorbatschev, An outline of the Precambrian evolution of the Baltic Shield, *Precam. Res.*, 35, 15–52, 1987.

Gorman, A. R., R. M. Clowes, R. M. Ellis, T. J. Henstock, G. D. Spence, G. R. Keller, A. Levander, C. M. Snelson, M. J. Burianyk, E. R. Kanasewich, I. Asudeh, Z. Hajnal, and K. Miller, Deep Probe: imaging the roots of western North America, *Can. J. Earth Sci.*, 39, 375–398, 2002.

Grand, S. P., R. D. van der Hilst, and S. Widiyantoro, Global seismic tomography: A snapshot of convection in the earth, *GSA Today*, 7, 1–7, 1997.

Gries, R. R., and R. C. Dyer (Eds*.), Seismic Exploration of the Rocky Mountain Region, Rocky Mountain Association of Geologists and Denver Geophysical Society*, 300 p., Denver, Colorado, 1985.

Henstock, T. J., A. Levander, C. M. Snelson, G. R. Keller, K. C. Miller, S. H. Harder, A. R. Gorman, R. M. Clowes, M. J. A. Burianyk, and E. D. Humphreys, Probing the Archean and Proterozoic lithosphere of western North America, *GSA Today*, 8, 1–5 and 16–17, 1998.

Hole, J. A., Nonlinear high-resolution three-dimensional seismic travel time Tomography, *J. Geophys. Res.*, 97, 6553–6562, 1992.

Hole, J. A., and B. C. Zelt, 3-D finite-difference reflection travel times, *Geophy. J. Int.*, 121, 427–434, 1995.

Humphreys, E. D., E. Erslev, and G. L. Farmer, A metasomatic origin for Laramide tectonism and magmatism, *Abstracts with Programs, GSA*, 33, 50, 2001.

Isaacson, L. B. and S. B. Smithson, Gravity anomalies and granite emplacement in west-central Colorado, *Geol. Soc. Amer. Bull.*, 87, 22–28, 1976.

Johnson, R. A., K. E. Karlstrom, S. B. Smithson, R. S. Houston, Gravity profiles across the Cheyenne belt: A Precambrian crustal suture in southeastern Wyoming, *J. Geodyn.*, 1, 445–472, 1984.

Kaplan, S. S., and R. C. Skeen, North-south regional seismic profile of the Hanna basin, in *Seismic Exploration of the Rocky Mountains*, edited by R. R. Gries and R. C. Dyer, *Rocky Mountain Association of Geologists and Denver Geophysical Society*, pp. 219–224, Denver, Colorado, 1985.

Karlstrom, K. E., S. S. Harlan, M. L. Williams, J. McLelland, J. W. Geissman, and K-I. Ahall, Refining Rodinia: Geologic evidence for the Australia-Western U.S. connection in the Proterozoic, *GSA Today*, 9, 1–7, 1999.

Karlstrom, K. E., and E. D. Humphreys, Persistent influence of Proterozoic accretionary boundaries in the tectonic evolution of southwestern North America: Interaction of cratonic grain and mantle modification events, *Rocky Mt. Geol.*, 33, 161–180, 1998.

Keller, G. R., K. E. Karlstrom, M. L. Williams, K. C. Miller, C. Andronicos, A. Levander, C. Snelson, and C. Prodehl, The dynamic nature of the continental crust-mantle boundary: Crustal evolution in the Southern Rocky Mountain region as an example, this volume.

Keller, G. R., and W. S. Baldridge, The Southern Oklahoma Aulacogen, in *Olsen, K. H. (ed.), Continental Rifts: Evolution, Structure, Tectonics*, Elsevier, Amsterdam, 427–435, 1995.

Keller, G. R., and W. S. Baldridge, The Rio Grande rift: A geological and geophysical review, *Rocky Mt. Geol.,* 34, 131–148, 1999.

Kluth, C. F., and P. J. Coney, Plate tectonics of the Ancestral Rocky Mountains, *Geology*, 9, 10–15, 1981.

Lana, C., R. L. Gibson, A. F. M. Kisters, and W. U. Reimold, Archean crustal structure of the Kaapvaal craton, South Africa – evidence from the Vredefort dome, *Earth Planet. Sci. Lett.*, 206, 133–144, 2003.

Lange, J. K., and R. E. Wellborn, Seismic profile North Park basin, in *Seismic Exploration of the Rocky Mountains*, edited by R. R. Gries and R. C. Dyer, *Rocky Mountain Association of Geologists and Denver Geophysical Society*, pp. 239–245, Denver, Colorado, 1985

Levander, A., C. Zelt, and M. B. Magnani, Crust and upper mantle velocity structure of the southern Rocky Mountains from the Jemez lineament to the Cheyenne belt, this volume.

Li, A., D. W. Forsyth, and K. M. Fischer, Evidence for shallow isostatic compensation of the Southern Rocky Mountains from Rayleigh wave tomography, *Geology*, 30, 683–686, 2002.

Litak, R. K., and L. D., Brown, A modern perspective on the Conrad discontinuity, *EOS*, 70, 713,722–724, 1989.

Luetgert, J. H., MacRay: interactive two-dimensional seismic raytracing for the Macintosh, *U. S. Geol. Surv. Open-File Report 92–356*, 1–2, 1992.

MacLachlan, J. C., H. C. Bemis, R. S. Bryson, R. D. Holt, C. J. Lewis, and D. E. Wilde, Configuration of the Precambrian rock surface, in *Geologic Atlas of the Rocky Mountain region*, pp. 53, *Rocky Mt. Assoc. Geol.*, Colo., 1972.

Magnani, B., A. R. Levander, K. C. Miller, T. Eshete, and K. Karlstrom, Seismic investigations of the Yavapai-Mazatzal transition zone and the Jemez lineament in northeastern New Mexico, this volume

Magnani, B. M., A. R. Levander, E. A. Erslev, N. Bolay-Koenig, and K. Karlstrom, Listric thrust faulting in the Laramide from north-central New Mexico guided by Precambrian basement anisotropies, this volume

McCoy, A., M. Roy, L. Treviño, and G. R. Keller, Gravity modeling of the Colorado mineral belt, this volume.

Morozova, E., X. Wan, K. R. Chamberlain, S. B. Smithson, R. Johnson, and K. E. Karlstrom, Inter-wedging nature of the Cheyenne belt – Archean-Proterozoic suture defined by seismic reflection data, this volume.

Mosher, S., Tectonic evolution of, the southern Laurentian Grenville orogenic belt, *Geol. Soc. Am. Bull.*, 110, 1357–1375, 1998.

Nelson, K. D., A unified view of craton evolution motivated by recent deep seismic reflection and refraction results, *Geophys. J. Int.*, 105, 25–35, 1991.

Prodehl, C., G. R. Keller, R. A. Johnson, C. Snelson, and H.-M. Rumpel, Background and overview of previous controlled source seismic studies, this volume.

Prodehl, C., and P. Lipman, Crustal structure of the Rocky Mountain region, in *Geophysical framework of the continental United States*, edited by L. C. Pakiser and W. D. Mooney, *Geological Society of America Memoir 172*, pp. 249–284, Geological Society of America, Boulder, Colo., 1989.

Rumpel, H.-M., C. M. Snelson, C. Prodehl, and G. R., Keller, Refraction/Wide-angle reflection experiment: The upper crust, this volume.

Sachs, P. M., and T. H. Hansteen, Pleistocene underplating and metasomatism of the lower continental crust: a xenolith study, *J. Petrology,* 41, 331–356, 2000.

Salisbury, M. H. and D. M. Fountain, The seismic velocity and Poisson's ratio structure of the Kapuskasing uplift from laboratory measurements, *J. Can. Earth. Sci.*, 31, 1052–1063, 1994.

Sandmeier, K.-J., Untersuchung der Ausbreitungseigenschaften seismischer Wellen in geschichteten und streuenden Medien. *Ph.D. Thesis*, Karlsruhe University, 1990,

Sheehan, A. F., G. A. Abers, C. H. Jones, A. L. and Lerner-Lam, Crustal thickness variations across the Colorado Rocky Mountains from teleseismic receiver functions, *J. Geophys. Res.*, 100, 20391–20404, 1995.

Sheehan, A.F., V. Schulte-Pelkum, O. Boyd, and C. Wilson, Passive source seismology of the Rocky Mountain region, this volume.

Sloss, L. L., Tectonic evolution of the craton in Phanerozoic time, in *The Geology of North America*, D-2, *Sedimentary Cover - North America Craton; U. S.*, edited by L. L. Sloss, Geological Society of America, Boulder, Colo., 1988.

Snelson, C. M., T. J. Henstock, G. R. Keller, K. C. Miller, and A. R. Levander, Crustal and uppermost mantle structure along the Deep Probe seismic profile, *Rocky Mt. Geol.*, 33, 181–198, 1998.

Stein, H. J., and J. G. Crock, Late Cretaceous-Tertiary magmatism in the Colorado mineral belt; rare earth element and samarium-neodymium isotopic studies, in *The nature & origin of Cordilleran magmatism*, edited by J. L. Anderson, Geological Society of America, Boulder, Colo., 174, 195–223, 1990.

Suleiman, A.S. and G. R. Keller, A geophysical study of basement structure in northeastern New Mexico: New Mexico. Geol. Soc., *36th Field Conf. Guidebook*, 153–159, 1985.

Treviño, L. and G. R. Keller, Structure of the North Park and South Park basins, Colorado: An integrated geophysical study, this volume.

Van der Lee, S., and G. Nolet, Upper mantle S velocity structure of North America: *J. Geophys. Res.*, 102, 22,815–22,838. 1997.

Vidale, J., Finite-difference calculation for travel times, *Bull. Seis. Soc. Am.*, 78, 2062–2076, 1988.

Vidale, J. E., Finite-difference calculation of traveltimes in three dimensions, *Geophysics*, 55, 521–526, 1990.

Wellborn, R. E., Structural style in relation to oil and gas exploration in North Park-Middle Park basin, Colorado, Colorado, in *Exploration Frontiers of the Central and Southern Rockies,* edited by H. K. Veal, *Rocky Mountain Association of Geologists*, pp. 41–60, Denver, Colorado, 1977.

Wessel, P. and W. H. F. Smith, New, improved version of the Generic Mapping Tools released, *EOS Trans. AGU*, 79, 579, 1998.

Wolf, L. W., and J. J. Cipar, Through thick and thin: A new model for the Colorado Plateau from seismic refraction data from Pacific to Arizona crustal experiment, *J. Geophys. Res.*, 98, 19,881–19,894, 1993.

Woodward, L. A., Tectonic map of the Rocky Mountain region of the United States, in *The Geology of North America, D-2, Sedimentary Cover - North American Craton: U. S., plate 2,* edited by L. L. Sloss, Geological Society of America, Boulder, Colo., 1988.

Zelt, C. A., and R. B. Smith, Seismic traveltime inversion for 2-D crustal velocity structure, *Geophys. J. Int.*, 108, 16–34, 1992.

Zelt, B. C., R. M. Ellis, R. M. Clowes, and J. A. Hole, Inversion of three-dimensional wide-angle seismic data from the southwestern Canadian Cordillera, *J. Geophys. Res.*, 101, 8503–8529, 1996.

Zelt, C.A. and P. J. Barton, Three-dimensional seismic refraction tomography; a comparison of two methods applied to data from the Faeroe Basin, *J. Geophys. Res., 103*, 7187–7210, 1998.

Zurek, B., and K. Dueker, Lithospheric stratigraphy beneath the southern Rocky Mountains, this volume.

G. Randy Keller, Department of Geological Sciences, University of Texas at El Paso, 500 University Ave., El Paso, Texas 79968, keller@geo.utep.edu

Kate C. Miller, Department of Geological Sciences, University of Texas at El Paso, El Paso, 500 University Ave., Texas 79968, miller@geo.utep.edu

Claus Prodehl, Geophysikalisches Institut, University of Karlsruhe, Hertzstr.16, Karlsruhe, Germany 76187, claus.prodehl@epost.de

Hanna-Maria Rumpel, Geophysikalisches Institut, University of Karlsruhe, Hertzstr.16, Karlsruhe, Germany 76187, Hanna-Maria.Rumpel@gpi.uni-karlsruhe.de

Catherine M. Snelson, Department of Geoscience, University of Nevada Las Vegas, 4505 Maryland Parkway, MS 4010, Las Vegas, Nevada 89154-4010, csnelson@unlv.nevada.edu

Crust and Upper Mantle Velocity Structure of the Southern Rocky Mountains from the Jemez Lineament to the Cheyenne Belt

Alan Levander, Colin Zelt, and Maria Beatrice Magnani

Department of Earth Science, Rice University, Houston, TX 77005

We have interpreted the refraction/wide-angle reflection seismic profile acquired as part of the Continental Dynamics of the Rocky Mountains project. The profile extends ~955km from northern New Mexico to southern Wyoming, crossing the Tertiary-Quaternary volcanics of the Jemez lineament, the Paleoproterozoic Mazatzal and Yavapai terranes, the Cheyenne belt, and the southern Archean Wyoming Province. We inverted the travel-time data from the 10 shot profile with both a layer based inversion method and a tomographic method. The two techniques yield comparable upper and middle crustal velocity structures. Lower crustal velocities are well constrained in the layer based model but are not in the tomographic model. From the layer based model, velocities in the crystalline crust and the upper mantle are lower than typical for continents and for modern orogens. Lower crustal velocities rarely exceed 7.00 km/s, likely due to the regionally high heat flow. We infer that the low upper mantle velocities beneath the Jemez lineament (7.70–7.76 km/s) are indicative of upper mantle partial melt. Crustal thickness increases from south to north, with thinner crust under the Jemez lineament (40–42 km), and thicker crust under northern Colorado, the Cheyenne belt, and southern Wyoming (51–53 km). Although the Cheyenne belt outcrops as a narrow zone separating Paleoproterozoic and Archean terranes, the seismic model shows broad lateral variation in crustal velocity near the suture, and a thick crust in the northern half of the profile. Part or all of the crustal thickening is likely to have occurred subsequent to continental accretion.

INTRODUCTION

As discussed in several papers in this volume, the southwestern United States formed during the Proterozoic (~1.1–1.8 Ga) by northward accretion of a series of island arc terranes to Archean protocontinents, notably the Wyoming Province. The northernmost Proterozoic accretionary boundary, the Cheyenne Belt, strikes northeast, forming a distinct geological boundary [*Karlstrom and Houston*, 1984; *Chamberlain*, 1998; *Henstock et al.*, 1998, *Snelson et al.*, 1998] and marking profound regional crust and mantle seismic velocity differences between the Archean Wyoming Province, and the Proterozoic Yavapai-Mazatzal terranes. Geological and geochemical evidence indicates that this boundary resulted from island arc collision with a northeast oriented passive margin at the southern edge of the Wyoming province [e.g., *Chamberlain*, 1998]. The Yavapai-Mazatzal terranes were accreted to the Wyoming craton in the Paleoproterozoic (1.65–1.8 Ga), and extend to the south some ~1100 km to the Grenville Front. Several internal boundaries within the Yavapai and Mazatzal terranes have been identified geochemically, and structurally

The Rocky Mountain Region: An Evolving Lithosphere
Geophysical Monograph Series 154

10.1029/154GM22

appear as broad shear zones [*Karlstrom and Bowring*, 1988, 1993]. Subsequent to Paleoproterozoic island arc accretion, this large area experienced a period of anorogenic intrusion of granites (1.3–1.4 Ga), and its southern edge was the site of the complex Grenville continent-continent collision forming the Rodinian supercontinent.

The Phanerozoic history is punctuated by tectonic activity within long periods of quiescence including development of the ancestral Rockies in the Pennsylvanian, Mesozoic-early Cenozoic Laramide uplifts, and late Cenzoic-Recent Rio Grande rift and Basin and Range extension. *Karlstrom and Humphreys* [1998] noted that although tectonic activity in the Phanerozoic has been oriented along a north-south axis, upper mantle low velocity anomalies identified in regional tomography studies roughly parallel the northeast striking accretionary boundaries such as Cheyenne Belt, the less distinct geochemical boundaries within the Yavapai-Mazatzal terrane, and several spatially associated modern trends, including the Jemez lineament. The latter is a northeast striking zone of Tertiary-Quaternary felsic to mafic volcanic centers extending from the southwestern edge of the Colorado plateau in Arizona to the Great Plains near the New Mexico-Colorado border. This coincidence of modern lineaments and mantle anomalies with ancient accretionary boundary locations led *Karlstrom and Humphreys* [1998] to speculate that the accretionary boundaries are aligned zones of lithospheric weakness that have provided structural control on subsequent tectonic activity.

As part of the Continental Dynamics of the Rocky Mountains project (CD-ROM) we acquired a ~955km long seismic refraction/wide-angle reflection profile extending across the Jemez lineament, much of the Mazatzal-Yavapai terranes, the Cheyenne Belt, and the southernmost Wyoming Province. This experiment complemented coincident reflection profiling and a broadband passive array deployment in north central New Mexico across the Jemez Lineament and Yavapai-Mazatzal boundary [*Magnani et al.,* 2003, *Magnani et al.*, 2005, this volume], and similar reflection profiling and a passive array deployment across the Cheyenne Belt in northern Colorado and southern Wyoming [*Morozova et al.*, this volume; *Dueker*, 2002; *Yuan and Dueker*, 2005, this volume, *Zurek and Dueker*, 2005 this volume]. Together these experiments constitute a multi-band look at upper mantle and crustal structure from the surface to the base of the mantle transition at various resolutions from a few hundred meters in the crust, to a few tens of kilometers in the mantle.

In this paper we describe the analysis and interpretation of the wide-angle seismic profile crossing virtually the entire CD-ROM project area. These data have vertical resolution as good as a few kilometers, and variable horizontal resolution of 20–100km. Two other papers in this volume present alternate interpretations of the seismic refraction data [*Snelson et al.*, 2005, this volume, *Rumpel et al.*, 2005, this volume].

PREVIOUS SEISMIC STUDIES

The number of crustal-scale active source seismic studies in the Rocky Mountain region is relatively modest considering geologic interest in this area as a modern orogenic plateau, and its mineral and hydrocarbon resources. The geophysical framework of the region is reviewed by *Prodehl et al. [2005* this volume], and will only be touched on here. *Prodehl and Lipman* [1989] and more recently *Keller et al.* [1998] have analyzed and synthesized the existing data and provided a map of the thickness of the crust in the Rocky Mountain region [*Keller et al.*, 1998]. The summaries show crustal thickening of up to 12 km across the Colorado-Wyoming border. The Moho discontinuity deepens from 40–44 km under southern Wyoming to 44–52 km under Colorado with mean velocity in the crust and the upper mantle decreasing in the same direction. Crustal thickness remains constant across the Rocky Mountain front into the Great Plains, whereas the mean velocity of the upper crust and lower crust are lower under the mountains than the plains. An average lower crustal velocity of 6.6 km/s and velocities up to 7.2 km/s at the base of the crust are observed on a N-S seismic refraction profile that extends across the Southern Rockies west of the Rio Grande rift from Sinclair, Wyoming to Lumberton, New Mexico [*Prodehl and Pakisier*, 1980]. Along the CD-ROM profile (Figure 1), *Snelson* [2001], and *Snelson et al.* [2005, this volume] have interpreted the same data that we model here. They find a relatively thick crust (45–60km) with a high velocity lower crustal layer ($Vp > 7.0$ km/s).

Sheehan et al. [1995] used Rocky Mountain Front (RMF) experiment teleseismic receiver functions to investigate the crustal thickness variations from the Great Plains in Kansas across the Colorado Rocky Mountains to the Colorado Plateau [see *Sheehan*, 2004, this volume]. The poor correlation between topography and crustal thickness, i.e. lack of a distinct crustal root associated with Rocky Mountain topography led [*Sheehan et al.*, 1995] to show that neither an Airy-type crustal root nor a Pratt variable density model can explain the difference in elevations between the Rockies and the Great Plains. The implication is that the upper mantle provides part of the isostatic support for the Rocky Mountain elevation, by density variations in the mantle inferred from low velocity mantle anomalies. Tomographic models from the Rocky Mountain Front experiment [*Lee and Grand*, 1996] corroborate such a hypothesis, detecting low mantle velocities under the high elevations of Colorado and New Mexico Rocky Mountains [*Lerner-Lam et al.,* 1998].

The Deep Probe experiment, a continuous 2400 km refraction transect from Canada to New Mexico (Figure 1), was collected with a seismograph spacing of 1.25 km and a small

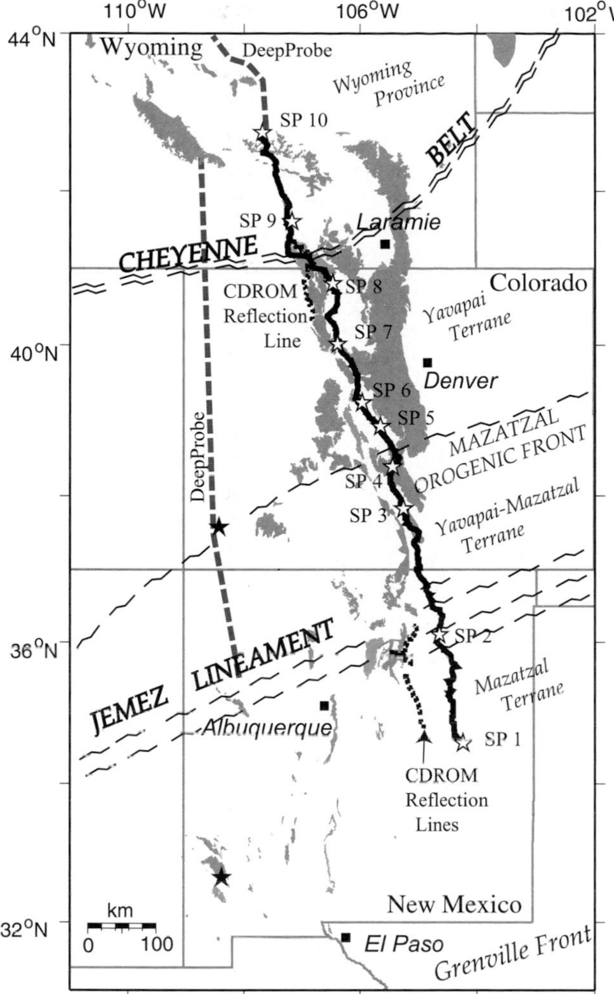

Figure 1. Location map showing the CD-ROM refraction profile (solid black line) with CD-ROM shotpoints (open stars). The CD-ROM reflection profiles (short dashes) are near either end of the refraction profile. The location of the DeepProbe experiment (long dashes) and shotpoints (solid stars) are also shown. Outcrops of Precambrian basement are shown as grey. The Wyoming Province, the Proterozoic terranes and their structural boundaries, and the Grenville Front are indicated.

number of large, widely spaced shots [*Henstock et al.,* 1998; *Snelson et al*., 1998]. Although the focus of the experiment was the mantle structure of the Archean Hearne and Wyoming provinces and the Proterozoic accreted terranes in the U.S. southwest, the data also provide insights on crustal structure. Each province is characterized by a distinct crustal type: the Proterozoic accreted terranes have low average crustal velocity, 6.3 km/s, and thickness of 40–45 km, less than expected for its high elevations. Crustal thickness in northwestern Colorado at the Cheyenne belt is 50–52 km, a measurement confirmed by the receiver function studies in the area [*Sheehan et al*., 1995, *Crosswhite and Humphreys*, 2003]. Most of the Archean province has a fast and thick crust (6.6 km/s and 50 km or greater) that is thicker than the average for Archean cratons [*e.g., Durrheim and Mooney*, 1991]. The northern two thirds of the Archean Wyoming Province is also characterized by a ~20–25 km thick high velocity lower crust (Vp~7.05 to > 7.50 km/s) that extends from ~150 km north of the Cheyenne belt and crosses the Great Falls Tectonic Zone and Medicine Hat Block, to the southern boundary of the Hearne province in Canada [*Gorman et al*, 2002].

A P-wave low velocity zone was inferred in the upper mantle just beneath the Moho at 50–80 km depth under the Proterozoic accreted terranes based on a Pn shadow zone and traveltime modeling [*Henstock et al,* 1998]. The upper mantle low velocity zone extends north under Colorado to the Cheyenne belt in Wyoming, where it terminates over a distance of less than 250 km. Upper mantle low velocity zones have also been observed in the Rio Grande rift [*Keller et al.,* 1990], and the Basin and Range transition zone in Arizona [*Benz and McCarthy*, 1994]. Together the data suggest that low velocity zones in the shallow upper mantle are widespread under the Proterozoic terranes, but are likely not contiguous features.

Teleseismic datasets collected for both the RMF and CD-ROM experiments have been used to make upper mantle velocity tomograms under the Southern Rocky Mountains and the surrounding areas [*Sheehan*, 2005, this volume; *Lee and Grand,* 1996; *Lerner-Lam et al,* 1998; *Dueker et al.*, 2001; *Yuan and Dueker*, 2005, this volume]. Both P and S wave data show that the mantle under the Rocky Mountain area is characterized by very low velocities relative to the Great Plains to a depth of 200 km, and that slow velocities extend some 200 km east of the Rocky Mountain Front as far as the Colorado-Kansas border [*Lee and Grand*, 1996]. Lee and Grand identify the onset of high velocities in western Kansas as the western edge of thick, high velocity North American cratonic mantle [*Grand*, 1987]. *Dueker et al.* [2001] and *Yuan and Dueker* [2005, this volume] identify low velocity mantle anomalies along the axis of the Rio Grande rift and southern Rockies as far north as the Cheyenne belt. Beneath the Cheyenne belt they identify an isolated north dipping high velocity anomaly. To the north, continent-wide S wave tomographic investigations by [*van der Lee and Nolet*, 1997] show that the high velocity continental root of the North American craton does not extend under the Archean Wyoming province, implying that the mantle in this area has undergone modification since its formation in the Archean, or originally formed without a tectospheric root.

Lastly, we note that heat flow measurements are often highly correlated with mantle seismic velocity anomalies [e.g., *Goes and van der Lee*, 2002]. In the CD-ROM region [*Morgan and Gosnold*, 1989] characterize the Rockies south of the Cheyenne belt as a high heat flow regime (63 to >

105 mWm^{-2}, in places exceeding 150 mWm^{-2}; see Figure 5D). They note that *no conductive model of steady-state heat flow* in the Southern Rockies is reasonable without melting large percentages of crustal rocks. In contrast central Wyoming north of the Cheyenne Belt is an average heat flow regime (54–67 mWm^{-2}; [*Decker et al.,* 1980]. The U.S. heat flow map from the Global Heat Flow Database [*Pollack et al.*, 1993] shows a relatively abrupt transition from heat flow values of 71–200 mWm^{-2} in the southern Rockies to 41–81 in Wyoming and the Great Plains to the east and south. Heat flow in the Jemez Lineament and surrounding regions in the CD-ROM study area can be very high, in places 145–180 mWm-^{2} [*Rieter et al.,* 1979].

SEISMIC DATA

The CD-ROM refraction/wide-angle reflection profile extends ~955km from Fort Sumner New Mexico, in the northern Proterozoic Mazatzal Province to Day Loma Wyoming, in the southern Archean Wyoming Province (Figure 1). The profile consists of 10 shots, varying in separation from 40.8 km to 198.7 km, with a mean spacing of 105.8 km [*Snelson*, 2001]. Shot spacing in the center third of the profile is considerably closer, ~60 km. Shot size varied from 167 kg to 2727 kg. The shot near Fairplay, CO, was substantially smaller than the other shots (167 kg), but produced detectable signals to ~100 km. Approximately 600 seismographs (see Figure 1) were deployed twice along the profile, for an average receiver spacing of 800 m. The shot spacing is large by modern refraction profiling standards, however the data provide reversed coverage of Pg, and PmP, and some reversed coverage of PcP (a mid-crustal reflection), and Pn.

Traveltimes were picked on records filtered with minimum phase bandpass filters of variable bands (1–6Hz, 2–8Hz, 4–12Hz). The phases used for the inversion (Figure 2), and the assigned travel time picking errors, are listed in Tables 1 and 2. Reciprocal times between corresponding phases were within the picking error. The crooked line geometry was treated by projecting the shots perpendicularly onto a best-fit straight line, while maintaining the correct shot-receiver offsets, as described by [*Zelt,* 1999], thus allowing for conventional 2-D modeling.

Snelson et al. [2005, this volume] picked a refraction arrival from the lower crust that we feel is not warranted. As a result *Snelson et al.* [*2005,* this volume] interpret higher deep crustal velocities and a thicker crust in their model than in the preferred model we describe below.

TRAVELTIME INVERSION

We inverted the data for two different models using the layer based traveltime inversion method of [*Zelt and Smith*, 1992], and the first arrival traveltime tomography code of [*Zelt and Barton*, 1998]. Using the tomographic velocity model obtained from inverting the first arrivals, reflection tomography was applied to PcP and PmP. The tomographic model is viewed as an independent check on the more subjectively-derived layered model, similar to the examples presented by [*Zelt et al*., 2003].

Preferred Model M1

We prefer the layered model, designated M1 (Plate 1), obtained by inverting the traveltimes from all phases using the [*Zelt and Smith*, 1992] algorithm. This inversion is designed to yield the most simple model that is geologically reasonable and honors the traveltime data [*Zelt et al.* 2003]. We identified three separate arrivals making up the crustal refraction; a shallow (sedimentary layer) arrival, Ps, an upper-mid crustal phase, Pg, and a mid-lower crustal phase, Pc, that often extended well beyond the Pn crossover distance. The upper mantle refraction, Pn, was identified as a clear arrival on 6 of the 10 shot points. The reflected arrivals included a prominent mid-crustal reflection, PcP, on 8 of the shot records, and the Moho reflection, PmP, on 9 of the shot records. PmP was more prominent and displayed more uniform character across the profile than PcP. As a result, the picking uncertainty assigned to PcP was the largest of any phase (Table 1). The data were fit successively in inversions that started with the surface layer and proceeded by adding progressively deeper layers down to the upper mantle using the standard approach for the *Zelt and Smith* [1992] algorithm [e.g. *Zelt,* 1999; *Zelt et al.,* 2003]. Model M1 includes a surface layer about 1–2 km thick, an upper crustal layer to 15 km depth, a mid-crustal layer to 22–29 km depth, a lower crustal layer, and a mantle layer (Plate 1). The upper-to-mid crustal interface at 15 km is an artificial boundary in the sense that it does not represent a velocity discontinuity, but it allows for changes in the horizontal and vertical velocity gradients needed to model the Pg and Pc phases. The upper crust was first modeled using a half-space beneath the surface layer, followed by inversions in which the surface layer and the 15 km interface with a half-space beneath were all included.

M1 is described by 136 independent model parameters, 106 velocity nodes and 30 boundary nodes. The travel-time fits and normalized χ^2 values for the different phases are given in Table 1. Overall the 3617 data were fit with an RMS residual of 0.167s and a normalized χ^2 value of 1.51 (Figure 4). The RMS misfit of Ps, Pg and Pc is relatively high because the short-wavelength fluctuations in these phases would require short-wavelength variations in upper crustal velocity that would not be uniquely constrained by only 10 shots along a

955-km-long profile [e.g. *Zelt and White*, 1995]. Ray coverage throughout the crust, including PcP and PmP reflections, is relatively good (Figures 3 and 4), making the estimation of all crustal velocities reliable. Although Pn is only strictly reversed between two of the shots, there is overlapping Pn ray coverage across the whole model. In addition, the geometry of the Moho is well constrained by PmP reflections, thus reducing the ambiguity usually associated with unreversed Pn data. As a result we consider the upper mantle velocities to be constrained between X=125 to 750 km.

Tomography Model M2

The second velocity model, M2 (Plate 2), was derived using a modified version of the [*Zelt and Barton*, 1998] regularized tomographic approach that incorporates constraints on the second derivative and perturbation of the model from a simple laterally homogeneous starting model [e.g., *Zelt et al.,* 1999; 2003]. The first arriving crustal refractions and Pn were inverted for velocities on a uniform grid with a cell size of 2 km laterally by 1 km vertically. Traveltimes were calculated with an eikonal traveltime solver on a 1 km grid. The 2201 first arrivals were fit with a 0.099 s RMS residual and a normalized χ^2 of 1.01. Using the same approach applied by [*Zelt et al.,* 2003] to PmP, both the PcP and PmP reflections traveltimes were inverted to constrain the mid-crustal boundary and Moho using a first derivative flatness constraint and the velocity model, held fixed, derived by first-arrival tomography (Table 2; Plate 2). The crust-mantle boundary in M2 is a broad, smooth transition zone. This is a consequence of using a fine grid parameterization, smoothing regularization, and only first arrivals to constrain velocities in M2. Therefore, the velocities in M2 below about 25–30 km should not be used as either representative of lower crustal Earth velocities, or as a check on the velocity in M1 at a specific point. M2 can be used as a check on *relative* lateral variations in M1 below 25–30 km [*Zelt et al.,* 2003].

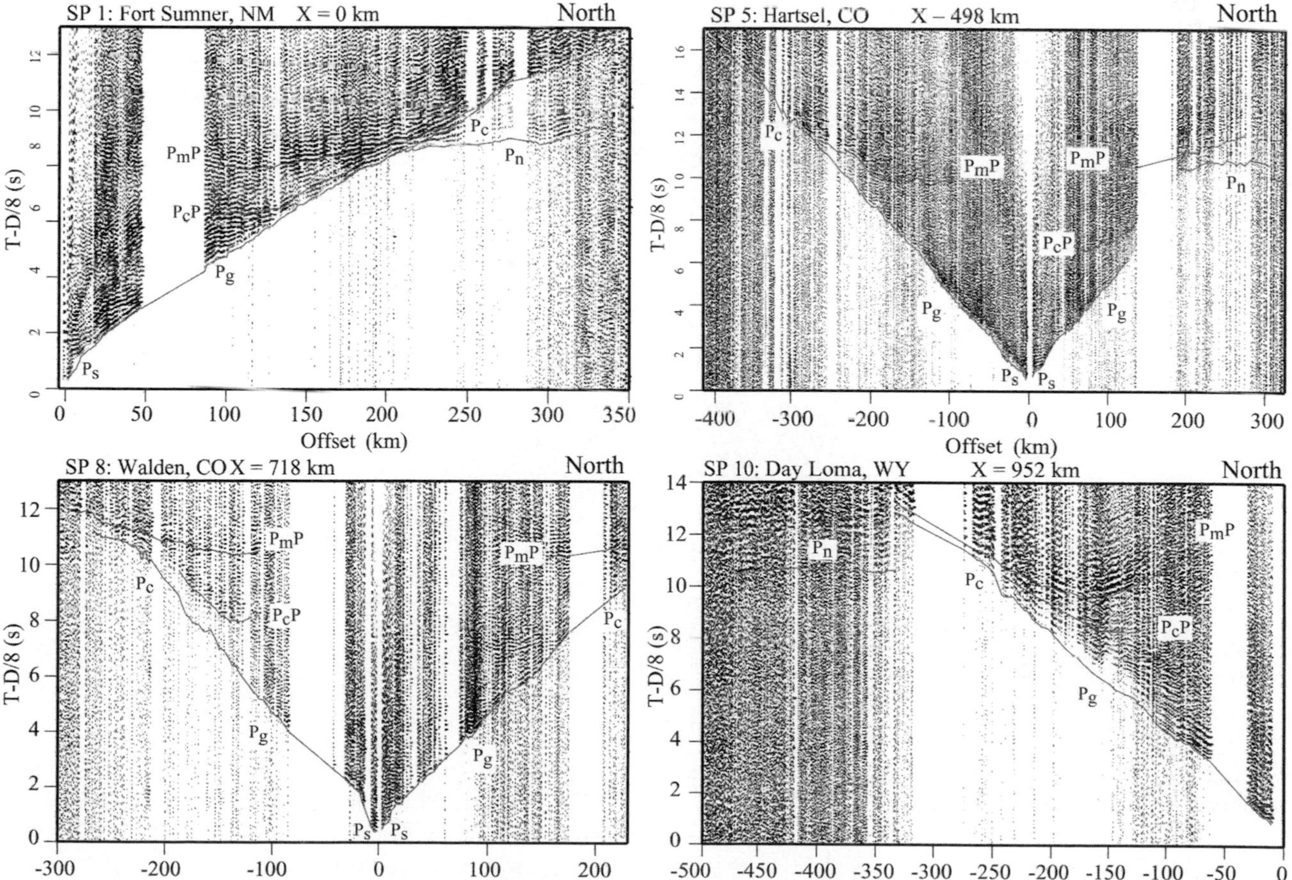

Figure 2. Seismic shot records from four of the CD-ROM shotpoints (SP1, SP5, SP8, and SP10). The traveltime picks are indicated as pink lines. The phases used in the modeling are labeled and described in the text. The data are reduced using a velocity of 8 km/s and have been bandpass filtered from 2–8 Hz.

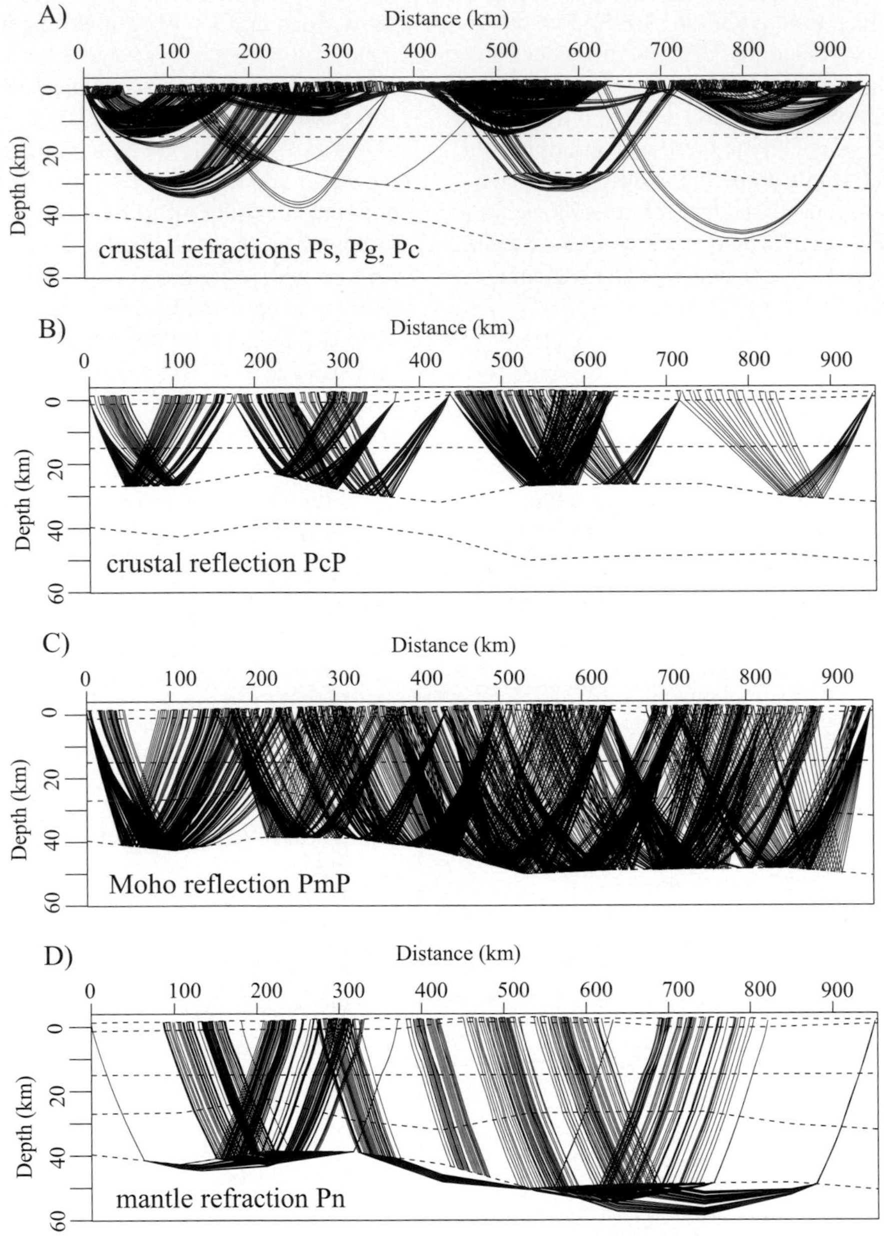

Figure 3. Ray coverage for all phases used to determine model M1. Every third point along each ray is plotted giving the rays a somewhat kinked appearance. A. The Ps, Pg, and Pc refractions sampled the sedimentary layer and the crystalline crust. B. The crustal reflection PcP was used to determine upper and middle crustal velocities and to constrain the top of the lower crustal layer. C. The PmP reflections were used to determine lower crustal velocities and the geometry of the Moho. D. Pn refractions determined upper mantle velocities and helped constrain the Moho geometry.

DISCUSSION OF THE MODELS

We describe locations in the models using the coordinate system X=0 to X=955, with X=0 corresponding to the southernmost shotpoint (SP 1), and X=955 slightly north of the northernmost shotpoint (SP 10). The two models M1 and M2 are very similar to one another in crustal structure to depths of ~25 km, with well correlated slow and fast regions above the lower crust (compare Plates 1 and 2). Both models have a low velocity surface layer (3.2–5.6 km/s) that is thickest in the northern third and southern half of the profile, and is somewhat attenuated in the center of the profile. Both mod-

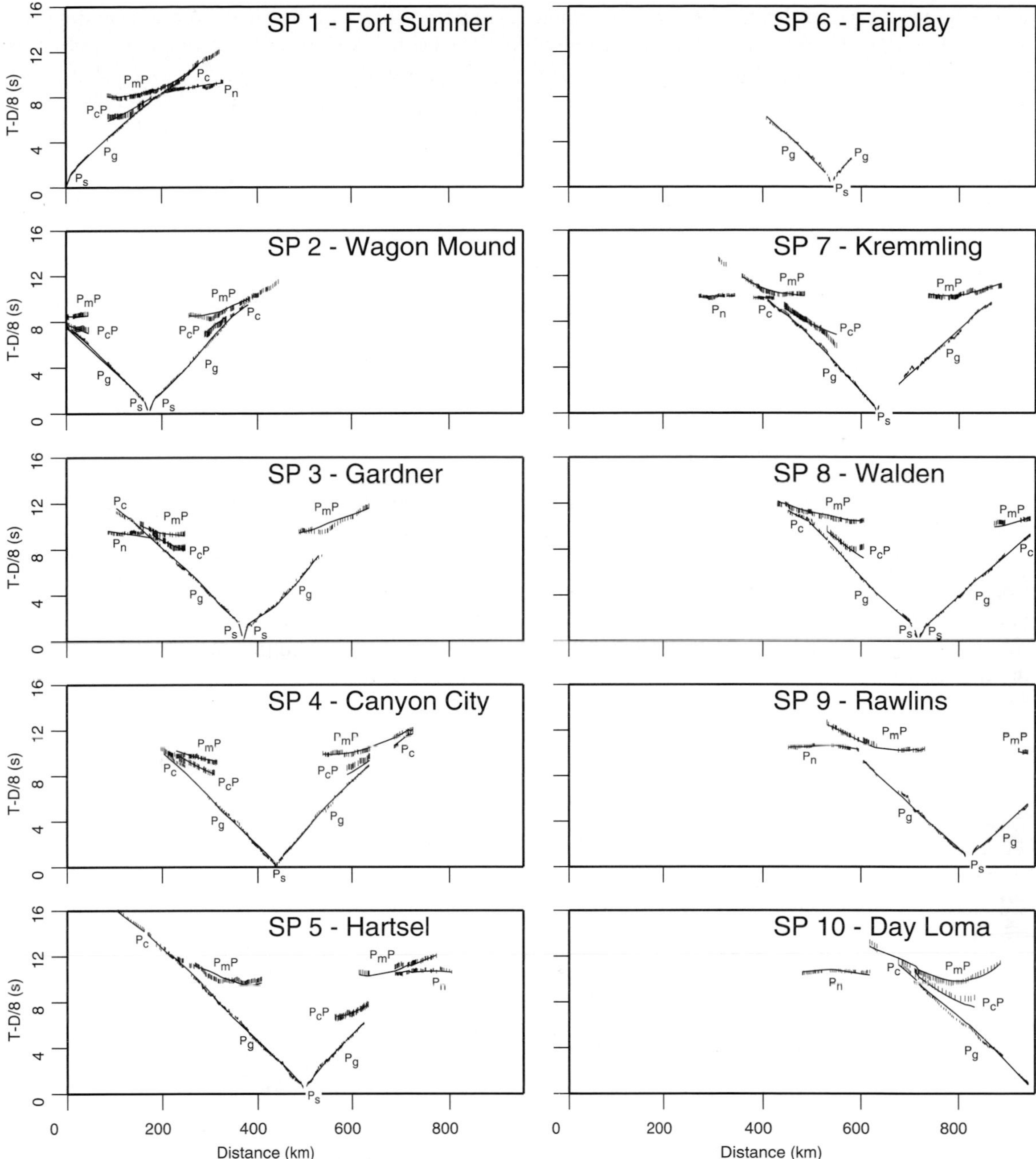

Figure 4. Calculated and observed traveltimes for each CD-ROM shotpoint relative to the 955km long model M1. Each panel shows the calculated times (solid lines) and the observed traveltime picks (bars showing travel time errors) for all phases as labeled. The data are reduced using a velocity of 8 km/s.

els show pronounced lower velocity (~6.0 km/s) regions in the upper crust below the surface layer at X=225-325 km, X=450-650 km, and at X=700-750 km. The average velocity of the lower crustal layer in M1 is 6.5 km/s. Velocities at the base of the crust range from 6.6 to 7.1 km/s, with an average velocity immediately above the Moho of 6.9 km/s.

Moho depths vary from near 40 km in the southern part of the profile (X=0–425 km) to more than 50 km in the northern half of the profile (X=500–955 km). Upper mantle velocities are on average low, being 7.85 km/s, and ranging from 7.7 to 8.0 km/s. The lowest velocities (7.70–7.76 km/s) are found beneath the thinnest crust (40–41 km), from X~150-300 km (Figure 5).

We note that using the layered model approach, which makes use of all first and later refracted and reflected traveltimes to estimate velocities and interface positions, we have developed a model (M1), that has about 5 km thinner crust, and significantly lower velocities in the lower crust than the model of *Snelson* [2001, and *Snelson et al.* [2005, this volume].

GRAVITY MODELING

We have modeled Bouguer gravity data from the National Geodetic Survey database along the CD-ROM refraction profile using a 2-D commercial gravity modeling package, GMSYS, based on the method of [*Talwani et al.,* 1959]. The model was constructed from the interfaces and velocities from M1 (Plate 3), with densities derived from a standard velocity-density relationship [see *Zelt,* 1989]. The surface topography was modified to correspond to the elevations of the gravity observations. The location of nodes in the shallowest layer, the upper 4 km of the model, were adjusted manually to account for basement uplifts and sedimentary basins as indicated by the tomography model M2. Density values in the crystalline crust and mantle were adjusted slightly (generally less than 25 kg/m^3). The largest errors occur in the southern end of the profile (~25-30 mgal), where control on the shallow crust from the seismic data is poor due to the large shot spacing. We attribute the mismatch to the effects of unmodeled sedimentary basins and shallow uplifts.

The large gravity negative in the center of the profile required reducing the density by 100 kg/m^3 to 2600 kg/m^3, and modifying the location of the nodes at X=500-600 km at a depth of 15 km (Plate 3). The data can be similarly fit by reducing the density only 50 kg/m^3 to 2650 kg/m^3, and extending the body depth to 22 km. In doing this we adopted the approximate geometry and density value of *McCoy and Roy* [2005, this volume], who have modeled this structure more extensively, interpreting it as a deep granitic/felsic batholith of the Colorado Mineral belt. Otherwise only slight adjustments to the density determined from the velocity-density relationship were required for the upper and middle crust, to ~27 km depth.

No modifications to the boundaries of the lower crust or Moho were required to fit the data. Lower crustal densities vary from 2850–2900 kg/m^3, and required almost no adjustment. Mantle densities also required little adjustment and range from 3230 kg/m^3 beneath the upwarped Moho under the Jemez lineament, to 3330 kg/m^3 on the northern side of the Cheyenne Belt.

Overall, after minor adjustments, the gravity calculated using the crust and Moho interfaces determined from M1 and densities determined from conversion of its velocity field agrees well with the observations, having an RMS misfit of less than 14.1 mgal. The somewhat poorer fit to the gravity in the southern part of the profile is a consequence of unmodeled shallow sedimentary basins for which we have little subsurface control, and is not relevant to the larger features of the model.

INTERPRETATION

We first compare the average one-dimensional CD-ROM velocity-depth function to global averages [*Christensen and Mooney,* 1995] for all continental crust and for orogens in Figure 6a. Except for the uppermost crust the average CD-ROM velocity function has lower values than the averages for either orogens or the average continent, although the CD-ROM velocities fall within the standard errors of the global compilations. Average upper mantle velocity along the profile is 7.85 km/s, also well below the average upper mantle velocity for orogens 8.01±0.22 km/s, or continents 8.09±0.20 km/s [*Christensen and Mooney*, 1995]. Average crustal thickness on the CD-ROM profile, 45.0 km, is well within the average of orogens of 46.3±9.5 km, but on the high side of average continental crust, 41.0±6.2 km.

We next compare one-dimensional profiles through M1 at various locations (X=250, 450, 550, 750, and 850 km) to the average profile for M1 (Figure 6b). The figure shows the northward increase in crustal thickness and in mantle velocities (also see Figure 5), as well as the relatively low upper and middle crustal velocities. In regions of Laramide uplifts (X=450 km at the Wet Mountains, and X=750 km at the Sierra Madre uplift) velocities in the upper crust exceed those immediately below it.

Surface Layer

The tomography model M2 has the best resolution of the near surface, with surface velocities that correspond to surface geology. Low velocities appear in the Raton, North Park, and Hanna Basins (X=300-350 km, X=625-725km, and X=900-955 km). Higher surface velocities are associated with the crystalline basement uplifts throughout the center of the profile (400–625 km) and at the northern end of the profile at the Sierra Madre and Rawlins uplifts (X=725-800 km and X=850-875 km, respectively).

Upper and Middle Crust

At the southern end of the profile, south of the Jemez Lineament, relatively high crustal velocity (6.6 km/s) appears as shallow as ~12.5 km depth. High velocities in the southern

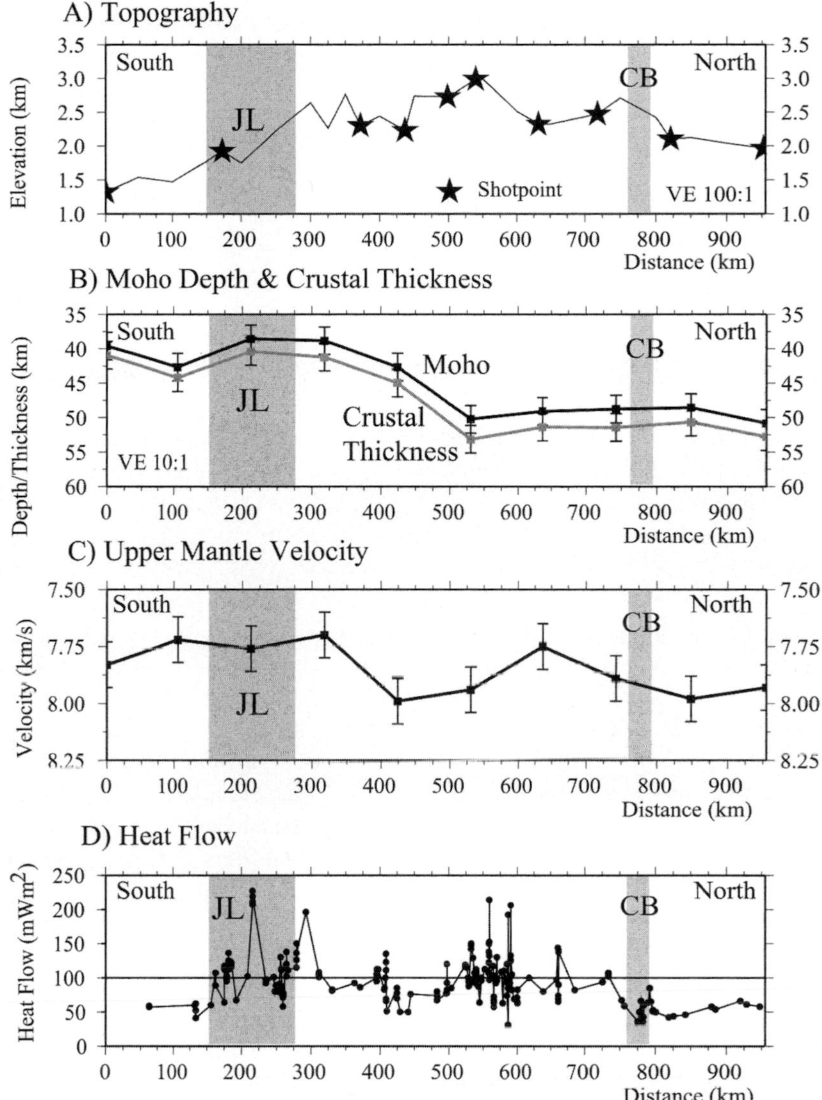

Figure 5. A. Topography, B. crustal thickness, C. upper mantle velocity, and D. heat flow. B. The depth to Moho (black) and the total crustal thickness (gray) from model M1. C. The Pn velocity from model M1. D. Heat flow along the CD-ROM corridor. Thin crust (41–44 km), low upper mantle velocities (7.7–7.8 km/sec) and high heat flow are associated with the Jemez Lineament (JL). Thicker crust (50–52 km), higher

part of the Jemez lineament and to the southern end of the CD-ROM profile are associated with a crustal duplex imaged on the CD-ROM reflection data [*Magnani et al.,* 2004 and this volume].

Low upper crustal velocities (6.0–6.2 km/s) overlap with and extend north of the Jemez lineament (X=225-325 km) to depths of 15–20 km. The low velocities in M2 have the same shape but are more localized than those in M1. These velocities are compatible with felsic lithologies, granite and granitic gneisses in a moderate to high heat flow regime. Heat flow measurements are locally high to very high (145–185 mWm^{-2}) in the region of the Raton volcanic field and the localities to the north and south [*Reiter et al.,* 1979]. We discuss this further below.

In the center of the profile, X=450-750 km, relatively low velocities, 6.2 km/s, extend to 25 km depth in a number of locations. Low velocities and densities in the Colorado mineral belt (X=500-600 km) are associated with massive, deep, Proterozoic (1.4 Ga) and Laramide felsic plutons. The low velocity anomaly in M1 has approximately the same shape at the low density anomaly (2600–2650 kg/m^3) required in the gravity model [see also *McCoy and*

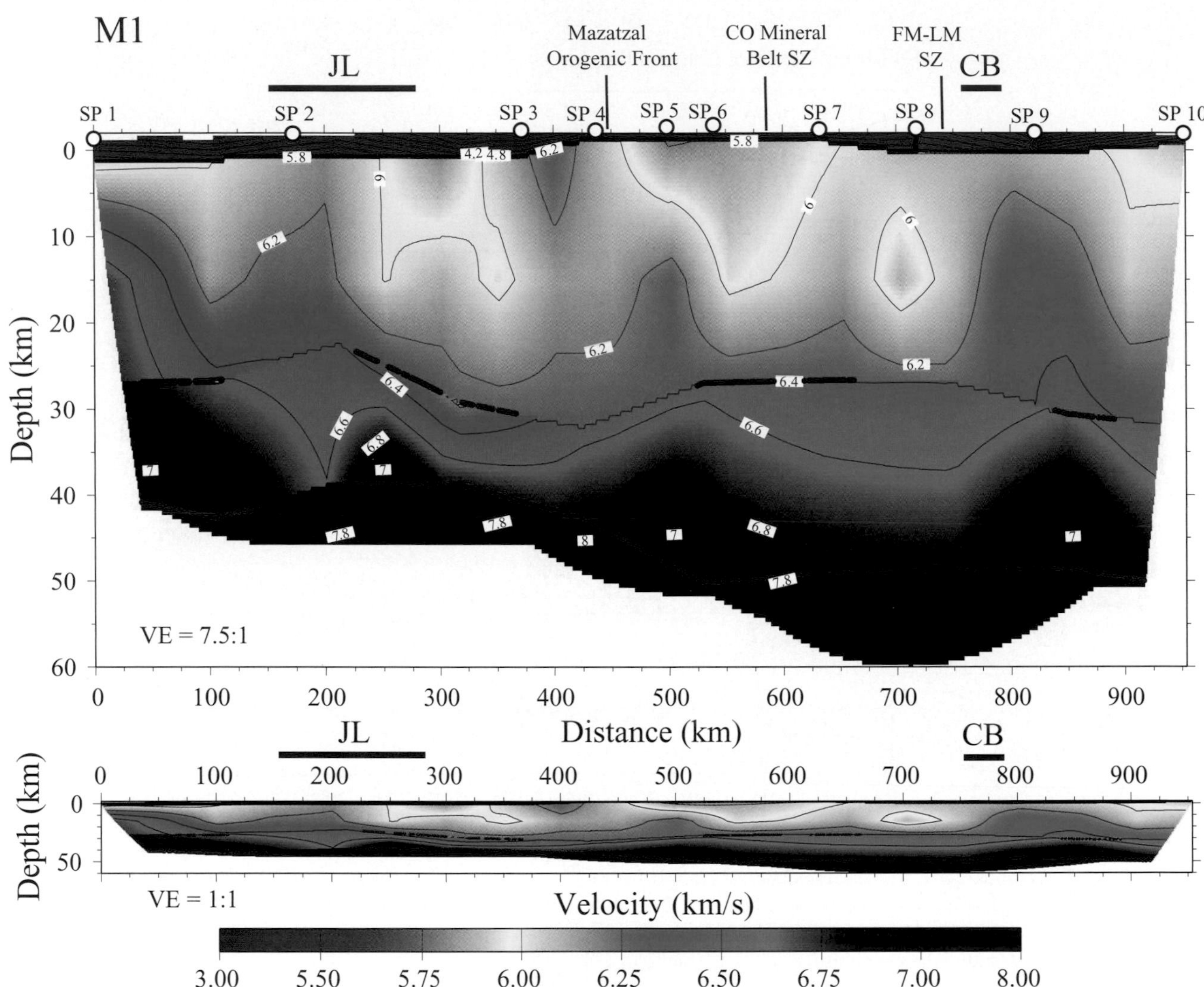

Plate 1. Top. Model M1 determined using the layer based inverse method [*Zelt and Smith*, 1992]. Reflection points for PcP and PmP are shown as pink dots. Shot points and geologic features are indicated at the top of the model. JL=Jemez lineament, FM-LM=Fairwell Mountain-Lester Mountain shear zone, CB=Cheyenne belt. Bottom. Model M1 shown with no vertical exaggeration.

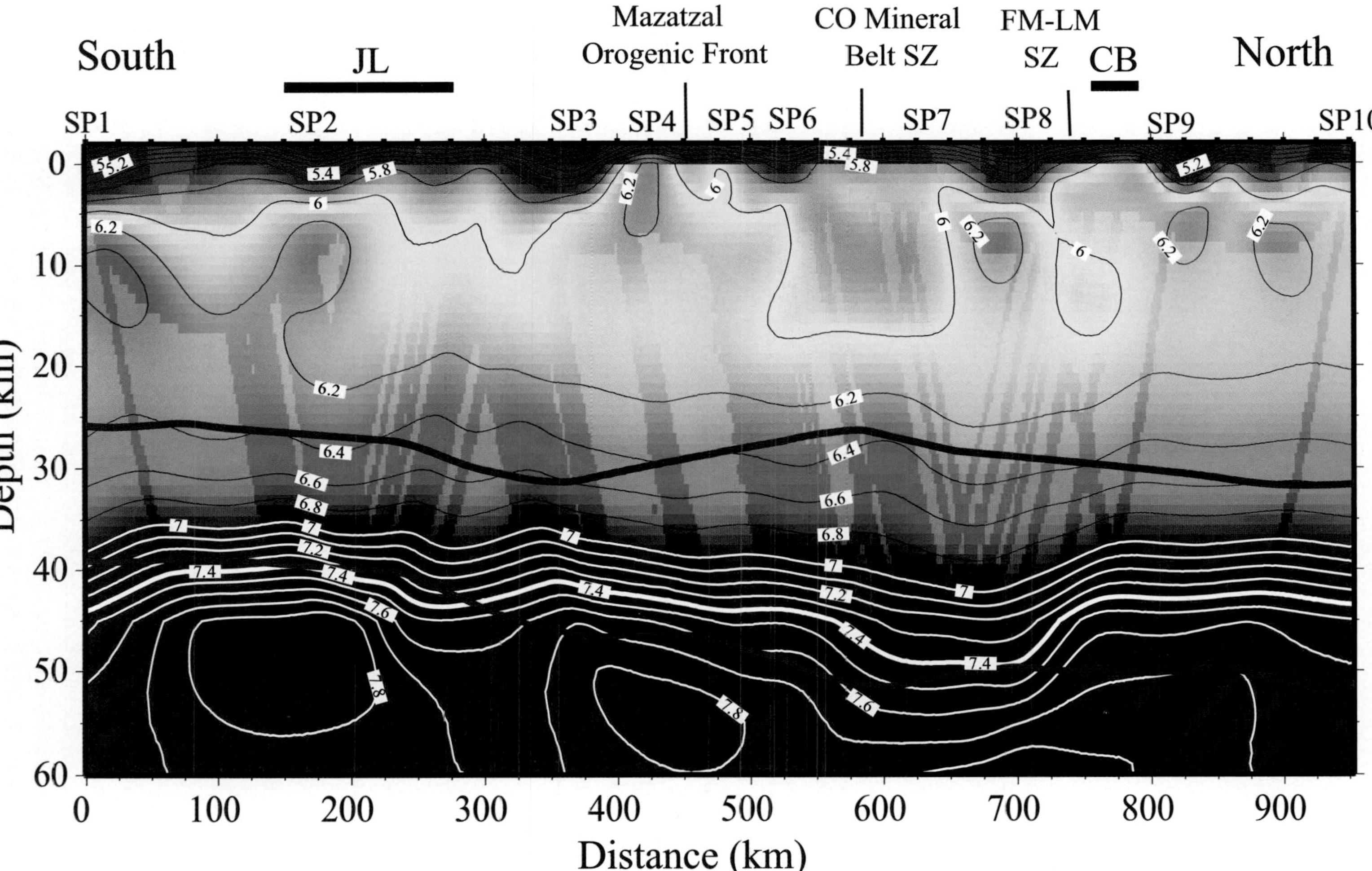

Plate 2. Model M2 determined using the first-arrival tomography method [*Zelt and Barton*, 1998]. The positions of the lower crustal and Moho boundaries were inverted for separately using PcP and PmP and are shown as pink lines. Compare to the preferred model M1 in Plate 1. Geologic features at the top of the model as in Plate 1. Those portions of the model not sampled by first arrival rays are indicated by reduced color intensity.

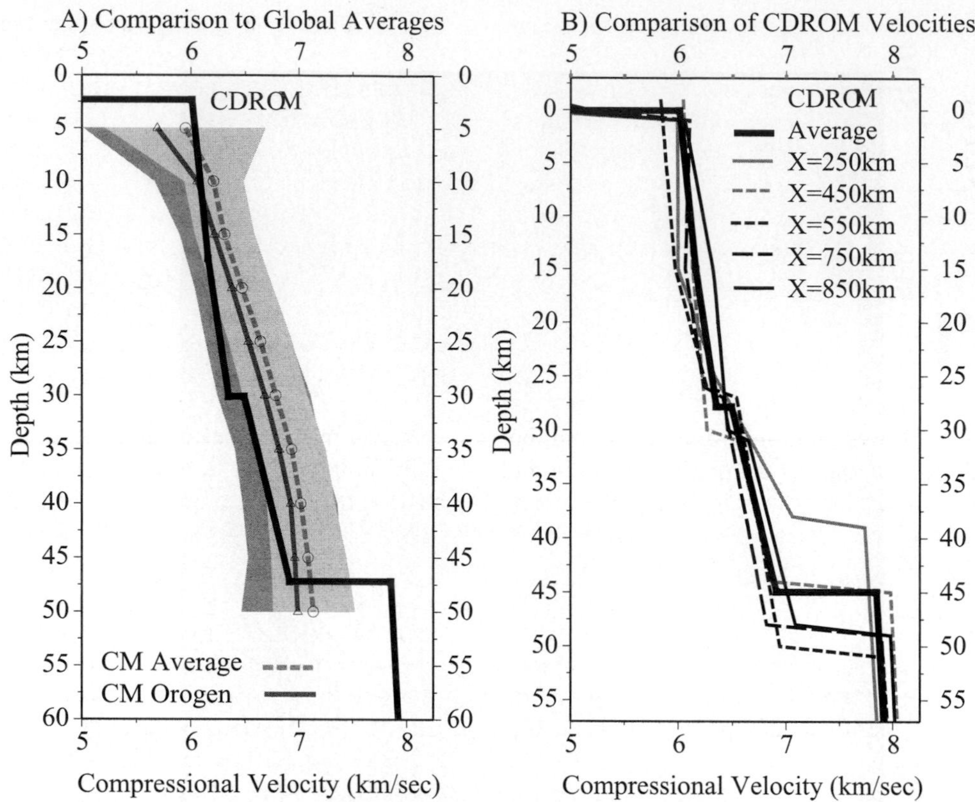

Figure 6. A. Comparison of the average one-dimensional CD-ROM velocity profile to global averages for continental crust (dashed gray line) and Phanerozoic orogens (solid gray line) [from *Christensen and Mooney*, 1995]. Standard errors for the average continent and orogens are shown as light and dark gray shading, respectively. Errors on orogens extend entirely beneath the errors for average continents. The CD-ROM curve has been shifted to sea level for the comparison. B. Comparison of average one-dimensional CD-ROM velocity profile to representative 1-D profiles extracted in the Jemez Lineament region (X=250 km), the Mazatzal orogenic front (X=450 km), the Colorado Mineral Belt (X=550 km), the Sierra Madre uplift (X=750 km), and southern Wyoming north of the Cheyenne Belt (X=850 km). Crustal velocity in the Wyoming Province is higher than the average and most values shown to the south. Crustal velocities at the southern end of the profile (not shown) are also higher than the average (see Plates 1 and 2).

Roy, 2005, this volume]. The low velocities south of the Fairwell Mountain-Lester Mountain shear zones are associated with exposed Paleo- and Mesoproterozoic felsic plutons.

Just north of the Cheyenne belt, upper and midcrustal velocities increase (6.2–6.5 km/s) in the region of the Sierra Madre uplift. *Morozova et al.* [2004, this volume] have interpreted CD-ROM deep reflection data as showing tectonic wedging with Archean rocks being displaced to the south both in the shallow (0–10 km) and deep (20–45 km) crust relative to the Proterozoic terranes and the surface expression of the Cheyenne belt. The high velocities in the upper and middle crust observed in M1 and M2 are compatible with the lithologies of the Archean granitic gneisses intruded by Paleoproterozoic (2.1–2.0 Ga) mafic sills and dikes [*Karlstrom and Houston*, 1984; *Chamberlain*, 1998].

Lower Crust

As noted above, M1 has a relatively low-velocity lower crust. Compressional velocities just above the Moho average 6.9 km/s and vary from $6.6 < Vp < 7.1$ km/s. The average velocity of the lower crust between the lower crustal reflections and the Moho is 6.7 km/s. Note that above the upwarped Moho, at X=150-200 and X=300-425, relatively low velocities (~6.6–6.8 km/s) extend through the lower crust to the Moho. Small regions of high velocity (7.0–7.1 km/s) appear at the southern end of the profile (X < 100km), beneath the Mazatzal orogenic front (X~500 km), and in the Archean terranes just to the north of the Cheyenne Belt (X=825-875 km). The reflection from the top of the lower crustal layer across the CD-ROM profile is intermittent, being imaged over less than half of the profile, suggesting either a compositionally or geometrically irregular boundary with variable impedance contrasts [e.g. *Holliger et al.,* 1994]. The lower crust is also highly variable in thickness (~10–22.5 km), with the thinnest section occurring just north of the Jemez Lineament (~10 km), and the thickest section being under the Colorado Mineral Belt (~22–23 km). Although the velocities are low, overall the velocities in the lower crust are compatible with mafic granulites, or gabbro compositions in a moderate to high heat flow regime [*Christensen and Mooney,* 1995].

Upper Mantle

As we noted above (see Plate 1 and Figures 5 and 6) upper mantle velocities are lowest, and the whole crust and the lower crust are thinnest along the CD-ROM profile in the distance range X=100-325 km, where it crossed the Jemez lineament (X~150-275 km). These features appear above south dipping negative P and S mantle velocity anomalies that extend from the base of the crust to depths of at least 150 km [*Dueker,* 2001; *CD-ROM Working Group,* 2002; *Yuan and Dueker*, 2005, this volume]. We interpret the low velocities in the mantle to be indicative of a partially melted upper mantle, and one that likely is supplying basalt to the crustal column.

The teleseismic tomography images also show positive P and S velocity anomalies dipping to the north at the Cheyenne belt, where we observe upper mantle velocities of 7.9–8.0 km/s. Crustal thickness beneath the Cheyenne belt region is ~52 km, similar to estimates beneath the Cheyenne belt further west along the Deep Probe profile [*Henstock et al.,* 1998; *Snelson et al.*, 1998; *Crosswhite and Humphreys*, 2003].

DISCUSSION

The overall low velocities in the entire crust beneath the CD-ROM profile from the center of the Jemez Lineament to the Cheyenne belt (X=225-750 km, Plates 1 and 2, Figure 6) are indicative of a largely felsic upper and middle crust, and a heterogeneous lower crust in a higher than average crustal heat flow regime. As described above, the CD-ROM refraction profile from its southern end to the Cheyenne belt lies in a high heat flow regime (63 to > 105 mWm^{-2}; [*Morgan and Gosnold*, 1989; *Pollack et al*., 1993], with highest heat flow in the Jemez lineament region and northern central Colorado. North of the Cheyenne belt in Wyoming heat flow drops to average levels (54–67 mWm^{-2}; [*Decker et al.,* 1980; *Morgan and Gosnold*, 1989; *Pollack et al*., 1993].

In and adjacent to the Jemez Lineament region near the CD-ROM profile, Quaternary-Recent basalt flows are found in the Ocate volcanic field (0.8–8.3 Ma) and the Raton volcanic field (0.03–8.77 Ma), both of which lie over the low mantle velocities and thin crust under the CD-ROM profile. In this region deep seismic reflection data (Figure 1); [see *Magnani et al.,* 2004, and this volume] show a series of bright sill-like reflections at upper-middle crustal depths across the Jemez region that we interpret to be modern mafic sills. The Jemez lineament region along the CD-ROM refraction profile has also experienced slow Late Cenozoic upwarping relative to surrounding regions, as determined from canyon incision data and other geomorphic evidence from the Canadian River [*Wisniewski and Pazzaglia,* 2002]. The coincidence of high heat flow, slow uplift, recent mafic volcanism and mafic sills intruded in the crust, accompanied with low velocities in the upper, middle, and lower crust suggest that the upper mantle beneath the Jemez lineament is hot and partially molten, and is heating the crust through movement of basaltic magmas into the crust. The 0.25 km/s reduction in upper mantle velocity beneath the Jemez Lineament from a reference of 8.0 km/s is consistent with 1% partial melt [*Hammond and Humphreys*, 2000]. The Jemez lineament, which marks the southern edge of the Mazatzal-Yavapai transition region, is a long lived zone of lithospheric weakness and therefore is a likely zone for magma ascension into the crust [*Karlstrom and Humphreys*, 1998].

The velocities of the lower crust beneath the CD-ROM profile are considerably lower than those beneath the northern two thirds of the Archean Wyoming Province and the Great Falls Tectonic Zone measured along the DeepProbe profile where a thick, ~20–30 km, high velocity (7.0–7.5 km/s) lower crustal layer has been identified [*Henstock et al.,* 1998; *Snelson, et al.,* 1998; *Gorman et al*, 2002]. *Gorman et al.* [2002] have interpreted this layer to be a heterogeneous mafic underplate added to the crust during the Proterozoic, however whether the underplating occurred in the Archean, Proterozoic or as a number of Precambrian events is unclear. The CD-ROM lower crustal velocities are on average about 0.25–0.30 km/s slower than those beneath the northern Wyoming Province [*Henstock et al.,* 1998; *Snelson et al.,* 1998; *Gorman et al.*, 2002]. If the lower crustal layers are composed of the same lithologies, this would require a temperature increase at the base of the crust of approximately 430–600°C to 1000–1200°C from the Wyoming Province into the Proterozoic terranes, a very large difference between average and high heat flow regimes [e.g. *Morgan and Gosnold*, 1989], and one likely to melt lower crustal rocks. To explain the large lower crustal velocity difference (~4%) between the Proterozoic and Archean lower crusts, we suggest that in addition to higher temperatures in the Proterozoic crust, the lithologies at the base of the crust are variable between the northern Wyoming Province and the Proterozoic terranes between the Jemez Lineament and the Cheyenne Belt, differing in bulk chemistry and/or mineralogy.

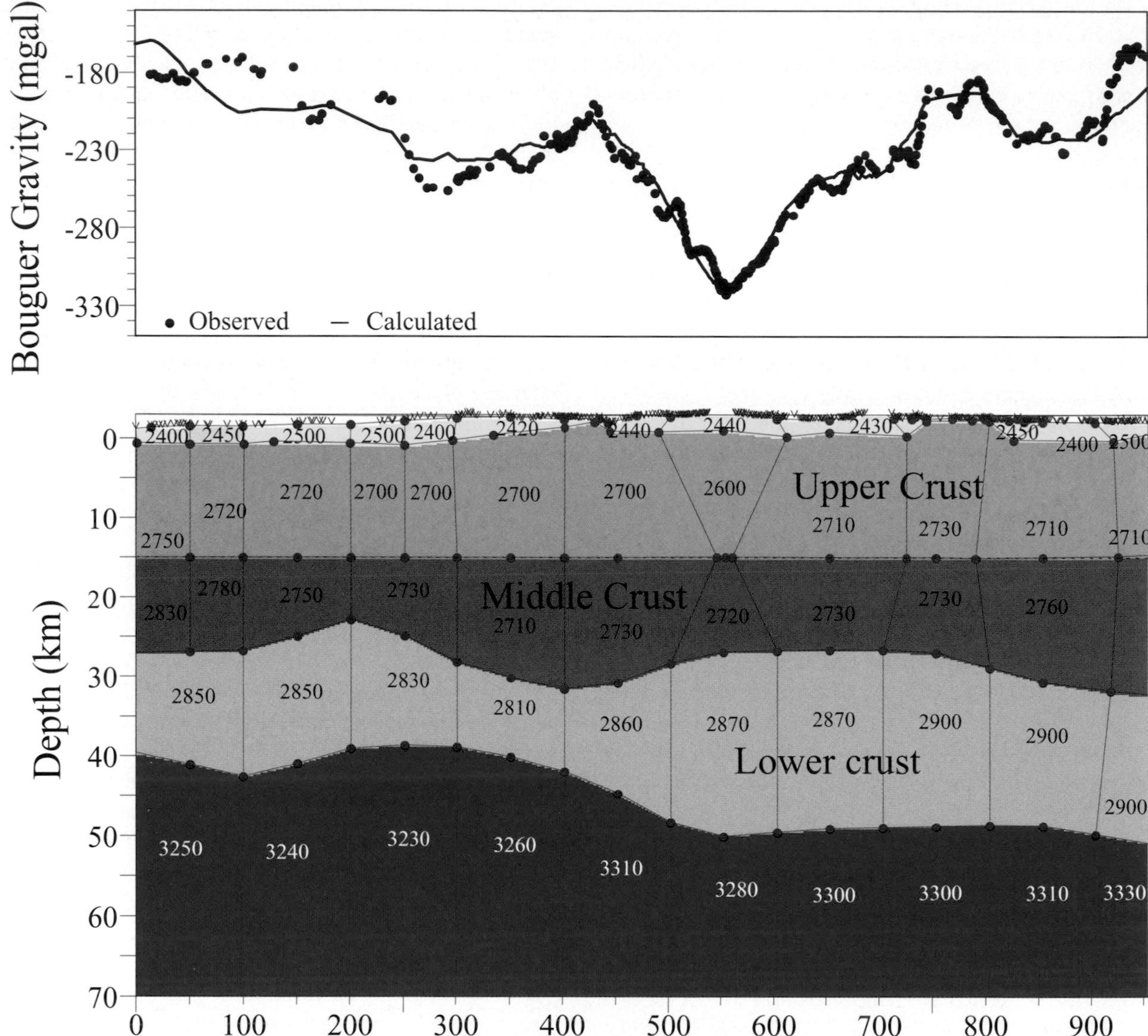

Plate 3. Top. Observed and calculated Bouguer gravity along the CD-ROM refraction profile. Bottom: Density model used to fit the gravity data. The model was developed directly from velocity model M1 with only minor changes to the shallow interfaces and to densities determined from an empirical velocity-density relationship [*Zelt*, 1989]. Densities in the polygons are given in kg/m^3.

CONCLUSIONS

The seismic velocity structure of the crust and upper mantle along the CD-ROM profile in the Southern Rocky Mountains from the Proterozoic Mazatzal-terrane south of the Jemez Lineament to the Archean Wyoming Province north of the Cheyenne Belt is characterized by low seismic velocities and a highly heterogeneous crust and upper mantle. Upper, middle and lower crustal velocities from the Jemez Lineament to the Cheyenne belt are low relative to global compilations for average continental crust and for Phanerozoic orogenic crust. Similarly, upper mantle velocities are low relative to those of the global compilations. The low velocities in the upper and middle crust are attributed to numerous Proterozoic and Phanerozoic felsic plutons, and a high geothermal gradient. The velocities in the lower crust, although low, are consistent with mafic lithologies provided that the temperature at the base of the crust is high. The low upper mantle velocities are indicative of high temperature mantle heating the crust at various locations along the profile, particularly beneath the Jemez lineament, but also beneath the Colorado Mineral Belt.

The northern half of the Jemez lineament appears as a crustal boundary between the higher velocities at the southern end of the profile in the Mazatzal terrane, and low seismic velocities in the crust of the mixed Mazatzal-Yavapai, and Yavapai terranes extending to the Cheyenne belt. The crust is thin, 40–42 km, and upper mantle velocities are low, 7.7–7.8 km/s, indicative of high mantle temperatures. A 0.25 km/s decrease in upper mantle velocity from a reference of 8.0 km/s is consistent with 1% partial melt [*Hammond and Humphreys*, 2000]. Recent basaltic flows, mafic sills in the crust, high heat flow, and recent upwarping support an interpretation that the upper mantle under the Jemez lineament is above its solidus temperature over a region almost 200 km wide and is supplying basaltic melt to the crust.

The region north and south of the Cheyenne belt is characterized by a thick crust, 50–52 km, with higher crustal velocities in the Wyoming Archean Province than in the adjacent Proterozoic terranes to the south. The northern half of the CD-ROM profile, including the Colorado Mineral Belt and the Cheyenne Belt, has a seismically distinct structure from the south, almost entirely in the thickness of the lower crust. Just north of the Cheyenne Belt, midcrustal velocities reach the highest values anywhere on the profile. Beneath the Cheyenne Belt upper mantle velocities are only slightly below normal for a continental region (~0.02–0.07km/s), in contrast to the lower mantle velocities under most of the upper mantle to the south. Although the Cheyenne Belt is a major geologic boundary, in the refraction models it does not stand out clearly as a major seismic velocity boundary. This is in contrast to *Morozova et al.'s* [2005, this volume] seismic reflection interpretation, in which the Cheyenne Belt is a distinct suture zone penetrating the entire crust. The disparity in the reflection and refraction images is consistent with the different imaging capabilities of the two types of seismic probes.

Acknowledgements. We would like to thank the UTEP seismology group for their organization and management of the field campaign, particularly Steve Harder, the many volunteers who acquired the seismic data, and the personnel at the PASSCAL Instrument Center. Randy Keller kindly gave us the gravity profile. Karl Karlstrom provided valuable input on the interpretation of the seismic velocity models. Cathy Snelson and Hannah Rumpel provided valuable insights in analysis of the seismic dataset, as the different subgroups working on the CD-ROM refraction dataset developed their models.

REFERENCES

Benz, H.M. and J. McCarthy, 1994, Evidence for an upper mantle low velocity zone beneath the southern Basin and Range-Colorado Plateau transition zone, *Geophys. Res. Let., 21*, 509–512, 1994.

CD-ROM Working Group, 2002, Structure and Evolution of the lithosphere beneath the Rocky Mountains: initial results from the CD-ROM experiment, *GSA Today*, 12, 4–10.

Chamberlain, K.R., 1998, Medicine Bow orogeny: Timing of deformation and model of crustal structure produced during continent-arc collision, ca 1.78 Ga, southeastern Wyoming, *Rocky Mountain Geology*, 33, 259–278.

Christensen, N.I., and W.D. Mooney, 1995, Seismic velocity structure and composition of the continental crust: A global view, *J. Geophys. Res.*, 100, 9761–9788.

Crosswhite, J. A., and E.D. Humphreys, 2003, Imaging the mountainless root of the 1.8 Ga Cheyenne Belt suture and clues to its tectonic stability, *Geology*, 31, 669–672.

Decker, E.R., K.R. Baker, G.J. Bucher, and H.P. Heasler, 1980, Preliminary heat flow and radioactive studies in Wyoming, *J. Geophy. Res.*, 85, 311–321.

Dueker, L.G., K.E. Karlstrom, S.A. Bowring, K.R. Chamberlain, K.R., T. Eshete, E.A. Erslev, G.L. Farmer, M. Heizler, E.D. Humphreys, R.A. Johnson, G.R. Keller, S.A. Kelley, A. Levander, M.B. Magnani, J.P. Matzel, A.M. McCoy, K.C. Miller, E.A. Morozova, F.J. Pazzaglia, C. Prodehl, H.M. Rumpel, C.A. Shaw, A.F. Sheehan, E. Shoshitaishvili, S.B. Smithson, C.M. Snelson, L.M. Stevens, A.R. Tyson, and M.L. William, 2002, Structure and evolution of the lithosphere beneath the Rocky Mountains; initial results from the CD-ROM experiment, *GSA Today*, 12, 4–10.

Dueker, K., X. Yuan and B. Zurek, 2001, Thick-structured Peoterozoic lithosphere of the Rocky Mountain region, *GSA Today*, 11, 4–9.

Durrheim, R.J. and W.D. Mooney, 1991, Archean and Proterozoic crustal evolution; evidence from crustal seismology, *Geology*, 19, 606–609.

Goes, S., and S. van der Lee, 2002, Thermal structure of North American upper mantle inferred from seismic tomography, *J. Geophys. Res.*, 107, 10,1029.

Gorman, A.R., R.M. Clowes, R.M. Ellis, T.J. Henstock, G.D. Spence, G.R. Keller, A. Levander, C.M. Snelson, M.J.A. Burianyk, E.R. Kaneswich, I. Asudeh, Z. Hajnal, and K.C. Miller, 2002, Deep Probe: Imaging the roots of western North America, *Canadian Journal of Earth Sciences*, 39, 375–398.

Grand, S.P., 1987, Tomographic inversion for shear velocity beneath the North American Plate, *Journal of Geophysical Research, B, Solid Earth and Planets*, 92, 14,065–14,090.

Hammond, W. C. and E.D. Humphreys, 2000, Upper mantle seismic wave velocity; effects of realistic partial melt geometries, *Journal of Geophysical Research, B, Solid Earth and Planets,* 105, 10,975–10,986.

Henstock, T.J., A. Levander, C.M. Snelson, G.R. Keller, K.C. Miller, S.H. Harder, A.R. Gorman, R.M. Clowes, M.J. Burianyk, and E.D.Humphrey,

E.D., 1998, Probing the Archean and Proterozoic Lithosphere of Western North American, *GSA Today*, 8, 1–29.

Humphreys, E.D., and K. Dueker, 1994, Western U.S. upper mantle structure, *J. Geophys. Res*., 99, 9615–9634.

Humphreys, E.D., and K. Dueker, 1994, Physical state of the western U.S. upper mantle, *J. Geopys. Res*., 99, 9635–9650.

Jackson, W.H., S.W. Steward, and L.E. Pakiser, 1963, crustal structure in eastern Colorado from seismic refraction measurements, *J. Geophys. Res*., 68, 5767–5776.

Karlstrom, K.E. and R.S. Houston, 1984, The Cheyenne Belt: Analysis of a Proterozoic suture in southern Wyoming, *Precambrian Research*, 25, 415–446.

Karlstrom, K. and S.A. Bowring, 1988, Early Proterozoic assembly of tectonostratigraphic terranes in southwestern North America, *Journal of Geology*, 96, 561–576.

Karlstrom, K. and S.A. Bowring, 1993, Evolution of Proterozoic lithosphere in the Southwestern U.S., Abstracts with Programs - *Geological Society of America – Annual Meeting*, 25, 237.

Karlstom. K.E. and E.D. Humphreys, 1998, Persistent influence of Proterozoic accretionary boundaries in the tectonic evolution of southwestern North America; interaction of cratonic grain and mantle modification events. Lithospheric structure and evolution of the Rocky Mountains; Part I, *Rocky Mountain Geology*, 33, 161– 180.

Karlstrom, K., and the CDROM Working Group, 2001, Structure and evolution of the lithosphere beneath the Rocky Mountains: Initial results from the CD-ROM experiment, *GSA Today*, 12, 4–10.

Keller, G R; P. Morgan, and W.R. Seager, 1990, Crustal structure, gravity anomalies and heat flow in the southern Rio Grande Rift and their relationship to extensional tectonics, *Tectonophysics*, 174, 21–37.

Keller, G.R., C.M. Snelson, A.F. Sheehan and K. Dueker, 1998, Geophysical studies of crustal structure in the Rocky Mountain region: A review, *Rocky Mountain Geology*, 33, 217–228.

Lee, D.K. and S.P. Grand, 1996 Upper mantle shear structure beneath the Colorado Rocky Mountains, *Journal of Geophysical Research*, 101, 22,233–22,244.

Lerner-Lam, A.L., A. F. Sheehan, S. Grand, E.D. Humphreys, K.G. Dueker, E. Hessler, H. Guo, D.K. Lee, M. and M. Savage, 1998, Deep structure beneath the Southern Rocky Mountains from the Rocky Mountain Front broadband seismic experiment, *Rocky Mountain Geology*, 33, 199–216,

Magnani, M.B., K.M. Miller, A. Levander, K. Karlstrom, 2003, The Yavapai-Mazatzal boundary: A long-lived assembly structure in the lithosphere of southwestern North America, submitted to *Geol. Soc. Am. Bull.*

Magnani, M.B., A. Levander, K.M. Miller, T. Eshete, K. Karlstrom, 2005, Seismic investigations target the Yavapai-Mazatzal Transition Zone and the Jemez Lineament, *this volume.*

McCoy, A., Roy, M., 2005, Gravity Modeling of the Colorado Mineral Belt, *this volume.*

Morozova, E.A., X. Wan, K. R. Chamberlain, S. B. Smithson, R. Johnson, and Karlstrom, K.E., Inter-Wedging Nature of the Cheyenne Belt–Archean-Proterozoic Suture Defined by Seismic Reflection Data, *this volume.*

Morgan, P., and W.D. Gosnold, 1989, Heat flow and thermal regimes in the continental United States, *Geol. Soc. Am. Memoir 172*, L.E. Pakiser and W.D. Mooney, editors, 493–522.

Olsen, K.H., G.R. Keller, and J.N. Stewart, 1979, Crustal structure along the Rio Grande rift from seismic refraction profiles, in *Rio Grande Rift: Tectonics and Magmatism*, Am. Geophys. Union Monograph, R.E. Reicker, editor, 127–143.

Pollack, H.N., Hurter, S.J., and Johnson, J.R., 1993, Heat flow from the earth's interior: analysis of the global data set, *Reviews of Geophysics,* 31, 267–-280.

Prodehl, C., and L.C. Pakiser, 1980, Crustal structure of the southern Rocky Mountains from seismic measurements, Geological *Society of America Bulletin*, 91, I 147– I 155.

Prodehl, C., and P.W. Lipman, 1989, Crustal structure of the Rocky Mountain region, Memoir, *Geol. Soc. Am. Memoir 172,* L.E. Pakiser and W.D. Mooney, editors, 249–284.

Reiter, M., A.J., Mansure, and C. Shearer, 1979, Geothermal characteristics of the Rio Grande rift within the southern Rocky Mountain complex, in *Rio Grande Rift: Tectonics and Magmatism,* Am. Geophys. Union Monograph, R.E. Reicker, editor, 253–267.

Sheehan, A.F., G.A. Abers, A.G. Jones and A.L. Lerner-Lam, A.L., 1995, Crustal thickness variations across the Colorado Rocky Mountains from teleseismic receiver functions, *Journal of Geophysical Research*, 100, 20,391–204,404.

Sinno, Y.A., P.H. Daggett, G.R. Keller, O. Morgan, and S.H. Harder, 1986, Crustal structure of the southern Rio Grande rift determined from seismic refraction profiling, *J. Geophys. Res*., 91, 6143–6156.

Sinno, Y.A. and G. R. Keller, 1986, A Rayleigh wave dispersion study between El Paso, Texas and Albuquerque, New Mexico: *J. of Geophys. Res*., 91, 6168–6174.

Snelson, C.M., T.J. Henstock, G.R. Keller, K.C. Miller and A. Levander, 1998, Crustal and uppermost mantle structure along the Deep Probe seismic profile, *Rocky Mountain Geology*, 33, 181–198.

Snelson, C.M., 2001, Investigating crustal structure in western Washington and in the Rocky Mountains: Implications for seismic hazards and crustal growth, Ph.D. thesis, *University of Texas at El Paso*, 234 pages.

Snelson, C.M., Keller, G.R., and Miller, K.C., Rumpel, H., and Prodehl, C., 2005, Regional Crustal Structure Derived from the CD-ROM 99 Seismic Refraction/Wide-Angle Reflection Profile: The Lower Crust and Upper Mantle, *this volume.*

Talwani, M., J.L. Worzel, and M. Landisman, 1959, Rapid gravity computations for two-dimensional bodies with application to the Mendocino submarine fracture zone, *J. Geopys. Res.,* 64, 49–59.

Toppozada, T.R. and A.R. Sanford, 1976, Crustal structure in central New Mexico interpreted from the Gasbuggy explosion, *Bulletin of the Seismological Society of America*, 66, 877–866.

Van der Lee, S. and G. Nolet, 1997, Upper mantle S velocity structure of North America, *Journal of Geophys Research*, 102, 22,815–22,838.

Wisiniewski, P.A. and Pazzaglia, F., 2002, Epeirogenic controls on Canadian River incision and landscape evolution, Great Plains of northeastern New Mexico, *Journal of Geology*, 110, 437–456.

Yuan, H. and Dueker, K., 2005, Upper Mantle Tomographic V_p and V_s Images of the Rocky Mountains in Wyoming, Colorado and New Mexico: Evidence for a Thick Heterogeneous Chemical Lithosphere, *this volume*.

Zelt, C. A., Seismic structure of the crust and upper mantle in the Peace River Arch region, *Ph.D. thesis, Univ. British Columbia, Vancouver, B.C.*, 1989.

Zelt, C. A. and R. B. Smith, 1992, Seismic traveltime inversion for 2-D crustal velocity structure, *Geophys. J. Int*., 108, 16–34.

Zelt, C.A. and D.J. White, 1995, Crustal structure and tectonics of the southeastern Canadian Cordillera, *Journal of Geophysical Research, B, Solid Earth and Planets,*100, 24,255–24,273.

Zelt, C.A. and P.J. Barton, 1998, Three-dimensional seismic refraction tomography; a comparison of two methods applied to data from the Faeroe Basin, *Journal of Geophysical Research, B, Solid Earth and Planets* 103, 7,187–7,-210.

Zelt, C. A., 1999, Modelling strategies and model assessment for wide-angle seismic traveltime data, *Geophys. J. Int*., 139, 183–204.

Zelt, C. A., A. M. Hojka, E. R. Flueh, and K. D. McIntosh, 1999, 3D simultaneous seismic refraction and reflection tomography of wide-angle data from the central Chilean margin, *Geophys. Res. Lett*., 26, 2577–2580.

Zelt, C. A., K. Sain, J. V. Naumenko, and D. S. Sawyer, 2003 Assessment of crustal velocity models using seismic refraction and reflection tomography, *Geophys. J. Int*., 153, 609–626.

Alan Levander, Department of Earth Science, Rice University MS-126 Houston, TX 77005 USA

Maria Beatrice Magnani, Department of Earth Science, Rice University MS-126 Houston, TX 77005 USA

Colin A. Zelt, Department of Earth Science, Rice University MS-126 Houston, TX 77005 USA

Passive Source Seismology of the Rocky Mountain Region

Anne Sheehan, Vera Schulte-Pelkum, Oliver Boyd, and Charles Wilson

CIRES and Department of Geological Sciences, University of Colorado at Boulder, Boulder, CO

Two recent passive source (earthquake) seismic experiments have produced a teleseismic and regional event data set which provides constraints on the structure of the crust and upper mantle beneath the Colorado Rocky Mountains and two major Precambrian province boundaries. The passive source component of the Continental Dynamics of the Rocky Mountains (CD-ROM) experiment included two dense north-south linear arrays of broad-band seismometers straddling Precambrian province boundaries in Colorado, Wyoming, and New Mexico. The Rocky Mountain Front (RMF) experiment included thirty broadband seismometers spaced uniformly throughout Colorado. Results from a spectrum of seismological imaging and inversion techniques indicate that the Cheyenne Belt Archean-Proterozoic boundary in southern Wyoming has signatures in both the crust and upper mantle, while the Yavapai-Mazatzal province boundary is less clearly defined. Studies of data from the RMF experiment show pronounced low seismic velocities in the crust and upper mantle beneath the Rocky Mountains, and high attenuation (low Q) in the mantle. Techniques of passive source seismology used with RMF and CD-ROM experiment data are described here, along with reference to corresponding studies.

1. INTRODUCTION

The Continental Dynamics of the Rocky Mountains (CD-ROM) experiment was designed to explore the geology and geophysics of the lithosphere along a transect from Wyoming to New Mexico. Particular emphasis was placed upon the study of proposed Precambrian suture zones which may document the growth and stabilization of the continent. The project included a coordinated set of seismic experiments, utilizing both active and passive source techniques. This paper provides an overview of the passive seismic experiments (those using naturally occurring earthquake sources), including the techniques, their limitations, resolution, and the models that result from their interpretation. Active source experiments (those utilizing energy actively put into the ground, through explosions or other means) are described in *Prodehl et al.* [this volume]. The Rocky Mountain Front (RMF) passive source seismic experiment provides further information on the lithospheric structure beneath a broad area of the southern Rocky Mountains, and two papers from that experiment are described here. Techniques in this volume include P-wave seismic travel-time tomography [*Yuan and Dueker*, this volume], receiver functions [*Zurek and Dueker*, this volume], shear wave splitting [*Fox and Sheehan*, this volume], Pn travel times [*Lastowka and Sheehan*, this volume], shear wave attenuation tomography [*Boyd and Sheehan*, this volume], and surface wave tomography [*Li et al.*, this volume].

The Rocky Mountain Region: An Evolving Lithosphere
Geophysical Monograph Series 154

10.1029/154GM23

2. PASSIVE SOURCE EXPERIMENTS IN THE ROCKIES

The Program for Array Seismological Studies of the Continental Lithosphere (PASSCAL) of the Incorporated Research Institutions of Seismology (IRIS), funded by the US National Science Foundation, provides individual investigators use of portable, autonomous, digital seismographs with sensitive broad-band seismometers. Instruments from this facility were used for both CD-ROM and RMF experiments. The broadband seismometers are capable of recording both teleseisms and regional earthquakes with very high fidelity. Relatively large numbers of instruments can be deployed in an array concentrated geographically over the structures of interest, and the array dimensions can be designed according to imaging density and resolution criteria.

The passive source component of the Continental Dynamics of the Rocky Mountains (CD-ROM) experiment consisted of two north-south lines of broadband seismometers crossing major Precambrian province boundaries in Colorado, New Mexico, and Wyoming. The seismometers were deployed starting in April 1999 and removed in June 2000. The northern seismic line extended from Rawlins, Wyoming to Steamboat Springs, Colorado and traversed the ancient suture zone between the Archean Wyoming province and the Proterozoic Yavapai province. The southern line extended from the San Luis Basin in Colorado to Las Vegas, New Mexico and crossed the suture zone between the Yavapai province to the north and the younger Mazatzal province to the south. The southern line is also very proximal to the Rio Grand rift and the recent tectonism in the Jemez lineament. The deployment consisted of 25 instruments across the northern suture zone and 23 across the southern suture zone. The spacing between stations was approximately 10 km. Three-component sensors capable of recording teleseisms across a broad frequency band were installed at all sites, including 27 STS-2 seismometers, 15 CMG3T seismometers and six CMG40T seismometers. The sampling rate of the experiment ranged from 10 to 20 samples/second. The sampling rate was reduced because inclement weather would prevent the sites from being visited in the winter; therefore a lower sampling rate was needed in order to preserve field disk space.

The CD-ROM line was supplemented by the Laramie real time array. The Laramie array consisted of 30 broadband seismometers at a station spacing of 1.6 kilometers, deployed from June 2000 to May 2001. The line crosses the Cheyenne Belt near Laramie, Wyoming (Figure 1). Each station in the Laramie Array consisted of Guralp 40-T seismometers recording at 1, 40 and 100 samples per second. The data were telemetered to the University of Wyoming in real time.

The Rocky Mountain Front experiment consisted of a deployment of thirty-five broadband (CMG3-ESP and STS2) seismometers distributed throughout Colorado and extending into eastern Utah and western Kansas (Figure 1). The stations were deployed from May through December of 1992. Station spacing was approximately 75 km. Data was collected from each station in both continuous (10 samples per second) and triggered (20 samples per second) data streams. Numerous studies utilizing data from the Rocky Mountain Front array have been published, including shear wave velocity tomography [*Lee and Grand*, 1996], surface wave tomography [*Li et al.*, 2002], receiver functions [*Sheehan et al.*, 1995], shear wave splitting [*Savage et al.*, 1996; *Savage and Sheehan*, 2000], and deep discontinuity structure [*Dueker and Sheehan*, 1998; *Gilbert and Sheehan*, 2004].

Other recent passive seismic source seismic experiments in the Rocky Mountain and Colorado Plateau regions include the Lodore array [*Crosswhite and Humphreys*, 2003], the *Deep Probe* experiment [*Crosswhite et al.*, 1999], the *Colorado Plateau – Great Basin Experiment* [*Sheehan et al.*, 1997], and the RISTRA experiment [*Wilson and Aster*, 2003] (Figure 1).

3. RECEIVER FUNCTIONS

The isolation of conversions from compressional to shear waves on teleseismic waveforms can be used to determine crust and mantle velocity discontinuity structure beneath seismic stations using a technique referred to as receiver function analysis [*Burdick and Langston*, 1977; *Langston*, 1977; *Langston*, 1979]. An incident P wave from a distant earthquake generates converted shear waves at boundaries with impedance contrasts, such as the Moho. The delay time between the direct P arrival and the converted S wave, called Ps, is related to the velocity and depth to the conversion point. The near vertical incidence of teleseismic P allows the separation of the P and converted S wave because the converted shear wave is large on the radial component of the seismogram and the vertical component is dominated by the incident P wave. The P wave record on the vertical component provides a reference for the earthquake source and path effects, and receiver function analysis involves the deconvolution of the vertical component from the radial in order to enhance the near-receiver mode conversions. Typically, the most significant converted arrivals are those from the Moho, but conversions from velocity discontinuities in the mid-crust and in the mantle transition zone are also common.

Basic receiver function analysis assumes that the seismic structure of the crust and upper mantle are isotropic and comprised of flat lying planar interfaces. In the presence of anisotropy and dipping layers these assumptions are not accurate [*Bostock*, 2003; *Levin and Park*, 1997; *Savage*, 1998]. With a good back-azimuth distribution of recorded events, the presence of these more complicated structures may be recognized by changes in timing and amplitude of the converted phases.

High density passive seismic arrays stretching tens to thousands of kilometers have become the foundation of many modern receiver function studies [e.g., *Dueker and Sheehan*, 1997; *Rondenay et al.*, 2000a, 2000b; *Zurek and Dueker,* this

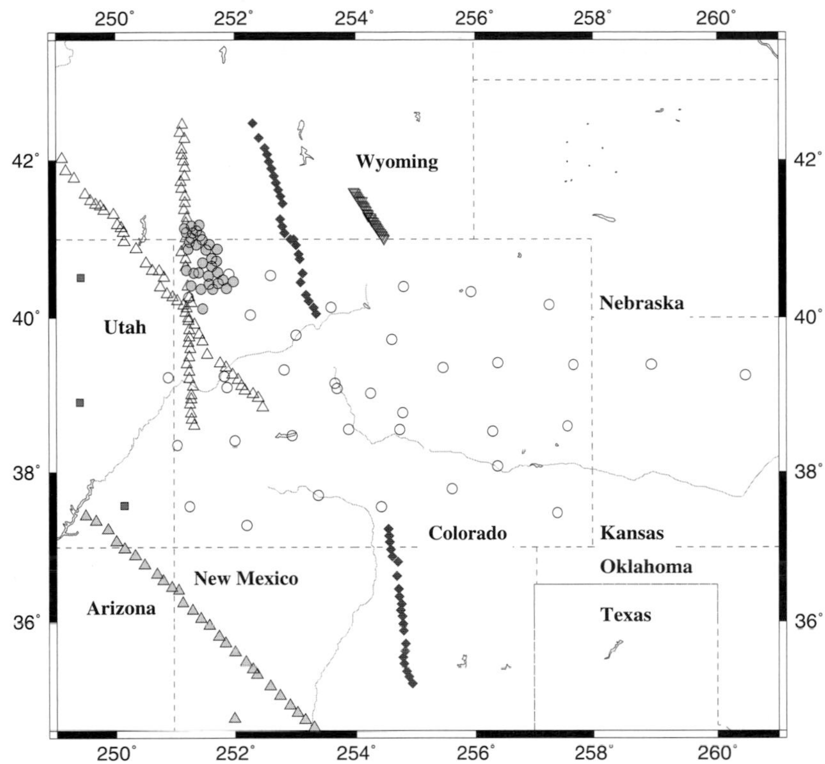

Figure 1. Seismograph stations of Rocky Mountain Front (RMF), Continental Dynamics of the Rocky Mountains (CD-ROM), Laramie Array, Deep Probe, RISTRA, and Colorado Plateau-Great Basin broadband passive source seismic PASSCAL experiments. Open circles indicate RMF station locations, diamonds denote CD-ROM station locations, inverted triangles Laramie Array stations, filled triangles RISTRA stations, squares Colorado-Plateau Great Basin stations, unfilled triangles Deep Probe stations, and filled circles Lodore stations.

volume] in order to combat signal generated noise and to determine high resolution lithospheric structure. Seismic arrays recording multiple observations of the same conversion point from many teleseismic events allows the employment of techniques such as common conversion point stacking or seismic migration [e.g., *Bank and Bostock*, 2003; *Wilson et al.*, 2003]. All of these seismic imaging approaches begin with the assumption of a seismic velocity model. The calculated receiver functions are then back projected along their theoretical ray paths to restore the converted energy to the point of conversion. Multiple samples from different directions of the same conversion point are required to recover the amplitude of the original conversion. In theory, the amplitude of the conversion should provide information about the variation in seismic properties across a conversion interface. The location restoration and amplitude recovery problem can be formulated as a forward problem (common conversion point stacking) or as an inverse problem (seismic migration) and the choice of technique depends upon factors including the geometry of the imaging target, the array geometry, and the signal quality.

With data from deployments such as the RMF and CD-ROM experiments, receiver function analysis provides a plan view of crustal thickness variations and cross sections of detailed discontinuity variations, which can then be interpreted in terms of the major structural trends exhibited in the geology. *Gilbert and Sheehan* [2004] use data from RMF, CD-ROM, Deep Probe, and other passive source studies for a crustal receiver function study throughout the western United States building upon previous studies of *Sheehan et al.* [1995] and others. They find a crustal thickness of 50 km beneath the Rockies in northern Colorado, thinning to 40 km in southern Colorado. Detailed crustal images from CD-ROM have been presented in *Dueker et al.* [2001] and *Karlstrom et al.* [2002]. The Moho along the southern CD-ROM line was found to be fairly flat, with a thickness of approximately 40 km, and a slight thickening to the north. The receiver function results from the CD-ROM north line are more complex. A thick crust of ~50 km is found in northern Colorado, thinning to 40–45 km just south of the Colorado/Wyoming border. The Moho appears to be imbricated about 30 km north of the Cheyenne belt suture. Crust to the north of the Cheyenne belt is of ~40 km thickness. In this volume, *Zurek and Dueker* present images of lithospheric receiver function discontinuities from the CD-ROM arrays. They argue for sub-crustal discontinuities to 150 km depth beneath the Proterozoic lithosphere immediately south of the Cheyenne belt and to 100 km beneath the Jemez lineament based upon layering observed to those depths. Mid-crustal features are also imaged under the Cheyenne belt, and correspond with CD-ROM reflection profiles [*Morozova et al.*, 2002; *Karlstrom et al.*, 2002].

4. TOMOGRAPHY: P, S, AND SURFACE WAVES

Three papers appearing in this volume use passive wave seismic tomography to examine crustal and upper mantle thermal and compositional structure. Tomographic problems are inversions of a large set of linear equations. An observation, such as P-wave travel time, is ideally the effect of a seismic attribute integrated over space, e.g. P-wave velocity integrated over the path the seismic wave has traveled. The equations are usually of the form $d=Gm$ where d is the data vector, such as a set of observations like P-wave travel times, m is the model vector which contains a set of model parameters such as P-wave slowness in the upper mantle, and G is the path kernel, a j by k matrix with, in this case, the distance traveled in k model bins for j observations. A solution for m may be obtained by various methods [*Iyer and Hirahara*, 1993]. The solution is usually determined in a least squares sense, e.g. the solution tries to minimize the squares of the residuals between predicted and observed data. The solution may be damped and/or weighted and solved by gaussian elimination, singular value decomposition, or other methods.

Some caution should be used when interpreting tomographic images. For example, variability in the measurements will affect the model regardless of whether the measured variation originated from within the model. A related problem is the assumption that the measured attribute is due to variations along the presumed seismic ray path [*Dahlen et al.*, 2000a; 2000b]. Finally, anomalies within the tomographic model tend to bleed along regions of poor ray coverage. Structures that dip away along ray paths at the edge of tomographic models should be viewed with skepticism.

Though dependence between model parameters can be observed in the resolution matrix, checkerboard tests are a useful guide to estimate the quality of features in the tomographic model. In this method, a synthetic model is generated and using the incomplete ray coverage of the real data, synthetic data are produced. The synthetic data are then inverted to see how well the input checkerboard model is recovered. Interdependence of model parameters and poorly resolved regions can be seen easily with this technique.

4.1. Body Wave Velocity Tomography

In *Yuan and Dueker* [this volume] upper mantle compressional and shear wave images from the CD-ROM experiment are presented. Two major anomalies are found. Beneath the Cheyenne belt along the CD-ROM North Line, a north dipping high velocity feature from the Moho to 200 km depth is found. Travel time modeling suggests that a significant portion of the dipping high velocity anomaly is due to anisotropy with a dipping fast axis. Shear wave splitting measurements [*Fox and Sheehan*, this volume] are also consistent with this dipping seismic anisotropic model. Beneath the Jemez lineament along the CD-ROM South Line, a 100-km wide low velocity anomaly is imaged between the Moho and 100 km depth.

4.2. Attenuation Tomography

Utilizing the Rocky Mountain Front (RMF) broadband seismic dataset, *Boyd and Sheehan* [this volume] derive the shear wave attenuation structure underlying the Southern Rocky Mountains and surrounding areas. Attenuation is measured using differential spectra of teleseismic S phase waveforms. Calculations of intrinsic attenuation coupled with current velocity models aids in the determination of temperature, partial melt distributions, and compositional variation [*Karato*, 1993]. A north-south zone of high shear wave attenuation (low Q) is found in the mantle in south central Colorado, along the northern reaches of the Rio Grande rift, and coincides with a region of low shear wave velocity [*Lee and Grand*, 1996]. The correlation between the velocity and attenuation models can be used to distinguish between thermal, compositional, and melt anomalies. Beneath the High Plains the combined velocity and attenuation variations are consistent with slight increases in temperature to the west, while beneath the southwest Rockies, the presence of compositional variations and partial melt are required to explain the observations.

4.3. Surface Wave Tomography

Li et al. [this volume] present a tomographic inversion of Rayleigh surface wave phase velocities from the RMF dataset across the southern Rocky Mountains. Previous surface wave work with the RMF data [*Lerner-Lam et al.*, 1998] showed large shear wave velocity variations across the Rocky Mountain Front, with lower velocities beneath the Rockies than beneath the High Plains. *Li et al.* [this volume] use the technique of *Forsyth et al.* [1998] and *Li* [2001], where the wavefield is represented as the sum of two interfering plane waves, and invert amplitude and phase data for phase velocity, then invert the phase velocity for shear velocity structure. They do not find a lithospheric lid beneath the Rocky Mountains, and find low mantle velocities beneath the Rockies to at least 150 km. Beneath the Great Plains a fast lithosphere is found to a depth of 150 km. The Colorado Plateau is found to have a fast seismic lid to 100 km depth, with low velocities below 100 km, similar to results reported by *Lastowka et al.* [2001]. The surface wave tomography results are broadly consistent with the RMF shear body wave tomography work of *Lee and Grand* [1996] but differ in some details, particularly with respect to the depth extent of the velocity anomalies.

There are some differences between the surface wave derived crustal thicknesses and the receiver function and refraction based crustal thicknesses [*Sheehan et al.*, 1995; *Keller et al., 1998*]. These are explained in part by the large lateral variations in crustal wave speeds allowed by the Li et al. model and differences in lateral and vertical resolution between surface wave and receiver function techniques. *Li et al.* [this volume] do not find low upper mantle velocities beneath west-central Colorado near Aspen to the extent reported by *Dueker et al.* [2001], but instead find unusually low crustal velocities there. A tomographic study with denser station distribution in central Colorado may be needed to resolve the depth extent and nature of the Aspen anomaly.

5. VELOCITIES OF UPPERMOST MANTLE (Pn)

Lastowka and Sheehan [this volume] report interstation Pn velocities from the CD-ROM array. The Pn portion of the waveform arrives first in the distance range 2 to 16 degrees and represents high frequency seismic energy propagating in the uppermost mantle. It has been modeled successfully both as a whispering gallery phase [*Menke and Richards*, 1980] and as a mantle-lid refraction [*Sereno and Orcutt*, 1985]. *Lastowka and Sheehan* [this volume] measure Pn velocities using an interstation method. The Pn arrival times of two stations along the same great circle path from an event are measured, differenced, and the distance between the stations is divided by the differential time. The interstation method reduces errors due to hypocenter mislocation. For the southern CDROM line, Pn is measured by constructing a travel time curve of all of the southern line stations that recorded a single earthquake in Mexico, which happened to be along the same great circle line as the stations. Such a measurement is more robust than individual interstation measurements.

Low Pn velocities of 7.8 ± 0.1 km/s were found beneath the southern CD-ROM array, consistent with low upper mantle velocities and the absence of a mantle lid. This low velocity is consistent with the high heat flow in the Rio Grande rift and the evidence for modern rifting. These data provide independent confirmation of the low upper mantle velocities from P wave tomography found along the CD-ROM south line by *Yuan and Dueker* [this volume]. Measurements of Pn made in northern Colorado and in Wyoming are near the global average Pn value of 8.1 km/s, suggesting an intact mantle lid.

6. MANTLE ANISOTROPY

Most passive source techniques, such as tomography and receiver function analysis, provide an image of the current structure of the crust and mantle. In contrast, the measurement of seismic anisotropy has the potential to give information on the dynamic state of the Earth. Olivine, the constituent mineral in the upper mantle, is highly anisotropic, and orients itself systematically during deformation under conditions that allow dislocation creep. Thus, mantle anisotropy can give images of strain in the mantle due to current processes, or provide a history of past deformation [*Blackman et al.*, 2002; *Karato et al.*, 1998; *Mainprice and Silver*, 1993].

The most popular method to measure seismic anisotropy is through shear wave splitting. The polarization of a shear wave is perpendicular to its direction of propagation. When a shear wave enters an anisotropic region, its initial polarization may be at an angle to the medium's fast and slow axes, and the component oscillating along the fast axis will propagate faster than the component polarized along the slow axis. Eventually, the fast component will separate from the slow component, so that the single incident shear phase is split into two phases with identical shape. In a medium with hexagonal or orthorhombic anisotropic symmetry, the fast and slow symmetry axes and therefore the corresponding split shear phases are orthogonal to each other. Shear wave splitting techniques measure the time delay between the fast and slow phase, delta t, and the polarization azimuth of the fast phase, ϕ [e.g., *Silver and Chan*, 1988, 1991]

The phases most commonly used for this process are the core phases SKS and SKKS. These phases convert from shear to compressional motion in the fluid outer core, and after conversion back to shear, travel steeply from the core-mantle boundary to the seismic station. The analysis is simpler than for teleseismic S since information about source-side anisotropy is removed on the core leg. However, measured SKS and SKKS splitting can still stem from anywhere between the core-mantle boundary and the station. By making measurements at several stations from a number of different source areas, some depth constraints can be obtained by determining the overlap of different ray paths [*Alsina and Snieder*, 1995].

Near the Rocky Mountain region, previous studies have found fast orientations which were interpreted in terms of asthenospheric flow beneath the Snake River Plain [*Schutt et a.l*, 1998] and the Rio Grande rift [*Sandvol et al.*, 1992]. Closer to the Rocky Mountain Front, *Schutt and Humphreys* [2001] and *Savage et al.* [1996] found more complex splitting signatures, with fast orientations dependent on backazimuth of the incident shear phase, rapid lateral variation of observations between stations, and many null measurements. A null measurement, i.e. no observation of splitting, occurs when the incident shear phase is already polarized along one of the symmetry axes of the medium, or if it is propagation along the single symmetry axis in a hexagonal medium, or if the medium is isotropic. *Savage et al.* [1996] favor a vertically aligned hexagonal symmetry to explain null measurements and inconsistent splitting results seen in the Rocky Mountain Front experiment. Other possible explanations given by *Savage et al.* [1996] are depth-dependent anisotropy, orthorhombic symmetry, plunging axes of anisotropy, and lateral variations.

In this volume, *Fox and Sheehan* analyze shear wave splitting using a larger data set from the CD-ROM and Laramie arrays. They find consistent fast orientations that correlate with absolute plate motion in southern Colorado and New Mexico, similar to that seen by *Li et al.* [this volume] using seismic surface waves. To the north of the Cheyenne belt, *Fox and Sheehan* [this volume] find a complex pattern with backazimuthal dependence along both the CD-ROM and Laramie arrays in southern Wyoming. With the large number of stations and events and good azimuthal coverage, they are able to perform hypothesis testing for layered anisotropy and plunging symmetry axes, and arrive at a preferred model with a steeply plunging axis of symmetry that may be related to a high-velocity anomaly seen in a tomographic inversion for the area [*Yuan and Dueker,* this volume].

7. CONCLUSION

Detailed passive seismic studies of the Rocky Mountains provide much information on the crust and mantle beneath this complex orogenic region. Data from the Rocky Mountain Front experiment have been used to determined large scale structure, and data from the CD-ROM experiment provide details of Precambrian province boundaries. Studies using the Rocky Mountain Front data show that the crust is thick beneath the Rocky Mountains and the western Great Plains, and that P and S wave velocities in the mantle beneath the Rockies are low. Seismic attenuation is high in this region, consistent with both compositional and thermal origins for the velocity and attenuation heterogeneity. Data from the CD-ROM array show a high velocity body in the mantle dipping to the north beneath the Cheyenne belt. The presence of this feature is also detected with seismic shear wave splitting, and it has a strong anisotropic signature with plunge to the northeast. Beneath the Jemez lineament low velocities are found in the upper mantle with both seismic body wave tomography and Pn travel times.

Acknowledgments: We thank Jason Crosswhite and Ken Dueker for their efforts with the CD-ROM field program and Lynda Lastowka and Otina Fox for CD-ROM data management. We thank Art Lerner-Lam for leading the Rocky Mountain Front (RMF) experiment. Both RMF and CD-ROM experiments utilized seismic equipment from the IRIS PASSCAL instrument pool and we thank staff from the Lamont and New Mexico PASSCAL instrument centers for their assistance.

REFERENCES

Alsina, D., and R. Snieder, Small-scale sublithospheric continental mantle deformation: Constraints from SKS splitting observations, *Geophys. J. Int., 123,* 431–448, 1995.

Bank, C.-G., and M. G. Bostock, Linearized inverse scattering of teleseismic waves for anisotropic crust and mantle structure: 2. Numerical examples and application to data from Canadian Stations, *J. Geophys. Res., 108*(B5), 2259, doi:10.1029/2002JB001951, 2003.

Blackman, D. K., H. R. Wenk, and J. M. Kendall, Seismic anisotropy in the upper mantle: 1. Factors that affect mineral texture and effective elastic properties, *G-Cubed*, 3(9), 8601, doi:10.1029/2001GC000247, 2002.

Bostock, M.G., Linearized inverse scattering of teleseismic waves for anisotropic crust and mantle structure: 1. Theory, *J. Geophys. Res., 108*(B5), doi:10.1029/2002JB001950, 2003.

Boyd, O. S. and A. F. Sheehan, Attenuation tomography beneath the Rocky Mountain Front: Implications for the physical state of the upper mantle, *in* AGU Monograph on the Lithospheric Structure of the Rocky Mountains, K. Karlstrom and G. R. Keller, eds., *in press (this volume),* 2004.

Burdick, L. J., and C. A. Langston, Modeling crustal structure through the use of converted phases in teleseismic body waveforms, *Bull. Seismol. Soc. Am., 67,* 677–692, 1977.

Crosswhite, J. A., and E. D. Humphreys, Imaging the mountainless root of the 1.8 Ga Cheyenne belt suture, *Geology*, *31*, 2003.

Crosswhite, J. A., K. Dueker, and G. Humphreys, The Lodore, Deep Probe and CDROM Teleseismic Arrays; Imaging and Archean-Proterozoic Suture Using Receiver Function Stacking, *Geological Society of America, 1999 annual meeting, Abstracts with Programs - Geological Society of America, 31*, 129, 1999.

Dahlen, F.A., S.-H. Hung, and G. Nolet, Frechet kernels for finite-frequency traveltimes – I. Theory, *Geophys. Jour. Int., 141,* 157–174, 2000a.

Dahlen, F.A., S.-H. Hung, and G. Nolet, Frechet kernels for finite-frequency traveltimes – II. Examples, *Geophys. Jour. Int., 141,* 175–203, 2000b.

Dueker, K. G. and A. F. Sheehan, Mantle discontinuity structure from midpoint stacks of converted P and S waves across the Yellowstone hotspot track, *J. of Geophys. Res., 102*(B4), 8313–8327, 1997.

Dueker, K. G., and A. F. Sheehan, Mantle discontinuity structure beneath the Colorado Rocky Mountains and High Plains, *J. Geophys. Res., 103,* 7153–7169, 1998.

Dueker, K., H. Yuan, and B. Zurek, Thick Proterozoic lithosphere of the Rocky Mountain region, *GSA Today, 11,* 4–9, 2001.

Forsyth, D. W., S. Webb, L. Dormann, and Y. Shen, Phase velocities of Rayleigh waves in the MELT experiment on the East Pacific Rise, *Science, 280,* 1235–1238, 1998.

Fox, O. and A. F. Sheehan, Shear wave splitting beneath the CD-ROM transects, *in* AGU Monograph on the Lithospheric Structure of the Rocky Mountains, K. Karlstrom and G. R. Keller, eds., *in press (this volume),* 2004.

Gilbert, H. J., and A. F. Sheehan, Images of crustal thickness variations in the intermountain west, *J. Geophys. Res., in press*, 2004.

Gilbert, H. J., A. F. Sheehan, K. G. Dueker, and P. Molnar, Receiver functions in the western United States, with implications for upper mantle structure and dynamics, *J. Geophys. Res., 108*(B5), doi:10.1029/2001JB001194, 2003.

Iyer, H. M., and K. Hirahara, Tomography using both local earthquakes and teleseisms; velocity and anisotropy; theory, in *Seismic tomography; Theory and practice,* Chapman & Hall, London, 493–518, 1993.

Karato, S., Importance of anelasticity in the interpretation of seismic tomography, *Geophys. Res. Lett., 20,* 1623–1626, 1993.

Karato, S., S. Zhang, M. E. Zimmerman, M. J. Daines, D. L. Kohlstedt, Experimental studies of shear deformation of mantle materials: Towards structural geology of the mantle, *Pageoph., 151,* 589–603, 1998.

Karlstrom, K. E., S. A. Bowring, K. R. Chamberlain, K. G. Dueker, T. Eshete, E. A. Erslev, G. L. Farmer, M. Heizler, E. D. Humphreys, R. A. Johnson, G. R. Keller, S. A. Kelley, A. Levander, M. B. Magnani, J. P. Matzel, A. M. McCoy, K. C. Miller, E. A. Morozova, F. J. Pazzaglia, C. Prodehl, H. M. Rumpel, C. A. Shaw, A. F. Sheehan, E. Shoshitaishvili, S. B. Smithson, C. M. Snelson, L. M. Stevens, A. R. Tyson, and M. L. Williams, Structure and evolution of the lithosphere beneath the Rocky Mountains: Initial results from the CD-ROM experiment, *GSA Today, v. 12,* no. 3, p. 4–10, March 2002.

Keller, G. R., C. M. Snelson, A. F. Sheehan, and K. G. Dueker, Geophysical studies of crustal structure in the Rocky Mountain region: a review, *Rocky Mtn. Geol., 33,*217–228,1998.

Langston, C.A., Corvalis, Oregon, crustal and upper mantle receiver structure from teleseismic P and S waves, *Bull. Seismol. Soc. Am., 67,* 713–724, 1977.

Langston, C. A., Structure under Mount Rainier, Washington, inferred from teleseismic bodywaves, *J. Geophys. Res., 84,* 4749–4762, 1979.

Lastowka, L. A., and A. F. Sheehan, CDROM interstation Pn study across the Rio Grande Rift, *in* AGU Monograph on the Lithospheric Structure of the Rocky Mountains, K. Karlstrom and G. R. Keller, eds., *in press (this volume),* 2004.

Lastowka, L. A., A. F. Sheehan, and J. M. Schneider, Seismic evidence for partial lithospheric delamination model of Colorado Plateau uplift, *Geophys. Res. Lett., 28,* 1319–1322, 2001.

Lee, D. K., and S. P. Grand, Upper mantle shear structure beneath the Colorado Rocky Mountains, *J. Geophys. Res., 101,* 22,233–22,244, 1996.

Lerner-Lam, A. L., A. F. Sheehan, S. Grand, E. Humphreys, K. Dueker, E. Hessler, H. Guo, D. Lee, M. Savage, Deep Structure beneath the Southern Rocky Mountains from the Rocky Mountain Front Broadband Seismic Experiement, *Rocky Mountain Geology, 33,* 199–216, 1998.

Levin, V., and J. Park, Crustal anisotropy in the Ural Mountains from teleseismic receiver functions, *Geophys. Res. Lett., 24;11,* 1283–1286, 1997.

Li, A., Crust and mantle discontinuities, shear wave velocity structure, and azimuthal anisotropy beneath North America, Ph. D. dissertation, Brown University, 2001.

Li, A., D. W. Forsyth, and K. M. Fischer, Rayleigh wave constraints on shear-wave structure and Azimuthal Anisotropy Beneath the Colorado Rocky Mountains, *in* AGU Monograph on the Lithospheric Structure of the Rocky Mountains, K. Karlstom and G. R. Keller, eds., *in press (this volume),* 2004.

Li, A., D. W. Forsyth, and K. M. Fischer, Evidence for shallow isotastic compensation of the southern Rocky Mountains from Rayleigh wave tomography, *Geology, 30,* 683–686, 2002.

Mainprice, D. and P. G. Silver, Constraints on the interpretation of telesiesmic SKS observations from kimberlite nodules from the subcontinental mantle, *Physics of the Earth and Planetary Interiors, 78,* 257–280, 1993.

Menke, W. H. and P. G. Richards, Crust mantle whispering gallery phases: a deterministic model of teleseismic Pn wave propagation, *J. Geophys. Res., 85,* 5416–5422, 1980.

Morozova, E. A., Wan, X., Chamberlain, K. R., Smithson, S. B., Morozova, I. B., Boyd, N. K., Johnson, R. A., Karlstrom, K. E., Tyson, A. R., and Foster,C. T., Geometry of Proterozoic sutures in the central Rocky Mountains from seismic reflection data: Cheyenne belt and Farwell Mountain structures, *Geophys. Res. Lett., 29,* 1639, 10.1029/2001GL013819, 2002.

Prodehl, C., G. R. Keller, R. A. Johnson, C. Snelson, and H. M. Rumpel, Background and overview of previous controlled source seismic studies, *in* AGU Monograph on the Lithospheric Structure of the Rocky Mountains, K. Karlstrom and G. R. Keller, eds., *in press (this volume),* 2004.

Rondenay, S., M. Bostock, T. Hearn, D. White, H. Wu, G. Senechal, S. Ji, and M. Mareschal, Teleseismic studies of the lithosphere below the Abitibi-Grenville Lithoprobe transect, *Can. J. Earth Sci., 37*(2–3), 415–426, 2000a.

Rondenay, S. G., M. G. Bostock, T. M. Hearn, D. J. White, and R. M. Ellis, Lithospheric assembly and modification of the SE Canadian Sheild: Abitibi-Grenville teleseismic experiment, *J. Geophys. Res., 105,* 13,735–13,754, 2000b.

Sandvol, E., J. Ni, S. Ozalaybey, and J. Schlue, Shear wave splitting in the Rio Grande rift, *Geophys. Res. Lett., 19,* 2337–2340, 1992.

Savage, M. K., Lower crustal anistropy or dipping boundaries? Effects on receiver functions and a case study in New Zealand, *J. Geophys. Res., 103,* 15,069–15,087, 1998.

Savage, M. K., and A. F. Sheehan, Seismic anisotropy and mantle flow from the Great Basin to the Great Plains, western United States, *J. Geophys. Res., 105,* 13,715–13,734, 2000.

Savage, M. K., A. F. Sheehan, and A. Lerner-Lam, Shear wave splitting across the Rocky Mountain Front, *Geophysical Research Letters*, I (B17), 2267–2270, 1996.

Schutt, D. L. and E. D. Humphreys, Evidence for a deep asthenosphere beneath North America from western United States SKS splits, *Geology, 29,* 291–294, 2001.

Schutt, D., E. D. Humphreys, and K. Dueker, Anisotropy of the Yellowstone hot spot wake, eastern Snake River Plan, Idaho, *Pure Appl. Geophys., 151,* 443–462, 1998.

Sereno, T.J. and J. Orcutt, Synthesis of realistic oceanic Pn waves trains, *J. Geophys. Res., 90,* 12,755–12,776,1985.

Sheehan, A. F., G. A. Abers, C. H. Jones, A. L. Lerner-Lam, Crustal thickness variations across the Colorado Rocky Mountains from teleseismic receiver functions, *J. Geophys. Res., 100,* 20,391–20,404, 1995.

Sheehan, A. F., C. H. Jones, M. K. Savage, S. Ozalaybey, and J. M. Schneider, Contrasting lithospheric structure beneath the Colorado Plateau and Great Basin: Initial results from Colorado Plateau - Great Basin PASSCAL experiment, *Geophys. Res. Lett., 24,* 2609–2612, 1997.

Silver, P. G., and W. W. Chan, Shear Wave Splitting and Subcontinental Mantle Deformation, *J Geophys. Res., 96* (B10), 16,429–16,454, 1991.

Silver, P. G., and W. W. Chan, Implications for continental structure and evolution from seismic anisotropy, *Nature, 335,* 34–39, 1988.

Wilson, D. and Aster, R., Imaging crust and upper mantle seismic structure in the southwestern United States using teleseismic receiver functions, *The Leading Edge, 22,* 232–237, 2003.

Wilson, C. K., C. H. Jones, and H. J. Gilbert, Single chamber silicic magma system inferred from shear wave discontinuities of the crust and uppermost mantle, Coso geothermal area, California, *J. Geophys. Res., 108,* 1–16, 2003.

Yuan, H., and K. Dueker, Upper mantle tomographic V_p and V_s images of the Middle Rocky Mountains in Wyoming, Colorado, and New Mexico: Evidence for a thick heterogeneous chemical lithosphere, *in* AGU Monograph on the Lithospheric Structure of the Rocky Mountains, K. Karlstrom and G. R. Keller, eds., *in press (this volume),* 2004.

Zurek, B. and K. Dueker, Lithospheric Stratigraphy beneath the Southern Rocky Mountains, USA, *in* AGU Monograph on the Lithospheric Structure of the Rocky Mountains, K. Karlstrom and G. R. Keller, eds., *in press (this volume),* 2004.

Oliver Boyd, CIRES and Department of Geological Sciences, University of Colorado at Boulder, Boulder, CO.

Vera Schulte-Pelkum, CIRES and Department of Geological Sciences, University of Colorado at Boulder, Boulder, CO.

Anne Sheehan, CIRES and Department of Geological Sciences, University of Colorado at Boulder, Boulder, CO.

Charles Wilson, CIRES and Department of Geological Sciences, University of Colorado at Boulder, Boulder, CO.

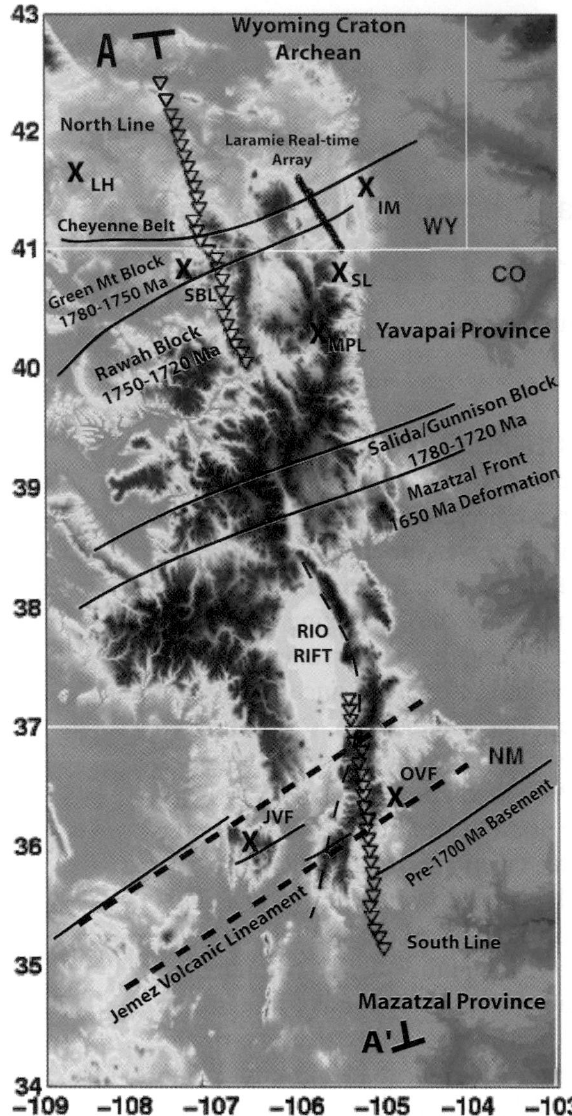

Plate 1. Continental Dynamics – Rocky Mountains (CD–ROM) PASSCAL passive broad-band deployment. Base map of the Rocky Mountain region encompassing the states of Wyoming (WY), Colorado (CO) and New Mexico (NM). Major geological boundaries are shown as black lines [*CD–ROM working group*, 2002]. Seismic stations are denoted by triangles. Seismic transect is indicated A–A'. Volcanic fields of interested are marked by 'X' and their geographic labels are denoted as: LH – Lucite Hills, IM – Iron Mountain, SL – State Line, SBL – Steamboat lamproites, MPL – Middle Park lamproite, OVF – Ocate volcanic field, JVF – Jemez volcanic field.

aries created by island arc accretion (Plate 1): 1) The Archean–Proterozoic boundary known as the Cheyenne belt, which separates the 2.4–3.9 Ga Wyoming craton from the 1.78–1.74 Ga Yavapai Proterozoic province [*Frost et al.*, 1993]; 2) the Jemez suture, a Proterozoic–Proterozoic boundary, which separates the 1.73 Ga Yavapai province from the 1.65 Ga Mazatzal province [*Karlstrom and Bowring*, 1988].

Today, this ancient lithosphere is being thermally and chemically reworked, as evidenced by xenolith P-T-t studies [*Smith*, 2000], high heat flow [*Morgan and Gosnold*, 1989], late-Cenozoic volcanism [*Thompson et al.*, 1997] and post-Miocene uplift of the southern Rocky Mountains and High Plains [*Thompson and Zoback*, 1979; *Heller* et al., 2003]. Interestingly, this region of lithospheric reworking correlates well with low upper mantle velocities [*Deep Probe working group*, 1998; *Goes and van der Lee*, 2002]. Conversion of these low velocities to thermal structure suggests that most of this regions sub-crustal mantle is at or near the dry peridotite solidus [*Goes and van der Lee*, 2002]. This suggests a casual link between the low velocities and the high heat flow, volcanism and possible late Cenozoic uplift in this region.

Constraining the thickness of the lithosphere is fundamental to understanding the origin of this low velocity region. One popular model suggests that this region is high standing and warm because the lithosphere has been thinned either by: the Laramide slab advecting much of the sub-crustal lithosphere eastward beneath the High Plains [*Bird*, 1988] or delamination of the sub-crustal lithosphere [*Bird*, 1979]. However, this thin lithosphere model does not predict post-Miocene uplift and is inconsistent with thick lithosphere observations: xenolith P-T-t studies [*Smith*, 2000], late-Cenozoic high-K lamprolitic lavas [*Thompson et al.*, 1997] and diamonds from the Stateline Devonian kimberlite pipes [*McCallum et al.*, 1975]. Thus, we propose a thick lithosphere model in which the modern day lithosphere is chemically thick, perhaps hydrated along its sutures, and is presently low velocity because the lithosphere is being reheated by mantle processes [*Dueker et al.*, 2001]. To test this model, new detailed images of mantle layering are presented and synthesized with new tomographic velocity results [*Yuan and Dueker*, this volume].

2. METHOD

Receiver functions are used to isolate receiver side P to S conversions (denoted as a $P_d s$ phase where d is depth to the interface generating the S-wave) created by significant velocity contrasts (Figure 1). A $P_d s$ conversion is generated when a teleseismic P-wave partitions a small fraction of its amplitude (i.e., <10%) to a S-wave at a relatively sharp velocity gradient. The Moho is usually the largest lithospheric velocity contrast with a 20% change in P-velocity. Receiver functions have been primarily been used to measure crustal thickness by identifying the Moho $P_d s$ arrival. The use of receiver functions to study sub-crustal layering is hindered by: 1) inadequate spatial sampling with broad-band seismometers; 2) signal generated noise: especially, the interfer-

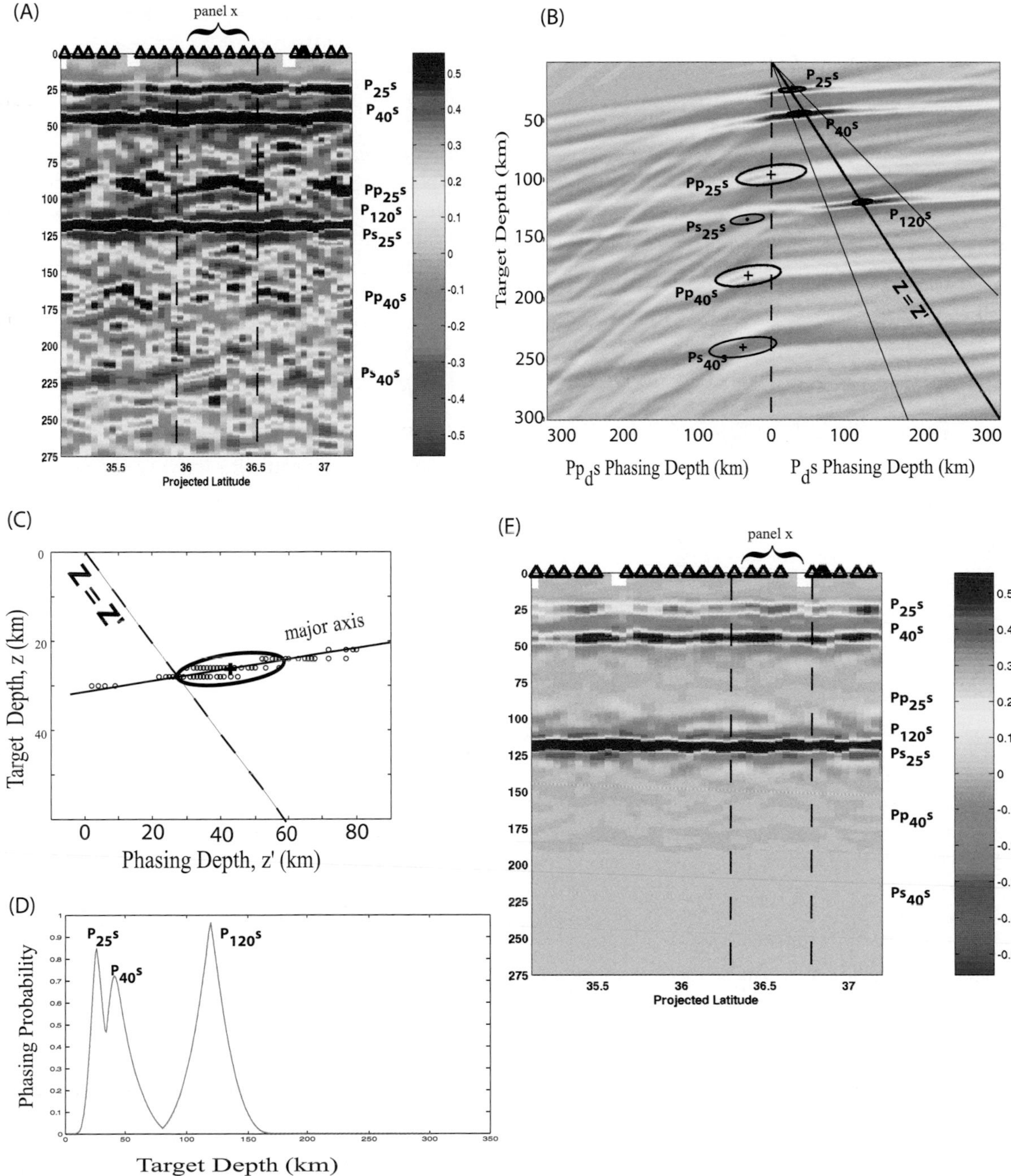

Plate 2. Phasing analysis of synthetic receiver function stacks using CD–ROM south line ray set. Velocity discontinuities have been placed at 25, 40 and 120 km. P_ds, Pp_ds and Ps_ds phases are included in the synthetic receiver functions. Noise was added using randomly selected waveform segments. (A) Common conversion point image. Notice that the deeper the discontinuity, the more incoherent its reverberations become because the reverberations are being mis-stacked using the P_ds move-out curves. (B) Phasing diagram for P_ds (right half) and Pp_ds arrivals (left half). The Pp_ds target depth axis has been mapped to P_ds depth to facilitate comparison between direct and reverberated converted arrivals. The peak values at which arrivals are phased are marked by a '+' and the associated standard deviations are indicated by the ovals. (C) Bootstrapped mean phasing location. Circles represent the phasing location for each bootstrapped iteration. The oval represents the 2-D Gaussian distribution fitted to the bootstrapped phasing peaks. (D) Muting function derived from evaluating phasing probabilities along the line of optimal phasing (z=z') in (b). (E) Muted image derived by multiplying each data panel by its corresponding mute function.

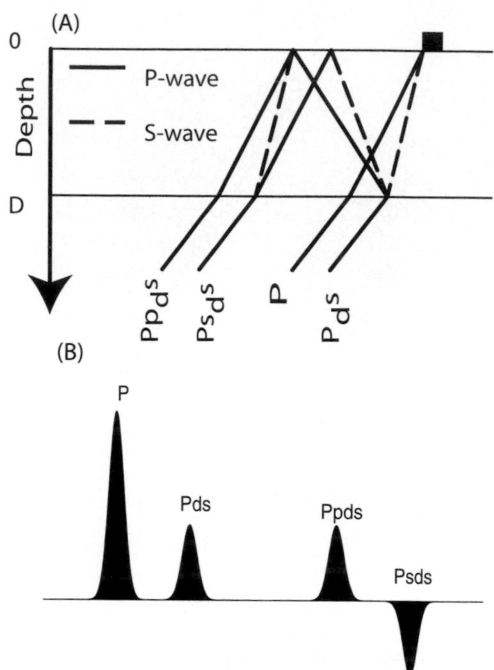

Figure 1. Ray diagram showing seismic response from one discontinuity. (A) Ray paths for the three primary converted phases shown: the direct arrival (P_ds) and two free-surface reverberations (Pp_ds and Ps_ds). (B) The relative arrival times, amplitudes and polarities predicted for these phases.

ence of free surface multiples [*Bostock*, 1996], complex P-S resonance in sedimentary basins [*Levander and Hill*, 1985] and free surface P to Rg topographic scattering [*Clauser and Langston*, 1995]. However, we believe by taking the necessary steps, receiver functions can be used to identify structure in the mantle lithosphere. The problems associated with sub-crustal imaging are mitigated by: 1) deploying arrays with dense 10 km station spacing; 2) stacking the data across many stations via common conversion points stacking of receiver functions (converted-wave equivalent of common mid-point stacks in reflection seismology); 3) the suppression of reverberations by the application of phasing analysis.

2.1 Data Set

The CD–ROM natural source seismic experiment consisted of the deployment of 47 PASSCAL broad-band sensors along two transects: CD–ROM North and CD–ROM South (Plate 1). These lines operated for a full year, from June 1999 to June 2000. The data comes from direct P waves with source magnitudes >5.5 and a distance of 20–95((Figure 2). The initial set of 2100 waveforms is culled by: 1) requiring the zero-lag P arrival to be the largest arrival on the receiver function, and 2) using receiver functions with RMS values < 20% of the P-arrival. This culling produces a final data set of ~1400 receiver functions.

2.2 Deconvolution

Receiver functions are made via deconvolution of the vertical component from the radial component of the seismogram using water-level spectral division [*Langston*, 1977] with a water-level of 0.1 [*Clayton and Wiggins*, 1976]. The data is windowed from 10 s before to 120 s after the P arrival. To maximize the signal to noise ratio, a zero-phase Butterworth filter with a band-pass of 1.5–0.3 Hz is used to isolate P_ds arrivals.

2.3 Mapping Time to Depth

To make a depth image of discontinuity structure in the lithosphere, the time of a P_ds arrival must be mapped to depth. For a flat Earth, this time is simply a function of ray parameter and depth to interface (Figure 1)[*Gurrola et al.*, 1994]:

$$T_{P_ds}(p, D) = \int_{-D}^{0} \left[\sqrt{V_S(z)^{-2} - p^2} - \sqrt{V_P(z)^{-2} - p^2} \right] dz \quad (1)$$

where D is the depth to interface, V_S is the S-velocity, V_P is the P-velocity and p is the Cartesian ray parameter (s/km). The tectonic 1-D North American shear wave model (TNA) [*Grand and Helmberger*, 1984] is used for $V_S(z)$. $V_P(z)$ is calculated assuming the IASP upper mantle average V_P/V_S of 1.84 [*Kennett and Engdahl*, 1991]. Ray parameter is calculated using the source to receiver distance and the IASPI velocity model. An Earth flattening transformation is applied to correct for the Earth's sphericity.

Note that the relative depth variations are better constrained than the absolute depths. The absolute depths are contingent upon the P- and S- velocity model used for mapping time to depth. False relative depth variation would arise from large, lateral bulk-crustal velocity variations. However, no large-scale P-wave velocity variations are observed in the CD–ROM refraction model [*Snelson and Keller*, this volume]. Furthermore, it is encouraging that the crustal velocity model based on the P-wave refraction model and assuming a Vp/Vs of 1.8 maps the P_ds Moho to within the errors of the refraction model. Therefore, we believe that 1-D velocity model used is an acceptable approximation for mapping time to depth.

2.4 Forward Formulation of the Model Space

To identify velocity discontinuities in the lithosphere, a 2-D model is constructed using the common conversion point

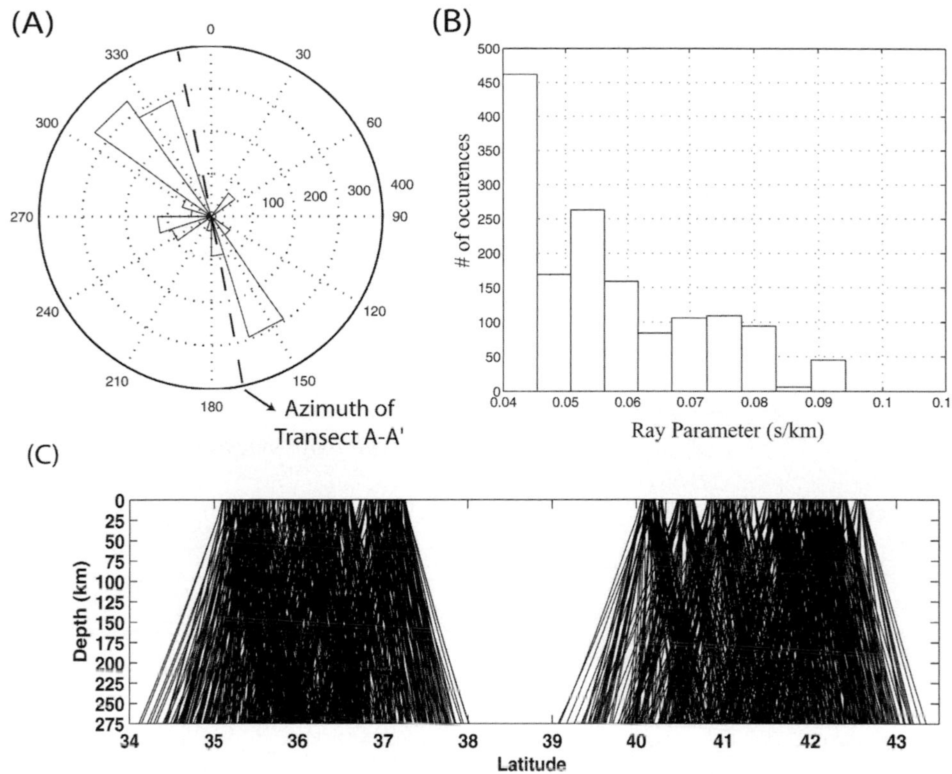

Figure 2. CD–ROM ray-set characteristics. (A) The distribution of earthquake back-azimuths for all data. (B) The distribution of direct P ray parameters, a wide distribution is needed to properly phase arrivals as outlined in section 3.6. (C) Converted S-wave ray coverage.

(CCP) method [*Dueker and Sheehan*, 1997]. The CCP image is constructed by solving the linear matrix equation:

$$\mathbf{Am} = \mathbf{d}, \tag{2}$$

where **A** is the scattering matrix, **m** is the model vector and **d** is the data vector. The model vector is a 2-D grid of scattering points with 2 by 1 km spacing. The data vector is the concatenation of the radial receiver functions. The scattering matrix is a matrix of zeros and ones that specify which grid points are sampled by the receiver functions. The construction of the scattering matrix is accomplished via 1-D ray tracing downward through the velocity model. To properly account for variations in model sampling, each column of the scattering matrix is normalized by its' column sum. Given that the slowness of a typical $P_d s$ ray path is <4% smaller than the P-wave slowness [*Gurrola et al.*, 1994], it is a good approximation to use the IASPI P-wave slowness value for ray tracing. Note that this formulation ignores diffractions and ray bending. These effects are not considered important to the image quality because 3-D Kirchoff migration of the CD–ROM data set has shown little improvement in image quality [*Morozova and Dueker*, 2002; *Morozova and Dueker*, 2003].

To find our model vector, this matrix formulation is iteratively solved via re-weighted back-projection (Eqn. 3 and 4). To down weight the effects of noisy data, the data misfit is calculated and used to re-weight the system of equations using the data covariance matrix $\mathbf{C_d}$. In practice, only two iterations are used, which has the effect of down weighting about 30% of the dataset. The net effect is that the model amplitudes are increased and the variance reduction increases from 30% to 60%; however, no new structure is created in the process. The equation solved is:

$$\left[\mathbf{C_d}^{-1/2}\mathbf{A\,S}\right]\left(\mathbf{S}^{-1}\mathbf{m}\right) = \mathbf{C_d}^{-1/2}\mathbf{d} \tag{3a}$$

where **S** is a nearest neighbor smoothing matrix. This can be rewritten as

$$\mathbf{A'} = \mathbf{C_d}^{-1/2}\mathbf{A\,S},\ \ \mathbf{d'} = \mathbf{C_d}^{-1/2}\mathbf{d},\ \ \mathbf{m'} = \mathbf{S}^{-1}\mathbf{m} \tag{3b}$$

hence:

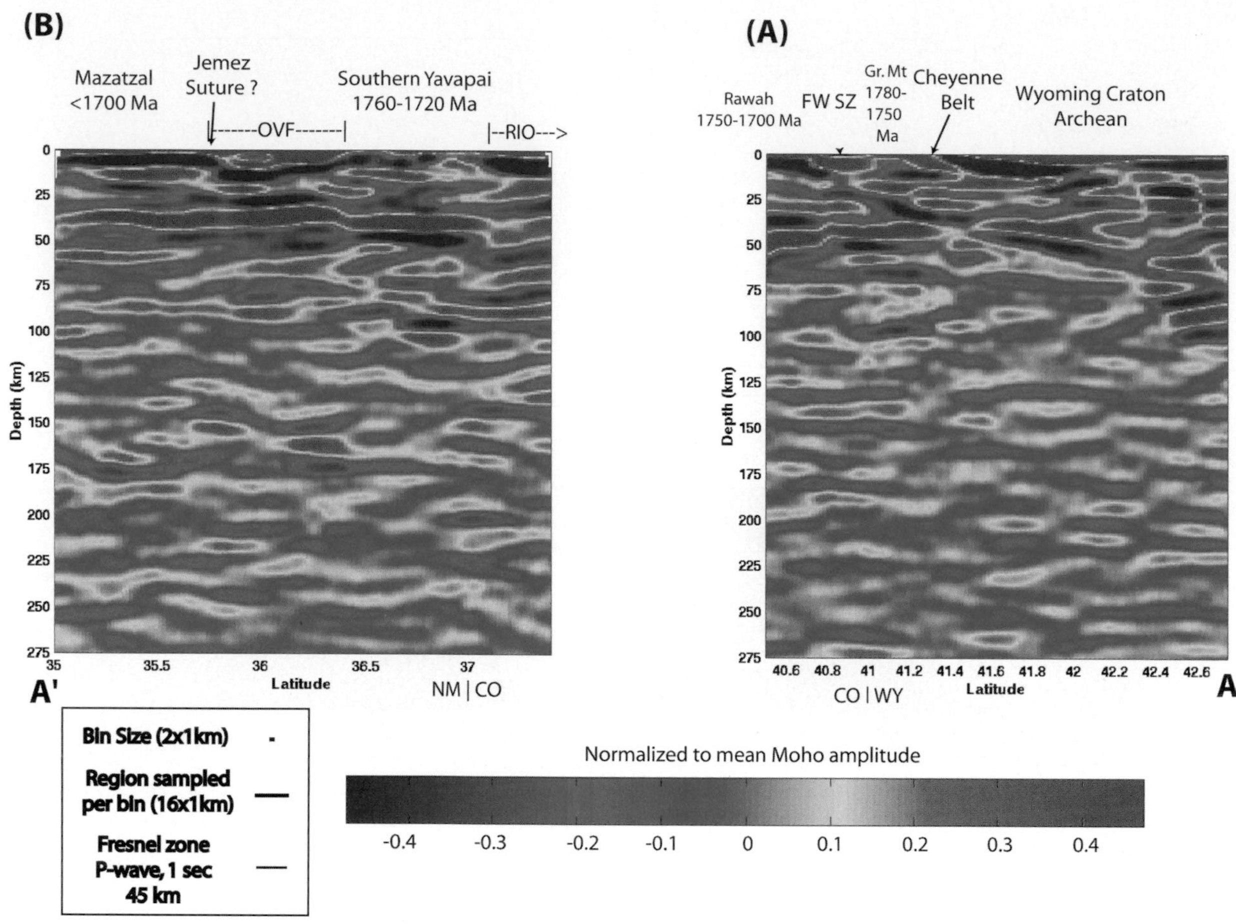

Plate 3. Crust and mantle layering images. (A) North transect. (B) South transect. Amplitudes are normalized with respect to the mean Moho amplitude. The variance reduction of the model was 69%. No vertical exaggeration is applied. Geological surface boundaries are labeled: OVF—Ocate volcanic field, RIO—eastern edge of Rio Grande rift; FW SZ—Farwell mountain shear zone.

$$\mathbf{A'}\,\mathbf{m'} = \mathbf{d'}. \tag{3c}$$

Direct inversion of the **A'** matrix has been done with a LSQR matrix solver, however the solution requires strong regularization and shows no improvement over simple back-projection [*Shearer et al.*, 1999]. Therefore, our final solution is constructed via simple back projection:

$$\mathbf{m'} = \mathbf{A'}^{t}\,\mathbf{d'}. \tag{4a}$$

The final model solution **m** is obtained by smoothing the "roughened model" vector **m'** (Eqn. 3b):

$$\mathbf{m} = \mathbf{S}\,\mathbf{m'}. \tag{4b}$$

To summarize, each model parameter estimate is the mean of the data points "sampling" a given scattering point. The data covariance matrix $\mathbf{C}_d$ is recalculated based upon the residuals of the data to converge towards a more robust solution. A smoothing matrix **S** imposes a horizontal correlation between the three nearest neighbors to either side of each scattering point. This imposes a lateral correlation length of 14 km upon the image. Structures imaged that have length scales less than this correlation length should not be interpreted as real structure.

2.5 Bootstrapping and Error Estimation

An estimate of the model errors can be found by calculating the variance of data at each model point. However, the variance is only a reasonable proxy for error if the underlying probability distributions are Gaussian-like. Therefore, to assess the underlying distributions of the model parameters, bootstrapping with replacement [*Efron and Tibshitani*, 1986] is

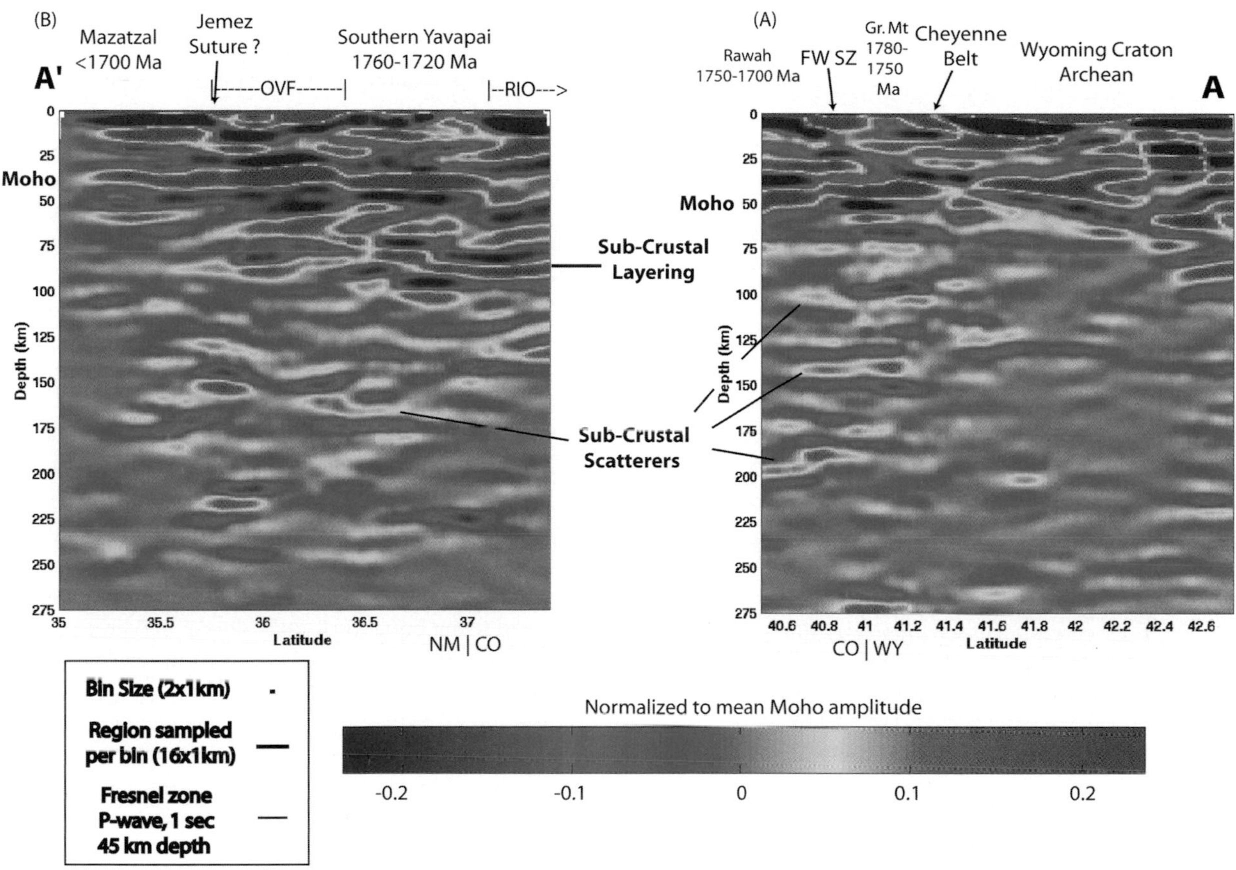

Plate 4. Muted crust and mantle layering images. (A) North transect. (B) South transect. Phasing panels are 16 km wide with 50% sharing between panels. The same amplitude scale is used as Plate 3. No vertical exaggeration is applied. Geological surface boundaries are labeled as Plate 3.

used. This is done by constructing 100 realizations of the model where each model realization is constructed from a randomly chosen set of waveforms. Then, histograms of each model parameter are constructed to visually assess the Gaussian-like nature of the distributions. This procedure shows that the model distributions are generally bell-shaped in form and not multi-modal. This suggests that the standard errors are a reasonable estimate of the model errors.

2.6 Phasing Analysis and Reverberation Suppression

A primary source of image degradation is the contamination of the image by Pp_ds and Ps_ds crustal reverberations (Figure 1); thus, a goal of receiver function imaging should be the identification and suppression of these reverberations. The strongest reverberation is the Pp_ds arrival, whereas the Ps_ds reverberation is the superposition of two arrivals (Figure 1) that can be "split" by lateral velocity heterogeneity, interface dip, or anisotropy making this phase less visible in our images. Reverberation contamination in the images depends on several factors: 1) the dominant period of the data; 2) the depth of the interface; 3) interface dip; 4) lateral crustal velocity variations (both isotropic and anisotropic); 5) the ray parameter distribution of the data.

To suppress reverberations, the mode of the arrival (i.e., P_ds, Pp_ds or Ps_ds) is determined via phasing analysis. The CCP image can then be muted to suppress arrivals that do not display the moveout associated with a direct P_ds arrival. Move-out is the timing difference between a converted phase arrival and its equivalent vertically incident arrival (i.e., p=0). The move-out for a P_ds arrival is:

$$\delta T_{P_ds}(p,z) = T_{P_ds}(p,z) - T_{P_ds}(0,z) \tag{5}$$

where z is depth of discontinuity, p is ray parameter, $T_{pds}(0,z)$ (Eqn. 1a) is the zero incident angle arrival time, and $T_{pds}(p,z)$ is the P_ds arrival time. To form a phasing diagram, the move-out for a range of phasing depths is calculated

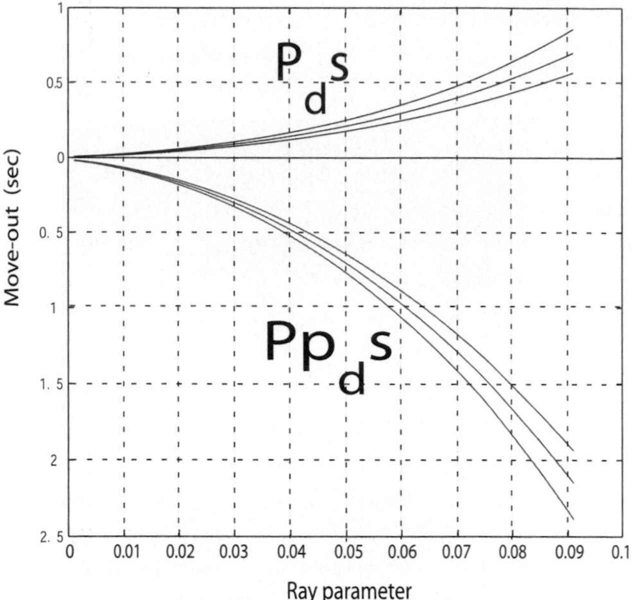

Figure 3. Move-out curves for discontinuity at 40 km depth. The curves are for P_ds and Pp_ds move-out. The three curves for each phase represent the change in move-out with a +/- 5% anti-correlated change in Vp and Vp/Vs with respect to our 1-D velocity model. Note the maximum differential move-out between P_ds and Pp_ds is 2.5 s, which is distinguishable in our phasing analysis using data with a 1–2 s dominant period.

$$T_{phase}(p,z,z') = \delta T(p,z')_{phase} + T_{phase}(0,z) \quad (6)$$

where phase is P_ds, Pp_ds or Ps_ds, p is the ray parameter for a given ray, z' is the trial phasing depth and z is the depth. A mode is said to phase when the maximum amplitude in the phasing plot resides close to the z=z' line.

To apply this phasing algorithm to our 2-D spatial image, the raw image is divided into overlapping spatial panels (Plate 2a). These panels are 16 km wide that corresponds to the lateral correlation scale length imposed by our S matrix (Eqn. 3a). A 50% spatial overlap between panels is used to smooth the results. In each data panel, a phasing plot is constructed (Plate 2b). To select which arrivals to further analyze, the standard deviation of each phasing panel is calculated. Any arrivals with an amplitude greater than two standard deviations above the average are identified as candidate arrivals.

To determine the phasing errors of these arrivals, each candidate arrival is bootstrapped 100 times (Plate 2c) and the coordinates (z and z') of the peak value are recorded. Then, each arrival that is close to the line of perfect phasing (z=z') are selected (Plate 2b). A two dimensional Guassian is fitted to the bootstrapped distribution of peak phasing values (Plate 2c). To construct the 1-D muting function, the ensemble of 2-D Guassian distributions are evaluated along the line of optimal phasing (Plate 2d). Finally, the 1-D muting functions for each panel are interpolated to generate a 2-D weighting function that is then applied to the raw image to create the phase muted image (Plate 2e).

The phasing analysis ability to resolve the moveout of our data depends on the dominant period of the data (i.e. the move-out needs to be > 1/4 the dominant period of the waveforms). Equation 1 predicts that an interface at 40 km has up to 2.0 s of differential move-out between P_ds and Pp_ds arrivals (Figure 3). The 2.0 s of move-out between P_ds and Pp_ds is resolvable with respect to the 1–2 s dominant period of the data. The caveats are that the interface is nearly flat (i.e., <10 % dip) and that a wide distribution of ray parameters (p=0.04 to 0.1 s/km) is available.

To test the robustness of the phasing analysis, a suite of synthetic receiver functions have been made. Plate 2 shows the results of one such test with discontinuities placed at 25, 40 and 120 km. The signals for the receiver functions are constructed using equation 1 and contain the direct and two primary reverberations. To simulate realistic velocity heterogeneity, a random time picked from a Gaussian distribution with a half width of 0.5 s was added to the timing of each arrival. The deconvolution and signal generated noise are simulated by randomly selecting rescaled receiver functions and adding them to each synthetic receiver function. This noise simulation consists of taking 80 s segments with a random start time between 2–22 s of the P-arrival and scaling the RMS value to 1/3 the amplitude of the synthetic arrivals to approximate the signal to noise ratio of the data [Morozov and Dueker, 2003]. The same filter used for real data is applied to the synthetics. This synthetic receiver function dataset was constructed using the CD–ROM south ray paths (Figure 2). The results of this test indicate that this technique can distinguish layers that are separated by >10 km with 1–2 s dominant period data.

2.7 Identification of Lithosphere

A fundamental question is whether our mantle layering resides in the lithosphere or asthenosphere. We propose that the creation of strong, horizontal layering in the asthenosphere is difficult and therefore any observed layering resides in the lithosphere. The only well accepted solid-state phase transition between the Moho and 200 km depth range is the spinel to garnet transition [*Anderson*, 1989]. However, this transition is both small in amplitude and has a broad phase loop incapable of producing the amplitudes observed in our images. Chemical layering in the asthenosphere can be ruled out as most researchers believe the asthenosphere is convectively mixed [*Helffrich and Wood*, 2001] and chemical layering in the asthenosphere could only be maintained if the asthenosphere is not convecting. Another mechanism capable of producing velocity layering in the asthenosphere would be the proposed change in the dominant creep mechanism from dislocation to diffusion creep [*Karato and Wu*, 1993]. However, this creep transition is thought to only occur in cool, high stress regions of the mantle, not the warmer, lower stress asthenosphere. Thus, lithospheric layering seems more plausible especially given the observation of strong chemical layering in xenolith columns [*Griffin et al.*, 1999], flat slabs juxtaposing oceanic crust against peridotite mantle [*Bostock*, 1998] and mantle shear zones associated with the collisional tectonics [*Levin and Park*, 2000].

3. RESULTS

For comparison the raw and reverberation suppressed images of lithosphere layering are shown in Plate 3 and 4, albeit interpretation will primarily be focused on the reverberation suppressed image. Note the Moho arrival phases well as a P_ds arrival and is not suppressed by the muting function (Plate 4). The Moho Pp_ds reverberation underneath the south line (reverberation falsely mapped to ~175 km depth) is muted because it did not phase as a P_ds arrival. For the north line, the reverberations from the Moho are not clearly visible (Plate 4), most likely due to the large amount of Moho topography and a smaller data set.

3.1 North Line

Differences between the Archean and Proterozoic lithospheres are observed in crustal thickness and sub-crustal reflectivity. Underneath the Cheyenne belt the crust is ~40 km thick, at the Proterozoic south end of the line the crust is ~50 km thick and at the Archean north end of the line the crust is ~40 km thick. Evidence for suturing tectonics is suggested by the image of an imbricated Moho just north of the Cheyenne suture, where the Proterozoic crust appears thrust under the Archean crust. The spatial density and lateral continuity of the sub-crustal structure varies north and south of the Cheyenne belt. The mantle layering in the Proterozoic sub-crustal region truncates approximately below the Cheyenne Belt. This subtle change in structure across the Cheyenne suture supports the notion that the lithosphere on either side is different. There is a lack of sub-crustal structure underneath the rifted 2.1 Ga rifted margin Archean crust. As discussed previously, the sub-crustal layering is interpreted as lithospheric in origin, which implies a lithosphere that is at least 150 km thick. Interestingly, the region in-between the Proterozoic lithosphere and the Archean rifted margin is devoid of sub-crustal reflectivity. At the northern end of this line, Archean sub-crustal layering is observed. The depth of this layering deepens to the north and is interpreted to manifest the southern margin of the 2.1 Ga rifted lithosphere.

Mid-crustal features imaged under the Cheyenne belt, correspond well with CD–ROM seismic reflection profiles [*Morozova et al*, this volume; *Tyson et al*, 2002]. Downward projection of the Cheyenne suture reveals mid-crustal features that correspond to the reflectivity "wedge" imaged in the reflection profile.

3.2 South Line

The southern line crosses four physiographic provinces: Mazatzal province to the south of the Jemez volcanic lineament; the 100 km wide Jemez volcanic lineament; southern Yavapai province between the Jemez and Rio Grande rift; the Rio Grande rift. The Moho is relatively flat across the transect at a depth of ~ 40 km, with small changes in the thickness of the crust occurring at the northern edge of the Jemez volcanic lineament and southern end of the Rio Grande rift. Underneath the Jemez volcanic lineament structural variations are observed in the crust. To the south and to the north of the Jemez volcanic lineament a strong mid-crustal discontinuity is observed at ~15 km depth. Below the Jemez volcanic lineament a weaker discontinuity is observed at ~20 km depth. A series of S-dipping mid-crustal reflections observed in the CD–ROM seismic reflection experiment suggests that a paleo-subduction exist to the north of the Jemez volcanic lineament [*Maganai and Levander*, this volume]. The south dipping paleo-subduction is broadly consistent with a step in the Moho and truncation of the crust and mantle layering.

The subcrustal lithosphere in this region is defined by two laterally coherent layers at ~60 and 85 km. The most profound sub-crustal layering is present beneath the Jemez volcanic lineament where coherent layering extends from the Moho to 100 km. The sub-crustal discontinuities show structural changes to the north and south of the Jemez volcanic lineament Below 100 km, there is structure that is comparable to that observed in the north, therefore the lithosphere is interpreted to be at least 100 km thick and perhaps thicker.

3.3 Integrated Tomography – Receiver Functions

Shear-wave splitting, tomographic velocity images and our mantle layering results suggest a complex history of continental–island arc collisions and lithosphere development and modification. To elucidate the origin of the observed mantle layering, the correlation between the tomographic velocity image [*Yuan and Dueker*, this volume] and mantle layering are assessed (Plate 5).

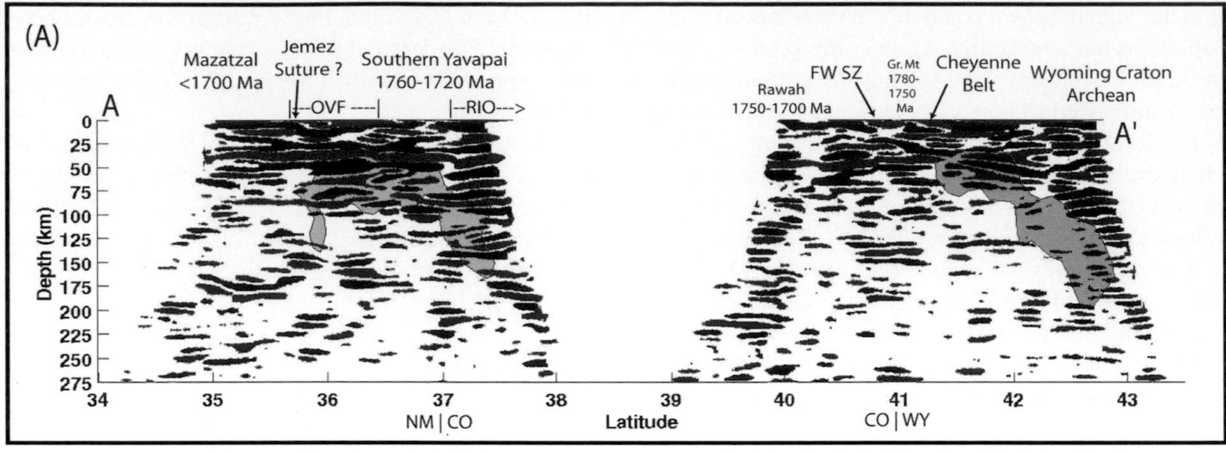

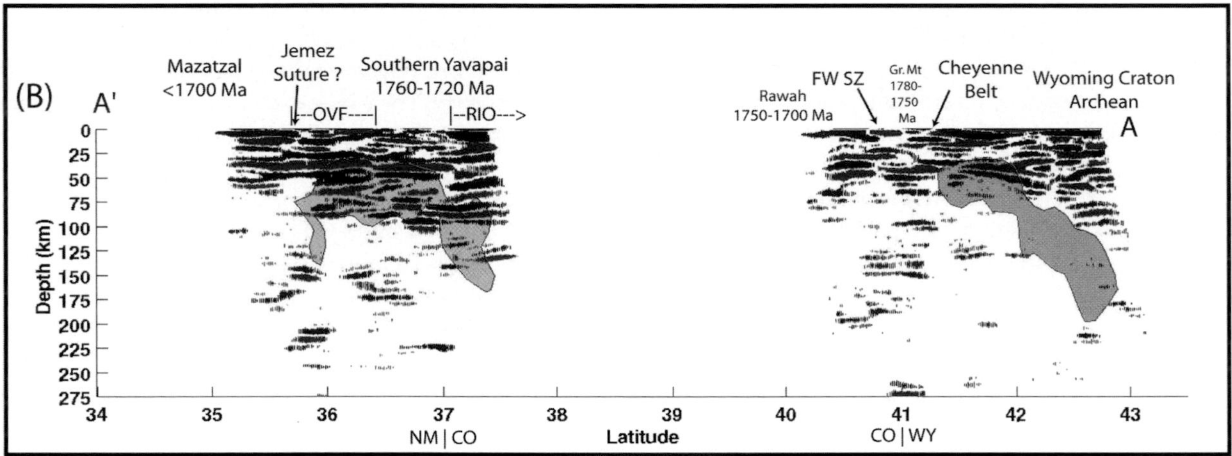

Plate 5. Lithospheric layering and P-wave tomographic image. (A) The non-muted model (Plate 3) (B) the muted image (Plate 4). Blue discontinuity shading denotes a positive shear wave con-trasts and red denotes a negative shear wave contrast. The tomo-graphic velocity anomalies are denoted by the transparent over-lays. The red overlay denotes the region with >2% low velocity anomaly and the blue overlay is the >2% high velocity anomaly. [Yuan and Dueker, this volume]. Transect is shown in Figure 1 as A-A". Geological surface boundaries are labeled as following: CB-Cheyenne belt, FW-SZ-Farwell Mountain shear zone, RIO-eastern edge of Rio Grande rift, OVF-Ocate volcanic field. No vertical exaggeration is applied.

When comparing the receiver functions with the regional tomography, an understanding of what information is contained in these techniques and how this information is complementary is required. Receiver functions image sharp vertical gradient of velocity and tomography images volume velocity variations. These two methods provide complementary information.

Comparison of the two images (Plate 5) reveals several correlations. In the northern line, the fast "slab" anomaly is in a region largely devoid of sub-crustal layering. The upper surface of this velocity anomaly truncates layering in the Archean lithosphere to the north. On the southern line, a slow velocity anomaly is observed between 40–100 km depth beneath the Jemez lineament. This is where the strongest sub-Moho lithospheric layering is observed.

4. DISCUSSION

Beneath the North line, the truncation of Archean mantle layering against the top of the high-velocity "slab" is consistent with the slab's tectonic emplacement after the creation of the 2.1 Ga rifted margin. The slab's upper surface, beneath the imbricated Moho, is consistent with the slab emplacement perhaps "driving" the crustal imbrication. We propose that the imbricated Moho (Plate 5), and north dipping high-velocity anomaly [*Yuan and Dueker*, this volume], are the remnants of a short-lived subduction zone that flipped polarity during the Proterozoic accretion events [*Tyson et al.*, 2002]. The evolutionary scenario is: 1) layered Archean mantle exists pre-rifting; 2) rifting reshaped the edge of continental lithosphere into a sloping boundary [*Buck*, 1991]; 3) the slab fragment is emplaced against this lithospheric boundary via a post-collision polarity flip [*Tyson et al.*, 2003]. Subduction zone polarization reversal are observed today in Taiwan [*Chemenda et al.*, 2001; *Teng et al.*, 2000] and at the Banda arc [*Snyder and et al.*, 1996].

The New Mexico mantle lithosphere is best defined by its layered structure and high heat flow. Under the Jemez volcanic lineament, a low velocity body is imaged down to 100 km depth (Plate 5)[Yuan and Dueker, this volume]. Models to explain this low velocity body are: 1) active upwelling from below (i.e. a plume) eroding the lithosphere; 2) passive asthenospheric upwelling from rifting lithosphere; 3) low-solidus heterogeneities in the lithosphere associated with hydrated regions and/or basaltic dykes [*Carlson and Irving*, 1994]. Major passive upwelling is unlikely because the Jemez lineament has not undergone any major extension. The Jemez low velocity anomaly is 50–100 km off axis from the Rio Grande Rift that has a total net extension of <20 km [*Chapin and Cather*, 1994] and is currently extending at <1 mm/yr [*Thatcher et al.*, 1999]. This small amount of crustal thinning cannot fully explain the high heat flow, indicating that another mechanism is needed to generate this heat [*Keller et al.*, 1990].

Our preferred model is that the Jemez low velocity anomaly and mantle layering define the minimum lithospheric thickness. If the mantle beneath the Jemez volcanic lineament is low-solidus mantle, then this mantle could reduced the velocities due to small melt fractions. Given that the Jemez lineament is a suture between the Yavapai and Mazatzal provinces, this region may have been hydrated by Proterozoic subduction [*Selverstone et al.*, 1999]. In this case, the Jemez volcanism would be an indicator of the "fertility" of the Mazatzal – Yavapai suture zone. The high heat flow in this region suggests that the base of the lithosphere is being heated.

5. CONCLUSION

Our results indicate structural differences preserved in the mantle lithosphere between the Wyoming Archean province and Proterozoic island arc terranes. This result supports the proposal that the lithosphere underneath the north line is thick, old and stable. On the other hand, continental lithosphere under the south line has been modified in recent times by rifting and magmatic events and is 100 km thick. The Jemez volcanic lineament may mark a hydrated Proterozoic–Proterozoic suture zone. The Wyoming Archean lithosphere is cold and at least 150 km thick (strong and stable) and is relatively unaffected by the warm asthenosphere to the south.

Acknowledgements. This work was made possible by the IRIS/PASSCAL instrumentation pool and by a grant from the National Science Foundation Continental Dynamics Program (EAR-9614862).

REFERENCES

Anderson, D. L., *Theory of the Earth*, 367 pp., Blackwell Scientific Publications, Boston, 1989.

Balling, N., Deep seismic reflection evidence for ancient subduction and collision zones within the continental lithosphere of northwestern Europe, *Tectonophysics*, *329* (1–4), 269–300, 2000.

Bank, C. G., M. G. Bostock, R. M. Ellis, and J. F. Cassidy, A reconnaissance teleseismic study of the upper mantle and transition zone beneath the Archean Slave Craton in NW Canada, *Tectonophysics*, *319* (3), Pages 151–166, 2000.

Bird, P., Continental delamination and the Colorado Plateau, *Journal of Geophysical Research*, *84* (B13), 7561–7571, 1979.

Bird, P., Formation of the Rocky Mountains, western United States: a continuum computer model, *Science*, *239* (4847), 1501–1507, 1988.

Bostock, M. E., Anisotropic upper-mantle stratigraphy and architecture of the Slave Craton, *Nature (London)*, *390* (6658), Pages 392–395, 1997.

Bostock, M. G., Ps conversions from the upper mantle transition zone beneath the Canadian landmass, *Journal of Geophysical Research, B, Solid Earth and Planets*, *101* (4), 8383–8402, 1996.

Bostock, M. G., Mantle stratigraphy and evolution of the Slave Province, *Journal of Geophysical Research B: Solid Earth*, *103* (B9), 21,183–21,200, 1998.

Buck, W. R., Modes of continental lithospheric extension, *Journal of Geophysical Research*, *96* (B12), 20,161–20,178, 1991.

Carlson, R. W., and A. J. Irving, Depletion and enrichment history of subcontinental lithospheric mantle: an Os, Sr, Nd and Pb isotopic study of ultramafic xenoliths from the northwestern Wyoming Craton, *Earth & Planetary Science Letters*, *126* (4), 457–472, 1994.

CD–ROM working group, Structure and Evolution of the Lithosphere Beneath the Rocky Mountains: initial results from the CD–ROM experiment, *GSA Today*, *12* (3), 4–10, 2002.

Chapin, C. E., and S. M. Cather, Tectonic setting of the axial basins of the northern and central Rio Grande rift, in *Basins of the Rio Grande Rift: Structure, Stratigraphy and Tectonic Setting*, edited by G.R. Keller, and S.M. Cather, pp. 5–26, Geological Society of America, Boulder, 1994.

Chemenda, A. I., E. A. Konstantinvskaya, G. M. Ivanov, R. K. Yang, and J. F. Stephan, New results from physical modeling or arc-continent collision in Taiwan: Evolutionary model, *Tectonophysics*, *333* (1–2), 159–178, 2001.

Clauser, R. H., and C. A. Langston, Modeling P–Rg coversions from isolated topographic features near the NORESS array, *Bulletion Seismological Society America*, *85*, 859–873, 1995.

Clayton, R. W., and R. A. Wiggins, Source shape estimation and deconvolution of teleseismic body waves, *Geophys. J. R. Astron. Soc.*, *47*, 151–177, 1976.

Deep Probe working group, Probing the Archean and Proterozoic Lithosphere of Western North America, *GSA Today*, *8* (7), 1–5, 1998.

Dueker, K., H. Yuan, and B. Zurek, Thick-structured Proterozoic lithosphere of the Rocky Mountain region, *GSA Today*, *11* (12), 4–9, 2001.

Dueker, K. G., and A. F. Sheehan, Mantle discontinuity structure from midpoint stacks of converted P and S waves across the Yellowstone hotspot track, *Journal of Geophysical Research*, *102* (B4), 8313–8327, 1997.

Efron, B., and R. Tibshitani, Bootstrap methods for standard errors, confidence intervals, and other measures of statistical accuracy, *Stat. Sci*, *1*, 54–77, 1986.

Frost, B. R., D. Frost Carol, H. Lindsley Donald, S. Scoates James, and N. Mitchell Jeremy, The Laramie anorthosite complex and Sherman Batholith; geology, evolution, and theories of origin, *Memoir Geological Survey of Wyoming*, *5*, 118–161, 1993.

Goes, S., and S. van der Lee, Thermal structure of the North American uppermost mantle inferred from seismic tomography, *Journal of Geophysical Research B: Solid Earth*, *107* (B3), ETG 2, 2002.

Grand, S. P., and D. V. Helmberger, Upper mantle shear structure of North America, *Geophysical Journal-Royal Astronomical, Society*, 399–438, 1984.

Griffin, W. L., S. Y. O'Reilly, R. Davies, K. Kivi, E. van Achterbergh, L. M. Natapov, B. J. Doyle, C. G. Ryan, and N. J. Pearson, Layered mantle lithosphere in the Lac de Gras area, Slave craton: Composition, structure and origin, *Journal of Petrology*, *40* (5), 705–727, 1999.

Gurrola, H., J. B. Minster, and T. Owens, The use of velocity spectrum for stacking receiver, *Geophysical Journal International*, *117* (2), 1994.

Helffrich, G. R., and B. J. Wood, The Earth's mantle, *Nature*, *412* (6846), 501–507, 2001.

Heller, P., K. Dueker, and M. McMillan, Post-Paleozoic alluvial gravel transport as evidence of continental tilting in the U.S. Cordillera, *GSAB*, *115* (9), 1122–1132, 2003.

Jones, C. H., J. R. Unruh, and L. J. Sonder, The role of gravitational potential energy in active deformation in the southwestern United States, *Nature*, *381* (6577), 37–41, 1996.

Karato, S., and P. Wu, Rheology of the upper mantle: a synthesis, *Science*, *260* (5109), 771–778, 1993.

Karlstrom, K. E., and S. A. Bowring, Early Proterozoic assembly of tectonostratigraphic terranes in southwestern North America, *Journal of Geology*, *96* (5), 561–576, 1988.

Karlstrom, K. E., and R. S. Houston, The Cheyenne belt: analysis of a Proterozoic suture in southern Wyoming, *Precambrian Research*, *25* (4), 415–446, 1984.

Keller, G. R., P. Morgan, and W. R. Seager, Crustal structure, gravity anomalies and heat flow in the southern Rio Grande rift and their relationship to extensional tectonics, *Tectonophysics*, *174* (1–2), 21–37, 1990.

Kennett, B. L. N., and E. R. Engdahl, Traveltimes for global earthquake location and phase identification, *Geophysical Journal International*, *105* (2), 429–465, 1991.

Langston, C. A., Corvallis, Oregon, crustal and upper mantle receiver structure from telesismic P and S waves, *Bulletion Seismological Society America*, *67*, 713–724, 1977.

Levander, A., and N. R. Hill, P–SV resonances in irregular low-velocity surface layers, *Bulletion Seismological Society America*, *75*, 847–864, 1985.

Levin, V., and J. Park, Shear zones in the Proterozoic lithosphere of the Arabian Shield and the nature of the Hales discontinuity, *Tectonophysics*, *323* (3–4), Pages 131–148, 2000.

McCallum, M. E., D. H. Eggler, and L. K. Burns, Kimberlitic diatremes in northern Colorado and southern Wyoming, *Physics and Chemistry of the Earth*, *9*, 149–161, 1975.

Morgan, P., and W. D. Gosnold, Heat flow and thermal regimes in the continental United States, in *Geophysical Framework of the Continental United States*, edited by L.C. Pakiser, and W.D. Mooney, pp. 493–522, Geol. Soc. Am., 1989.

O'Reilly, S. Y., W. L. Griffin, Y. H. Poudjom Djomani, and P. Morgan, Are lithospheres forever? Tracking changes in subcontinental lithospheric mantle through time, *GSA Today*, *11* (4), 4–10, 2001.

Rondenay, S., G. Bostock Michael, M. Hearn Thomas, J. White Donald, H. Wu, G. Senechal, S. Ji, and M. Mareschal, Teleseismic studies of the lithosphere below the Abitibi–Grenville Lithoprobe transect, *Canadian Journal of Earth Sciences = Revue Canadienne des Sciences de la Terre*, *37* (2–3), 415–426, 2000.

Selverstone, J., A. Pun, and C. Condie Kent, Xenolithic evidence for Proterozoic crustal evolution beneath the Colorado Plateau, *Geological Society of America Bulletin*, *111* (4), 590–606, 1999.

Shearer, P. M., M. P. Flanagan, and M. A. H. Hedlin, Experiments in migration processing of SS precursor data to image upper mantle discontinuity structure, *Journal of Geophysical Research B: Solid Earth*, *104* (4), 7229–7242, 1999.

Smith, D., Insights into the evolution of the uppermost continental mantle from xenolith localities on and near the Colorado Plateau and regional comparisons, *Journal of Geophysical Research B: Solid Earth*, *105* (7), 16,769–16,781, 2000.

Snyder, D. B., and et al., A dual doubly vergent orogen in the Banda Arc continent-arc collision zone as observed on deep seismic reflection profiles, *Tectonics*, *15* (1), 34–53, 1996.

Teng, L. S., C. T. Lee, Y. B. Tsai, and L. Y. Hsiao, Slab breakoff as a mechanism for flipping of subduction polarity in Taiwan, *Geology*, *28* (2), 155–158, 2000.

Thatcher, W., E. Quilty, G. W. Bawden, G. R. Foulger, B. R. Julian, and J. Svarc, Present-day deformation across the Basin and Range Province, western United States, *Science*, *283* (5408), 1714–1718, 1999.

Thompson, G. A., and M. L. Zoback, Regional geophysics of the Colorado Plateau, *Tectonophysics*, *61* (1–3), 149–181, 1979.

Thompson, R. N., J. G. Mitchell, A. P. Dickin, S. A. Gibson, D. Velde, P. T. Leat, and M. A. Morrison, Oligocene lamproite containing an Al-poor, Ti-rich biotite, Middle Park, northwest Colorado, USA, *Mineralogical Magazine*, *61* (4), 557–572, 1997.

Warner, M., J. Morgan, P. Barton, P. Morgan, C. Price, and K. Jones, Seismic reflections from the mantle represent relict subduction zones within the continental lithosphere, *Geology (Boulder)*, *24* (1), 39–42, 1996.

Upper Mantle Tomographic V_p and V_s Images of the Rocky Mountains in Wyoming, Colorado and New Mexico: Evidence for a Thick Heterogeneous Chemical Lithosphere

Huaiyu Yuan and Ken Dueker

Department of Geology and Geophysics, University of Wyoming, Laramie, Wyoming

Upper mantle tomographic body wave images from the CD-ROM deployment reveal two major lithospheric anomalies across two primary structural boundaries in the southern Rocky Mountains: a ~200 km deep high velocity north-dipping "Cheyenne slab" beneath the Archean-Proterozoic Cheyenne belt, and a 100 km deep low velocity "Jemez body" beneath the Proterozoic Proterozoic Jemez suture. The Cheyenne slab is most likely a slab fragment accreted against the Archean Wyoming during the Proterozoic arc collision processes. This interpretation suggests that the ancient slab's thermal signature has been diffused away and non-thermal explanations for the high velocity slab are required. Tomographic modeling of possible chemical and anisotropic velocity variations associated with the slab shows that our isotropic velocity images can be explained via non-thermal models. In addition, the de-correlation of the P- and S-velocity images and the CD-ROM shear-wave splitting modeling are consistent with a dipping slab. The Jemez body plausibly results from the combination of low-solidus materials in the suture lithosphere and the late Cenozoic regional heating of the lithosphere. The 100 km deep lithospheric layering and the uniform shear-wave splitting measurements support our contention that the Jemez body is a lithospheric anomaly. A third low velocity structure extends beneath the middle Rio Grande Rift to 300 km depth. This anomaly may manifest a thermal upwelling that could increase heat flow into the lithosphere. Our results suggest that lithospheric heterogeneities related to fossil accretionary processes have been preserved in the Precambrian sutures, and are preferentially affecting the subsequent tectonism in this region.

1. INTRODUCTION

From a global perspective, natural and controlled source seismic imaging of Precambrian suture zones has shown that the sub-crustal lithosphere beneath sutures is often seismically, chemically and anisotropically heterogeneous [*Shragge et al.*, 2002; *Judenherc et al.*, 2002; *Shomali et al.*, 2002; *Babuska and Cara*, 1991; *Balling*, 2000; *Snyder*, 2002; *Poupinet et al.*, 2003; *Sandoval et al.*, 2003a]. Tomographic velocity images reveal sharp lateral variations that often demarcate distinct lithospheric blocks separated by suture zones. Given the old age of suture zones and no obvious variations in modern day heat flow, compositional and anisotropic velocity heterogeneities are required to explain these velocity anomalies. Sub-Moho dipping reflectors observed from the controlled-source imaging studies under suture zones are often explained as fossil slabs [*Warner et al.*, 1996; *Eaton and Cassidy*, 1996; *Pharaoh*, 1999; *Abramovitz and Thybo*, 2000; *Gorman et al.*, 2002; *White et al.*, 2002]. For instance, two uppermost mantle

The Rocky Mountain Region: An Evolving Lithosphere
Geophysical Monograph Series 154

10.1029/154GM25

dipping reflectors beneath the Great Fall/Vulcan suture at the northern end of the Wyoming craton have been imaged [*Gorman et al.*, 2002]. Interpretation of the seismic signature and tectonic history of this region suggests that the reflectors are best explained as fossil slabs related to suturing between the Wyoming Province, the Medicine Hat Block, and the Hearne Province. Fossil slabs, whose associated oceanic crust resides in eclogitic facies, are consistent with the observed impedance contrast where detailed petrophysical and seismic modeling has been done [*Morgan et al.*, 2000].

South of the Wyoming craton, good basement exposures exist along a 1000 km wide sequence of Proterozoic oceanic terranes [*Karlstrom and Houston*, 1984]. Detailed structural, basement age, geochemical, and pressure-temperature time histories allow this region to be divided into a sequence of distinct blocks separated by sutures [*Karlstrom and Bowring*, 1988; *Condie*, 1992]. Key to constraining how these blocks accreted and transformed into stable continental lithosphere is accurate imaging of the sub-crustal structures. A large teleseismic experiment was deployed across two primary suture zones in this region: the Archean-Proterozoic Cheyenne belt (imaged by the North line), and the Proterozoic Jemez suture (imaged by the South line). The main target is constraining the seismic lithospheric heterogeneities related to the Proterozoic suturing processes in the southern Rocky Mountains.

Our results show a high velocity north dipping slab-like anomaly (referred to as "the Cheyenne slab" herein) beneath the Cheyenne belt, and a low velocity body (referred to as "the Jemez body") extending to 100 km beneath the Jemez suture. We interpret the Cheyenne slab as a fossil Proterozoic slab fragment accreted against the Archean Wyoming margin. The Jemez body is most likely preferentially molten low-solidus materials trapped in the suture lithosphere. At the same time, upper mantle convective rolls [*Richter*, 1973] may be focusing heat along the NE-trending Jemez suture [*Dueker et al.*, 2001].

2. TECTONIC SETTING

The relevant geologic history of the southern Rocky Mountains in the western United States begins with rifting of the southern margin of the Archean Wyoming province at 2.1 Ga [*Karlstrom and Houston*, 1984], followed by 300 Ma of passive margin sediment accumulation. Proterozoic arc accretion began when the Green Mountain arc accreted to the Wyoming craton at 1.78 Ga [*Chamberlain*, 1998] to create the Archean-Proterozoic suture known as the Cheyenne belt. Accretion of Proterozoic island arcs continued to the south until 1.65 Ga when the Mazatzal terrane accreted to the southern margin of the Yavapai terrane to form the Jemez suture [*Karlstrom and Bowring*, 1988; *Wooden and DeWitt*, 1991; *CD-ROM Working Group*, 2002]. Post 1.65 Ga, the primary magmatic event preserved in the basement rocks is the pervasive exposure of 1.4 Ga granitic batholiths whose petrogenesis most plausibly requires the emplacement of a large volume of mantle derived magma into the lithosphere [*Williams et al.*, 1999]. In addition, Laramide aged calc-alkaline magmatism modestly affected this region along with a significant peak in volcanism in Colorado and New Mexico around 35 Ma [*Armstrong and Ward*, 1991]. Two compressional events, the Pennsylvanian Ancestral Rockies [*Kluth and Coney*, 1981] and the late Cretaceous Laramide [*Hamilton and Myers*, 1966] orogenies, significantly deformed the Proterozoic and Archean aged lithosphere. While debated, it appears that the later shortening event did reactivate the Precambrian structures, and produced basement uplifts and deep sediment basins in the southern Rockies. Rifting of the lithosphere along the N-S trending Rio Grande rift [*Chapin*, 1979] started in the late Oligocene to early Miocene time. Associated rift-related magmatic activity is relatively minor, except along the late-Tertiary Jemez volcanic lineament. This volcanic lineament appears to be roughly following the trend of the Proterozoic Jemez suture [*CD-ROM Working Group*, 2002; *Wooden and DeWitt*, 1991; *Shaw and Karlstrom*, 1999].

The modern day geophysical characteristics of the Proterozoic Rocky Mountains (i.e., south of the Wyoming/Colorado border) are distinct with respect to the Archean Wyoming province as evidenced by its higher elevations, higher heat flow [*Decker et al.*, 1988], on-going rifting [*Ingersoll et al.*, 1990], and volcanism [*Aldrich and Laughlin*, 1984; *Baldridge et al.*, 1995]. In the upper mantle, the seismic shear wave velocity increases from very slow in southern New Mexico to slightly slower than the global average beneath the Wyoming province [*Deep Probe working group*, 1998; *Goes and van der Lee*, 2002]. Thermal calculations from the surface wave tomographic images suggest that the upper mantle beneath this region is at or near the dry peridotite solidus [*Goes and van der Lee*, 2002]. An interesting observation in the intermountain western U.S. is the spatial correlation between the young volcanic fields and Proterozoic suture zones [*Dueker et al.*, 2001; *Karlstrom and Humphreys*, 1998; *CD-ROM Working Group*, 2002]. These suture zones are generally underlain by low velocity upper mantle anomalies. Whether the low velocity anomalies represent compositional variations in the lithosphere or thermal effects of upwelling asthenosphere or some combination of both is still poorly constrained [*Dueker et al.*, 2001].

3. TELESEISMIC TRAVEL-TIME PROCESSING

3.1 Travel-Time Residual Measurement

The CD-ROM teleseismic experiment consists of a deployment of two dense (average 12 km station spacing) broadband arrays of 25 sensors each across the Cheyenne belt and the Jemez suture (Figure 1). The one-year deployment from May 1999 to June 2000 creates a dataset with good earthquake azimuthal and ray parameter coverage. Arrivals of P, S, PP, SS, PcP, ScS, PKiKP and SKSac phases from Aleutian, west Pacific, Tonga and south America are measured to obtain a diversity of ray sampling for the tomographic inversion. Both P- and S-waveforms are band-pass filtered (0.05 Hz – 0.3 Hz) to minimize the background noise, and to avoid the frequency dependent attenuation effects that may affect the scaling between P- and S-velocities [*Warren and Shearer*, 2000]. The radially symmetric IASPEI-91 velocity model [*Kennett and Engdahl*, 1991] is used to calculate the predicted moveout associated with the arrivals for each event. A time window of 4- and 6-s for the P- and S-waveform cross-correlation window is found to give robust residual estimates.

The travel-time residuals are measured using the multichannel cross-correlation technique, which gives robust residual measurements and standard errors [*VanDecar and Crosson*, 1990; *Allen et al.*, 2002]. For each event, cross-correlation functions for all waveform pairs are computed, and the time lag associated with the peak value of each cross-correlation function is calculated. Given n waveforms, this process results in a data vector of $\frac{1}{2}n(n-1)$ time shifts between station pairs. By adding a constraint equation that the mean residual associated for an event is zero, this system of equations becomes over-determined and is solved via a least-squares algorithm. For the P- and S-waves, the number of measured relative residuals is 4032 and 2058 with average standard error of 0.05 and 0.1 s, respectively. Noteworthy is that by forcing the mean of each event's residuals to zero, it creates a relative travel-time residual dataset that can only recover relative velocity variations, not absolute velocities.

Simple statistical analysis is performed to identify bad residuals with large error and spurious time shift. The primary cause of the bad residuals is due to measurement of low signal to noise ratio seismograms. These data are easily identified and removed by their large standard error from the cross-correlation solution. Study of the relation between the wave shape coherency and the standard error suggests that 0.15 and 0.3 s are reasonable thresholds for rejecting data for the P- and S-residuals, respectively. The secondary cause of the bad data is due to shifted waveforms by faulty station clocks that resulted in a few spurious travel-time measurements. These types of outliers are easily identified via spatial residual analysis (residual vs. station latitude plot). For each event, the misfits between the event data and a robust least-square quadratic polynomial fit are calculated. Then spurious travel-times are found and removed when their value is greater than twice the standard variation (>2σ) of the polynomial fit. The last step to

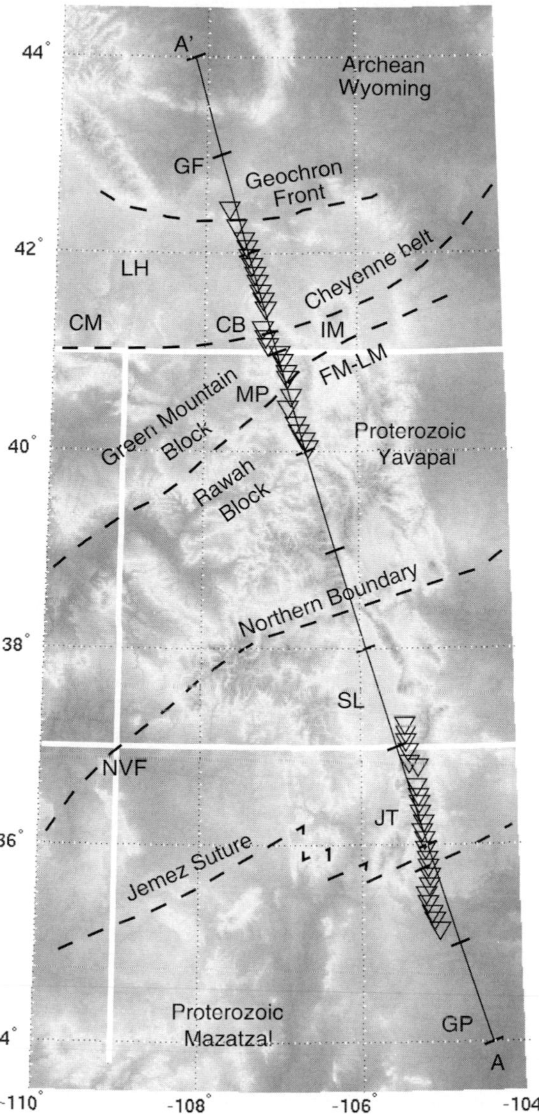

Figure 1. Crustal provinces of the southern Rockies and CD-ROM transects (black triangles). Geographic labels are: GF, Geochron Front [Chamberlain, 1998]; CB, Cheyenne belt; FM-LM, Farewell Mountain-Lester Mountain shear zone that separates the Green mountain and Rawah arcs; SL, San Luis basin; JT, the Jemez volcanic trend/suture; GP, the Great Plains; CM, Cedar Mountains; LH, Leucite Hills; NVF, Navajo Volcanic Field; and MP, Middle Park. Tomographic images are presented along A-A'.

identify bad data is via slowness residual analysis (polar plot). The residuals for each station are binned into non-overlapping grids in 2-D slowness space and the differences between the bin median and individual residuals are calculated. Then, individual residuals with large time differences (>2σ) are removed.

3.2 Choice of an S-Wave Particle Motion Coordinate System

Unlike for the P-waves, where the residuals are measured on the vertical channel of the seismometer, it has to be decided which particle motion system to use in measuring the travel-time residuals for the S-waves. Typical choices include the SH component, the average fast or slow anisotropy axis detected from shear wave splitting, or the maximum particle motion direction. The SH component waveform is least contaminated by the scattering from near station structures, and is generally the "cleanest" coordinate system. However, exclusive use of SH results in a significant reduction in available data. In passing through an anisotropic volume, the SV and SH components of ground motion become coupled, causing their waveforms to become distorted which can bias the residuals measured via waveform cross-correlation. If a uniform anisotropic layer is present beneath the seismic stations, the waveform distortion and time shift are uniform for each event. The time shifts imparted to the residuals therefore can be removed via demeaning of the residuals. The average fast or slow anisotropic velocity axis is commonly used [e.g., *Toomey et al.,* 1998] when a uniform anisotropic layer exists. If the anisotropic domain varies along the array, demeaning of the residuals cannot remove the travel-time effect of the seismic anisotropy due to the non-uniform distortion of the waveforms, and the biased residuals can create imaging artifacts. In the maximum particle motion coordinate system, the horizontal waveforms are projected onto the first principal component [e.g., Jackson, 1991] of ground motion. This produces a shear waveform dataset that possesses the maximum signal amplitude and hence the greatest number of measurable shear waves travel-times.

Given that the shear wave splitting results shows that the seismic anisotropy domain is uniform beneath the south line and changes to back-azimuthal dependent across the Cheyenne belt [Fox and Sheehan, this volume], the S-wave travel-time residuals are measured in both the maximum particle motion direction, and the average fast velocity axis orientation (N50(E approximately parallel to the North America plate motion direction). Inversion results of both datasets are compared and discussed in section 4.6.

4. TOMOGRAPHIC INVERSION

4.1 Summary Ray Dataset

Obtaining comparable V_p and V_s images is important to address the thermal and/or non-thermal causes of the observed upper mantle velocity anomalies [*Sobolev et al.*, 1999; *Goes and van der Lee*, 2002; *Schutt and Humphreys*, in press]. Making a summary ray dataset from the raw dataset allows very similar ray sets and hence very similar resolution matrices for both the P- and S-wave inversions [e.g., *Schutt and Humphreys*, in press]. To make our summary ray datasets for each station, all residuals from all back-azimuth are binned in 2-D horizontal slowness space. The summary ray bin size is an adjustable parameter. It is found in this study that 0.01 s/km and 0.015 s/km for the P- and S-waves summary ray bin size provides a reasonable trade-off between smoothing of the dataset and still having enough residuals within each bin to get a robust mean residual and standard error estimate. A summary ray residual is the weighted mean of the raw residuals at each summary ray point. The weighting function is a Gaussian function with a half width equal to the bin size. This procedure produced nearly identical datasets of ~1200 P- and S-wave summary rays from the original P- and S-wave residuals.

4.2 P- and S-Wave Residual Patterns

The P- and S-wave summary ray residuals show large spatial and back-azimuth dependent variations along the CD-ROM transect (Figure 2). The north line has early residuals (0.27 s and 0.63 s average for the P- and S-waves, respectively), and the south line has late residuals (-0.20 s and –0.85 s average, respectively). The peak-to-peak magnitude in the P- and S-residuals is 1.4 s and 3.7 s, accordingly. These large variations in residual patterns require large velocity anomalies in the upper mantle. A simple model to match the magnitude of these residuals would require a 5% (V_p) and 10% (V_s) velocity variation over a 200 km thick layer. If these residuals were purely crustal, implausibly large velocity variations (i.e., ~15% and 30% for P- and S-waves, respectively) would be required. The large azimuthal variation in residuals across the Cheyenne belt and the Jemez suture (CB and JT, respectively in Figure 2a and b) fundamentally require sub-crustal velocity anomalies. Crustal velocity variations cannot produce such large azimuthal variations due to the shallow depth.

4.3 Inversion Method

A linearized inversion for a 2-D velocity model is used to invert the travel-time residuals for structure. The dimension of our model space is 1600 km long by 300 km in depth. Each model block is 10 km (horizontal) by 15 km (depth). The Tectonic North American (TNA) shear wave model [*Grand and Helmberger*, 1984] is used as the S-wave background velocity model, and the P-wave model is computed assuming a 1.84 V_p/V_s ratio. Rays are traced from each event-receiver pair using the slowness computed from the IASPEI-91 velocity model [*Kennett and Engdahl*, 1991].

To find an optimal model m, the following matrix equation is solved via full matrix inversion

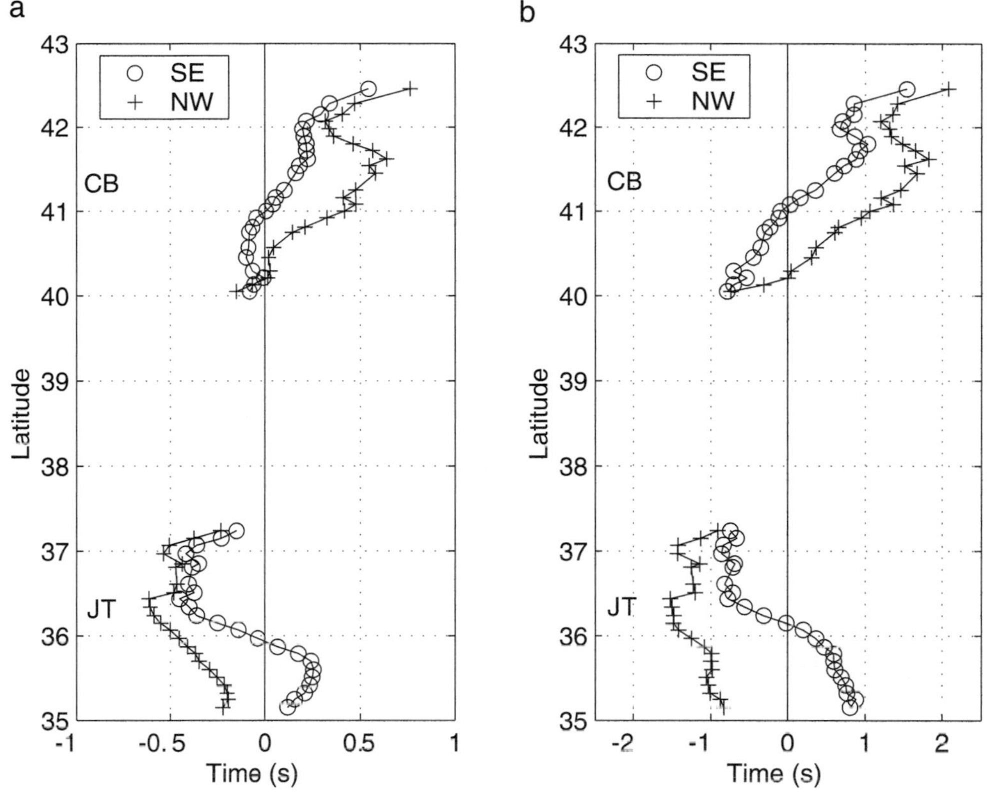

Figure 2. Station median residuals for two back-azimuth quadrants, the southeast (open circles) and the northwest (crosses). (a) P-wave residuals. (b) S-waves residual. Note the large peak-to-peak residual amplitudes (1.4 and 3.7 sec for P and S, respectively), and strong azimuthal variations across the Cheyenne belt (CB) and the Jemez volcanic trend/suture (JT).

$$C_d^{-1/2} ASS^{-1} m = C_d^{-1/2} d \ , \qquad (1)$$

where A is the data kernel containing the ray path length in each model block, m the slowness perturbations, d the travel-time residuals, C_d the diagonal a priori data covariance matrix, and S a 3 by 3 nearest neighbor smoothing matrix [*Meyerholtz et al.*, 1989]. The full model resolution and covariance matrices are calculated from the inverse matrix. A range of damping parameters is used to study the trade-off between resolution and model variance reduction. Empirically, the optimal model is at the "elbow" of the model energy versus the resolution spread curve, which is controlled by the damping parameter. The preferred inversion results are slightly over-damped to minimize artifacts in the images.

4.4 Resolution of Crustal Versus Mantle Velocity Anomalies

An important concern in teleseismic tomography is the separation of the crust and mantle velocity anomalies. The degree of concern is conditioned by the resolving power of the seismic ray set and the relative contribution of the travel-time anomalies from the crust versus the mantle. Given the large station spacing in most teleseismic tomography studies, two approaches have been proposed to address the crustal effects: either by adding station static term to absorb shallow structures [*Dziewonski and Anderson*, 1983], or by removing the crustal travel-times using an a priori crustal model [e.g., *Sandoval et al.*, 2003b]. With large station spacing (i.e. > 20 km), the station static term absorbs not only the crustal travel-time variations, but also mantle structures, and therefore is not used in some studies [*Dueker et al.*, 1993; *Sandoval et al.*, 2003b].

The CD-ROM stations are densely spaced (average 12 km spacing), indicating that crustal scale velocity anomalies can be well resolved. To examine the resolving power of our ray set, a suite of tests on models with various combinations of crust and upper mantle velocity heterogeneity patterns is performed. Two end-member models are presented that are parameterized by the amplitude ratio (θ) of the crust to mantle anomalies, and spatial wavelength (and for crust and mantle, respectively) of sinusoidal velocity anomaly variations. The crustal model has θ=3, λ_c=30 km and λ_m=90 km (Plate 1a). The mantle model has θ=1/2, λ_c=60 km and λ_m=90 km (Plate

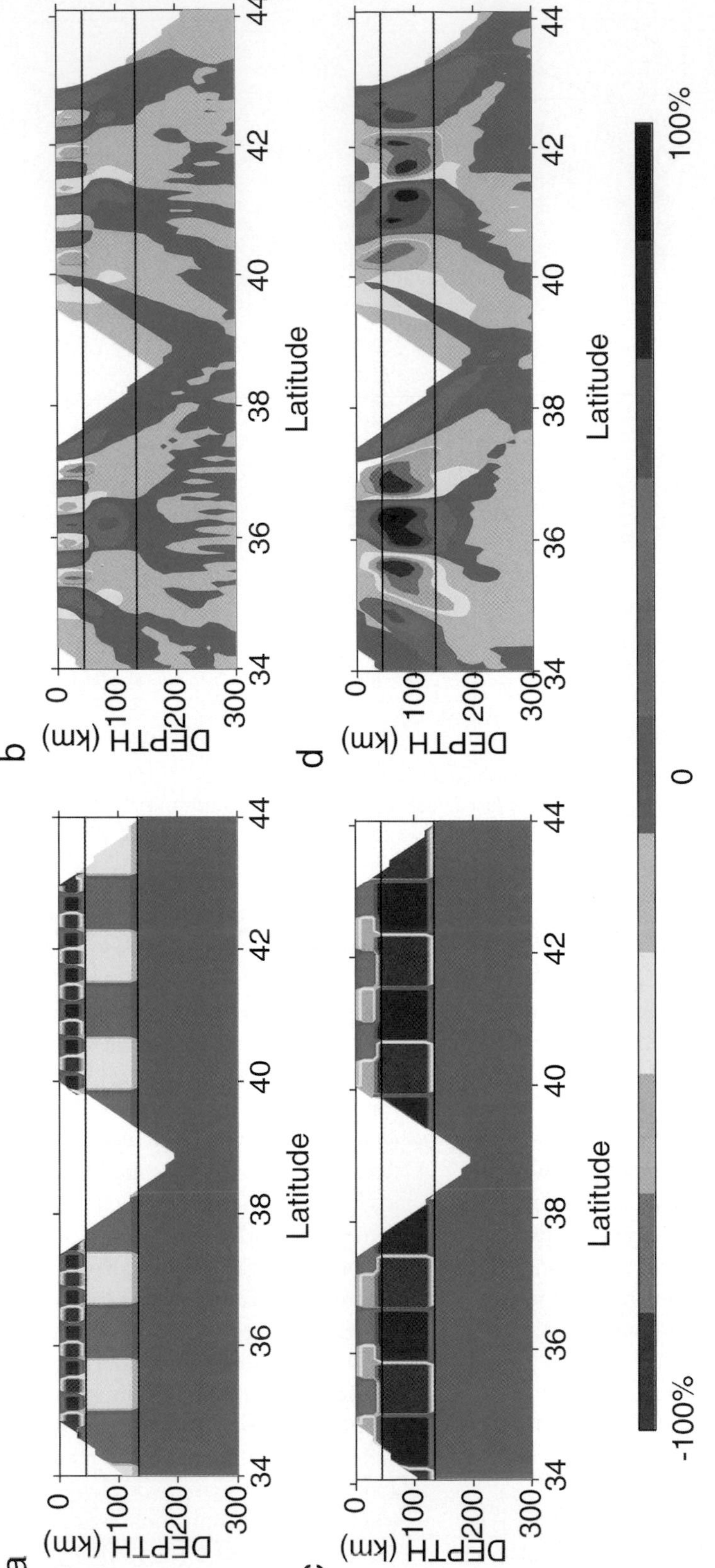

Plate 1. Crust and mantle resolution tests. Input models are parameterized by amplitude ratio θ between the crust and mantle, and spatial wavelength λ_c (for the crust) and λ_m (for the mantle) of the synthetic velocity variations. (a) Crust model with θ=3, λ_c=30 km and λ_m=90 km. (b) Inverted crust model. (c) Mantle model θ=1/2, λ_c=60 km and λ_m=90 km. d) Inverted mantle model.

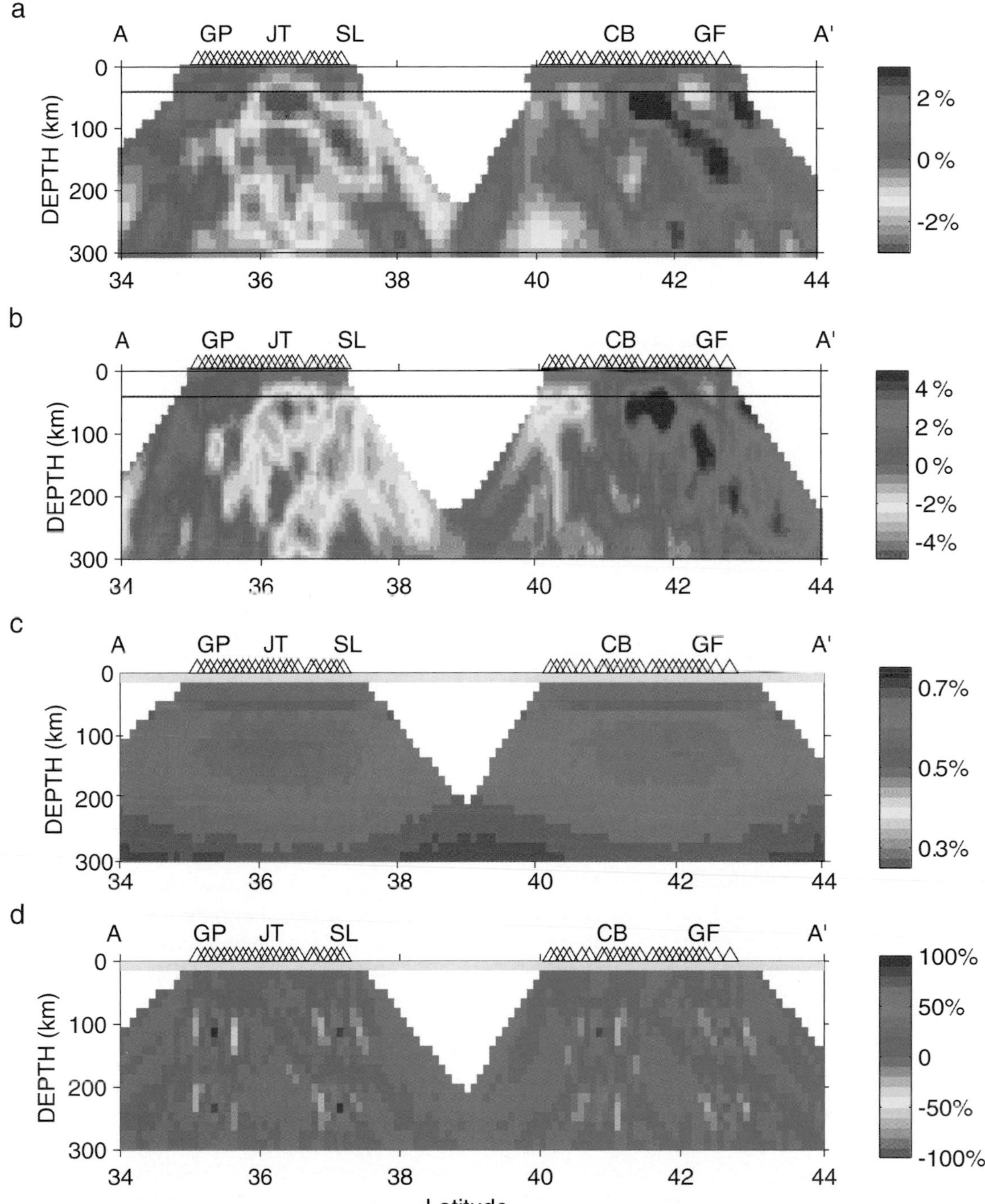

Plate 2. Tomographic inversion results. (a) The P-wave image. (b) The S-wave image inverted using the dataset measured in the maximum particle motion coordinate system. Blue and red denote high and low velocity variations. Triangles at zero depth are stations. Variance reduction is 84% and 81% for the P and S image, respectively. Note the 3 major anomalies, the Cheyenne slab, the Jemez body and the low velocity pipe (more obvious in the S-wave image). (c) Standard errors of the velocity models. (d) Selected resolution kernels for the P-wave inversion. The Cheyenne slab, the Jemez body and the low velocity pipe are in regions with low standard error and compact resolution kernels. The standard error and resolution kernels for the S-wave image are similar to this figure, except that the standard error amplitudes range from 0.5% – 0.9%.

1c). Gaussian distributed noise is added that is scaled to 10% RMS (root-mean-square) amplitude of the synthetic residuals. This 10% value is estimated from the noise to signal ratio of the CD-ROM dataset. All synthetic models are inverted with the same damping parameter as in the inversions presented in Plate 2. The inverted crustal model image (Plate 1b) reconstructs an average 70% of the input crustal velocity structure. The inverted mantle model image (Plate 1d) reconstructs 60-70% of the input mantle structure. The amplitudes of the recovered crustal anomalies vary with respect to the amplitude variation of the mantle anomalies. For instance, the reconstructed crust amplitude is higher than the input where the crustal and mantle anomalies are in-phase (the signs for the crustal and mantle anomalies are the same), and the amplitude vanishes where the crustal and mantle anomalies are out of phase. The constructive and destructive effects indicate that vertical smearing of the long wavelength mantle anomalies into the crust is significant.

These synthetic tests demonstrate that when the crustal anomalies dominate, the crustal structure is well reconstructed. On the other hand, if large mantle anomalies exist, false crustal anomalies can be generated. Given that the integrated travel-times from the crustal velocity heterogeneity [*Magnani and Levander*, this volume; *Snelson et al.*, this volume] is much smaller (~10% of the residual RMS amplitude) with respect to the mantle, the mantle structures are readily smeared into the crust in our inversions. Therefore, we choose to weight the crust and mantle model parameters differently to minimize the mantle smearing in our inversions. This is done by re-weighting the crustal part of the smoothing matrix S in equation (1) to 10% of the mantle parameter value. Comparison of the inversions with and without the crust down-weighted shows insignificant difference in the mantle. The difference in data variance reduction is only 2%, supporting our contention that the contribution of crustal velocity anomalies to our dataset is minor.

4.5 P- and S-Wave Inversion Results

The two primary anomalies in the V_p and V_s images (Plate 2a and b) are the high velocity Cheyenne slab beneath the Cheyenne belt, and the low velocity Jemez body beneath the Jemez suture. The Cheyenne slab anomaly extends from the Moho to ~200 km at a 45° dip. In the V_s image, the slab anomaly is not as continuous as in the V_p image. In section 4.6, we show that this de-correlation of the P- and S-images may result from the effect of dipping anisotropy. The Jemez low velocity anomaly is 100 km wide and extends to 100 km depth. A third velocity anomaly is the low velocity "pipe" beneath the south line, which is more continuous in the V_s image. This pipe extends to the bottom of our model space, with its surface projection beneath the San Luis basin of the middle Rio Grande rift.

The standard model error and the resolution matrix (Plate 2c and d) indicate that the velocity anomalies discussed above reside in a region of low standard model error and high resolution. Because the V_p and V_s images are constructed with the summary ray sets, the resolution kernels are very similar for both images. Selected P-image resolution kernels (Plate 2c) demonstrate that the lateral and vertical resolution is good with kernel amplitudes decaying to 40% in 30 km horizontal and 50 km vertical distances. The average standard errors associated with the Cheyenne slab and the Jemez body are ~0.6% and ~0.8% in the V_p and V_s images. The low velocity pipe anomaly extends to the base of the model where the error is larger (0.9% in the V_s), yet still much smaller than the pipe's 2% S-wave anomaly. The south-dipping direction is opposite to the smearing predicted by the resolution kernel, indicating that the pipe is not a resolution artifact.

4.6 Comparison of S-Wave Images From Different Datasets

At sub-solidus absolute temperatures, a thermal anomaly is generally expected to affect the V_p and V_s variations in a linearly scaled manner [*Karato*, 1993]. Hence, the de-correlation between the V_p and V_s images of the Cheyenne slab (Plate 2a and b) may indicate non-thermal mantle velocity variation. Melts, fluids and seismic anisotropy variations in the mantle can decorrelate the V_p and V_s variations [*Karato and Jung*, 1998; *Sobolev et al.*, 1999; *Goes and van der Lee*, 2002]. Given that the CD-ROM shear wave splitting modeling is consistent with a north dipping fast axis anisotropy across the Cheyenne belt [*Fox and Sheehan*, this volume], the de-correlation of the Cheyenne slab V_p and V_s anomaly may largely result from dipping seismic anisotropy.

One way to assess this dipping anisotropy's effects on our tomographic images is to invert the two S-wave residual datasets, measured in the average fast axis direction and the maximum polarization direction coordinate system. Due to the use of summary ray technique, the two inversions have similar variance reductions of 79% and 81%, respectively, and similar model standard errors and resolution kernels. However, the two S-wave velocity images are significantly different with respect to each other and the P-wave image of the Cheyenne slab (Figure 3). In the two different S-wave images, the slab is imaged as a continuous feature in the average fast axis coordinate system, while it breaks into segments in the maximum particle motion direction system. For an isotropic velocity anomaly, the two S-wave images are expected to be identical. The de-correlation of the two S-images would be consistent with non-isotropic velocity variation such as a dipping anisotropic slab across the Cheyenne belt.

4.7 Synthetic Forward Modeling of Dipping Anisotropy

When anisotropic velocity variations are present, large velocity artifacts are possible in isotropic tomographic inversion. A dipping anisotropic velocity structure results in inconsistent residuals as rays sample the fast and slow velocity axis. As an result, fast and slow velocity artifacts that are perpendicular to each other are present in the inverted P-wave image [*Sobolev et al.*, 1999].

To assess the effects of a dipping anisotropic velocity structure upon our P-wave isotropic inversion, residuals from a set of synthetic slab models are calculated and inverted. The synthetic slab coincides with the 2% contoured region of the inverted P-wave Cheyenne slab (thick contour lines in Plate 3a – d). The three input models are: an isotropic slab, an anisotropic slab, and a third model which is the combination of the first two models. The isotropic slab has a 2% high velocity anomaly. The anisotropic slab has an 8% peak-to-peak velocity variation, with its fast velocity axis fixed at the values from the shear wave splitting modeling [*Fox and Sheehan*, this volume]: the fast axis strikes *N45°W* and dips at 45°. The 8% anisotropy specified for the slab is in the range of the computed and measured peridotite aggregates in the mantle [*Soedjatmiko and Christensen*, 2000; *Saruwatari et al.*, 2001; *Babuska and Cara*, 1991]. Gaussian noise scaled to 10% RMS amplitude of the synthetic travel-times is added. For simplicity, only the Cheyenne slab is modeled in our synthetic tests.

The inversion well reconstructs the synthetic isotropic slab (Plate 3a). Vertical smearing occurs slightly at the bottom of the slab. As expected, the isotropic inversion cannot recreate the input anisotropic slab (Plate 3b). The inverted image shows two high velocity bodies in-line with the fast symmetry axis (the labeled F axis of the ellipsoid velocity tensor in Plate 3b), and two slow velocity bodies (labeled as 1 and 2) along the slow symmetry axis (the S axis in Plate 3b). A third slow velocity anomaly is present along the slow axis near the bottom of the model space). These three low velocity bodies are consistent with the ones in the observed P-wave image (Plate 3d). The inverted image for the third model (Plate 3c) gives the best match to the observed images: the Cheyenne slab and the associated low velocity bodies (1,2 and 3 in Plate 3d). Given that the other small velocity anomalies are not in the synthetic model (e.g., 5 and 6 in Plate 3d), the inversion with a combined anisotropic and isotropic velocity model produces an acceptable fit to the observed P-wave slab image.

5. DISCUSSION

5.1 The Cheyenne Slab

Temperature variation in the upper mantle has been suggested as the main factor affecting seismic velocities [e.g., *Sobolev et al.*, 1996; *Goes and van der Lee*, 2002]. However, it is unreasonable that the high velocity of the Cheyenne slab

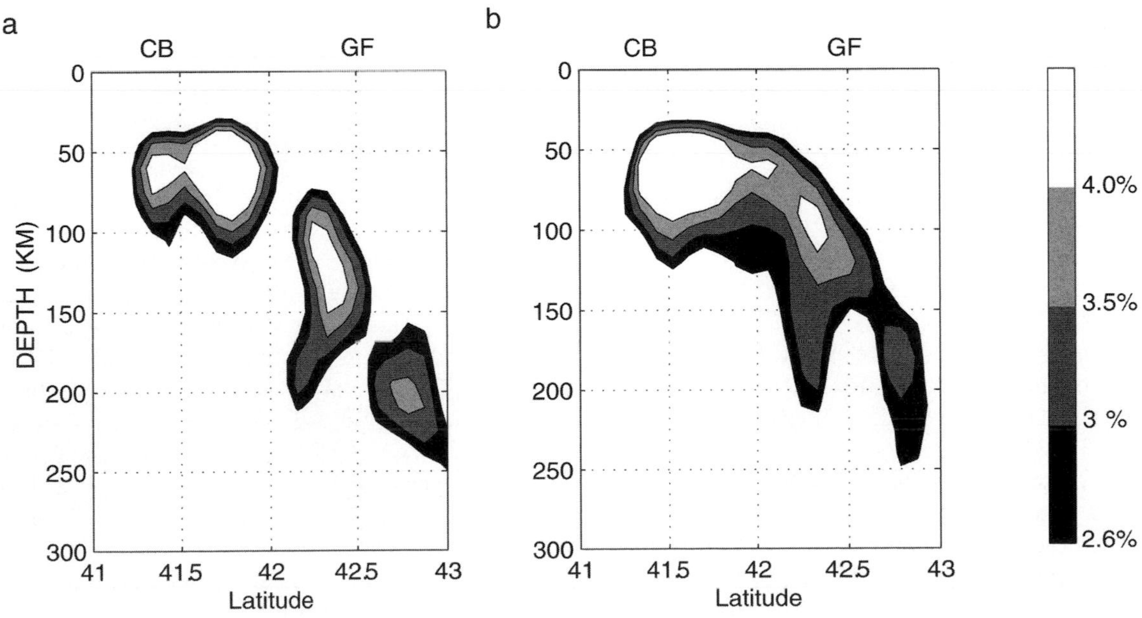

Figure 3. S-wave velocity variations of the Cheyenne slab for the two different S-wave dataset. a) For the maximum particle motion direction. b) For the average fast velocity direction. Note the Cheyenne slab breaks at 2.6% velocity variation level in a).

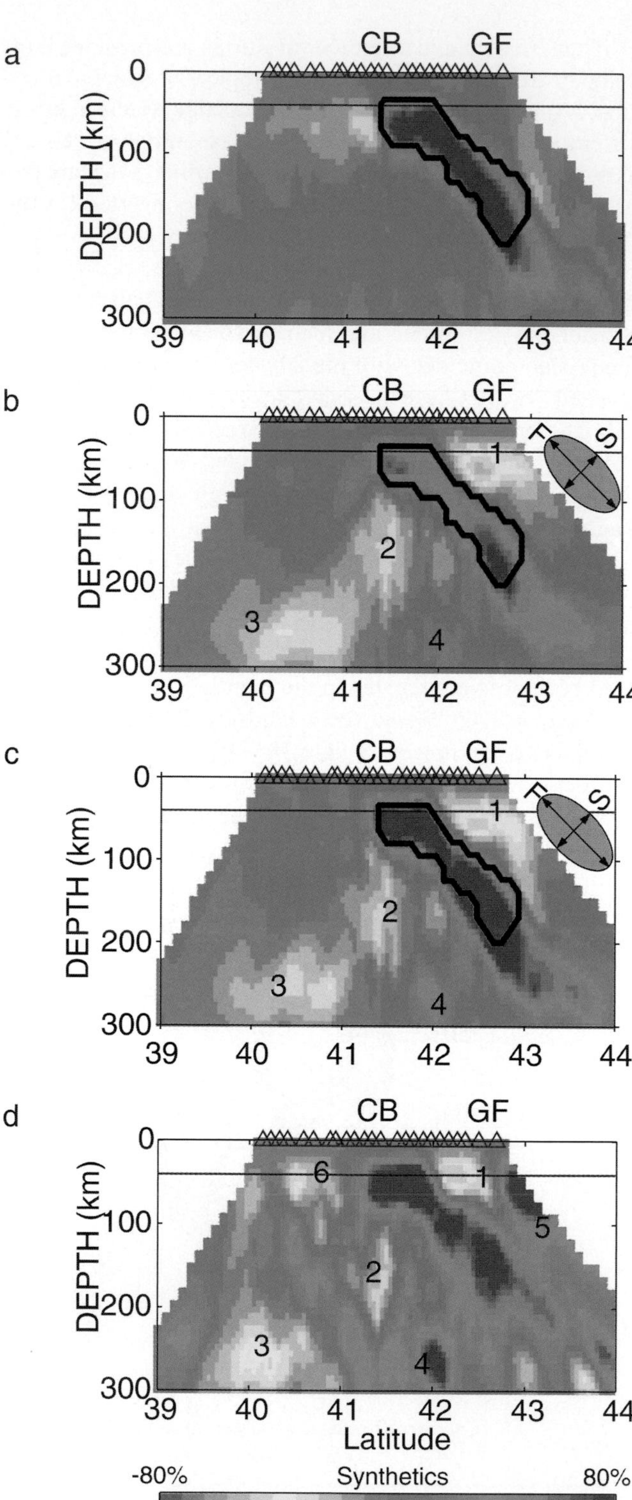

Plate 3. P-wave anisotropic synthetic inversions. The input model is defined as the 2% contour region of the P-wave Cheyenne slab. (a) Image for the isotropic model with 2% isotropic velocity variation within the slab. (b) Image for the anisotropic model. An 8% dipping fast velocity anisotropy is assigned to the slab. The ellipse to the upper right corner illustrates the simplified velocity tensor for the assigned anisotropy with its fast and slow axes labeled as F and S, respectively. (c) Inverted image for the combined anisotropic and isotropic model. (d) Observed P-wave image. Triangles at zero depth represent stations. Colorbar for the synthetics (a, b, and c) is scaled to 80% of the observed image (d). Visual comparisons with the observed P-wave image suggest that the image of the combined model best matches the observed image best (regions 1 – 4). Note in the input synthetic models other small anomalies (5 and 6) are not modeled.

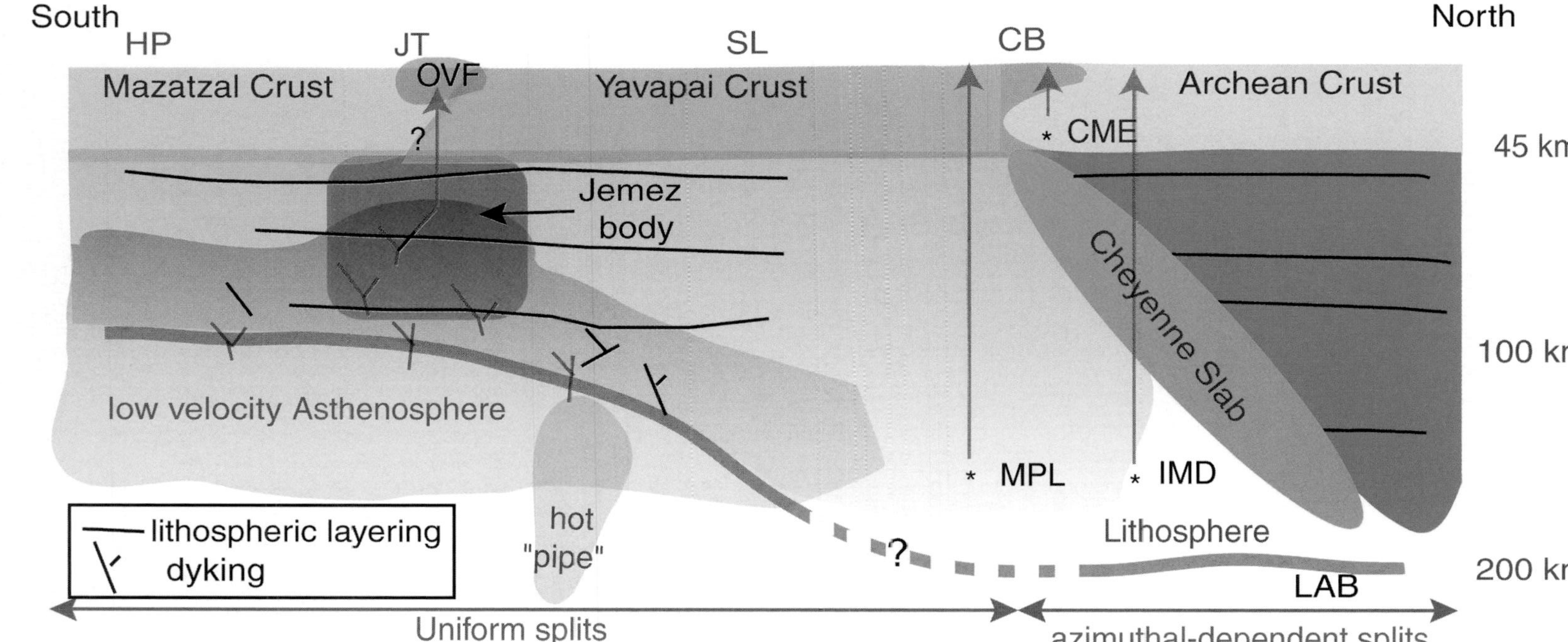

Plate 4. Cartoon of the southern Rocky Mountains lithosphere beneath the CD-ROM transect. Labels are: HP, High Plains; SL, San Luis basin, and OVF, Ocate volcanic field. The northern extent of the lithosphere/asthenosphere boundary (LAB) is drawn based upon the following structural elements: the depth extent of Cheyenne slab velocity image; receiver function layering [Zurek and Dueker, this volume]; the Stateline kimberlite pipes [Lester and Farmer, 1998]; the Middle Park Lamproites (MPL), and the Cedar Mountain (CM) eclogites and metabasites. Variations in shear wave splitting measurements occur north of the Cheyenne belt, while the shear wave measurements are uniform to the south. Receiver functions [*Zurek and Dueker*, this volume] show imbricated Moho beneath the Cheyenne belt (CB). The southern position of the LAB is based upon the receiver function layering within the Jemez low velocity anomaly to 100 km. The low velocity asthenosphere [Goes and van der Lee, 2002; West et al., 2002] probably was formed after the impact of upwelling thermals, and provides heat input to the southern line lithosphere.

is thermally controlled if the slab is related to the Cheyenne suture formation at 1.78 – 1.75 Ga. Thermal diffusion should have erased any temperature variation within the Cheyenne slab long ago. It is also improbable for the Cheyenne slab to be a young thermal anomaly, e.g., a subducting slab or an on-going delaminating/convectively downwelling lithosphere. The Cheyenne slab is more than 1000 km inboard of the North America and Pacific plate boundary where subduction is on going. The 45° dip and the 80 km width of the Cheyenne slab is inconsistent with the predictions of most delamination models and the delamination anomalies proposed in other studies [*Houseman and Molnar*, 1997; *Schott and Schmelling*, 1998].

Our synthetic tomographic modeling (Plate 3) demonstrates a possible combined chemical and anisotropic origin for the Cheyenne slab velocity anomaly. Seismic properties of this model are consistent with the isotropically high velocity eclogitic oceanic crust and velocity anisotropy resulting from the sub-crustal olivine lattice preferred orientation (LPO) generally observed in sub-crustal oceanic lithosphere [*Babuska and Cara*, 1991]. Problematic is that the oceanic eclogitic crust is too thin (~8 km) to match an 80 km wide isotropic high velocity anomaly image. Thus, we propose that the Cheyenne slab's sub-crustal lithosphere is also isotropically fast with respect to the surrounding mantle.

An isotropically fast Cheyenne slab may result from the Cheyenne slab being less hydrated compared to its surroundings. This is possibly due to water removal via melting during the formation of the slab's oceanic lithosphere [*Hirth and Kohlstedt*, 1996]. North of the Cheyenne belt, shear wave images show that the cratonic Wyoming lithosphere is low in shear wave velocity [*Goes and van der Lee*, 2002; *Frederiksen et al.*, 2001]. This observation is contrary to conventional wisdom that expects high velocity mantle beneath most Archean lithosphere [*van der Lee and Nolet*, 1997; *Grand*, 1994; *James and Fouch*, 2002; *Freybourger et al.*, 2001]. The low shear wave velocities seem anomalous given the low reduced heat flow of ~27 mWm^{-2} [*Decker et al.*, 1988] of the non-volcanic portions of the Wyoming Province. The Archean lower crust xenoliths found in the Leucite Hills volcanic field contain an abundance of hydrous minerals (amphibole, biotite) [*Farmer et al.*, this volume]. If hydration occurred via a volatile flux from the mantle, it would be reasonable to assume that the Wyoming mantle lithosphere is also hydrated. South of the Cheyenne belt, post-Miocene volcanism is occurring around Steamboat Springs, Colorado, where upper mantle P- and S-velocities are low in an absolute sense [*Deep Probe working group*, 1998; *Goes and van der Lee*, 2002]. Lamproite lavas found around Steamboat Springs and Middle Park, Colorado [*Thompson et al.*, 1997] indicate a thick (150 –200 km) and hydrated lithosphere south of the Cheyenne belt. Overall, it appears plausible that a dry Cheyenne slab is isotropically fast with respect to the wetter surrounding lithospheric mantle.

Our interpretation of the high velocity anomaly as the fossil image of the north dipping Cheyenne slab is also supported by sub-crustal Eocene eruption aged eclogites and metabasites found near Cedar Mountain, Wyoming [*Kuehner and Irving*, 1999]. Cedar Mountain is located ~250 km to the west of where our seismic line crosses the Cheyenne belt (Figure 1). The geotherms calculated from the mineral assemblages suggest these rocks were erupted from 50 – 80 km depths. While it is difficult to constrain the emplacement history of these ultra-mafic xenoliths (i.e., fossil oceanic slab versus magmatic eclogite), this xenolith assemblage is consistent with emplacement of metamorphosed oceanic crust [*Kuehner and Irving*, 1999].

Shear wave splitting and receiver function studies from the CD-ROM experiment [*Fox and Sheehan*, this volume; *Zurek and Dueker*, this volume] provide more support for the north dipping Cheyenne slab anomaly. The splitting fast axis change from approximately parallel to the absolute North American plate motion direction beneath the south line to back-azimuthal-dependent north of the Cheyenne belt [*Fox and Sheehan*, this volume]. The forward modeling favors either a two-layer anisotropic domain or a dipping fast axis north of the Cheyenne belt. The latter is consistent with our anisotropic modeling. Receiver function analysis for the CD-ROM north line [*Zurek and Dueker*, this volume] shows lithospheric layering to ~200 km in the Archean upper mantle. Interestingly, this layering is truncated near the top of the Cheyenne slab [Plate 5 in Zurek and Dueker, this volume], and the layering is non-existent within the slab region. The slab is steeply dipping and therefore cannot be imaged with our receiver function imaging technique, but this truncation of mantle layering would be consistent with emplacement of a slab against the Wyoming cratonic margin.

5.2 Tectonic Model for Emplacement of Cheyenne Slab

Between the 2.1 Ga age of passive rift margin formation and the 1.78 Ga accretion of the Green Mountain arc, subduction polarity was south directed (i.e., outboard of the Wyoming craton) as evidenced by the lack of 2.1–1.78 Ga arc magmatic rocks north of the Cheyenne belt [*Karlstrom and Houston*, 1984]. If this scenario is true, then the emplacement of our north dipping Cheyenne slab demands a flip in subduction polarity post 1.78 Ga [*Dueker et al.*, 2001; *Tyson et al.*, 2002].

A previous tectonic model to explain the origin of the Cheyenne slab [*Tyson et al.*, 2002; *CD-ROM Working Group*, 2002] suggests that the Cheyenne slab resulted from underthrusting of a ~200 km wide segment of the 2.1 Ga passive margin oceanic crust immediately offshore of the Wyoming passive margin. The south-directed subduction that formed the Green Mountain arc stopped at 1.78 Ga when the arc accreted to the Wyoming margin. Then, the remaining 2.1 Ga lithosphere was underthrust northward to emplace the Cheyenne slab. A concern with the Tyson model is why the 200 km oceanic lithosphere immediately south of the Wyoming passive margin would subduct beneath the Wyoming craton while the Green Mountain arc was still 200 km offshore. In the reference frame of the Wyoming craton, the force driving the closing of the ocean basin between the Green Mountain arc and the Wyoming craton is most likely the rollback of the subducting slab. If true, it is unclear how this rollback force would instigate the subduction of the last 200 km of passive margin oceanic lithosphere.

An alternate tectonic model is that the Cheyenne slab is subducted back-arc basin lithosphere, which formed in between the Green Mountain arc and the 1.74 Ga Rawah arc accretionary event (Figure 4). Our model begins with the Green Mountain and the Rawah arcs accreted to the Wyoming craton via south dipping subduction around 1.78 Ga and 1.74 Ga, respectively (Figure 4a). After the two accretion events, a back-arc basin opened between the two arcs, due to northward-directed subduction (Figure 4b and 4c). This back-arc basin subsequently "collapsed" via a short episode of north-directed subduction beneath the Wyoming margin (Figure 4d). The small amount of subduction (~150 km) could result in no arc magmatism beneath the Green Mountain arc. Subduction flip and lack of related magmatism are commonly seen at other cratonic margins [*Snyder et al.*, 1996; *Teng et al.*, 2000; *Snyder*, 2002]. The ongoing collision at the North Banda Sea of the Banda arc to Australia is a modern-day example of an arc-continent collision in the process of flipping subduction polarity [*Snyder et al.*, 1996].

5.3 The Jemez Body

Given the lack of significant crustal dilatation along the Jemez suture zone, a model whereby the asthenosphere is passively pulled-up via lithospheric extension can be ruled out. Geodynamic models to explain the origin of the low velocity Jemez body thus can be separated into end-member asthenospheric and lithospheric controlled models [*Dueker et al.*, 2001]. While reviewing each model, our conclusion is that end-member models alone are not fully consistent with the seismic petrologic and geochemical observations in the study area. An interaction between the lithosphere and asthenosphere seems mostly required to explain the Jemez velocity anomaly.

An asthenospheric controlled model suggests the Jemez velocity anomaly resulting from active upwelling from upper mantle convective rolls, and/or the impact and spreading of upwelling thermals (plume). In the regional P-wave tomographic images of the western U.S [*Dueker et al.*, 2001], one of three northeast oriented low velocity trends spatially co-exists with the Jemez lineament. The alignment of these low velocity bodies with the current plate motion, and the

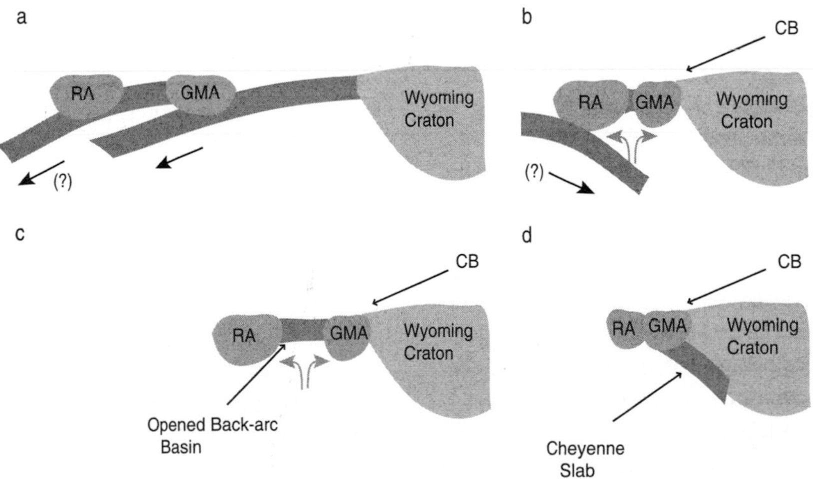

Figure 4. Cartoon of subduction polarity flip to emplace the Cheyenne slab. (a) Pre- 1.78 Ga while the Green Mountain Arc (GMA) and the Rawah Arc (RA) were being formed offshore via south dipping subduction. Arrows show the subduction direction. The Cheyenne belt (CB) formed at ~1.78 Ga after the Green Mountain Arc accreted to the Wyoming passive margin. The accretion of the Rawah arc occurred at 1.74 Ga [Tyson et al., 2002]. (b) After the Rawah Arc accreted to the Green Mountain Arc, a back-arc basin is opened, due to later north-directed subduction. (c) The opening of the back-arc basin formed a ~200 km wide section of oceanic lithosphere. (d) The closure of the basin resulted via north-directed subduction emplaced the Cheyenne slab against the Archean mantle keel.

400–500 km spacing of these anomalies indicate that they may manifest the upwelling limbs of upper mantle convective rolls [*Richter*, 1973; *Schmelling*, 1985; *Dueker et al.*, 2001]. Upwelling thermals, as indicated by the south dipping low velocity pipe in our S-wave image, can diapiricly invade and eventually move aside the lithosphere. In the convecting asthenospheric flows, it is difficult to remain any pre-existing sharp mantle layering due to the presence of chemical layers [*Griffin et al.*, 1999], juxtaposed flat slabs [*Bostock*, 1998], or mantle shear zones [*Levin and Park*, 2000]. Nevertheless, *Zurek and Dueker* [this volume] observed strong seismic layering co-residing with the Jemez low velocity body beneath the south line. The coexistence of the Jemez velocity anomaly with the sharp seismic layering thus strongly suggests that the Jemez anomaly is unlikely to be upwelling asthenosphere.

A lithospheric-controlled model requires that the Jemez suture contains lower-solidus materials (fertile and/or hydrated rocks in a fossil subduction zone) with respect to its surrounding mantle, and these low-solidus rocks preferentially melt during a regional heating event [Dueker et al., 2001]. Low solidus mantle rocks have been reported from late Cenozoic aged mantle-derived magmas at the Navajo Volcanic Field in the Four-Corner region [*Carlson and Nowell*, 2001] that straddles a proposed suture zone and the northern boundary of the Proterozoic Mazatzal province [*Condie*, 1992]. It is possible that the hydrated materials were preserved in the Jemez suture from its formation. The most favorable evidence for the Jemez velocity anomaly being lithospheric comes from the lithospheric mantle layering [*Zurek and Dueker*, this volume] and the sharp lateral velocity gradient in our tomographic images. A disadvantage of the lithospheric controlled model is that the low solidus rocks need heat input to melt at this moment in time. If early heat events initiated the melting process within the Jemez suture, the low-solidus rocks would be "depleted" long ago via melt removal.

Given that Jemez volcanism is young (the Ocate Volcanic Field above our low velocity anomaly is ~6–8 Ma old [*Wood and Kienle*, 1992]), it would be more reasonable to suggest that the Jemez low velocity anomaly originates from a coupled system in which the low solidus lithosphere preferentially melts in response to rising thermal currents (i.e., the low velocity pipe in out S-wave image) beneath southern Colorado and New Mexico in late Cenozoic time [*Dueker et al.*, 2001]. The impact of upwelling asthenosphere agrees with the low surface wave velocity (4.1–4.3 km/s) imaged at 100–150 km depth in this region [*Goes and van der Lee*, 2002; *West et al.*, 2002; *Godey et al.*, 2003]. In addition, late Cenozoic uplift [*Eaton*, 1982; *Heller et al.*, 2003] is also consistent with emplacement of warm buoyant mantle beneath the lithosphere.

When ascending thermals impinge upon the lithosphere, flow-induced olivine LPO aligns olivine's fast velocity axis. As a result, the shear wave splits are expected to be in complicated patterns [*Rümpker and Silver*, 2000; *Savage and Sheehan*, 2000; *Park and Levin*, 2002]. Non-uniform splitting parameters have been reported in regions thought to be undergoing small-scale convection [e.g., *Gao et al.*, 2003]. However, beneath the CD-ROM south line [*Fox and Sheehan*, this volume] and the RISTRA transects [*Gok et al*, 2003], no sharp variations in the splitting parameters are observed. Given the generally poor back-azimuth sampling of the shear wave splitting results, it seems that the constraints provided from shear wave studies on lithospheric and asthenospheric fabric beneath the CD-ROM south line are rather non-unique.

A concern with our model is the heat conduction would be too slow [*Turcotte and Schubert*, 1982] to form the 60 km thick Jemez mantle anomaly. Advective heat transfer must have enhanced the heat transport from the asthenosphere into the lithosphere. A similar heat transport problem occurs beneath Hawaii where a broad lithospheric low P-wave velocity region (up to 5%) is observed in the oceanic lithosphere between 40 and 80 km depth [*Tilmann et al.*, 2001]. Calculation shows that conductive heating by one primary conduit cannot explain the broad elongated low velocity zone. Complicated melt "pathways" are thus suggested to advect heat into a much wider region to create the broad velocity anomaly [*Tilmann et al.*, 2001]. This scenario may also apply to the Jemez suture lithosphere. Asthenospheric melts are transferred into the lithosphere via extensive dyking, which significantly enhances the advective heat transport into the lithosphere, and enhances melting of low solidus minerals trapped in the Jemez suture zone.

Another puzzle with the shear wave splitting observation is the lack of variation (both azimuthal and split time) in the splitting parameters across the partially molten Jemez body. Studies of the re-heated and slightly molten peridotite samples from the Ronda massif (Spain) show that the olivine LPO remains unperturbed even after the heating and partial melting event [*Vauchez and Carlos*, 2001]. The Jemez body may represent a region where a small amount of the melt (i.e., 0.5 %) is present, which would cause large velocity reductions [*Hammond and Humphreys*, 2000], yet still retain an undisturbed olivine LPO.

6. CONCLUSION

The CD-ROM passive seismic results suggest that fossil lithospheric heterogeneities have been preserved beneath the two primary suture zones in the southern Rocky Mountains lithosphere (Plate 4). A high velocity Proterozoic slab fragment exists beneath the Cheyenne belt. The slab is high velocity probably due to some combination of eclogitic facies oceanic crust, frozen-in oceanic olivine LPO, and/or the juxtaposition of this fossil slab against hydrated mantle. A low velocity Jemez body resides beneath the Jemez suture. The low velocity anomaly most probably represents low solidus materials preserved in the Jemez suture lithosphere, and is now molten due to a late Cenozoic regional heating event. These observations suggest that a thick chemical lithosphere is present in the Rock Mountain regions, down to 200 km beneath the Cheyenne belt, and 100 km under the Jemez suture.

Acknowledgments. We thank Matthew Fouch, Aibing Li and Michael West for valuable comments and help for improving the manuscript. We also thank the CD-ROM group led by Karl Karlstrom, the IRIS PASSCAL Instrument Center, and the NSF Continental Dynamics Program. The work was funded by NSF Continental Dynamics grant 0614410.

REFERENCE

Abramovitz, T., and H. Thybo, Seismic images of Caledonian, lithosphere-scale collision structures in the southeaster North Sea along MONA LISA Profile 2, *Tectonophysics*, *317*, 27–54, 2000.

Aldrich, M.J., and A.W. Laughlin, A model for the tectonic development of the southeastern Colorado Plateau boundary, *J. Geophys. Res.*, *89*, 10207–10218, 1984.

Allen, R.M., G. Nolet, W.J. Morgan, K. Vogfiord, B.H. Bergsson, P. Erlendsson, G.R. Foulger, S. Jakobsdottir, B.R. Julian, M. Pritchard, S. Ragnarsson, and R. Stefansson, Imaging the mantle beneath Iceland using integrated seismological techniques, *J. Geophys. Res.*, *107*, 2001JB000595, 2002.

Armstrong, R.L., and P. Ward, Evolving geographic patterns of Cenozoic magmatism in the North American cordillera: the temporal and spatial association of magmatism and metamorphic core complexes, *J. Geophys. Res.*, *34*, 149–164, 1991.

Babuska, V., and M. Cara, *Seismic anisotropy in the Earth*, Kluwer Academic, Dordrecht, 1991.

Baldridge, W.S., G.R. Keller, V. Haak, E. Wendlandt, G.R. Jiracek, and K.H. Olsen, *The Rio Grande Rift*, Elsevier, 1995.

Balling, N., Deep seismic reflection evidence for ancient subduction and collision zones within the continental lithosphere of northwestern Europe, *Tectonophysics*, *329*, 269–300, 2000.

Bostock, M.G., Mantle stratigraphy and evolution of the Slave province, *J. Geophys. Res.*, *103*, 21183–21200, 1998.

Carlson, R.W., and G.M. Nowell, Olivine-poor sources for mantle-derived magmas: O_s and H_f isotopic evidence from potassic magmas of the Colorado Plateau, *Geochem. Geophys. Geosyst.*, *2*, 2000GC000128, 2001.

CD-ROM Working Group, Structure and evolution of the lithosphere beneath the Rocky Mountains: initial results from the CD-ROM experiment, *GSA Today*, *10*, 4–10, 2002.

Chamberlain, K.R., Medicine Bow orogeny: Timing of deformation and model of crustal structure produced during continent-arc collision, ca 1.78 Ga, southeastern Wyoming, *Rocky Mountain Geology*, *33*, 259–277, 1998.

Chapin, C.E., Evolution of the Rio Grande rift - A summary, in *Rio Grande Rift: Tectonics and Magmatism*, edited by R.E. Riecker, American Geophysical Union, Washington D.C, 1979.

Condie, K.C., Proterozoic terranes and continental accretion in southwestern North America, in *Proterozoic crustal evolution*, edited by K.C. Condie, pp. 447–480, Elsevier, 1992.

Decker, E.R., H.P. Heasler, K.L. Buelow, K.H. Baker, and J.S. Hallin, Significance of past and recent heat flow and radioactivity studies in the southern Rocky Mountains region, *Geological Society of America Bulletin*, *100*, 1851–1885, 1988.

Deep Probe working group, Probing the Archean and Proterozoic lithosphere of western North America, *GSA Today*, *8*, 1–7, 1998.

Dueker, K., E.D. Humphreys, and G. Biasi, Teleseismic imaging of the western United States upper mantle structure using the simultaneous iterative reconstruction technique, in *Seismic Tomography*, edited by K. Hirahara, pp. 265–298, Chapman & Hall, London, 1993.

Dueker, K., H. Yuan, and B. Zurek, Thick Proterozoic lithosphere of the Rocky Mountain region, *GSA Today*, *11*, 4–9, 2001.

Dziewonski, A.M., and D.L. Anderson, Travel-time and station corrections for P-waves at teleseismic distances, *J. Geophys. Res.*, *88*, 722–743, 1983.

Eaton, D.W., and J.F. Cassidy, A relic Proterozoic subduction zone in western Canada: New evidence from seismic reflection and receiver function data, *Geophys. Res. Lett.*, *23*, 3791–3794, 1996.

Eaton, G.P., The basin and Range province: Origin and tectonic significance, *Annual Review of earth and Planetary Sciences*, *10*, 409–440, 1982.

Farmer, L., S. Bowring, N. Christensen, M. Williams, J. Matzel, and L. Stevens, Contrasting lower crustal evolution across an Archean-Proterozoic suture: Physical, Chemical and Geochromologic studies of lower crustal xenoliths in southern Wyoming and northern Colorado, this volume.

Fox, O., and A.F. Sheehan, Shear wave splitting beneath the CD-ROM transects, this volume.

Frederiksen, A.W., M.G. Bostock, and J.F. Cassidy, S-wave velocity structure of the Canadian upper mantle, *Physics of the Earth and Planetary Interiors*, *124*, 175–191, 2001.

Freybourger, M.J., J.B. Gaherty, and T. Jordan, Structure of the Kaapvaal craton from surface waves, *Geophys. Res. Lett.*, *28*, 2489–2492, 2001.

Gao, S.S., K.H. Liu, P.M. Davis, P.D. Slack, Y.A. Zorin, V.V. Mordvinova, and V.M. Kozhevnikov, Evidence for small-scale mantle convection in the upper mantle beneath the Baikal rift zone, *J. Geophys. Res.*, *108*, 10.1029/2002JB002039, 2003.

Godey, S., R. Snieder, A. Villasenor, and H.M. Benz, Surface wave tomography of North America and the Caribbean using global and regional broad-band networks: Phase velocity maps and limitations of ray theory, *Geophys. J. Int.*, *152*, 620–632, 2003.

Goes, S., and S. van der Lee, Thermal structure of the North American uppermost mantle, *J. Geophys. Res.*, *107*, 2000JB000049, 2002.

Gorman, A.R., R.M. Clowes, R.M. Ellis, T.J. Henstock, G.D. Spence, G.R. Keller, A. Levander, C.M. Snelson, M.J.A. Burianyk, E.R. Kanasewich, I. Asudeh, Z. Hajnal, and K.C. Miller, Deep Probe: Imaging the roots of western North America, *Canadian Journal of Earth Sciences*, *39*, 375–398, 2002.

Grand, S., Mantle shear structure beneath the Americas and surrounding oceans, *J. Geophys. Res.*, *99*, 11591–11621, 1994.

Grand, S., and D.V. Helmberger, Upper mantle shear structure of North America, *Geophys. J. R. astr. Soc.*, *76*, 399–438, 1984.

Griffin, W.L., S.Y. O'Reilly, R. Davies, K. Kivi, E. van Achterbergh, L.M. Natapov, B.J. Doyle, C.G. Ryan, and N.J. Pearson, Layered mantle lithosphere in the Lac de Gras area, Slave craton: Composition, structure and origin, *Journal of Petrology*, *40*, 705–727, 1999.

Hamilton, W., and W.B. Myers, Cenozoic tectonics of the western United States, *Reviews of Geophysics*, *4*, 509–549, 1966.

Hammond, W.C., and E.D. Humphreys, Upper mantle seismic wave velocity: effects of realistic partial melt geometries, *J. Geophys. Res.*, *105*, 10975–10986, 2000.

Heller, P.L., K. Dueker, and M.E. McMillan, Post-Paleozoic alluvial gravel transport as evidence of continental tilting in the U.S. Cordillear, *Geological Society of America Bulletin*, *115*, 1122–1132, 2003.

Hirth, G., and D.L. Kohlstedt, Water in the oceanic upper mantle: Implications for rheology, melt extraction and the evolution of lithosphere, *Earth Planet. Sci. Lett.*, *144*, 93–108, 1996.

Houseman, G.A., and P. Molnar, Gravitational (Rayleigh-Taylor) instability of a layer with non-linear viscosity and convective thinning of continental lithosphere, *Geophysical Journal International*, *128*, 125–150, 1997.

Ingersoll, R.V., W. Cavazza, W.S. Baldridge, and M. Shafiqullah, Cenozoic sedimentation and paleotectonics of north-central New Mexico: Implications for initiation and evolution of the Rio Grande rift, *Geological Society of America Bulletin*, *102*, 1280–1296, 1990.

James, D.E., and M.J. Fouch, Formation and evolution of Archaean cratons: insights from southern Africa, in *The Early Earth: Physical, Chemical and biological Development*, edited by C.J. Hawkesworth, pp. 1–26, Spec. Pub., London, 2002.

Judenherc, S., M. Granet, J.P. Brun, G. Poupinet, J. Plomerova, A. Mocquet, and U. Achauer, Images of lithospheric heterogeneities in the armorican segment of the Hercynian Range in France, *Tectonophysics*, *358*, 121–134, 2002.

Karato, S., Importance of anelasticity in the interpretation of seismic tomography, *Geophys. Res. Lett.*, *20*, 1623–1626, 1993.

Karato, S., and H. Jung, Water, partial melting and the origin of the seismic low velocity and high attenuation zone in the upper mantle, *Earth Planet. Sci. Lett.*, *157*, 193–207, 1998.

Karlstrom, K.E., and S.A. Bowring, Early Proterozoic assembly of tectonostratigraphic terranes in southwestern North America, *Journal of Geology*, *96*, 561–576, 1988.

Karlstrom, K.E., and R.S. Houston, The Cheyenne belt: analysis of a Proterozoic suture in southern Wyoming, *Precambrian research*, *25*, 415–446, 1984.

Karlstrom, K.E., and E.D. Humphreys, Persistent influence of Proterozoic accretionary boundaries in the tectonic evolution of southwestern North America: Interaction of cratonic grain and mantle modification events, *Rocky Mountain Geology*, *33*, 161–179, 1998.

Kennett, B.L.N., and E.R. Engdahl, Traveltimes for global earthquake location and phase identification, *Geophys. J. Int.*, *105*, 429–465, 1991.

Kluth, C.F., and P.J. Coney, Plate tectonics of the Ancestral Rocky Mountains, *Geology*, *9*, 10–15, 1981.

Kuehner, S.M., and A.J. Irving, Eclogite and metabasite xenoliths of subducted slab origin from the Paleogene Cedar Mountain diatremes, southwestern Wyoming, USA, *Proceedings of the International Kimberlite Conference*, *1*, 485–493, 1999.

Levin, V., and J. Park, Shear zones in the Proterozoic lithosphere of the Arabian shield and the nature of the Hales discontinuity, *Tectonophysics*, *323*,131–148, 2000.

Magnani, M.B. and A. Levander, Listric thrust faulting in the Laramide front of north-central New Mexico guided by Precambrian basement anisotropies, this volume.

Meyerholtz, K.A., G.L. Pavlis, and S.A. Szpakowski, Convolutional quelling in seismic tomography, *Geophysics*, *54*, 570–580, 1989.

Morgan, R.P.L., P.J. Barton, M. Warner, J. Morgan, C. Price, and K. Jones, Lithospheric structure north of Scotland-I.P-wave modeling, deep reflection profiles and gravity, *Geophys. J. Int.*, *142*, 716–736, 2000.

Park, J., and V. Levin, Seismic anisotropy: Tracing plate dynamics in the mantle, *Science*, *296*, 485–489, 2002.

Pharaoh, T.C., Paleozoic terranes and their lithospheric boundaries within the Trans-European suture zone (TESZ): a Review, *Tectonophysics*, *314*, 17–41, 1999.

Poupinet, G., N. Arndt, and P. Vacher, Seismic tomography beneath stable tectonic regions and the origin and composition of the continental lithospheric mantle, *Earth Planet. Sci. Lett.*, *212*, 89–101, 2003.

Rümpker, G., and P. G. Silver, Calculating splitting parameters for plume-type anisotropic structures of the upper mantle, *Geophys. J. Int. 143,* 507–520, 2000

Richter, F.M., Dynamical models for sea floor spreading, *Reviews of Geophysics and Space Physics*, *11*, 223–287, 1973.

Sandoval, S., E. Kissling, J. Ansorge, and the SVEKALAPKO Seismic Tomography Working Group, High-resolution body wave tomography beneath the SVEKALAPLO array: II. Anomalous upper mantle structure beneath the central Baltic Shield, *Geophys. J. Int.*, *submitted*, 2003a.

Sandoval, S., E. Kissling, J. Ansorge, and "the SVEKALAPKO Seismic Tomography Working Group", High-resolution body wave tomography beneath the SVEKALAPLO array: I. A priori three-dimensional crustal model and associated traveltime effects on teleseismic wave fronts, *Geophys. J. Int.*, *153*, 75–87, 2003b.

Saruwatari, K., S. Ji, C. Long, and M. Salisbury, Seismic anisotropy of mantle xenoliths and constraints on upper mantle structure beneath the southern Canadian Cordillera, *Tectonophysics*, *339*, 403–426, 2001.

Savage, M.K., and A.F. Sheehan, Seismic anisotropy and mantle flow from the Great Basin to the Great Plains, western United States, *J. Geophys. Res.*, *105*, 13715–13734, 2000.

Schmelling, H., Numerical models on the influence of partial melt on elastic, anelastic and electric properties of rocks, Part I: elasticity and anelasticity, *Physics of the Earth and Planetary Interiors*, *41*, 34–57, 1985.

Schott, B., and H. Schmelling, Delamination and detachment of a lithospheric root, *Tectonophysics*, *296*, 225–247, 1998.

Schutt, D.L., and E.D. Humphreys, P and S wave velocity and Vp/Vs in the wake of the Yellowstone Hotspot, *J. Geophys. Res.*, *in press*.

Shaw, C.A., and K.E. Karlstrom, The Yavapai-Mazatzal crustal boundary in the southern Rocky Mountains, *Rocky Mountain Geology*, *34*, 37–52, 1999.

Shearer, P.M., *Introduction to Seismology*, Cambridge University Press, New York, 1999.

Shomali, Z.H., R.G. Roberts, and the TOR Working Group, Non-linear body wave teleseismic tomography along the TOR array, *Geophys. J. Int.*, *148*, 562–574, 2002.

Shragge, J., M.G. Bostock, C.G. Bank, and R.M. Ellis, Integrated teleseismic studies of the southern Alberta upper mantle, *Canadian Journal of Earth Sciences*, *39*, 399–411, 2002.

Snelson, C.M., H.M. Rumpel, G.R. Keller, K.C. Miller, and C. Prodehl, Regional crustal structure derived from the CD-ROM seismic refraction-/wide-angle reflection experiment: The lower crustal and upper mantle, this volume.

Snyder, D.B., Lithospheric growth at margins of cratons, *Tectonophysics*, *355*, 7–22, 2002.

Snyder, D.B., H. Prasetyo, Blundell, D. J., C.J. Pigram, A.J. Barber, A. Richardson, and S. Tjokosaproetro, A dual doubly-vergent orogen in the Banda Arc continent-arc collision zone as observed on deep seismic reflection profiles, *Tectonics*, *15*, 34–53, 1996.

Sobolev, S.V., A. Gresillaud, and M. Cara, How robust is isotropic delay time tomography for anisotropic mantle?, *Geophys. Res. Lett.*, *24*, 509–512, 1999.

Soedjatmiko, B., and N.I. Christensen, Seismic anisotropy under extended crust: evidence from upper mantle xenoliths, Cima volcanic field, California, *Tectonophysics*, *321*, 279–296, 2000.

Teng, L., S., C.T. Lee, Y.B. Tsai, and L.Y. Hsiao, Slab breakoff as a mechanism for flipping of subduction polarity in Taiwan, *Geology*, *28*, 155–158, 2000.

Thompson, R.N., J.G. Mitchell, A.P. Dickin, S.A. Gibson, D. Velde, P.T. Leat, and M.A. Morrison, Oligocene lamproites containing an Al-poor, Ti-rich biotite, Middle Park, northwest Colorado, USA, *Mineralogical Magazine*, *61*, 557–572, 1997.

Tilmann, F.J., H.M. Benz, K.F. Priestley, and P.G. Okubo, P wave velocity structure of the uppermost mantle beneath Hawaii from travel time tomography, *Geophys. J. Int.*, *146*, 594–606, 2001.

Turcotte, D., and G. Schubert, *Geodynamics: Applications of continuum physics to geological problems*, 450 pp., John Wiley and Sons, New York, 1982.

Tyson, E., K.E. Morozova, K.E. Karlstrom, K.R. Chamberlain, S.B. Smithson, K.G. Dueker, and C.T. Foster, Proterozoic Farwell Mountain-Lester Mountain suture zone, northern Colorado: Subduction flip and progressive assembly of arcs, *Geology*, *30*, 943–946, 2002.

van der Lee, S., and G. Nolet, Upper mantle S velocity structure of North America, *J. Geophys. Res.*, *102*, 22815–22838, 1997.

VanDecar, J.C., and R.S. Crosson, Determination of teleseismic relative phase arrival times using multi-channel cross correlation and least squares, *Bull. Seismol. Soc. Am.*, *80*, 150–159, 1990.

Vauchez, A., and J.G. Carlos, Seismic properties of an asthenospherized lithospheric mantle: constraints from lattice preferred orientations in peridotite from the Ronda Massif, *Earth Planet. Sci. Lett.*, *192*, 235–249, 2001.

Warner, M., J. Morgan, P. Barton, C. Price, and K. Jones, Seismic reflections from the mantle represent relict subduction zones within the continental lithosphere, *Geology*, *24*, 39–42, 1996.

Warren, L.M., and P.M. Shearer, Investigating the frequency dependence of mantle Q by stacking P and PP spectra, *J. Geophys. Res.*, *105*, 25391–25402, 2000.

West, M., J. Ni, D. Wilson, R. Aster, S. Grand, W. Gao, R. Gok, S. Baldridge, S. Semken, and J. Schlue, Structure of the uppermost mantle beneath the

RISTRA array from surface waves, *AGU 2002 Fall Meet. Suppl., Abstract S61A-1116*, 2002.

White, D.J., S.B. Luca, W. Bleeker, Z. Hajnal, J.F. Lewry, and H.V. Zwanzig, Suture-zone geometry along an irregular Paleoproterozoic margin: The Superior boundary zone, Manitoba, Canada, *Geology*, *30*, 735–738, 2002.

Williams, M.L., K.E. Karlstrom, A. Lanzirotti, A.S. Read, J.L. Bishop, C.E. Lombardi, J.N. Pedrick, and M.B. Wingsted, New Mexico middle crustal cross sections: 1.65 Ga macroscopic geometry, 1.4 Ga thermal structure and continued problems in understanding crustal evolution, *Rocky Mountain Geology*, *34*, 53–66, 1999.

Wood, C.A., and J. Kienle, *Volcanoes of North America: United States and Canada*, 354 pp., Cambridge University Press, 1992.

Wooden, J.L., and E. DeWitt, Pb isotopic evidence for a major early crustal boundary in western Arizona, in *Proterozoic geology and ore deposits of Arizona*, edited by K.E. Karlstrom, Arizona Geological Society Digest, 1991.

Zurek, B., and K. Dueker, Images of lithospheric stratigraphy beneath the CD-ROM lines, this volume.

K. G. Dueker, Department of Geology & Geophysics, University of Wyoming,, Laramie, WY 82071, USA.

H. Yuan, Department of Geology & Geophysics, University of Wyoming,, Laramie, WY 82071, USA.

Upper Mantle Anisotropy Beneath Precambrian Province Boundaries, Southern Rocky Mountains

Otina C. Fox[1] and Anne F. Sheehan

University of Colorado at Boulder, Boulder Colorado

Teleseismic shear wave splitting is used to estimate mantle anisotropy beneath Precambrian province boundaries in the southern Rocky Mountains of Wyoming, Colorado, and New Mexico. Data from the passive seismic experiment of the Continental Dynamics of the Rocky Mountains (CD-ROM) project and Laramie Seismic Array are used. Data from the south CD-ROM line in New Mexico show consistent northeast fast directed shear-wave splitting, correlating with absolute plate motion and parallel to the fabric of Proterozoic accretion structures. In Wyoming, a clear change in shear-wave splitting parameters is seen across the Cheyenne Belt Archean/Proterozoic suture zone. South of the suture zone, northeast directed fast directions are observed. North of the Cheyenne Belt, shear wave splitting parameters vary with backazimuth, with the same backazimuthal patterns observed with the CD-ROM and Laramie arrays, despite the arrays being separated by over 100 km distance. This similarity in backazimuthal variation suggests a common anisotropic structure beneath both regions rather than lateral variations in anisotropy. The azimuthal variations of both the north CD-ROM and Laramie arrays are best modeled with a plunging axis of anisotropy. Our preferred model of anisotropy has an axis of symmetry that plunges steeply to the northwest. CD-ROM tomography has shown a steeply plunging fast velocity anomaly that has been forward modeled as an anisotropic slab. This feature may also be the source of the observed pattern of shear wave splitting.

INTRODUCTION

The western United States has undergone multiple deformational events that have made modifications to the original Archean and Proterozoic lithospheric structure. The Continental Dynamics of the Rocky Mountains (CD-ROM) experiment was designed to explore whether one can image remnants of the ancient assembly structures preserved in the mantle or in the crust and whether these ancient structures control the style and locus of subsequent deformation [Karlstrom et al., 2002]. Archean and Proterozoic mantle may have been preserved and one can sample this ancient mantle structure through various geophysical and geochemical techniques. An alternative possibility is that the Archean and Proterozoic mantle may have been eroded away by means of more recent processes such as sub-horizontal subduction of the Farallon flat slab [Bird, 1988].

The Archean/Proterozoic suture called the Cheyenne belt in southeastern Wyoming shows zones of deformed rocks and changes in crustal structure across the suture, as inferred from seismic reflection and gravity studies [Johnson et al, 1984].

[1] Now at Alaska Earthquake Information Center, Fairbanks, Alaska.

The Rocky Mountain Region: An Evolving Lithosphere
Geophysical Monograph Series 154

10.1029/154GM26

South of the Cheyenne Belt the general structural trend of arches and folds represents Laramide north-south deformation, while north of the suture zone there is a decrease of Laramide faulting [Ehrlich and Erlsev, 1999]. The suture zone acts as a bounding fault to deformation whose main influence could reside in the lower crust and upper mantle [Ehrlich and Erlsev, 1999]. Strike-slip movements, oblique convergence, or transpression could have caused some of the discontinuities observed across the Cheyenne Belt [Karlstrom and Houston, 1984; Duebendorfer and Houston, 1987]. We seek to explore whether these surface structural trends continue into the lower crust and mantle.

In this paper we use teleseismic shear wave splitting to determine the patterns of mantle anisotropy beneath an Archean/Proterozoic suture zone at the Cheyenne Belt in southern Wyoming and a Proterozoic/Proterozoic suture zone near the Jemez lineament in New Mexico. Mantle anisotropy can be used to infer patterns of mantle flow or the deformation history of the mantle [Tanimoto and Anderson, 1984; Schutt et al., 1998]. Previous shear-wave splitting experiments in this region have shown various results. Results from the Rocky Mountain Front experiment in Colorado mainly show null measurements (absence of shear wave splitting) with a few positive shear wave splitting measurements that do not correlate well with surface geology [Savage et al., 1996; Savage and Sheehan, 2000]. Lodore and Deep Probe experiment (western Colorado and Wyoming) shear wave splitting results have considerable spatial variability [Schutt and Humpheys, 2001], while the nearby Snake River Plain experiment produces remarkably uniform shear wave splitting results consistent with anisotropy created by absolute plate motion or the flattening of the Yellowstone plume [Schutt and Humpheys, 2001; Schutt et al., 1998]. The complexity of the Lodore and Deep Probe shear wave splitting measurements may indicate a more complex anisotropy than a single horizontal layer. In the Rio Grand Rift, the pattern of shear wave splitting has been interpreted in terms of northward-directed asthenospheric flow [Sandvol et al., 1992].

Network Descriptions

The CD-ROM passive source seismic experiment consisted of 47 broadband seismometers deployed in two north-south linear arrays (Table 1; see CDROM in back cover sleeve). The southern line (Figure 1), with 23 seismometers, crosses an inferred Mazatzal-Yavapia Proterozoic suture zone and follows the trend of the Sangre de Cristo Mountains. The northern line, consisting of 24 broadband seismometers, crosses the Cheyenne Belt, an Archean-Proterozioc suture zone near the Colorado-Wyoming border (Figure 2). The seismometers recorded from April 1999 to June 2000. Each CD-ROM seismic station consisted of a Reftek digital data acquisition system that recorded continuously at 10 or 20 samples per second (reduced sample rate during winter months). Fifteen stations had Guralp CMG-3T seismometers; six stations used Guralp CMG-40T seismometers and 27 stations operated with Streckheisn STS-2 sensors.

The Laramie array consisted of 30 broadband seismometers (Table 2; see CDROM in back cover sleeve) at a station spacing of 1.6 kilometers, deployed from June 2000 to May 2001. The line crosses the Cheyenne Belt near Laramie, Wyoming (Figure 2). Each station in the Laramie Array consisted of Guralp 40-T seismometers recording at 1, 40 and 100 samples per second. The data were transferred by Passcal telemetry to the University of Wyoming in real time.

Shear Wave Splitting Measurements

Measurements of mantle anisotropy provide a glimpse of upper mantle structure and possible information on the past history of deformation. Mantle anisotropy is believed to be the result of lattice preferred orientation (LPO) of olivine [Nicolas and Christensen, 1987; Christensen, 1984; Ismail and Mainprice, 1998] due to finite strain. When a shear wave passes through an anisotropic region it will split into a fast and slow shear wave, with delay time between the fast and slow shear waves, δt, and will be polarized in the direction of the first arriving phase, ϕ, or fast polarization direction.

Proposed mechanisms for anisotropy in the mantle include mantle flow associated with absolute plate motion, as seen in oceanic upper mantle and certain continental areas [Tanimoto and Anderson, 1984; Vinnik et al., 1992; Schutt et al, 1998], crustal stress translating into the upper mantle [Helffrich et al 1994], and large-scale lithospheric deformation [Russo et al, 1996; Kay et al 1999]. The first two mechanisms reflect present day processes as the cause of anisotropy while the last would reflect "fossil" anisotropy or crust and mantle deforming coherently [Silver and Chan, 1988; Silver and Chan, 1991; Savage et al, 1990; Russo et al 1996]. Small-scale convection has also been proposed [Sandlov et al., 1992; Savage and Sheehan, 2000]. At shallow crustal depths, other mechanisms may be present, such as alignment of cracks, presence of fluids, fractures and fissures [Crampin, S., 1994; Evans et al, 1996; Leary et al, 1990; Kuo et al, 1994; Levin and Park, 1997]. For a more complete review of anisotropy mechanisms, see the reviews by Savage [1999] and Silver [1996].

SKS waveforms were examined for earthquakes with $M_b >$ 5.7 and epicentral distances > 85°, and S waveforms were analyzed from earthquakes with depths greater than 150 km and epicentral distances of 60°–87° (Table 3 and Table 4; see CDROM in back cover sleeve). The method of Silver and

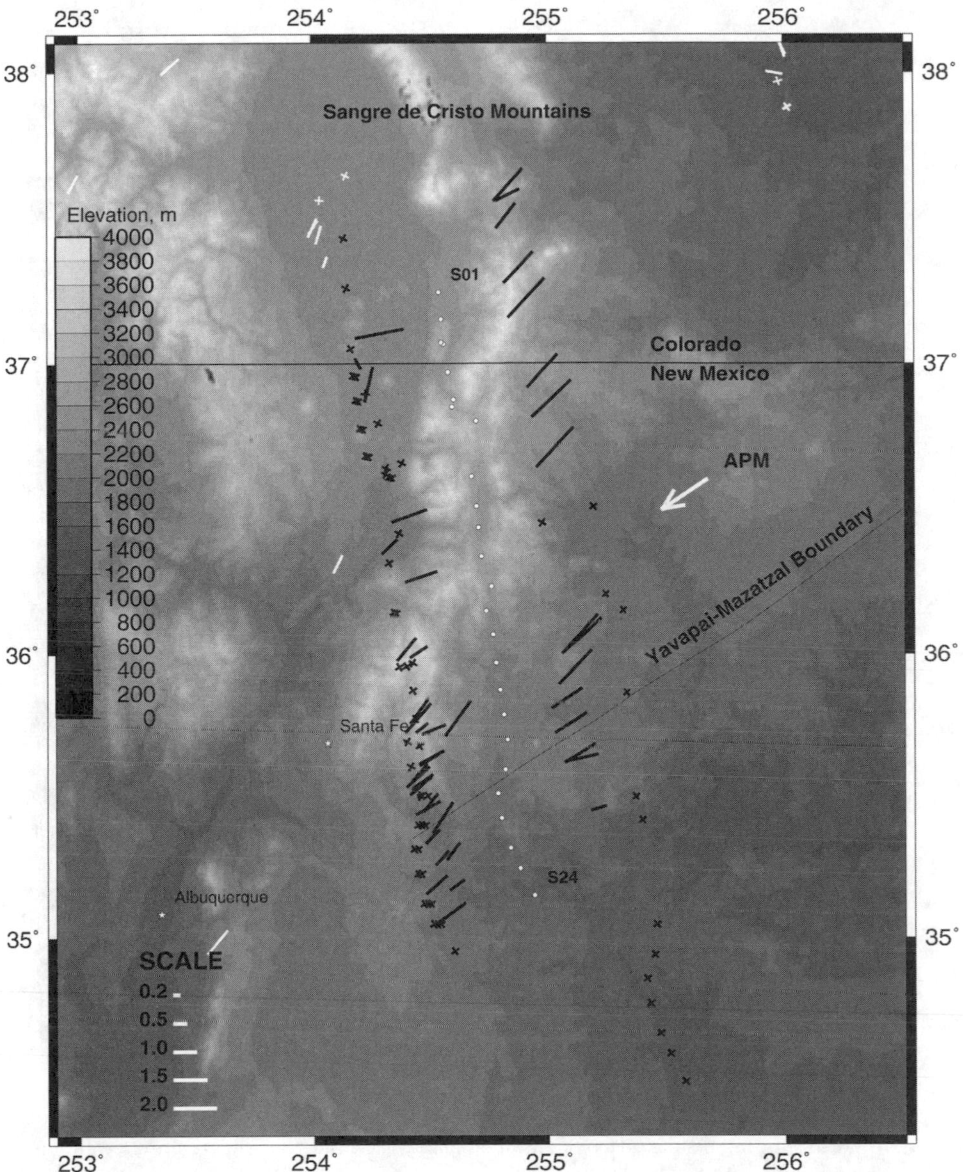

Figure 1. CD-ROM south splitting results on top of elevation map. Black lines represent splitting measurements from this study. Angle from north is ϕ and the length of the line represents δt. Black crosses represent null measurements. White lines and crosses are splitting results from past experiments. Black splits and nulls are plotted to their 220 km piercing points. S01 is in the north and S24 is in the south.

Chan [1991] is used to obtain shear wave splitting parameters. The method uses a grid search to find the values of fast polarization direction ϕ and delay time δt that best represent the splitting.

Individual splitting measurements were rated by the similarity in waveforms upon correction, the amount of reduction in tangential energy, the change from elliptical particle motion to linear particle motion, and the error plot of the grid search. Splitting measurements were rated as "great", "good", "ok", "weak" or "poor" (Table 5 and Table 6; see CDROM in back cover sleeve). In general the "weak" and "ok" splits follow the same pattern as the "good" splitting results in fast direction ϕ (Figure 3a). More scatter is seen in values of δt with the "weak" and "ok" splits (Figure 3b).

Stations with very little or no splitting are termed "null" measurements, and the incoming polarization direction is noted (dotted lines in Figure 5). Nulls are characterized by little to no tangential energy and particle motion that is not lin-

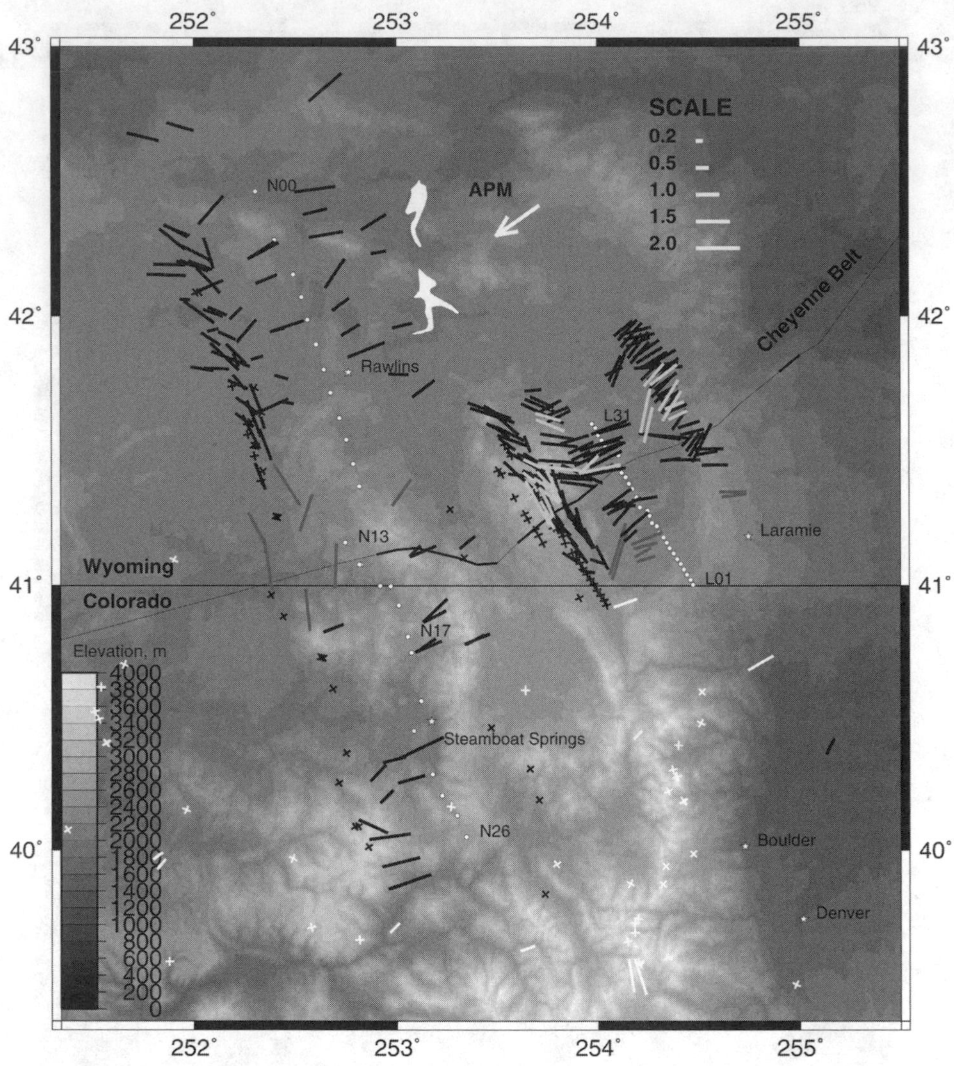

Figure 2. CD-ROM north splitting results on top of elevation map. Black and gray lines represent splitting measurements from CD-ROM and the Laramie Array. Angle from north is ϕ and the length of the line represents δt. Black crosses represent null measurements. White lines and crosses are splitting results from past experiments. Black and grey splits and nulls are plotted to their 220 km piercing points. N00 is in the north, N26 is to the south, and L01 is to the south and L31 is to the north. The Cheyenne Belt is expressed as a thick line where there is surface expression, and as a thin line where the contact is inferred. CD-ROM stations N14-N16 are medium gray. Laramie Array L01 and L02 are medium gray and stations L16, L18 and L22 are light gray.

ear. Null measurements may be due to an absence of anisotropy (isotropic conditions), vertical anisotropy with the a-axis of the olivine crystal aligned in the direction of SKS propagation, or incoming wave polarization along the fast or slow azimuth of ϕ [Savage et al, 1996]. Thus the grouping of results was in terms of positive results (great, good, ok, etc.), null results, and non-results (data neither fall under the positive or null splitting definition).

All events were uniformly filtered with a 4-pole Butterworth bandpass filter from 1–30 seconds. Thirteen events were improved (errors decreased to an acceptable range, as well as improved signal to noise ratio) by a narrower filter from 2–30s and a filter from 3–30s was used for one event. Sensitivity tests with various filters suggest that it is best to use the broadest filter possible for a noisy event. Narrower band filters may help reduce noise, but they also can increase the

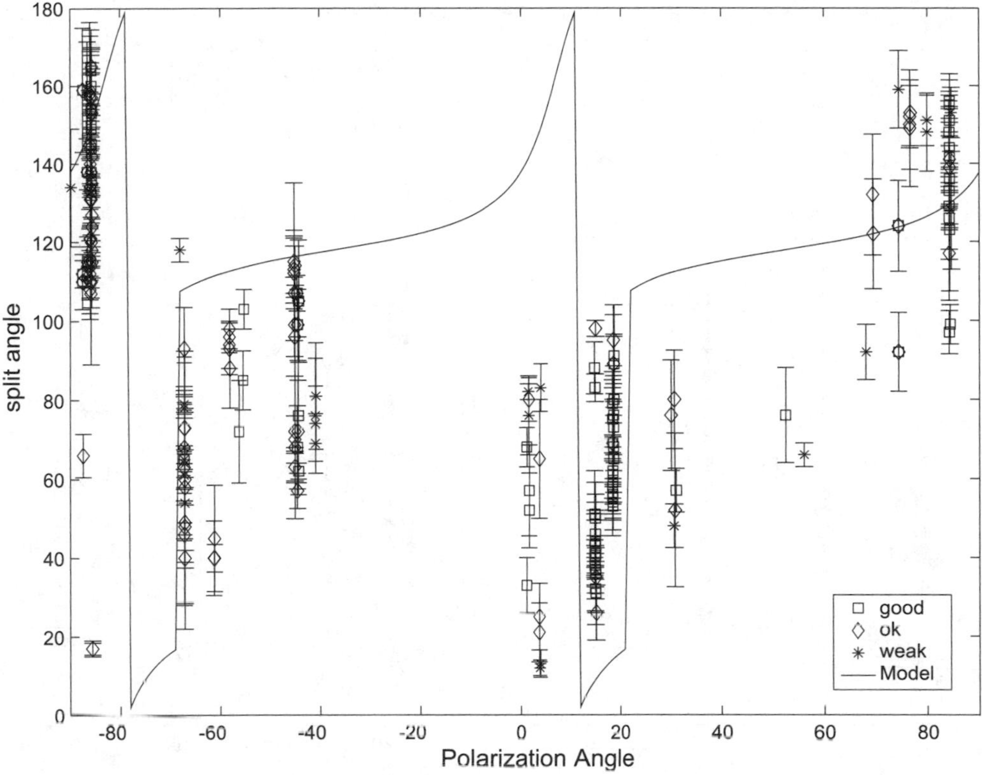

Figure 3. Two-layer model 1. $\phi1=62°$, $\phi2=120°$, $\delta t1=0.7s$, $\delta t2=1.5s$. Black line is 2-layer model. Dotted lines are null directions from observations. Squares represent good splitting measurements, diamonds are ok measurements and asterisks represent poor measurements. A. ϕ fit. B. δt fit.

error bars on the splitting parameters. Narrower band filters occasionally produce ok results, but the waveforms become more oscillatory.

Selection of a time window on the seismogram for analysis depends upon the noise character of the waveform as well as which window selection provides the best constrained splitting parameters. For example, windows can span the SKS, SKKS or both SKS and SKKS phases. Tests were performed using narrow, medium, and wide window length. Including both SKS and SKKS phases tends to produce more tightly constrained splitting results, in part due to the increased number of cycles that must be fit in the grid search. For a low noise event window selection can cause varying results with error bars, but little variation with ϕ and δt. In all cases the best results are achieved when the fast and slow waveforms are almost identical. With the wider windows, noise can be added into the analysis and increase the error. The narrowest windows may not include all the energy of the phase and therefore a well-constrained measurement may not be achieved. A noisy event was tested with a range of window sizes at several different stations, with variable results. On two occasions the smaller window had the smaller error, two times the larger window had the smaller error, and once the errors were the same for both window sizes. In all cases the fast polarization directions and delay times were similar, but the size of the error bar varied with window size, in no discernable pattern.

From these examples, it can be seen that there is some analyst discretion that must be exercised to determine optimal window size. With certain stations and events there was no window that would produce an acceptable split, and therefore labeled a non-result or a poor result.

Shear Wave Splitting Results

Uniform NE-SW fast polarization directions of seismic anisotropy are with data from the CD-ROM southern line in New Mexico, while splitting from the CD-ROM northern line (Table 5; see CDROM in back cover sleeve) and the Laramie array (Table 6; see CDROM in back cover sleeve) in Wyoming show a complicated pattern that depends upon incident ray back azimuth (Figure 5; see CDROM in back cover sleeve). The shear wave splitting parameters are plotted at the pro-

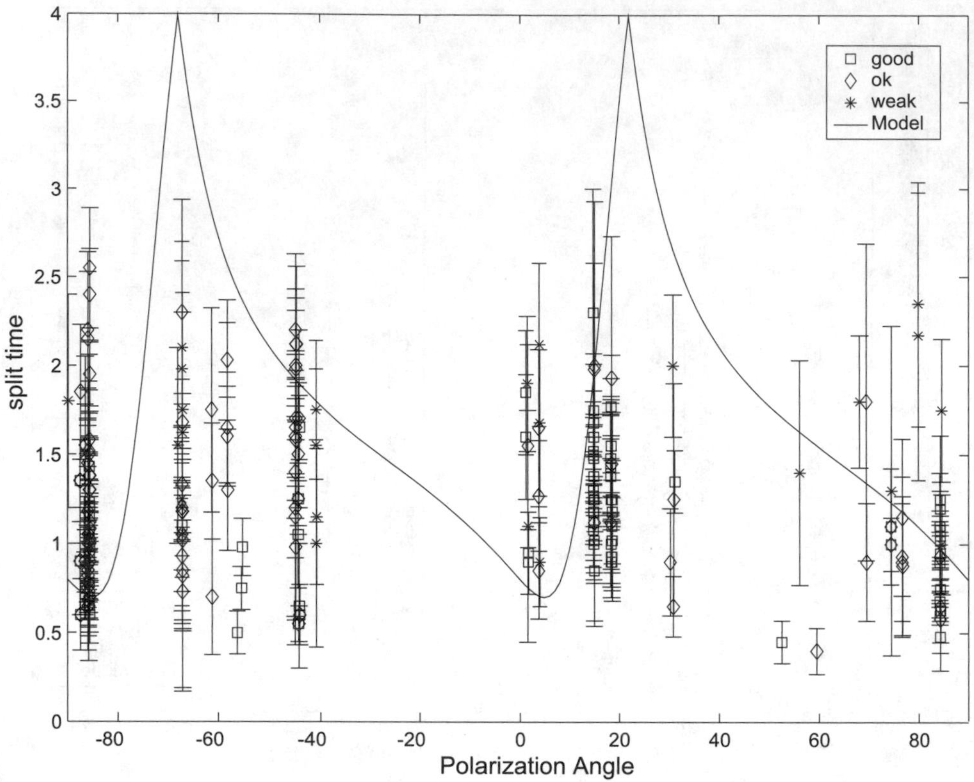

Figure 3B.

jection of their 220 km piercing points to separate back-azimuths visually (Figure 1 and Figure 2; see CDROM in back cover sleeve).

CD-ROM South Stations. The CD-ROM southern stations have a distinct 50-degree average fast polarization direction, ϕ, that correlates well with the absolute plate motion direction (Figure 1, Table 5; see CDROM in back cover sleeve). Absolute plate motion across the southern array varies between 241.7° (61.7°) in the northern part of the array to 242.4° (62.4°) in the south [DeMets et al., 1990]. Events from western back-azimuths produce shear wave splitting measurements smaller in magnitude (smaller δt) than the single good event from an eastern back-azimuth. This might suggest that the anisotropic layer is thicker to the east. Erosion from the Rio Grande Rift might be the mechanism for thinning on the western edge. Null measurements correlate well with the observed splits. Nulls from back-azimuths 236° and 237° would be coming along the fast axis and nulls from back-azimuth 141° would be arriving along the slow axis of splitting.

There is no clear shear-wave splitting evidence delineating the inferred Yavapai-Mazatzal Proterozoic- Proterozoic boundary in northern New Mexico (Figure 1). Stations throughout the array show similar splitting parameters, and no backazimuthal variation in ϕ is observed. The Yavapai-Mazatzal paleo-accretion zone is northeasterly directed [Karlstrom and Humphreys, 1998]. Such a geometry could produce northeast directed anisotropy, thus it is uncertain whether the splitting observed is from fossil accretion or current mantle flow associated with motion of the North American plate [Vinnik et al., 1992], as both are consistent with northeast directed splitting. Splitting parameters from past experiments in the area [Sandvol et al., 1992; Savage et al., 1996] do not compare well with CD-ROM south splits in all areas, which may suggest a more complicated flow pattern or different causes of anisotropy (Figure 1). The results from the Rio Grande rift were proposed to be the result of small scale asthenospheric flow within the rift [Sandvol et al., 1992]. The extent of these northeast directed splits and consistency with APM and paleoaccretion directions leads us to a different interpretation than small scale rift flow.

CD-ROM North Stations and Laramie Array. Northern CD-ROM station splitting parameters show more variation than those from the southern array, but can be separated into distinct areas (Figure 2, Figure 5, Table 5 and Table 6; see

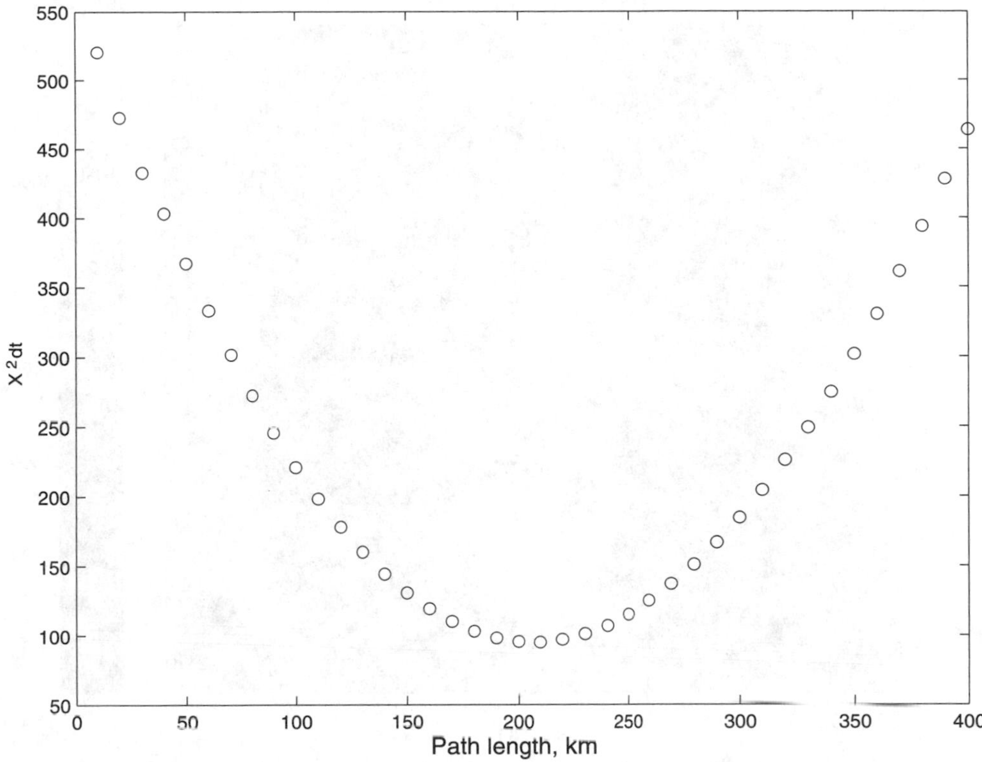

Figure 4 a. $X^2_{\delta t}$ as a function of path length for the best fitting plunging symmetry axis parameters, for CDROM stations N00 – N13.

CDROM in back cover sleeve). The southernmost stations of the north line (N17-N26), south of the Cheyenne Belt, show an average fast direction of 63°, roughly parallel with absolute plate motion direction in this region. Absolute plate motion across the northern array varies between 239.4° (59.4°) in the northern part of the array to 240.47° (60.5°) in the south part of the northern array [DeMets et al., 1990]. CD-ROM stations N14-N16 have variations different than the stations to the north or south of them (Figure 2 and Figure 4). The values of fast direction for these stations vary from 20° to –25°. These 3 stations could be imaging the anisotropic regions to the north and south, instead of being a separate anisotropic zone. CD-ROM stations N00 to N10 have a consistent back-azimuthal variation that is also observed at the Laramie array stations. At these stations events from the north to northeast and northwest back-azimuths range from 20° to 80° in fast polarization direction, while events from the west trend from 100° to 170° in fast polarization direction (Figure 4a).

The Laramie Array crosses the Cheyenne Belt north of Laramie, Wyoming (Figure 2). Variations in shear wave splitting parameters from stations within the Laramie array were examined closely to search for any variation associated with the Cheyenne Belt. The only stations with shear wave splitting measurements with different back-azimuthal variation than the majority of the stations are the southernmost stations L01, L02 and one result from L03 (Table 6; see CDROM in back cover sleeve). When plotting the variations of CD-ROM stations N00-N13 and Laramie array stations by polarization angle, a mode-180 variation is seen with polarization angle (Figure 3). Plots of δt versus polarization angle show a wide scatter from 0.5 seconds to 2.5 seconds (Figure 3b). Models with two layers of anisotropy predict variation with a mode-90 pattern (solid black line in Figure 3) with polarization angle and back-azimuth [Ozalaybey and Savage, 1994]. This mode-90 pattern can help distinguish between a 2-layer anisotropic model and a model with a plunging symmetry axis, as the latter has a mode-360 variation with back-azimuth. These possibilities will be discussed more fully in a later section.

The backazimuthal variations in shear wave splitting polarization directions of the Laramie Array and the northern CD-ROM stations (N00-N13), are remarkably similar (Figure 4). Events from the north to northeast and northwest back-azimuths range from 20° to 80° in fast polarization direction, while events from the west trend from 100° to 170° in fast polarization direction. The consistent variation in shear-wave

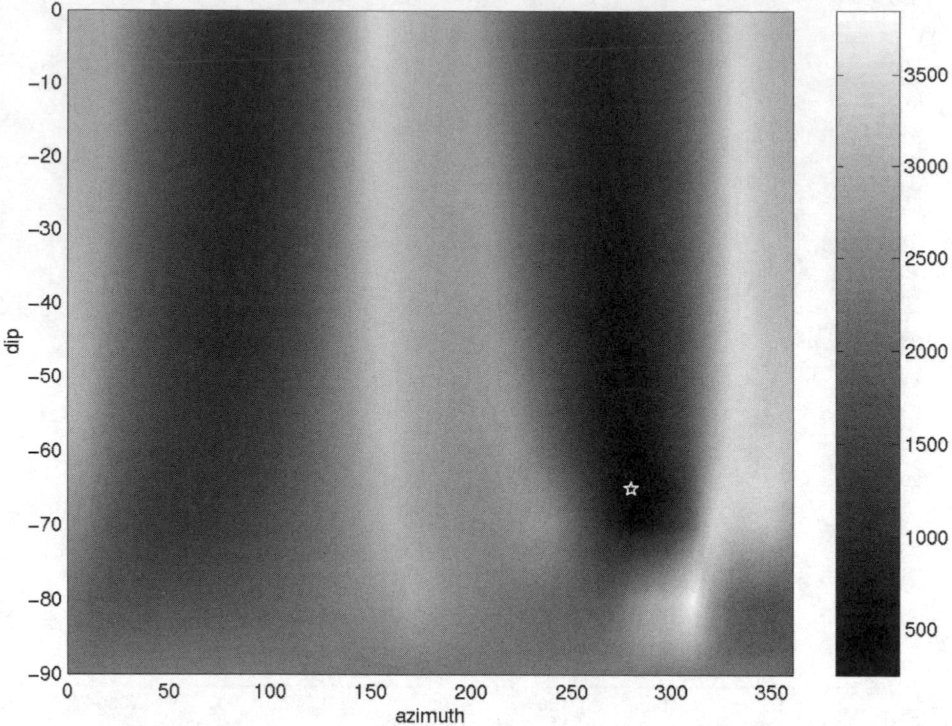

Figure 4 b. Grid search over anisotropic plunge and plunge azimuth for CD-ROM stations N00-N13. Contours represent ($X^2_{\delta t} + X^2_{\phi}$) for range of a-axis of orientation of olivine. White colors represent large misfit, while black represents good fit to the data. The white star is the best fitting model at plunge = 65 degrees, azimuth = 280 degrees.

splitting parameters with respect to back-azimuth between the CD-ROM and Laramie lines, despite their east-west separation by 100 km, strongly suggests a common regional mechanism for the anisotropy rather than geographically heterogeneous anisotropy.

Using fresnel zones, we can arrive at a rough depth to anisotropy estimate [Alsina and Snieder, 1995]. As stations N13 and N17 are about 40 km apart and show distinctly different shear wave splitting, it can be assumed that these two stations represent different anisotropic regions. For 8 s period waves (similar to the periods recorded in this study), Fresnel zones at 80, 200, and 440 km depth have diameters of approximately 60 km, 100 km, and 140 km, respectively. Therefore the anisotropy seen should be no deeper than 80 – 140km in depth. This would put the anisotropy in the lithosphere.

Two-layer Modeling. Two-layer modeling was performed to try and model the back azimuthal variations in shear wave splitting parameters observed at the northern CD-ROM and Laramie array stations. The forward modeling algorithm described in Ozalaybey and Savage [1994] was used in these analyses. The input parameters are the ϕ and δt for each layer.

A grid search was performed over the four splitting parameters (ϕ_1, ϕ_2, δt_1 and δt_2) for two layers of anisotropy, where layer 1 is the bottom layer and layer 2 is the top layer. From these grid searches, a model with a low root mean squared (rms) value was used as starting model to plot against the data to search for acceptable models. Root mean squared values are the sum of the square root of the difference of the observed and predicted values, divided by the number of values:

$$rms = \Sigma_i^m [(| \phi_{obs} - \phi_{pred} |^2)/ n^2]^{1/2} \quad (1)$$

where ϕ_{obs} are the observed data (both ϕ and δt), ϕ_{pred} are the model predicted values (both ϕ and δt) and n is the number of input observed values.

The best fitting two-layer model is shown in Figure 3. The fast polarization for the lower layer, ϕ_1, is 62°, parallel to absolute plate motion. The delay time, δt_1, for the lower layer is 0.7 seconds. The upper layer has fast polarization direction, ϕ_2, of 120° and delay time δt_2 of 1.5 seconds. This upper layer might represent some fossil lithospheric deformation. The two-layer model shown in Figure 3 is plotted versus polarization angle to show the observed data in a more detailed

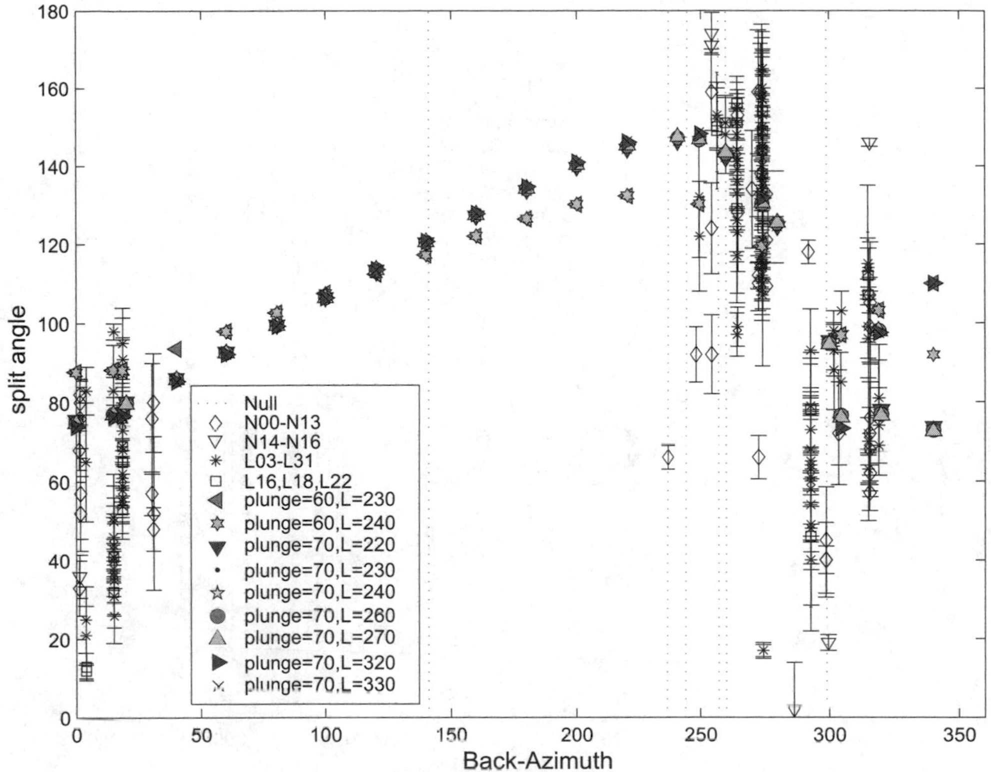

Figure 5. Models of plunging axis of symmetry. Azimuth is set at 290 degrees, plunge ranges from 60 to 70 degrees, and path length, L, varies from 220 to 330 km. Open symbols represent stations (diamonds: N00-N13, triangle: N14-N16, asterisks: L03-L31, squares: L16,L18,L22) , while filled symbols are the models. Dashed lines are observed null directions. A. ϕ versus back-azimuth. Note that models for a given plunge overlap in many cases. B. δt versus back-azimuth.

way, as many back-azimuths are poorly represented. The preferred model fits the large ϕ change observed near polarization –70°(back-azimuth 280°) well as well as the ϕ-observed between polarization 70° to 90° and –90 to –95° (240° and 274° backazimuth). However the ϕ data between polarization 0° to 60° (back-azimuths 0° to 32°) are not fit well by the model. Predicted ϕ's from this model are much larger than those observed between polarizations –70° to –40° (back-azimuths 290° to 320°). In the plot of observed δt and model δt (Figure 3b), the model does not fit the minimal variations in observed δt, nor does the model line up well with observed null measurements.

Many two-layer models fit the observed data in certain places, but none fit all the variations. There is little variation in the observed δt's with back-azimuth or polarization angle, so it is more difficult to constrain a single model. Any anisotropic two-layer model will show 90-degree variation in both back-azimuth and polarization angle. More complete back azimuthal coverage would be needed to further distinguish between possible models. In the plot of back azimuth there is not a mode-90 variation seen clearly (Figure 4). With a mode-90 variation there should be observed-ϕ of about 120–160° between polarization angles –20 to at least 5 (Figure 3). There are no observed splits between –20° to 0° polarization (Figure 3), making it difficult to constrain the model. Where there are observed shear wave splitting measurements, around polarization angle 4, the values of observed ϕ are between 20° and 90°, which do not correlate with the observed ϕ from polarization angle –86° (110° to 165° observed ϕ).

With many models fitting the data equally well it is difficult to select a final a 2-layer model and assign meaningful uncertainties to it. One advantage of the grid search is that many poorly fitting models can be rejected, and trade-offs between parameters can be visualized. The broad class of 2 layer models that fit the data best have a lower layer of anisotropy aligned northeast to east-northeast and an upper layer of northwest to north-northwest.

Plunging Symmetry Axis Modeling. Since the stereographic projections of the data do not show 90° or 180° symmetry of splitting with backazimuth, models with a plunging symmetry axis of anisotropy are tested. Hartog and Schwartz [2000]

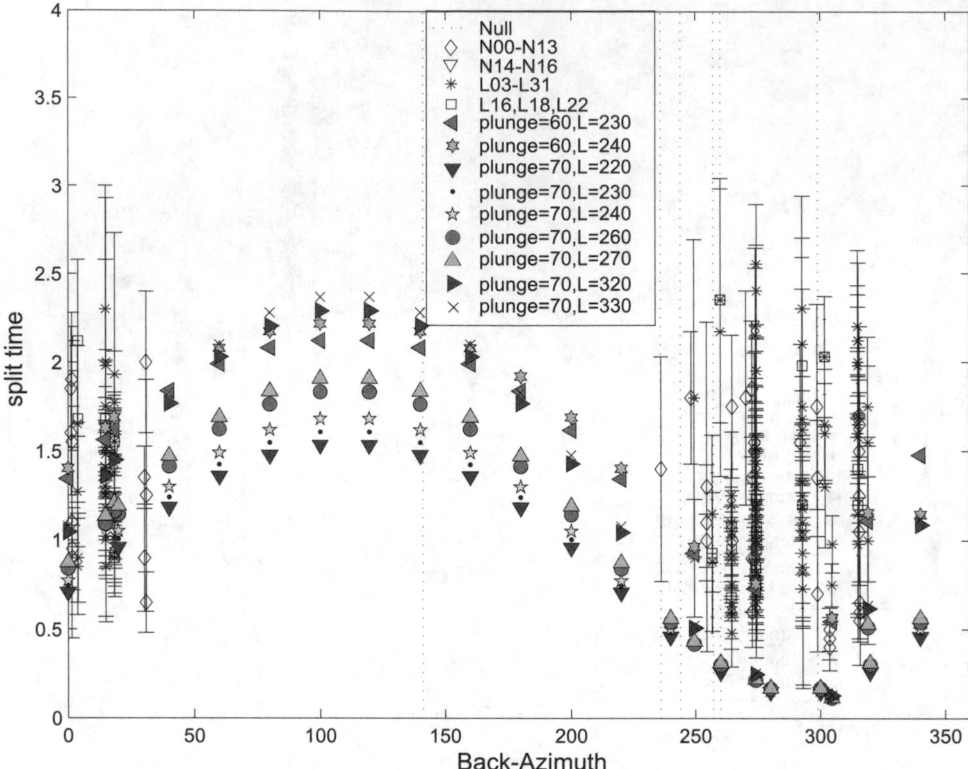

Figure 5b.

found evidence for a plunging axis of symmetry in anisotropy north of the Mendecino triple junction. They suggest that the subduction of the Juan de Fuca plate has caused the plunging anisotropy.

The codes used to perform this modeling were acquired from Renate Hartog [Hartog and Schwartz, 2000]. They perform a grid search over a suite of symmetry axis azimuths, plunges, and path lengths for 4% anisotropy for a single anisotropic layer, comparing splitting parameters predicted from each model with those observed. Models with either hexagonal or orthorhombic symmetry can be explored. With hexagonal symmetry there are two symmetry values (a, and b/c), while orthorhombic models have three values (for the a, b, and c axes). As olivine anisotropy is mainly dependent upon the a-axis orientation [Christensen 1984], the hexagonal model is used in this study. The code also allows for input of specific olivine or petrological information if it is available. Though at present, no such data is available in the CD-ROM study region. An end-member model, as used in Hartog and Schwartz [2000], where olivine is perfectly aligned, is used. The percent weight of the olivine is 25%, with 75% being an isotropic medium. This composition produces elastic properties similar to those of the upper mantle values from IASPEI91 model (Vp = 8 km/s, Vs = 4.47, ρ = 3.32 gm/cc), resulting in a percent anisotropy of 4% [Hartog and Schwartz, 2000]. As the choice of 4% anisotropy is due to lack of physical data, the acceptable path lengths would be larger.

The grid search over shear wave splitting measurements from stations Northern CD-ROM stations N00 to N13 (using only "good" or better measurements) results in an inclined axis model of azimuth 280°, plunge 65° (from horizontal) and a path length of 220 km or 1.5 seconds (Figure 4 and 5). The path length is converted to δt by the following equation [Hartog and Schwartz, 2000]

$$\delta t = L\,(V_f - V_s)/(V_f * V_s) \qquad (2)$$

Where L is path length, Vf is fast S-wave velocity, and Vs is slow S-wave velocity. As can be seen by the plot of error estimates ($X^2_{\delta t}$ and X^2_{ϕ}) in Figure 4a, the model is more dependent on azimuth than on the plunge angle. This is due to the model is more dependant on azimuth than plunge. The azimuth is mainly controlled by the fast direction of shear wave splitting and the plunge is dependent on the delay times. The shear wave splitting data shows more variation in fast directions than in delay times, so such a dependency is expected. For a given azimuth, a large range of plunge angles fit the data with similar misfit, thus the plunge angle is not well resolved. The path length also has a range of acceptable values. The model does show significant sensitivity to plunge azimuth, which is thus better resolved than the actual plunge inclination.

The misfit values were calculated by summing the squared differences of the observed and predicted values and dividing by the squared measured errors [Hartog and Schwartz, 2000]:

$$X_{\phi}^{2} = \left(\Sigma\left(\phi_{pred} - \phi_{obs}\right)^{2}\right) / \left(\sigma_{\phi}\right)^{2}, \text{ and}$$
$$X_{\delta t}^{2} = \left(\Sigma\left(\delta t_{pred} - \delta t_{obs}\right)^{2}\right) / \left(\sigma_{\delta t}\right)^{2}, \tag{3}$$

where ϕ_{pred} and δt_{pred} are the predicted model values, the ϕ_{obs} and δt_{obs} are the observed values and $\sigma_{\delta t}$ are the measured misfits. The smaller the chi-squared value, the better the model fits the data. Both the ϕ and δt observed values are given equal weighting in the model.

The Laramie Array data were split into three groupings based upon similarity of splitting measurements and geographic proximity to invert for models. For Laramie array southern stations L03 to L12, the resulting best fitting model has azimuth=290° (ie. plunges to the west northwest), plunge=65°, L=220 km, and δt=1.5s. Stations L13 to L22 (without L16, L18, L22) produced a best fitting model with azimuth=290°, plunge=70°, L=340 km and δt=2.3. Stations L23 to L31 resulted in a model of azimuth=270°, plunge=70°, L=270 km and δt=1.3s. Stations L16, L18 and L22 produced anomalous results when they were included with the other Laramie Array stations. Eliminating these stations from the modeling produced acceptable results. These stations have similar results to the near-by stations (Table 6; see CDROM in back cover sleeve, Figure 2, Figure 4), so the anomalous model results are difficult to explain.

Models were determined for four subsets of the data, and produced similar results. Azimuths vary between 270° and 290°, plunges from 60 to 80 and the δt's from 1.3 to 2.3 seconds. It might be expected that there would be a variation between north CD-ROM and Laramie arrays, given that they are separated by 110 km east-west, as well as within the individual arrays, but it is striking that they still come up with similar models.

Comparing a suite of plunging symmetry axis models with a range of plunge azimuths from 270° to 300°, plunges between 50° and 70° and varying path lengths, the models in the 290°-azimuth suite best fit the observed data (shown in Figure 5). Models with plunge azimuths from 270° to 280° have lower predicted phi values in the 264° to 274° backazimuthal range than is observed. The amplitude of predicted ϕ variations is smaller in amplitude than the observed ϕ variations, except between back-azimuth 0° to 32° and 280° to 320°. Models with plunge azimuths of 300° over estimate values of ϕ between back-azimuths 0° to 30°.

Azimuth-290° models (plunge = 60° to 70°) closely approximate the majority of the observed splitting parameter amplitude variations (Figure 5). The plunge=70°, δt-model does not fit as well as the plunge = 60° in the west and northwest back-azimuths, while the plunge=70°, ϕ-model fits better than the plunge = 60° in amplitude of variations. Choosing a path length (L) is complicated by there not being a distinct variation in split time, δt, with ϕ. The δt observed values also do not vary greatly with back-azimuth, further complicating the choice. Therefore the best fitting model has azimuth 290° +/- 10°, plunge =65° +/- 5° and δt = 1.9 +/- 0.5 seconds.

The observed null directions cluster around back-azimuth 250°, with outliers at 299° and 141° (shown as dotted lines in Figure 5). If there were a plunging symmetry axis, one would expect the nulls and small split values to be in the azimuth of plunge, as there would be no splitting observed along the axis of symmetry (if the plunge angle were the same as the incidence angle). With a plunge of 65°, incidence angles of around 25° would show nulls along back-azimuth 290°, as this would be the axis of plunging anisotropy. S-waves would have an incidence angle this large. One S-wave event recorded on the CD ROM array, with back-azimuth 303.7 and incidence angle of about 17°, had results for two CD-ROM stations with very small δt values (Table 3 and 5; see CDROM in back cover sleeve).

Laramie southernmost stations L01 and L02 were modeled using all splitting results (good, ok and weak) and the resulting plunging symmetry axis model has symmetry axis azimuth=340°, plunge=70°, L=150 km and δt =0.9. The L01 and L02 plunge azimuth is more north-directed than the plunge axis found with the northern Laramie array stations, and points towards the other models (L01 and L02) plunging axis. These 2 stations may be influenced by the back azimuthal variations to the north as well as from different anisotropy variations to the south. As the Laramie Array did not continue further south it is difficult to discern whether this is a separate anisotropy zone, or a freznel zone complication. A similarity happens in the CD-ROM line with N14, N15 and N16.

Travel time tomography using data from the CD-ROM north array shows a high P-wave velocity anomaly dipping steeply (approximately 50 degree plunge) north [Dueker et al, 2001]. This may be indicative of a down welling piece of lithosphere that has caused a thermal anomaly or of a fossil slab with a fast velocity anomaly in the down-dip direction.

A possibility for the change between L01 and L02 to the rest of the array (with respect to ϕ and backazimuth), is that it may indicate the location of the Cheyenne Belt (Figure 2). As the Laramie array is only 50 km long, it may be difficult to locate the Cheyenne Belt through shear-wave splitting.

While SKS phases have a steep incidence angle (5° to 12°), many ray paths could go through areas north and south of the Cheyenne Belt, therefore complicating the splitting parameters. As most of the splitting events have come from the northwest and western back-azimuths, few rays sample the mantle south of the Cheyenne Belt. This may be the reason for not seeing a change in splitting parameters between L24-L25, which is north of the surface expression of the Cheyenne Belt, and L23 to the south.

Discussion

A single layer of anisotropy is an appropriate model for the southern CD-ROM stations as well as for the region between N17 to N26. North of the Cheyenne Belt a single layer is not viable, as the average fast axis for the observed data is 110° and does not intersect any data.

Neither the two layer anisotropic models nor the plunging anisotropic symmetry axis model explain all of the shear wave splitting observations perfectly. Introduction of additional free parameters, such as more than two anisotropic layers or more than one plunging symmetry axis, may fit the observed data better, but such complex modeling may not be warranted without better back-azimuthal data coverage.

With the two anisotropic layer best fitting model, the upper layer has fast axis at 120°–160°, roughly perpendicular to the Cheyenne Belt which trends ENE-WSW, while the lower layer (layer 1) is roughly parallel to the Cheyenne Belt. As the back-azimuthal variations are noticed north of the suture zone, the variations in the upper layer may be indicative of fossil strain in the lithosphere. Layer 2 may correlate to surface deformation during orogenesis. Dikes mapped north of the Cheyenne Belt could be indicative of the anisotropy in the upper layer [Duebendorfer and Houston, 1987]. Layer 1, roughly parallel to the Cheyenne Belt, could represent the accretion event, strain during wrench fault deformation, or it could be current strain due to absolute plate motion deformation.

The plunging anisotropic symmetry axis model (Figure 5) is our preferred model. Viewed back-azimuthally a mode-90 variation could still be possible with the observed ϕ, as there are some missing polarization angles and backazimuths, but there is no mode-90 pattern with the observed δt values. As the δt values only vary minimally, small at the western and northwestern back-azimuths and larger at the northern and northeastern back-azimuths, a model that reflects this very minimal variation is preferable. Two-layer models have rapidly varying δt values, but with the plunging symmetry axis models the variation is much more gentle, as we observe.

The plunging symmetry axis model is difficult to explain by surface geology. The plunging axis might be due to asthenospheric flow interacting with strong lithosphere-asthenosphere boundary topography associated with the Archean-Proterozoic transition. Such a model was suggested by Bormann et al. [1996] to explain splitting measurements in Europe. Another hypothesis is that the splitting patterns could be related to the high seismic velocity feature seen in the teleseismic travel time tomography of Dueker et al. [2001]. Down-welling associated with such a feature could cause considerable strain in the surrounding area and result in alignment of the a-axis of olivine, although one might expect more of a lateral change in inclined axis values. For the number of stations and spatial size of the consistent inclined axis, the downwelling would need to be rather large and consistent over this large area. Other splitting results in the extended region show a consistent NE-SW orientation well west of the CD-ROM array at the Snake River Plain [Schutt et al. 1998], but complex and inconsistent orientations closer to the CD-ROM array from the Deep Probe and Lodore arrays [Humphreys et al., 2001]. Without additional measurements it is difficult to assess the lateral extent of such a downwelling feature.

Another possible explanation for the northwest plunging axis of symmetry would be the combination of compressional and shear motion across the Cheyenne Belt. North-south compression combined with left-lateral shear across the Cheyenne Belt could produce the NW directed plunge axis modeled. Shear determined from surface outcrops of rocks at the Cheyenne Belt is thought to consist mostly of right-lateral motion [Duebendorfer and Houston, 1987], opposite of what we are seeing at depth.

A structural possibility, and our preferred model, as proposed by Dueker et al. [2001] would involve a "fossil slab" accreted to the bottom of the lithosphere during the Proterozoic accretion event (Figure 6). This slab could be composed of either eclogite or some anisotropically fast structure, as would be suggested by the shear wave splitting results. A fossil slab would show consistent results over a large lateral area. The tomographic results show a fast velocity feature plunging 50° to the north. Without depth constraints on the shear-wave splitting it is difficult to discern between a dipping slab with slab-parallel anisotropy and a slab with plunging anisotropy.

Conclusions

Shear wave splitting results from South CD-ROM show consistent northeast directed shear-wave splitting, indicative of either absolute plate motion (APM) and/or fossil accretion. North CD-ROM back-azimuthal results compare well with the Laramie Array splitting parameters, suggesting a common anisotropic fabric rather than laterally heterogeneous anisotropy. The azimuthal variations of splitting from both north CD-ROM

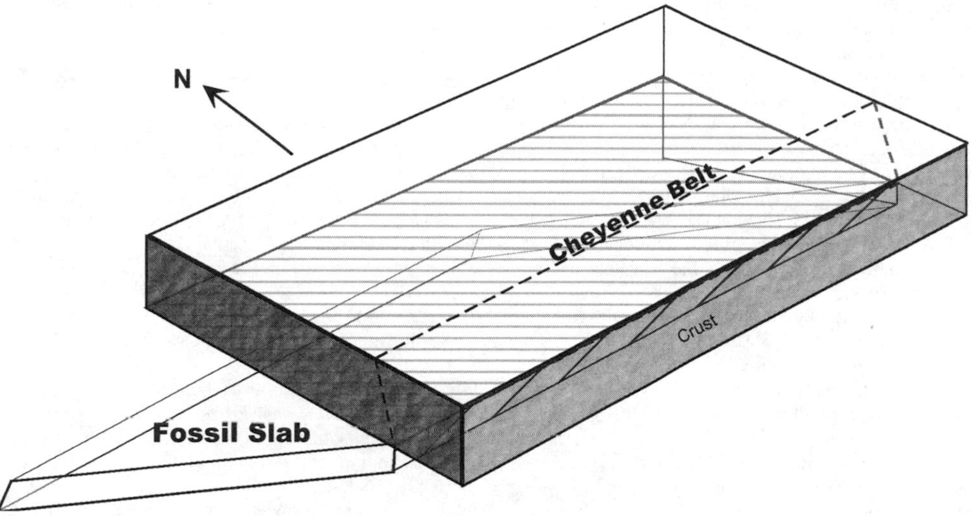

Figure 6. Block model of shear-wave splitting interpretation, showing the plunging symmetry axis as a north-dipping slab. The plunging anisotropy is only seen north of the Cheyenne Belt in the CD-ROM north data. The Laramie array tomography shows a high velocity feature that plunges to the north (Huan and Dueker, this issue). The slab may or may not touch the base of the crust, as tomography shows it at about 200km depth.

and Laramie are best modeled with a plunging axis of anisotropy. Our preferred model has azimuth 290° +/- 10°, plunge =65° +5° / -25° (see Figure 4b) and δt = 1.9 +/- 0.5. As there is little depth constraint on SKS shear-wave splitting measurements, it is difficult to choose between a structural feature flush with the bottom of the crust or one that plunges. From fresnel zone arguments, the depth to the anisotropic feature, should be within the lithosphere, below the Cheyenne Belt. A lower range for anisotropy would be between 80 and 140km depth, with the majority of the anisotropy occurring above this value.

Tomography [Huan and Dueker, this issue] has shown that the best model is the one with a slab beneath the crust that plunges at high angle to the base of the crust (Figure 6). The tomography has been forward modeled as an anisotropic slab that steeply dips to the north [Huan and Dueker, this issue]. As the variations start north of the Cheyenne Belt on the CD-ROM north line, this could be the slab from a paleo-accretion event [Karlstrom et al., 2002].

Acknowledgments. We thank Renate Hartog for sharing her computer programs for modeling anisotropy with a plunging axis of symmetry, and Martha Savage for advice and discussion. Ken Dueker generously shared the Laramie array data and provided useful comments on the paper. Some figures in this paper were made with GMT [Wessel and Smith, 1998]. IRIS Passcal seismic instruments were used in this project and the Passcal Instrument Center provided assistance. This work was funded by NSF grants EAR-9614410 and EAR-0003747.

REFERENCES

Alsina, D., and Snieder, R., Small-scale sublithospheric continental mantle deformation: constraints from SKS splitting observations, *Geophysical Journal International*, 123, 431–448, 1995.

Bormann,P., Grunthal,G., Kind,R., and Montag,H., Upper mantle anisotropy beneath Central Europe from SKS wave splitting: effects of absolute plate motion and lithosphere- asthenosphere boundary topography?, *Journal of Geodynamics*, 22;1–2, 11–32, 1996.

Christensen, N.I., The magnitude, symmetry and origin of upper mantle anisotropy based on fabric analysis of ultramafic tectonites, *Geophys.J.R. Atron. Soc.*, 76, 89–111, 1984.

Crampin, S., The fracture criticality of crustal rocks, *Geophysical Journal International*. 118, 428–438, 1994.

DeMets, C., Gordon, R.G., Argus, D.F., and Stein, S., Current plate motions, *Geophysical Journal International*, 101, 425–478, 1990.

Duebendorfer, E.M., Houston, R.S., Proterozoic accretionary tectonics at the southern margin of the Archean Wyoming craton, *GSA Bulletin*, 98, 554–568, 1987.

Dueker, K., Zurek, B., and Yuan, H., Western U.S. mantle; old structured lithosphere over young restless asthenosphere, *GSA Today,* 2001.

Ehrlich, T.K. and Erlsev, E.A., The influence of the Cheyenne Belt on Cenozoic deformation in northern Colorado and southern Wyoming, *Geological Society of America*. 31; 7, 130, 1999.

Evans, J.R., Foulger, G.R., Julian, B.R., and Miller, A.D., Crustal shear-wave splitting from local earthquakes in the Hengill triple junction, southwest Iceland, *Geophysical Research Letters*, 23, No 5, 455–458, 1996.

Hartog, R. and Schwartz, S.Y., Subduction-induced strain in the upper mantle east of the Mendocino triple junction, California, *Journal of Geophysical Research*, 105, No B4, 7909–7930, 2000.

Huan Y. and Dueker, K., Upper mantle tomographic Vp and Vs images of the Middle Rocky Mountains in Wyoming, Colorado and New Mexico: Evidence for a thick heterogeneous chemical lithosphere, AGU monograph on Evolution of Rocky Mountain Lithosphere, this issue, 2002.

Ismail, W.D.,and Mainprice, D., An olivine fabric database: an overview of upper mantle fabrics and seismic anisotropy, *Tectonophysics*, 296,145–157, 1998.

Johnson, R.A., Karlstrom, K.E., Smithson, S.B., and Houston, R.S., Gravity profiles across the Cheyenne Belt, a Precambrian crustal suture in Southeastern Wyoming, *Journal of Geodynamics*, 445–472, 984.

Karlstrom, K.E. and Humphreys, E.D., Persistent influence of Proterozoic accretionary boundaries in the tectonic evolution of southwestern North America: Interaction of cratonic grain and mantle modification events, *Rocky Mountain Geology*, 33, no. 2, 161–179, 1998.

Karlstrom, K. E., S. A. Bowring, K. R. Chamberlain, K. G. Dueker, T. Eshete, E. A. Erslev, G. L. Farmer, M. Heizler, E. D. Humphreys, R. A. Johnson, G. R. Keller, S. A. Kelley, A. Levander, M. B. Magnani, J. P. Matzel, A. M. McCoy, K. C. Miller, E. A. Morozova, F. J. Pazzaglia, C. Prodehl, H. M. Rumpel, C. A. Shaw, A. F. Sheehan, E. Shoshitaishvili, S. B. Smithson, C. M. Sneson, L. M. Stevens, A. R. Tyson, and M. L. Williams, Structure and evolution of the lithosphere beneath the Rocky Mountains: Initial results from the CD-ROM experiment, *GSA Today, 12*, no. 3, p. 4–10, 2002.

Kay, I., Sol, S., Kendall, J-M., Thompson, C., White, D., Asude, I., Roberts, B., and Francis, D., Shear wave splitting observations in the Archean craton of Western Superior, *Geophysical Research Letters*, 26, No 17, 2669–2672, 1999.

Keller, G.R., Snelson, C.M., Sheehan, A.F., Dueker, K.G., Geophysical studies of crustal structure in the Rocky Mountain region: A review, *Rocky Mountain Geology*, 33, no. 2, 217–228, 1998.

Kou, B-Y, Chen, C-C., and Shin, T-C, Split S waveforms observed in northern Taiwan: Implications for crustal anisotropy, *Geophysical Research Letters*, 21, No 14, 1491–1494, 1994.

Leary, P.C., Crampin, S., and McEvilly, T.V., Seismic fracture anisotropy in the Earth's crust: An overview, *Journal of Geophysical Research*, 95, No B7, p 11,105–11,114, 1990.

Levin, V., Park, J., Crustal anisotropy in the Ural Mountains foredeep from teleseismic receiver functions. *Geophysical Research Letters*, 24, No 11, 1283–1286, 1997.

Nicolas A. and Christensen N.I., Formation of anisotropy in upper mantle peridotites-A review, *American Geophysical Union*, 111–123, 1987.

Ozalaybey, S. and Savage, M.K., Double-layer anisotropy resolved from S phases, *Geophysical Journal International*. 117, 653–664, 1994.

Russo, R.M., Silver, P.G., Franke, M. Ambeh, W.B., and James, D.E., Shear-wave splitting in northeast Venezuela, Trinidad and the eastern Carribbean, *Physics of the Earth and Planetary Interiors*, 95, 251–275, 1996.

Sandvol, E., Ni, J., Ozalaybey, S., and Shule, J., Shear-Wave splitting in the Rio Grande Rift, *Geophysical Research Letters*, 19, 2337–2340, 1992.

Savage, M.K., Sheehan, A.F., and Lerner-Lam, A., Shear wave splitting across the Rocky Mountain Front, *Geophysical Research Letters*, 23, No B17, 2267–2270, 1996.

Savage, M.K., Seismic anisotropy and mantle deformation: What have we learned from shear wave splitting? *Reviews of Geophysics*, 37, 65–106, 1999.

Savage, M.K. and Sheehan, A.F., Seismic anisotropy and mantle flow from the Great Basin to the Great Plains, western United States, *Journal of Geophysical Research*, 105, No B6, 13,715–13,734, 2000.

Schutt, D., Humphreys, E.D., and Dueker, K., Anisotropy of the Yellowstone hot spot wake, eastern Snake River Plain, Idaho, *Pure and Applied Geophysics*, 151, 443–462, 1998.

Schutt, D. and Humpreys, E.D., Evidence for a deep asthenosphere beneath North America from western United States SKS splits, *Geology*, 29, No 4, 291–294, 2001.

Silver, P.G., Seismic anisotropy beneath the continents: probing the depths of geology, *Annual Review Earth and Planetary Science*, 24, 385–432, 1996.

Silver, P.G. and Chan, W.W., Shear Wave Splitting and Subcontinental Mantle Deformation, *Journal of Geophysical Research*, 96, No B10, 16,429–16,454, 1991.

Silver, P.G. and Chan W.W., Implications for continental structure and evolution from seismic anisotropy, *Nature*, 335, 34–39, 1988.

Tanimoto, T. and Anderson, D.L., Mapping convection in the mantle, *Geophysical Research Letters*, 11, No. 4, 287–290, 1984.

Wessel, P. and W. H. F. Smith, New, improved version of the Generic Mapping Tools released, EOS Trans. AGU, 79, p. 579, 1998.

Vinnik, L.P., Makeyeva, L.I., Milev, A., and Usenko, A.Yu., Global patterns of azimuthal anisotropy and deformations in the continental mantle, *Geophysical Journal International*, 111, 433–447, 1992.

Otina C. Fox, Geophysical Institute, PO Box 757320, Fairbanks, Alaska 99775, Voice: (907) 474–5481; Fax: (907) 474–5618, otina@giseis.alaska.edu

Anne Sheehan, Campus Box 399, Department of Geological Sciences, University of Colorado at Boulder, Boulder, Colorado 80301, Voice: (303) 492–4597; Fax: (303) 492–2606, afs@terra.colorado.edu

Attenuation Tomography Beneath the Rocky Mountain Front: Implications for the Physical State of the Upper Mantle

Oliver S. Boyd and Anne F. Sheehan

Dept. of Geological Sciences and CIRES, University of Colorado, Boulder, CO 80309

Utilizing the Rocky Mountain Front (RMF) broadband seismic dataset acquired in 1992, this study has derived the seismic attenuation structure underlying part of the Southern Rocky Mountains and surrounding areas through measurements of differential t^* of S-phase waveforms. Previous studies of the area include P, S and surface wave travel time tomography, and all indicate low upper mantle velocities below the Rocky Mountain region. Calculations of intrinsic attenuation coupled with current velocity models aid in the determination of temperature, partial melt distributions, and compositional variation. A N-S zone of high shear wave attenuation ($Q_s \approx 30$) is found in the mantle beneath the Rocky Mountains and lies east of the region of lowest shear wave velocity. Relationships between shear wave attenuation and shear wave velocity are consistent with both thermal and compositional variability. Along the eastern Colorado Rockies and due north of the Rio Grande Rift, the relationships are consistent with an interpretation of elevated temperatures by up to 50 K at 125 km depth. West of this region low velocities and low attenuation suggest either unusual composition or very high temperatures. The low density mantle material beneath the Colorado Rocky Mountains in addition to increased crustal thickness and low density crustal intrusions provides a density contrast sufficient to support its overburden.

1. INTRODUCTION

For decades, geologists have questioned the mechanisms responsible for the high topography of the Rocky Mountains. Are the mountains supported by crustal thickening due to shortening (Airy root) or are there lateral density contrasts in the crust (Pratt compensation)? Is the lithosphere sufficiently rigid to support the topography? Does support come from the mantle rather than the crust? Measurements of crustal thickness by Sheehan et al. [1995] and Li et al. [2002] indicate an increase in crustal thickness that is not entirely able to support the overlying Rocky Mountains. Models of flexural support indicate that little of the observed topography is supported by bending of the elastic plate [*Sheehan et al.*, 1995]. These observations imply that regions of the crust and/or mantle have a reduced relative density.

Previous measurements that have explored the role of reduced crust/mantle density centered on measuring the change in seismic velocity across the region. Lee and Grand [1996], using the Rocky Mountain Front seismic dataset, measured shear wave travel times from teleseismic S-phase arrivals beneath the Colorado Rocky Mountains. The resulting shear wave velocity structures indicate reduced upper mantle velocity directly beneath the Colorado Rocky Mountains (Figure 1). The reduced upper mantle velocities trend north-south beneath the Colorado Rocky Mountains where the minimum shear wave velocity, -9% relative to western Kansas,

The Rocky Mountain Region: An Evolving Lithosphere
Geophysical Monograph Series 154

10.1029/154GM27

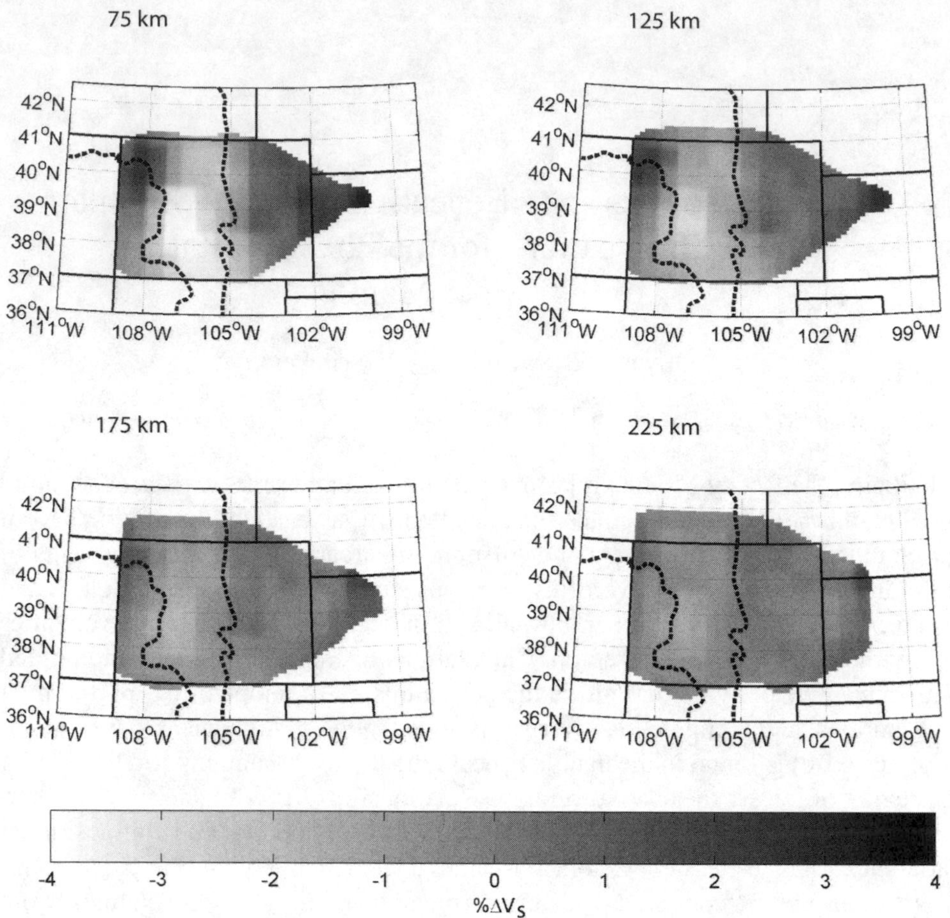

Figure 1. Horizontal slices of shear velocity variation [Lee and Grand, 1996] at four depths. Solid black lines denote state boundaries. Thick dashed solid lines delineate the Rocky Mountain region. Negative values (white) indicate slower velocities. Shear velocities beneath the Colorado Rocky Mountains are reduced by up to 9% relative to western Kansas. The transition to higher velocities in the east is believed to signify the transition to the stable craton.

occurs along this trend northwest of the Rio Grande Rift. Hessler [1997] produced a similar map from P wave travel time tomography but with smaller variations in P wave velocity, -4% (also summarized by Lerner-Lam et al. [1998]). Li et al. [2002] measured surface wave dispersion and inverted for crustal and upper mantle shear wave velocity. The greatest difference between the models of Lee and Grand and Li et al. is from 50 to 100 km depth beneath southwest Colorado where Lee and Grand's greatest decrease in shear velocity, -9%, corresponds to a decrease of only -6% from the model of Li et al. In addition, this position correlates with Li et al.'s thickest crust. Li et al. also find slow crustal shear velocities beneath the Sawatch Range in central Colorado which is correlated with high heat flow [*Decker et al.*, 1988]. Decker et al. used heat flow measurements and Bouguer gravity anomalies to infer the presence of granitic intrusions that were emplaced in the Miocene through the Quaternary. Li et al. point out that this compositional variation could explain the reduced shear velocities.

Changes to seismic velocity can result from changes to several material properties including temperature, composition and partial melt. If Lee and Grand's measured 9% lateral decrease in shear velocity in the upper mantle were due solely to increased temperature, using the scaling relations of Nataf and Ricard [1996] for changes in velocity with respect to temperature,

$$\frac{\partial \ln V_S}{\partial T} = -1.1\times10^{-4} K^{-1}, \tag{1}$$

changes in temperature would be over 800 K. Lateral changes of this magnitude are unlikely. Such changes would produce density contrasts far exceeding those required to support the Rocky Mountains as well as produce wide spread melting.

Karato [1993] shows that velocity-temperature scaling relationships are dependent on attenuation (Q^{-1}), a fact that Nataf and Ricard briefly considered but did not fully exploit. Using Karato's relations with a Q of 50, a 9% S-wave velocity contrast will predict a change in temperature of almost 400 K. If only half of the measured velocity contrast is attributable to changes in temperature, a Q of 50 coupled with ΔV_S of 4.5% implies a change in temperature of only 200 K.

Attenuation models coupled with velocity models can also lead to models of the unrelaxed velocity [*Minster and Anderson*, 1981] which is sensitive to composition including the effects of partial melt [*Duffy and Anderson*, 1989; *Hammond and Humphreys*, 2000b]. If half of the measured velocity contrast were due to partial melt, Hammond and Humphreys [2000b] predicts that a 5% decrease in shear velocity would accompany a 0.6% increase in partial melt.

Previous studies of the attenuation structure beneath western North America indicate broad regional variations. Lay and Wallace [1988] examined multiple ScS wave attenuation west of the Rocky Mountains and found the highest attenuation values, $Q_{ScS} = 95 \pm 4$, beneath the Basin and Range. The lowest attenuation values, $Q_{ScS} = 344 \pm 88$, were measured beneath the Pacific Northwest. Al-Khatib and Mitchell [1991] measured Rayleigh wave attenuation coefficients across several regions of western North America. One measurement traversed the eastern Colorado Rocky Mountains on its way from southern New Mexico to Edmonton, Alberta. Average Q_β along this path reaches a minimum of 35 at 150 km depth. A higher resolution study [*Slack et al.*, 1996] measured P and S wave travel time delays and differential t^* across the Rio Grande Rift. Differential t^* spans a range of 3 seconds for S-waves and 2 seconds for P-waves, but there is significant scatter in their data with respect to the relationship between delay time and differential t^*. Possibly because of this scatter, they do not attempt to solve for Q. Global models [*Romanowicz*, 1995; *Bhattacharyya et al.*, 1996; *Reid et al.*, 2001] are consistent with the above studies and also lack the resolution to quantify changes in temperature, partial melt and composition in the upper mantle beneath the Rocky Mountains.

In this paper, we measure the integrated differential attenuation of teleseismic S-phases, δt^* [*Sheehan and Solomon*, 1992], and correct for sedimentary basin reverberations. This measurement along with an a priori velocity model and ray path modeling is used to derive the shear wave attenuation structure in the upper mantle. This information is coupled with the Lee and Grand shear wave velocity model [1996] to identify regions of compositional and thermal variability. Karato's shear velocity - temperature derivative relationship [1993] is applied to estimate the mantle thermal component of isostatic compensation while an analysis is performed using the residual topography to estimate the best relationship between density and unrelaxed shear velocity, $\partial \ln\rho / \partial \ln V_S$, and better constrain the likely compositions. The mantle component of isostatic compensation is combined with estimates of compensation due to changes in crustal thickness using the model of Li et al. [2002] to evaluate overall isostatic compensation.

2. δt^* MEASUREMENT

The RMF seismic dataset was acquired by a two dimensional array of 37 broadband, multicomponent seismometers, 26 of which were used for this study (Figure 2, Table 1; see CDROM in back cover sleeve), deployed for a period of nearly seven months during 1992 [*Sheehan et al.*, 1995; *Lerner-Lam et al.*, 1998]. Twenty-seven stations were Guralp CMG3-ESP seismometers with a 30 second corner period, two were Guralp CMG3 seismometers with a 10 second corner period, and eight were Streckeisen STS-2 seismometers with a 120 second corner period. The instrument response of the seismometers is flat below the corner period. Events are initially extracted based on the following criteria:

$$65° < \Delta < 85° \ \&\ m_b > 5.6$$
$$45° < \Delta < 65° \ \&\ m_b > 5.4$$
$$30° < \Delta < 45° \ \&\ m_b > 5.3$$

where Δ is the epicentral distance in degrees and m_b is the body wave magnitude. All waveforms are visually inspected before picking windows for signal and noise (Figure 3a). Waveforms with no recognizable S phase are removed. The resulting dataset consists of 380 S-wave traces from 37 events (Figure 4, Table 2; see CDROM in back cover sleeve).

Each waveform undergoes conditioning before the calculation of δt^*. The channels are rotated to obtain the transverse component so as to isolate SH motion and minimize S to P conversions. The waveforms are band pass filtered to between 0.005 and 0.4 Hz and cropped to a 200 second window surrounding the hand picked S arrival. The waveform's trend and mean are removed, after which a 10% cosine taper is applied. The station's instrument response is removed and then the waveform is reconvolved with a common instrument response, that of the CMG-3 ESP broadband seismometer. The waveform is further reduced to a 30 second window centered on the S-pulse and multiplied by a gaussian taper. Larger windows and less severe tapers result in lower signal to noise ratios which degrades the measurement of δt^*. A multi-taper spectral analysis [*Percival and Walden*, 1993] with a time-bandwidth product of 3 is performed on the prepared time series to obtain the station spectra (Figure 3b).

For our measurements, t^* for a given signal is measured relative to a reference spectrum. The reference spectrum is derived from a pseudo-source which is the alignment and

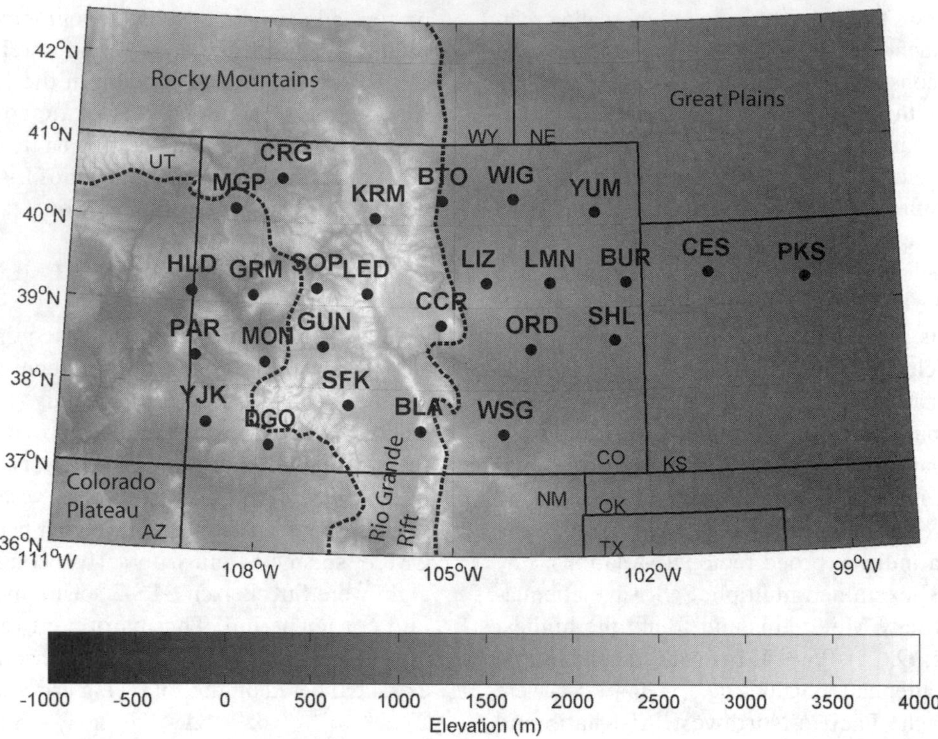

Figure 2. Seismic stations of the Rocky Mountain Front seismic experiment (RMF) used in this study on gray-scale topographic relief map. Stations are denoted by filled circles and a three letter station code. Geologic provinces are separated by thick dashed lines.

stack of all signals for a given event. The spectra for the station trace and the reference trace (Figure 5a) are given by:

$$A_S(f) = G_S S(f) I_S(f) R_S(f) e^{-\pi f t_S^*} \quad (2)$$

$$A_R(f) = G_R S(f) I_R(f) R_R(f) e^{-\pi f t_R^*} \quad (3)$$

where f is the frequency, G is the geometrical spreading, S is the source function, I is the instrument response, R is the crustal response, and $e^{-\pi f t^*}$ results from the energy loss along the ray's path from source to receiver. The subscripts S and R refer to the station and reference respectively. To find δt^*, we look at the ratio of spectral amplitudes, $A_S(f)/A_R(f)$. Relative t* is calculated for a given event to facilitate the cancellation of the source term in the spectral ratio. If the geometrical spreading is frequency independent, δt^* will not be a function of G. The instrument response cancels because we have removed the station instrument response, $I_S(f)$, and reconvolved the waveform with a common instrument response, $I_R(f)$. At this point, the crustal response terms are removed. Synthetic tests are performed to further examine the contribution of crustal response terms (Appendix 1.1; see CDROM in back cover sleeve) and a correction to δt^* is made using estimated basin thicknesses, shear velocities and impedance contrasts. The remaining spectral ratio is

$$\frac{A_S(f)}{A_R(f)} = e^{-\pi f (t_S^* - t_R^*)}. \quad (4)$$

Solving for δt^*, we have

$$\delta t^* = t_S^* - t_R^* = \partial \left[\ln\left(\frac{A_S(f)}{A_R(f)} \right) \Big/ -\pi \right] \Big/ \partial f \quad (5)$$

The slope of the curve between $\ln(A_S(f)/A_R(f))/(-\pi)$ and f is δt^*. We make δt^* measurements by finding the slope of the straight line that is fit to the curve of $\ln(A_S(f)/A_R(f))/(-\pi)$ vs. f over the frequency range 0.02 to 0.1 Hz (Figure 5b). This frequency range is used in the measurements for two reasons. The spectral energy at these frequencies is high, the spectral decay at higher frequencies can be dominated by scattering attenuation, and the spectral energy at lower frequencies can be contaminated by decreased sensitivity of the sensors.

The total number of traces for which δt^* is calculated is 380 (Table 3; see CDROM in back cover sleeve). The number of traces originating from the northwest is 280, the northeast, 25, the southeast, 75, and southwest, 0. After removing apparent δt^* due to basin reflections (Appendix 1.1; see CDROM in back cover sleeve), the variance of δt^* is 0.63 seconds. Appendix 1.2 (see CDROM in back cover sleeve) uses synthetics to determine that 75% of this variance could be due to normally distributed random noise. This may seem excessive but Appendix 2 (see CDROM in back cover sleeve) shows that a properly weighted inversion can produce a reasonable model.

Figure 6 is an average δt^* map for events originating from the northwest (top) and southeast (bottom) for uncorrected (left) and basin corrected (right) δt^* values. All δt^* values for events originating from a given direction for a given station are weighted by the inverse of the standard deviation of δt^* error due to normally distributed random noise (Appendix 1.2; see CDROM in back cover sleeve) and averaged, and the resulting weighted average is placed at the station location. The basin corrections tend to be negative resulting in an increase in δt^* in the basins after the correction is applied. Observations of the spatial distribution of δt^* in dependence on backazimuth are inconclusive. It appears possible that high δt^* moves to the northwest when examining events from the southeast and to the southeast for events from the northwest. The lack of an obvious pattern may mean that the measurements have considerable error or that the distribution of attenuation in the upper mantle is sufficiently complex to make the examination of these δt^* maps inconclusive.

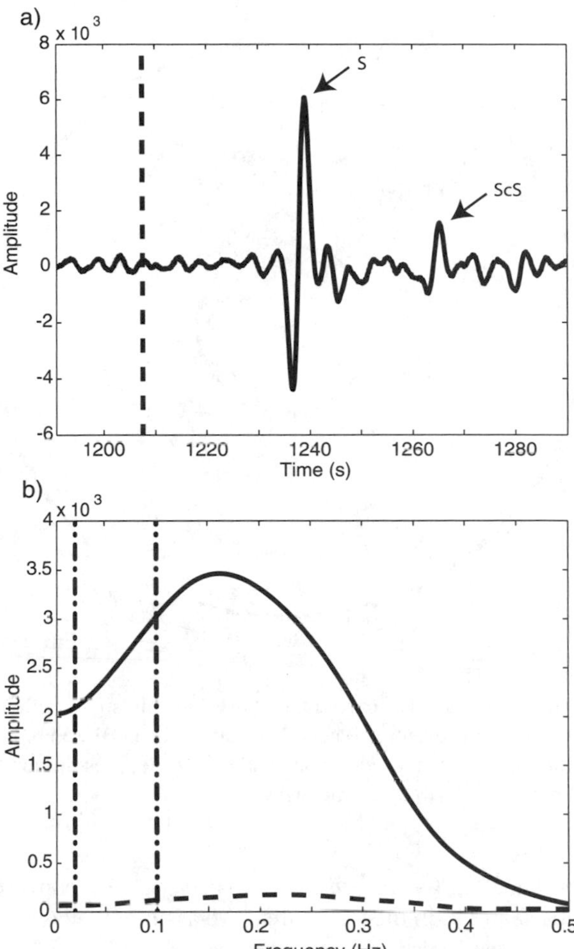

Figure 3. Example time series (a) and associated spectra for signal and noise (b). (a) Seismic phases S and ScS are denoted for reference. The dashed vertical line is the location of the center of the 30 second time window for the noise. This time series was recorded at station DGO and is from a magnitude 5.7 earthquake in northern Japan (backazimuth of 318 degrees, distance of 76 degrees, and depth of 317 km). (b) Signal spectra (solid line) and pre-event noise spectra (dashed line). Signal to noise ratio is 28.5. The vertical dashed dot lines span the frequency range over which the δt^* measurement is made.

3. δQ^{-1} INVERSION

Solving for differential attenuation is a relatively straight forward inverse problem. Relative t^* is related to attenuation, Q^{-1}, through the following expression:

$$\delta t^* = \int_S Q_R^{-1} d\tau_R - \int_S Q_S^{-1} d\tau_S \tag{6}$$

where τ is the travel time. As before, the subscripts S and R refer to the station and reference respectively. δt^* is the difference in the integrated effect of energy loss along the ray path between a reference trace and a station trace. In order to turn (6) into a tractable inverse problem, we assume that $d\tau_R = d\tau_S = d\tau$ such that

$$\delta t^* = \int \Delta Q^{-1} d\tau \tag{7}$$

where

$$\Delta Q^{-1} = Q_R^{-1} - Q_S^{-1} \tag{8}$$

We use the δt^* measurements, the ray path derived from 2-D ray tracing and the western U.S. one dimensional veloc-

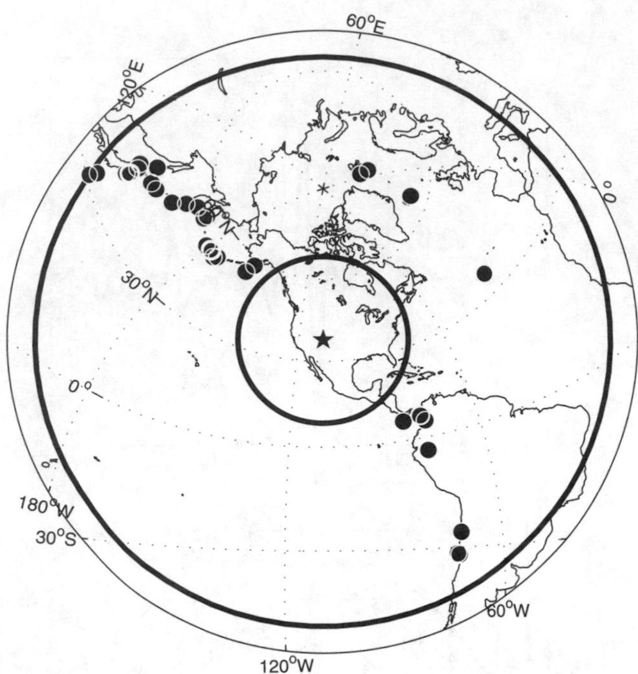

Figure 4. Locations of earthquakes used in this study (filled circles). Most of the events originate from either the northwest or southeast. Thick solid black circles are 30 and 80 degrees distant from the center of the RMF seismic array.

ity model, TNA [*Grand and Helmberger*, 1984], to invert for variations in Q^{-1} in the study area. We assume that variations in Q^{-1} between seismic rays for a given event are confined to the upper 400 km of the mantle.

Singular value decomposition [*Menke*, 1984] is used to do the inversion. Equation 7 can be written in the form

$$\delta t_i^* = \sum_M d\tau_{ij} \Delta Q_j^{-1} \tag{9}$$

where δt_i^* is the data vector consisting of N δt^* measurements, ΔQ_{j-1} is the model vector of differential attenuation in each of M blocks, and $d\tau_{ij}$ is the data kernel matrix with ray path travel time information. Singular value decomposition seeks a solution to

$$\Delta Q^{-1} = d\tau^{-1} \delta t^* \tag{10}$$

where

$$d\tau^{-1} = D_p \Lambda_p M_p^T \, . \tag{11}$$

Equation 11 is referred to as the natural generalized inverse. It is composed of D, a matrix of eigenvectors that span the data space, M, a matrix of eigenvectors that span the model space, and Λ, a matrix of eigenvalues whose diagonal elements are the singular values. The parameter p corresponds to the number of largest singular values kept when calculating $d\tau^{-1}$. In our case we have chosen to use all singular values within two orders of magnitude of the largest. Using only the largest p singular values is referred to as a truncated singular value decomposition (TSVD) and is equivalent to damping the solution, reducing model variance at the expense of model resolution.

In order to reduce the adverse effects of significant random noise, we use a weighted TSVD where the weights, W, are the inverse of the standard deviation of δt^* error due to normally distributed random noise (Appendix 1.1; see CDROM in back cover sleeve) [*Meju*, 1994]. Multiplying both sides of (9) by W, we have

$$W \delta t^* = W d\tau \Delta Q^{-1} \, . \tag{12}$$

We now seek a solution to

$$\Delta Q^{-1} = \left(W d\tau\right)^{-1} W \delta t^* \, . \tag{13}$$

The eigenvectors and eigenvalues of equation (11) are now found for the quantity $(Wd\tau)^{-1}$.

The inversion is solved for average attenuation values in 100 x 100 x 100 km bins extending from 113.7°W to 96°W, 33.7°N to 43.6°N, and 0 to 400 km depth. The bins are offset by 10 km in each direction and the inversion is repeated, producing a total of 1000 inversions. This procedure is done twice. In the first iteration, a solution for the event mean is included. In the second, the 1000 estimates of the event mean from the first iteration are averaged, removed from the δt^* data, and the inversion is performed again without the solution for the event mean. The inversions are sequentially combined, generating a psuedo-effective resolution of 10 km, and smoothed by convolving the model with a 200 x 200 x 50 km unit box.

For regions in the model with a resolution derived from the resolution matrix (A2.1) greater than 0.3, the range in ΔQ^{-1} is 0.04. The variance reduction of the corrected δt^* data due to the resulting model is 10%, a value that is expected given the amount of random error in δt^* (Appendix 2; see CDROM in back cover sleeve). Romanowicz [1995] reports a variance reduction of 49% for her global model of upper mantle shear wave attenuation, a value approaching the variance reduction achieved in velocity tomography. Reid et al. [2001] generate a variance reduction of 23% in their upper mantle shear wave

Q model while their velocity model is able to produce a variance reduction of 67%.

The ray path travel times are calculated using a one-dimensional velocity model. Because of this assumption, the resulting Q^{-1} model is incorrect by an amount approximately equal to the fractional difference between the one-dimensional velocity model and the true Earth velocity structure. For example, where the Earth velocities are 5% lower than the one-dimensional velocities, the actual Q^{-1} model should also be reduced by 5%. A competing effect not accounted for is that due to focusing and defocusing of the seismic energy from lateral velocity heterogeneities. Our analysis of this phenomenon suggests that when using our technique for measuring t^* and the velocity perturbations of Lee and Grand, Q^{-1} will decrease by less than 5% in regions of low velocity (focusing) and increase by less than 5% in regions of high velocity (defocusing). These regions will be flanked by decaying oscillations of high and low Q^{-1}. Work by Allen et al. [1999] indicates that focusing/defocusing effects on t^* in the vicinity of the Icelandic plume could be more than two orders of magnitude larger than those determined here. The difference is primarily due to the potentially small radius of the Iceland plume, 100 km, its vertical orientation, and its large shear velocity anomaly, -12%. We calculate that the combined errors due to the focusing/defocusing and ray path travel time effects are minimal.

Figure 7 provides horizontal slices of the resulting differential shear wave attenuation model. When basin corrections are not applied, the negative attenuation anomalies correlate very well with the positions of the basins. The primary feature of this corrected model is the large relative attenuation trending north-northwest directly beneath the eastern Colorado Rockies. This feature extends to around 150 km depth and coincides with a modest decrease in shear wave velocity (Figure 1). An anomalous region exists just to the west where a decrease in attenuation coincides with the lowest velocities. Is this relationship a consequence of the high temperature side of an attenuation peak or compositional variability? At greater depths, the correlation to features at shallow levels disappears.

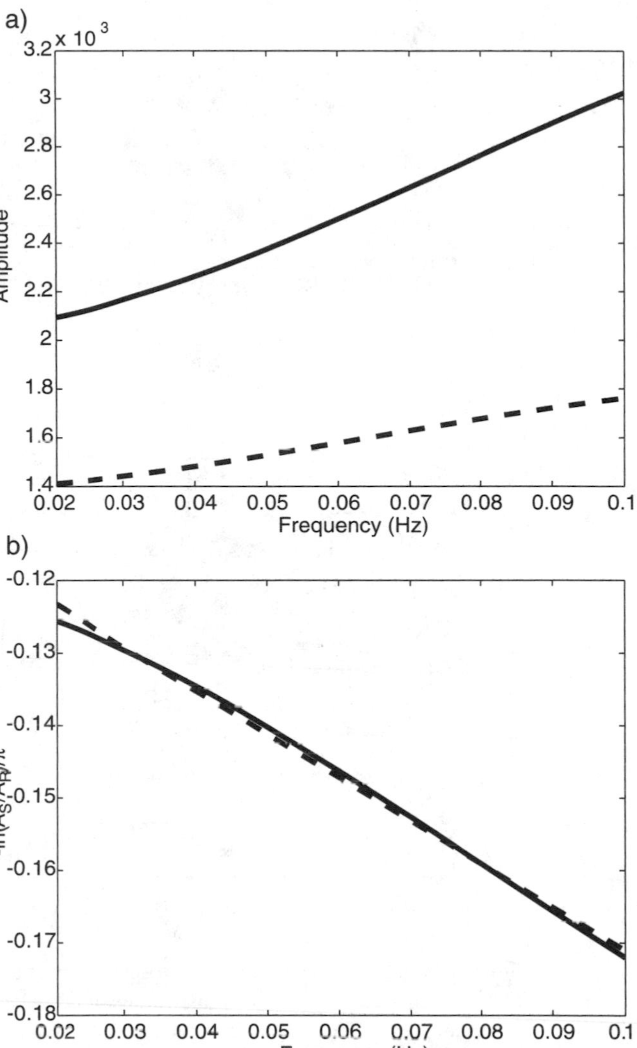

Figure 5. Example spectra for the signal (solid line) and reference (dashed line) (a) and spectral ratio (solid line) with linear fit (dashed line) (b). (a) The signal spectrum is that shown in Figure 3b limited to the frequency range 0.02 to 0.1 Hz. The reference spectrum is the average of spectra from all stations present for a given event. (b) The solid line is the natural log of the ratio of amplitudes between the station and the reference divided by $-\pi$. The dashed line is the linear fit to the solid line over the frequency range 0.02 to 0.1 Hz. For this example, δt^* is –0.59. The difference to the value given in Table 3 (see CDROM in back cover sleeve) is due to the removal of the event mean.

4. THE PHYSICAL STATE OF THE UPPER MANTLE

To fully appreciate these results, attenuation measurements are combined with velocity measurements [*Roth et al.*, 2000; *Anderson*, 1989; *Nowick and Berry*, 1972]. This analysis can be used to estimate the temperature [*Karato*, 1993; *Sato and Ryan*, 1994; *Goes et al.*, 2000], percent partial melt [*Hammond and Humphreys*, 2000b; *Hammond and Humphreys*, 2000a; *Sato and Ryan*, 1994], and composition [*Karato and Jung*, 1998; *Duffy and Anderson*, 1989] of the material through which the seismic wave has passed. Most regional velocity studies have either assumed a constant one dimensional Q profile with depth or assumed a simple relationship between velocity and Q [*Goes et al.*, 2000; *Sobolev et al.*, 1996]. Not knowing the true value of Q or its relationship to temperature, frequency, or velocity can be detrimental in the interpretation of sub-surface anomalies.

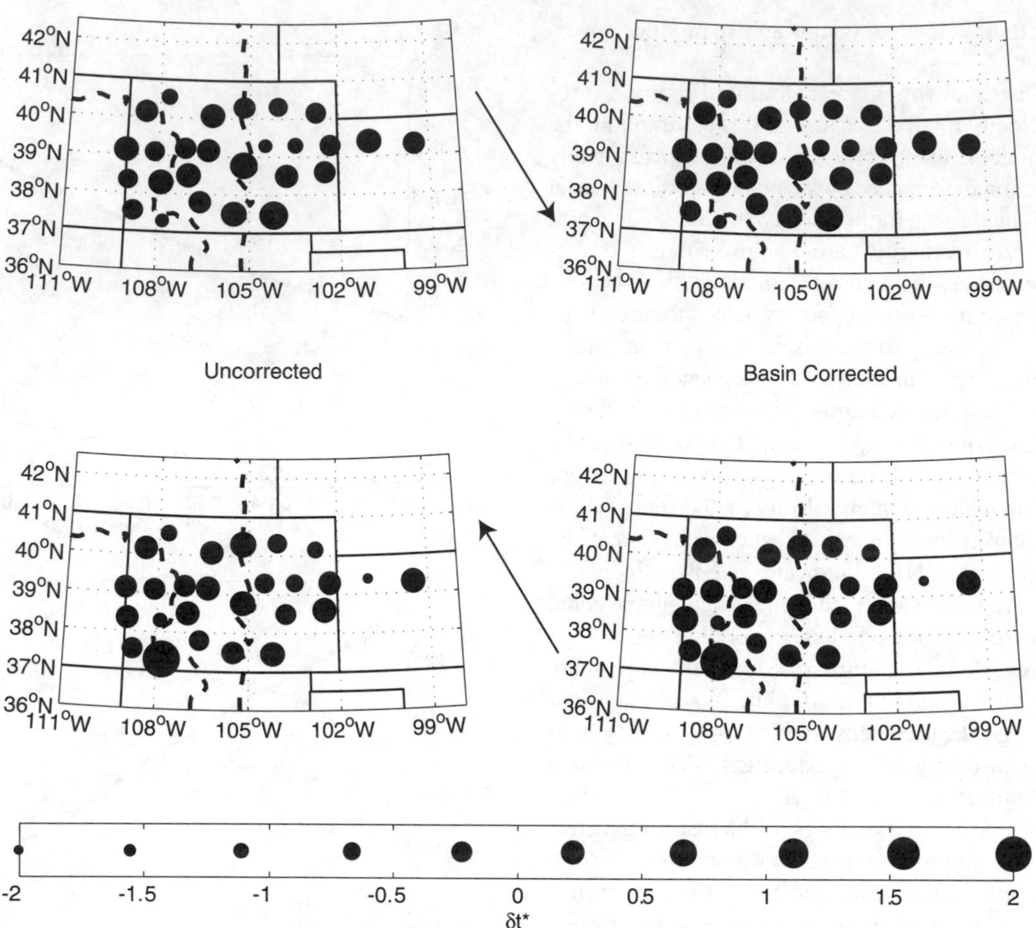

Figure 6. Weighted average δt^* maps for events originating from the northwest (top) and southeast (bottom) for uncorrected (left) and sediment corrected (right) values of δt^*. The weight is the inverse of the expected standard deviation of δt^* due to random noise (Appendix 1.2; see CDROM in back cover sleeve). Symbols are placed at station locations. Larger symbols indicate greater δt^*.

Based on laboratory experiments [*Jackson*, 1993], attenuation is commonly observed to follow a power-law relationship where Q is proportional to angular frequency, w, raised to some positive power α. This model is believed to be due to the movement of dislocations having a range of thermally activated strengths and relaxation times and/or activation energies encompassing periods of at least 10^{-1} to 10^2 seconds. Minster and Anderson [1981] proposed that there should be a minimum and maximum relaxation time spanning two to three decades in frequency. Above and below these values, attenuation is predicted to drop off to zero and produce an attenuation band or peak [*Nowick and Berry*, 1972]. Anderson and Given [1982] presented frequency dependent attenuation results in the mantle to which they ascribed an attenuation band, but their results as to the maximum and minimum relaxation times was inconclusive. From frequency dependent t^* measurements, Warren and Shearer [2000] are unable to conclusively observe an attenuation band but are able to state that attenuation decreases with increasing frequency.

4.1 Effects of Composition

The relationships between the shear attenuation model and shear velocity model are presented in Figure 8. The relative shear velocity model is converted to absolute velocities by assuming the values given by Lee and Grand [1996] are relative to PREM and that the average shear velocities in the model below 250 km depth are equal to PREM. To make this comparison and perform the analysis below, we must also assume that the frequencies used to measure velocity and attenuation are the same, ~0.06 Hz.

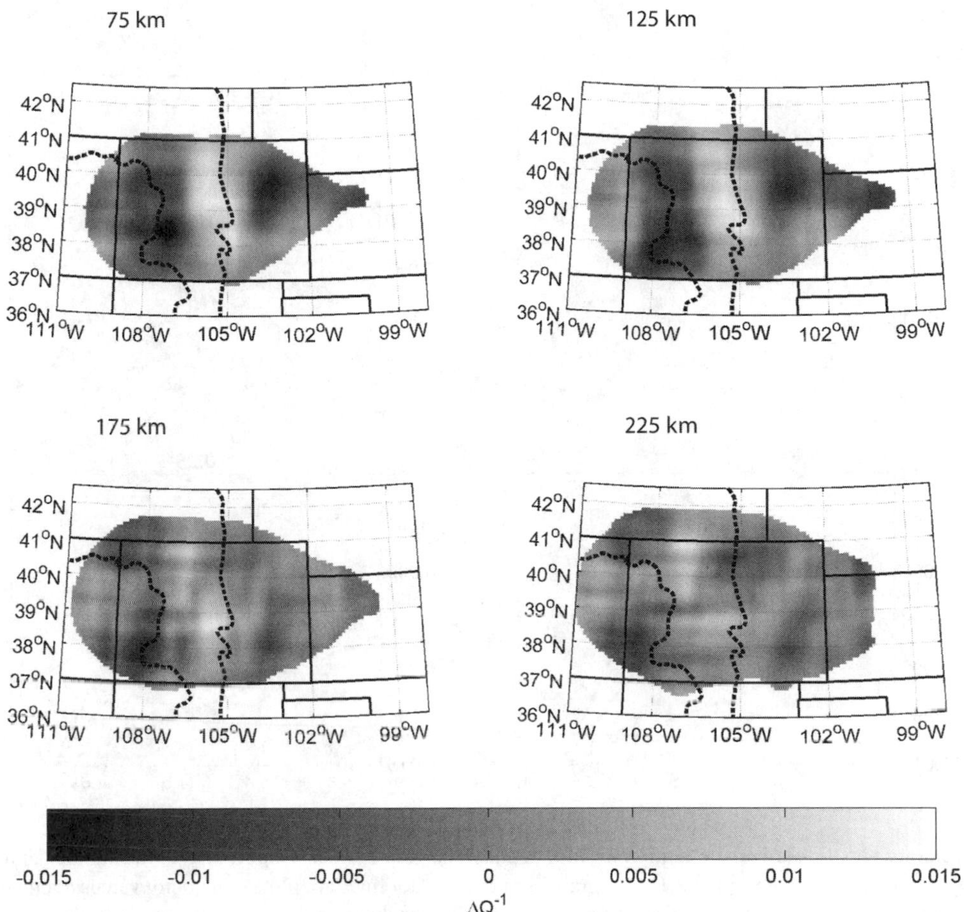

Figure 7. Differential attenuation tomography results at four depths. Relatively strong attenuation occurs beneath the eastern Rocky Mountains to depths of ~150 km and trends north from the Rio Grande Rift.

Compositional and thermal effects can result in a given value of attenuation and velocity. If we assume that Q has a positive power law dependence with respect to frequency ($\alpha > 0$), Minster and Anderson [1981] provide the following relationship between velocity and attenuation

$$V_S(w) \approx V_U\left[1-\frac{1}{2}\cot\frac{\alpha\pi}{2}Q_S^{-1}(w)\right]. \quad (14)$$

V_U is the unrelaxed shear velocity due either to sufficiently high frequency or low temperature where anelastic mechanisms do not have time or energy to operate. The unrelaxed velocity depends primarily on composition. For example, the fractional change in the unrelaxed shear velocity relative to the fractional change in the iron content for Proterozoic sub continental lithospheric mantle (SCLM) [*Griffin et al.*, 1999], $\partial \ln V_S/\delta X_{Fe}$, is approximately -0.3 (Table 4; see CDROM in back cover sleeve). A 1% decrease in iron content increases the shear velocity by 0.3%. Decreasing iron content, an effect that accompanies older and more mature SCLM and the melting of mantle peridotite, increases the seismic velocity. Another compositional effect that can change shear velocity is simply the bulk mineralogy. More mature SCLM typically has less garnet and more olivine. In going from the Phanerozoic to Archean compositions reported by Griffin et al. [1999] where garnet decreases by 4% and olivine increases by 3%, shear velocity increases by 5%.

If the relationship provided by Minster and Anderson is correct and we know α, then we can use velocity and attenuation models to solve for changes in V_U, an indicator of compositional variation. Values for α measured in the laboratory on polycrystalline aggregates under upper mantle conditions at seismic frequencies range from 0.2 to 0.3 [Karato and Spetzler, 1990]. Seismological estimates for α are on the order of 0.15 [Warren and Shearer, 2000; Sobolev et al., 1996].

The straight lines in Figure 8 represent different values of a in equation 14. For the upper layers at 75 and 125 km depth,

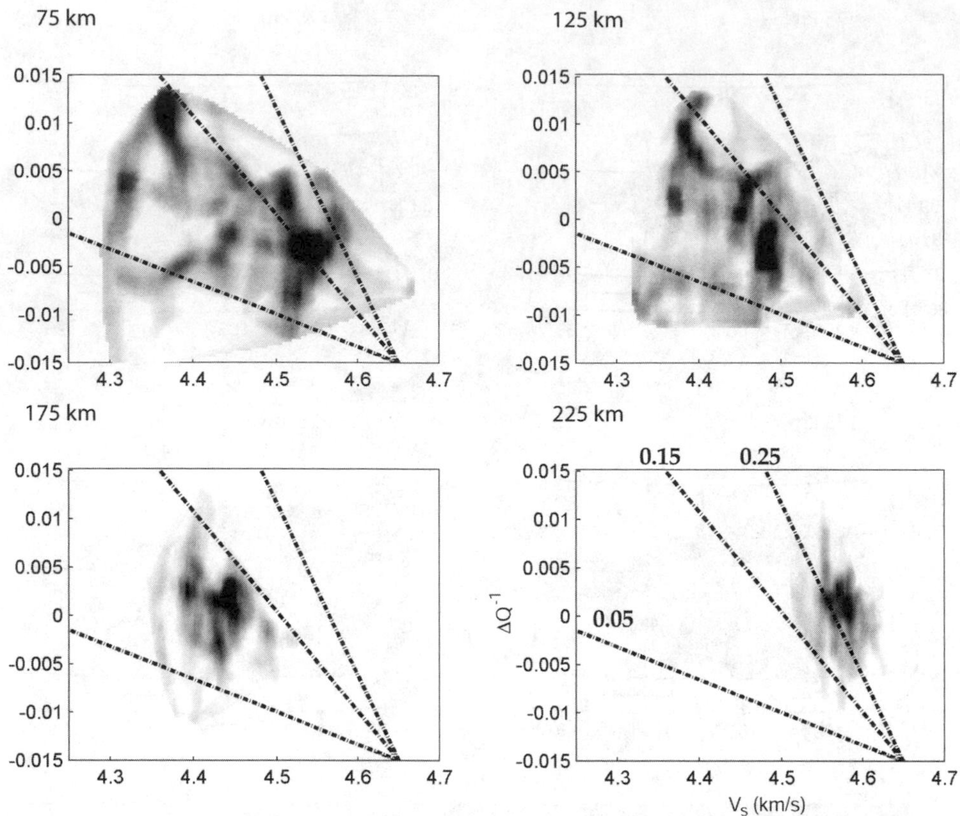

Figure 8. Velocity versus relative attenuation for four depths. The gray scale is proportional to the number of model points having that velocity and attenuation, a 2-D histogram. The dashed-dot lines are plots of velocity versus attenuation according to equation 14 for different values of *a*. Movement along a line is due to changing temperature while movement across a line is due to changing V_U or composition. Seismological studies have found α equal to 0.15 [*Warren and Shearer*, 2000; *Sobolev et al.*, 1996] while laboratory experiments have found α between 0.2 and 0.3 [*Karato and Spetzler*, 1990].

the overall trend in the data shows low values of α, on the order of 0.15. If we examine the slope of the smaller scale trends and trends at greater depth, 175 and 225 km, we see that the slope increases to values of α closer to 0.25. We therefore examine compositional variability by rearranging equation 14, using the V_S and Q^{-1} models and assuming α equals 0.25, and solving for the percent change in unrelaxed shear velocity, $\%\Delta V_U$. The resulting compositional models presented below do not change significantly for α equal to 0.15.

Figure 9 contains horizontal slices of the relative unrelaxed shear velocity. Unrelaxed shear velocities tend to increase gradually to the east and may reflect temperature dependent phase changes such as the reactions in Table 4 (see CDROM in back cover sleeve). Application of equation 14 predicts very low unrelaxed shear velocities beneath the southwestern Colorado Rocky Mountains. If this were partial melt and the local fluid flow attenuation mechanism were operating [*Hammond and Humphreys*, 2000b; *Hammond and Humphreys*, 2000a], we would see relatively low unrelaxed velocities. A shear velocity reduction of 6.5% would indicate 0.8% partial melt (Table 4; see CDROM in back cover sleeve). But conventional models of attenuation in which attenuation simply increases with temperature suggest that attenuation would remain high since high temperatures would presumably be needed to produce partial melt. Since the shear wave attenuation has dropped considerably, we have either observed a very unusual composition or the high temperature side of an attenuation peak [*Anderson and Given*, 1982]. If we are in fact on the high temperature side of an attenuation peak, equation 14 no longer holds in this region and unrelaxed velocities are greater than presented in Figure 9. After assessing the possible changes in temperature, we will return to the question of composition.

4.2 Effects of Temperature

We can calculate changes in temperature for the case of either unusual composition or the high temperature side of an attenuation peak. We adopt the relation for the temperature

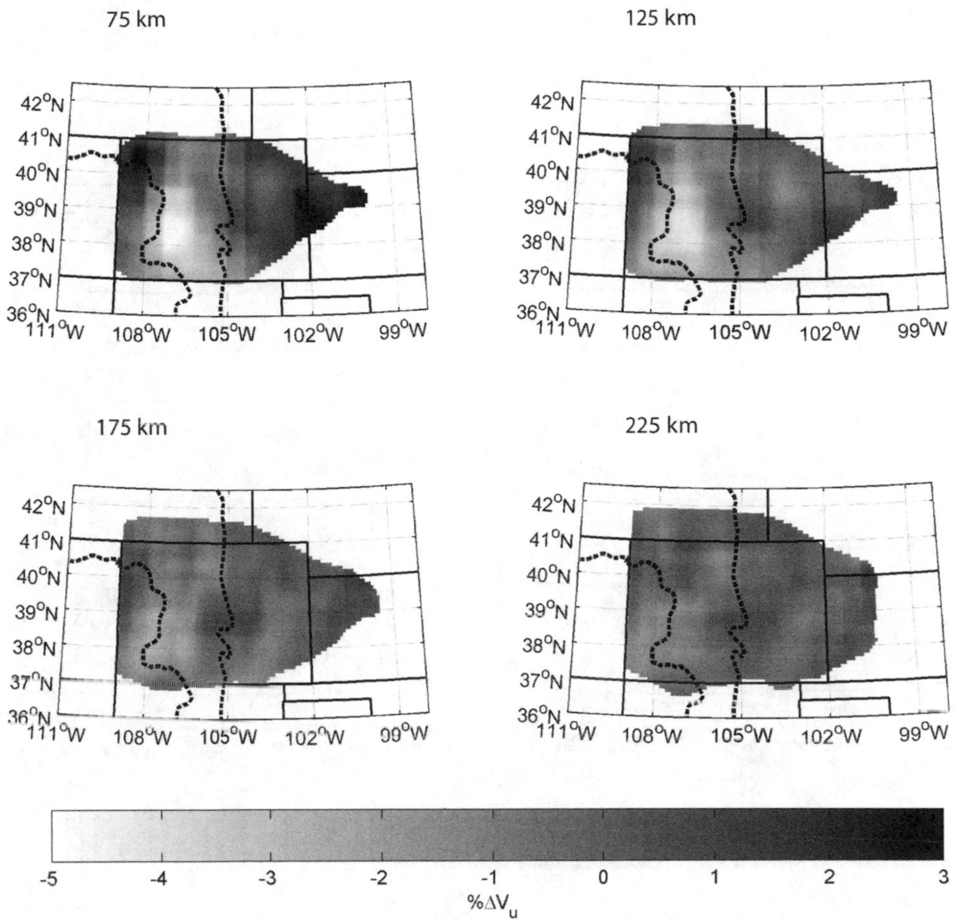

Figure 9. Unrelaxed shear velocity at four depths. Unrelaxed shear velocity gradually increases to the east from the Colorado Rockies. A significant drop in unrelaxed shear velocity occurs beneath the southwest Colorado Rocky Mountains and may indicate either high temperatures and an attenuation peak or unusual compositions.

derivative with respect to shear velocity given by Karato [1993] including effects of pressure,

$$\frac{\partial \ln V_S(\omega)}{\partial T} = \frac{\partial \ln V_U}{\partial T} - F(\alpha)\frac{Q^{-1}(\omega)}{\pi}\frac{H^* + PV^*}{RT^2} \ . \quad (15)$$

$\partial \ln V_U / \partial T$ is the normalized derivative of the shear velocity with respect to temperature for the unrelaxed state. From 50 to 400 km depth, $\partial \ln V_U / \partial T$ for olivine, assumed to be the dominant mineral in the upper mantle for this region, is approximately -0.76 x 10^{-4} K^{-1} [*Karato*, 1993]. ω is the angular frequency. $F(\alpha)$ is a constant related to the frequency dependence of attenuation where Q is proportional to ω^{α}. $F(\alpha)$ is given by

$$F(\alpha) = \frac{\pi\alpha}{2}\cot\frac{\pi\alpha}{2} \ . \quad (16)$$

Again we choose $\alpha = 0.25$ which gives $F(\alpha) = 0.95$. As α goes to 0, $F(\alpha)$ goes to 1. A larger value of a reduces the dependence of (15) on Q. H^* is the activation energy of the thermally activated process, V^* is the activation volume, P is the pressure, R is the gas constant and T is the reference temperature. We use H^* = 500 kJ/mol, the activation energy for the diffusion of oxygen through olivine [*Karato and Spetzler*, 1990]. The activation volume, V^*, controls how pressure affects the activation energy. We use V^* = 4 x 10^{-6} m^3/mol [*Anderson*, 1989]. The pressure in the mantle is taken from PREM [*Dziewonski and Anderson*, 1981] and the reference temperature is taken from the geotherm reported by Stacey [1992], ~1200 to 1800 K from 50 to 400 km depth.

This formulation requires absolute attenuation. In addition, to account for composition, we must consider $\partial \ln V_S$ for velocities relative to the unrelaxed shear velocity, $(V_S - V_U)/V_U$, for which the calculation of V_U also requires absolute attenuation. We determine absolute attenuation by assuming that the

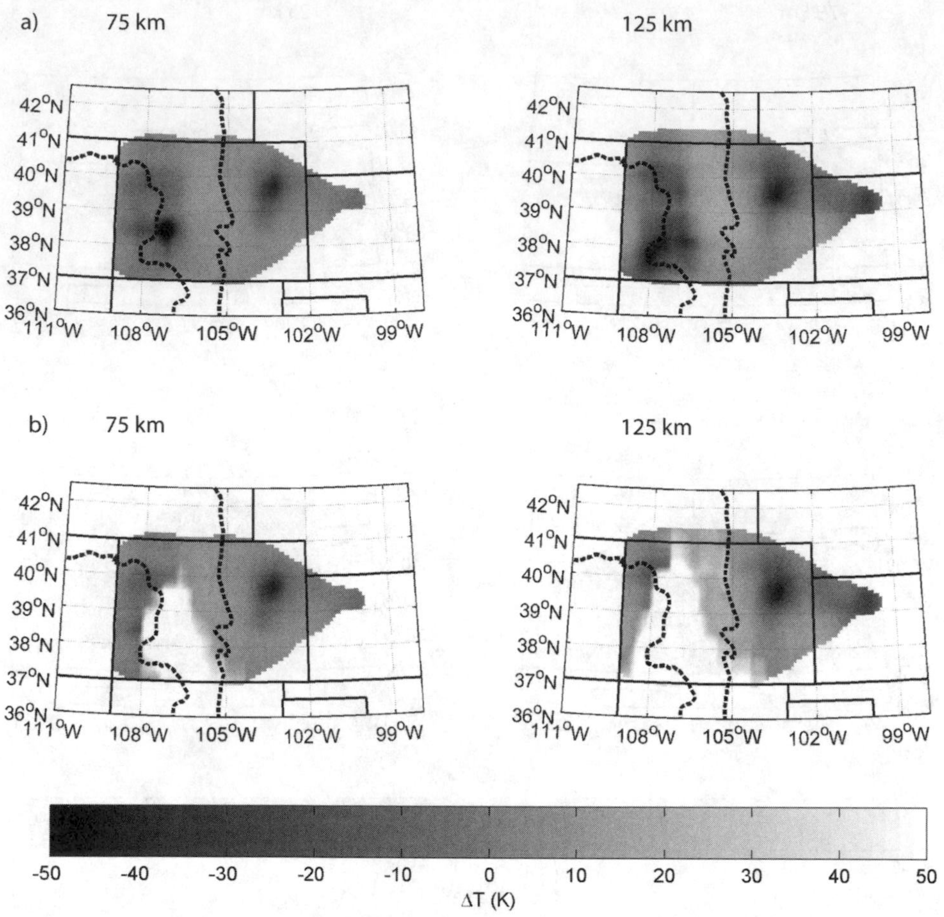

Figure 10. Temperature contrast at 75 and 125 km depth from the use of equation 15. In a) equation 14 holds in all regions, i.e. there is no attenuation peak. Maximum temperature contrasts are limited to elevated values of 80 K at 125 km beneath the eastern Colorado Rockies. In b) equation 14 holds everywhere except in the region of very low unrelaxed shear velocity, i.e. the low velocities beneath southwest Colorado are due to an attenuation peak resulting from very high temperatures. Unrelaxed shear velocity is not allowed to drop below the values on the edge of the anomalously slow region, -2%ΔV_U. Maximum temperature contrasts increase to elevated values of over 300 K at 75 km beneath the southwestern Colorado Rockies.

smallest absolute attenuation in our model is close to zero and add a constant value to our differential attenuation model. We add 0.015 to our relative attenuation model and require attenuation to be positive and greater than 0.005. The requirement that absolute attenuation be greater than 0.005 is done only to inhibit extreme changes in temperature.

We must now consider two cases when applying (15) to derive changes in temperature. The first case assumes that the power-law relation between velocity and attenuation, equation 14, holds in all regions and that the composition of the upper mantle beneath the southwest Colorado Rocky Mountains is unusual. In this case, changes in temperature derived from equation 15 are independent of velocity since the fractional change in velocity becomes only a function of attenuation,

$$\Delta \ln V_S = -\frac{1}{2}\cot\frac{\alpha\pi}{2}Q_S^{-1}\,. \qquad (17)$$

Changes in temperature are then proportional to attenuation with the greatest increase in temperature beneath the eastern Colorado Rocky Mountains (Figure 10a). At 125 km depth, temperatures are elevated to about 80 K greater than the anomalous region to the southwest and only 50 K greater than the Great Plains. For the second case, where the anomalous region is due to an attenuation peak, the unrelaxed shear velocity is not allowed to fall below the highest unrelaxed shear velocity along the boundaries of this anomalous region. Equation 15 for velocity-temperature relations still holds, though for a different value of α as it goes to zero and becomes negative on the high temperature side of the attenuation peak. We continue to use α equals 0.25 and incur a small penalty. For this case, temperatures at 125 km depth beneath the southwest Colorado Rocky Mountains increase by over 300 K (Figure 10b).

4.3 Upper Mantle Composition and Isostatic Compensation

An additional constraint to help us differentiate the possible thermal and compositional models is the subsurface density distribution required to support the Rocky Mountains and surrounding regions. We first account for thermal contributions to upper mantle isostacy for the two temperature models and isostatic compensation due to thickened crust [*Li et al.*, 2002] assuming a constant density contrast across the Moho of 400 kg/m^3. We then look for the relationship between density and velocity in the upper mantle that produces the best variance reduction of the residual topography. This relationship is compared to predicted values to estimate the compositional heterogeneity in the upper mantle. The remaining topography, after accounting for the above factors, is due to density anomalies in the crust.

Fractional changes of density with respect to velocity, $\partial\ln\rho/\partial\ln V_S$, are estimated for several different mineral reactions, changes in temperature, and partial melt (Table 4; see CDROM in back cover sleeve). Density, velocity and thermal expansivity (~ 3.3 x 10^{-5} K^{-1}) are calculated using the techniques of Bina and Helffrich [1992] and Holland and Powell [1998] and the mineral data compiled by Hacker et al. [2003]. A starting mineral assemblage of preferred Proterozoic SCLM [*Griffin et al.*, 1999] is used and the above values are averaged from 50 to 200 km depth. We compute the mantle thermal component of isostatic compensation using the two differential temperature models described previously while assuming that mass anomalies affecting compensation are confined to between 50 and 200 km depth. The mass needed to compensate the mass deficit in the mantle results in topography having a density of 2700 kg/m^3 [*Snelson et al.*, 1998]. We also calculate the crustal component of isostatic compensation using the crustal thickness map of Li et al. [2002] and a density contrast across the Moho of 400 kg/m^3.

Figure 11 shows the variance reduction of smoothed topography in dependence on $\partial\ln\rho/\partial\ln V_S$ for the upper mantle due to changes in composition while accounting for upper mantle thermal and crustal thickness contributions. Thickened crust alone accounts for 57% of the topographic variance (solid horizontal line). Up to an additional 7% of the variance can be explained by the thermal variations in the upper mantle (short horizontal lines spanning their associated values of $\partial\ln\rho/\partial\ln V_S$). Two cases are presented representing the two possible relationships between velocity and attenuation.

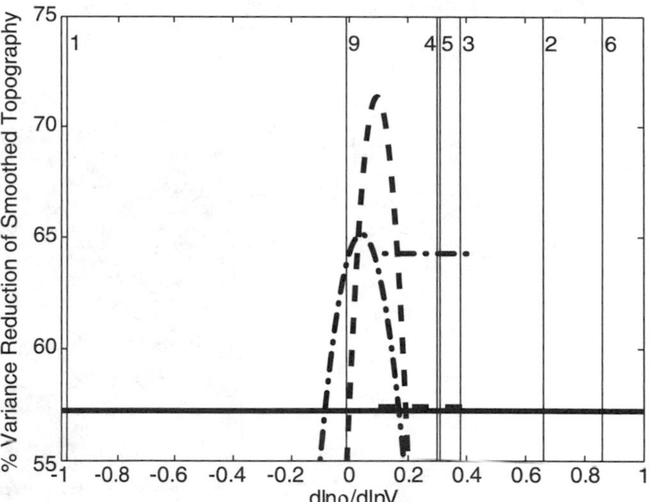

Figure 11. Variance reduction of elevation data versus fractional derivative of density with respect to shear velocity. The solid horizontal line is the variance reduction for thickened crust. The dashed line represents the low temperature model while the dashed-dot line represents the high temperature model. The short horizontal lines indicate the variance reduction due to the addition of thermal effects and span the appropriate range of $\partial\ln\rho/\partial\ln V_S$ for that phenomenon. The curved lines represent the addition of the compositional component of mantle compensation for a range of $\partial\ln\rho/\partial\ln V_S$. The vertical lines are values of $\partial\ln\rho/\partial\ln V_S$ for the different compositional effects in Table 4 (see CDROM in back cover sleeve).

Each case results in a thermal model and compositional model. The dashed-dot line represents mantle contributions to isostacy when there is an attenuation peak (case 1) while the dashed line represents contributions when attenuation has a simple linear relationship to velocity (case 2). The vertical lines are values of $\partial\ln\rho/\partial\ln V_S$ for the different compositional effects in Table 4 (see CDROM in back cover sleeve). For case 1 where high temperatures coincide with the lowest shear velocities, the thermal model increases the variance reduction by 7%. The compositional counterpart for this case adds little improvement. For case 2 where unusual composition coincides with the lowest shear velocities, the thermal model increases the variance reduction by less than 1% but the corresponding compositional model improves the variance reduction by 14%.

For both relationships between velocity and attenuation, the best $\partial\ln\rho/\partial\ln V_S$ lies somewhere between melt with basalt depletion and the other mineral reactions. Assuming that no Fe/Mg fractionation occurs for reactions 1-6 (Table 4; see CDROM in back cover sleeve), the measured value of $\partial\ln\rho/\partial\ln V_S$ implies that some melt is present. The high temperature case would make hydrous phases unstable. Antigorite is stable below 1000 K [*Bromiley and Pawley*, 2003] while phlogopite can remain stable up to 1600 K [*Sato et*

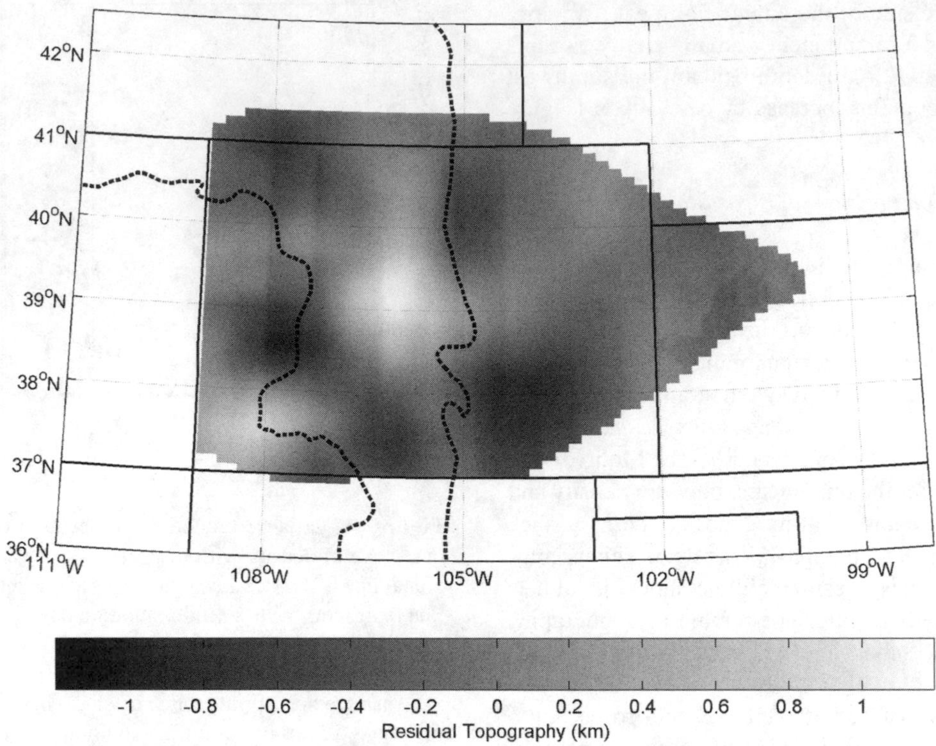

Figure 12. Residual topography after the compensation of thickened crust and upper mantle density contrasts. The remaining uncompensated topography is in the central Colorado Rockies, a position coinciding with low crustal shear velocities and the inferred presence of a large low density granitic intrusion.

al., 1997]. Increasing temperature is likely to drive garnet to orthopyroxene and corundum (reaction 2) [*Aranovich and Berman*, 1997] or garnet and olivine to orthopyroxene and spinel (reaction 6) [*Danckwerth and Newton*, 1978]. These reactions have the correct sense in that higher temperatures produce compositions that are slower and lighter, however reaction 2 is unlikely to produce a large signal because of its small values of $\partial \ln V_S / \partial R$ (Table 4; see CDROM in back cover sleeve) and $\partial R / \partial T$ [*Aranovich and Berman*, 1997]. Though the high temperature model with its' associated compositional variations produces a consistent interpretation, the second case in which velocity has a simple linear relationship with respect to attenuation produces a better variance reduction. For such a model, elevated temperatures are slight and melt will only be present if the melting temperature has been reduced by the presence of hydrous phases. Antigorite is unlikely to be stable very far below the Moho while phlogopite could be stable throughout most of the mantle lithosphere. The presence of phlogopite in the upper mantle beneath southwest Colorado is consistent with its presence in lamproite dikes in the Colorado Plateau [*Wannamaker et al.*, 2000] and highly potassic magmatism in southwest Colorado [*Mutschler et al.*, 1987].

Researchers have speculated on the presence of phlogopite in the upper mantle beneath southwest Colorado but have not been able to determine its subsurface extent. If we assume that our data is explained by the presence of phlogopite and partial melt, we can derive their proportions from the estimated $\partial \ln \rho / \partial \ln V_S$ in Figure 11, 0.096, and the relations

$$\partial \ln \rho = \{\partial \ln \rho\}_A + \{\partial \ln \rho\}_B \tag{18}$$

$$\partial \ln V_S = \{\partial \ln V_S\}_A + \{\partial \ln V_S\}_B \tag{19}$$

which give

$$\frac{\partial \ln \rho}{\partial \ln V_S} = \left\{\frac{\partial \ln \rho}{\partial \ln V_S}\right\}_A \frac{\{\partial \ln V_S\}_A}{\partial \ln V_S} + \left\{\frac{\partial \ln \rho}{\partial \ln V_S}\right\}_B \frac{\partial \ln V_S - \{\partial \ln V_S\}_A}{\partial \ln V_S} \cdot \tag{20}$$

The subscript A denotes reaction 9, melt with depletion, and subscript B denotes reaction 4, substitution of olivine with phlogopite. In solving for $(\partial \ln V_S)_A / \partial \ln V_S$, the fraction of velocity anomaly due to partial melt, we find that 66% of the velocity anomaly is due to melt and the remainder, 34%, to phlogopite. Using the derivatives of velocity versus phlogopite or melt fraction (Table 4; see CDROM in back cover sleeve) and the fractional change in the unrelaxed velocity, we conclude that if up to 6% of the unrelaxed velocity anomaly is due to these processes, there is up to 0.5% partial melt and up to 3% substitution of olivine with phlogopite beneath the southwest Colorado Rockies. If the unrelaxed shear velocities beneath southwest Colorado were higher due either to greater anelastic shear velocities or greater attenuation, the value of $\partial \ln \rho / \partial \ln V_S$ for which there is a maximum variance reduction of residual topography would be greater. Because of this and the reduction in the velocity anomaly, there would be less partial melt.

The residual elevation anomaly after accounting for thickened crust and the thermal effects of the low temperature upper mantle model and its associated compositional effects as stated in the previous paragraph is shown in Figure 12. Nearly 1.5 km of elevation remains in central Colorado. Work by Li et al. [2002] have found slow shear wave velocities in the crust of central Colorado and attributed these slow velocities to low density granitic intrusions [*Decker et al.*, 1988]. To account for the residual elevations in central Colorado, the density of this material distributed over the full height of the crust, 45 km, should be reduced by about 3%. Decker et al. suggested 4.5% but distributed the anomaly over 30 km of the upper crust.

Some of the negative elevation residual anomalies coincide with the thickest crust (e.g. southwest Colorado). Li et al. point out that their crustal thickness model would allow a 5 km reduction in crustal thickness to be compensated by a 3% reduction in shear velocity. If limited to the negative anomalies in the southwest Colorado Rocky Mountains, this effect would reduce the variance of the elevation residual and provide a better correlation between the velocity models of Li et al. [2002] and that of Lee and Grand [1996].

5. CONCLUSION

Before too much weight is given to any interpretation, we must respect the amount of uncertainty. The primary problem with the results presented above and most field t^* measurements is the significant amount of noise expected for each measurement. Appendix 1 (see CDROM in back cover sleeve) sheds light on some of the possible errors present in the δt^* measurement while Appendix 2 (see CDROM in back cover sleeve) evaluates the effect of some of this error on model resolution. Errors considered for the δt^* measurement include normally distributed random noise and non random noise from basin reflections. Additional errors that are not considered include anisotropy, multipathing, improper ray tracing resulting from not using a three dimensional velocity model and frequency dependent variations in amplitudes across the array resulting from source radiation patterns. In addition to the uncertainty of the observations, the theoretical relationship between shear velocity and attenuation is uncertain. Rock physics experimentation has made progress but has had difficulty making absorption and dispersion measurements under mantle conditions at seismic frequencies with rocks of appropriate composition. Further, the temperature models presented above are relative and knowing the correct phase relationships between mineral assemblages requires knowing the absolute pressure and temperature.

Aside from the uncertainties, several results can be stated. High attenuation underlies the eastern Colorado Rockies and is likely due to elevated temperatures. East of the Rockies, attenuation gradually decreases as velocity gradually increases reflecting at least a simple decrease in temperature. To the west low attenuation coincides with the lowest velocities, and the interpretation is less conclusive. If a frequency limited attenuation band is active, the low velocities and low attenuation may be very hot mantle, over 300 K above its surroundings. If attenuation obeys a more simple relationship to velocity as in equation 14, this anomalous region is due to compositional variation. Our preferred model assumes a linear relationship between velocity and attenuation over the temperatures and frequencies encountered in this study and suggests that the southwest Colorado Rockies are underlain by up to 0.5% partial melt and the substitution of up to 3% olivine with phlogopite.

After the addition of various density anomalies, we find that the Colorado Rocky Mountains are supported by low density mantle, thick crust [*Li et al.*, 2002], and in some cases, low density crust [*Li et al.*, 2002]. The low density mantle may either be thermal and/or compositional in origin. The presence of recent and potassic magmatism above the anomalous region suggests that partial melt and the presence of phlogopite are responsible for at least some of the decrease in velocity and density.

Acknowledgements. We thank Karen Fischer and Steve Grand for thorough reviews that helped to substantially improve this manuscript and Randy Keller for his work as editor. Instruments used in the Rocky Mountain Front study were from the IRIS Passcal instrument pool. This work was supported in part by National Science Foundation grants EAR-9614410 and EAR-0003747.

REFERENCES

Al-Khatib, H.H., and B.J. Mitchell, Upper Mantle Anelasticity and Tectonic Evolution of the Western United-States from Surface-Wave Attenuation, Journal of Geophysical Research, 96 (B11), 18129–18146, 1991.

Allen, R.M., G. Nolet, W.J. Morgan, K. Vogfjord, B.H. Bergsson, P. Erlendsson, G.R. Foulger, S. Jakobsdottir, B.R. Julian, M. Pritchard, S. Ragnarsson, and R. Stefansson, The thin hot plume beneath iceland, Geophysical Journal International, 137, 51–63, 1999.

Anderson, D.L., Anelasticity, in Theory of the Earth, pp. 279–302, Blackwell Scientific Publications, Boston, 1989.

Anderson, D.L., and J.W. Given, Absorption Band Q Model for the Earth, Journal of Geophysical Research, 87 (B5), 3893–3904, 1982.

Aranovich, L.Y., and R.G. Berman, A new garnet-orthopyroxene thermometer based on reversed Al2O3 solubility in FeO-Al2O3-SiO2 orthopyroxene, American Mineralogist, 82, 345–353, 1997.

Bhattacharyya, J., G. Masters, and P. Shearer, Global lateral variations of shear wave attenuation in the upper mantle, Journal of Geophysical Research, 101 (B10), 22273–22289, 1996.

Bina, C.R., and G.R. Helffrich, Calculation of elastic properties from thermodynamic equation of state principles, Annu. Rev. Earth Planet. Sci., 20, 527–552, 1992.

Bromiley, G.D., and A.R. Pawley, The stability of antigorite in the systems MgO-SiO2-H2O (MSH) and MgO-Al2O3-SiO2-H2O (MASH): The effects of Al3+ substitution on high-pressure stability, American Mineralogist, 88 (1), 99–108, 2003.

Danckwerth, A., and R.C. Newton, Experimental determination of the spinel peridotite to garnet peridotite reaction in the system MgO-Al2O3-SiO2 in the range 900-1000o C and Al2O3 isopleths of enstatite in the spinel field, Contributions to Mineralogy and Petrology, 66, 189–201, 1978.

Decker, E.R., H.P. Heasler, K.L. Buelow, K.H. Baker, and J.S. Hallin, Significance of past and recent heat-flow and radioactivity studies in the Southern Rocky Mountain region, Geological Society of America Bulletin, 100, 1851–1885, 1988.

Duffy, T.S., and D.L. Anderson, Seismic Velocities in Mantle Minerals and the Mineralogy of the Upper Mantle, Journal of Geophysical Research, 94 (B2), 1895–1912, 1989.

Dziewonski, A.M., and D.L. Anderson, Preliminary reference Earth model, Physics of the Earth and Planetary Interiors, 25, 297–356, 1981.

Goes, S., R. Govers, and P. Vacher, Shallow mantle temperatures under Europe from P and S wave tomography, Journal of Geophysical Research, 105 (B5), 11153–11169, 2000.

Grand, S.P., and D. Helmberger, Upper mantle shear structure of North America, Geophysical Journal International, 76, 399–438, 1984.

Griffin, W.L., S.Y. O'Reilly, and C.G. Ryan, The composition and origin of sub-continental lithospheric mantle, in Mantle Petrology: Field observations and High Pressure Experimentation: A tribute to Francis R. (Joe) Boyd, edited by B.O. Mysen, pp. 13–46, The Geochemical Society, Houston, 1999.

Hacker, B.R., G.A. Abers, and S.M. Peacock, Subduction factory 1: Theoretical mineralogy, densities, seismic wave speeds, and H2O contents, Journal of Geophysical Research, 108 (B1), 2029, doi:10.1029/2001JB001127, 2003.

Hammond, W.C., and E.D. Humphreys, Upper mantle seismic wave attenuation: Effects of realistic partial melt distribution, Journal of Geophysical Research, 105 (B5), 10987–10999, 2000a.

Hammond, W.C., and E.D. Humphreys, Upper mantle seismic wave velocity: Effects of realistic partial melt geometries, Journal of Geophysical Research, 105 (B5), 10975–10986, 2000b.

Hessler, E.K., Upper mantle seismic structure of the Southern Rocky Mountains - Great Plains transition, M.S. thesis, University of Oregon, Eugene, 1997.

Holland, T.J.B., and R. Powell, An internally consistent thermodynamic data set for phases of petrological interest, Journal of Metamorphic Petrology, 16, 309–343, 1998.

Jackson, I., Progress in the Experimental Study of Seismic Wave Attenuation, Annu. Rev. Earth Planet. Sci., 21 (1), 375–406, 1993.

Karato, S., Importance of anelasticity in the interpretation of seismic tomography, Geophysical Research Letters, 20 (15), 1623–1626, 1993.

Karato, S., and H. Jung, Water, partial melting and the origin of the seismic low velocity and high attenuation zone in the upper mantle, Earth and Planetary Science Letters, 157 (3–4), 193–207, 1998.

Karato, S., and H.A. Spetzler, Defect microdynamics in minerals and solid state mechanisms of seismic wave attenuation and velocity dispersion in the mantle, Reviews of Geophysics, 28 (4), 399–421, 1990.

Lay, T., and T.C. Wallace, Multiple ScS Attenuation and Travel-Times beneath Western North-America, Bulletin of the Seismological Society of America, 78 (6), 2041–2061, 1988.

Lee, D.-K., and S.P. Grand, Upper mantle shear structure beneath the Colorado Rocky Mountains, Journal of Geophysical Research, 101 (B10), 22233–22244, 1996.

Lerner-Lam, A.L., A.F. Sheehan, S.P. Grand, E. Humphreys, K.G. Dueker, E. Hessler, H. Guo, D.-K. Lee, and M.K. Savage, Deep Structure Beneath the Southern Rocky Mountains from the Rocky Mountain Front Broadband Seismic Experiment, Rocky Mountain Geology, 33 (2), 199–216, 1998.

Li, A., D.W. Forsyth, and K.M. Fischer, Evidence for shallow isostatic compensation of the southern Rocky Mountains from Rayleigh wave tomography, Geology, 30 (8), 683–686, 2002.

Meju, M.A., Error analysis in linear inversion, in Geophysical data analysis: Understanding inverse problem theory and practice, pp. 294, Society for Exploration Geophysicists, Tulsa, 1994.

Menke, W., Singular Value Decomposition and the Natural Generalized Inverse, in Geophysical Data Analysis: Discrete Inverse Theory, pp. 119–124, Academic Press, Orlando, 1984.

Minster, J.B., and D.L. Anderson, A model of dislocation-controlled rheology for the mantle, Philosophical Transactions of the Royal Society of London Series a-Mathematical Physical and Engineering Sciences, 299 (1449), 319–356, 1981.

Mutschler, F.E., E.E. Larson, and R.M. Bruce, Laramide and younger magmatism in Colorado - new petrologic and tectonic variations on old themes, Colorado School on Mines Quarterly, 1, 1–47, 1987.

Nataf, H.C., and Y. Ricard, An a priori tomographic model of the upper mantle based on geophysical modeling, Physics of the Earth and Planetary Interiors, 95, 101–122, 1996.

Nowick, A.S., and B.S. Berry, Anelastic Relaxation in Crystalline Solids, 677 pp., Academic Press, New York, 1972.

Percival, D.B., and A.T. Walden, Spectral Analysis for Physical Applications: Multitaper and Conventional Univariate Techniques, Cambridge University Press, Cambridge, 1993.

Reid, F.J.L., J.H. Woodhouse, and H.J.v. Heijst, Upper mantle attenuation and velocity structure from measurements of differential S phases, Geophysical Journal International, 145, 615–630, 2001.

Romanowicz, B., A global tomographic model of shear attenuation in the upper mantle, Journal of Geophysical Research, 100, 12375–12394, 1995.

Roth, E.G., D.A. Wiens, and D.P. Zhao, An empirical relationship between seismic attenuation and velocity anomalies in the upper mantle, Geophysical Research Letters, 27 (5), 601–604, 2000.

Sato, H., and M.P. Ryan, Generalized Upper Mantle Thermal Structure, in Magmatic Systems, edited by M.P. Ryan, pp. 259–290, Academic Press, New York, 1994.

Sato, K., T. Katsura, and E. Ito, Phase relations of natural phlogopite with and without enstatite up to 8 GPa: implication for mantle metasomatism, Earth and Planetary Science Letters, 146, 511–526, 1997.

Sheehan, A.F., G.A. Abers, C.H. Jones, and A. Lerner-Lam, Crustal thickness variations across the Colorado Rocky Mountains from teleseismic receiver functions, Journal of Geophysical Research, 100 (B10), 20391–20404, 1995.

Sheehan, A.F., and S.C. Solomon, Differential Shear Wave Attenuation and Its Lateral Variation in the North Atlantic Region, Journal of Geophysical Research, 97 (B11), 15339–15350, 1992.

Slack, P.D., P.M. Davis, W.S. Baldridge, K.H. Olsen, A. Glahn, U. Achauer, and W. Spence, The upper mantle structure of the central Rio Grand rift region from teleseismic P and S wave travel time delays and attenuation, Journal of Geophysical Research, 101 (B7), 16003–16023, 1996.

Snelson, C.M., T.J. Henstock, G.R. Keller, K.C. Miller, and A. Lavender, Crustal and uppermost mantle structure along the Deep Probe seismic line, Rocky Mountain Geology, 33 (2), 181–198, 1998.

Sobolev, S.B., H. Zeyen, G. Stoll, F. Werling, R. Altherr, and K. Fuchs, Upper mantle temperatures from teleseismic tomography of French Massif Central including effects of composition, mineral reactions, anharmonicity, anelasticity, and partial melt, Earth and Planetary Science Letters, 139, 147–163, 1996.

Stacey, F.D., Thermal parameters of the Earth, in Physics of the Earth, pp. 457–460, Brookfield Press, Kenmore, Brisbane, 1992.

Wannamaker, P.E., J.B. Hulen, and M.T. Heizler, Early Miocene lamproite from the Colorado Plateau tectonic province, Southeastern Utah, USA, Journal of Volcanology and Geothermal Research, 96, 175–190, 2000.

Warren, L.M., and P.M. Shearer, Investigating the frequency dependence of mantle Q by stacking P and PP spectra, Journal of Geophysical Research-Solid Earth, 105 (B11), 25391–25402, 2000.

Oliver S. Boyd and Anne F. Sheehan, Dept. of Geological Sciences, University of Colorado at Boulder, Campus Box 399, 2200 Colorado Ave., Boulder CO 80309-0399 (email: Oliver.Boyd@Colorado.EDU, Anne.Sheehan@Colorado.EDU)

CDROM Interstation Pn Study Along the Rio Grande Rift

Lynda A. Lastowka[1] and Anne F. Sheehan

University of Colorado and CIRES, Boulder, Colorado

Upper mantle velocities in the regions of the Rio Grande rift and the Southern Rockies were investigated using Pn waves from the broadband seismic data obtained from the Continental Dynamics - Rocky Mountain (CDROM) experiment. A velocity of 7.8 +/- 0.1 km/s on the eastern flank of the Rio Grande rift in New Mexico was measured from epicentral distance vs. travel time data for ten Pn measurements from a south-western New Mexico earthquake that was in-plane with the southern CDROM line. Thirty-two interstation Pn measurements were made using eight western United States earthquakes. Using these measurements another bulk velocity estimate was made by solving for the best-fit velocity for interstation distances and travel times; this method also shows that the upper mantle beneath the Rio Grande rift has a slow velocity of 7.8 +/- 0.1 km/s. This low velocity is consistent with the high heat flow in the Rio Grande rift area and the evidence of modern rifting. Individual measurements made in the north end of the southern Rocky Mountains are near the global average Pn value of 8.1 km/s.

1. INTRODUCTION

The Continental Dynamics - Rocky Mountain (CDROM) passive source seismic experiment studies the effects of the assembly, stabilization, and reactivation of the western United States by providing constraints on the modern crust and mantle structure. The present day mantle structure may reflect past tectonism and/or a reactivated tectonic regime. More specifically, the seismic velocity of the upper mantle can help us determine the physical state of the upper mantle and the presence of current tectonic activity.

Two simultaneous deployments of 48 seismometers continuously recorded seismic activity from April of 1999 until June of 2000. The broadband instruments recorded local, regional and teleseismic earthquakes. The seismic stations were arranged in two NNE-SSW trending lines, which coincide with the active seismic source studies. Station locations are provided in Figure 1 and Table 1. The combination of the passive-source and active-source studies will provide us with both a detailed understanding of the shallow subsurface and a glimpse at the deeper structure of the mantle. The northern seismic line extended from Rawlins, Wyoming to Steamboat Springs, Colorado and traversed the ancient suture zone between the Archean Wyoming province and the Proterozoic Yavapai province. The southern line extended from the San Luis Basin in Colorado to Las Vegas, New Mexico and crossed the suture zone between the Yavapai province and the younger Mazatzal province. The southern line is also very proximal to the Rio Grand rift and the recent tectonism in the Jemez lineament.

The southern deployment was in place from April of 1999 until March of 2000 and the northern line was in place from June of 1999 until June 2000. Periods of bad weather and equipment malfunction caused some power outages, especially at the beginning of the experiment, but for the most part a good signal was continuously recorded throughout the experiment. The deployment consisted of 25 instruments

[1] Currently at United States Geological Survey, Golden, Colorado.

The Rocky Mountain Region: An Evolving Lithosphere
Geophysical Monograph Series 154

10.1029/154GM28

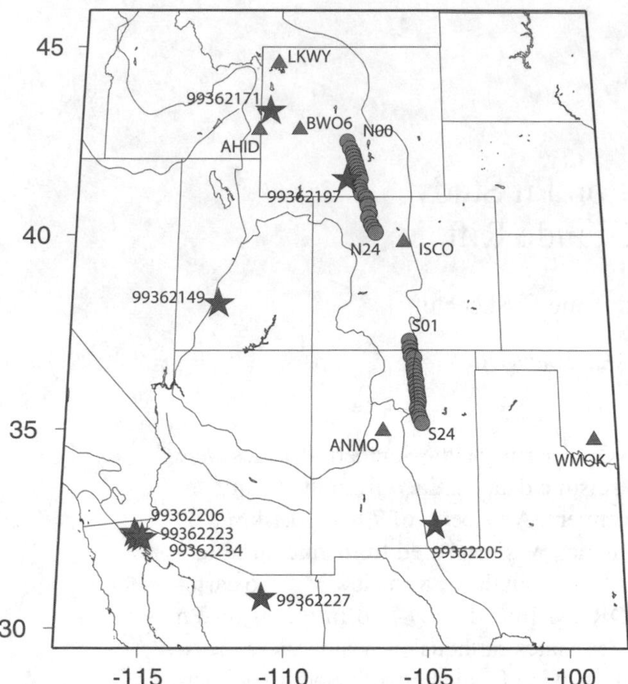

Figure 1. Map showing the location of the CDROM stations and the earthquakes used in this study. The stations are shown as circles and the earthquakes are shown as stars. The station spacing is approximately 10 km.

across the northern suture zone and 23 across the southern suture zone. The instruments included 27 STS-2 seismometers, 15 CMG3T seismometers and six CMG40T seismometers. The sampling rate of the experiment was 25 samples/sec for the beginning (April - October) of the experiment and then was reduced to 10 samples/sec. The sampling rate was reduced because inclement weather would prevent the sites from being visited in the winter; therefore a lower sampling rate was needed in order to preserve field disk space. The stations were arranged in a linear, north-northwest trending array with station spacing at approximately 10 km. This short interstation distance was necessary in order to document possible abrupt changes in the mantle that may correspond to terrane boundaries.

This study contributes to the spirit of the Continental Dynamics project by providing an independent data set that can be used to enhance the findings of researchers working in other areas of seismology and earth science. The upper mantle seismic velocities obtained in this study could potentially enhance the crustal and upper mantle seismic velocity models developed by the researchers working on the active source seismic data. The active source studies often use PmP, a reflected phase, to determine the depth to the Moho, but this phase offers little information on the upper mantle velocities. The results from this study will provide an additional constraint for the near-surface structural model.

2. INTERSTATION PN STUDY

The arrangement of the CDROM stations in a NNW-SSE linear array was a tradeoff between crossing the province boundaries in a normal sense and recording teleseismic events for in-plane tomography and other analysis. However, it severely limited the number of events that could be used for interstation velocity analysis. Interstation velocity analysis has the advantages of reducing error due to earthquake mislocation and constraining the measured velocities to a finite location. The interstation Pn analysis requires an earthquake to be within three degrees of the same great circle path as two seismic stations. In order to create more suitable interstation paths, the study was supplemented with data from IRIS Global Seismographic Network (GSN) station ANMO and United States Geological Survey - United States National Seismograph Network (USGS - USNSN) stations AHID, BW06, ISCO, LKWY and WMOK.

The locations of the earthquakes that were used in this study are shown in Figure 1 and Table 2. These locations were taken from the USGS Preliminary Determination of Epicenters (PDE) bulletin. The seismic phase of most interest in this study is Pn. Pn is a refracted body wave that travels in the uppermost mantle. It is the first arrival when station-earthquake distances are between 2 and 16 degrees [*Beghoul and Barazangi,* 1989]. When the source-station distance is less than 2 degrees the first arrival is Pg, when the distance is greater than 16 degrees the first arrival is P. The minimum distance where Pn becomes the first arrival is known as the crossover point. The crossover point depends on a variety of factors such as crustal thickness and velocity and mantle velocity. Pn arrivals were picked on stations that had a clear Pn arrival and that were on the same great circle path as another station for that same event. The stations were determined to be on the same great circle path as the events when the azimuth between the stations was within three degrees of the azimuth from each station to the event. A difference of three degrees provokes a velocity difference of less than 1% over our average interstation distance. There were eight earthquakes that fit the requirements for the study; these eight events produced 32 velocity measurements. The velocities were calculated by dividing the distance between the two stations by the difference in the Pn arrival time at the two stations (t_a and t_b) as shown in equation (1)

$$Velocity = distance\ between\ stations\ /(t_a - t_b) \qquad (1)$$

Calculated velocities are given in Table 3. All of the velocities in Table 3 fall within an acceptable range of Pn values and were picked from clear Pn arrivals. Velocities that were < 7.4 km/s or > 8.5 km/s were discarded due to an obvious error in selecting the phase. The northern CDROM stations produced velocities from 7.87 km/s to 7.99 km/s. The southern CDROM velocities ranged from 7.48 km/s to 8.4 km/s with a mean of 7.9 km/s. The uncertainties for this study were estimated by using the extrema of possible arrival time picks to find the maximum and minimum possible velocity. The average Pn velocity uncertainty was 0.04 km/s.

Due to a small number of events, it was difficult to determine variations in Pn velocity on similar interstation paths. There were three repeat paths in the study with a standard deviation of 0.02 km/s. Ten stations, and one event, 99362205, (Located in SE New Mexico, 75 miles SE of Roswell, New Mexico, May 30, 1999, ml=3.9, depth = 10 km) produced twenty-two measurements. Four of the eight events produced only one measurement each and three events produced two measurements.

Only three events were in locations suitable for the interstation analysis using the northern stations. Event 99362149 (located in SW Utah, January 2, 1998, ml = 4.5, depth = 5 km) gave a velocity of 7.94 km/s between AHID and LKWY. Event 99362171 (located in western Wyoming, June 20, 1998, ml = 4.4, depth = 0 km) provided a velocity of 7.99 km/s between BW06 and ISCO. Event 99362197 (located in southern Wyoming April 6, 1999, ml = 4.3, depth = 10 km) gave a velocity of 7.87 km/s between ISCO and WMOK.

Event 99362205, located in southern New Mexico, was nearly perfectly aligned with both the northern and southern CDROM lines. Unfortunately, the northern stations had not yet been deployed at the time of this event and the event was only clearly recorded at ten southern stations. The other events used for the southern station analysis (99362206, 99362223, 99362227, 99362234) were located in northern Mexico and all utilized station ANMO in the interstation measurements. Event 99362227, located in northern Mexico at a SW azimuth from the CDROM south line, gives a particularly low velocity of 7.6 km/s and is discussed further in the next section.

Since event 99362205 was recorded at the most stations and was aligned with the southern CDROM line we calculated the upper mantle velocity in the Rio Grande rift region with data from that event. We plotted epicentral distance vs. the total phase travel time for event 99362205, shown in Figure 2. The scatter in the data set is rather insignificant, +/- 0.2 s. This small amount of variation can be caused by lateral variations in earth structure such as topography on the moho, station elevation, and slight differences in the Pn picks.

We also calculated the upper mantle velocity using all of the southern interstation travel time and distance data from the

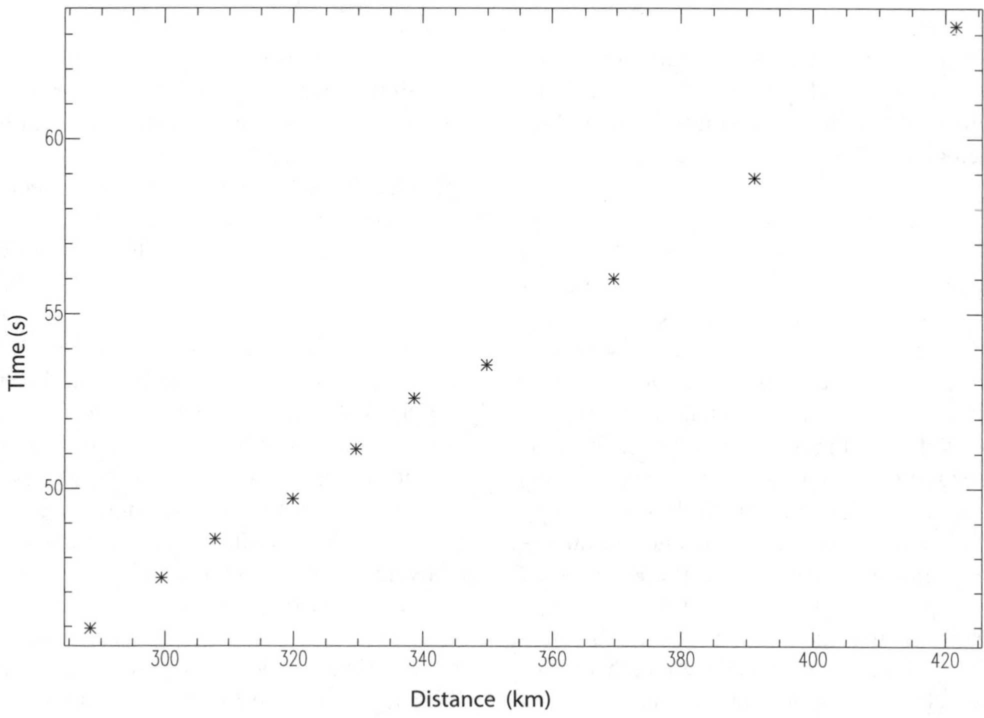

Figure 2a. Plot showing epicentral distance vs. Pn travel time for event 99362205. Ten stations recorded a clear Pn arrival for this event.

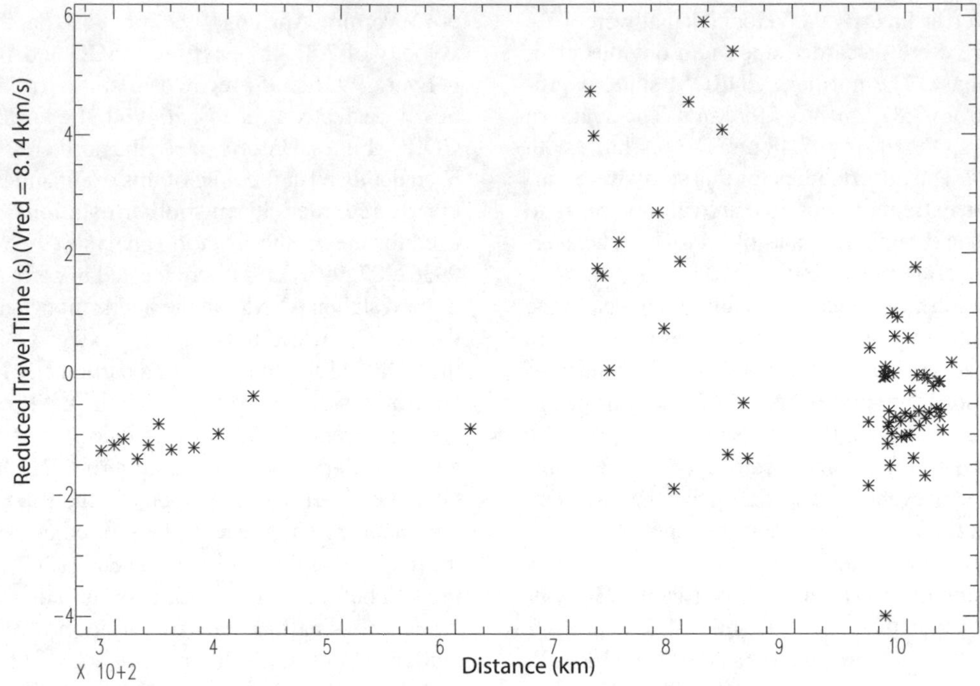

Figure 3. Reduced velocity plot showing data from the eight stations used in the interstation study. Each data point represents the distance between a two-station pair plotted against the interstation travel time. The best fit velocity for the data set is 7.83 +/- 0.12 km/s.

eight western US events. We plotted interstation distance vs. interstation travel time. A Pn velocity of 7.8 +/- 0.12 km/s best fit this data. The data is shown in a reduced velocity plot with a reduction velocity of 7.83 km/s in Figure 3.

3. DISCUSSION

We characterize the upper mantle in the Rio Grande rift area using the velocity of 7.8 +/- 0.1 km/s. This velocity was achieved through two methods: calculating the best-fit line through our distance vs. travel time data for one event (99362205) and by using all of the interstation travel time and distance data available for the southern stations. The individual velocity measurements can be seen plotted at the interstation midpoints in Figure 4. The particularly low velocity produced by event 99362227 may be due be due to an interstation path that lies almost entirely within the Rio Grande rift. The NE-SW alignment of the ray path from event 99362227 is in the direction of the dominant fast SKS polarization direction for the region (Fox and Sheehan, this issue) thus the slow Pn velocities are inconsistent with the regional mantle anisotropy. Robust Pn estimates for the region are not available. A Moho dipping down to the north could also produce slow Pn velocities along this path. However, the Moho imaged by CDROM receiver functions beneath the CDROM South line (where the majority of our interstation paths are) is nearly flat [Zurek and Dueker, 2001; Karlstrom et al., 2002]. The Rio Grande rift has a thin crust, slow Pn velocities, and high heat flow. Previous unreversed seismic profiles show the Pn velocity beneath the southern Rio Grande rift is about 7.7 km/s [*Olsen et al.*, 1979]. The northern portion of the rift exhibits a Pn velocity of 7.7–7.8 km/s, which is much slower than the velocity of 8.2 km/s reported for the Great Plains [*Stewart and Pakiser*, 1962] but is very similar to the values of 7.7 km/s reported in the Basin and Range [*Keller et al.*, 1976]. A Rayleigh wave dispersion study [*Sinno and Keller*, 1986] also shows low Pn velocities (7.7 km/s) in the Rio Grande rift. The low Pn velocity we observe is consistent with the Pn velocities in other tectonically active areas [*Hearn et al.*, 1991]. The Basin and Range province displays a similar Pn velocity of 7.8–8.0 km/s [*Hearn et al.*, 1991]. The Snake River Plain also displays a low Pn velocity (7.9 km/s), this is due to thermal heating. Heat flow measurements in the Rio Grande rift are varied and complex, but in general they are very high at about 75–125 mWm^{-2}. Near surface anomalies can locally elevate heat flow measurements up to 160 mWm^{-2} [*Keller et.al.*, 1990]. These high heat flow values are consistent with a low Pn velocity. The velocities measured adjacent

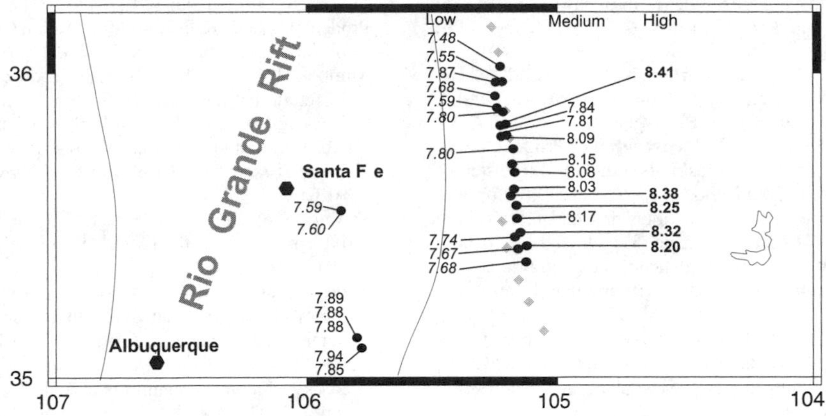

Figure 4. Pn velocities plotted at the midpoint of the stations used for calculation. Italicized numbers represent velocities < 7.8 km/s, regular numbers are velocities between 7.8 and 8.2 km/s, bold numbers are > 8.2 km/s.

to the Rio Grande rift are comparable to the velocities measured in the Eastern Great Basin making it hard to distinguish between the two provinces based on upper mantle Pn velocities. The presence of high mantle temperatures in a rifted area was also shown to exist beneath the Kenya rift [*Fuchs et al.*, 1997; *Prodehl et al.*, 1994]. The rift exhibits low velocities of 7.5–7.8 km/s while the adjoining unrifted craton has Pn velocities of 8.1–8.3 km/s. This is similar to the slow velocities of the Rio Grande rift and the adjoining Great Plains.

We have showed the upper mantle velocities surrounding the Rio Grande rift are slower than the surrounding Great Plains. The slower velocities can be attributed to elevated mantle temperatures and/or partial melt. *Davis et al.* [1993] find an 8% velocity reduction from the global average beneath the Rio Grande rift. This velocity reduction is attributed to 1% partial melt. We see only a 3–4% velocity reduction, at a shallower depth observed by *Davis et al.* [1993]. There is a strong correlation in between elevated upper mantle temperature and low Pn velocities [*Hearn et al.*, 1991; *Black and Braile*, 1982]. It seems clear that the low Pn velocities we are observing are also strongly correlated to a raised mantle temperature.

We also tested whether individual Pn velocity measurements could be interpreted or if they may be influenced by Moho topography. The individual velocities measured have a variance of 6%. These variations may be due to changes in the upper mantle, but they also might be due to variations in Moho topography. We are using relatively short interstation path lengths for this study, therefore a small variation in Moho topography may represent a considerable percentage of the total path length. The CDROM refraction/wide angle reflection study [*Rumpel,* 2001] and receiver function study [*Zurek*, 2001] show a maximum crustal thickness variation of 5 km over the area we are observing. Using forward modeling we measure the potential influence of Moho topography on velocity over our average interstation path length of 90 km. We see that 5 km of Moho topography can influence the apparent Pn velocity by up to 7% of 8.0 km/s. Our measured velocities vary 6% from 8.0 km/s therefore we cannot believe individual velocity measurements for they may reflect Moho topography and not the state of the upper mantle. The results of Zurek et al., [2001] show a coarse picture of Moho topography. Our undulating pattern of results may reflect this topography variation as well as variations on a smaller scale.

*Acknowledgments***.** We would like to acknowledge NSF grant *9614410*, IRIS for providing the instruments, and the personnel at the PASSCAL instrument center for their support in the field and with the data management associated with this experiment. We would like to acknowledge P. Wessel and W. H. F. Smith for the use of their GMT software. We thank editor Randy Keller, Tom Hearn and Ulrich Achauer for constructive reviews, and Craig Jones for comments on the paper.

REFERENCES

Beghoul, N., and M. Barazangi, Mapping high Pn velocity beneath the Colorado Plateau constrains uplift models, *J. Geophys. Res.*, 94, 7083–7104, 1989

Black, P.R. and L.W. Braile, Pn velocity and cooling of the mantle lithosphere, *J. Geophys. Res.*, 87, 10557–10568,1982.

Davis, P.M., P. Slack, H.A. Dahlheim, W.V. Green, R.P. Meyer, U. Achauer, A. Glahn, and M. Granet, Teleseismic Tomography of Continental Rift Zones, Chapter in book publ. Chapman and Hall, *Seismic Tomography, Theory and Practice*, edited by H.M. Iyer and K. Hirahara, 397–439, 1993.

Fox, O. C. and A. F. Sheehan, Upper mantle anisotropy beneath Precambrian province boundaries, southern Rocky Mountains, AGU monograph on Lithospheric Structure of the Rocky Mountains (this volume), 2003.

Fuchs, K., R. Altherr, B. Mueller and C. Prodehl, Structure and dynamic processes in the lithosphere of the Afro-Arabian rift system, *Tectonophysics*, 278, 1–352, 1997.

Hearn, T.M., M.N. Beghoul, and M. Barazangi, Tomography of the Western United States from regional arrival times *J. Geophys. Res.*, 96, 16,369–16,381, 1991.

Karlstrom, K. E., S. A. Bowring, K. R. Chamberlain, K. G. Dueker, T. Eshete, E. A. Erslev, G. L. Farmer, M. Heizler, E. D. Humphreys, R. A. Johnson, G. R. Keller, S. A. Kelley, A. Levander, M. B. Magnani, J. P. Matzel, A. M. McCoy, K. C. Miller, E. A. Morozova, F. J. Pazzaglia, C. Prodehl, H. M. Rumpel, C. A. Shaw, A. F. Sheehan, E. Shoshitaishvili, S. B. Smithson, C. M. Sneson, L. M. Stevens, A. R. Tyson, and M. L. Williams, Structure and evolution of the lithosphere beneath the Rocky Mountains: Initial results from the CD-ROM experiment, *GSA Today, v. 12*, no. 3, p. 4–10, March 2002.

Keller, G. R., P. Morgan, and W. R. Seager, Crustal structure, gravity anomalies and heat flow in the southern Rio Grande rift and their relationship to extensional tectonics, *Tectonophysics*, 174, 21–37, 1990.

Keller, G. R., R. B. Smith, L. W. Braile, R. Heaney, and D. H. Shubert, Upper crustal structure of the eastern Basin and Range, northern Colorado Plateau, and middle Rocky Mountains from Rayleigh wave dispersion, *Bull. Seism. Soc. Am.*, 66, 869–876, 1976.

Lastowka, L. A., A. F. Sheehan, and J. M. Schneider, Seismic evidence for partial delamination model for Colorado Plateau uplift, *Geophys. Res. Lett.*, 28, 1319–1322, 2001.

Olsen, K. H., G. R. Keller, and J. N. Stewart Crustal structure along the Rio Grande Rift from seismic refraction profiles. In Riecker R. E. ed., *Rio Grande Rift; tectonics and magmatism*. Am. Geophys. Union. Washington, DC, United States. 127–143, 1979.

Prodehl, C., G.R. Keller and M. A. Kahn, Crustal and upper mantle structure of the Kenya rift, *Tectonophysics*, 236, 1–483, 1994.

Rumpel, H. C. , C. M. Snelson, C. Prodehl, and G. R. Keller. Results of the Refraction/Wide Angle Reflection Seismic Experiment in the Southern Rocky Mountains (CDROM'99) AGU abstract, Fall Meeting, 2001.

Sinno, Y.A. and G.R. Keller, A Rayleigh wave dispersion study between El Paso, Texas and Albuquerque, New Mexico, *J. Geophys. Res.*, 91,6168–6174, 1986.

Stewart, S. W. and L. C. Pakiser, Crustal structure in eastern New Mexico interpreted from the GNOME explosion, *Bull. Seis. Soc. of Am.*, 52, 1017–1030, 1962.

Zurek, B., and K. Dueker, Observations of mantle discontinuities beneath the western United States: Identifying direct PdS arrivals in common conversion point images., AGU abstract, Fall meeting, 2001.

Zurek, B. and K. Dueker, Lithospheric stratigraphy beneath the southern Rocky Mountains, submitted to AGU special monograph 8/2002 (this volume), 2003.

L. A. Lastowka, USGS, PO Box 25046, DFC, MS 966, Lakewood, CO, 80226. (e-mail: llastowka@usgs.gov)

A. F. Sheehan, Dept. of Geological Sciences and CIRES University of Colorado, Boulder, CO, 80309. (email: afs@cires.colorado.edu)

Rayleigh Wave Constraints on Shear-Wave Structure and Azimuthal Anisotropy Beneath the Colorado Rocky Mountains

Aibing Li

Department of Geosciences, University of Houston, Houston, Texas

Donald W. Forsyth and Karen M. Fischer

Department of Geological Sciences, Brown University, Providence, Rhode Island

We inverted Rayleigh wave data recorded in the Rocky Mountain Front Broadband Seismic Experiment for shear-wave velocity structure and azimuthal anisotropy. Distinctive structures are imaged beneath the southern Rocky Mountains, the western Great Plains, and the eastern Colorado Plateau. Beneath the southern Rockies, shear velocities are anomalously low from the Moho to depths of 150 km or more, suggesting replacement or delamination of the mantle lithosphere. The lowest velocities are beneath the extension of the Rio Grande rift into southern Colorado and are probably associated with partial melt. Beneath the Colorado Plateau, a thin, high-velocity lid is underlain by a low velocity layer to a depth of at least 160 km. Under the high plains, the velocities are above average down to ~150 km depth, but not as fast as beneath the cratonic core of the continent. A crustal, low-velocity anomaly is observed beneath the high elevations of central Colorado. Elsewhere, inferred crustal thickness correlates with elevation, with the thickest crust beneath the San Juan Mountains in southwestern Colorado. These crustal anomalies suggest that much of the isostatic compensation for the high topography takes place within the crust. We observe a simple pattern of azimuthal anisotropy in the Rocky Mountain region with fast directions rotated slightly counterclockwise from the absolute plate motion of the North America plate and strength increasing with period. The observed anisotropy can be explained by deep asthenospheric flow dominated by current plate motion and shallower and perhaps laterally variable anisotropy in the upper lithosphere.

INTRODUCTION

The Colorado Rocky Mountains are located between the tectonically stable Great Plains to the east and the elevated, but stable Colorado plateau to the west. The Rio Grande rift extends into the Rockies in southern Colorado and associated normal faulting may extend northward through much of the state [*Keller and Baldridge*, 1999]. The southern Rockies have experienced a number of tectonic events, including the Laramide orogeny (~75–50 Ma), subsequent erosion, current regional uplift, and on-going extension [*Burchfiel et al.*, 1992; *Karlstrom and Humphreys*, 1998].

Seismic images beneath this region can provide observational constraints on variations of velocity and thickness of the crust and mantle lithosphere across different tectonic provinces. Seismic tomography can therefore help us to better under-

The Rocky Mountain Region: An Evolving Lithosphere
Geophysical Monograph Series 154

10.1029/154GM29

stand the formation of the southern Rockies and the stability and evolution of the Great Plains. For example, tomographic studies of North America have shown that the Rocky Mountain Front represents a transition from a seismically slow upper mantle in western United States to a fast, cratonic structure in central and eastern North America [*Grand,* 1994; *van der Lee and Nolet*, 1997]. This transition is imaged in more detail in regional body wave tomography in Colorado and western Kansas [*Lee and Grand*, 1996; *Lerner-Lam et al.*, 1998]. These studies reveal a low velocity volume in the mantle beneath the southern Rockies, which the authors interpret as indicating isostatic support of the Rockies through thermal buoyancy and partial melt in the upper mantle. Seismic anisotropy, which may be sensitive to deformation in the lithosphere and flow in the asthenosphere, also provides important constraints on regional dynamic processes. There are rapid lateral variations in the fast direction of shear wave splitting in Colorado and many null measurements [*Savage et al.*, 1996]. *Savage and Sheehan* [2000] suggested that this complicated pattern indicated a complex strain regime that might be consistent with asthenospheric upwelling sheared by plate motion.

In this paper, we use Rayleigh wave data recorded from the Rocky Mountain Front (RMF) Broadband Seismic Experiment (Figure 1) [*Lerner-Lam et al.*, 1998] to obtain spatial and azimuthal variations in phase velocities, which we interpret in terms of crustal and upper mantle shear-wave structure beneath the Colorado Rocky Mountain region and surrounding terrenes. Because there is a high density of stations in the RMF experiment, the lateral resolution in our study is greater than that in large-scale tomography studies [*Grand*, 1994; *van der Lee and Nolet,* 1997]. Because Rayleigh waves at different periods sample velocity structure in different depth ranges and the fundamental mode is insensitive to deep structure, measuring phase velocities from 20 to 100 s period provides better resolution for velocity structure in the upper 100 km than can be obtained in regional, body-wave tomographic studies with station spacing comparable to the RMF experiment. In

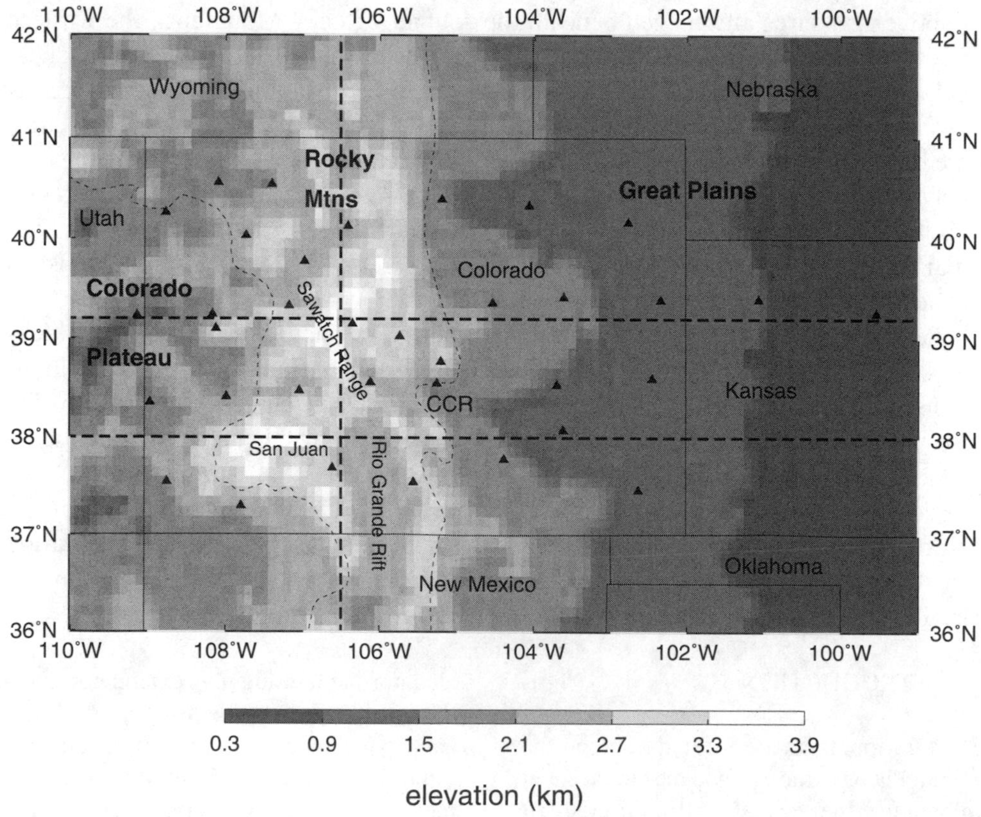

Figure 1. Station locations on a relief map of the Colorado Rocky Mountains and surrounding terrenes. The two areas with the highest elevations include the Sawatch Range and the San Juan Mountains, respectively. The black triangles represent seismic stations of the Rocky Mountain Front Seismic Experiment [*Lerner-lam et al.*, 1998]. Thin dashed lines mark tectonic boundaries and thin solid lines are state boundaries. Thick dashed lines are locations of profiles A-C in Plate 3.

addition, the frequency dependence of the azimuthal anisotropy of Rayleigh waves can provide constraints on variation of anisotropy with depth that complement the vertically averaged measure of anisotropy obtained from shear-wave splitting.

Our approach involves two steps. The first step is to invert Rayleigh wave amplitude and phase data for phase velocities of the waves. This gives a direct indication of the information contained in the data independent of any assumptions needed for a stable inversion for earth structure. The second step is to invert for shear-wave velocity structure from the phase velocities obtained in the first step. We construct models initially ignoring azimuthal anisotropy, seeking only to resolve lateral variations. Then we allow for the possibility of anisotropy and show the effects of this added model complexity on the estimates of lateral variations in phase velocity.

DATA SELECTION AND PROCESSING

Rayleigh wave data from 74 teleseismic earthquakes with body wave magnitudes larger than 5.0 and epicentral distances from 30° to 120° (Figure 2) were used in this study. The events were recorded by 35 RMF broadband seismic stations that operated during May to December in 1992 (Figure 1). Most of the stations were located in Colorado, with one in eastern Utah and two in western Kansas. The sites were not all occupied simultaneously, so the actual coverage is not as good as would be expected given the distribution of sources and receivers. Nevertheless, the coverage for surface waves is excellent (Figure 3), yielding many crossing paths that are needed for tomographic studies both inside and immediately outside the array.

Figure 2. Distribution of earthquake sources used in this study. The azimuthal equidistant projection is centered on the center of the RMF seismic array. Each event is connected to the array center by a straight line representing the great circle ray path. Note the good azimuthal coverage of the data, which is important in solving for azimuthal anisotropy.

Because several types of seismometers were used in the RMF stations, we corrected instrument responses to match a single type. We filtered vertical-component seismograms with a series of 10 mHz wide, zero-phase-shift, 4th-order Butterworth filters centered at frequencies of 50, 45, 40, 35, 30, 25, 20, 17, 15, 12, and 10 mHz. These frequencies sample velocity structure to depths of 300–400 km and provide good vertical resolution to about 150 km depth. Fundamental mode Rayleigh wave trains were isolated by windowing each filtered seismogram. Frequency bands with signal-to-noise ratio less than 3:1 were rejected. A careful comparison of Rayleigh wave trains at all possible stations for each frequency was important for identifying problems such as timing errors and anomalous instrument responses. We then converted the filtered and windowed seismograms to the frequency domain to obtain phase and amplitude data.

PHASE VELOCITY

Phase velocity c in a uniform slightly anisotropic medium varies as

$$c(\omega,\psi)=A_0(\omega)+A_1(\omega)\cos(2\psi)+A_2(\omega)\sin(2\psi) +A_3(\omega)\cos(4\psi) +A_4(\omega)\sin(4\psi) \quad (1)$$

where ω is frequency, ψ is the azimuth of propagation of the wave, and A_0 to A_4 are velocity coefficients [*Smith and Dahlen*, 1973]. We neglected A_3 and A_4 terms here because they should be small for Rayleigh waves [*Smith and Dahlen*, 1973]. Phase velocities are represented as weighted averages of values of A_0, A_1, and A_2 at neighboring points on a grid of nodes (Figure 4). The spatial resolution is controlled by adjusting the characteristic scale length of the 2-D Gaussian weighting function, which is supposed to be a function of wavelength. However, to simplify the calculation, we use 80 km, an intermediate value, for all frequencies in this study. To account for wave propagation effects such as focusing and multipathing that occur between the sources and the array, each incoming wavefield is represented as the sum of two interfering plane waves with amplitude, initial phase, and propagation direction for each wave to be determined in the inversion [*Forsyth et al.*, 1998; *Li,* 2001]. We solve simultaneously for these wavefield parameters and the velocity parameters in an iterative, least-squares inversion. Two stages are employed for each iteration: a simulated annealing method is used first to solve for the two-plane wave parameters; then a generalized linear inversion [*Tarantola and Valette*, 1982] with damping and

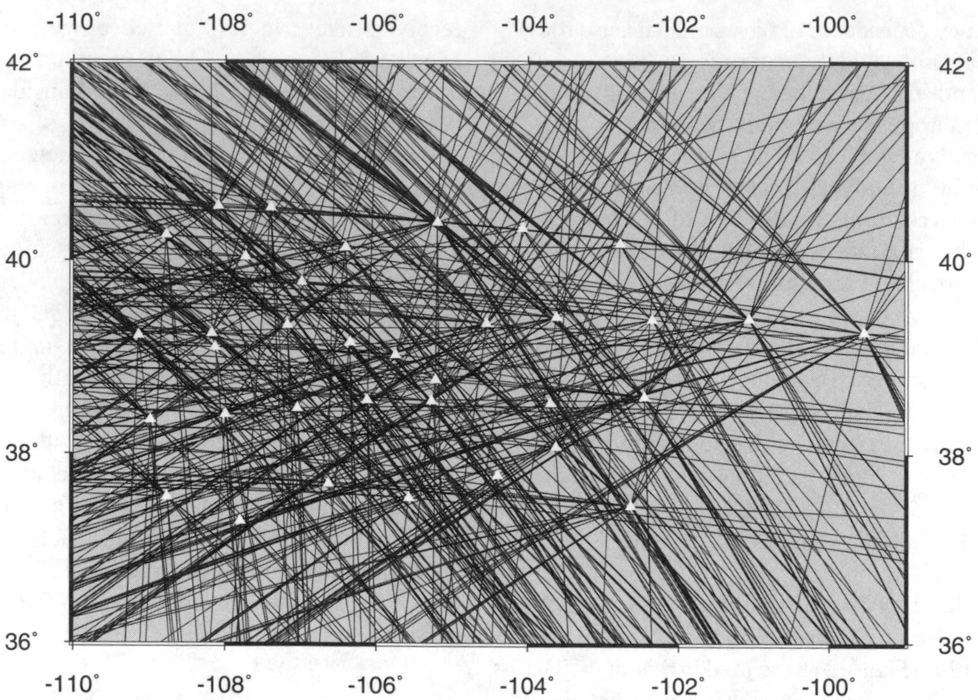

Figure 3. Great circle ray paths in the vicinity of the RMF array at frequency 0.017 Hz. Ray path population varies somewhat with frequency due to the change in number of events with acceptable noise levels. White triangles represent the RMF stations. Note good density of crossing paths both inside and immediately outside of the array.

smoothing is applied to find both phase velocity coefficients at each node and the wave field parameters.

Isotropic Phase Velocities

To generate a reference model that will serve as a starting model in inversions for lateral variations in phase velocity, we first assume that phase velocities are uniform in the whole study area to obtain an average phase velocity at each frequency (Figure 5). Because the study area is characterized by three distinct tectonic provinces, the Colorado Plateau, the Rocky Mountains, and the Great Plains, we also solved for average phase velocities in each province by grouping the nodes by tectonic province (Figure 4). The nodes outside the array are treated as another group, but this region is heterogeneous and less well-constrained, so the results are not shown in Figure 5. This regionalized inversion is equivalent to a classic "pure-path" inversion, except that we use the average velocity for the whole study area as a starting model. The damping in the version is light and has virtually no effect except at the longest periods and in the least-well-constrained province, the Colorado Plateau.

The average phase velocities from 20 s to 100 s beneath the Rocky Mountains are overall much lower than those beneath the Colorado Plateau and the Great Plains (Figure 5), indicating the presence of high temperatures and/or partial melt in the crust and upper mantle. Phase velocities in the Colorado Rockies and the Great Plains increase gradually and smoothly with increasing period. The largest differences between these two provinces are at periods of 25 to 40s, which are primarily sensitive to structure in the crust and uppermost mantle down to depths of about 100 km (Figure 6). Diminishing differences at longer periods indicate a decrease in velocity contrast at depths greater than 100 km. The dispersion for the eastern Colorado Plateau lies between that for the Plains and the Rockies, but the velocity does not increase as smoothly with increasing period. Part of this oscillatory character may be due to larger uncertainties for this region, but the general feature of a decreased slope around 50 to 60 s is similar to that reported in an independent experiment in the Colorado Plateau [*Lastowka et al.*, 2001] and is indicative of a pronounced low-velocity zone in the upper mantle underlying a high-velocity lithosphere.

Using the average phase velocity for the region as a whole as a starting value, we solved for 2-D phase velocity variations across the array without *a priori* regionalization. Phase velocities at the grid points (Figure 4) are used to generate maps on a finer grid of 0.1° by 0.1° using a smoothing length of 80

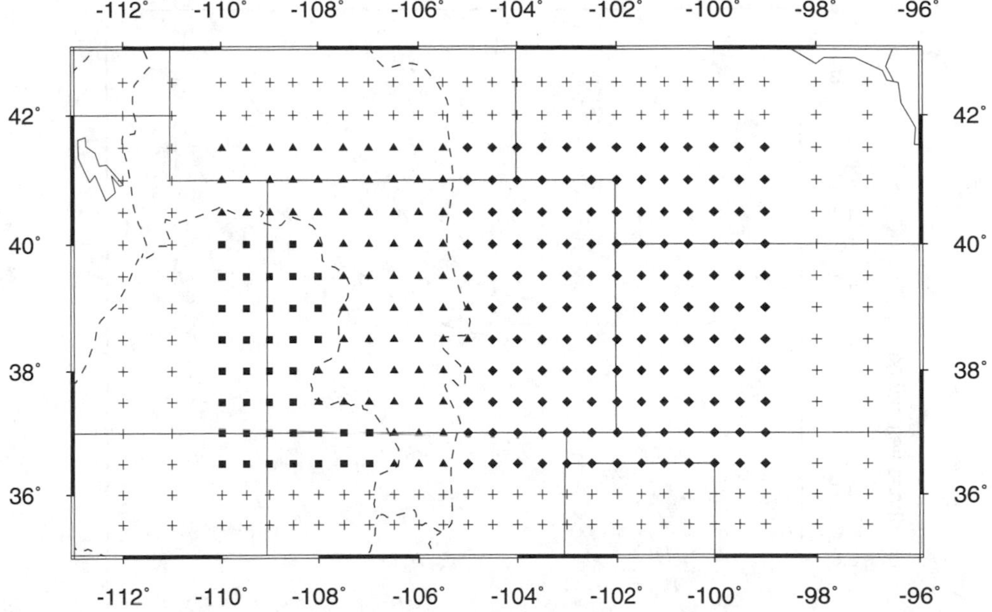

Figure 4. The grid used in phase velocity inversion. The nodes are grouped by tectonic province for calculating average phase velocity and azimuthal anisotropy of each province. Squares are nodes for the Colorado Plateau, triangles for the southern Rockies, and diamonds for the Great Plains. Crosses mark the nodes outside the array where model variance is large.

km. Plate 1 shows the maps of phase velocity anomalies at periods of 25, 33, 40, 50, 59, and 67 s. A striking and consistent feature in these maps is a band of low velocities confined to the Colorado Rockies region. Confirming the results of the *a priori* regionalizaton, average velocities are highest in the western Great Plains, lowest in the Rockies, and intermediate in the eastern Plateau. At 25 s, there is an almost circular pattern of low velocities in central Colorado. This pattern suggests that the crust must be thicker and/or slower in this region than elsewhere in the southern Rockies, because phase velocities at 25 s are primarily sensitive to the upper 50 km (Figure 6). At 40 s, there are low phase velocities in the northern extension of the Rio Grande rift into Colorado. Because this anomaly is absent at 25 s, it suggests that there must be a pronounced, low, shear-velocity anomaly in the uppermost mantle beneath the rift, probably indicating the presence of partial melt. Compared to the shorter periods, the lateral contrast in phase velocities at 50 to 67 s is reduced, but the contrast at 67 s is still about 3% from the slowest area beneath the Rockies to the fastest beneath the Great Plains. Because these periods are relatively insensitive to the crust, these maps provide a clear image of a transition from a tectonic to a cratonic upper mantle in the vicinity of the RMF.

Resolution

To understand the significance of lateral variations in the phase velocity maps presented in Plate 1, a description of the resolution is needed. In our approach, the value at each point in a map represents a Gaussian weighted average over adjacent node points, with the value of the weight decreasing to 1/e of the maximum at the characteristic distance of 80 km. Lateral variations in resolution are indicated by lateral variations in the standard error or uncertainty in these averages, taking into account the covariance between values of adjacent node points. In the generalized linear inversion for phase velocities, the *a posteriori* covariance matrix, $\mathbf{C}_{MM}$, for model parameters (phase velocity at each node) can be directly calculated from

$$\mathbf{C}_{MM} = (\mathbf{G}^T\mathbf{C}_{nn}^{-1}\mathbf{G} + \mathbf{C}_{mm}^{-1})^{-1} \quad (2)$$

where $\mathbf{G}$ is the partial derivative or sensitivity matrix relating predicted changes in phase and amplitude to perturbations in phase velocity, and $\mathbf{C}_{nn}$ and $\mathbf{C}_{mm}$ are the *a priori* data and model covariance matrices, respectively [*Tarantola and Valette*, 1982]. The *a priori* model covariance acts to damp the least squares solution by assigning an uncertainty to the starting model and retaining it as a constraint in the inversion. In an undamped least squares inversion, very large velocity anomalies tend to be assigned to poorly constrained regions of the model. In a highly damped solution, lateral velocity variations are usually underestimated and velocity variations in the poorly constrained regions are very small. We choose rel-

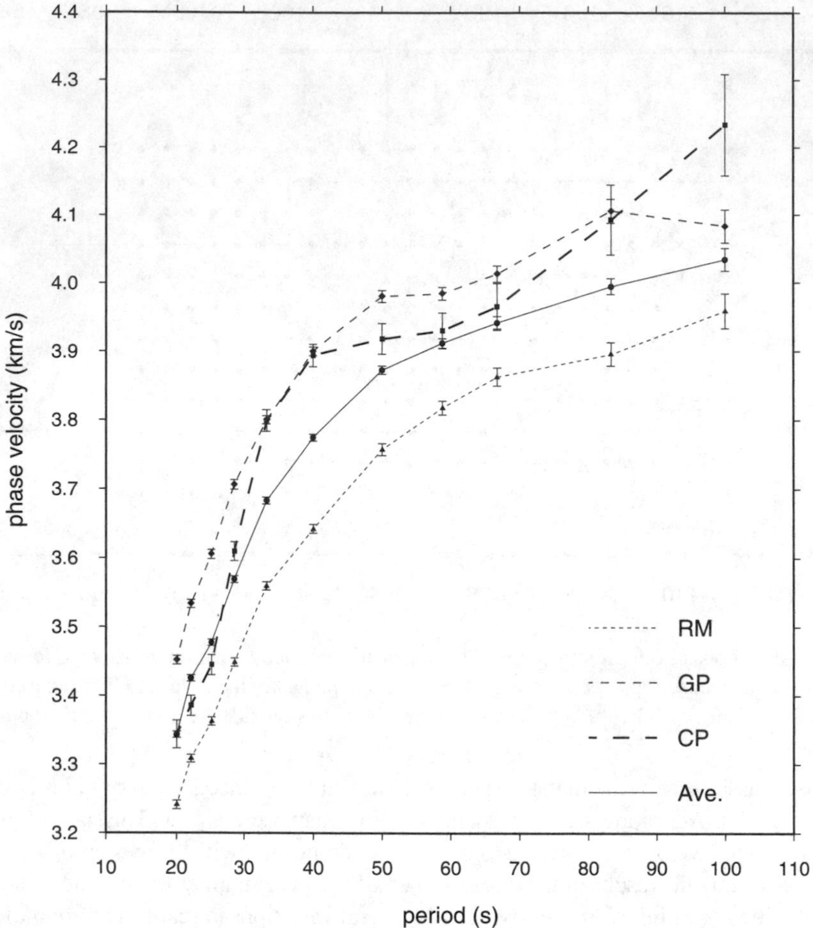

Figure 5. Average phase velocities in tectonic regions. Circles and the solid line are for the whole area (Ave.), squares and the thick dashed line for the Colorado Plateau (CP), triangles and the thin dashed line for the Rockies (RM), and diamonds and the intermediate dashed line for the Great Plains (GP). The bars represent one standard error on each measurement. Note that the errors are larger in the Colorado Plateau.

atively light damping that tends to leave the amplitude of the velocity variations roughly constant throughout the study area, but assigns larger *a posteriori* errors to the parts of the model that are poorly constrained.

We plot in Plate 1G twice the standard error of the weighted average velocities for a period of 33 s. Maps of the standard error at other periods are similar in form, but the errors increase for longer periods in the same manner as in Figure 5. The reason that variance increases with period is that the same magnitude error in Rayleigh wave phase produces larger uncertainty in travel time at longer periods and the signal-to-noise ratio and the number of acceptable signals also decrease. Not surprisingly, the errors are smallest where the density of crossing paths is greatest (compare to Figure 3). In the maps of velocity in Plate 1A-F, we use the 2.2% contour from 33 s as a mask, eliminating the illustration of velocity variations in regions outside this contour for being relatively poorly constrained. Twice the standard deviation yields a rough guide to the 95% confidence level for velocity anomalies; if the change in velocity from one point to another in the maps is greater than two standard errors, then there is only about a 5% likelihood that this difference would have arisen by chance. Thus, at 40 s, for example, the difference in phase velocity between the Rio Grande rift region and the central Rockies at 40° N is significant at the 95% confidence level.

SHEAR-WAVE STRUCTURE IN THE CRUST AND UPPER MANTLE

Methodology

Although the variations in Rayleigh wave phase velocity yield measures of the lateral and vertical variations in structure that are easily interpreted qualitatively, for geological

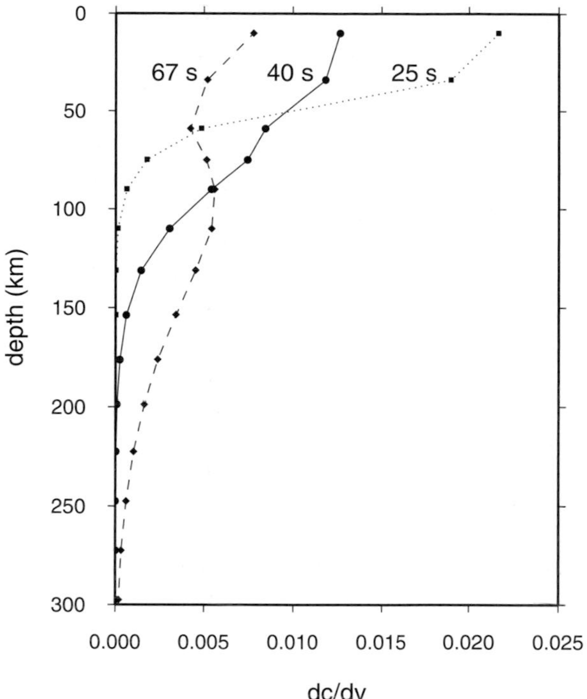

Figure 6. Rayleigh wave sensitivity kernels at periods of 25 (squares and the dotted line), 40 (circles and the solid line) and 67s (diamonds and the dashed line). The kernels are calculated from the reference 1-D velocity structure in the southern Rockies (solid line in Figure 7).

interpretation, we need to invert these observations for the velocity structure of the earth. Because the problem is non-linear and the structure is imperfectly resolved, the particular models we present will depend somewhat on the starting model. To obtain a reference model for this area, we started with the phase velocities near station CCR and an initial, average earth model (AK 135 from *Kennett et al.* [1995]) that was modified to have a 48 km crust as constrained from receiver functions at CCR [*Sheehan et al.,* 1995]. We chose CCR because it lies in the middle of the study area (Figure 1) where velocities are best constrained and the velocities are intermediate between that of the central Rockies and the Great Plains. Model parameters are shear velocities in approximately 20-km-thick layers extending to a depth of 200 km, with thicker but ultimately unresolvable layers extending to 600 km. P-wave velocities were coupled to S-wave variations with the same Poisson's ratio in each layer as in the starting model, so the sensitivity kernels shown in Figure 6 actually represents the combined response to coupled velocity changes, not just the shear velocity. In isotropic models, the dominant effect is S wave velocity, but P velocity is important near the surface. For example, much of the sensitivity to crustal velocity at 67 s (Figure 6) is due to P wave sensitivity, but the mantle response is almost entirely due to S wave sensitivity. There is not sufficient resolution to uncouple the P and S velocities in the inversion. We add crustal thickness as a parameter, preserving a velocity discontinuity at the Moho, by increasing the thickness of the lower crustal layer at the expense of the thickness of the uppermost mantle layer. We adjusted the relative damping of crustal thickness and velocity variations to maximize agreement between Rayleigh wave and receiver function data, although the latter were not explicitly employed as constraints in the inversion.

Synthetic phase velocities were computed using Thomson's algorithm [*Thomson,* 1997; *Martin et al.*, 1997] which allows for general anisotropy. Partial derivatives of phase velocity with respect to changes in model parameters were obtained by finite differences from a series of forward models in which individual parameters were perturbed. When inverting the maps of phase velocity for 3-D structure, we assumed for computational efficiency that as long as the crust is constant thickness, the partial derivatives are constant, because the primary non-linearity in the problem stems from variations in thickness of the layers. We tabulated partial derivatives for 17 models with crustal thickness ranging from 34 to 66 km, then after each iteration in an inversion, selected two sets of partial derivatives whose corresponding crustal thickness are most close to the crustal thickness in the current model, and obtained the most appropriate partial derivatives by interpolating the selected two sets. The shear velocity models are damped by assigning *a priori* standard deviations of 0.1 km/s to the starting model velocities and smoothed by introducing off-diagonal terms in the model covariance matrix that enforce a 0.4 correlation in the changes to adjacent layers.

In this reference model at CCR with crustal thickness fixed at 48 km, the S-wave velocity in the upper mantle is indistinguishable from the AK135 model, never differing by more than 0.01 km/s (Figure 7). The S-velocity in the crust is 3.32 km/s in the top layer (0–20 km) and 3.73 km/s in the bottom layer (20 km-Moho), about 0.13 km/s slower than in model AK135. The similarity of mantle structure is coincidental; there are substantial deviations from these reference models in other parts of the study area.

Average Structure in Tectonic Provinces

One-dimensional shear-wave models for the Colorado Plateau, southern Rockies, and the Great Plains are distinctly different (Figure 7). Shear velocity under the Rockies is remarkably low from the surface to ~140 km depth. Velocities in the model are also slightly slow compared to AK135 at greater depths where our data do not have good resolution; although the difference is probably real, the depth distribution

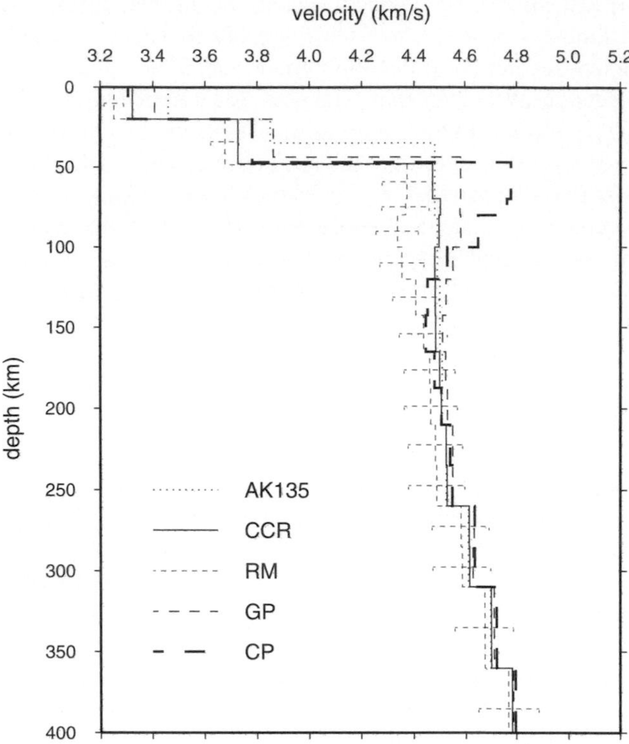

Figure 7. One-dimensional shear-wave velocity structures. The model of AK135 is plotted as a dotted line. The thick solid line represents our reference model at station CCR. The thin solid line is for the southern Rocky Mountains (RM). The thin and thick dashed lines correspond to the western Great Plains (GP) and the eastern Colorado Plateau (CP), respectively. These models are based on the phase velocities for average regions shown in Figure 5. One standard errors of shear velocity under the Rocky Mountains are plotted at the center of each layer, based on the *a posteriori* covariance of the damped inversion.

of the anomaly is not well constrained. If such differences do exist, as is suggested by body-wave tomography [*Lee and Grand*, 1996; *Lerner-Lam et al.*, 1998], the lack of resolution from surface waves coupled with the damping of the model inversions would lead to an underestimate of the magnitude of the anomaly. Our model agrees with the preferred pure-path model for the southern Rockies shown in the paper of *Lerner-Lam et al.* [1998]. There is no discernible, fast, lithospheric lid.

Vertical resolution depends on the period range sampled and the precision with which the phase velocities are measured. One measure of the resolution is the rank of the inverse matrix, which indicates the number of linearly independent combinations of model parameters that are resolved, or, equivalently, the number of pieces of information about the velocity structure that the data provides. For the southern Rockies and western Great Plains regions, the rank is 3.8; for the Colorado plateau, with larger errors on the velocities (Figure 5) the rank is 2.8; and for typical points in the maps of Plate 1, the rank is 2.9. The vertical distribution of the information is described fully by the resolution matrix. Crudely, three well-resolved, independent pieces of information about the vertical velocity structure are: the average velocity in the crust; the Moho depth and/or average velocity from 40 to 70 km; and average velocity from 70 to ~150 km. For the better-resolved, average structure of the Rockies and Great Plains, there is some additional information extending to depths greater than 150 km. *A posteriori* standard errors for individual layers are smaller where the information density is higher, like in the crust (Figure 7). Average velocities over resolvable depth ranges are better constrained than the standard errors indicated for individual layers. The typical standard error for crustal thickness is a few kilometers, but it should be recognized that standard errors in any inversion of this type do not provide extreme limits.

The model of the Colorado Plateau shows a fast lid from Moho to ~100 km and a strong low velocity layer underneath it. Velocities in the low-velocity zone approach those beneath the Rockies in the same depth range. This relatively thin lid beneath the Colorado Plateau supports the model that nearly horizontal subduction of the Farallon slab mechanically thinned the lithosphere [*Humphreys and Dueker*, 1994; *Spencer*, 1996], in agreement with the surface wave interpretation of *Lastowka et al.* [2001]. Our model velocity in the lid is unusually high; this may be caused partially by chance variations in the observed phase velocities within the bounds expected given the larger standard errors (Figure 5), but there is also some tradeoff possible between lid velocity and crustal thickness. Our best fitting crustal thickness for this average dispersion curve is about 47 km. If we constrained the thickness to be in the 40–44 km range typical of seismic refraction and receiver function estimates for the Colorado Plateau [*Sheehan et al.*, 1997; *Keller et al.*, 1998], the lid velocity could be reduced to 4.6 to 4.7 km/s.

In the model for the average western Great Plains, shear wave velocity is everywhere faster in the crust and upper mantle than in AK135 and the reference model, with a subtle low velocity zone, indicating a cratonic lithosphere. Defining a thickness of the lid or lithosphere is difficult because there is a relatively small velocity contrast between lid and low-velocity zone. The structure is quite similar to that found for the western Australian craton from Rayleigh wave tomography [*Simons et al.*, 1999], but the velocities are clearly lower than beneath the core of stable North America [*Brune and Dorman*, 1963]. The lower average velocities compared to the continental interior are not surprising considering that the Rayleigh wave phase velocities systematically decrease and the topography systematically increases from western Kansas to the Rocky Mountain Front (Plate 1 and Figure 1). The form of this variation is shown by our inversion for three-dimensional structure.

3-D Structure

Variations of crust thickness and velocity anomalies are shown in Plate 2 and 3. The crust is generally thick (48–52 km) beneath the Colorado Rockies and thins gradually to 44 km in the nearby Colorado Plateau and to 40 km in the Great Plains. There is a good, general correlation of crustal thickness with regional elevation. The thickest crust is beneath one of the two most elevated areas of the Rockies, the San Juan Mountains (Plate 2A and 3A). An exception to this correlation is local crustal thinning beneath the other highest area of the southern Rockies near the Sawatch Range (Plate 2A and 3B), but in this area there is a strong, low velocity (-3%) anomaly in the crust (Plate 2B, 2C, and 3B) that may indicate low crustal densities which could provide local isostatic compensation for the high topography [*Li et al.*, 2002]. Some trade-off between the low velocity anomaly and crustal thickness is possible, but the crustal anomaly is resolvably different than the effects of a change in crustal thickness, and the local thinning is confirmed by receiver function analysis beneath stations of the RMF experiment [*Sheehan et al.*, 1995; *Keller et al.*, 1998].

The variations of Moho depth in Plate 2A are significantly different than those reported in compilations of results from seismic reflection and refraction profiles and receiver function studies [*Prodehl and Lipman*, 1989; *Sheehan et al.*, 1995; *Keller et al.*, 1998]. Those compilations show the thickest band of crust to the east of the Rocky Mountain Front and a generally poor correlation of crustal thickness with elevation in Colorado. The combination of this lack of correlation with the discovery that P and S velocities beneath the southern Rockies are lower than beneath either the Great Plains or the Colorado Plateau led to the hypothesis that isostatic compensation of the southern Rockies takes place largely in the mantle [*Eaton,* 1987; *Sheehan et al.,* 1995; *Lee and Grand,* 1996; *Lerner-Lam et al.*, 1998; *Karlstrom and Humphreys,* 1998]. In contrast, *Li et al.* [2002] showed that the Bouguer gravity anomaly in Colorado could be matched by a combination of crustal thickness variations as mapped in Plate 2A and intracrustal density variations that are proportional to the velocity anomalies mapped in Plate 3B and 3C, indicating that no significant contribution of buoyancy from the mantle is required.

It is possible to reconcile the estimates of crustal thickness based on surface waves with the estimates from seismic reflection, refraction and receiver function analysis. First, there is some positive correlation between S-wave vertical travel times through the crust estimated from receiver function analysis and the times predicted from our crustal models. Second, the compilations include data from various types of studies with differing quality and there may be some ambiguous possible interpretations that are necessarily hardened into a definite value in compiling a map. The Rayleigh wave data set is more uniform. Third, and most importantly, Rayleigh waves do not directly detect seismic discontinuities. They are sensitive to average velocities over depth ranges, so a 5 km change in Moho depth with a 0.7 km/s velocity contrast at 50 km is roughly equivalent to a 0.1 km/s velocity change over a depth range of 35 to 70 km. The important point, however, is that the Rayleigh wave data require either substantial variations in crustal thickness that correlate with topography, as shown in Plate 2A, or substantial variations in velocity of the lowermost crust and uppermost mantle that correlate with topography. Either way, the combination of upper crustal velocity variations and variations near Moho depths suggest that much of the isostatic compensation of the southern Rocky Mountains takes place in the shallow lithosphere.

In the upper mantle from the Moho to 140 km depth, our velocity images reveal two primary features: a band of low velocities under the Colorado Rockies with a gradual transition to high velocities beneath eastern Colorado and western Kansas (Plate 2D to H, and 3C); and a local low velocity anomaly centered beneath the northern extension of the Rio Grande rift into southern Colorado that is most pronounced just below the Moho (Plate 2D to F, and 3A). Previous studies have found similar slow anomalies beneath the Rio Grande rift in New Mexico [*Parker et al.*, 1984; *Davis et al.*, 1993], although these anomalies have also been attributed to the northeast trending Jemez lineament in New Mexico [*Humphreys and Dueker*, 1994; *Dueker et al.*, 2001]. Our observations agree with the S-wave tomography model of *Lee and Grand* [1996] in terms of the overall pattern of lowest velocity beneath the Rockies, but differ from it in detail. For example, their lowest velocity anomaly in the upper 100 km is essentially uniform in amplitude extending from central Colorado southward into the Rio Grande rift. In our models, there are three components to this anomaly: the crustal velocity anomaly centered on the Sawatch range, the Moho depth anomaly near the San Juan mountains, and the sub-Moho anomaly in the Rio Grande rift.

Lee and Grand [1996] also show a pronounced low-velocity anomaly beneath the central Colorado Rockies extending from 100 to 300 km. Because the horizontal outline of this anomaly is remarkably similar to our crustal anomaly and they did not allow for any crustal velocity anomalies, we are concerned that some of the shallow structure may have "leaked" into the deeper parts of their model. In an inde-

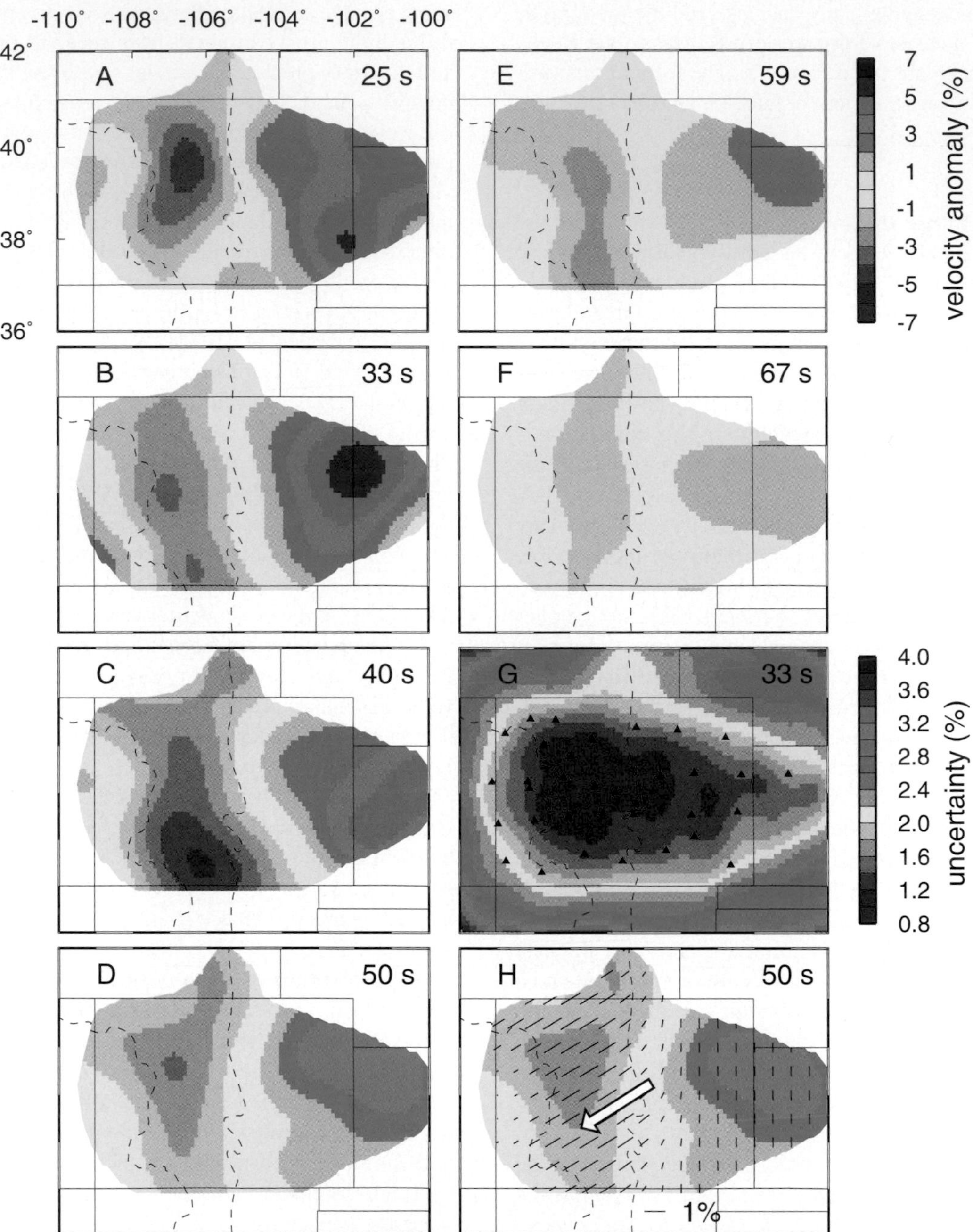

Plate 1. Variations of phase velocities and phase velocity uncertainties. Plate 1 A to E show phase velocity anomalies at 6 periods (25 s, 33 s, 40 s, 50 s, 59 s, 67 s) relative to the average phase velocities (solid line) in Figure 5. Although the inversion is performed in the whole study region, the shown area is outlined by the error contour of 2.2% in Plate 1G which shows twice the standard errors of phase velocity anomalies at 33 s. Plate 1H is a map of azimuthally anisotropic phase velocities at 50 s. Both the average velocity and azimuthal anisotropy were smoothed with a Gaussian characteristic length of 80 km. Azimuthal anisotropy is represented by the black bars. The orientation of the bars indicates the fast direction of azimuthal anisotropy and the length of the bars is proportional to the strength of anisotropy. The absolute plate motion of the North America Plate according to *Gripp and Gordon* [1990] is marked as the big white arrow. Black dashed lines indicate tectonic boundaries.

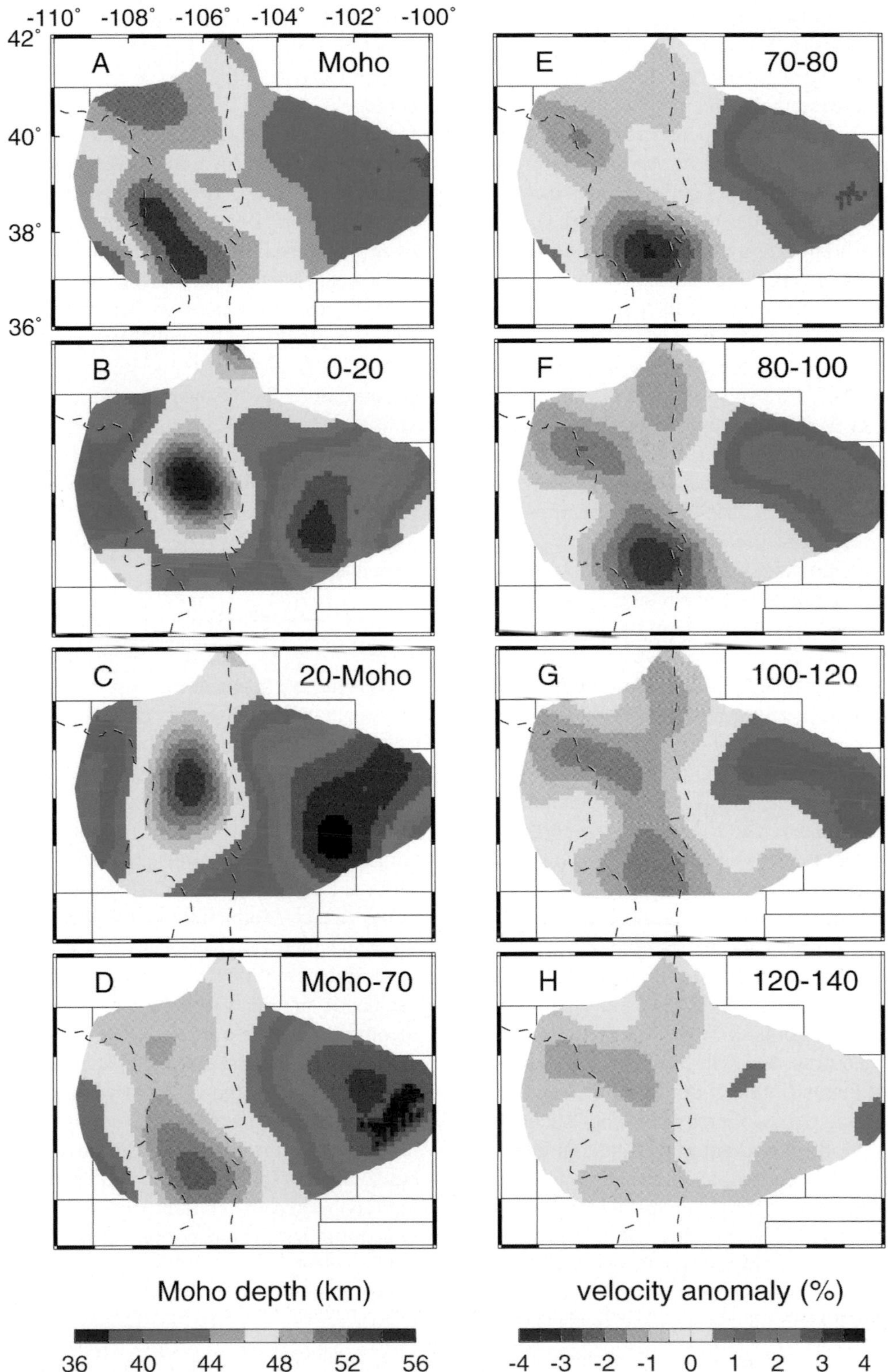

Plate 2. Variations of crustal thickness and shear-wave velocity anomalies in 7 layers from the surface to 140 km depth. The velocity anomalies are calculated relative to the 1-D reference model (CCR) in Figure 7.

pendent, P-wave tomography study that did allow for the possibility of shallow structure, however, *Dueker et al.* [2001] report the existence of a strong (~1.5%), slow, P-wave velocity anomaly extending from the Moho to about 250 km. They call this feature, which coincides in location roughly with the shallow structure (Plate 2B and C), the Aspen anomaly. We see hints of the Aspen anomaly in our deepest slices (Plate 2H), but our images suggest that it does not connect to the crustal anomaly. Body wave tomography with the existing station distribution provides little vertical resolution within the upper 100 km. A tomographic study with denser station distribution in central Colorado is needed to resolve this question.

The transition from a slow tectonic mantle in the southern Rockies to a fast cratonic lithosphere in the Great Plains is imaged near the Rocky Mountain Front, consistent with large-scale tomography studies [*Grand*, 1994; *van der Lee and Nolet*, 1997] and regional body-wave tomography [*Lee and Grand*, 1996; *Lerner-Lam et al.*, 1998]. This transition represents the western edge of the North American craton. Although this study cannot image a sharp lateral boundary, it is clear that in the mantle the transition has significant breadth and extends well into the High Plains east of the Rocky Mountain Front. At depths of 80 km and more in southeastern Colorado, east of the Rio Grande rift, the cratonic lithosphere has been eroded or is absent, suggesting that the old, cold, and presumably more rigid tectosphere can be modified by small-scale convection or rifting events beyond the surface expression of extension. Similar modification of cratonic lithosphere has also been observed in Brazil, and the northeastern U.S./Canadian shield [*VanDecar et al.*, 1995; *van der Lee and Nolet*, 1997; *Rondenay et al.*, 2000]. *Dueker et al.* [2001] suggest that the initial compositional lithosphere may be intact and that localized low velocity anomalies may be created by partial melting of hydrated, olivine-poor lithologies embedded in the lithosphere during earlier, Proterozoic suturing events.

In the upper mantle at depths of 50 to 80 km, the average shear-wave velocity contrast is 4.5 to 5% from the southern Rockies to the Great Plains (Figure 7 and Plate 2). The range decreases with increasing depth, but may be somewhat underestimated because smoothing and damping tend to minimize the variation. Body wave tomography shows a ~9% S-wave velocity contrast across the region [*Lee and Grand*, 1996], but this contrast includes a ~3% variation within Kansas, an area which is not well-resolved by surface waves, and, as discussed above, the contrast may be exaggerated by leaking of crustal anomalies into the mantle part of the model. According to *Nataf and Ricard* [1996], a 100 °C increase in temperature decreases the shear velocity in the shallow mantle by about 1.1%. Our result thus suggests a minimum of 400°C variation in temperature from the Colorado Rocky Mountains to the Great Plains, which is compatible to the temperature contrast in this area obtained by *Goes and van der Lee* [2002], although *Karato* [1993] suggests that thermal coefficients of velocity may be larger if Q is low. The required temperature contrast is reduced if partial melt is present. *Faul et al.* [1994], for example, estimated that 1% melt distributed in inclusions with a range of aspect ratios would decrease shear-wave velocity by about 3.3%. The large, lateral velocity contrasts led *Lee and Grand* [1996] and *Lerner-Lam et al.* [1998] to hypothesize that upwelling and mantle melting is occurring beneath the Rockies, perhaps associated with the northern continuation of rifting and extension from the Rio Grande rift into the Colorado Rockies. Although the absence of a high-velocity lid is consistent with such a scenario, the minimum, absolute, shear velocity for the average southern Rockies is about 4.33 km/s (Figure 7), much faster than the ~4.0 km/s found beneath the East Pacific Rise at comparable depths [*Nishimura and Forsyth*, 1989]. Rather than melt being widely distributed beneath the Rockies, it is more likely that it exists in only a few localities, like beneath the Rio Grande rift or possibly in the Aspen anomaly.

AZIMUHTAL ANISOTROPY

Shear wave splitting measurements [*Savage et al.*, 1996] demonstrate that anisotropy exists beneath at least some parts of our study area in a form that should cause azimuthal variations in Rayleigh wave phase velocities. Although our data have good azimuthal coverage, they are not sufficient for resolving continuous 2-D variations in azimuthal anisotropy. Consequently, we required anisotropy to be constant in each of three tectonic regions, the eastern Colorado Plateau, Colorado Rocky Mountains, and western Great Plains (Figure 4). We inverted simultaneously for the A_1 and A_2 terms needed to describe azimuthal anisotropy along with the A_0 terms at each node and the two-plane-wave parameters describing the incoming wavefield from each earthquake. Our first inversions showed that the Colorado Plateau terms were very poorly constrained, so in the inversions presented here, those terms are fixed at zero. Including anisotropy in the inversions makes only minor improvements in the least-squares fit to the data, but adding the four anisotropic parameters (A_1 and A_2 for two regions) increases the rank by 3 to 4, depending on period, indicating that the anisotropic terms are well resolved.

Observations

Anisotropy varies laterally between the tectonic regions and is also frequency dependent (Plate 1H and Figure 8). Anisotropy under the Rockies shows a simple pattern. It increases with period from near zero at 40 s to over 7% at 100 s with a nearly constant NE-SW fast direction. At periods less than 40 s, the strength is less than 2% and the fast direction progressively rotates to nearly N-S as the period decreases and the waves become primarily sensitive to crustal structure. The strikes of young normal faults in the Rockies [*Eaton*, 1987] are also nearly N-S, suggesting that this apparent crustal anisotropy may be caused by the alignment of cracks. In the Great Plains, anisotropy is generally weaker at most periods than in the Rockies, the 95% confidence limits at several periods include zero anisotropy, and the directions are inconsistent from one period to the next. We do not consider the existence of azimuthal anisotropy in this region to be resolved convincingly; the average effect is likely to be less than 1%. In both regions, the uncertainty increases with increasing period as the velocities decrease in accuracy, the noise increases, and the azimuthal coverage worsens.

Trade-offs

There is always a possible trade-off between velocity and anisotropy when they are jointly inverted. Travel-time variations caused by true azimuthal anisotropy can always be perfectly mimicked by allowing in the model sufficiently strong lateral variations in isotropic velocity on sufficiently short distance scales. Generally, however, the resulting models are highly heterogeneous with large velocity variations over distances comparable to the station separation. With the length scale of velocity variations allowed by the smoothing in this study, there is no significant tradeoff between lateral velocity variations and regional anisotropy. This is illustrated in Plate 1D and 1H at a period of 50s. Adding the anisotropic terms causes only a slight diminution of the amplitude of the low velocity

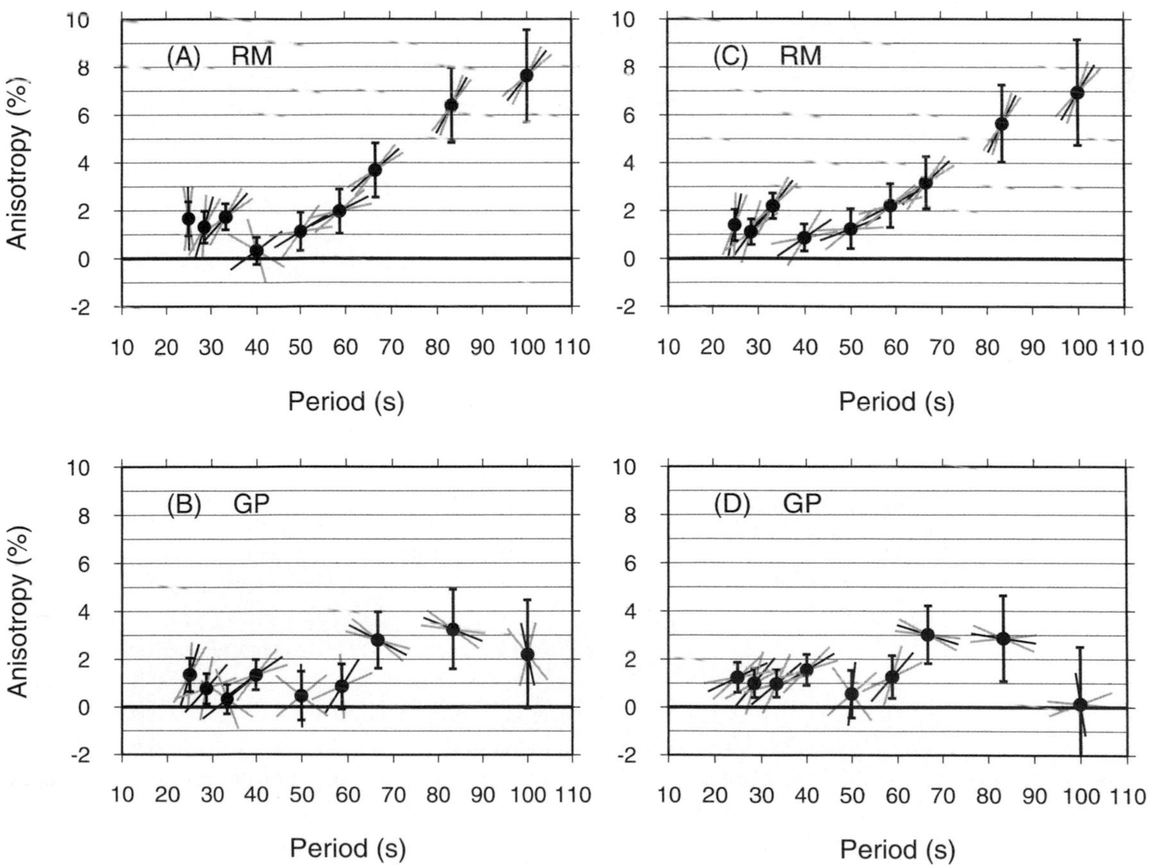

Figure 8. Variations of azimuthal anisotropy with period beneath the Rocky Mountain region (A and C) and the Great Plains (B and D). Black circles indicate the strength of anisotropy in percent. Vertical bars represent one standard error in strength. Fast directions of azimuthal anisotropy are indicated by the orientations of black bars as if in a map view with north up on the diagram. Gray bars show one standard error of the fast directions. For a given frequency, anisotropy in A and B is solved simultaneously in the inversion. For C or D, the anisotropy in that region is allowed to vary, but in the rest of the study area, it is fixed to zero.

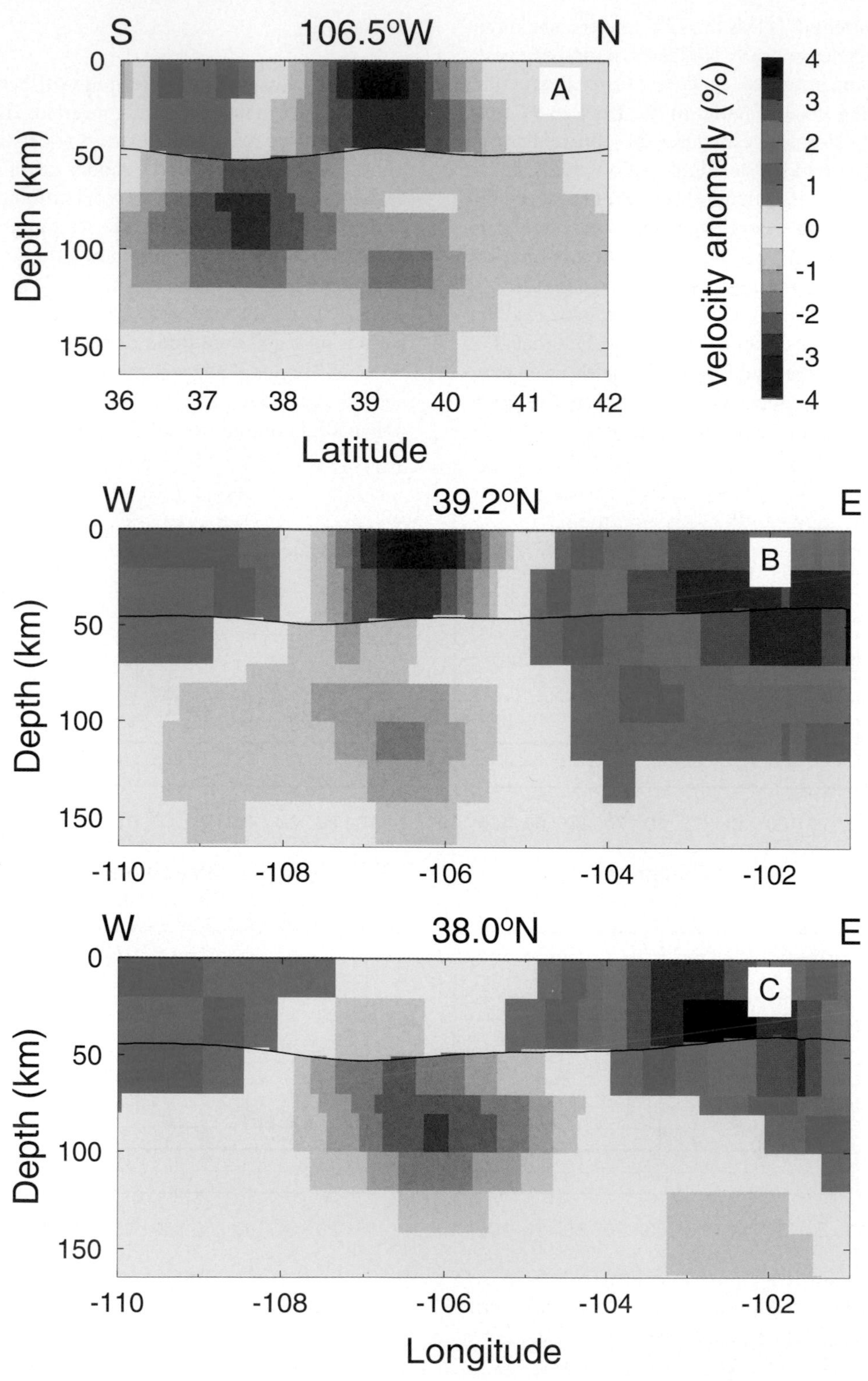

Plate 3. Profiles of share-wave velocity anomalies (similar to Figure 2D-F in [*Li et al.*, 2002]). (A) Profile along 38°N. (B) Profile along 39.2°N. (C) Profile along 106.5°W. Thick black line denotes crustal thickness from Plate 2A.

anomalies in the Rockies. Even though the amplitude of the anisotropic terms is larger at some other periods, the effects are still small. Similarly, the estimated azimuthal anisotropy is not due to the trade-off with true velocity anomalies, as can be seen by the fact that the anisotropy is in general stronger at longer periods even as the velocity anomalies become weaker.

A trade-off of anisotropy between different tectonic provinces is also possible. We were concerned because the anisotropy is unexpectedly large in both the Rockies and Great Plains at long periods (Figure 8A and B), but their fast directions are nearly orthogonal to each other. Could this be due to the trade-off of anisotropy between the two regions? To answer this question, we solved for anisotropy in the Rockies and the Great Plains independently. In these inversions, we allowed uniform anisotropy in just one tectonic province and kept other areas isotropic. The complete results from this experiment are displayed in Figures 8C and D. The trade-off does reduce the anisotropy slightly, by less than 1%, but the patterns of anisotropy in both the Rockies and the Plains are unchanged.

Discussion

The small and variable anisotropy in the Great Plains could indicate that the crust and uppermost mantle are nearly azimuthally isotropic, but our estimates might be biased by the assumption of uniform anisotropy in each tectonic province that artificially averages out lateral variations of anisotropy within the province. Significantly different anisotropy within the high plains is indeed revealed in shear-wave splitting analyses [*Savage et al.*, 1996; *Savage and Sheehan*, 2000], changing from a NE fast direction in northeastern Colorado to a SSE direction in southeastern Colorado. Therefore, the apparent abrupt change in direction with frequency that we observe could be due to a changing population of paths sampling the two areas differently at different frequencies. Unfortunately, there is simply not enough resolution to productively break down the study area into smaller subregions.

The most interesting observation is the simple pattern of azimuthal anisotropy in the Rocky Mountain region (Figure 8A) at periods longer than 40 s. A remarkably consistent fast direction is oriented NE-SW. The observed fast directions on average are rotated 10 to 20 degrees counterclockwise from the direction of the absolute plate motion of the North American plate [*Gripp and Gordon*, 1990] and agree well with the fast direction of shear wave splitting found in the northern Rio Grande rift area of northern New Mexico and southern Colorado [*Sandvol et al.,* 1992; *Savage and Sheehan*, 2000]. Although the absolute values are less believable at the longest periods due to relatively large uncertainties, the trend of anisotropy increasing with period is significant. Therefore, the pattern strongly suggests that anisotropy beneath the Rocky Mountains is dominated by a source at depths of 100 km or greater, probably a highly anisotropic asthenosphere strongly sheared by the North America plate. Beneath the Plains, asthenospheric shearing may occur at greater depths, beyond the range of detection of Rayleigh waves in this study.

Our results of strong and consistent anisotropy at longer periods appear to contradict the observations of shear-wave splitting at stations in the northern Colorado Rockies [*Savage et al.*, 1996], which are characterized by variable direction and many null measurements. The strength of apparent azimuthal anisotropy we observe should produce strong and consistent shear wave splitting. Although the SKS and SKKS phases employed in the splitting studies sample velocity structure from the core-mantle boundary to the earth's surface, analytic studies [*Saltzer et al.,* 2000] suggest that shear-wave splitting measurements tend to be dominated by the structure at shallow depth. If this is true, then the null and variable measurements of shear-wave splitting in the northern Colorado Rockies could be consistent with rapid lateral and vertical changes in anisotropy in the upper lithosphere, although no consistent evidence of a two-layer structure has been reported for the splitting.

The large apparent anisotropy at longer periods is not yet well understood. Maximum anisotropy is ~3% at 83 s and 100 s assuming that the upper mantle is 70%olivine and 30% orthopyroxene and their fast axes are perfectly aligned horizontally, and only ~1% anisotropy distributed throughout the upper mantle (0–400 km) is needed to obtain about 1 s observed delay times of shear-wave splitting. There are several possible solutions to this problem. First, with large uncertainties at 83 and 100s, it could just be a fortuitous accident that the estimated values continue the trend established at intermediate periods. The maximum could be no more than 3% without violating any of the Rayleigh wave observations at the 95% confidence level. However, even at 3%, predicted shear-wave splitting would still exceed observed delay times, unless mitigated by variable anisotropy at shallower depths not resolved by the Rayleigh waves. Second, problems with resolving lateral heterogeneity at finite wavelengths may exist. Except for our weighting function, we implicitly assume ray theory in our interpretation, which is strictly valid only when the wavelength is much smaller than the heterogeneity. At long periods, the wavelength, ~400 km at 100s, exceeds the width of the Rocky Mountain region and the actual sensitivity to structure is spread over a broad region around the ray path [*Marquering et al.*, 1998, 1999] much like a Fresnel zone [*Gudmundsson*, 1996], so that banded lateral heterogeneity with a scale less than a wavelength may be more effectively represented as azimuthal anisotropy. But if the large anisotropy comes from a low velocity anomaly beneath the Rocky Mountain region, the average fast direction is expected to be perpendicular to the band in the E-W direction, not NE as observed. Third, the anisotropy could be due to dipping fast axes that yield greater variation in shear velocity if the propagation is not in symmetry planes [*Babuska and Cara*, 1991] and affect Rayleigh waves and shear waves differently. For instance, if the upper mantle is pure olivine, the strongest azimuthal anisotropy at 83 s is 5.6% when the fast axis is 35° dipping from horizontal. Dipping fast axes should be observable from the variation of splitting with azimuth if SKS phases propagate slightly off a vertical ray path. Further synthetic calculations are needed in order to better understand the large amount of anisotropy from the Rayleigh waves and to compare this result with shear-wave splitting observations.

CONCLUSIONS

Using Rayleigh waves propagating across the Rocky Mountain Front seismic array, we observed distinct structures beneath the western Great Plains, southern Rocky Mountains and the eastern Colorado Plateau. There is no high velocity lid beneath the Rockies and anomalously low velocities continue to depths of 150 km and greater. Beneath the Plateau, there is a high velocity lid and pronounced low velocity zone and beneath the western Plains, velocities are high and the low-velocity zone is almost absent. The largest velocity contrast across the region is in the crust, with the lowest velocities beneath the elevated region around the Sawatch Range, suggesting that, in addition to crustal thickening, density variations within the crust play an important role in compensating the high topography of the Rocky Mountains. In the shallow upper mantle, the strongest low-velocity anomaly is imaged in southern Colorado, near the Rio Grande rift, indicating the presence of partial melt. The transition from cratonic mantle in the eastern part of the study area to tectonic mantle beneath the Rockies is gradual and not confined to the vicinity of the Rocky Mountain Front.

We constrained azimuthal anisotropy in a joint inversion that included spatial variations in the azimuthally averaged isotropic phase velocities. With the scale length of spatial variations allowed in this study, there is little trade-off between azimuthal and spatial variations in velocity. Azimuthal anisotropy beneath the Rocky Mountains increases with period and its fast direction at depth is close to the absolute plate motion of the North American plate. This pattern is consistent with deep asthenospheric flow dominated by the current plate motion.

Acknowledgments. We thank Kenneth Dueker and Anne Sheehan for providing information about station characteristics and instrument responses, Aaron Velasco and Vladislav Babuska for their constructive comments. Seismic data were obtained from the IRIS Data Management Center. This research was supported by NSF grants EAR-9614705 and EAR-9903026.

REFERENCES

Babuska, V., and M. Cara, *Seismic Anisotropy in the Earth*, pp.145, Kluwer Academic Publishers, Dordrecht, Netherlands, 1991.

Brune, J. and J. Dorman, Seismic waves and Earth structure in the Canadian Shield, *Bull. Seis. Soc. Am., 53,* 167–209, 1963.

Burchfiel, B. C., D. S. Cowan, and G. A. Davis, Tectonic overview of the Cordilleran orogen the western United States, in *The Cordilleran Orogen: Conterminous U.S.*, edited by B. C. Burchfiel, P.W. Lipman, and M. L. Zoback, pp. 407–479, Boulder, Colorado, Geol. Soc. Am., 1992.

Davis, P. M., P. Slack, H. A. Dahlheim, W. V. Green, R. P. Meyer, U. Achauer, A. Glahn, and M. Granet, Teleseismic tomography: Theory and Practice, edited by H. M. Iyer and K. Hirahara, pp. 397–439, Blackwell, Cambridge, Mass., 1993.

Dueker, K., H. Yuan, and B. Zurek, Thick-structured Proterozoic lithosphere of the Rocky Mountain region, *GSA Today, 11*, no. 12, 4–9, 2001.

Eaton, G. P., Topography and origin of southern Rocky Mountains and Alvarado Ridge, in *Continental Extension Tectonics*, *Geol. Soc. Spec. Publ.* London, 28, pp.335–369, 1987.

Faul, U. H., D. R. Toomy, and H. S. Waff, Intergranular basaltic melt is distributed in thin, elongated inclusions, *Geophys. Res. Lett.*, *21*, 29–32,1994.

Forsyth, D. W., S. Webb, L. Dorman, and Y. Shen, Phase velocities of Rayleigh waves in the MELT experiment on the East Pacific Rise, *Science*, *280*, 1235–1238, 1998.

Grand, S. P., Mantle shear structure beneath the Americas and surrounding oceans, *J. Geophys. Res.*, *99*, 11,591–11,621, 1994.

Gripp, A. E., and R. G. Gordon, Current plate velocities relative to the hotspots incorporating the NUVEL-1 global plate motion model, *Geophys. Res. Lett.*, *17*, 1109–1112, 1990.

Goes, S. and S. van der Lee, Thermal structure of the North American uppermost matnle inferred from seismic tomography, *J. Geophys. Res.*,10.1029/2000JB000049, March, 2002.

Gudmundsson, O., On the effect of diffraction on traveltime measurements, *Geophys. J. Int.*, *124*, 304–314, 1996.

Humphreys, E. D., and K.G. Dueker, Physical state of the western U.S. upper mantle, *J. Geophys. Res.*, *99*, 9635–9650, 1994.

Karato, S., Importance of anelasticity in the interpretation of seismic tomography, *Geophys. Res. Lett., 20,* 1623–1626, 1993.

Karlstrom K. E., and E. D. Humphreys, Persistent influence of Proterozoic accretionary boundaries in the tectonic evolution of southwestern North America: Interaction of cratonic grain and mantle modification events, *Rocky Mountain Geology*, *33*, 161–179, 1998.

Kennett, B. L., N., E. R. Engdahl, and R. Buland, Constraints on seismic velocities in the Earth from traveltimes, *Geophys. J. Int.*, *122*, 108–124, 1995.

Keller, G. R., C. M. Snelson, A. F. Sheehan, and K. G. Dueker, Geophysical studies of crustal structure in the Rocky Mountain region: A review, *Rocky Mnt. Geol.*, *33*, 217–228, 1998.

Keller, G.R., and Baldridge, W.S., The Rio Grande rift: A geological and geophysical overview: *Rocky Mountain Geology, 34*, 121–130, 1999.

Lastowka, L.A., A.F. Sheehan, and J.M. Schneider, Seismic evidence for partial lithospheric delamination model of Colorado Plateau uplift, *Geophys. Res. Lett., 28*, 1319–1322, 2001.

Lee, D.-K., and S. P. Grand, Upper mantle shear structure beneath the Colorado Rocky Mountains, *J. Geophys. Res.*, *101*, 22,233–22,244, 1996.

Lerner-Lam, A. L., A. F. Sheehan, S. P. Grand, E. D. Humphreys, K. G. Dueker, E. Hessler, H. Gao, D.-K. Lee, and M. Savage, Deep structure beneath the southern Rocky Mountains from Rocky Mountain Front Broadband Seismic Experiment, *Rocky. Mnt. Geol.*, *33*, 199–216, 1998.

Li, A., Crust and mantle discontinuities, shear wave velocity structure, and azimuthal anisotropy beneath North America, Ph. D. dissertation, Brown University, 2001.

Li, A., D. W. Forsyth, and K. M. Fischer, Evidence for shallow isostatic compensation of the southern Rocky Mountains from Rayleigh wave tomography, *Geology, 30*, 683–686, 2002.

Marquering, H., G. Nolet, and F. A. Dahlen, Three-dimensional waveform sensitivity kernels, *Geophys. J. Int.*, *132*, 521–534, 1998.

Marquering, H., F. A. Dahlen, and G. Nolet, Three-dimensional sensitivity kernels for finite frequency traveltimes: the banana-doughnut paradox, *Geophys. J. Int.*, *137*, 805–815, 1999.

Martin, B.E., and C. J. Thomson, Modeling surface waves in anisotropic structures II: Examples, *Phys. Earth Planet. Int.. 103*, 253–279, 1997.

Nataf, H. C., and Y. Ricard, 3SMAC: An *a priori* tomographic model of the upper mantle based on geophysical modeling, *Phys. Earth Planet. Int.*, *95*, 101–102, 1996.

Nishimura, C.E., and D.W. Forsyth, The anisotropic structure of the upper mantle in the Pacific, *Geophys. J., 96*, 203–229, 1989.

Parker, E. C., P. M. Davis, J. R. Evans, H. M. Iyer, and K. H. Olsen, Upwarp of anomalous asthenosphere beneath the Rio Grande rift, *Nature*, *312*, 354–356, 1984.

Prodehl, C., and P. W. Lipman, Crustal structure of the Rocky Mountain region, in *Geophysical framework of the continental United States*, edited by L. C. Pakiser and W. D. Mooney, pp.249–284, Geol. Soc. Am. Mem. 172, 1989.

Rondenay, S., M. G. Bostock, T. M. Hearn, D. J. White, and R. M. Ellis, Lithospheric assembly and modification of the SE Canadian Shield: Abitibi-Grenville teleseismic experiment, *J. Geophys. Res.*,*105*,13,735–13,754, 2000.

Saltzer, R. L., J. B. Gaherty, and T. H. Jordan, How are vertical shear wave splitting measurements affected by variation in the orientation of azimuthal anisotropy with depth?, *Geophy. J. Int., 141*, 374–390, 2000.

Sandvol, E., J. Ni, S. Ozalaybey, and J. Schlue, Shear-wave splitting in the Rio Grande rift, *Geophys. Res. Lett.*, *19*, 2337–2340, 1992.

Savage, M. K., A. F. Sheehan, and A. Lerner-Lam, Shear wave splitting across the Rocky Mountain Front, *Geophys. Res. Lett.*, *23*, 2267–2270, 1996.

Savage, M. K., and A. F. Sheehan, Seismic anisotropy and mantle flow from the Great Basin to the Great Plains, western United States, *J. Geophys. Res.*, *105*, 13,715–13,734, 2000.

Sheehan, A. F., G. A. Abers, A. L. Lerner-Lam, and C. H. Jones, Crustal thickness variations across the Rocky Mountain Front from teleseismic receiver functions, *J. Geophys. Res.*, *100*, 20,291–20,304, 1995.

Sheehan, A.F., C.H. Jones, M.K. Savage, S. Ozalabey, and J.M. Schneider, Contrasting lithosphere structure between the Colorado Plateau and the Great Basin: Initial results from Colorado Plateau - Great Basin PASSCAL experiment, *Geophys. Res. Lett., 24,* 2609–2612, 1997.

Simons, F.K., A. Zielhuis, and R.D. van der Hilst, The deep structure of the Australian continent from surface wave tomography, *Lithos., 48,* 17–43, 1999.

Smith, M. L., and F. A. Dahlen, The azimuthal dependence of Love and Rayleigh wave propagation in a slightly anisotropic medium, *J. Geophys. Res.*, *78*, 3321–3333, 1973.

Spencer, J.E., Uplift of the Colorado Plateau due to lithosphere attenuation during Laramide low-angle subduction, *J. Geophys. Res., 101*, 13595–13609, 1996.

Tarantola, A., and B. Valette, Generalized non-linear problems solved using the least-squares criterion, *Rev. Geophys. Sp. Phys., 20*, 219–232, 1982.

Thomson, C. J., Modeling surface waves in anisotropic structures I. Theory, *Phys. Earth Planet.. Int.*, *103*, 195–206, 1997.

VanDecar, J. C., D. E. James, and M. Assumpcao, Seismic evidence for a fossil mantle plume beneath South America and implications for plate driving forces, *Nature*, *378*, 25–31, 1995.

Van der Lee, S., and G. Nolet, Upper mantle *S*-velocity structure of North America, *J. Geophys. Res.*, *102*, 22,815–22,838, 1997.

The Dynamic Nature of the Continental Crust-Mantle Boundary: Crustal Evolution in the Southern Rocky Mountain Region as an Example

G. Randy Keller[1], Karl E. Karlstrom[2], Michael L. Williams[3], Kate C. Miller[1], Christopher Andronicos[1], Alan R. Levander[4], Catherine M. Snelson[5], Claus Prodehl[6]

The evolution of continents involves modification of the lithosphere through time, including changes in crustal thickness and composition that create a dynamic crust-mantle boundary (Moho). The geological history of the southern Rocky Mountain region is relatively well understood and recent additions of modern seismic data provide an ideal opportunity to investigate the evolution of the crust. The results presented in this volume show that crust in the southern Rocky Mountains is relatively thick compared to the global average for the continents. The mafic lower crust and crust-mantle boundary of the Proterozoic provinces of the southwestern U.S. likely formed, and reformed, in several stages. Initial formation of juvenile continental crust took place by development and assembly of magmatic arcs between 1.8 and 1.6 Ga. Volcanic and plutonic rocks of this age record whole-crust differentiation and probably resulted in a mafic lower crustal residue. From 1.45 to 1.35 Ga, the crust underwent another period of differentiation leading to emplacement of A-type granites in the middle crust across southern Laurentia. Voluminous granitoid emplacement ca. 1.4 Ga, petrology of granitoids, widespread metamorphism, and 1.4 Ga lower crustal xenoliths are best explained by mafic underplating. Subsequent mafic additions to the lower crust likely took place at each of the times when basalts were emplaced in the Rocky Mountain region (1.1 Ga Grenville orogeny, Laramide orogeny, Oligocene ignimbrite flare-up and Neogene extensional tectonism), but these events were more local in distribution and not widespread enough to produce the thick, mafic crust observed over an extensive area.

1. INTRODUCTION

Understanding the crust-mantle boundary (Moho) is a key aspect of understanding the evolution of continents. Exchange of material across the Moho is the primary process by which continental crust is created and recycled back into the mantle [*Ardnt and Goldstein*, 1989]. The depth of this boundary can be changed by tectonic thickening and thinning, magmatic

[1]Department of Geological Sciences, University of Texas at El Paso, El Paso, Texas

[2]Department of Earth and Planetary Sciences, University of New Mexico, Albuquerque, New Mexico

The Rocky Mountain Region: An Evolving Lithosphere
Geophysical Monograph Series 154

10.1029/154GM30

[3]Department of Geosciences, University of Massachusetts, Amherst, Massachusetts

[4]Department of Geology and Geophysics, Rice University Houston, Texas

[5]Department of Geoscience, University of Nevada Las Vegas, Las Vegas, Nevada

[6]Geophysikalisches Institut, University of Karlsruhe, Karlsruhe, Germany

underplating (an array of processes whereby mafic melts derived from the mantle are added to lowermost portion of the crust), and lithospheric delamination. Additionally, some studies have concluded that the Moho represents the base of the lithosphere during crustal formation [*Roy Chowdhury and Hargraves*, 1981]. Therefore the evolution of this boundary has profound implications for the mechanical and petrologic evolution of the continents.

The long-term evolution of continents typically involves modifications of the lithosphere by plate margin tectonism (orogeny) and by intraplate lithosphere-asthenosphere interactions in the "interior" (plume/hot spot impacts and rifting). Inevitably in continents that remain coherent over billions of years, this evolution involves changes in the thickness and composition of the crustal column through time that can lead to "cratonization" (increased stability), or incipient rifting (decreased stability). The southern Rocky Mountain region records both a major period of crustal growth and cratonization at 1.8–1.6 Ga and repeated reactivation events due to subsequent intracratonic tectonism [Karlstrom and the CD-ROM *Working Group*, 2002].

The goal of this paper is to present our analysis of the evolution of the crust of the southern Rocky Mountain region based on the integration of geologic data with recent seismic results. The lithosphere of this region has been investigated by diverse seismic methods that can be compared, contrasted, and integrated. These include: 1) Broad-band teleseismic data recorded at a grid of stations in the Rocky Mountain Front experiment [*Sheehan et al.*, 1995; *Lerner-Lam et al.*, 1998; *Li et al.*, this volume]; 2) The Deep Probe experiment, the longest modern refraction line in North America, which extended from the Canadian Arctic through the Rockies to southern New Mexico [*Henstock et al.*, 1998; *Snelson et al.*, 1998; *Gorman et al.*, 2002] and also included a teleseismic recording effort [*Crosswhite and Humphreys*, 2003]; and 3) The Continental Dynamics of the Rocky Mountain Project (CD-ROM) seismic experiments that included a combination of high resolution, and approximately co-located, refraction, teleseismic, and reflection studies (as presented in a series of papers in this volume). We use the CD-ROM results from the southern Rocky Mountain region combined with earlier studies to discuss the evolution of the crust and to argue that the crust-mantle boundary in this region has been a dynamic feature throughout the past 1.8 Ga.

2. NATURE OF THE CRUST-MANTLE BOUNDARY

A discussion of the crust-mantle boundary is complicated and clouded by semantics and differing interpretations of the composition and evolution of the lower crust and mantle lithosphere. The crust-mantle boundary is commonly taken to be identical to the Moho, which is traditionally defined [e.g., *Jarchow and Thompson,* 1989] based on seismic refraction measurements as the deep horizon (usually a zone of finite thickness) in the Earth where seismic compressional wave velocity (Vp) exceeds 7.6 km/s. However, *Nelson* [1991] argued that the actual crust-mantle boundary does not have a prominent seismic expression and is the contact between cumulate ultramafic rocks and the residuum below and may lie well below the Moho. *Furlong and Fountain* [1986] also argued that underplating of the continental crust is a common phenomena that adds material to the base of the crust that can have velocities greater than 7.6 km/s. Additionally, *Griffin and O'Reilly* [1987] pointed out that mafic lower crustal rocks can be transformed into eclogites, which are seismically identical to typical mantle peridotites, but which represent crustal material. Thus in some places, there is likely to exist a seismic Moho (Vp > 7.6 km/s) that resides at a shallower depth than the petrologic Moho, which can be defined as the base of rocks with mafic compositions and could be considered the crust-mantle boundary.

In addition to seismic refraction determinations of P-wave velocity, a seismic discontinuity referred to as the Moho is also mapped by deep seismic reflection profiling and receiver function studies. As discussed below, the Moho as defined by each of these techniques is based on a different seismic response (frequency and wave type) to a zone representing a major velocity discontinuity or series of discontinuities. The seismic studies reported in this volume and previous studies in the region [*Prodehl et al.*, this volume] have detected a strong discontinuity at a depth of 40–55 km, and considering the uncertainties in each method, nearby results agree well on the depth to the Moho. We have taken the new results reported in this volume and combined them with the older refraction and receiver function results and gridded and contoured them to produce Figure 1. We have not significantly smoothed the data before contouring, and the map is quite irregular in places. This irregularity is partly due to the use of different types of data with different resolutions. One must remember that +/- 10% is a reasonable rule of thumb for estimating the uncertainty of Moho depth determinations [*Mooney*, 1989] for all but the high-resolution studies that have been possible in recent years as large numbers of instruments have become available. On the other hand, the crust in the Rocky Mountain region has experienced a complex evolution that probably played a role in producing some of the undulations shown on the map (Fig. 1). As discussed below, at least the main features on this map represent the relief on a major seismic discontinuity underlain by P-wave velocities >7.6 km/s, which we conclude is the Moho today. Thus, the observed 15+ km of relief on the Moho shown in this image represents a first-order signature of the lithosphere of the Southern Rocky

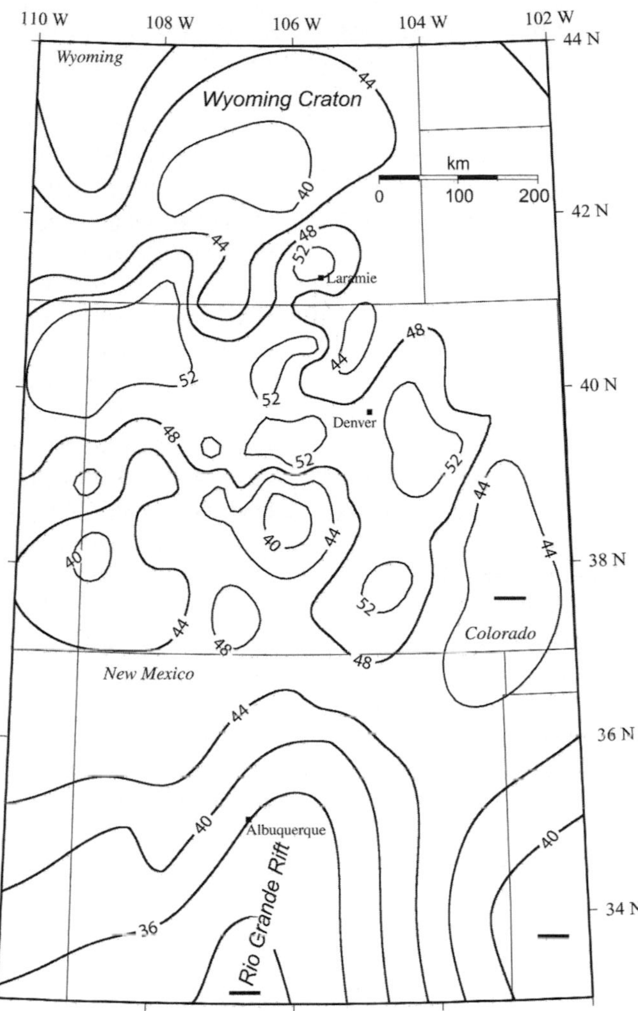

Figure 1. Map of the thickness of the crust today derived by gridding and contouring the results of previous studies in the area [seismic refraction – Prodehl et al., this volume; receiver functions - Sheehan et al., this volume] with the seismic results presented in this volume. Minus signs (-) indicate minima in crustal thickest that may not be obvious from the labeling. Contour interval – 4 km.

Mountains. As we discuss the processes at work during the evolution of the crust, we will focus on possible past configurations of the crust-mantle boundary and speculate on the position of the paleo-Moho that seismologists would have mapped had they had been able to conduct experiments long ago.

A different approach to understanding the crust-mantle boundary utilizes petrologic data from xenoliths and volcanic/plutonic/metamorphic rocks, in order to add a time dimension to the discussion. For example *Griffin and O'Reilly* [1987] pointed out that mixed "crustal" (i.e. mafic) and "mantle" (i.e. ultramafic) lithologies could result in average velocities that would be interpreted as "crust" or "mantle", depending on the proportions and physical state (degree of differentiation or conditions of metamorphism) of interlayered materials. For example, "crustal" lithologies in granulite facies (garnet granulites) can have seismic velocities that would be identified as upper mantle (i.e., Vp >7.6 km/s), eclogites (Vp ~8.4 km/sec) can be present in the crust (derived from metamorphism of crustal basalts) or mantle, and hydrated mantle rocks or mantle rocks at elevated temperatures can have velocities that are transitional between crust and mantle (7.6 km/sec or lower). Hence, the Moho that is imaged seismically in many continental regions is usually a complex zone (typically ~5 km thick) rather than a sharp interface and is thus capable of producing diverse seismic signatures that are subject to differing interpretations. In addition, underplating may result in complex layering and a range of seismic velocities that fall above and below 7.6 km/s [e. g., *Furlong and Fountain*, 1986]. Transitional structures such as these may, for example, produce different interpretations of refraction/wide angle-reflection data [e.g., *Snelson et al.*, this volume; *Levander et al.*, this volume]. In spite of these semantic and physical complications, we believe that the different ways to define and study the crust-mantle boundary have each provided important insight in our effort to answer questions concerning its origin and history.

3. SEISMIC CHARACTERISTICS OF THE CRUST-MANTLE BOUNDARY

As discussed above, the classic definition of the Moho is based on seismic refraction data, and refraction waveforms can commonly be modeled in continental and oceanic areas as arising from a smooth or complex velocity increase from crust to mantle velocities in a thin (several kilometers thick) layer at the base of the crust [e.g., *Braile and Chang*, 1986; *Wolf and Cipar*, 1993]. This same waveform character has also been modeled as resulting from a Moho with laterally variable topography [e.g., *Larkin et al.*, 1997]. Other types of seismic measurements provide information about the geometry of the change from crustal to mantle velocities. On seismic reflection images, the Moho can appear as a single reflection, a zone of multicyclic reflections, or a gradual loss of reflections [e.g. *Sadowiak et al.*, 1991; *Mooney and Meissner*, 1992]. In addition, receiver functions have, for over a decade, provided useful measurements of the travel time of S-waves through the crust. These measurements are based on the P to S conversion at the Moho, because this phase is usually manifested as a single, easily recognized pulse on the receiver function [e.g., *Sheehan et al.*, 1995; *Sheehan et al.*, this volume; *Zurek and Dueker*, this volume] with a dominant wavelength on the order of 5 km. This pulse is interpreted as being due to an interface (a zone ~5km thick), and if the velocity of the crust is estimated, these travel times can be converted to depths. Finally, surface wave dispersion studies can also define the Moho based on S-wave velocity variations. Although this method determines only average velocity across regions with dimensions of ~100 km or more, it does offer complementary information in Colorado because of the 3-D grid of stations of the Rocky Mountain Front experiment [*Sheehan et al.*, 1995; *Li et al.*, 2002, and this volume].

These different seismic techniques can produce conflicting values for the depth of the Moho, but at least some of the apparent differences can result from the different, commonly non-overlapping, bandwidths used in these techniques. For example, near vertical-incidence reflection data generally contain higher frequencies (10–50 Hz) and will commonly respond to a highly layered Moho differently than lower frequency (2–10Hz) refraction data, which in turn respond differently than the 0.3–0.5 Hz teleseismic receiver function. We have found that, in the southern Rocky Mountain region, differences in reported Moho depth are typically within the resolution limits of the different data sets. However the more vertically and laterally complicated the Moho is in a particular region, the greater is the potential for differences between techniques to arise.

Another key to understanding the crust-mantle boundary is knowledge of the uppermost mantle. The velocity of the uppermost portion of the mantle is best measured by the refracted/diving P-wave (Pn phase) that travels just below the Moho. The distance at which this phase becomes a first arrival (the Pn crossover distance) is a key observation in determining Moho depth (i.e., the greater the crossover distance, the thicker the crust). The global average uppermost mantle velocity (~8.1 km/s) is very consistent in tectonically stable areas [e.g., *Christensen and Mooney*, 1995], but in areas of high heat flow such as rift zones, the velocity of the uppermost mantle may be as low as ~7.5 km/s [e.g., *Mechie et al.*, 1994]. Velocities greater than ~8.2 km/s suggest low heat flow and rocks that are rich in olivine and/or in eclogite facies [e.g., *Christensen and Mooney*, 1995].

4. THE PRESENT CRUST-MANTLE INTERFACE IN THE SOUTHERN ROCKY MOUNTAINS

Analysis of the CD-ROM seismic refraction [*Snelson et al.*, this volume; *Levander et al.*, this volume], deep reflection [*Morozova et al.*, this volume, *Magnani et al.*, this volume, a], and teleseismic [*Dueker et al.*, 2001; *Zurek and Dueker*, this volume] data show that the crustal thickness varies from 40 km to greater than 50 km along the CD-ROM seismic transect. These results are in good agreement with previous studies in the region [summarized in *Sheehan et al.*, this volume; *Prodehl et al.*, this volume]. When viewed regionally (Fig. 1), the crustal thickness shows less correlation with modern topography than simple isostatic models would predict [*Sheehan et al.*, 1995; *Snelson et al.*, this volume]. In addition, the crust is generally thinner than predicted by these models, and thus, some of the compensating mass must lie in the upper mantle. However, the crust is considerably thicker than the ~40 km thick global average for continental crust [*Christensen and Mooney*, 1995], and as shown by *Li et al.* [2002] and *Snelson et al.* [this volume], the variations in crustal thickness and composition, and therefore density, contribute substantially to isostatic balance.

Underplating is a major process in the growth and evolution of continental crust and directly changes the position and composition of the crust-mantle boundary [*Furlong and Fountain*, 1986; *Ardnt and Goldstien, 1989; Nelson*, 1991]. Thus, it is important to discuss seismic evidence for underplating in the Rocky Mountain region. However, this is difficult for several reasons. First, the term underplating has been used to describe a variety of magmatic processes including intrusion (sills and dikes) and true ponding of mafic magma in the lower crust and uppermost mantle. Secondly, in areas with relatively thick (>40 km) crust, the presence of lower crustal layers with P-wave velocity >7.0 km/s have commonly been interpreted as evidence for underplating. However, if underplating is widespread, the use of average crustal properties, such as derived by *Christensen and Mooney* [1995], to search for crustal structure anomalies indicative of underplating is doomed in most cases because seismic measurements in many areas that are underplated have been used in the process of determining average values for the crustal thickness (~40 km) and Vp (~6.45 km/s). Finally in terms of velocity, seismic techniques have their lowest resolution in the lower crust. Receiver functions are seldom able to detect interfaces within the crust, and in particular, to determine the velocity layers within the crust. Reflection data have very low resolution of velocity in the lower crust. In refraction studies, unless a clear first arrival is observed from the lowermost crust [e.g. the DEEP PROBE experiment, *Gorman et al*., 2002], different interpretations of the lower crust are commonly possible based on the subjectivity of picking and correlating phases that follow the first arrival in time. This is the case in the studies of *Snelson et al*. [this volume] and *Levander et al*. [this volume]. In these two models, the crustal thicknesses are comparable, but the velocity of the lowermost crust is different by about 0.2 km/s (~7.0 to ~7.2). Similar but greater differences exist (for similar reasons) in interpretations of the PACE experiment data on the Colorado Plateau [e.g., *Wolf and Cipar*, 1993; *Parsons et al*., 1996].

With these complicating factors in mind, it is nonetheless appropriate to compare the southern Rocky Mountain region (Fig. 1), which includes the eastern Colorado Plateau and the western Great Plains, to global averages of crustal thickness (Tc) and P-wave velocity (Vp) [*Christensen and Mooney*, 1995] for orogens (Tc~46 km, Vp~6.4 km/s) and shields and platforms (Tc~41 km, Vp~6.4). In orogens, the average crustal velocity in the lower crust near the Moho (45–50 km) is ~7.0 km/s and in shields and platforms (40–45 km) this value is 7.1–7.2 km/s. Thus, the CD-ROM determinations of crustal thickness are thick relative to the average for orogens within the Rocky Mountains and about average in the Great Plains. In terms of crustal evolution, the differences in the CD-ROM refraction models for the lower crustal velocity (7.0–7.2 km/s) are not very significant because virtually all of the values fall within the expected range for mafic material. Thus, the relatively thick crust in the region (Fig. 1) argues most strongly for considerable underplating during the evolution of the crust in the Southern Rocky Mountains. Further east across the Great Plains and Midcontinent region, existing refraction results were synthesized by *Braile* [1989] who constructed a profile along 37ºN extending from 95º to100º W. All along this profile, the crust is thicker than 40 km and the lower crust is ~20 km thick with an average Vp of 7.1–7.2 km/s. The combined results of the CD-ROM project and *Braile's* [1989] synthesis show that a relatively thick portion of the lower crust beneath most of the Rocky Mountain region has a P-wave velocity high enough to be consistent with it being formed by mafic underplating.

An interesting result shown in Figure 1 is the large crustal thickness variations (>15 km) in the southern Rocky Mountain region. For example, there is at least 10 km of relief on the Moho along the CD-ROM transect that begins along the southern margin of the Archean Wyoming craton. The connection of this transect to the DEEP PROBE profile [*e.g., Gorman et al*., 2002] provides another image of deep structure from the early Proterozoic southern Rocky Mountain terranes, across the margin of this craton, into its core. Detailed geologic studies of Archean exposures in southernmost Wyoming provide convincing evidence that the southern margin of the Wyoming craton was rifted to form a passive continental margin at about 2.1 Ga [*Karlstrom et al*., 1983; *Karlstrom and Houston*, 1984]. A north to south thinning of today's crust by 10 km in this region follows this ancient plate boundary suggesting that the relative crustal thickness change may in part still reflect this ancient event [*Prodehl and Lipman*, 1989; *Johnson et al*., 1994; ; *Henstock et al*., 1998; *Snelson et al*., 1998; this volume]. Other thickness variations such as the area of thick crust in northwest Colorado along the Cheyenne belt imaged by the DEEP PROBE teleseismic array [*Crosswhite and Humphreys*, 2003] could also result from ancient tectonics. The CD-ROM seismic reflection data also show that the structure of the Cheyenne belt suture is very complex suggesting that the Proterozoic crust of the Colorado provinces was interleaved with Archean crust of the Wyoming province during suturing [*Morozova et al*., this volume]. In addition, some of the localized areas of thick crust lie in areas that were significantly affected by the Ancestral Rocky Mountain and Laramide orogenies (e.g., Front Range area, Fig.1). Finally, the crustal thinning associated with the Rio Grande rift in New Mexico (Figure 1) is well documented and correlates well with estimates of extension that show it increasing southward [*Cordell*, 1982; *Keller and Baldridge*, 1999]

The upper mantle velocities determined in the CD-ROM refraction analyses agree [*Snelson et al*., this volume; *Levander et al*., this volume] and are low (~7.9 km/s) compared typical continental values (8.0–8.1 km/s). In addition, teleseismic tomography [*Dueker et al.,* 2001; *Yuan and Dueker*, this volume] and surface wave tomography [*Li et al*., 2002, and this volume] show many low velocity regions in the uppermost mantle with Rio Grande rift/Jemez lineament being the most prominent. The origin of these values may be due to the high regional heat flow, but the heat flow anomaly associated with the northern Rio Grande rift is narrow [e.g., *Reiter et al.*, 1975]. Compositional variations due to subduction of the Farallon plate have been invoked to explain low velocities in the upper mantle of the Rocky Mountain region [e.g., *Dueker et al.*, 2001]. However, it is also possible that the anomalies represent an older compositional anomaly [*Karlstrom et al.*, this volume].

5. GEOLOGIC EVOLUTION OF THE CRUST-MANTLE BOUNDARY IN THE SOUTHERN ROCKY MOUNTAIN REGION

Notwithstanding the difficulties in defining and mapping the crust-mantle boundary today, the real challenge is trying to construct a credible interpretation of the evolution of this boundary through time. Modern exposures of Precambrian rocks over a large region of Colorado and northern New Mexico were at middle crustal depths of 10 to 20 km at about 1.4 Ga [*Grambling*, 1986; Williams and Karlstrom, 1996; *Farmer et al.*, this volume]. Because the crust is ~45 km thick today and the exposures are too extensive to be due to Laramide thrusting, the crust was either ~65 km thick at 1.4 Ga or 20 km or more of material has been added to the crust since 1.4 Ga. This does not include the likelihood that Neoproterozoic and/or Phanerozoic sediments and volcanics were deposited and subsequently removed. Thus, important questions are how and when did the crust attain its present thickness, and how did its thickness change through time? There are several mechanisms (e.g., shortening, extension, magmatism, migration of metamorphic fronts, and lower crust and upper mantle delamination) by which the crust-mantle boundary can "migrate" through time. We will discuss them where they apply below.

If we use modern island arcs as a reference [e.g., *Holbrook et al.*, 1999; *Fliedner and Klemperer*, 1999], then building today's crust in the southern Rocky Mountain region from the accretion of island arcs required a major chemical refining process to increase the silica content of mafic island arc crust to that of crust typical of cratonal areas. Today, about 70% of the exposed Precambrian rocks are 1.6 Ga or greater in age. Although there are many lithologies present, these rocks are dominantly felsic intrusions. About 25% of the exposed Precambrian rocks are ~1.4 Ga in age and are almost entirely felsic. From a geophysical perspective, several large gravity anomalies in the region help delineate the large extent of both mafic and felsic material added to what is now the upper crust over the past 1.6 Ga [e.g., *Plouff and Pakiser*, 1972; *Schneider and Keller*, 1994; *Adams and Keller*, 1996; *McCoy et al.*, this volume; *Quezada et al.*, this volume].

Geological and geophysical data presented in this volume indicate that the crust we see today in southwestern North America is the result of at least seven major tectonic/magmatic events (Figs. 2–5) including: A) crustal accretion at 1.8–1.6 Ga; B) intracratonic magmatism at 1.4 Ga; C) final assembly of Rodinia at 1.1–1.0 Ga; D) breakup of Rodinia at 0.8–0.6 Ga; E) assembly of Pangea and the Ancestral Rocky Mountain orogeny at ~300 Ma; F) the Laramide orogeny at 70–40 Ma; and G) ongoing extension and thinning of continental lithosphere from 35–0 Ma. Arguably, the first two and last two events had the most regional and profound effect on the lithospheric structure of the southwestern U. S. [*Karlstrom and Humphreys*, 1998; *Karlstrom et al.*, this volume]. Figures 2–5 show the nature and spatial extent of each event from the perspective of crustal modification and crustal columns in which the type of modification of the lithosphere that we have deduced is shown.

Step 1- Initial Formation of the Crust

Proterozoic juvenile (at 1.8 Ga) lithosphere extends in a ~1500 km-wide belt south from the Cheyenne belt (along the Wyoming-Colorado border) and extends to the rifted margin of Laurentia in west Texas and northern Mexico (Fig. 2). The formation of this lithosphere in the Proterozoic involved the accretion of a number of island arc/oceanic terranes culminating with the Neoproterozoic Grenville orogeny at about 1.1 Ga [e.g., *Hoffman*, 1988; *Karlstrom and Bowring*, 1988; *Mosher*, 1998; *Karlstrom et al.,* 2001; *Karlstrom et al.*, this volume]. *Bowring and Karlstrom* [1990] argued that the crust in New Mexico and Colorado was assembled from "thin" (25–30 km) arcs. These arcs were shortened and thickened during convergence to "normal" thickness, isostatically-stable continental crust by about 1.6 Ga. This interpretation is based on thermochronologic and metamorphic data that suggest slow cooling and an absence of regional denudation from 1.6 to 1.4 Ga.

The velocity structure of this early crust is critical to determining the evolution of the crust-mantle boundary. The velocity models derived from recent seismic studies in the Aleutian arc [*Holbrook et al.*, 1999; *Fliedner and Klemperer*, 1999] would be the starting point if the assembly of the Rocky Mountain region was purely due to oceanic arc assembly. Other recent seismic models that provide some insights come from the Cascades arc in Washington [*Miller et al.*, 1997] that has continental arc affinities and the coastal region of Canada and southeastern Alaska (ACCRETE project) where the continent formed by the assembly of exotic terranes and arc magmatism [*Morozov et al.*, 2001, 2003].

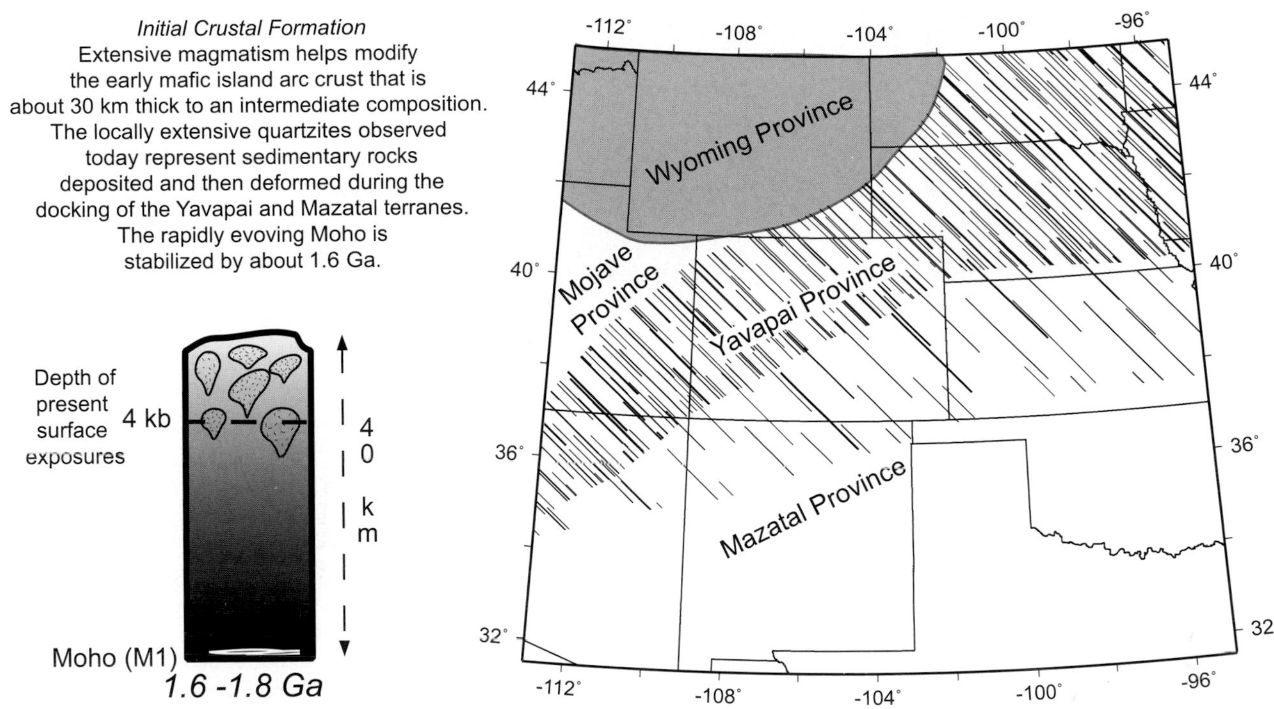

Initial Crustal Formation
Extensive magmatism helps modify the early mafic island arc crust that is about 30 km thick to an intermediate composition. The locally extensive quartzites observed today represent sedimentary rocks deposited and then deformed during the docking of the Yavapai and Mazatal terranes. The rapidly evoving Moho is stabilized by about 1.6 Ga.
Depth of present surface exposures
4 kb
40 km
Moho (M1)
1.6 -1.8 Ga
-112°
-108°
-104°
-100°
-96°
44°
40°
36°
32°
Wyoming Province
Mojave Province
Yavapai Province
Mazatal Province

Figure 2. Map of the southern Rocky Mountain region showing Precambrian terranes that formed southwestern Laurentia and our interpretation of crustal evolution during this period of time. The 4 kb isobar revealed by studies of outcrops is shown as a datum. The original Moho is labeled as M1.

The crust in the Aleutian arc is only ~30 thick, but its high average velocity (6.8 km/s) indicates a bulk composition consisting of basalt or basaltic andesite [*Holbrook et al.,* 1999]. In contrast, typical continental crust has lower seismic velocity consistent with an andesitic bulk composition [e.g., *Christensen and Mooney*, 1995; *Rudnick and Fountain*, 1995]. This observation suggests that typical continental crust cannot be formed from island arc material alone and that a combination of chemical and mechanical processes are needed to modify the composition to the average intermediate composition of continental crust [*Taylor and McClennan*, 1985; *Christensen and Mooney*, 1995; *Rudnick and Fountain*, 1995; *Rudnick*, 1995]. Some of the compositional modification needed likely took place during the arc collisions [*Williams and Karlstrom*, 1996], collision-related crustal differentiation, and granitic magmatism, and protracted convergence [*Shaw et al.*, 2001]. However, magmatism had to play the dominant role because mechanical and metamorphic processes in general do not change bulk composition. The magmatism likely resulted in the production of mafic restite in the lower crust. By 1.6 Ga, the crust had become stabilized, as demonstrated by the lack of U-Pb zircon dates for magmatic rocks or known tectonism from 1.6 to 1.5 Ga [*Bowring and Karlstrom*, 1990], and we infer that it looked approximately as shown in Figure 2.

One modern analog is the crust of the western flank of the Cascade Range of Washington that is about 45 km thick, and based on the rather high average P-wave velocity (~6.7 km/s) of the crystalline portion of the crust [*Miller et al.*, 1997], it is more mafic than typical continental crust. This crustal structure would need to be further modified by addition of felsic material via processes such as magmatism or imbrication of the upper crust to be similar to that observed in the southern Rocky Mountain region today, but could be analogous to the Rocky Mountain crust after the effects of the ~1.4 Ga magmatic event discussed below. Another analog comes from the ACCRETE project whose results indicate that the crust of the Coast Plutonic Complex has properties of average continental crust that has had its upper ~20 km removed by extension early after it formed [*Andronicos et al.*, 2003]. Because the Coast Plutonic complex was constructed out of mafic island arc crust, it is likely that the generation of the Coast Plutonic Complex batholiths was the critical step in differentiating an andesitic crust from the older arcs [*Andronicos et al.*, 2003].

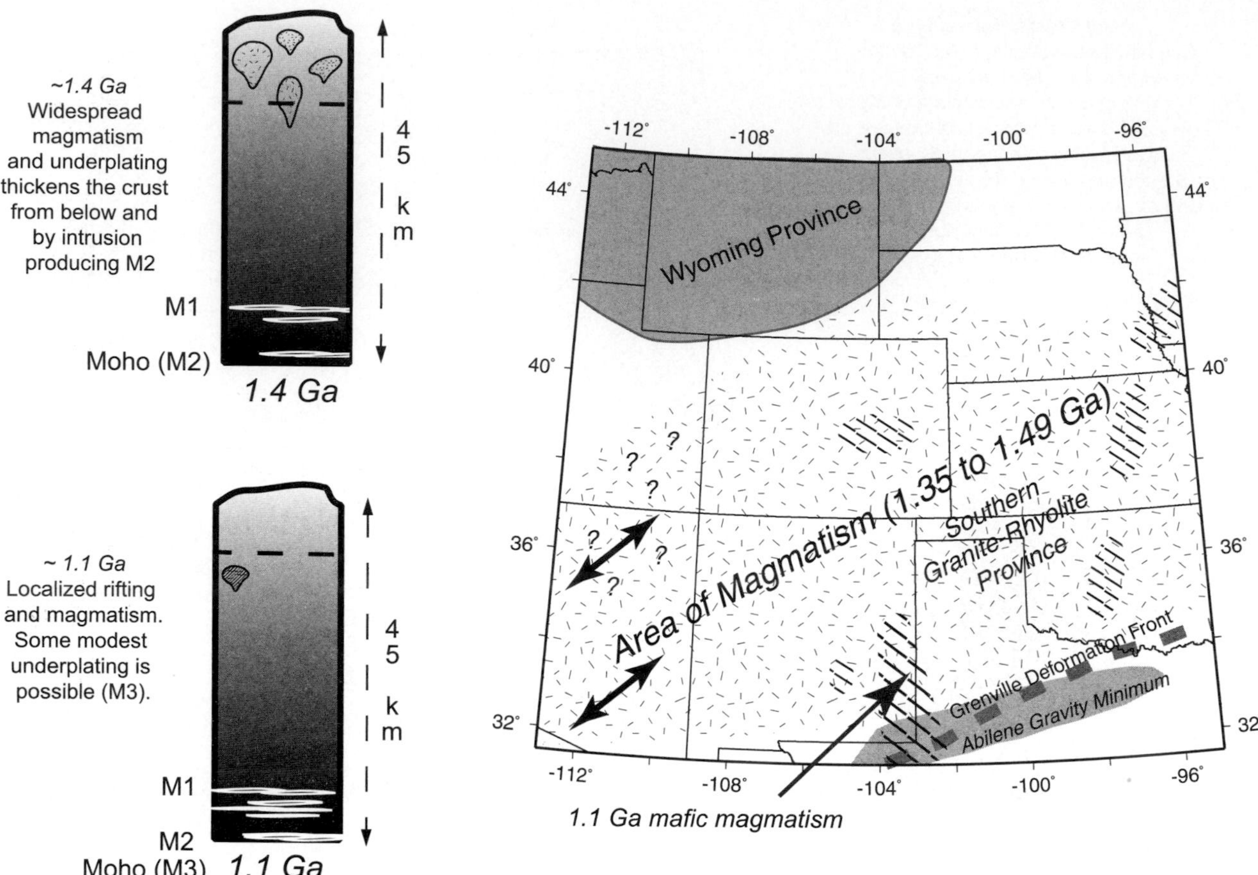

Figure 3. Map showing the extent of 1.4 Ga magmatism and 1.1 Ga magmatism (Midcontinent rift, Pecos mafic igneous suite – PMIC). Arrows indicate areas of localized Neoproterozoic extension, basin formation, and magmatism. The crustal columns show our interpretation of crustal evolution during this period of time. The Abilene gravity minimum is very similar to the gravity low associated with the Sierra Nevada [*Adams and Keller*, 1994]. The 4 kb isobar revealed by studies of outcrops is shown as a datum. The original Moho is labeled as M1.

This suggests that the crustal column shown in Figure 2 was likely generated by processes similar to those that formed the crust of the Coast Plutonic Complex, but without as much of the upper crust being stripped off by erosion and extension.

In the Southern Rocky Mountains, evidence for isobaric heating paths followed by slow cooling indicate that the crust was of "normal" thickness following initial crustal assembly [*Bowring and Karlstrom*, 1990; *Williams and Karlstrom*, 1996]. Thus, we suggest that a fairly typical crustal column on the order of 40 km thick formed early in the assembly of the southern Rocky Mountain region. The lower portion of this crust was probably mafic due to differentiation and intrusion of mafic melts into sedimentary wedges that were in turn melted, as proposed in the ACCRETE project area [*Hollister and Andronicos*, 1997]. Another way of justify this starting crustal column is that it is very similar to the present day crust in the Midcontinent area [e.g., *Braile*, 1989] that shares a similar Precambrian tectonic history with the southern Rocky Mountain region. However, the crust of most of the Midcontinent region has been stable since about 1.4 Ga. The use of a single crustal column does not imply that we believe the crustal structure was simple at this time. In fact, many of the lateral thickness variations in Figure 1 may date from ~1.6 Ga. For example, the crustal root for the Cheyenne belt that *Crosswhite and Humphreys* [2003] interpreted as the cause for thick crust in northwestern Colorado is reminiscent of the thick (~55 km) crust in the southern Baltic shield that also appears to date from a Proterozoic orogenic event [*Luotso et al.,* 1990].

Following crust stabilization at 1.6 Ga, *Bowring and Karlstrom* [1990] argued that there was no major extensional unroofing event as one might expect if crust had been significantly overthickened. The evidence for this interpretation

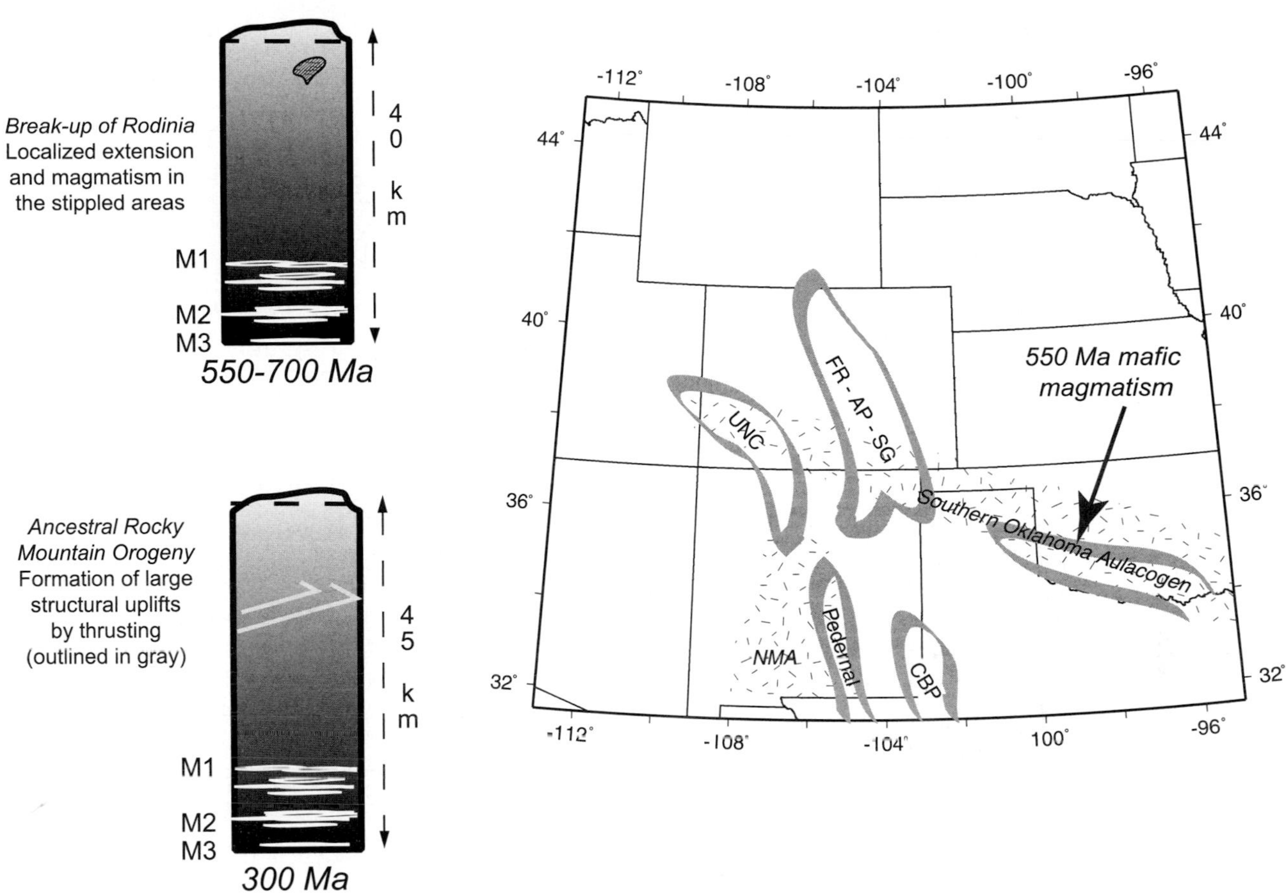

Figure 4. Map of features associated with the break-up of Rodinia (Southern Oklahoma aulacogen; New Mexico aulacogen – NMA [*McLemore et al.*, 2000]) and the Ancestral Rocky Mountains (Front Range - Apishapa-Sierra Grande uplift - FR-AP-SG; Uncompahgre uplift - UNC; Central basin platform – CBP). The 4 kb isobar revealed by studies of outcrops is shown as a datum. The original Moho is labeled as M1.

comes from studies of Proterozoic rocks exposed today in northern New Mexico, Colorado, and Arizona that show evidence for: 1) slow cooling and long-term residence at mid-crustal depths; and 2) these rocks still being at 10–20 km depths at 1.65 and 1.4 Ga over wide regions. This interpretation is primarily based on thermobarometry studies for depth of metamorphism at the different times [*Williams and Karlstrom*, 1996; *Williams et al.*, 1999].

Step 2 - Intracratonic Magmatism (1.4 Ga)

The first major crustal modification event following crust stabilization occurred at ca. 1.45–1.35 Ga when felsic magmatism was extensive along a broad belt that extended from the southwestern U.S. across the midcontinent region and beyond (Fig. 3) [*Anderson*, 1989; *Van Schmus et al.*, 1993; *Karlstrom et al.*, 2001]. This has been termed the "anorogenic event" [*Windley*, 1993], but evidence from numerous regions in New Mexico, Colorado, the Midcontinent, and northeastern Laurentia also show evidence of intracontinental deformation and metamorphism at this time [e. g., *Nyman et al.*, 1994]. Recent petrologic models, combined with the nearly ubiquitous mafic enclaves associated with the A-type granites, suggest that granites were formed via differentiation of tholeiitic, mantle-derived basalt and partial melting of lower crust by these basaltic intrusions [*Frost and Frost*, 1997]. The huge volume of A-type granite (and rarer gabbro and anorthosite) exposed at the surface (up to 25% of exposed crust) implies major mantle magmatism, and we infer that this event involved a significant amount of mafic underplating to thicken the crust, as shown in Figure 2 [*Karlstrom and Humphreys*, 1998]. Although the seismic models are consistent with underplating having played a role in the formation of the lower crust, direct geologic evidence for such an underplate is sparse, except

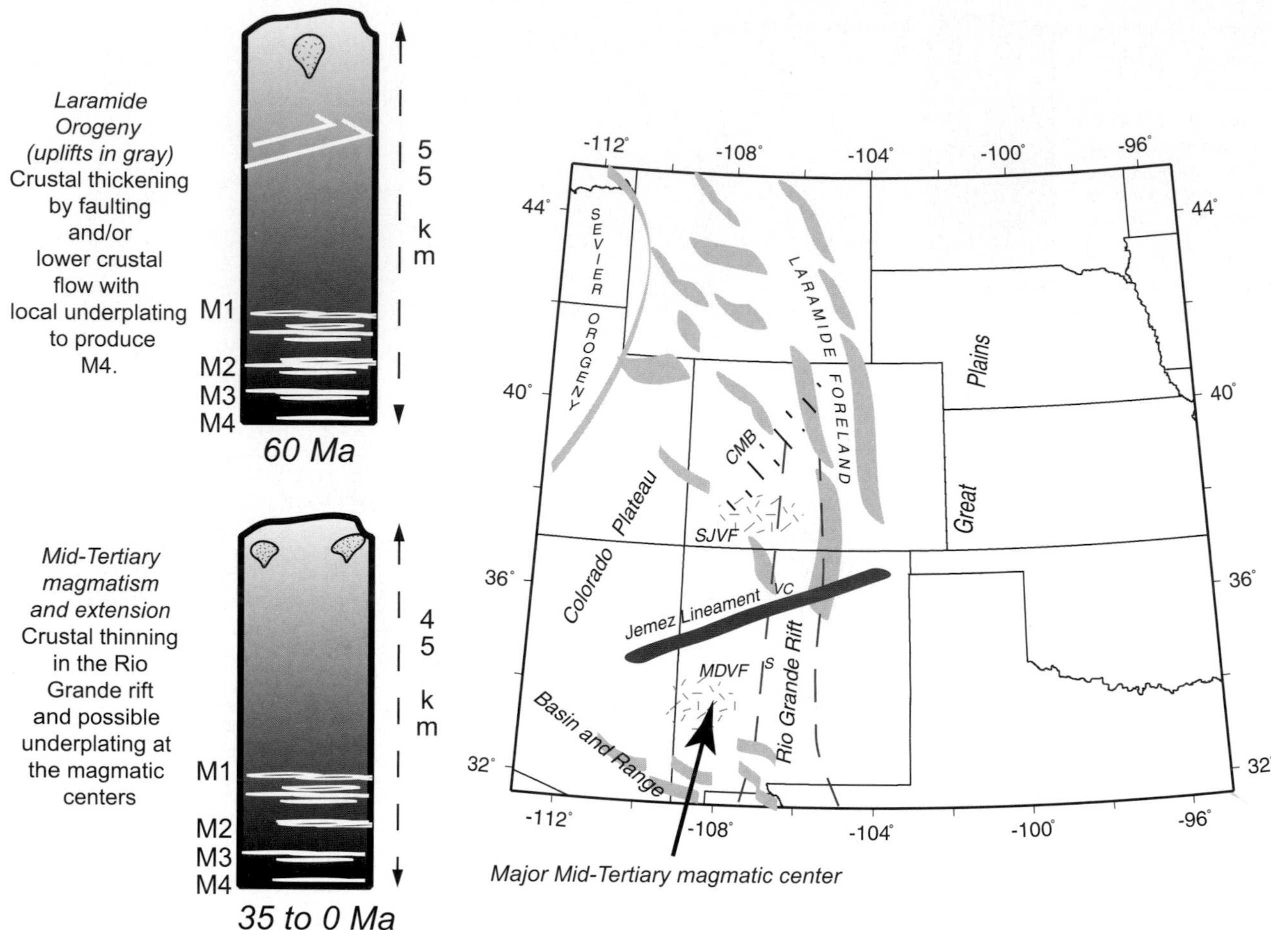

Figure 5. Regional map of Laramide structures (gray areas), the Colorado mineral belt (CMB), major mid-Tertiary volcanic centers (Mogollon-Datil volcanic field – MDVF; San Juan volcanic field – SJVF), and Rio Grande rift / Basin and Range. V – Valles caldera; S – Socorro magma chamber. The 4 kb isobar revealed by studies of outcrops is shown as a datum. The original Moho is labeled as M1. After several events that produced at least local underplating and 2 major orogenic events that produced crustal shortening, the Moho mapped today is M4.

perhaps for a component of 1.4 Ga mafic xenoliths from the Stateline area in northern Colorado [*Farmer et al.*, this volume] and strong evidence for underplating in the Four Corners area of northeast Arizona and northwest New Mexico [*Crowley et al.*, 2003]. The extensive 1.4 Ga magmatic event could have been coincident with delamination of the early lower crust. However based on geochemical arguments, *Frost and Frost* [1997] also argue that the 1.4 Ga event involved significant remelting of mafic lower crustal rocks that were already in place, as indicated by 1.8–1.6 Ga Nd model ages from many 1.4 Ga granites. Thus in addition, the process of producing the extensive 1.4 Ga felsic rocks must have also created a significant amount of mafic restite.

Our crustal column in Figure 2 suggests that true underplating of mantle-derived basalts thickened the crust from below at about 1.4 Ga. Because of the extensive occurrence of the ~1.4 Ga event (Fig. 3 and beyond), this same argument applies to the formation of the Midcontinent and Colorado Plateau crust. We cannot exclude the possibility that some mafic material added to the crust at this time was recycled into the mantle via delamination, but the geochemical data preclude the delamination of the entire lower crust. Rather, the geologic evidence is that this crust was relatively thick and stable like observed in large areas of eastern Baltica [e.g., *EUROBRIDGE Seismic Working Group*, 1999; *Grad et al.*, 2003].

Step 3 - Assembly of Rodinia at 1.1–1.0 Ga

The next events, related to the assembly and breakup of Rodinia (Fig. 3), were globally significant, but were expressed in the Rocky Mountain region as relatively localized intracratonic events. In the Grand Canyon region and in other scattered outcrops extending into far west Texas, there is evidence for extensional basins that appear to have been near sea level at 1.25 Ga [*Seeley*, 1999; *Timmons et al.*, 2001; *Amarante et al.*, this volume]. The record of the spatial extent of this period of extension is sketchy, but the fact that many Proterozoic metamorphic rocks, for example in northern New Mexico, do not record widespread exhumation at this time suggests that the event was rather localized as are most continental rifting events. Thus, this event is unlikely to have modified the regional crustal thickness of 40–45 km over the entire Southwest.

From ~1.1 to 1.0 Ga, the Grenville orogeny affected all of eastern Laurentia and extension affected a major portion of the Midcontinent. The Grenville orogeny marked the final event of the Precambrian assembly of southwestern North America [e.g., *Mosher*, 1998], and from a larger perspective, it completed the formation of the supercontinent Rodinia in this region [e.g., *Dalziel*, 1997]. The crust was undoubtedly thickened along the Grenville orogen, but there is no evidence suggesting significant compressional deformation and crustal thickening in intracratonic areas such as the southern Rocky Mountain region. However, geophysical data indicate that there was a significant amount of extension and magmatism to the south and east of the southern Rocky Mountain region at about the same time as the Grenville orogeny [*Adams and Keller*, 1994; *Adams and Miller*, 1995]. For example, the upper crust in southeastern New Mexico was extensively intruded by mafic material [*Adams and Miller*, 1995; *Amarante et al.*, this volume], the Pikes Peak batholith was emplaced, and NW-trending extensional faults formed across the region [*Timmons* et al., 2001]. The CD-ROM seismic reflection profile in New Mexico also shows evidence of sills that may be of this age [*Magnani et al.*, this volume]. Because of uncertainties in age relationships, this extension could be either related to the Midcontinent rift zone [e.g., *Hinze and Braile*, 1988] or the Grenville orogeny.

Large linear gravity highs delineate the upper crustal mafic intrusions that have been used to define the regional extent of rifting, but there is no evidence to suggest that significant magmatic activity extended far beyond the limits of the rifted regions shown in Figure 3. This fact and the lack of evidence for deformation related to the Grenville orogeny lead us to conclude that the crust in the southern Rocky Mountain region was mostly unaffected by this event, leaving the 1.4 Ga crust largely intact.

Step 4 - Breakup of Rodinia - 0.8–0.6 Ga

After ~1.0 Ga, the southern Rocky Mountain region was stable for a relatively short period until Rodinia began to break-up at ~700 Ma [e.g., *Torsvik et al.*, 1996; *Karlstrom et al.*, 2001]. This period of continental break-up involved several phases of rifting that extended into the Cambrian and created a passive continental margin that wrapped completely around southwestern North America [e. g., *Stewart*, 1976]. In the Grand Canyon region, intraplate extension also occurred (0.9 to 0.7 Ga) as shown by *Timmons et al.* [2001] and *Seeley and Keller* [2003]. The main effect in the southern Rocky Mountain region was formation of a failed rift zone that extends from the rifted margin in Oklahoma to the Uncompahgre uplift [*Hoffman et al.*, 1974; *Larson et al.*, 1985] (Fig. 4). Gravity [*Jenkins and Keller*, 1989] and seismic [*Snelson et al.*, 1998] data show that mafic rocks reside in the upper crust of the Uncompahgre uplift area (Fig. 4). In addition to considerable mapped extrusive activity, deep seismic reflection, seismic refraction/wide-angle reflection, and gravity data show that ~10 km of mafic material was emplaced into the upper crust in the Wichita Mountains region of southern Oklahoma during the formation of this feature [550 Ma; e.g., *Keller and Baldridge*, 1995]. In the southern Rocky Mountain region, this rifting may have been more widespread than portrayed by *Larson et al.* [1985]. For example, Cambrian mafic igneous rocks are found in the Wet Mountains of Colorado and are associated with large positive velocity and gravity anomalies [*Rumpel et al.*, this volume], and Cambrian igneous rocks occur in a number of localities in New Mexico [*McLemore et al.*, 1999].

In addition to this intraplate extension, the break-up of Rodinia established the structural framework of the southwestern margin of Laurentia. Ensuing formation of the Cordilleran passive margin resulted in dramatic thinning of the continental lithosphere west of the Wasatch line and western edge of the Colorado Plateau. The combination of intraplate extension and formation of this margin probably produced some thinning of the Rocky Mountain crust and lithosphere (Fig. 4) in areas of local extension, but again, we must conclude that the crustal column established at ~1.4 Ga in the southern Rocky Mountain region was not significantly altered except by erosion due to regional uplift. However after the formation of the Great Unconformity, the period from the late Cambrian to the Pennsylvanian, stratigraphic evidence [*Baars et al.*, 1988] indicates that the southern Rocky Mountain region was either near sea level or covered by shallow seas. The Trans-Continental arch did extend through much of Colorado resulting in little sedimentation during this period.

Step 5 - Assembly of Pangea and the (300Ma) Ancestral Rocky Mountain orogeny

Until the end of the Paleozoic, there was little opportunity for crustal modification because the Rocky Mountain region experienced virtually no deformation or magmatism. However at the end of the Paleozoic, continents again collected to form the supercontinent Pangea. The southern part of the Laurentian margin interacted with Gondwandan terranes during the Ouachita orogeny, and the continental interior was deformed to form the Ancestral Rocky Mountains. The Ancestral Rockies orogeny (Fig. 4) was a major but enigmatic tectonic event [*Kluth*, 1986; *Ye et al.*, 1996]. The formation of the Ancestral Rocky Mountains involved the development of several major structures in the middle and upper crust. For example, the Wichita uplift overthrusted the Anadarko basin producing 10–15 km of structural relief [e.g., *Brewer et al.*, 1983; *Keller and Baldridge*, 1995]. This feature extends across the Texas Panhandle (Amarillo uplift) and may extend through New Mexico and Colorado to the Uncompahgre uplift [*Larson et al.*, 1985]. There is also strong evidence for lateral movements along the northern front of the Wichita and Amarillo uplifts (Fig. 3) [*McConnell*, 1989]. In addition, recent studies in New Mexico have provided evidence for the formation of uplifts and pull-apart basins due to transpression and transtension [*Barrow and Keller*, 1994; *Broadhead*, 2003] suggesting lateral movement between crustal scale blocks. However, the magmatic modification of the crust at this time is negligible, and any crustal thickening due to large-scale block movements was apparently localized. The correlation of several of the areas in thick crust in Figure 1 with Ancestral Rocky Mountains uplifts (Fig. 4) is intriguing, but many of these same uplifts correlate with Laramide structures. Except on the uplifts, stratigraphic evidence shows that the surface was near sea level [e.g. *Baars*, 1988] suggesting that the crust was ~40 km thick in most areas but could have been thickened in the areas of the uplifts.

Step 6 - Laramide orogeny - 70–40 Ma

Since the break-up of Pangea in the Triassic, southwestern North America has experienced a complex series of tectonic events related to subduction and transform faulting accompanied by terrane accretion, magmatism, regional uplift, and extension. Delineating the crustal effects of specific events and differentiating their effects from those of older tectonic events is a challenge. The Laramide orogeny had a major effect on the Rocky Mountain region (Fig. 5) [e.g., *Bird*, 1984; *Erslev*, this volume], and Laramide structures in the upper crust are well mapped geologically and geophysically. For example, a deep seismic reflection profile across the Wind River Mountains provides a good picture of the deeper effects on the crust there [e.g., *Smithson et al.*, 1978]. These data show that the Wind River thrust penetrates into the middle crust (~27 km) at a dip of ~30° [*Sharry et al.*, 1986], a dip also observed on Laramide faults in the Sangre de Cristo Mountains area [*Magnani et al.*, this volume]. From a regional perspective, the amount and extent of crustal thickening in the Southern Rockies is a very important unknown. *Bird* [1984] has argued that the crustal thickening was widespread. On the other hand, *Coney and Harms* [1984] argue that an orogenic welt was centered along the Cordilleran foreland thrust belt and collapsed during the Cenozoic. Locally, it is likely that mechanical thickening of the crust occurred under major structural uplifts [*Erslev*, 1993; this volume], and crustal scale buckling may also have taken place [*Tikoff and Maxson*, 2001]. During the Laramide orogeny in the Southern Rocky Mountains, magmatic activity was modest except along the Colorado mineral belt [*McCoy et al.*, this volume]. The crust was almost certainly thickened to its present thickness or greater (>50 km) at this time as shown by paleoelevation studies [*Wolfe et al.*, 1998], and regional deformation models [e.g., *Spencer*, 1996]. Estimates of crustal shortening in the Laramide Rockies are typically on the order of 10 to 15% [e.g., *Erslev*, 1993] and are consistent with this estimate.

Thus, we conclude that the Laramide magmatism in central Colorado [*McCoy et al.*, this volume] and mechanical processes significantly thickened the crust in large areas of the Southern Rockies. Felsic magmatism in the upper crust was probably accompanied by underplating and formation of restites in the lower crust along the Colorado mineral belt. Outside the Rio Grande rift region, the end result was the Moho that is observed today in most places.

Step 7 - Extension and thinning of continental lithosphere from 35–0 Ma

The mid-Tertiary was a time of pervasive local magmatic modification of the crust in the San Juan and Mogollon-Datil volcanic fields (Fig. 5). Geophysical data provide a fairly well documented crustal model for the Mogollon-Datil volcanic field [*Schneider and Keller*, 1994] that seems applicable to the San Juan volcanic field [*Plouff and Pakiser*, 1972]. These data indicate that the volcanic fields are underlain by felsic batholiths whose thickness exceeds 5 km and whose emplacement would be accompanied by the formation of significant volumes of mafic residuum in the lower crust/uppermost mantle. However, existing geophysical data cannot resolve the extent of this residuum. Isotopic data also indicate that the crustal modification was locally significant [*Riciputi et al.*, 1995]. The Mogollon-Datil volcanic field could be an example of the crustal thickness remaining nearly constant as the combined effects of extension (thinning the crust) and magmatic additions to the crust (upper crustal batholith and underplating near the Moho) balanced. However, extension was minimal in the San Juan volcanic field.

The lithosphere began to extend along the Rio Grande rift (Fig. 4) at ~35 Ma, and activity peaked at about 10 Ma [e.g., *Chapin and Cather*, 1994]. Earthquake activity and Holocene volcanism show that the extension is still active today. The timing of tectonism associated with this rift is well established by studies of exposed basin fill and of the modest amount of accompanying volcanism. The crustal structure of the rifted area south of Santa Fe, New Mexico is fairly well known and shows crustal thinning in a zone that increases in width southward (Fig. 1), correlating with the surface expression of the rift and the southward increase in extension [*Cordell*, 1982; *Keller and Baldridge*, 1999; *Wilson and Aster*, 2003]. However, the deep structure from Santa Fe, New Mexico northward is poorly known. The San Luis Valley is the primary expression of the rift in Colorado, and a minor amount of extension is indicated by upper crustal structure [e.g., *Kluth and Schaftenaar*, 1994; *Brister and Gries*, 1994]. There are many small volcanic fields associated with the rift, but the volcanism along Jemez lineament [*Aldrich*, 1986] (Fig. 5), especially at the Valles caldera, is the only evidence for magmatism that significantly modified the crust. The Valles caldera is, in fact, associated with considerable crustal modification at all crustal levels [*Steck et al.*, 1998]. The CD-ROM refraction data show minor crustal thinning and lowered crust and mantle velocities across the Jemez lineament east of the Rockies (*Levander et al.*, this volume; *Yuan and Dueker*, this volume)

Given that the effects of the Laramide and especially the rifting significantly varied across the region, it is hard to generalize about the effects on the crust and Moho. In contrast to the period from the late Cambrian to the Laramide, the regional uplift in the Cenozoic produced a situation in which significant erosion could occur. Thus, the crust could be thinned from the top during this time by erosion, and this is shown in Figure 5.

6. SPECULATIONS ON EVOLUTION OF SURFACE ELEVATION IN THE ROCKY MOUNTAIN REGION.

Plate 1 shows a summary of inferred changes in surface elevation that may have accompanied and followed the various tectonic events and modifications of the crustal thickness. Appreciable topography may have existed during crustal assembly, as is common in Indonesian or Aleutian-style orogenic belts (Plate 1A), but the slow cooling and long-term residence of middle crustal rocks from 1.6 to 1.4 Ga would seem to imply rather modest average surface topography and an isostatically stable lithospheric column (Plate 1B; *Bowring and Karlstrom* [1990]). The inferred mafic underplating and accompanying crustal deformation and differentiation at ~1.4 Ga may have resulted in regional uplift and formation of an orogenic plateau, somewhat similar to the Colorado Plateau today (Plate 1C). Using isostatic arguments [*Morgan and Karlstrom*, 1995], a surface elevation of approximately 1.5 km would be required to explain the ensuing erosional removal of ~ 10 km of crust during the ~1 Ga of the "Great Denudation" [*Dutton*, 1882] to bring middle crustal basement rocks to sea level by Cambrian time (Plate 1D; Great Unconformity of *Powell* [1876]). The next major change in surface elevation was addition of 1.5 km of surface elevation due to a combination of enhanced mantle buoyancy and crustal thickening in the Laramide [*Humphreys et al.*, 2003], perhaps with addition of Cenozoic epeirogeny due to continued magmatic processes [Plate 1E; *Karlstrom et al.*, this volume].

7. DISCUSSION AND CONCLUSIONS A SCENARIO OF CRUSTAL MODIFICATION

The theme of Figures 2–5 is one of repeated changes (of varying magnitudes) in crustal thickness that involve interplay of magmatism (upper and lower crust) and tectonism (both extension and shortening). Distinctive regional attributes of the crust appear to have been that it has been at least average in thickness for continental areas since 1.4 Ga and average to slightly above average in terms of velocity. The area was in fact mostly cratonal between 1.4 Ga and the Ancestral Rocky Mountains orogeny, a period of over 1 Ga.

Although studies of modern island arcs show that we could expect the original crust to have high velocities relative to typical continental crust, each subsequent igneous event involving in situ melting and differentiation in the lower crust would have helped to compositionally differentiate the crust, producing a high velocity lower crust and a typical felsic upper crust. We postulate repeated, but modest, magmatic events that progressively modified the lower crust by a combination of underplating and melting that left mafic residue behind. There were also times at which lower crustal flow and mechanical thickening of the upper crust occurred. The best time to form the mafic lower crust in most of the region was during initial formation of the crust between 1.8 and 1.6 Ga and at ~1.4 Ga. Neoproterozoic and pre-Laramide Phanerozoic events produced only locally important changes in the crustal structure of the Southern Rocky Mountains, but ~10 km of erosion occurred between 1.4 Ga and approximately the beginning of the Phanerozoic. Consequently to attain the crustal thicknesses observed today, the Laramide had to be a time of major mechanical thickening of the crust. Thus over the last 1.8 Ga, the crustal thickness has changed numerous times by the interplay of magmatism and tectonism, but the main crust shaping events were early, and late. Thus, we conclude that the crust was mainly constructed during two events in the Precambrian and has recently been undergoing major modification beginning with the Laramide orogeny.

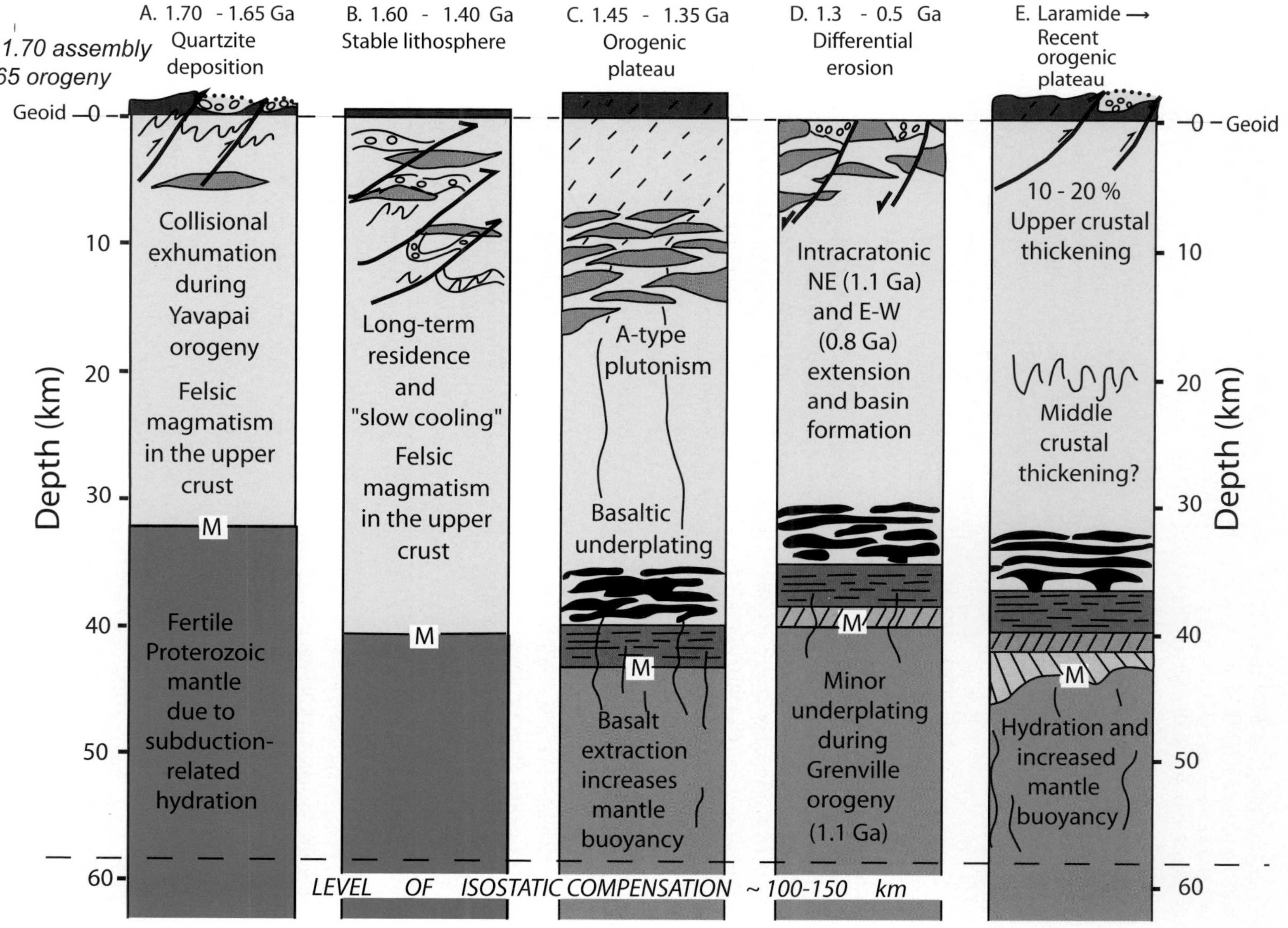

Plate 1. Lithospheric columns for the southern Rocky Mountain region showing the surface elevation evolution from the assembly of juvenile crust to Laramide and post-Laramide reactivation. A: Assembly of initial crust from collision of arc during the Yavapi orogeny (Fig. 2) [Bowring and Karlstrom, 1990]. B: Continued thickening and burial of quartzites to 10-15 km followed by stabilization [Bowring and Karlstrom, 1990; Heizler, 2002]. C: Uplift of an orogenic plateau due to mafic underplating and related 1.4 Ga A-type magmatism (Fig. 3) [Karlstrom and Humphries, 1998]. D: The "Great Denudation' [Dutton, 1882] and diachronous erosion to expose 10-15 km deep rocks at sea level. E: Laramide (and Neogene) uplift combine to create the present elevated plateau.

An important implication of the history summarized above is that the crust has been modified from below numerous times in its history. We have stressed the crust because we have so few geologic constraints on which to base an interpretation of the mantle lithosphere. We have used the word "underplating" to refer to an array of processes whereby mafic melts derived from the mantle are added to lowermost portion of the crust. These processes represent mantle tectonic events resulting from differentiation of ultramafic mantle rocks and upward movement, into the crust, of the basaltic differentiates. Decompression melting accompanies asthenospheric upwelling, and these processes usually are the trigger for contemporaneous crustal tectonic processes that differentiate the lower crust. Thus, it is difficult to determine the relative abundances of lower crustal mafic rocks that may represent new additions (underplate) versus reworked older crust (restite) for any given event. For example, melting of amphibolite at 30 to 40 km will make a mafic garnet granulite restite with a velocity of ~7.00 km/s once cooled. However, relatively 'small' additions of mantle melts injected into hot lower crust (underplate) may be enough to push the rocks into a thermal region where widespread lower crustal melting occurs. The mafic rocks that are the result of this process can be subsequently remelted during later heating events leaving behind a very complex mafic residuum.

The lower crust in most of the southern Rocky Mountain region was likely mainly in place by the end of the 1.4 Ga magmatic event. This event stabilized the crust creating a lithosphere that was stable until the Ancestral Rocky Mountain orogeny, and not profoundly modified until the Cenozoic. This lithosphere is still stable from the Great Plains eastward to the Appalachians. Stability does not imply simplicity in terms of structure, particularly near the crust-mantle interface. The lower crust and uppermost mantle contain important keys to understanding the evolution of the lithosphere, but they have been difficult to interpret seismically and require continued attempts to combine different seismic techniques with geological and complementary geophysical data.

Acknowledgments. Financial support was provided by the National Science Foundation - Continental Dynamics Program (EAR-9614269) and the Deutsche Forschungsgemeinschaft (DFG). The data compilation and integration effort was part of the GEON (Geoscience Network) project that is supported by the National Science Foundation. We thank Walter Mooney for a particularly helpful review.

REFERENCES

Adams, D. C., and Keller, G. R., Possible extension of the Midcontinent Rift in west Texas and eastern New Mexico, *Canadian Journal of Earth Science, 31*, 709–720, 1994.

Adams, D. C. and Keller, G. R., Precambrian basement geology of the Permian Basin region of West Texas and eastern New Mexico: A geophysical perspective, *Am. Assoc. Petrol. Geol. Bull., 80*, 410–431, 1996.

Adams, D. C., and Miller, K. C., Evidence for late Middle Proterozoic extension in the Precambrian basement beneath the Permian basin, *Tectonics, 14*, 1263–1272. 1995.

Aldrich, M. L., Jr., Tectonics of the Jemez lineament in the Jemez Mountains and Rio Grande rift, *J. Geophys. Res., 91*, 1753–1762, 1986.

Anderson, J. L., Proterozoic anorogenic granites of the southwestern U. S., in *J. P. Jenny and S. J. Reynolds (eds.), Arizona Geological Digest, 19*, 211–238, 1989.

Arndt, N. T., and Goldstein, S. L., An open boundary between lower continental mantle: its role in crust formation and crustal and recycling, *Tectonophysics, 161*, 201–212, 1989.

Baars, D. L., Triassic and older stratigraphy: Southern Rocky Mountains and Colorado Plateau, in *L. L. Sloss (ed.), Sedimentary Cover – North American Craton*, Geological Society of America, Boulder, CO, The Geology of North America, *D-2*, 53–64, 1988.

Barrow, R., and Keller, G. R., An integrated study of the Estancia basin, central New Mexico, in *G. R. Keller and S. M. Cather (eds.), Basins of the Rio Grande Rift: Structure, Stratigraphy, and Tectonic Setting*, Geological Society of America, *Special Paper 291*, 171–186, 1994.

Bird, P., Laramide crustal thickening event in the Rocky Mountain foreland and Great Plains, *Tectonics, 3*, 741–758, 1984.

Braile, L. W., Crustal structure of the continental interior, in *Pakiser, L. C. and Mooney W. D. (eds.), Geophysical Framework of the Continental United States*, Geological Society of America, *Mem. 172*, 285–315, 1989.

Braile, L. W., Hinze, W. J., von Frese, R. R. B., and Keller, G. R., Seismic properties of the crust and uppermost mantle of the conterminous United States and adjacent Canada, in *Pakiser, L. C. and Mooney W. D. (eds.), Geophysical Framework of the Continental United States*, Geological Society of America, *Mem. 172*, 655–680, 1989.

Brewer, J. A., Good, R., Oliver, J. E., Brown L. D., and Kaufman, S., COCORP profiling across the Southern Oklahoma Aulacogen; overthrusting of the Wichita Mountains and compression within the Anadarko basin, *Geology, 11*, 109–114, 1983.

Brister, B. S., and Gries, R. R., Tertiary stratigraphy and tectonic development of the Alamosa basin (northern San Luis Basin), Rio Grande rift, south-central Colorado, in *G. R. Keller and S. M. Cather (eds.), Basins of the Rio Grande Rift: Structure, Stratigraphy, and Tectonic Setting*, Geological Society of America, *Special Paper 291*, 39–58, 1994.

Broadhead, R. F., Petroleum source rocks in late Paleozoic elevator basins of the southern Ancestral Rockies: The Tucumcari, Estancia, Vaughn, and Carrizozo Basins, *American Association of Petroleum Geologists Bulletin*, in review, 2003.

Bowring, S. A., and Karlstrom, K. E., Growth, stabilization and reactivation of Proterozoic lithosphere in the southwestern United States, *Geology, 18*, 1203–1206, 1990.

Chamberlain, K. R., Medicine Bow orogeny: Timing of deformation and model of crustal structure produced during continent-arc collision, ca.1.78 Ga, southwestern Wyoming, *Rocky Mountain Geology, 33*, 259–277, 1988.

Chapin, C. E., and Cather, S. M., Tectonic setting of the axial basins of the northern and central Rio Grande rift, in *G. R. Keller and S. M. Cather (eds.), Basins of the Rio Grande Rift: Structure, Stratigraphy, and Tectonic Setting*, Geological Society of America, *Special Paper 291*, 5–25, 1994.

Christensen, N. I., and Mooney, W. D., Seismic velocity structure and composition of the continental crust: A global review, *J. Geophys. Res.*, 100, 9761–9788, 1995.

Coney, P. J., and Harms, T. A., Cordilleran metamorphic core complexes: Cenozoic extensional relics of Mesozoic compression, *Geology, 12*, 550–554, 1984.

Continental Dynamics of the Rocky Mountains Working Group, Structure and evolution of the lithosphere beneath the Rocky Mountains: Initial results from the CD-ROM experiment, *GSA Today, 12*, 4–10, 2002.

Cordell, L, Extension in the Rio Grande rift, *J. Geophys. Res., 87*, 8561–8569, 1982.

Crosswhite, J. A., and Humphreys, E. D., Imaging the mountainless root of the 1.8 Ga Cheyenne belt suture and clues to its tectonic stability, Geology, 31, 669–672. 2003.

Crowley, J. L., Schmitz, M. D., Bowring, S. A., Williams, M. L., and Karlstrom, K. E., Timing of magmatism and metamorphism in the lower crust beneath the Colorado Plateau from coupled U-Pb and Lu-Hf isotopic analysis of Zircon from xenoliths, *EOS Trans. AGU, 84*, Fall Meet. Suppl., Abstract V32C–1036, 2003

Dalziel, I. W. D., Neoproterozoic – Paleozoic geography and tectonics: review, hypothesis, environmental speculation, *Geol. Soc. Am. Bull., 109*, 16–42, 1997.

Dueker, K., Yuan, H. and Zurek, B., Thick Proterozoic lithosphere of the Rocky Mountain region, *GSA Today, 11*, 4–9, 2001.

Dutton, C.E., Tertiary History of the Grand Canyon District: *U.S. Geological Survey, Monograph 2*, 264, 1882.

Erslev, E. A., 2D Laramide geometries and kinematics in the Rocky Mountains of the western United States, this volume.

Erslev, E. A., Thrusts, back-thrusts, and detachment of Rocky Mountain foreland arches, in Schmidt, C. J. Chase, R. B. and Erslev, E. A. (eds.), *Laramide basement deformation in the Rocky Mountain foreland of the western United States*, Geological Society of America, *Spec. Paper 280*, 339–358, 1993.

EUROBRIDGE Seismic Working Group, Seismic velocity structure across the Fennoscandia-Sarmatia suture of the East European Craton beneath the EUROBRIDGE profile through Lithuania and Belarus, *Tectonophysics*, *314*, 193–217, 1999.

Fliedner, M. M., and Klemperer, S. L., Structure of an island-arc: Wide-angle seismic studies in the eastern Aleutian Islands, Alaska, *J. Geophys. Res., 104*, 10,667– 10,694, 1999,

Frost, C. D., and Frost, B. R., Reduced rapakivi-type granites: The tholeiite connection, *Geology, 25*, 647–650, 1997.

Furlong, K. P., and Fountain, D. M., Continental crustal underplating: Thermal considerations and seismic-petrologic consequences, *J. Geophys. Res.*, *91*, 8285–8294. 1986.

Gorman, A.R., Clowes, R. M., Ellis, M., Henstock, T. J., Spence, G.D., Keller, G. R. Levander, A. R., Snelson, C. M. Burianyk, M., Kanasewich, E. R., Asudeh, E., Hajnal, Z. and Miller, K. C., Deep Probe, imaging the roots of western North America, *Canadian Journal of Earth Sciences*, *39*, 375–398, 2002.

Grad, M., Jensen, S. L., Keller, G. R., Guterch, A., Thybo, H., Janik, T., Tiira, T., Yliniemi, J., Luosto, U., Motuza, G., Nasedkin, V., Czuba, W., Gaczynski, E., Sroda, P., Miller, K. C., Wilde-Piorko, M., Komminaho, K., Jacyna, J., and Korabliova, L., Crustal structure of the Trans-European suture zone region along POLONAISE'97 seismic profile P4: Journal of Geophysical Research, v. 108, doi:10.1029/2003JB002426, 2003.

Grambling, J. A., Crustal thickening during Proterozoic metamorphism and deformation in New Mexico, *Geology*, *14*, 149–152, 1986.

Griffin, W. L., and O'Reilly, S. Y., Is the continental Moho the crust-mantle boundary?, *Geology*, *15*, 241–244, 1987.

Henstock, T. J., Levander, A., Snelson, C. M., Keller, G. R., Miller, K. C., Harder, S. H., Gorman, A. R., Clowes, R. M., Burianyk, M. J. A., and Humphreys, E. D., Probing the Archean and Proterozoic lithosphere of western North America, *GSA Today*, *8*, 1–5 and 16–17, 1998.

Hinze, W. J., and Braile, L. W., Geophysical aspects of the craton, in *L. L. Sloss (ed.), Sedimentary Cover – North American Craton*, Geological Society of America, Boulder, CO, The Geology of North America, *D-2*, 5–24, 1988.

Hoffman, P. F., Dewey, J. F. and Burke, K., Aulacogens and their genetic relationship to geosynclines, with a Proterozoic example from the Great Slave Lake, Canada, *Soc. Econ. Paleontologists and Mineralogists, Spec. Pub. 19*, 38–55, 1974.

Hoffman, P. F., United Plates of America, birth of a craton: Early Proterozoic assembly and growth of Laurentia, *Annual Rev. Earth Planet. Sci.*, *16*, 543–603, 1988.

Holbrook, W. S., Lizarralde, D., McGeary, S., Bangs, N., and Diebold, J., Structure and composition of the Aleutian arc and implications for continental crustal growth, *Geology*, *27*, 31–34, 1999.

Hollister, L. S., and Andronicos, C. L., A candidate for the Baja British Columbia fault system in the Coast Plutonic Complex, *GSA Today*, *7*, 1–7, 1997.

Humphreys, E., Hessler, E. Dueker, K. Erslev, E. Farmer, G.L. and Atwater, T., How Laramide-age hydration of North America by the Farallon slab controlled subsequent activity in the western U.S., *Inter. Geol. Rev.*, 45, 575–595, 2003.

Jarchow, C. M., and Thompson, G. A., The nature of the Mohorovicic discontinuity, *Annual Rev. Earth Planetary Sci.*, *17*, 475–506, 1989.

Jenkins, R. D., and Keller, G. R., Interpretation of basement structures and geophysical anomalies in the southeastern Colorado Plateau, *New Mexico Geological Society*, *Guidebook 40*, 135–142, 1989.

Johnson, R. A., Karlstrom, K. E., Smithson, S. B., Houston, R. S., Gravity profiles across the Cheyenne belt: A Precambrian crustal suture in southeastern Wyoming, *J. Geodynamics*, *1*, 445–472, 1984.

Karlstrom, K. E., Flurkey, A. J., and Houston, R. S., Stratigraphy and depositional setting of the Proterozoic Snowy Pass Supergroup, southeastern Wyoming: record of an early Proterozoic Atlantic-type cratonic margin, *Geol. Soc. Am. Bull.*, *94*, 1257–1274, 1983.

Karlstrom, K. E., and Houston, R. S., The Cheyenne Belt: analysis of a Proterozoic suture in southern Wyoming, *Precambrian Res.*, *25*, 415–446, 1984.

Karlstrom, K. E., and Bowring, S. A., Early Proterozoic assembly of tectonostratigraphic terranes in southwestern North America, *Geology*, *96*, 561–576, 1988.

Karlstrom, K. E., and Humphreys, E. D., Persistent influence of Proterozoic accretionary boundaries in the tectonic evolution of southwestern North America: Interaction of cratonic grain and mantle modification events, *Rocky Mt. Geol.*, *33*, 161–180, 1998.

Karlstrom, K.E., and Morgan, P., Comparison of Mesoproterozoic and Cenozoic plateau uplift in the Southwest - Isostatic response to thermal/magmatic event, *American Geophysical Union*, *76*, no. 46, p. F619, 1995.

Karlstrom, K. E., Ahall, Karl-Inge, Harlan, S. S., Williams, M. L., McLelland, J., and Geissman, J. W., Long-lived (1.8–1.0 Ga) convergent orogen in southern Laurentia, its extensions to Australia and Baltica, and implications for refining Rodinia, *Precambrian Res.*, *111*, 5–30, 2001.

Keller, G. R., and Baldridge, W. S., The southern Oklahoma aulacogen, in *Olsen, K. H. (ed.), Continental Rifts: Evolution, Structure, Tectonics*, Elsevier, Amsterdam, 427–435, 1995.

Keller, G. R., and Baldridge, W. S., The Rio Grande rift: A geological and geophysical review, *Rocky Mt. Geol.*, *34*, 131–148, 1999.

Kluth, C. F., Plate tectonics of the Ancestral Rocky Mountains, *American Association of Petroleum Geologists, Memoir 41*, 353–369, Tulsa, Oklahoma, 1986.

Kluth, C. F., and Schaftenaar, C. H., Depth and geometry of the Rio Grande rift in the San Luis Basin, south-central Colorado, in *G. R. Keller and S. M. Cather (eds.), Basins of the Rio Grande Rift: Structure, Stratigraphy, and Tectonic Setting*, Geological Society of America, *Special Paper 291*, 27–37, 1994.

Larkin, S. P., Levander, A., Henstock, T. J., and Pullammanappalli, S., Is the Moho flat? Seismic evidence for a rough crust-mantle interface beneath the northern Basin and Range, *Geology, 25*, 451–454, 1997

Larson E. E., Patterson, P. E., Curtis, G., Drake, R. and Mutschler, F. E., Petrologic, paleomagnetic, and structural evidence of a Paleozoic rift system in Oklahoma, New Mexico, Colorado, and Utah, *Geol. Soc. Am. Bull.*, *96*, 1364–1372, 1985.

Lerner-Lam, A., Sheehan, A., Grand, S., Humphreys, E., Dueker, K., Hessler, E., Guo, H., Lee, D. K., and Savage, M., Deep structure beneath the Southern Rocky Mountains from the Rocky Mountain front broadband seismic experiment, *Rocky Mt. Geol.*, *33*, 199–216, 1998.

Li, A., D. W. Forsyth, and K. M. Fischer, Evidence for shallow isostatic compensation of the Southern Rocky Mountains from Rayleigh wave tomography, *Geology, 30*, 683–686, 2002.

Luosto, U., Tiira, T., Korhonen, H., Azbel, I., Burmin, V., Buyanov, A., Kosminskaya, I., Ionkis, V., and Sharov, N., Crust and upper mantle structure along the DSS Baltic profile in SE Finland, *Geophys. J. Int., 101,* 89–110, 1990.

McConnell, D. A., Determination of offset across the northern margin of the Wichita uplift, southwestern Oklahoma, *Geol. Soc. Am. Bull.*, *101*, 1317–1332, 1989.

McLemore, V. T., McMillan, N. J., Heizler, M., and McKee, C., Cambrian alkaline rocks at Lobo Hill, Torrance County, New Mexico: More evidence for a Cambrian-Ordovician aulacogen, *New Mexico Geological Society, Guidebook 50*, 247–253, 1999.

Mechie, J., Keller, G. R., Prodehl, C., Gaciri, S., Braile, L. W., Mooney, W. D., Gajewski, D., Sandmeier, K.–J., Crustal structure beneath the Kenya Rift from axial profile data, in *Prodehl, C., Keller, G. R., and Khan, M. A. (eds.), Crustal and Upper Mantle Structure of the Kenya Rift, Tectonophysics, 236*, 179–200, 1994.

Miller, K. C., Keller, G. R., Gridley, J. M., Luetgert, J. H., Mooney, W. D., and Thybo, H., Crustal structure along the west flank of the Cascades, western Washington, *J. Geophys. Res., 102*, 17,857–17,873, 1997.

Mooney, W.D., and Meissner, R., Multi-genetic origin of crustal reflectivity: a review of seismic reflection profiling of the continental lower crust and Moho, in *Fountain, D.M., Arculus, R., and Kay, R.W. (eds.), Continental Lower Crust*, Developments in Geotectonics, vol. 23, Elsevier, Amsterdam, 45–79, 1992.

Morozov, I. B., Smithson, S. B., Chen, J. and Hollister, L. S. Generation of new continental crust and terrane accretion in southeastern Alaska and western British Columbia: constraints from P- and S-wave wide-Angle seismic data (ACCRETE), *Tectonophysics*, *341*, 49–67, 2001.

Morozov, I. B., Christensen, N. I., Smithson, S. B., and Hollister, L. S., 2003, Seismic and laboratory constraints on crustal formation in a former continental arc (ACCRETE, southeastern Alaska and western British Columbia), *J. Geophys. Res., 108(B1)*, 2041, doi:10.1029/2001JB001740, 2003.

Mosher, S., Tectonic evolution of, the southern Laurentian Grenville orogenic belt, *Geol. Soc. Am. Bull., 110*, 1357–1375, 1998.

Nelson, K. D., A unified view of craton evolution motivated by recent deep seismic reflection and refraction results, *Geophys. J. Int., 105*, 25–25, 1991.

Nyman, M., Karlstrom, K. E., Graubard, C. and Kirby, E., Mesozoic contractional orogeny in western North America: Evidence from ca. 1.4 Ga plutons, *Geology*, 22, 901–904, 1994.

Parsons, T., McCarthy, J., Kohler, W. M., Ammon, C. J., Benz, H. M., Hole, J. A., and Criley, E. E., Crustal structure of the Colorado Plateau, Arizona: Application of new long-offset seismic data analysis techniques, *J. Geophys. Res., 101*, 11,173 –11,194, 1996.

Plouff, D., and Pakiser, L. C., Gravity study of the San Juan Mountains, Colorado, *U.S. Geological Survey*, *Prof. Paper 800-B*, B183–B190, 1972.

Powell, J.W., Exploration of the Colorado River of the West, Washington D.C., *Smithsonian Institution,* 291 p., 1876.

Prodehl, C. P., and Lipman, P. W., Crustal structure of the Rocky Mountain region, in *Pakiser, L. C. and Mooney W. D. (eds.), Geophysical Framework of the Continental United States*, Geological Society of America, *Mem. 172*, 249–284, 1989.

Reiter, M., Edwards, C. L., Hartman, H. and Weidman, C., Terrestrial heat flow along the Rio Grande rift in New Mexico and southern Colorado, *Geol. Soc. Am. Bull., 86*, 811–818, 1975.

Riciputi, L. R., Johnson, C. M., Sawyer, D. A., and Lipman, P. W., Crustal and magmatic evolution in a large multicyclic caldera complex: Isotopic evidence from the central San Juan volcanic field, *J. Volcanol. Geothermal Res., 67*, 1–28, 1995.

Roy Chowdhury, K., and Hargraves, R. B., Deep seismic soundings in India and the origin of continental crust, *Nature, 291*, 648–650, 1981.

Rudnick, R. L., Making continental crust, *Nature, 378*, 571– 578, 1995.

Rudnick, R. L. and Fountain, D. M., Nature and composition of the continental crust: A lower crustal perspective, *Reviews Geophys., 33*, 267–309, 1995.

Sadowiak, P., Meissner, R., and Brown, L., Seismic reflectivity patterns: comparative investigations of Europe and North America, in *Meissner, R., Brown, L., Dürbaum, H.-J., Franke, W., Fuchs, K., and Seifert, F. (eds.), Continental Lithosphere: Deep Seismic Reflections,* Geodynam. Ser. vol. 22, Am. Geophys. Un., Washington, D.C., 363–369, 1991.

Schneider, R. V. and Keller, G. R., Crustal structure of the western margin of the Rio Grande rift and Mogollon-Datil volcanic field, southwestern New Mexico and southeastern Arizona, in *G. R. Keller and S. M. Cather (eds.), Basins of the Rio Grande Rift: Structure, Stratigraphy, and Tectonic Setting*, Geological Society of America, *Special Paper 291*, 207–226, 1994.

Seeley, J., Studies of the Proterozoic tectonic evolution of the southwestern United States, *PhD Dissertation*, University of Texas at El Paso, 262 pp., 1999

Seeley, J. M., and Keller, G. R., Delineation of subsurface Proterozoic Unkar and Chuar Group sedimentary basins in northern Arizona utilizing gravity and magnetics: Implications for hydrocarbon source potential: *AAPG Bulletin*, *87*, 1299–1321, 2003.

Sharry, J., Langan, R. T., Jovanovich, D. B., Jones, G. M., Hill, N. R., and Guidish, T. M., Enhanced imaging of the COCORP seismic line, Wind River Mountains, in *Reflection Seismology: A Global Perspective*, edited by M. Barazangi and L. Brown, AGU, *Geodynamic Series, 13*, pp. 223–236, Washington, D. C., 1986.

Shaw, C. A., and Karlstrom, K. E., The Yavapai Mazatzal crustal boundary in the Southern Rocky Mountains, *Rocky Mt. Geol., 34*, 37–52, 1999.

Shaw, C. A., Karlstrom, K. E., Williams, M. L., Jercinovik, M. J., and McCoy A., Electron microprobe monazite dating of ca. 1710–1630 Ma and ca. 1380 Ma deformation in the Homestake shear zone, Colorado: Origin and early evolution of a persistent intracontinental tectonic zone, *Geology, 29*, 739–742, 2001.

Sheehan, A. F., Abers, G. A., Lerner-Lam, A. L., and Jones, C. H., Crustal thickness variations across the Colorado Rocky Mountains from teleseismic receiver functions, *J. Geophys. Res., 100*, 20,391–20,404, 1995.

Smithson, S. B., Brewer, J. A., Kaufman, S., Oliver, J. E., Hurich, C. A., Nature of the Wind River thrust, Wyoming, from COCORP deep reflection data and from gravity data, *Geology, 6*, 648–652, 1978.

Snelson, C. M., Henstock, T. J., Keller, G. R., Miller, K. C., and Levander, A. R., Crustal and uppermost mantle structure along the Deep Probe seismic profile, *Rocky Mt. Geol., 33*, 181–198. 1998.

Spencer, J. E., Uplift of the Colorado Plateau due to lithospheric attenuation during Laramide low-angle subduction, *J. Geophys. Res, 101*, 13595–13609, 1996.

Steck, L. K., Lutter, W. J., Fehler, M. C., Thurber, C. H., Roberts, P. M., Baldridge, W. S., Stafford, D. G., and Sessions, R., Crust and upper mantle P-wave velocity structure beneath the Valles caldera, New Mexico: Results from the JTEX teleseismic experiment, *J. Geophys. Res., 103*, 24,301–24,312, 1998.

Stewart, J. H., Late Precambrian evolution of North America: Plate tectonics implication, *Geology*, *4*, 11–15. 1976.

Taylor, S. R. and McClennan, S. M., The geochemical evolution of the continental crust, *Rev. Geophys., 33*, 241–265, 1985.

Tikoff, B. and Maxson, J., Lithospheric buckling of the Laramide foreland during Late Cretaceous and Paleogene, western United States, *Rocky Mt. Geol., 36,* 13–35, 2001.

Timmons, J. M., Karlstrom, K. E., Dehler, C. M., Geissman, J. W. and Heizler, M. T., Proterozoic multistage (ca. 1.1 and 0.8 Ga) extension recorded in the Grand Canyon Supergroup and establishment of northwest- and north-trending grains in the southwestern United States, *Geol. Soc. Am. Bull., 113*, 163–180, 2001.

Torsvik, T. H., Smethurst, M. A., Meert, J. G., Van der Voo, R., McKerrow, W. S., Brasier, M.D., Sturt, B. A. and Walderhaug, H. J., Continental break-up and collision in the Neoproterozoic and Palaeozoic: a tale of Baltica and Laurentia, *Earth Science Reviews, 40*, 229–258. 1996.

Van Schmus, W. R., Bickford, M. E., Sims, P. K., Anderson, R. R., Shearer, C. K., and Trever, S. B., Proterozoic geology of the western midcontinent basement, in *Reed, J. C., Bickford, M. E., Houston, R. S., Link, P. K., Rankin, D. W., Sims, P. K., and Van Schmus, W. R. (eds.): Precambrian Conterminous U.S.*, Geology Society of America, The Geology of North America, *C-2*, 239 – 259. 1993.

Williams, M. L., Karlstrom, K. E., Lanzirotti, A., Read, A. S., Bishop, J. L., Lombardi, C. E., Pedrick, J. N., and Wingsted, M. B., New Mexico middle-crustal cross sections: 1.65-Ga macroscopic geometry, 1.4-Ga thermal structure, and continued problems in understanding crustal evolution, *Rocky Mt. Geol., 34*, 53–66, 1999.

Williams, M. L., and Karlstrom, K. E., Looping P-T paths, high-T, low-P middle crustal metamorphism: Proterozoic evolution of the southwestern United States, *Geology, 24*, 1119–1122, 1996.

Wilson, D. and Aster, R, Imaging crust and upper mantle seismic structure in the southwestern United States using teleseismic receiver functions, *The Leading Edge, 11*, 232–237, 2003.

Windley, B. F., Proterozoic anorogenic magmatism and its orogenic connection, *Journal of the Geological Society, 150*, 39–50, 1993.

Wolf, L. W., and Cipar, J. J., Through thick and thin: A new model for the Colorado Plateau from seismic refraction data from Pacific to Arizona crustal experiment, *J. Geophys. Res., 98*, 19,881–19,894, 1993.

Wolfe, J. A., Forest, C. E., and P. Molnar, Paleobotanical evidence of Eocene and Oligocene paleoaltitudes in midlatitude western North America, *Geol. Soc. Am. Bull., 110*, 664–678, 1998.

Ye, H., Royden, L. Burchfiel, C., and Schuepbach M., Late Paleozoic deformation of interior North America: The greater Ancestral Rocky Mountains, *Am. Assoc. Petrol. Geol. Bull., 80*, 1397–1432, 1996.

Ye, H., Royden, L. Burchfiel, C., and Schuepbach M., Late Paleozoic deformation of interior North America: The greater Ancestral Rocky Mountains, *Am. Assoc. Petrol. Geol. Bull., 80*, 1397–1432, 1996.

G. Randy Keller, Department of Geological Sciences, University of Texas at El Paso, 500 University Ave., El Paso, Texas 79968, keller@geo.utep.edu

Karl E. Karlstrom, Department of Earth and Planetary Sciences, University of New Mexico, Albuquerque, New Mexico, 87131, kek1@unm.edu

Michael L. Williams, University of Massachusetts, Department of Geosciences, Amherst, MA 01003, mlw@geo.umass.edu

Kate C. Miller, Department of Geological Sciences, University of Texas at El Paso, 500 University Ave., El Paso, Texas 79968, miller@geo.utep.edu

Christopher Andronicos, Department of Geological Sciences, University of Texas at El Paso, 500 University Ave., El Paso, Texas 79968, chris@geo.utep.edu

Alan Levander, Rice University, Department of Geology and Geophysics, Houston, TX 77005, alan@geophysics.rice.edu

Catherine M. Snelson, Department of Geoscience, University of Nevada Las Vegas, 4505 Maryland Parkway, MS 4010, Las Vegas, Nevada 89154-4010, csnelson@unlv.nevada.edu

Claus Prodehl, Geophysikalisches Institut, University of Karlsruhe, Hertzstr.16, Karlsruhe, Germany 76187, claus.prodehl@epost.de

Synthesis of Results From the CD-ROM Experiment: 4-D Image of the Lithosphere Beneath the Rocky Mountains and Implications for Understanding the Evolution of Continental Lithosphere

Karl E. Karlstrom[1], Steven J. Whitmeyer[2], Ken Dueker[3], Michael L. Williams[4], Samuel A. Bowring[5], Alan Levander[6], E. D. Humphreys[7], G. Randy Keller[8], and the CD-ROM Working Group[9]

The CD-ROM experiment has produced a new 4-D understanding of the structure and evolution of the lithosphere of the southern Rocky Mountain region. We identify relicts of at least four subduction zones that were formed during assembly of dominantly oceanic terranes in the Paleoproterozoic. Crustal provinces with different geologic histories correspond to distinct mantle velocity domains, with profound mantle velocity contrasts associated with the ancient sutures. Typically, the transitions between the velocity domains are tabular, dipping, extend from the base of the crust to depths of 150–200 km, and some contain dipping mantle anisotropy. The present day heterogeneous mantle structure, although strongly influenced by ancient compositional variations, has undergone different degrees of partial melting due to Cenozoic heating and/or hydration caused by transient plumes or asthenospheric convection within the wide western U.S. active plate margin. A high-velocity mafic lower crust is present throughout the Rocky Mountains, and there is ~10-km-scale Moho topography. Both are interpreted to record progressive and ongoing differentiation of lithosphere, and a Moho that has changed position due to flux of basalt from the mantle to the crust. The mafic lower crust evolved diachronously via concentration of mafic restite during arc formation (pre-1.70 Ga), collision-related differentiation and granite genesis (1.70–1.62 Ga), and several episodes of basaltic underplating (1.45–1.35 Ga, ~1.1 Ga, and Cenozoic). Epeirogenic uplift of the western U.S. and Rocky Mountain regions, driven by mantle magmatism, continues to cause reactivation of the heterogeneous lithosphere in the Cenozoic, resulting in differential uplift of the Rocky Mountains.

[1]Department of Earth and Planetary Sciences, University of New Mexico, Albuquerque, NM 87108

[2]Department of Earth and Planetary Sciences, University of New Mexico, Albuquerque, NM 87108; currently at: Department of Earth and Planetary Sciences, University of Tennessee, Knoxville, TN 37996

The Rocky Mountain Region: An Evolving Lithosphere
Geophysical Monograph Series 154

10.1029/154GM31

[3]Department of Geology and Geophysics, University of Wyoming, Laramie, WY 82071

[4]Department of Earth Atmospheric & Planetary Sciences, Massachusetts Institute of Technology, Cambridge, MA 02139

[5]Department of Geosciences, University of Massachusetts, Amherst, MA 01003

[6]Department of Earth Science, Rice University, Houston, TX, 77005

[7]Department of Geological Sciences, University of Oregon, Eugene, OR 97403

[8]Department of Geological Sciences, University of Texas at El Paso, El Paso, TX 79968

INTRODUCTION

The Continental Dynamics of the Rocky Mountains (CD-ROM) geophysical and geological transect from Wyoming to New Mexico (Plate 1) obliquely crosses numerous Phanerozoic tectonic provinces (Southern Rocky Mountains, Rio Grande rift, Great Plains) and orthogonally crosses northeast-striking structures related to Proterozoic assembly of the crust. The oldest tectonic features, formed during assembly of the continent, are at high angles to the younger features, related to the Phanerozoic plate margin. Our goal was to study the present-day deep continental structure and compare it to the well-understood geological history deduced from exposed rocks in order to differentiate the components of the present lithospheric structure that reflect Precambrian growth and stabilization from those that reflect Cenozoic tectonic events. This paper presents an integration of the CD-ROM seismic experiments, xenolith studies, and geological studies of surface rocks in order to delineate crust and upper mantle structure and provide a better understanding of lithospheric evolution and geodynamical processes. The goal of the paper is to present a synthesis of some of the important and provocative results of the project, citing detailed papers in this volume and other recent contributions. We also present composite block diagrams that integrate surface and lithospheric structure using results from other teleseismic experiments that have been conducted in the region in the last decade.

GEOLOGIC AND SEISMIC EVIDENCE FOR PROTEROZOIC SUBDUCTION SCARS

Plate 1A shows the complex distribution of Proterozoic crustal provinces and younger physiographic/tectonic elements in the southern Rocky Mountains, including Precambrian crustal provinces, Laramide basement uplifts, the Colorado Plateau, Cenozoic volcanic belts, and physiographic provinces. As shown in tomographic images of the western U.S. (Plate 1B; Plate 2 of Karlstrom and Keller, this volume), this complex surface geology overlies a heterogeneous mantle that exhibits an overall regional trend where the stable core of the Laurentian craton (from the Great Plains northeastwards) is underlain by high velocity (cold, old, thick) lithospheric mantle, the Rocky Mountain/Colorado Plateau region is underlain by a zone of mixed, but intermediate velocity mantle, and the active western U.S. is underlain by low velocity (warm, young, and thinner) lithospheric mantle (Plate 2; Humphreys and Dueker, 1994; Grand, 1994; van der Lee and Nolet, 1997; Henstock et al., 1998; Godey et al., 2003). The overall shear wave velocity contrast is one of the largest velocity gradients resolved by surface wave analysis on Earth (van der Lee and Nolet, 1997; Godey et al., 2003). The boundary zone or transition between the continental-scale mantle velocity domains (red versus blue in the tomographic image) is a wide zone that includes the Great Plains-Rocky Mountain region and, when viewed at continental scale, has an overall north to northwest trend, parallel to the Cenozoic plate margin.

Within this zone are northeast-trending zones of low velocity mantle in the Rocky Mountain region, the Snake River Plain, Saint George lineament, and Jemez lineament (Karlstrom and Keller, this volume), that are subparallel to both NE-SW Proterozoic province boundaries, and the SW- directed absolute motion of the North American plate. These 30- to 100-km-scale velocity variations in the Rocky Mountain region are profound, being nearly the same magnitude as the continental scale variation. These velocity differences might be interpreted as primarily reflecting temperature differences, perhaps as much as 700 °C (100 °C per 2% velocity variation; Cammanaro et al., 2004), with low velocity mantle at 100 km depths close to 1350 °C, with partial melt present, and high velocity domains near 650 °C, compatible with shield geotherms. To explain these large velocity variations, one hypothesis is that, even though the crust is predominantly Proterozoic, the mantle under the Rocky Mountains might be essentially Cenozoic in age (e.g. Goes and van der Lee, 2002; Wilson, 2004), with low velocity domains as upwelling of asthenosphere and high velocity domains as intact, or downwelling, lithosphere.

An alternative hypothesis, presented in this paper, is that the lithospheric mantle under the Rocky Mountains, although extensively modified and reactivated by younger events, is primarily Proterozoic in age, and that Proterozoic structures are controlling some of the major velocity contrasts in the mantle. If this hypothesis is correct, large temperature, compositional, and rheology variations are present, but the low velocity domains may still be traveling with North American lithosphere because of their buoyancy. Even though these domains are weak and hot and rheologically similar to asthenosphere, they may not yet be entrained in the asthenospheric flow.

---

[9]CD-ROM Working Group (* denotes graduate students):
Chris Andronicos, Nicholas Bolay*, Oliver Boyd*, Sam Bowring, Kevin Chamberlain, Nick Christensen, Jim Crowley, Jason Crosswhite*, David Coblentz, Ken Dueker, Tefera Eshete*, Eric Erslev, Lang Farmer, Rebecca Flowers*, Otina Fox*, Matt Heizler, Gene Humphreys, Micah Jessup*, Roy Johnson, Karl Karlstrom, Randy Keller, Shari A. Kelley, Eric Kirby, Alan Levander, M. Beatrice Magnani, Kevin Mahan*, Jennie Matzal*, Annie McCoy*, Grant Meyer, Kate Miller, Elena Morozova, Frank Pazzaglia, Claus Prodehl, Adam Read*, Oscar Quezada*, Mousumi Roy, Hanna-Maria Rumpel, Jane Selverstone, Anne Sheehan, Liane Stevens*, Colin A. Shaw*, Elena Shoshitaishvili*, Scott Smithson, Cathy Snelson*, Mike Timmons*, Leandro Trevino*, Amanda Tyson*, Stacy Wagner*, Xin Wan*, Paul Wisniewski*, Michael Williams, Huaiyu Yuan*, Brian Zurek*

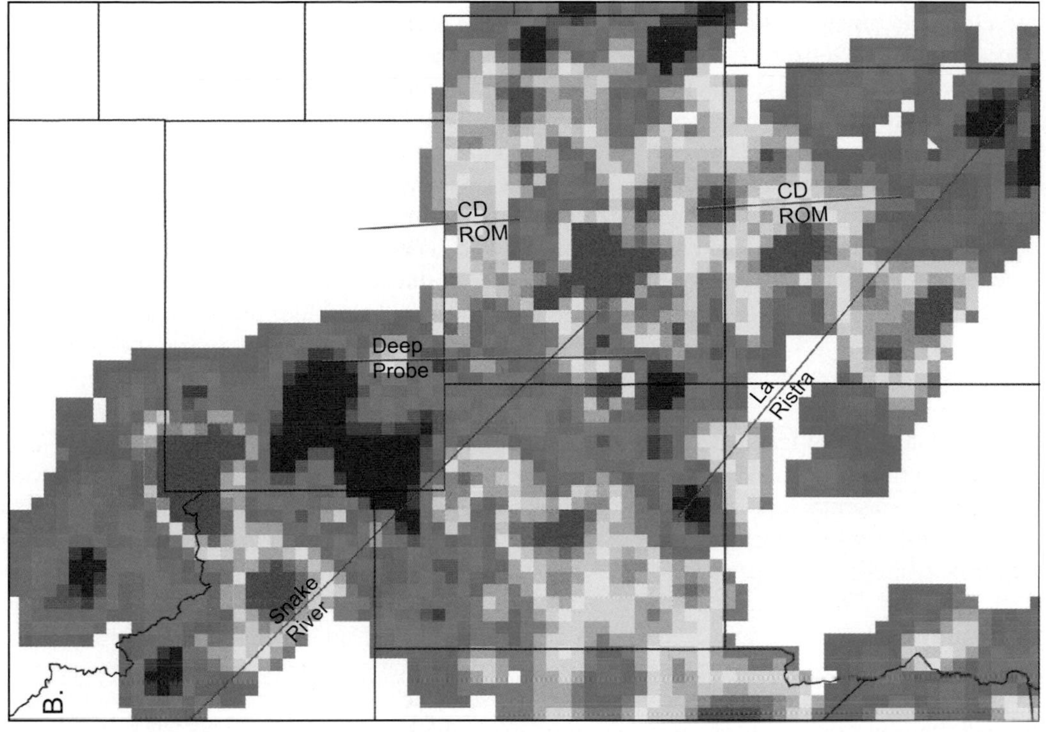

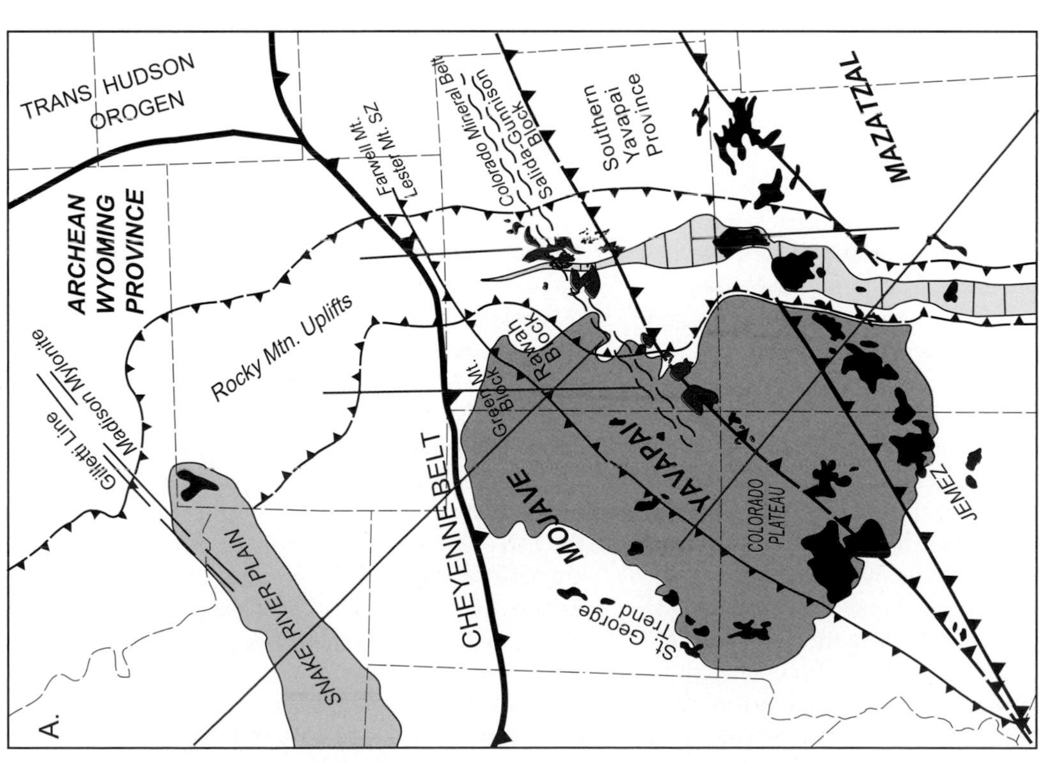

Plate 1. a) Geologic elements of southwestern North America showing locations of teleseismic lines in red. Precambrian provinces strike northeast, Laramide uplifts strike north-south, Cenozoic volcanic fields (red= Laramide; black= Neogene) strike northeast. b) Tomographic image of southwestern North America at 100 km depth; Snake River Plain, Deep Probe, CD-ROM, and La Ristra teleseismic lines indicated.

Arguments in favor of interpreting the low velocity domains to be lithosphere, as discussed in this paper, include the dipping nature of the velocity boundaries, their extent from near the base of the crust to >150–200 km, their correspondence with major Proterozoic geologic province boundaries at the surface, and the presence of horizontal layering imaged by receiver functions that is more compatible with structured lithosphere than with convecting asthenosphere. Thus, we explore the hypothesis that the observed seismic velocity variations may reflect initial compositional and structural heterogeneities that have helped control present day temperature variation.

CD-ROM Lithospheric Cross Section

Plate 3 shows a synthesis of the large-scale features of the Rocky Mountain lithospheric cross section based on CD-ROM data. The present lithospheric structure represents the cumulative effects of nearly two billion years of tectonic evolution. A continuing challenge is to interpret the age and history of major features on the map and cross-section to gain insight into processes of lithospheric evolution. Some of the most notable features on the lithospheric cross section are the significant lateral velocity gradients in the mantle, as mentioned above. The following discussion of Plate 3 proceeds from north to south.

Cheyenne Belt

The Cheyenne belt (Plate 1) is one of the most profound and long-lived tectonic boundaries in the Rocky Mountain region (Karlstrom and Humphreys, 1998). It is the crustal manifestation of the suture between Archean crust to the north and juvenile 1.8–1.7 Ga Proterozoic island arc crust to the south (Hills and Houston, 1979; Karlstrom and Houston, 1984). New seismic reflection images of the crust (depicted schematically in Plate 3 and based on Morozova et al., this volume) confirm that, in the upper crust, the Cheyenne belt dips south under the Proterozoic Green Mountain arc (Condie and Shadel, 1984), consistent with north-verging thrusting of Proterozoic rocks over Archean crust (Karlstrom and Houston, 1984; Duebendorfer and Houston, 1987; Chamberlain, 1998). However, reflection data show that the deeper crust is characterized by tectonic interwedging similar to other sutures between old continents and younger arcs (Cook et al., 1998; Snyder et al., 1996; Snyder, 2002; Morozova et al., this volume). In addition, north-dipping reflections from the Farwell Mountain area of the Park Range (Plate 1) can be projected through generally unreflective lower crust to coincide with a thrust-offset Moho (Plate 3), and with the top of a high velocity tomographic anomaly (blue slab of Plate 3) that dips north under the Archean (Dueker et al., 2001; Yuan and Dueker, this volume; Levander and Niu, 2004).

The most plausible explanation for the north-dipping, high velocity, mantle anomaly is that it is an ancient slab fragment, for example a remnant of the rifted margin of the Wyoming Province (Tyson et al., 2002) or a collapsed back-arc basin (Yuan and Dueker, this volume; Zurek and Dueker, this volume). Our interpretation is that Proterozoic oceanic lithosphere was underthrust beneath Archean crust during late stages of accretion of the Green Mountain arc, but this structure never developed into a self-sustaining subduction system, as shown by the absence of an associated magmatic arc. Eclogite xenoliths (Kuehner and Irving, 1999) and evidence for significant hydration of Archean lower crustal xenoliths (Farmer et al., this volume) are also compatible with this interpretation, as is dipping mantle anisotropy (Yuan and Dueker, this volume; Fox and Sheehan, this volume) that is more consistent with an old structure rather than active lithospheric mantle downwelling. The co-location of the blue slab with the Cheyenne belt in three dimensions may be supported by a marked velocity contrast in the southeastern extension of the Snake River Plain and Deep Probe teleseismic lines (Plate 4). CD-ROM receiver function migrations (Levander and Niu, 2004; Levander et al., this volume) suggest that the top of the northwest dipping slab marks the southern edge of the North American Archean cratonic lithosphere. Crustal thickness variations across the Cheyenne belt and a 100-km-wide crustal welt of 50–60 km thick crust also suggest preserved remnants of Proterozoic shortening across this zone (Johnson et al., 1984; Crosswhite and Humphreys, 2003). Refraction data show this thickening as a broad feature (Snelson et al., Rumpel et al., and Levander et al., this volume).

Farwell and Lester Mountain Structures

A related paleosuture zone is interpreted from combined geologic and geophysical results in the area near the Farwell and Lester Mountain structures just south of the Cheyenne belt in the Park Range (Tyson et al., 2002, Morozova et al., this volume). Here, a set of oppositely dipping reflectors project to the surface and form a "tee-pee" shaped structure that is interpreted to be part of the bivergent suture zone between the 1.78 – 1.76 Green Mountain arc and the 1.75 – 1.72 Rawah arc/back arc complex (Tyson et al., 2002; Plate 3). A zone of ultramafic fragments, pillow basalt, tectonized marble and cherts, and sillimanite pod rocks crop out along this boundary zone and may be remnants of an accretionary prism with dismembered ophiolitic fragments. Deformation in the zone took place at 1746–1740 Ma, with reactivation at ~1680 Ma (Tyson et al., 2002). Zircon of similar age is present along strike in lower crustal xenoliths in the Stateline diatremes

(Farmer et al., this volume) and hence are interpreted to reflect regional whole-crustal tectonism. Mantle structure also shows oppositely dipping sharp velocity gradients that seem to be downward continuations of crustal reflections (Plate 3). The N-dipping structure is contiguous with the "blue slab" discussed above. There is also a south-dipping velocity gradient that surfaces near the Lester Mountain zone (slower velocities to the south). It is best seen in the CD-ROM line, but can also be seen along strike in the Deep Probe and Snake River Plain teleseismic lines (Plate 4). This velocity gradient is adjacent to the upward projection of the Aspen anomaly so it may be associated either with the Lester Mountain zone (Tyson et al., 2002) or the Aspen anomaly. In either case, the regional continuity between teleseismic lines reinforces the interpretation that the gradient is genetically linked to ENE-trending Proterozoic crustal boundaries.

Aspen Anomaly

The Aspen anomaly (Dueker et al, 2001; Dueker and Yuan, 2004) is a large velocity anomaly in the CD-ROM cross section that is comparable in amplitude to the Snake River Plain low velocity anomaly. It is a distinct low velocity mantle domain that is located in central Colorado. This zone is still not completely imaged in central Colorado and may have a complex geometry related both to the N-S Rio Grande rift and the NE-trending Colorado mineral belt. But low velocity mantle is seen in the southern CD-ROM line (Plate 3) that projects towards the surface trace of the Colorado mineral belt and dips to the south to great depth (> 200 km). A similar narrow dipping anomaly appears along strike in the La Ristra line (Plate 4) reinforcing the interpretation that, in three dimensions, this may be a south-dipping tabular anomaly that correlates with the Colorado mineral belt.

The Colorado mineral belt is a northeast-striking zone defined by: a Proterozoic shear zone system (Tweto and Sims, 1963; McCoy et al., this volume), a suite of Laramide-aged plutons and related ore deposits (Plates 1 and 2), a major gravity low (Isaachson and Smithson, 1976; McCoy and Roy, this volume), low crustal velocities (Li et al., 2001; and this volume), and high heat flow (Decker et al., 1988). The presence of Laramide plutons suggests that the mantle in this region was modified during the early Cenozoic but the northeast trend of Laramide magmatic rocks is puzzling. The dipping geometry of the Aspen anomaly, its association with the Proterozoic shear zones, and its great depth extent, as well as the northeast trend of plutons, are most compatible with a Proterozoic heritage that helped guide the Cenozoic tectonic activity. As mentioned above, a similar and apparently correlative tabular low velocity anomaly is also portrayed in the La Ristra line (Plate 4), along the southwest continuation of the Colorado mineral belt where it crosses the Four Corners region, and a similar gradient may also be evident near the south end of the Deep Probe-Snake River Plain line (Plate 4). These anomalies appear to correlate along strike and, if so, the 3-D aspect of the anomaly would be a ~80 km wide, south-dipping tabular zone that extends from the base of the crust (more completely imaged near Four Corners) to depths >200–300 km (in both Colorado and near Four Corners). The aspect ratio seems far too narrow and tabular to be an upwelling of the lithosphere/asthenosphere boundary (c.f. Yuan and Dueker, this volume), but may be explained as a paleosubduction scar that is being reactivated in the Cenozoic. The depth extent of the structure seems to require that part of it is in the lithospheric (the low velocity material near the base of the crust), and part in the asthenosphere (the parts below 250–300 km).

The simplest way to explain the low velocity of the Aspen-Four Corners lithospheric zone is that it contains low solidus material with a small partial melt fraction (e.g. 0.1 to 1%) relative to adjacent sub-solidus zones. The age and character of the thermal perturbation that (perhaps cumulatively) gave rise to the partial melt can perhaps be correlated with the 50–25 Ma ages of magmatism above the zone and continued active hot springs and high heat flow (Decker et al., 1988). Given the volatile-rich magmas that erupted diachronously above the zone, with kimberlitic pipes in the Four Corners (Semken, 2003) and Laramide plutons in Colorado (Mutchler et al., 1987), our preferred model is that low-solidus Proterozoic lithospheric mantle within a fossil subduction zone was further hydrated during Laramide flat slab subduction (Humphreys et al., 2003), then heated from below and partially melted during the Oligocene ignimbrite flare-up, and again during Neogene basaltic volcanism. In this interpretation, the ancient structure may have guided asthenospheric upwelling and/or localization of partial melting in the lithosphere. Another interpretation involves upwelling of asthenosphere in small scale convection systems that are decoupled from and independent of the lithospheric structure (Grand et al., 2004).

Current interpretations support the interpretation that the lithosphere in the Four Corners region has been re-heated and xenoliths re-equilibrated in the late Cenozoic. A 13% reduction in shear wave velocities between the 32 Ma mantle xenoliths and modern day velocities, as constrained by surface waves, is consistent with reheating of the lithosphere since 32 Ma (K. Priestley, personal communication). This 13% velocity reduction would require a 300–400° temperature increase (Cammarano et al., 2003). Similarly, P-T work suggests that mantle xenoliths at 140 km were relatively cold (900° C) when sampled by diatremes in the Oligocene (Riter and Smith, 1996; Smith, 2000). While the scaling relationships between velocity and partial melt content are poorly constrained (Schmelling, 1985; Humphreys and Hammond,

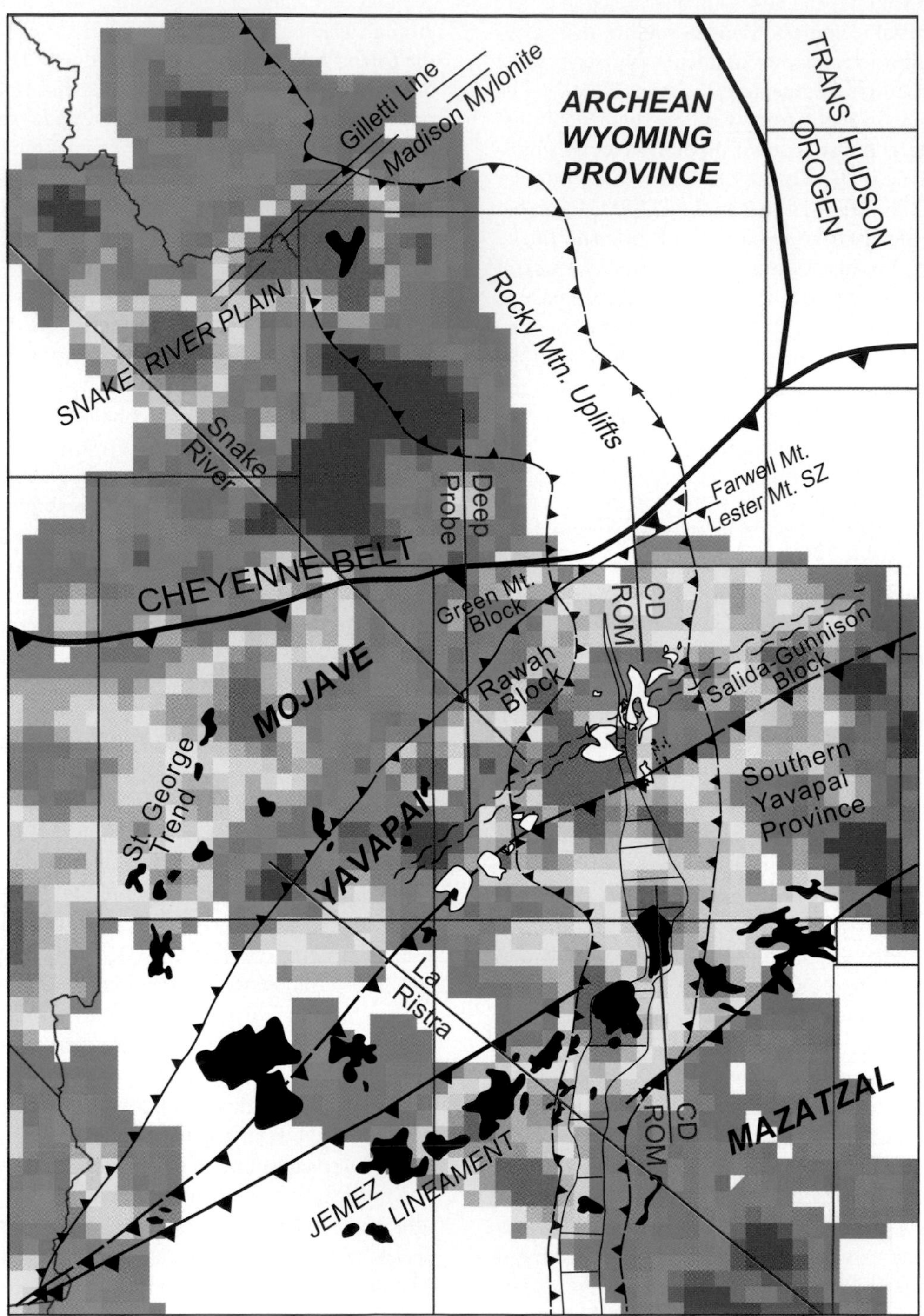

Plate 2. Geologic elements of southwestern North America (Plate 1a) superimposed on 100 km-depth tomography (Plate 1b). In the Rocky Mountain-Colorado Plateau region, "fingers" of hot mantle penetrate older lithosphere along northeast- striking zones. Young basalts (< 10 Ma) are present along the Snake River- Yellowstone (Y) trend, St. George trend, and Jemez lineament suggesting that these mantle domains are hot and producing basaltic melts.

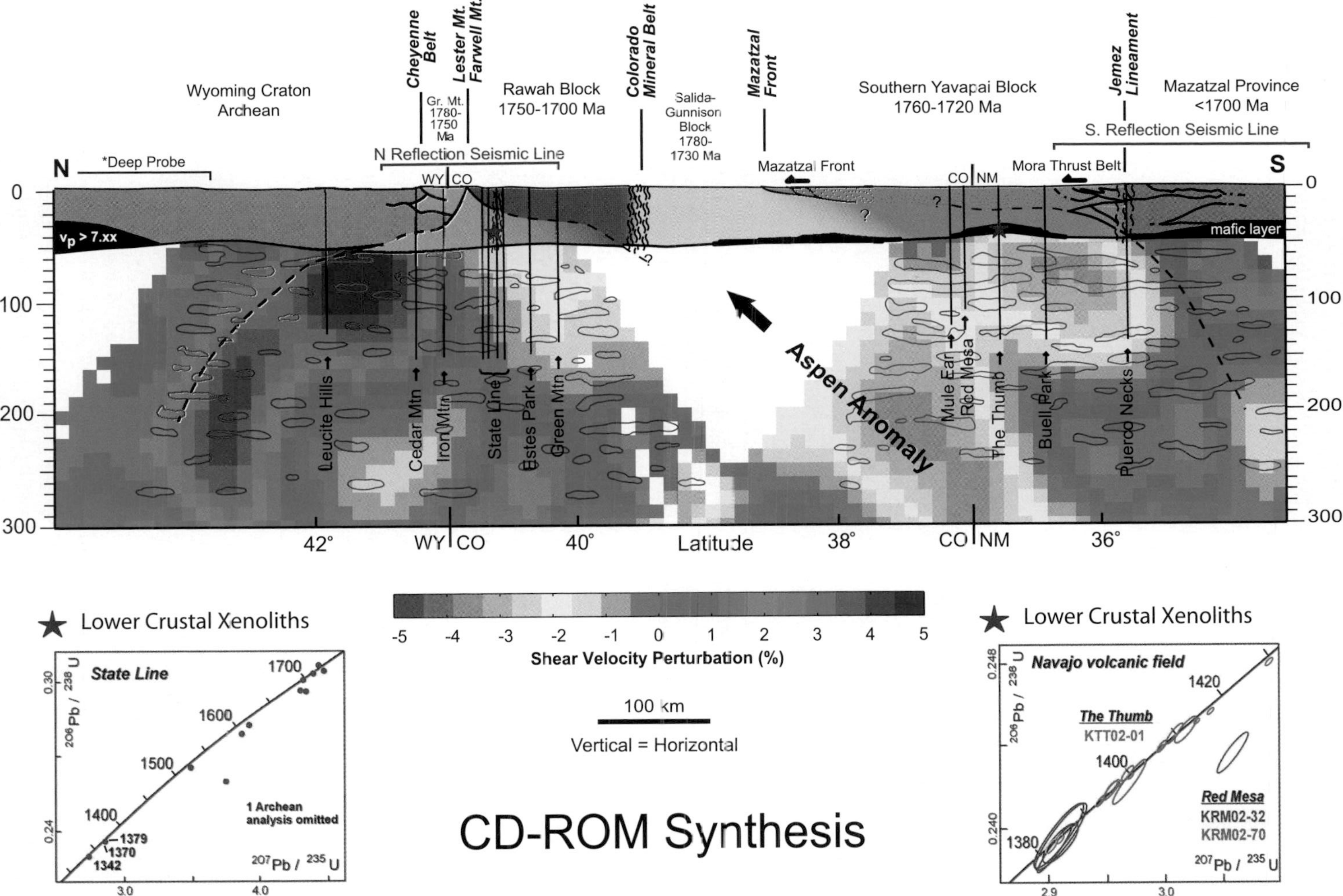

Plate 3. Cross-sectional synthesis of the CD-ROM transect (modified from Karlstrom et al., 2002). Generalized geologic cross section merged with S-wave tomographic image of Dueker et al. (2001). Crustal structures in the Cheyenne belt and Jemez lineament areas are generalized from seismic reflection data (Morozova et al., this volume; Magnani et al., this volume), with black solid lines in the crust representing well-defined reflections. Locations of xenolith pipes are shown as vertical lines. Dipping elements in the tomographic image, combined with overlying crustal structures are interpreted to be Proterozoic subduction scars, and North American lithosphere is interpreted to extend to > 200 km depth (the depth extent of the tomographic image is not well resolved below 200–250 km). Crustal thickness and lower crustal mafic layer from the CD-ROM refraction experiment (Levander et al., this volume, but c.f. Snelson et al., this volume and Rumpel et al., this volume for alternate interpretations). Receiver function images (Zurek and Dueker, this volume) are superimposed in red on the tomographic image and show significant upper mantle (lithospheric) layering. U-Pb zircon ages from lower crustal xenoliths show a predominance of 1.7–1.6 Ga ages in the State Line district and ~1.4 Ga in the Navajo volcanic field.

2000; Faul and Jackson, 2004), it is nonetheless clear that a mean equilibrium melt porosity of 0.01–0.1% can create substantial velocity reductions. Zircon dates and Nd studies (Smith et al., 2004) show that eclogites beneath the Four Corners are probably Proterozoic, and that younger dates (interpreted by Helmstaedt and Schulze (1991) and Usui et al (2003) to be from the Farallon slab) are more likely to be reset ages. Thus, we do not believe that available evidence strongly supports the alternative model, that the low velocity anomaly is a present-day volatile-rich plume impacting the crust.

Jemez Lineament

The Jemez lineament (Plate 1) is the farthest south of the proposed paleosuture zones in the CD-ROM cross section. This zone is defined by an alignment of Miocene to Recent volcanic rocks (Aldridge et al., 1986), and it is also interpreted to mark the surface boundary between 1.8–1.7 Ga crust of the Yavapai province (to the north) and ~1.65 Ga crust of the Mazatzal province (Wooden and DeWitt, 1991; Karlstrom et al., 2004). New reflection data (depicted schematically in Plate 3 based on Magnani et al., 2004, and Magnani et al., this volume) show another bi-vergent "teepee" structure, with south-dipping middle crustal reflections that seem to coincide with both a possible step in the Moho and a south-dipping mantle velocity boundary between fast (south) and slow (north) mantle that extends to great depth (> 200 km; Plate 3; also Dueker et al., 2001). North-dipping and south-dipping reflections merge in the lower middle crust below the Jemez lineament suggesting a bi-vergent orogenic belt similar to that seen in the Alps (Schmid et al., 1996) and similar to geometries seen in finite element geodynamic models (Beaumont et al., 1996). Based on these relationships, we interpret the Jemez lineament to mark a Proterozoic lithospheric suture zone that also localized Cenozoic magmatism. Refraction velocities in the uppermost mantle suggest that partial melt is present beneath the southern Jemez lineament, in the root zone of the bivergent orogen (Levander et al., this volume), at the top of the low velocity mantle anomaly (Yuan and Dueker, this volume). The same south-dipping mantle velocity gradient is observed in the La Ristra line (Plate 4), some 300 km to the southwest along the Jemez lineament (Plate 2), again providing evidence that, in three dimensions, the mantle velocity domain is a planar, south-dipping zone that is more closely co-located with the Jemez lineament than with the Rio Grande rift (cf. Wilson, 2004). The low velocity Jemez anomaly is unlikely to be completely explained as upwelling asthenosphere, as shown by the pronounced horizontal layered structure revealed by receiver function analysis (Plate 3; Zurek and Dueker, this volume). This layering extends across the boundary between both fast and slow mantle, suggesting that both are part of the lithosphere. We infer that the 3–5% shear wave discontinuities represented by the receiver function interfaces would be less likely to be present in asthenosphere because convective motion would disrupt the apparent horizontal layered structure (Zurek and Dueker, this volume).

TOWARDS A 3-D VIEW OF LITHOSPHERIC BOUNDARIES

One of the successes of the CD-ROM experiment was to develop a depth perspective of the subsurface nature of Proterozoic province boundaries. Important results (Figure 3) include the correspondence between crustal age provinces, changes in mantle velocity structure, and dipping nature of velocity transitions. A three dimensional view of the lithospheric structure is possible by trying to correlate important lithospheric boundaries between cross sections developed from independent geophysical experiments. Plate 4 is a set of tomographic cross sections at a uniform scale, comprising the Snake River Plain (Schutt and Humphreys, 2004), Deep Probe (Dueker and Yuan, 2004), CD-ROM and La Ristra (Gao et al., 2004) teleseismic lines. Plate 5 compiles these vertical cross sections with plan-view tomography at 100 km and surface geology into an exploded block diagram. From these figures it is apparent that the Proterozoic boundaries described in the previous section can be traced in pseudo 3-D through several of the seismic cross sections. The Cheyenne belt is a northwest dipping boundary that is imaged on the Snake River Plain and CD-ROM cross sections; the Farwell "teepee" structure apparently trends NNE, as seen on the Snake River Plain and CD-ROM; the Colorado mineral belt projects to depth along the southeast-dipping Aspen and Four Corners anomalies in the CD-ROM and La Ristra sections, respectively; and the Jemez lineament follows a NE trend above low velocity features on the La Ristra and CD-ROM sections (Plate 5). Most of these features are also apparent on the Deep Probe cross section (Plate 4).

ORIGIN OF THE MANTLE VELOCITY GRADIENTS, ANISOTROPY, THICKNESS OF THE LITHOSPHERE

While the interpreted Proterozoic subduction scars are inferred on the basis of sharp velocity contrasts, it is interesting that some of the slab-like features are high velocity (Cheyenne belt) and some low velocity (Aspen/Four Corners anomaly). Probably a mixture of causative processes need to be considered, but the common theme is that variable composition Proterozoic mantle responded in different ways to mantle heating such that low-solidus materials melted preferentially in response to increased late Cenozoic mantle heat flux (Karlstrom et al., 2002). The origin of the increased mantle heat flux is thought to result from upwelling asthenosphere associated with sinking of the lower lithosphere isotherms (Schott et al., 2000) or perhaps Neogene development of upper mantle small-scale free convection (Korenga and Jordan, 2003). This could be explained if superadiabatic mantle has flowed into this region in the last 50–10 Ma. The high velocity Cheyenne belt slab is interpreted to be a trapped remnant of an underthrust oceanic slab and hence its high velocity and anisotropy may be due to the presence of eclogite and lineated olivine (Yuan and Dueker, this volume). In contrast, the low velocity Aspen anomaly slab and Jemez anomaly are interpreted to be zones of Proterozoic hydration resulting from long-lived subduction of oceanic lithosphere, but without lithospheric capture of the final remnants of the downgoing slab, leaving hydrated mantle that has a lower solidus than adjacent regions.

Anisotropy From Surface and Body Wave Constraints

As discussed by Li et al. (this volume), there is a significant discrepancy between their Colorado surface wave anisotropy results and the Colorado SKS splitting results (Savage and Sheehan, 2000; Fox and Sheehan, this volume). The Colorado shear wave splitting results show a complicated anisotropy with many null measurements whereas the surface wave anisotropy analysis provides good evidence for a strong and uniform anisotropy, oriented northeast, dominantly below 100 km depth beneath the Rocky Mountains (Li et al., this volume). This anisotropic complexity is observed more clearly in northern CD-ROM and Laramie array stations that showed clear shear wave splits that are dependent on back-azimuth, requiring either a dipping fast velocity axis or two anisotropic layers to explain (Fox and Sheehan, this volume). Furthermore, shear wave splitting from the Deep Probe and southeastern Snake River plain arrays show a very complicated pattern of apparent splits inconsistent with any simple one or two layer anisotropy (Schutt and Humphreys, 2001). In contrast, the southernmost Rocky Mountains shear wave splitting results from the CD-ROM-south and La Ristra arrays appear to be adequately explained by a single layer with a horizontal northeast-trending fast axis (Gok et al., 2003; Fox and Sheehan, this volume). This anisotropy could be the manifestation of a plate-sheared asthenosphere (as North American absolute plate motion is within a few degrees of the fast axis), or Proterozoic fabrics in the lithosphere (as this is the dominant orientation of Proterozoic foliation). There are significant variations in the orientation and splitting time along both arrays that are easier to explain with a shallow (lithospheric) source. In neither the CD-ROM nor La Ristra lines is their evidence for dipping fast axis anisotropy such as might be created by asthenospheric small-scale convection (Ribe, 1989; Blackman et al., 1993; Kaminski and Ribe, 2001; Blackman et al., 2002; Kaminski and Ribe, 2002).

Thickness of the Lithosphere

Tomographic cross sections like those of Plates 3 and 4 do not readily show the base of the lithosphere because velocity contrasts are relative values for a particular depth slice, and hence vertical variations are inaccessible. However, there are several ways to interpret the images and infer the depth of today's lithosphere. Receiver function analysis in the CD-ROM and La Ristra lines (Dueker et al., 2001; Wilson and Aster, 2003; Zurek and Dueker, this volume) show that the upper mantle contains subhorizontal impedance contrasts ('reflectors'; Plate 3). The La Ristra line shows good evidence for a coherent reflector at 80 km depth beneath the Colorado Plateau (Wilson and Aster, 2003). Likewise, the CD-ROM lines show coherent layering between the Moho and 100 km with subhorizontal reflectors indicating coherent structures beneath the Cheyenne belt (Plate 3). While the genetic origin of these reflectors is poorly constrained, they do not appear to be crustal multiples (Zurek and Dueker, this volume); inference would suggest that they are chemical layers and/or anisotropic shear zones (Levin and Park, 2000; Sobolev et al., 1997; Sobolev and Babeyko, 1994; Thybo and Perchuc, 1997; Zurek and Dueker, this volume). The most robust conclusion from these images is that the chemical lithosphere (i.e., North American lithospheric mantle) is thicker than simple estimates of thermal (and rheological) lithosphere. Thus, upper mantle low velocity zones down to 200 km or more are interpreted by us to be part of the North American plate in terms of age (Proterozoic), composition (depleted Proterozoic mantle), and tectonic evolution, rather than asthenosphere (Godey et al., 2003; Gorman et al., 2002; Hicks et al., 2001; Lastowka et al., this volume; Li et al., this volume; van der Lee, 2001; van der Lee and Nolet, 1997).

Present mantle rheology within low velocity domains may be similar to asthenosphere (< 1 % partial melt and resulting weakness). But this weak low velocity lithosphere is buoyantly attached to North America and, given that the North American plate is drifting to the southwest relative to the deep mantle reference frame, may be confined by 200 km thick stronger high-velocity cratonic mantle to the east, and hence has not yet been entrained in the convecting mantle. Recent experimental studies (e.g. Karato, 2003 for a review) have shown that a small amount of water dissolved in olivine and other nominally anhydrous mantle minerals can significantly enhance the ductility of the mantle. These results suggest that perhaps the lithosphere-asthenosphere boundary could be a

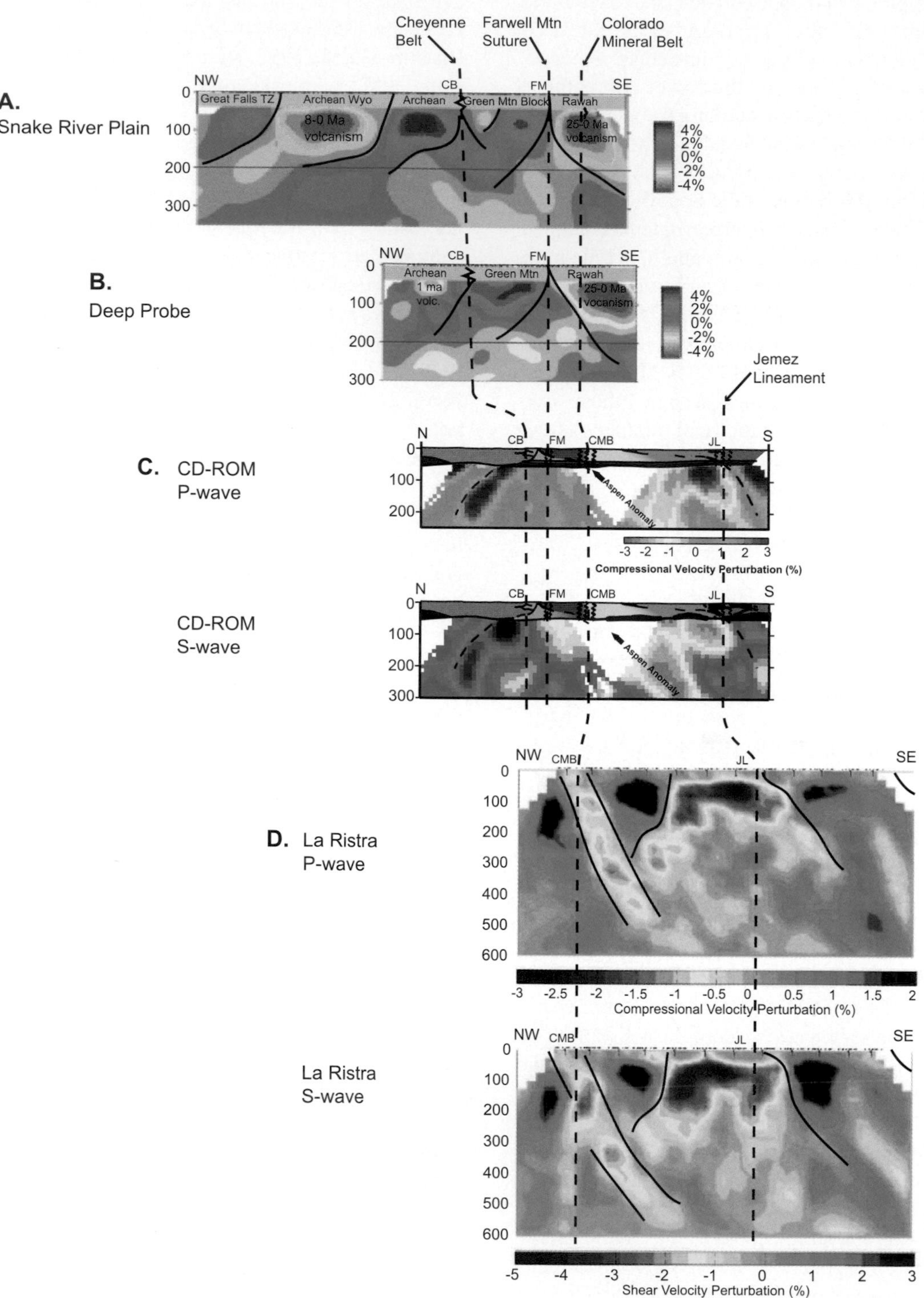

Plate 4. Cross-sectional tomographic images along lines indicated in Plate 1b: A= Snake River Plain line (Schutt and Humphreys, 2004), B= Deep Probe line (Dueker and Yuan, 2004), C= CD-ROM line (Deuker et al., 2001), D= La Ristra line (Gao et al., 2004). Note presence of Cheyenne belt (CB), Farwell Mountain (FM), Colorado mineral belt (CMB) and Jemez lineament (JL) sutures as sharp velocity gradients in multiple lines; also note dipping character of mantle velocity domains.

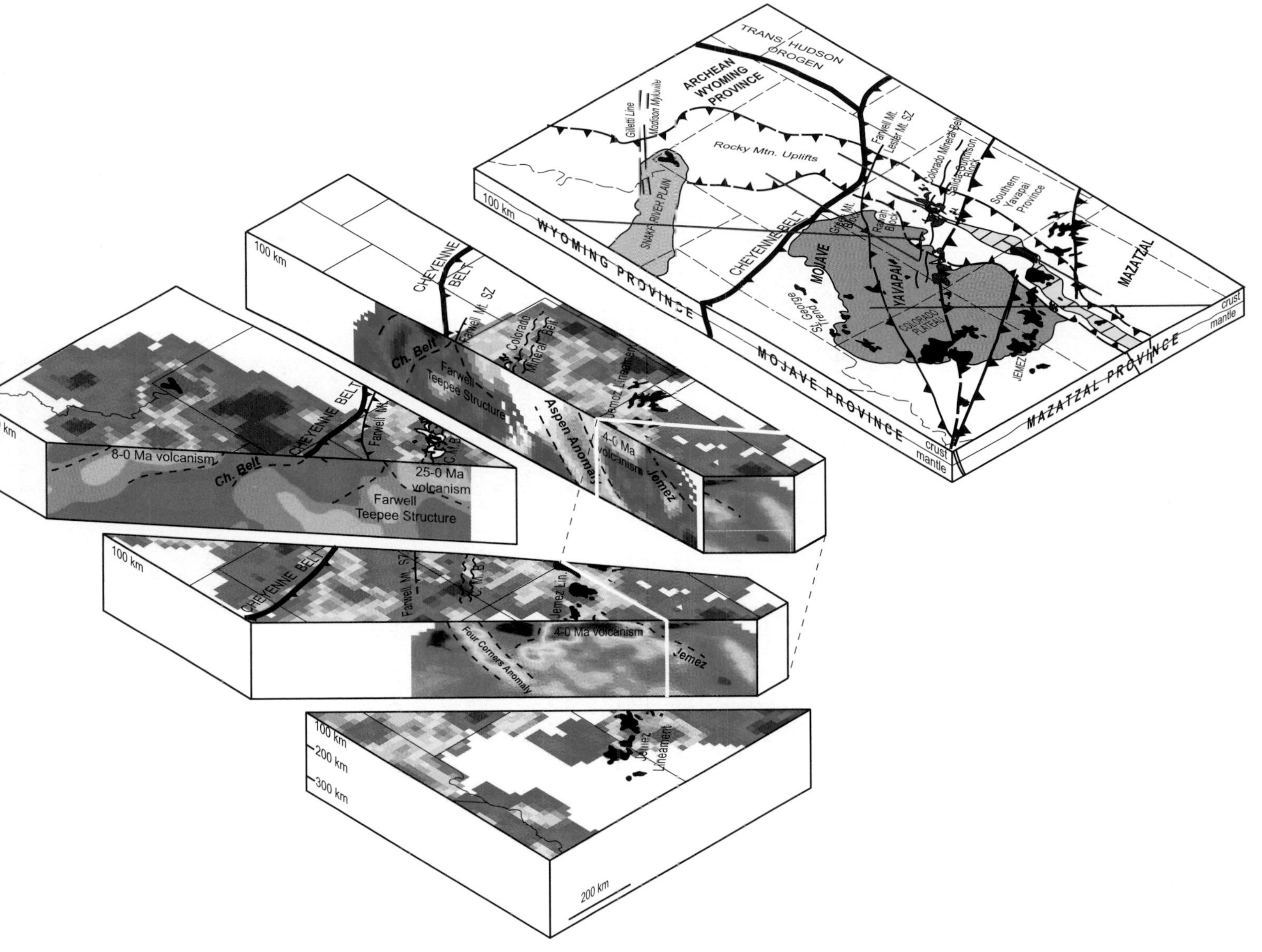

Plate 5. Block diagram showing results from the CD-ROM, Deep Probe, Snake River Plain, and La Ristra teleseismic lines. Major mantle velocity contrasts can be identified in multiple lines and seem to correspond (in pseudo 3-D) with surface Proterozoic province boundaries. The Aspen anomaly and Four Corners anomaly, especially, seem to correspond with the NE- trending Colorado mineral belt that has a Proterozoic ancestry.

transition from dry to wet olivine, and hence a thick but weak hydrous Proterozoic lithosphere is plausible. Other evidence for a relatively thick mantle lithosphere under the Rocky Mountains comes from geochemistry of basalts, which suggests depleted mantle to depths exceeding 80 km (Livicarri and Perry, 1993), and from xenoliths (Ritter and Smith, 1996).

The alternative explanation, that the low velocity domains represents upwelling asthenosphere (West et al., 2004), is supported by surface wave images from the La Ristra experiment which shows velocities down to 4.2 to 4.35 km/s in a 100–200 km wide region along the Jemez lineament and Rio Grande rift. These velocities are consistent with the 1300 °C mantle extending to depths as shallow as about 50 km. The La Ristra line crossed the Rio Grande rift near its intersection with the Jemez lineament, such that the low velocity anomaly coincides with both in this line of section. In contrast, apparently the same low velocity anomaly seen in the CD-ROM line is more clearly associated with the Jemez lineament and not the Rio Grande rift. Thus, we prefer the interpretation that the low velocity anomaly is associated with the northeast-trending Jemez lineament and is associated with hydrated and low solidus Proterozoic lithosphere (such as eclogite or pyroxenite dikes). The geometry of the combined CD-ROM/ La Ristra anomaly seems less readily explained as the product of passive asthenospheric upwelling because of the low extension (<20%) and low strain rates in the Rio Grande rift.

LOWER CRUSTAL RESPONSE: UNDERPLATING AND A DYNAMIC MOHO

Seismic and geologic data from the crust, including new seismic refraction, receiver function, and xenolith data highlight interactions between the crust and mantle through time. Crustal velocity models derived from refraction and wide-angle reflection data (Keller et al., 1998; Snelson et al., this volume; Rumple et al. this volume, Levander et al., this volume) differ somewhat, but all show appreciable topography on the Moho and a crust that varies from ~40 to 55 km thick (Plate 3). A high velocity lower crustal layer exists beneath the Proterozoic terranes, although its thickness, continuity and velocity are not uniquely resolved. Snelson et al. (this volume) and Rumpel et al. (this volume) show a continuous, 5–10 km thick layer with apparent velocity ranging from 7.0 to 7.5 km/s. Levander et al. (this volume) show the layer to be more discontinuous and to have an apparent velocity of 6.8–7.1 km/sec. Given the high heat flow and presumed high temperatures at the base of today's crust, both models are consistent with a mafic composition lower crust (Plate 3).

Xenoliths have been recovered from three key areas in proximity to the CD-ROM transect (Plate 1), the Leucite Hills of the Wyoming province, the Stateline diatremes just south of the Cheyenne belt in Colorado (Farmer et al., this volume), and the Navajo volcanic field in the Four Corners region (Crowley et al., 2003; in prep.). These sets of diatremes have been projected into the CD-ROM section in Plate 3. U-Pb geochronology provides insight into the age and complex tectonic evolution of the mafic lower crust, and shows different histories in the different regions. Xenoliths from the Stateline diatremes in the Proterozoic crust of northern Colorado differ markedly from the adjacent highly potassic lavas from the Leucite Hills in the Archean lithosphere of southern Wyoming (Farmer et al, this volume). Lower crustal xenoliths from the Archean lithosphere (0.8–1.0 GPa) consist of relatively felsic hornblende-pyroxene gneisses (without garnet); they typically display a weak to strong foliation primarily defined by amphibole. These, and mantle xenoliths from this locality, are more hydrated than the Proterozoic xenoliths to the south, compatible with a position above an underthrust oceanic slab that was dewatered during underthrusting. In contrast, the lower crustal xenoliths from the Proterozoic lithosphere (1.2 GPa) contain little fabric and include garnet, two-pyroxene granulites and rare eclogites, consistent with derivation from the thick, relatively dry, high-velocity mafic layer (Plate 3). U-Pb zircon geochronology of Archean xenoliths yields dates that are similar to the crystallization ages of rocks exposed at the surface (ca. 2.6–2.7 Ga).

In contrast, xenoliths from the Proterozoic Stateline side are metaigneous rocks that contain inherited grains as old as 3.1 Ga, Paleoproterozoic igneous grains, and metamorphic zircon that yields a complex range of ages: Devonian (the age of kimberlite eruption), ca. 500 Ma, 1370–1420 Ma, and 1640–1750 Ma (the dominant population). The volumetrically minor metamorphic zircon of ca. 500 Ma and 1370–1420 Ma is interpreted to record magmatic and/or tectonic events that affected the base of the crust, including mafic underplating, but there is no direct evidence in these diatremes for igneous zircon crystallized from underplated mafic magmas. The important conclusions from the Stateline district is that upper crust, lower crust, and mantle provinces across the Cheyenne belt are distinct lithospheric entities whose essential geometries, to > 200 km depths, date back to the time of assembly, and there was apparently not appreciable (> 10 km-scale) decoupling of lithospheric layers during thrust-related crustal assembly.

Xenoliths from diatremes of the Navajo volcanic field reveal a similar story in that the lower crust records the same range of ages and tectonic events as the nearby exposed (middle crustal) rocks (Karlstrom et al., 2004). However, here there is clear evidence for ca. 1.4 Ga igneous zircon that formed in mafic magmas (Crowley et al., in prep.). Xenoliths of mafic and felsic granulite are interpreted as being derived from lower crustal depths (at ~1.4 Ga) of 40 45 km based on the primary mineralogy of Cpx-Grt-Pl ± Qtz and thermobarometry estimates of ~1.3 GPa and 800°C. Mafic xenoliths contain a dominant population of weakly zoned metamorphic grains with U-Pb dates of 1420–1414, 1410–1395 (most prevalent), 1385, and 1360 Ma. Igneous cores in some of the mafic xenoliths have U-Pb dates of ~1435 Ma and nonradiogenic isotopic compositions (Hf = +4.1–7.8, T_{DM} = 1.7–1.6 Ga) consistent with ~1435 Ma mafic magmas having interacted with older crust. Zircons from felsic granulites contain igneous zircons cores and whole grains with U-Pb dates of ~1710 and 1640 Ma. On the Colorado Plateau west of the Four Corners area, seismic refraction data (e.g. Wolfe and Cipar, 1993) indicate that a distinct mafic layer is present at least in some areas, and the recent La Ristra receiver function results (Wilson and Aster, 2003) indicate that the crust is 45– 50 km thick, indicating that underplating has probably occurred. Thus, the combined data indicate a complex history of Proterozoic zircon growth and a polygenetic lower crust, including evidence for mafic magmatism and associated metamorphism from 1.42 to 1.36 Ga.

These xenolith data provide a new view of the seismically defined lower crustal mafic layer. Age distributions and the mix of igneous and metamorphic zircon populations indicate a complex origin involving several processes. These include segregation of crustal cumulates during the juvenile arc phase, refractory residues of partial melting during crustal stabilization, and underplated material. Each of these stages can also be identified in the exposed middle crustal rocks. For example, the early arc phase plutons are represented by 1.80 to 1.72 Ga calc alkaline plutons that commonly have mingled gabbro/diorite/granodiorite compositions (Jessup et al., this volume; Ilg et al., 1996; Hawkins et al., 1996). Lower crustal melting is recorded just after crust assembly by 1.70–1.68 Ga collision-related granites (Jessup et al., this volume; Ilg et al., 1996) that commonly have A-type compositions (Anderson and Bender, 1989) similar to later 1.4 Ga plutons. These granites are probably related to 1.75– 1.65 Ga metamorphic zircon left in refractory residues in both the Stateline and Four Corners areas.

The 1.45–1.35 Ga zircon is especially important in documenting igneous additions to the crust and the impressive thermal affect of these intrusions for whole crust metamorphism (also discussed below). We interpret them to be the first direct evidence for a mafic underplate of 1.45–1.35 Ga, but such an underplate has also been postulated based on: 1) the widespread appearance of middle crustal A-type plutons of bimodal composition (Emslie, 1978), 2) petrogenetic models suggesting that a large volume of ~1.4 Ga granitic magmatism in the middle crust was related to melting of tholeiitic basalt (Frost and Frost, 1997), 3) Ar-Ar themochronology (Karlstrom and Dallmeyer, 1997; Shaw et al., this volume), and 4) monazite zoning studies and the extent of 1.4 Ga middle crustal metamorphism and deformation (Williams et al., 1999; McCoy et al, this volume). The combined seismic and geologic data suggest that a large volume of basaltic rocks may reside in the lower crust.

The interpretation that the mafic lower crust is in part due to mafic underplating has important implications for crustal stabilization and models for a dynamic Moho and emphasizes an important interplay between the crust, mantle lithosphere, and asthenosphere during growth, stabilization, and reactivation of continental lithosphere. The model presented here (and in Keller et al., this volume) is that the Moho under the Rocky Mountains has been dynamically reshaped by mafic magmatism at several times such that its present complexity (10-km-scale topography, and variations in lower crustal and upper mantle velocity) is related to lithosphere stabilization events. In particular, episodic addition of basaltic magmas served to stabilize the continental lithosphere via several processes: it thickened the crust, provided thermal energy for episodic crustal differentiation and metamorphism, and was a manifestation of partial melting of the mantle, a process that reduces the density of and hence further stabilizes the continental lithospheric mantle.

REACTIVATION AND DIFFERENTIAL UPLIFT OF PROTEROZOIC LITHOSPHERE

Geologic studies indicate that the Proterozoic crust south of the Cheyenne belt was repeatedly reactivated, whereas the Archean lithosphere has been relatively stable (Karlstrom and Humphreys, 1998). Following protracted assembly of the lithosphere from 1.78 to 1.65 Ga, the first major reactivation event took place ~1.4 Ga and involved widespread bimodal magmatism and intracratonic transpressional deformation (Nyman et al., 1994). This event pervasively affected the Proterozoic lithosphere, but essentially terminated or was substantially subdued at the Cheyenne belt, most likely due to an existing Archean-Paleoproterozoic mantle tectosphere (Karlstrom and Humphreys, 1998; Levander and Niu, 2004; Levander et al., this volume).

In situ U-Pb dating of monazite using the electron microprobe (Williams et al., 1999; Williams and Jercinovic, 2002) documents the importance of recurrent movements, and hence persistent weakness, within the Colorado mineral belt (Shaw et al., this volume; McCoy et al., this volume). Monazite geochronology from shear zones, combined with microstructural studies of the relationship of the monazite to fabrics, indicate two protracted, ca. 100-m.y.-long, orogenic episodes (1.72–1.62 Ga and 1.45–1.35 Ga), each consisting of numerous pulses of deformation, plus 1.1 Ga, Paleozoic and Laramide movements (Allen, 1994). Ar-Ar data (Karlstrom et al, 1997; Shaw et al., this volume), and thermal modeling (Flowers et al., this volume) corroborate previous documentation (e.g. Chamberlain and Bowring, 1990; Bowring and Karlstrom, 1990; Hodges and Bowring, 1995) that discrete crustal blocks throughout the southwestern U.S. show very different cooling histories. Different cooling histories and differential uplift/exhumation histories are interpreted to result from different heat generation characteristics of different blocks, by reactivation of accretionary structures, and by variations in the thickness and nature of a developing mafic underplate.

In the Phanerozoic, fission track studies demonstrate post-Laramide differential uplift across the Colorado mineral belt (Kelley and Chapin, 2002). These data confirm and extend the hypothesis of Tweto and Sims (1963) that the Colorado mineral belt was a long-lived zone of weakness in the lithosphere. However, Phanerozoic reactivation did not take place across all Proterozoic structures (e.g., Cheyenne belt; Kelley at al., this volume), and reactivation styles were different between the Proterozoic and Archean lithospheric sections. For example, Ancestral Rocky Mountain uplifts formed almost exclusively south of the Cheyenne belt. Laramide deformation partially reactivated older boundaries in both areas, but analysis of minor faults shows a more complicated history of reactivations south of the Cheyenne Belt (Koenig and Erslev, in press). These observations suggest that lateral crustal rheologic differences (compositional inheritance) and the presence of weak structures (interface inheritance) influenced the expression of Cenozoic tectonism in the heterogeneous older lithosphere (Karlstrom and Humphreys, 1998).

LARAMIDE TECTONICS

The Laramide orogeny was characterized by tectonism, regional uplift, and magmatism at surprisingly great distances from the plate margin. This event had a major but still incompletely understood effect on both crust and upper mantle of the western U.S. Accumulating data for the deep crustal structure under the Rocky Mountains suggest that uplifts formed because of significant components of horizontal shortening. Basement-cored asymmetrical uplifts are commonly bounded by listric thrust and reverse faults that flatten at middle crustal depths (~12 km in the CD-ROM Sangre de Cristo line; Magnani et al., this volume), or penetrate to lower middle crustal depths (Lynn et al., 1983; Sharry et al., 1986; Cline and Keller, this volume). In Colorado and northern New Mexico, geometric and surface kinematic studies, combined with seismic and gravity data, seem most consistent with models involving upper crustal detachment on subhorizontal thrusts that root to the west, with possible distributed lower crustal thickening (Erslev, this volume). These faults were commonly localized along (and reactivated) a complex network of faults and shear zones that formed during Proterozoic ductile deformation, late Proterozoic normal faulting, and Ancestral Rocky Mountain thrusting (Marshak et al., 2001; Timmons et al., 2002).

Stresses that caused Laramide deformation were derived from the subducting Farallon plate system and were either transmitted in from the plate margin (Livaccari, 1991) and/or transmitted up from the interface with the subducting slab (Bird, 1984, 1988; Hamilton, 1988; Varga, 1993). Continuing controversies surround the interactions of driving forces and rock strengths in controlling the resulting deformation of the western U.S., including: 1) the influence of a relatively weak lower crust on the transfer of basal tractions to the upper crust (Bird, 1988, 1989; Schmidt and Perry, 1988; Urquhart, 1994; McQuarrie and Chase, 2000; Erslev, 1993; Leeman et al., 1992); 2) the strength (degree of coupling) across the Farallon-North America interface (Bird, 1984, 1988), and 3) how loading at the western margin is transferred across the weak but high-potential energy Sevier Mountains (Coney and Harms, 1984; Patino-Douce et al., 1990) to the Rocky Mountains.

The nature of mantle modifications during the Laramide orogeny is one key component for explaining both Laramide tectonism and today's low velocity mantle in the western U.S. There are three potential times when the major continental scale east to west mantle compositional variation (see Plate 2 of Karlstrom and Keller, this volume) may have been "set in": late Precambrian, Laramide, or post-Laramide, and each likely played a role. Some component of the regional-scale NW-trending velocity transition was perhaps already in place because of lithospheric thinning during late Precambrian rifting west of the Cordilleran hingeline (Karlstrom and Humphreys, 1998). Superimposed on this, Humphreys et al., (2003) suggested that the shallow angle of subduction of the Farallon slab caused pervasive hydration of the North American lithospheric mantle, but only limited magmatism (e.g. in the Colorado mineral belt), and Neogene tectonism and mantle reorganization related to asthenospheric processes under the active western U.S. is exploiting older preconditioned mantle to give modern magmatic and velocity structure.

Similar to the regional velocity variation, the 10- to 100-km-scale velocity variations also seems best explained as a combination of compositional inheritance and active processes. As suggested above, variations in mantle hydration associated with the Proterozoic assembly structures may have played an important role. Karlstrom et al. (2002) suggested that there was a distinctive Proterozoic lithosphere that was assembled via complex subduction-accretion processes, hence had long-term fertility, and responded very differently than Archean mantle to various reactivation events. For example, the parallelism of the magmatic trends in the Colorado mineral belt (early Tertiary), Jemez lineament (Neogene), St. George Lineament (Neogene), and Snake River Plain (Neogene) with old shear zone systems, the long duration of magmatism in the zones, and the lack of systematic distribution of ages (except for the Snake River Plain) may be simpler to explain in terms of reactivated Proterozoic boundaries that were compositionally distinct and hence fertile for melt production during Cenozoic tectonism, rather than due to flat slab subduction (for the Colorado mineral belt), focused slab removal (Humphreys, 1995), or asthenospheric upwelling in zones subparallel to North American plate motion (Goes and van der Lee, 2002; Wilson, 2004).

Neogene Tectonics and Regional Denudation

Impressive post-Laramide magmatism and tectonism continued to modify the western U.S. over the last 50 Ma. In the Rockies, major magmatic events include the San Juan and Jemez volcanism, and Rio Grande rift extension. Additionally, evidence is increasing for significant post-Laramide surface uplift in the Rockies (e.g., Heller et al., 2002; McMillan and Heller, 2002; Leonard, 2002) which diminishes the amount of surface uplift attributed to the Laramide orogeny. This prolonged tectonic activity probably was initiated by delamination of the Farallon slab from the base of the North American plate (e.g., Coney and Reynolds, 1977; Humphreys, 1995), and further activity resulted from subsequent lithospheric heating by conduction and melt advection.

Early Tertiary differential uplift is recorded by apatite fission track (AFT) data from the Southern Rocky Mountains. The northern to central Front Range data record a Laramide cooling history, but data from the southern Front Range and Wet Mountains of Colorado and the Sangre de Cristo Mountains and High Plains of New Mexico record a ~25–30 Ma thermal event that becomes progressively stronger to the south (Pazzaglia and Kelley, 1998; Kelley and Chapin, 1995; 2004; House et al., 2003). Sonic log data have been used to independently document ~ 2 km of middle Cenozoic exhumation on the High Plains that must have occurred prior to deposition of the 12 Ma Ogallala Formation (Kelley, 2002). These data are best explained in terms of linked processes that include removal of basalt from the mantle (e.g., Johnson, 1991), underplating, modification of the Moho, crust and mantle buoyancy changes (Jones et al., 1996; 1998; Humphreys, 1995), and isostatically driven rock and surface uplift (England and Molnar, 1990).

More regionally, stratigraphic and geomorphic studies suggest that the Great Plains region has undergone east-down tilting since deposition of the middle to upper Miocene Ogallala Formation (McMillian and Heller, 2002; Leonard, 2002; Anderson et al., 2002). Although the relative contribution of an isostatic response to fluvial incision and mass removal is debated, it appears that some measure of tectonic/epierogenic tilting is required, the influence of which appears to increase southward along the Front Range (Leonard, 2002). This interval of time between the Oligocene and the middle Miocene was one of significant surface modification in the southern Great Plains. First, the base of the Ogallala Formation is time transgressive from north to south. The Ogallala is nearly conformable with Oligocene-lower Miocene White River and Arikaree groups in the eastern Wyoming area, and the base of the Ogallala is progressively younger to the south. Second, the Ogallala was deposited on an erosion surface that progressively exhumed deeper/older Mesozoic units to the south. Third, recent AFT determinations from drill holes and surface outcrops indicate that the middle Cenozoic partial annealing zone (PAZ) slopes eastward, implying either tilting and/or spatially non-uniform thermal regimes (House et al., 2003). Ages preserved beneath this paleo-PAZ indicate that 2–4 km of material was eroded between Oligocene and the mid-late Miocene.

The spatial variations in the depth of erosion from north to south along the Rocky Mountain front corresponds to the present-day extent of anomalously slow mantle (Aspen anomaly), extensional deformation in the Rio Grande rift, and the degree of tectonically-induced tilting of the Ogallala formation (Leonard, 2002). These correlations suggest to us a possible genetic link, and we hypothesize that upward transfer of heat and melt during the Oligocene (coincident with voluminous ignimbrite volcanism) was responsible for significant regional surface uplift, the expression of which is recorded by long-wavelength tilting on the Great Plains. Increases in regional topographic slope changed the erosive power of fluvial systems and enhanced erosion on the Great Plains.

Similar processes may be ongoing. A provocative hypothesis is that the mantle structures that we have imaged seismically may have distinct topographic manifestations. A combined topographic-thermochronologic study by Pazzaglia and Kelley (1998) demonstrated that the mean local relief, mean elevation, and thermochronologically-determined exhumation history vary systematically across both the Cheyenne belt and Jemez lineament. Furthermore, there is good reason to believe that contemporary uplift is associated with the youthful magmatism concentrated along the Jemez lineament (Wisniewsky and Pazzaglia, 2002; Magnani et al., this volume). For example, the Canadian River has a distinct convexity or bulge in both its long profile and terrace profiles where it crosses the Jemez lineament. Here the river has been incising at a constant rate of 0.06 mm/yr since at least the middle Pliocene; this rate of incision is up to two times greater than similar reaches up or down stream of the lineament. These results are consistent with rock uplift above the anomalously low velocity mantle along the Jemez lineament (Yuan and Dueker, this volume; Dueker et al. 2001; Levander et al., this volume). Thus, in spite of the numerous complex processes that combine to shape landscapes, correlations such as this suggest that deep lithospheric structure and mantle-driven regional epeirogenic uplift exert important controls in producing present-day high elevation and relief (e.g., Pierce and Morgan, 1992; Wisniewski and Pazzaglia, 2002).

This is counter to recent studies that infer that the origin of present-day topographic relief in the Rocky Mountains is largely a function of glacial erosion and river incision accelerated by Pleistocene climate change (e.g., Small and Anderson, 1998; Zhang et al., 2001). These studies cite reconstructions of the paleoelevation history of the Laramide and post-Laramide Rocky Mountains using paleobotanical studies that suggest only minor differences between late Eocene and present elevations in Colorado (e.g. Gregory and Chase, 1994; Chase et al., 1998) and throughout much of the intermountain west (e.g., Wolfe et al., 1998). We do not discount the importance of climate change in increasing relief, but view this as a secondary factor superimposed on active epeirogenic surface uplift (e.g. Epis and Chapin, 1975; Eaton, 1986; Unruh, 1991; Sahagian, 2002). The driving mechanism for epeirogenic uplift is interpreted to be addition of buoyancy to the lithosphere by basalt extraction from the mantle and crustal thickening by addition of basalt to the crust. It becomes focused by lithospheric heterogeneity both because of the presence of more fertile compositions of lithosphere (volumetric inheritance) and melt pathways along zones of weakness (interface inheritance; Karlstrom and Humphreys, 1998).

DISCUSSION OF PROCESSES OF STABILIZATION AND EVOLUTION OF CONTINENTAL LITHOSPHERE

We envision the stable core of the North American continent, stabilized by thick lithospheric mantle that extends to depths of > 200 km in the northern part and > 150 km in the southern part of the CD-ROM line. The North American plate is moving southwest (relative to the mantle reference frame) through weaker asthenosphere. The lithospheric "keel" is cold (and strong) and/or buoyant due to the presence of strongly melt-depleted peridotite (Jordan, 1988). The Rocky Mountain lithosphere is where the leading edge of this mantle keel is being heated, chemically modified, and progressively incorporated into the asthenosphere. This region exhibits the largest mantle velocity gradient on Earth, from fast (cratonic) mantle to slow (orogenic) mantle over a remarkably short distance even at shallow mantle levels (Grand, 1994; van der Lee and Nolet, 1997; Henstock et al., 1997). The keel of Laurentia is the net result of Archean growth plus Paleoproterozoic collision of Archean microcontinents and stabilization during the Paleoproterozoic (1.9–1.8 Ga, Hoffman, 1988). High-velocity lithospheric mantle also extends beneath the Proterozoic provinces of the mid-continent (Van der Lee and Nolet, 1997), which can be viewed as an extension of the orogenic belts in the Southwest (Van Schmus et al., 1993).

The modification and disassembly of this lithospheric keel in the western U.S. is a Cenozoic phenomenon. Some workers have postulated that the mantle lithosphere was largely removed (Bird, 1988), or preserved to moderate depths (100 km; Livicarri and Perry, 1993), by shallow angle subduction of the Farallon slab in the Laramide. Other workers have postulated an upwelling of asthenosphere to shallow depth during removal of the Farallon slab mainly during the Miocene ignimbrite flare up (e.g. Humphreys, 1995). However, if the Proterozoic lithosphere was originally thicker, as we suggest, another possibility is that rather than removal, the mantle was modified in several steps: hydration above a Laramide flat slab, Oligocene ignimbrite flare-up, and Neogene magmatism. If the low-velocity mantle in the Southern Rocky Mountain region is old and essentially intact (e.g. below the Colorado mineral belt and Jemez lineament), then this mantle, although hot and weak, has not yet been entrained in the convecting asthenosphere, perhaps because it is buoyantly trapped between cold keels to the east (craton) and to some extent the west (portions of the Colorado Plateau) and north (Wyoming Province). The low velocity domains may also reflect zones of asthenospheric upwelling utilizing Proterozoic structures and domain boundaries.

The crustal record for lithospheric mantle melt-depletion events is inferred to include modification of the mafic lower crust and surface mafic magmatism. We infer that the Moho and the lower crustal layer are in part younger than the assembly structures and provide a record of changing crustal thickness and development of a mafic underplate. The lower crustal mafic layer is remarkably featureless on regional reflection profiles (Eshete et al., 2002, Magnani et al., 2004; this volume) and lies below well-developed bright reflectivity that we interpret to be a record of Proterozoic horizontal tectonics. Furthermore, the Precambrian rocks exposed at the surface were at depths of 10–20 km at 1.70 Ga, but we do not believe that the crust was 60–70 km thick at the end of the orogenic cycle because thermochronologic data indicate that rocks remained deep and relatively hot and were not quickly unroofed. This would suggest an isostatically adjusted lithosphere and relatively "normal" thickness crust (Bowring and Karlstrom, 1990). Thus, our hypothesis is that today's thick crust grew in part by underplating and addition of mafic intrusive bodies of a variety of ages. Based on thinning of the lower crustal mafic layer just north of the Cheyenne belt, the lack of evidence for Proterozoic overprinting of Archean lower crust to the north, and volumetrically minor Phanerozoic magmatism in the Archean lithosphere, this process seems to have preferentially affected the Proterozoic lithosphere. Geologic data suggest there was major underplating of mafic magma ca. 1.4 Ga, with additional magmatic additions to the crust ca 1.1 Ga, and during the Laramide and Cenozoic (Keller et al., this volume).

It is useful to examine the ongoing and incipient mantle modification in the eastern Rockies as a proxy for early stages of the overall plate-scale mantle modification event. Our recent teleseismic findings indicate that North American chemical lithosphere appears to extend to depths of at least 150–200 km beneath the Rocky Mountains, which is inconsistent with removal of North American mantle via shallow subduction of the Farallon plate (and is in agreement with Livicarri and Perry, 1993). North American lithosphere in this region has large velocity contrasts that reflect partial melt and a mantle that has gained buoyancy and lost strength beneath the Rocky Mountains (Humphreys et al., 2003), probably in several stages during Laramide and post-Laramide events (Karlstrom et al., 2002). Today's mantle in the low velocity domains is thus interpreted to be chemically part of the North American lithosphere in terms of composition, but it may be more similar to asthenosphere in terms of rheology. Its viscosity was apparently decreased by Farallon hydration and Neogene heating. It remains part of the North American plate only by virtue of its buoyancy and perhaps because it is confined between colder, stronger mantle domains.

SUMMARY

The Continental Dynamics of the Rocky Mountains (CD-ROM) experiment was a set of coordinated seismic and geologic investigations of the deep structure and tectonic evolution of the Rocky Mountain lithosphere in a region where there is an ongoing interplay between old structures and active processes. The geological and geophysical data combine to build a new 4-D image of the crust and upper mantle of the Rocky Mountain region in a swath extending from south-central Wyoming through Colorado to northern New Mexico, and they provide new models for the structure and evolution of the continental lithosphere. Crustal studies suggest Proterozoic paleosubduction zones were active at the Cheyenne belt (Archean-Proterozoic boundary), Lester-Farwell Mountain area of the Park Range (arc-arc suture), Colorado mineral belt (arc-arc suture), and Jemez lineament (arc-arc suture and Yavapai-Mazatzal province boundary). Mantle tomography indicates different mantle velocity domains that correspond to the Proterozoic crustal provinces, with profound velocity contrasts that are co-located with the province boundaries. Several of these velocity domain boundaries are dipping, contain dipping mantle anisotropy, and extend to depths of > 150–200 km. The appearance of similar velocity domain boundaries in multiple 2-D teleseismic lines (CD-ROM, Snake River Plain, Deep Probe, and La Ristra) provides support that, in three dimensions, they define tabular dipping mantle provinces whose sharp boundaries coincide with Proterozoic sutures developed during lithospheric assembly. We interpret them to have originated as paleosubduction zones that initially developed in the Proterozoic and were then frozen into the lithosphere following collisions of oceanic terranes. Remarkably, even after nearly 2 billion years, they have retained compositional variations in terms of degree of hydration and depletion. This heterogeneous mantle structure, although primarily due to ancient compositional variations, is presently responding in distinctly different ways (e.g. experiencing different degrees of partial melting) to Cenozoic thermal and fluid perturbations caused by transient plumes or small scale asthenospheric convection related to restructuring of the western plate margin of North America. The thick Proterozoic lithospheric mantle was part of North America by 1.6 Ga, was thickened at 1.4 Ga, and has remained both fertile and weak as shown by repeated intracratonic deformational and magmatic reactivations of ancient boundary zones from 1.4 Ga to the present.

The process of growth, stabilization, and reactivation of continental lithosphere involves an important interplay between the crust, mantle lithosphere, and asthenosphere that is expressed as a dynamic Moho (Keller et al., this volume). Throughout much of the southern Rocky Mountains, seismic refraction data delineate a high velocity (6.8 to 7.3 km/sec) lower crustal layer of variable thickness. The base of this layer is interpreted to be the Moho, and it varies from 40 to 55 km in depth. New geochronology from the xenoliths of the Stateline diatremes (northern Colorado) shows zircon ages of ~ 1720, 1620, and 1380 Ma from lower crustal (1.1–1.2 GPa) xenoliths. In contrast, Navajo volcanic field (Four Corners region) xenoliths have ages of 1710–1620 Ga, 1420–1414, 1410–1395 (most prevalent), 1385, and 1360 Ma. Taken together we infer that Proterozoic lower crust has a complex tectonic history that is strongly linked to that of the middle crust. The mafic lower crust evolved diachronously via two main processes. First, concentration of mafic restite took place during original arc formation (pre 1.70 Ga in both areas) and subsequent collision-related differentiation (1.70–1.62 Ga in both areas). Second, lower crust was also added during later basaltic underplating (as best recorded by Navajo volcanic field xenoliths). Episodic addition of basaltic magmas served to stabilize the continental lithosphere via several processes: it thickened the crust, provided thermal energy for episodic crustal differentiation and metamorphism, and was a manifestation of partial melting of the mantle, a process that served to increase the buoyancy (and hence further stabilize) the lithospheric mantle.

Proterozoic crust and lithospheric mantle of Colorado and New Mexico differ from lithosphere under the Archean core of the continent in terms of long-term strength, expressed as a tendency for the Proterozoic lithosphere to be more readily reactivated by intracratonic tectonic and magmatic events (Karlstrom and Humphreys, 1998). Both types of lithosphere are strongly segmented (Gorman et al., 2002; Cook et al., 1999; Bostock, 1998; Chamberlain, 1998), but the difference may be that the Archean lithosphere is more strongly depleted due to multiple high temperature Archean events and/or a longer history of depletion events. The juvenile Proterozoic lithosphere of the southwestern U.S. has been fertile, weak, yet fairly stable throughout its history. It originated by rapid subduction-accretion processes and was characterized by marked heterogeneity of mantle hydration because of Indonesian-style crustal growth by accretion of dominantly oceanic tectonic elements. Its original hydrous character and heterogeneity, combined with marked reactivation by hydration during flat slab Laramide subduction, have influenced later intracratonic tectonism and may provide an explanation of the anomalously wide zone of Laramide to Cenozoic tectonic activity in the Cordillera of southwestern North America.

Acknowledgements. The CD-ROM (Continental Dynamics of the Rocky Mountains) experiment was funded by the NSF Continental Dynamics Program starting with a workshop in 1995 (EAR-9506627, 9614787, 0003540, 0208473, 0310324). We thank Leonard Johnson for his support of the project. The refraction experiment was co-funded by the *Deutsche Forschungsgemeinschaft.* We thank Larry Brown and Steve Grand for reviews that helped improve the paper. We thank Patti Halcli for technical help in producing this volume.

REFERENCES CITED

Adams, D.C., and Keller, G.R., Possible extension of the Midcontinent Rift in west Texas and eastern New Mexico, *Can. J. Earth Sci.*, 31, 709–720, 1994.

Aldrich, M.L., Jr., Tectonics of the Jemez lineament in the Jemez Mountains and Rio Grande rift, *J. Geophys. Res.*, 91, 1753–1762, 1986.

Anderson, J.L., and Bender, E.E., Nature and origin of Proterozoic A-type granitic magmatism in the southwestern United States: *Lithos*, 23, 19–52, 1989.

Anderson, R.S., Riihimaki, C.A., Safran, E.B., and Stock, G., 2002, Late Cenozoic glacial and fluvial incision of Laramide ranges, *Geol. Soc. Am. Abstr. Prog.*, 34, 407, 2002.

Allen, J.L., III, Stratigraphic variations, fault rocks, and tectonics associated with brittle reactivation of the Homestake shear zone, central Colorado, PhD thesis, 321 pp., University of Kentucky, Lexington, 1994.

Barnes, M.A., Rohs, C.R., Anthony, E.Y., Van Schmus, W.R., and Denison, R.E., Isotopic and elemental chemistry of subsurface Precambrian igneous rocks, west Texas and eastern New Mexico, *Rocky Mt. Geol.*, 34, 245–262, 1999.

Beaumont, C., Ellis, S., Hamilton, J., and Fullsack, P., Mechanical model for Subduction-collision tectonics of Alpine-type compressional orogens, *Geology*, 24, 675–678, 1996.

Bird, P., Laramide crustal thickening event in the Rocky Mountain foreland and Great Plains, *Tectonics*, 3, 741–758, 1984.

Bird, P., Formation of the Rocky Mountains, western United States: a continuum computer model, *Science*, 239, 1501–1507, 1988.

Bird, P., Kinematic history of the Laramide orogeny in latitudes 35°–49° N, western United States, *Tectonics*, 17, 780–801, 1998.

Blackman, D.K., Orcutt, J.A., Forsyth, D.W., and Kendall, J.M., Seismic anisotropy in the mantle beneath an oceanic spreading centre, *Nature*, 366, 675–677, 1993.

Blackman, D.K., Wenk, H.R., and Kendall, J.M., Seismic anisotropy of the upper mantle; 1, Factors that affect mineral texture and effective elastic properties, *Geochemistry, Geophysics, Geosystems,* 9, pp. 18, 2002.

Bostock, M.G., Mantle stratigraphy and evolution of the Slave province, *J. Geophys. Res., 103*, 21193–21200, 1998.

Bowring, S.A., and Karlstrom, K.E., Growth, stabilization and reactivation of Proterozoic lithosphere in the southwestern United States, *Geology*, 18, 1203–1206, 1990.

Cammarano, F., Goes, S., Vacher, P., and Girardini, D., Inferring upper-mantle temperatures from seismic velocities, *Earth & Planetary Science Letters*, *138*, 197–222, 2003.

Korenaga, J., and Jordan, T.H., Physics of multi-scale convection in the Earth's mantle 1. Onset of sublithospheric convection, *submitted to Journal of Geophysical Research*, 2003.

Schott, B., Yuen, D.A., and Schmeling, H., The diversity of tectonics from fluid-dynamical modeling of the lithosphere-mantle system, *Tectonophysics*, *322* (1–2), Pages 35–51, 2000.

Carlson, R.W., Boyd, S.B., Shirley, S.B., Janny, P.E., Grove, T.L., Bowring, S.A., Schmitz, M.D., Bell, D.R., Gurney, J.J., Richardson, S.H., Tredoux, M., Menzies, A.H., Pearson, D.G., Hart, R.J., Wilson, A.H., and Moser, D., Continental growth, preservation, and modification in southern Africa, *GSA Today*, 10, 1–7, 2000.

Chamberlain, K.R., Medicine Bow orogeny: Timing of deformation and model of crustal structure produced during continent-arc collision, ca.1.78 Ga, southwestern Wyoming, *Rocky Mt. Geol.*, 33, 259–277, 1998.

Chamberlain, K.R. and Bowring, S.A., Proterozoic geochronologic and isotopic boundary in northwest Arizona, *J. Geol.*, 98, 399–416, 1990.

Chamberlain, K.R., Frost, C.D., and Frost, B.R., 2000, Late Archean to Mid-Proterozoic geologic evolution of the south-central Wyoming Province: Implications for crustal architecture and Deep-Probe seismic images, *Geol. Soc. Am. Abstr. Prog.*, 32, A-165, 2000.

Chase, C.G., Gregory-Wodzicki, K.M., Parrish, J.T., and DeCelles, P.G., Topographic history of the western cordillera of North America and controls on climate, in Tectonic boundary conditions for climate model simulation, edited by Crowley, T. J. and Burke, K., Oxford Monographs on geology and geophysics, Oxford University Press, 73–99, 1998.

Condie, K.C. and Shadel, C.A., An Early Proterozoic arc succession in southeastern Wyoming, *Can. J. Earth Sci.*, 21, 415–427, 1984.

Coney, P. and Harms, T., Cordilleran metamorphic core complexes: Cenozoic extensional relics of Mesozoic compression, *Geology*, 12, 550–554, 1984.

Coney, P.J. and Reynolds, S.J., Flattening of the Farallon slab, *Nature*, 270, 403–406, 1977.

Cook, F.A., van der Velden, A.J., Hall, K.W., and Roberts, B.J., Tectonic delamination and subcrustal imbrication of the Precambrian lithosphere in northwestern Canada mapped by Lithoprobe, *Geology*, 26, 839–842, 1998.

Cook, F.A., van der Velden, A.J., Hall, K.W., and Roberts, B.J., Frozen subduction in Canada's Northwest Territories; Lithoprobe deep lithospheric reflection profiling of the western Canadian Shield, *Tectonics, 18*, 1–24, 1999.

Crosswhite, J.A., and Humphreys, E.D., Imaging the mountainless root of the 1.8 Ga Cheyenne belt suture and clues to its tectonic stability, *Geology*, 31, 669–672. 2003.

Crowley, J. L., Schmitz, M.D., Bowring, S.A., Williams, M.L., and Karlstrom, K.E., U-Pb and Lu-Hf isotopic analysis of zircon in lower crustal xenoliths from the Navajo volcanic field : 1.4 Ga mafic magmatism and metamorphism beneath the Colorado Plateau: *J. of Geology*, in press., 2005.

Davis, W.J. and Ross, G.M., Proterozoic underplating of the Archean Wyoming craton and Medicine Hat block during assembly of western Laurentia, in Ninth Annual V.M. Goldschmidt Conference LPI Contribution No. 971, Lunar and Planetary Institute, Cambridge, 68–69, 1999.

Decker, E.R., Heasler, H.P., Buelow, K.L., Baker, K.H., and Hallin, J.S., Significance of past and recent heat flow and radioactivity studies in the Southern Rocky Mountains region, *Geol. Soc. Amer. Bull.*, 100, 1851–1885, 1988.

Ducea, M., The California Arc: Thick granitic batholiths, eclogitic residues, lithospheric-scale thrusting, and magmatic flare-ups, *GSA Today*, 11, 4–10, 2001.

Duebendorfer, E.M. and Houston, R.S., Proterozoic accretionary tectonics at the southern margin of the Archean Wyoming craton, *Geol. Soc. Amer. Bull.*, 98, 554–568, 1987.

Dueker, K.G., Yuan, H., and Zurek, B., Thick Proterozoic lithosphere of the Rocky Mountain region, *GSA Today*, 11, 4–9, 2001.

Dueker, K.G., and Yuan, H., Upper mantle P-wave velocity structure from PASSCAL teleseismic transects across Idaho, Wyoming and Colorado, *Geophys. Res. Lett., 34*, doi:10.1029/2004GL019476, 2004.

Eaton, G., A tectonic redefinition of the southern Rocky Mountains, *Tectonophysics*, 132, 163–193, 1986.

Emslie, R.F., Anorthosite massifs, rapakivi granites, and late Proterozoic rifting of North America: *Precambrian Res.*, 7: 61–98, 1978.

England P., and Molnar, P., Surface uplift, uplift of rocks, and exhumation of rocks, *Geology*, 18, 1173–1177, 1990.

Epis, R.C. and Chapin, C.E., Geomorphic and tectonic implications of the post-Laramide, late Eocene erosion surface in the southern Rocky Mountains, in Cenozoic history for the southern Rocky Mountains, edited by Curtis, B. F., *Geol. Soc. Am. Memoir*, 144, 45–74, 1975.

Erslev, E.A., Thrusts, backthrusts and detachment of Laramide foreland arches, in Laramide basement deformation in the Rocky Mountain foreland of the western United States, edited by Schmidt, C.J., Chase, R., and Erslev, E.A., *Geol. Soc. Am. Special Paper*, 280, 125–146, 1993.

Frost, C.D., and Frost, B.R., Reduced rapakivi-type granites: The tholeiite connection, *Geology*, 25, 647–650, 1997.

Gao, W., Grand, S., Baldridge, S., Wilson, D., West, M., Ni, J., and Aster, R., Upper Mantle Convection Beneath the Central Rio Grande Rift Imaged by P and S Wave Tomography, *J. Geophy. Res*, 109, B03305, doi:10.1029/2003JB002743, 2004.

Godey, S., Snieder, R., Villasenor, A., and Benz, H.M., Surface wave tomography of North America and the Caribbean using global and regional broad-band networks: Phase velocity maps and limitations of ray theory, *Geophys. J. Inter.*, 152, 620–632, 2003.

Gorman, A.R., Clowes, R.M., Ellis, R.M., Henstock, T.J., Spence, G.D., Keller, R., Levander, A.R., Snelson, C.M., Burianyk, M.J.A., Kanasewich, E.R., Asudeh, I., Hajnal, Z., and Miller, K.C., Deep Probe, imaging the roots of western North America, *Can. J. Earth Sci.*, 39, 375–398, 2002.

Goes, S. and van der Lee, S., Thermal structure of the North American uppermost mantle, *J. Geophys. Res.*, 107, 2000JB000049, 2002.

Gok, R., Ni, J.F., Sandvol, E., Wilson, D., Baldridge, W.S., Aster, R., West, M., Grand, S., Gao, W., Tilmann, F., and Semken, S., Shear wave splitting and mantle flow beneath LA RISTRA, *Geophys. Res. Lett.*, 2003.

Grand, S.P., Mantle shear structure beneath the Americas and surrounding oceans, *J. Geophys. Res.*, 99, 11591–11621, 1994.

Gregory, K.M. and Chase, C.G., Tectonic and climatic influences of a late Eocene low-relief, high-level geomorphic surface, Colorado, *J. Geophys. Res.*, 99, 20141–20160, 1994.

Hamilton, W., Tectonic settings and variations with depth of some Cretaceous and Cenozoic structural and magmatic systems of the western United States, in Metamorphism and crustal evolution of the western United States, Rubey Volume 7, edited by Ernst, W.B., Englewood Cliffs, New Jersey, pp. 1–40, 1988.

Hawkins, D.P., Bowring, S.A., Ilg, B.R., Karlstrom, K.E., and Williams, M.L., U-Pb geochronologic constraints on Proterozoic crustal evolution: *Geol. Soc. of Am. Bull*, 108, 1167–1181, 1996.

Heller, P.L., McMillan, M.E., Dueker, K., and Paola, C., Alluvial gravel transport as evidence of continental tilting and dynamic uplift of the U.S. Cordillera, *Geol. Soc. Am. Abstr. Prog.*, 33, 221, 2002.

Helmstaedt, H.H., and Schulze, D.J., Early to mid-Tertiary inverted metamorphic gradient under the Colorado Plateau; evidence from eclogite xenoliths in ultramafic microbreccias, Navajo volcanic field, *J. Geophys. Res.*, 96, 13, 1991.

Henstock, T.J., Levander, A., Snelson, C.M., Keller, G.R., Miller, K.C., Harder, S.H., Gorman, A.R., Clowes, R.M., Burianyk, M.J.A., and Humphreys, E.D., Probing the Archean and Proterozoic lithosphere of western North America, *GSA Today*, 8, 1–5 and 16–17, 1998.

Hicks, N.O., Keller, G.R., Miller, K.C., Montana, C., Duran, A., and Mickus, K.L., Comparisons between crustal structure in the Basin and Range Province of southern Nevada, California, and western Arizona and the Rio Grande and Kenya rifts, *Geol. Soc. Am. Abstr. Prog.*, 33, 62, 2001.

Hills, F.A., and Houston, R.S., Early Proterozoic tectonics of the central Rocky Mountains, North America, Contrib. Geol., University of Wyoming, 17, 89–109, 1979.

Hodges, K.V., and Bowring S.A., 40Ar/39Ar thermochronology of isotopically zoned micas: Insights from the southwestern USA Proterozoic orogen, *Geochem. Cos. Acta*, 59, 3205–3220, 1995.

Hoffman, P.F., United plates of America, the birth of a craton; Early Proterozoic assembly and growth of North America, *Annual Rev. Earth Planet. Sci.*, 16, 543–603, 1988.

House, M.A., Kelley, S.A., and Roy, M., Refining the footwall cooling history of a rift flank uplift, Rio Grande rift, New Mexico, *Tectonics*, 22, 1060, doi:10.1029/2002TC001418, 2003.

Humphreys, E.D., and Dueker, K.G., Western U.S. upper mantle structure, *J. Geophys. Res.*, 99, 9615–9634, 1994.

Humphreys, E.D., Post-Laramide removal of the Farallon slab, western United States, *Geology*, 23, 987–990, 1995.

Humphreys, E., Hessler, E., Dueker, K., Erslev, E., Farmer, G.L., and Atwater, T., How Laramide-age hydration of North America by the Farallon slab controlled subsequent activity in the western U.S., *Inter. Geol. Rev.*, 45, 575–595, 2003.

Ilg, B., Karlstrom, K.E., Hawkins, D., and Williams, M.L., Paleoproterozoic rocks of the Upper and Middle Granite Gorges, Grand Canyon, Arizona: New stratigraphic nomenclature, structural geometry, and insights into middle crustal processes, *Geol. Soc. Am. Bull.*, 108, 1148–1166, 1996.

Isaacson, L.B. and Smithson, S.B., Gravity anomalies and granite emplacement in west-central Colorado. *Geol. Soc. Amer. Bull.*, 87, 22–28, 1976.

Johnson, R.A., Karlstrom, K.E., Smithson, S.B., Houston, R.S., Gravity profiles across the Cheyenne belt: A Precambrian crustal suture in southeastern Wyoming, *J. Geodynamics*, 1, 445–472, 1984.

Johnson, C.M., Large-scale crust formation and lithosphere modification beneath middle to late Cenozoic calderas and volcanic fields, western North America, *J. Geophys. Res.*, 96, 13485–13507, 1991.

Jones, C.H., Unruh, J.R., and Sonder, L.J., The role of gravitational potential energy in active deformation in the southwestern United States, *Nature*, 381, p. 37–41, 1996.

Jones, C.H., Sonder, L.J., and Unruh, J.R., Lithospheric gravitational potential energy and past orogenesis; implications for conditions of initial Basin and Range and Laramide deformation, *Geology*, 26, 639–642, 1998.

Jordan, T.H., Structure and formation of the continental tectosphere, *J. Pet.*, 29, 11–37, 1988.

Kaminski, E., and Ribe, N.M., A kinematic model for recrystallization and texture development in olivine polycrystals, *Earth Planet. Sci. Lett.*, 189, 253–267, 2001.

Kaminski, E., and Ribe, N.M., 2002, Timescales for the evolution of seismic anisotropy in mantle flow, Geochemistry, Geophysics, *Geosystems*, 8, p. 17, 2002.

Karlstrom, K.E., and Houston, R.S., The Cheyenne belt: analysis of a Proterozoic suture in southern Wyoming, *Precambrian Res.*, 25, 415–446, 1984.

Karlstrom, K.E., and Humphreys, E.D., Persistent influence of Proterozoic accretionary boundaries in the tectonic evolution of southwestern North America: Interaction of cratonic grain and mantle modification events, *Rocky Mt. Geol.*, 33, 161–180, 1998.

Karlstrom, K.E., Dallmeyer, D.A., and Grambling, J.A., $^{40}Ar/^{39}Ar$ evidence for 1.4 Ga regional metamorphism in northern New Mexico: Implications for thermal evolution of the lithosphere in the southwestern U.S.A., *J. Geol.*, 105, 205–223, 1997.

Karlstrom, K.E., Amato, J.M., Williams, M.L., Heizler, M., Shaw, C.A., Read, A.S., and Bauer, P., Proterozoic tectonic evolution of the New Mexico region, in Mack, G.H. and Giles, K.A., eds., The Geology of New Mexico: A Geologic History: *New Mexico Geological Society Special Publication* 11, 1–34, 2004.

Karato, S., The dynamic structure of the deep Earth- An interdisciplinary approach, Princeton University Press, pp. 241, 2003.

Keller, G.R., Snelson, C.M., Sheehan, A.F., and Dueker, K.G., Geophysical studies of the crustal structure in the Rocky Mountain Region, A Review, *Rocky Mt. Geol.*, 33, 217–228, 1998.

Kelley, S.A., 2002, Evidence for elevated regional heat flow during Late Oligocene time on the Southern High Plains, *Geol. Soc. Am. Abstr. Prog.*, 34, 322, 2002.

Kelley, S.A., and Chapin, C.E., Apatite fission-track thermochronology of Southern Rocky Mountain-Rio Grande rift- western High Plains provinces, *New Mexico Geol. Soc. Guidebook*, 46, 87–96, 1995.

Kelley, S.A., and Chapin, C.E., Denudational histories of the Front Range and Wet Mountains, Colorado, based on apatite fission-track thermochronology, in Tectonics, geochronology and volcanism in the Southern Rocky Mountains and Rio Grande rift, edited by Cather, S.M., McIntosh, W., and Kelley, S.A., *New Mexico Bureau of Mines and Mineral Res. Bull.*, 2002.

Koenig, N.B., and Erslev, E.A., Internal and external controls on Phanerozoic Rocky Mountain structures, U.S.A.: Insights from GIS-enhanced tectonic maps, in SEPM volume on the Cenozoic Rocks of the Rocky Mountains, edited by Raynolds, R., 2002.

Kuehner, S.M., and Irving, A.J., Eclogite and metabasite xenoliths of subducted slab origin from the Paleogene Cedar Mountain diatremes, southwestern Wyoming, USA, Proceedings of the International Kimberlite Conference, I, 485–493, 1999.

Leeman, W.P., Oldow, J.S., and Hart, W.K., Lithosphere-scale thrusting in the western U.S. Cordillera as constrained by Sr and Nd isotopic transitions in Neogene volcanic rocks, *Geology*, 20, 63–66, 1992.

Leonard, E.M., Geomorphic and tectonic forcing of late Cenozoic warping of the Colorado piedmont, *Geology*, 30, 595–598, 2002.

Levander, A. and Niu, F., Imaging teleseismic P to S scattered waves using the Kirchhoff integral, in *Array Analysis of Broadband Seismograms*, edited by Levander, A and Nolet, G., *AGU Monograph*, accepted, 2004.

Levin, V., and Park, J., Shear zones in the Proterozoic lithosphere of the Arabian Shield and the nature of the Hales discontinuity, *Tectonophysics*, 323, 131–148, 2000.

Li, A., Forsyth, D.W., and Fischer, K.M., Evidence for shallow isostatic compensation of the southern Rocky Mountains from Rayleigh wave tomography, *Geology*, 30, 683–686, 2002.

Livaccari, R.F., Role of crustal thickening and extensional collapse in the tectonic evolution of the Sevier-Laramide orogeny, western United States, *Geology*, 19, 1104–1107, 1991.

Livaccari, R.F., and Perry, F.V., Isotopic evidence for preservation of Cordilleran lithospheric mantle during the Sevier-Laramide orogeny, western United States, *Geology*, 21, 719–722, 1993.

Lynn, H.B., Quam, S., and Thompson, G.A., Depth migration of the COCORP Wind River, Wyoming, seismic reflection data, *Geology*, 28, 462–469, 1983.

Magnani, M.B., Miller, K.M., Levander, A., and Karlstrom, K., Ancient to modern history of the Jemez Lineament of north central New Mexico, *Geol. Soc. Am. Bull.*, in press, 2004.

Marshak, S., Karlstrom, K.E., and Timmons, J.M., Inversion of Proterozoic extensional faults: An explanation for the pattern of Laramide and Ancestral Rockies intracratonic deformation, United States, *Geology*, 28, 735–738, 2000.

McCoy, A., The Proterozoic ancestry of the Colorado Mineral Belt: ca. 1.4 Ga shear zone system in central Colorado, MS thesis, 160 pp., University of New Mexico, Albuquerque, 2001.

McMillan, M.E., Angevine, C.L., and Heller, P.H., Postdepositional tilt of the Miocene-Pliocene Ogallala Group on the western Great Plains: Evidence of late Cenozoic uplift of the Rocky Mountains, *Geology*, 30, 63–66, 2002.

McQuarrie, N., and Chase, C.G., Raising the Colorado Plateau, *Geology*, 28, 91–94, 2000.

Morozova, E.A., Wan, X., Chamberlain, K.R., Smithson, S.B., Johnson, R.A., Karlstrom, K.E., Tyson, A.R., Morozov, I.B., Boyd, N.K., and Foster, C.T., Geometry of Proterozoic sutures in the central Rocky Mountains from seismic reflection data: Cheyenne belt and Farwell Mountain structures, *Geophys. Res. Lett.*, 2001.

Mutschler, F.E., Larson, E.E., and Bruce, R., Laramide and younger magmatism in Colorado, *Col. School Mines Quart.*, 82, 1–45, 1987.

Nyman, M., Karlstrom, K.E., Graubard, C. and Kirby, E., Mesozoic contractional orogeny in western North America: Evidence from ca. 1.4 Ga plutons, Geology, 22, 901–904, 1994.

O'Reilly, S.Y., W.L. Griffin, Y.H. Poudjom Djomani, and Morgan, P., Are lithospheres forever? Tracking changes in subcontinental lithospheric mantle through time, *GSA Today*, 11, 4–10, 2001.

Patino-Douce, A., Humphreys, E.D. and Johnston, A.D., Anatexis and metamorphism in tectonically thickened continental crust exemplified by the Sevier hinterland, western North America, *Earth Planet. Sci. Lett.*, 97, 290–315, 1990.

Pazzaglia, F.J., and Kelley, S.A., Large-scale geomorphology and fission-track thermochronology in topographic and exhumation reconstructions of the Southern Rocky Mountains, *Rocky Mt. Geol.*, 33, 229–257, 1998.

Pierce, K.L., and Morgan, L.A., The track of the Yellowstone hotspot: volcanism, faulting, and uplift, in Regional Geology of Eastern Idaho and Western Wyoming, Link, P.K., Kuntz, M.A., and Platt, L.B., *Geol. Soc. Am. Memoir 179*, 1–53. 1992.

Prodehl, C.P., and Lipman, P.W., Crustal structure of the Rocky Mountain region, in Geophysical Framework of the Continental United States, Pakiser, L. C. and Mooney W. D., *Geol. Soc. Am. Memoir 172*, 249–284, 1989.

Ribe, N.M., Seismic anisotropy and mantle flow, *J. Geophys*. Res., 94, 4213–4223, 1989.

Riter, J.C.A., and Smith, D., Xenolith constraints on the thermal history of the mantle below the Colorado Plateau, *Geology*, 24, 267–270, 1996.

Sahagian, D.A., Proussevitch, A., and Carlson, W., Timing of Colorado Plateau uplift: Initial constraints from vesicular basalt-derived paleoelevations, *Geology*, 30, 807–810, 2002.

Schmid, S.M., Pfiffner, O.A., Froitzheim, N., Schoenborn, G., and Kissing, E., Geophysical-geological transect and tectonic evolution of the Swiss-Italian Alps, *Tectonics*, 15, 1036–1064, 1996.

Schmidt, C.J., and Perry, W.J., Jr., Interaction of the Rocky Mountain Foreland and the Cordilleran Thrust Belt, *Geol. Soc. Am. Memoir, 171*, 582 p., 1988,

Schutt, D.L., and Humphreys, E.D., Evidence for a deep asthenosphere beneath North America from western United States SKS splits, *Geology*, 29, 291–294, 2001.

Schutt, D.L., and Humphreys, E.D., P and S Wave velocity and VP/VS in the wake of the Yellowstone hotspot, *J. Geophy. Res.*, in press, 2004.

Semken, S., Black rocks protruding up: The Navajo volcanic field, in Geology of the Zuni Plateau, edited by Lucas, S.G., Semken, S.C., Berglof, W.R., and Ulmer-Scholle, D.S., *New Mexico Geol. Soc. Field Guide*, 54, 133–138, 2003.

Sharry, J., Langan, R.T., Jovanovich, D.B., Jones, G.M., Hill, N.R., and Guidish, T.M., Enhanced imaging of the COCORP seismic line, Wind River Mountains, in *Geodynamics Series,* edited by Barazangi, M., and Brown, L.D., *13*, 223–236, 1986.

Shaw, C.A., Snee, L.W., Selverstone, J., and Reed, J.C., Jr., $^{40}Ar/^{39}Ar$ thermochronology of Mesoproterozoic metamorphism in the Colorado Front Range, *J. Geol.*, 107, 49–67, 1999.

Shaw, C.A., Karlstrom, K.E., Williams, M.L., Jercinovik, M.J., and McCoy A., Electron microprobe monazite dating of ca. 1710–1630 Ma and ca. 1380 Ma deformation in the Homestake shear zone, Colorado: Origin and early evolution of a persistent intracontinental tectonic zone, *Geology*, 29, 739–742, 2001.

Sheehan, A.F., Abers, G.A., Jones, C.H., Lerner-Lam, A.L., Crustal thickness variations across the Colorado Rocky Mountains from teleseismic receiver functions, *J. Geophys. Res.*, 100, 20391–20404, 1995.

Smith, D., Insights into the evolution of the uppermost continental mantle from xenolith localities on and near the Colorado Plateau and regional comparisons, *J. Geophys. Res.*, 105, 16769–16781, 2000.

Snelson, C.M., Henstock, T.J., Keller, G.R., Miller, K.C., and Levander, A.R., Crustal and uppermost mantle structure along the Deep Probe seismic profile, *Rocky Mt. Geol*., 33, 181–198. 1998.

Snyder, G.L., Preliminary geologic maps of the central Laramie Mountains, Albany and Platte counties, Wyoming, *U.S. Geological Survey Open File report,* 84–358, 1984.

Snyder, D.B., Prasetyo, H., Blundell, D.J., Pigram, C.J., Barber, A.J., Richarson, A., and Tjokosaproetro, S., A dual doubly vergent orogen in the Banda Arc continent-arc collision zone as observed on deep seismic reflection profiles, *Tectonics*, 15, 34–53, 1996.

Snyder, D.B., Lithospheric growth at margins of cratons, Tectonophysics, 2002.

Sobolev, S.V., Zeyen, H., Stoll, G., Werling, F., Altherr, R. and Fuchs, K., Upper mantle temperatures from teleseismic tomography of French Massif Central including effects of composition, mineral reactions, anharmonicity, anelasticity and partial melt, *Earth Planet. Sci. Lett*.,147, 147–163, 1996.

Sobolev, S.V. and Babeyko, A.Yu., Modeling of mineralogical composition, density and elastic wave velocities in the unhydrous rocks, *Surveys Geophys*., 15, 515–544., 1994.

Spence, W., and Gross, R.S., A tomographic glimpse of the upper mantle source of magmas of the Jemez lineament, J. Geophys. Res., 95, 10829–10849. 1990,

Thybo H., Perchuc E., The seismic 8o discontinuity and partial melting in continental mantle, *Science*, 275, 1626–1629, 1997.

Timmons, J.M., Karlstrom, K.E., Heizler, M., Bowring, S.A., and Crossey, L.C., Intracratonic deformation and basin formation during protracted Grenville orogenesis: regional inferences from the ca. 1254- ~1100 Ma Unkar Group, Grand Canyon, 2002.

Tweto, O. and Sims, P.K., 1963, Precambrian ancestry of the Colorado Mineral Belt, *Geol. Soc. Am. Bull.*, 74, 991–1014, 1963.

Tyson, A.R., Morozona, E.A., Karlstrom, K.E., Chamberlain, K.R., Smith son, S.B., Dueker, K.G., and Foster, C.T., Proterozoic Farwell Mountain-Lester Mountain suture zone, northern Colorado: Subduction flip and progressive assembly of arcs, *Geology,* 30, 943–946, 2002.

Unruh, J.R., The uplift of the Sierra Nevada and implications for late Cenozoic epeirogeny in the western Cordillera, *Geol. Soc. Am. Bull.*, 103, 1395–1404, 1991.

Urquhart, J.M., Laramide deformation, *Terra Abstracts*, 3, 231, 1991.

Usui, T., Nakamura, E., Kobayashi, K., Maruyama, S., and Helmstaedt, H., Fate of the subducted Farallon plate inferred from eclogite xenoliths in the Colorado Plateau. *Geology*, 31, 589–592, 2003.

van der Lee, S., and Nolet, G., Upper mantle S-velocity structure of North America, J. Geophys. Res., 102, 22815–22838, 1997.

van der Lee, S., Deep below North America, *Science*, 294, 1297–1298, 2001.

Varga, R.J., Rocky Mountain foreland uplifts: products of a rotating stress-field or strain partitioning, *Geology*, 21, 1115–1118, 1993.

Williams. M.L., Jercinovic, M.J., and Terry, M.P., Age mapping and dating of monazite on the electron microprobe: Deconvoluting multistage tectonic histories, *Geology*, 27, 1023–1026, 1999.

Wilson, D., Aster, R., and the RISTA Team, Imaging crust and upper mantle seismic structure in the southwestern United States using teleseismic receiver functions, *The Leading Edge*, 232, 234, 236–237, 2003.

Wisniewski, P.A. and Pazzaglia, F.J., Epeirogenic controls on the Canadian River incision and landscape evolution, High Plains of northeastern New Mexico, *J. Geol.*, 2002.

Wolf, L.W., and Cipar, J.J., Through thick and thin: A new model for the Colorado Plateau from seismic refraction data from Pacific to Arizona crustal experiment, *J. Geophys. Res.*, 98, 19881–19894, 1993.

Wolfe, J.A., Forest, C.E., and Molnar, P., Paleobotanical evidence of Eocene and Oligocene paleoaltitudes in midlatitude western North America, *Geol. Soc. Am. Bull.*, 110, 664–678, 1998.

Wooden, J.L., and DeWitt, E., Pb isotopic evidence for the boundary between the early Proterozoic Mojave and central Arizona crustal provinces in western Arizona, in Proterozoic geology and ore deposits of Arizona, edited by Karlstrom, K. E., *Arizona Geol. Soc. Digest*, 19, 27–50, 1991.

Samual A. Bowring, Department of Earth Atmospheric & Planetary Sciences, Massachusetts Institute of Technology, Cambridge, MA, 02139

Ken Dueker, Department of Geology and Geophysics, University of Wyoming, Laramie, WY 82071

E. D. Humphreys, Department of Geological Sciences, University of Oregon, Eugene, OR 97403

Karl E. Karlstrom, Department of Earth and Planetary Sciences, University of New Mexico, Albuquerque, NM 87108

G. Randy Keller, Department of Geological Sciences, University of Texas at El Paso, El Paso, TX 79968

Alan Levander, Department of Earth Science, Rice University, Houston, TX 77005

Steven J. Whitmeyer, Department of Earth and Planetary Sciences, University of Tennessee, Knoxville, TN 37996

Michael L. Williams, Department of Geosciences, University of Massachusetts, Amherst, MA 01003